Southeastern Section
of the
Geological Society of America

SOUTHEASTERN SECTION
OF THE
GEOLOGICAL SOCIETY OF AMERICA
(NORTHERN SECTOR)

① – CENTENNIAL FIELD GUIDE SITES

0 ‖‖‖‖‖‖ 50 MILES

0 ‖‖‖‖‖‖ 50 KILOMETERS

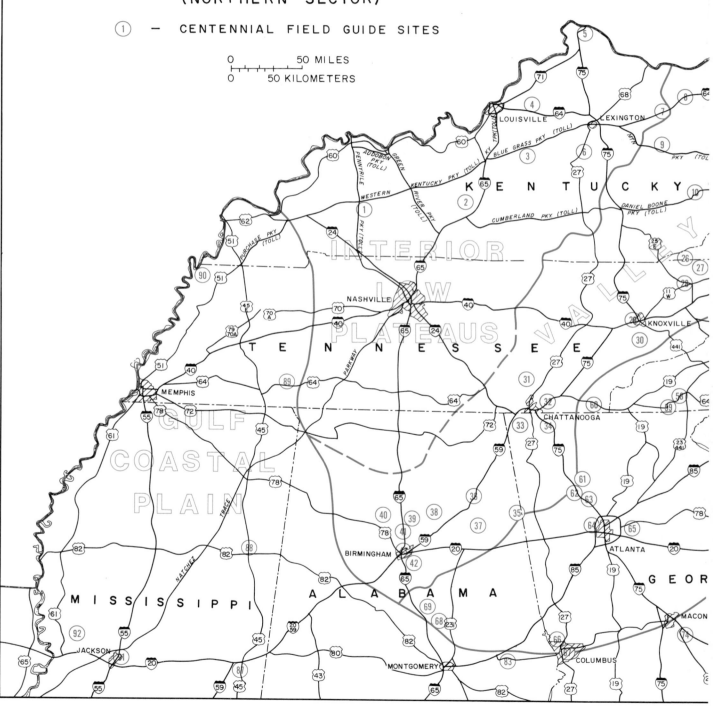

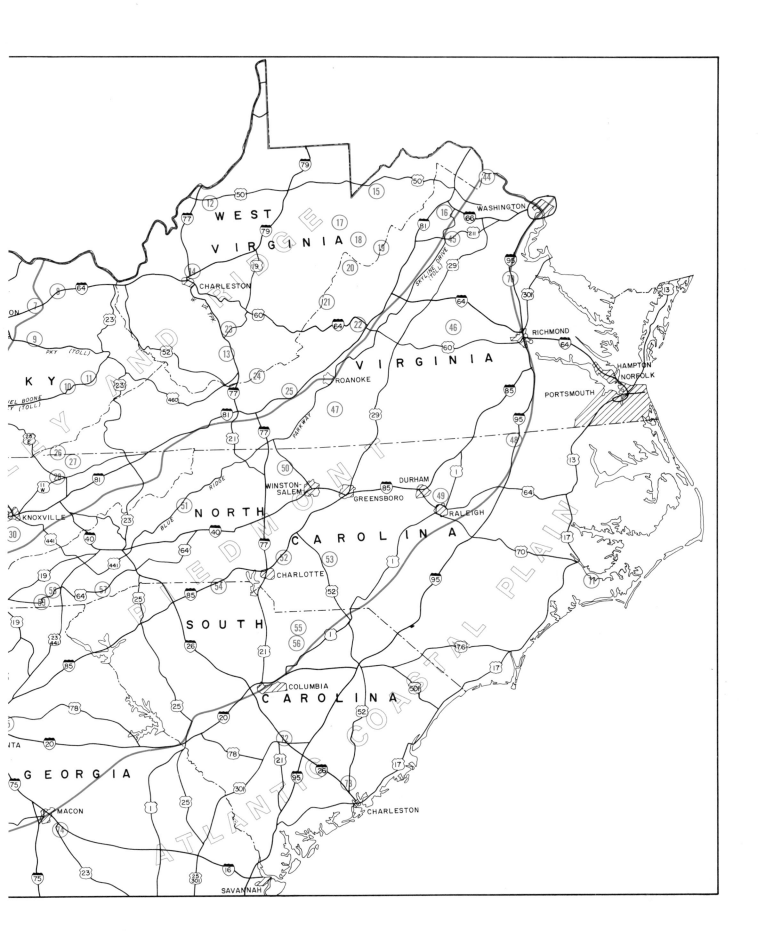

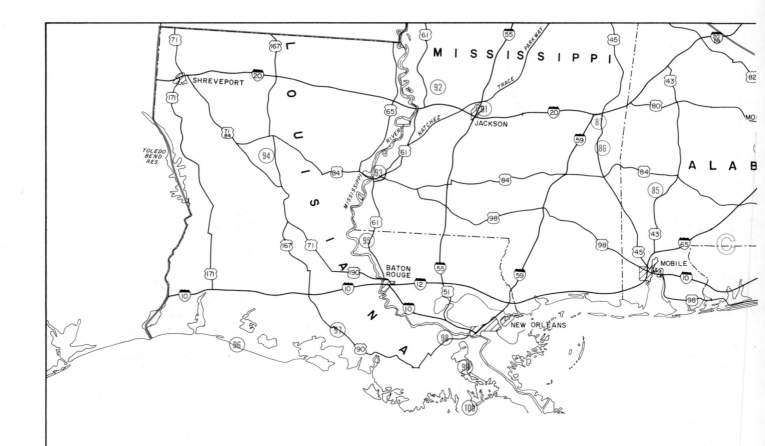

SOUTHEASTERN SECTION
OF THE
GEOLOGICAL SOCIETY OF AMERICA
(SOUTHERN SECTOR)

1 — CENTENNIAL FIELD GUIDE SITES

0   50 MILES

0   50 KILOMETERS

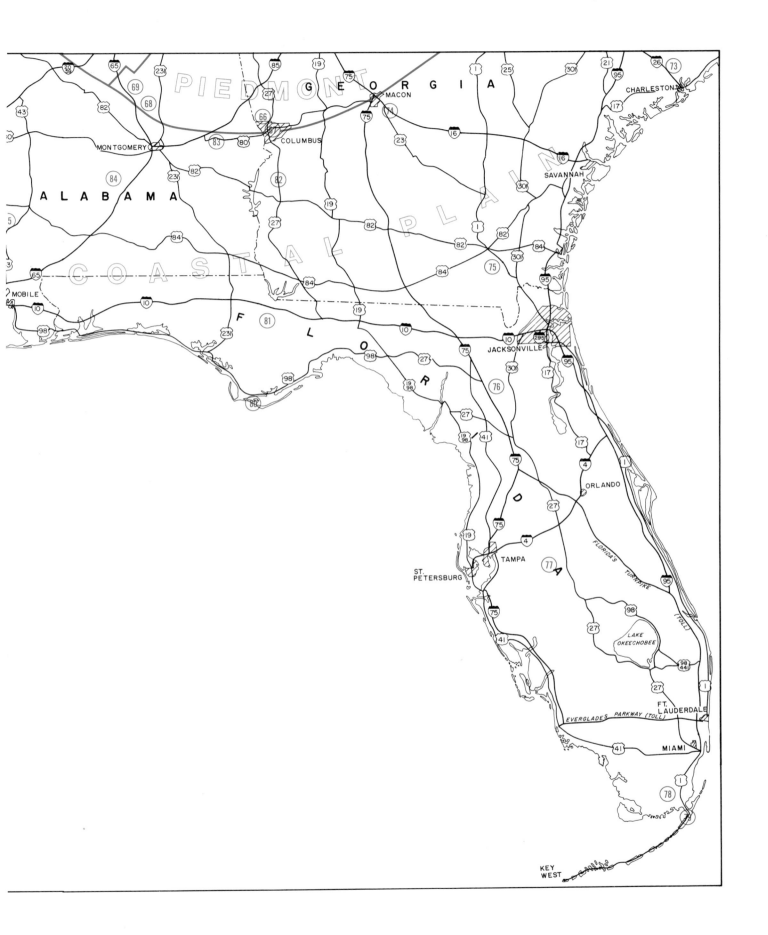

Centennial Field Guide Volume 6

Southeastern Section
of the
Geological Society of America

Edited by

Thornton L. Neathery
Geological Survey of Alabama
P.O. Drawer O
University, Alabama 35486

1986

Acknowledgment

Publication of this volume, one of the Centennial Field Guide Volumes of *The Decade of North American Geology Project* series, has been made possible by members and friends of the Geological Society of America, corporations, and government agencies through contributions to the Decade of North American Geology fund of the Geological Society of America Foundation.

Following is a list of individuals, corporations, and government agencies giving and/or pledging more than $50,000 in support of the DNAG Project:

ARCO Exploration Company
Chevron Corporation
Cities Service Company
Conoco, Inc.
Diamond Shamrock Exploration
 Corporation
Exxon Production Research Company
Getty Oil Company
Gulf Oil Exploration and Production
 Company
Paul V. Hoovler
Kennecott Minerals Company
Kerr McGee Corporation
Marathon Oil Company
McMoRan Oil and Gas Company
Mobil Oil Corporation
Pennzoil Exploration and Production
 Company

Phillips Petroleum Company
Shell Oil Company
Caswell Silver
Sohio Petroleum Corporation
Standard Oil Company of Indiana
Sun Exploration and Production Company
Superior Oil Company
Tenneco Oil Company
Texaco, Inc.
Union Oil Company of California
Union Pacific Corporation and
 its operating companies:
  Champlin Petroleum Company
  Missouri Pacific Railroad Companies
  Rocky Mountain Energy Company
  Union Pacific Railroad Companies
  Upland Industries Corporation
U.S. Department of Energy

Published by the Geological Society of America, Inc.
3300 Penrose Place, P.O. Box 9140, Boulder, Colorado 80301

Printed in U.S.A.

Library of Congress Cataloging-in-Publication Data

Centennial field guide—Southeastern Section of the
 Geological Society of America.

 (Centennial field guide ; v. 6)
 Includes bibliographies and index.
 1.Geology—Southern States—Guide-books.
2. Southern States—Description and travel—1981-
Guide-books. I. Neathery, Thornton Lee. II. Geological
Society of America. Southeastern Section.
III. Series.
QE77.C46 vol. 6 [QE78.5]   557.3 s [557.5] 86-11987
ISBN 0-8137-5406-2

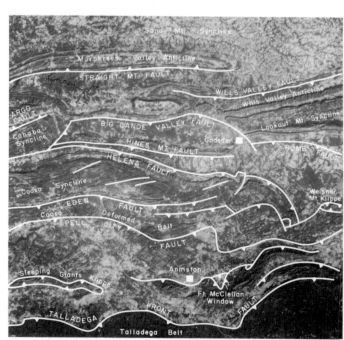

Front Cover: NASA Skylab 3-color infrared photograph taken September 1973 from an altitude of 234 nautical miles showing the central part of the Alabama Valley and Ridge structural belt. Sketch map (above) indicates various features that may be seen in the photograph.

Contents

BLUE RIDGE AND PIEDMONT

ATLANTIC AND GULF COASTAL PLAINS

TOPICAL CROSS-REFERENCES FOR FIELD GUIDE SITES

Preface

This volume is one of a six-volume set of Centennial Field Guides prepared under the auspices of the regional Sections of the Geological Society of America as a part of the Decade of North American Geology (DNAG) Project. The intent of this volume is to highlight the best and most accessible geologic localities of the southeastern United States for the geologic traveler and for students and professional geologists interested in major geologic features of regional significance. The leadership provided by the editor, Thornton L. Neathery; the support he received from his employer, the Geological Survey of Alabama; and from the members of the Southeastern Section of The Geological Society of America, are greatly appreciated.

Drafting services were offered by the DNAG Project to those authors of field guide texts who did not have access to drafting facilities. Particular thanks are given here to Ms. Karen Canfield of Lafayette, Colorado, who prepared final drafted copy of many figures from copy provided by the authors. Ms. Carolyn Thomas of Tuscaloosa, Alabama also provided drafting for some figures.

In addition to Centennial Field Guides, the DNAG Project includes a 29-volume set of syntheses that constitute *The Geology of North America,* and 8 wall maps at a scale of 1:5,000,000 that summarize the geology, tectonics, magnetic and gravity anomaly patterns, regional stress fields, thermal aspects, seismicity, and neotectonics of North America and its surroundings. Together, the synthesis volumes and maps are the first coordinated effort to integrate all available knowledge about the geology and geophysics of a crustal plate on a regional scale. They are supplemented, as a part of the DNAG project, by 23 Continent-Ocean Transects providing strip maps and both geologic and tectonic cross-sections strategically sited around the margins of the continent, and by several related topical volumes.

The products of the DNAG Project have been prepared as a part of the celebration of the Centennial of the Geological Society of America. They present the state of knowledge of the geology and geophysics of North America in the 1980s, and they point the way toward work to be done in the decades ahead.

Allison R. Palmer
Centennial Science Program Coordinator

Foreword

The southeastern region of the United States has been the source for observations fundamental to many modern geologic concepts. Generations of geologists have studied the sedimentary sequences of the Interior Lowlands, structure and stratigraphy of the Valley and Ridge, metamorphic and igneous rocks and sequences of the Blue Ridge and Piedmont, and the complex stratigraphic relationships and biostratigraphy of the Atlantic and Gulf Coastal plains. During the last 140 years, the rocks of the region have yielded important information about the evolution of eastern North America. Much additional information still remains to be discovered, and the rocks and structures of the southeast offer challenging opportunities for future geologic research.

This Field Guide is intended as a companion to the synthesis volumes of *The Geology of North America* that cover the area represented by the Southeastern Section of the Geological Society of America (The Sedimentary Cover of the Craton: U.S.; The Gulf of Mexico Basin; The Appalachian/Ouachita Regions: U.S.; The Atlantic Continental Margin: U.S.; Quaternary Non-glacial Geology: U.S.; and Precambrian: U.S.). The Field Guide contains descriptions of 100 outstanding geologic localities scattered throughout the 11-state region represented by the Southeastern Section. Each locality illustrates an important geologic relation of regional significance or an outstanding example of basic geology.

Because this book is deliberately limited to 100 localities, it is impossible to include every classical locale or everyone's favorite locale; however, this collection of sites will provide the student and professional alike with a good overview of the stratigraphy and structure of the southeastern United States. Used in conjunction with the regional synthesis volumes and maps of *The Geology of North America,* these sites should provide the interested geologist with "ground-truth" for many of the concepts presented in the syntheses.

The Field Guide sites in this book are grouped geographically into three sections, generally coinciding with the boundaries of the major physiographic divisions represented in the southeastern United States: Interior Low Plateaus and Appalachian Valley and Ridge; Blue Ridge and Piedmont; and the Atlantic and Gulf Coastal Plains. Forty-three papers describe stratigraphy and structure in the Interior Low Plateaus and Appalachian Valley and Ridge; 26 papers describe metamorphic and igneous rocks and structures in the Blue Ridge and Piedmont; and 31 papers describe stratigraphy, geomorphology, and modern depositional environments in the Atlantic and Gulf Coastal Plains. Some contributions describe a single outcrop or locale; others describe a series of related outcrops within a general area. An attempt has been made to include contributions on a variety of geologic topics, including geomorphology, hydrology, stratigraphy and sedimentology, and structural relations of sedimentary, metamorphic, and igneous rocks. By intent, localities that specifically involve collecting fossil or mineral specimens have not been included. Most localities or exposures are open to visitation without special permission. However, some specific localities do require permission for entry. Where permission is required, please make sure that you follow the visitation and safety rules so that everyone can enjoy the site.

Each site description includes a statement on location with information in the form of an index map, sometimes augmented by additional text; comments on accessibility of the site, if restricted or difficult; geologic background information including the regional significance of the locality, or geologic phenomena to be observed; a comprehensive site description, or several descriptions if the site is a cluster of related stops; and a list of key references for further study. Two contributions with related components that could not be contained within the page limits imposed on single sites, one by Englund and others (sites 13/14) and another by Thomas and Bearce (sites 42/43), are presented as double sites. Several papers are presented on similar stratigraphic sections but in different parts of the region. Individuals interested in these specific stratigraphic intervals will find these contributions to be valuable for regional analysis.

The help of the following individuals who provided guidance and support to develop this volume during the early planning and project formulation stage is gratefully acknowledged: Charlotte E. Abrams, Georgia Geological Survey; Katharine Lee Avary, West Virginia Geological Survey; J. Robert Butler and John M. Dennison, University of North Carolina; Robert D. Hatcher, Jr., University of South Carolina; Kenneth O. Hasson, East Tennessee University; J. Wright Horton, U.S. Geological Survey; Shea Penland, Louisiana Geological Survey; Juergen Reinhardt, U.S. Geological Survey; Ernest E. Russell, University of Mississippi; Thomas M. Scott, Florida Geological Survey; and William A. Thomas, University of Alabama.

The Southeastern Section of the Geological Society of America provided financial support for miscellaneous expenses related to this project, and the Geological Survey of Alabama provided logistical support. Secretarial assistance by Audrey T. Hartley is appreciated. Completion of this volume depended on the patience and cooperation of the many unnamed people who served as critical reviewers for the site description texts. To them, thank you for a job well done. Special acknowledgment of appreciation goes to my wife, Patricia, for her patience and understanding during the many months that manuscripts and work materials littered the house. Finally, to all of the authors who have contributed their time so generously to this Field Guide project, I tender my sincere thanks and appreciation.

Thornton L. Neathery, Editor
March, 1986

1

Selected exposures of Pennsylvanian rocks in western Kentucky

Peter W. Whaley Department of Geology, Murray State University, Murray, Kentucky 42071

LOCATION

Pennsylvanian rocks of western Kentucky are exposed in road cuts along Pennyrile and Kentucky Parkways north of Hopkinsville, Kentucky (Fig. 1). All stops are very accessible, but utmost caution should be exercised since these stops are on highways that carry moderate to heavy traffic. Year-round camping facilities with showers are available at John J. Audubon State Park near Henderson, Kentucky, and seasonal camping facilities with showers are available at Pennyrile Forest State Park near Dawson Springs, Kentucky (Fig. 1).

INTRODUCTION

Since the report of the First Geological Survey of Kentucky (Owen, 1856), the Pennsylvanian rocks of western Kentucky have been the subject of intense geological investigation because they have yielded substantial amounts of coal, oil, and natural gas. The lateral variation of rock units, vertical succession of rock types, and sedimentary structures of rocks from the Caseyville Formation, the Tradewater Formation, and the Carbondale Formation (Fig. 2) can be related to modern depositional environments. The most important coal in western Kentucky, W. Kentucky No. 9 or Springfield Coal (Williams and others, 1982), is completely exposed, and an underground mine collapse associated with this coal bed is visible. A large Pennsylvanian-age slump block preserved in the Caseyville Formation attests to past dynamic processes that can cause presentday safety problems in coal mine roof stabilization. A part of the Pennyrile Fault System displays differential movement along a fault, and the influence of lithology due to depositional environments on the location of some faults in the system can be observed.

Two field guides, Whaley and others (1979) and Whaley and others (1980), provide a detailed description of the geology exposed between the three stops that compose this cluster. Additional stops from the central and eastern portion of the Western Kentucky Coal Field are contained in Whaley and others (1980) and in Palmer and Dutcher (1979).

SITE DESCRIPTION

Stop 1

The three exposures that constitute Stop 1 are located at mile marker 23.7 on the Pennyrile Parkway (Fig. 1). On the Crofton 7½-minute geologic quadrangle, this stop is north of the parkway exit for Crofton, Kentucky, and south of McFarland Creek. The exposure is a fairly continuous 0.5-mile (0.9-km) long box cut on both sides of the road. Trace fossils, lithologic variation, and sedimentological structures found in the Caseyville Formation allow one to reconstruct the environments in which

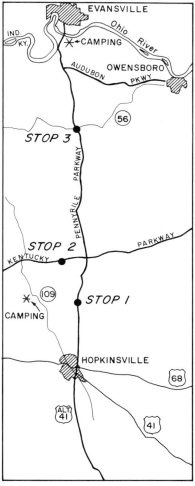

Figure 1. Location of cluster stops and adjacent camping facilities.

these rocks were deposited. The boundary between Pennsylvanian- and Mississippian-age rocks in this area is the Pennyrile Fault System. Six faults associated with this system are well exposed in this stop. Displacement along the major fault is a minimum of 200 ft (60.8 m). Differential movement and the influence of lithology on the location of minor faults within this cut is also evident.

Stop 1 A. Section A of Figure 3 is a diagram of the east side of the south end of the box cut. Near the south end of the cut, a fault with minor displacement occurs within the Mississippian-age rocks (Fig. 3, Section A, Point A). At the top of the cut, on the upthrown side of the fault, an orthoquartzitic, fine-grained sandstone is present. Kehn (1977) mapped this exposure as Menard Limestone and noted that sandstones are not seen in the Menard Limestone at the surface but are known to occur in the subsurface. It is most probable that this orthoquartzitic sandstone is a surface exposure of sandstone in the Menard Limestone. A

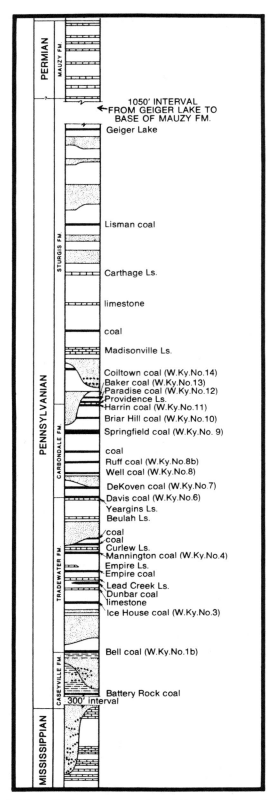

Figure 2. Generalized columnar section of Pennsylvanian units in the western Kentucky coal field (Williams et al., 1982).

second fault with a much larger displacement has placed the shale and limestone of the Menard Limestone in juxtaposition with the alternating siltstones and fine-grained, thin-bedded sandstones of the Caseyville Formation (Fig. 3, Section A, Point B). The strike of this major fault is N.72°E. This fault is the systemic boundary between the Mississippian and the Pennsylvanian in this area.

The upper part of the Caseyville Formation, between points B and C in Section A of Figure 3, contains thicker and more numerous beds of sandstone than the lower part. These sandstone beds exhibit ripple marks on the bedding surfaces and small-scale crossbeds. The sandstones contain trace fossils such as *Cruziana sp.* The majority of the trace fossils seem to occur parallel to the bedding; however, disturbed zones also intersect beds at various angles. A sandstone slab collected from this interval contained the thoracic segments of a trilobite. Crimes (1975) has related trilobites to the trace fossil *Cruziana sp.* The uppermost sandstone layer in this interval has been burrowed and contains shale clasts (Fig. 3, Section A, Point C).

The position and unique texture of a burrowed sandstone bed provide stratigraphic correlation across a minor fault that offsets this bed at Point C. Stratigraphic displacement indicates the south side of the fault went down; however, drag folding indicates that the south side went up (Fig. 3, Section A, Point C, and Fig. 4). Clearly, differential movement has taken place along this fault.

Above the faulted burrowed sandstone and below the interbedded siltstone and silty shale, a sequence of black siltstone beds strikes N.75°W. and dips 19°NE. About 2 in (5.0 cm) above the burrowed zone, microfossils recovered from a thin zone in the siltstones include the following pyritized forms: high- and low-spired gastropods, ostracods, brachiopods, and one cephalopod fragment. Also present are unaltered conodont fragments and foraminifera. An upward coarsening of the black siltstone sequence is evidenced by the occurrence of 0.1-in (0.3-cm) thick ripples of sand in the upper part of this unit. A well-developed joint set in the black siltstone has an average strike of N.54°E. Between pair of some of these joints a less steeply dipping joint set can be seen. This set is probably related to compressional forces that were generated during the second period of movement along the adjacent faults.

Above the black siltstone is a thicker, more variable, upward-coarsening sequence that is initiated by two interbedded silty shale and siltstone units and is terminated by a coal bloom that rests on a well-rooted rippled sandstone. The central part of this unit (Fig. 3, Section A, Point E) also exhibits well-developed, closely spaced joints. The average strike for these joints is N.75°E.

Above the coal bloom, the top of a silty shale and the lower 6 in (15.0 cm) of a rippled sandstone and siltstone have been disturbed by rooting. This rooted zone (Fig. 5) is different from the most common type of rooted zone in that the orientation of these roots is parallel rather than perpendicular to bedding. These horizontal roots are preserved as sandstone casts rather than the compressed vertical carbonized exteriors of the root.

The final upward-coarsening sequence in this cut starts with

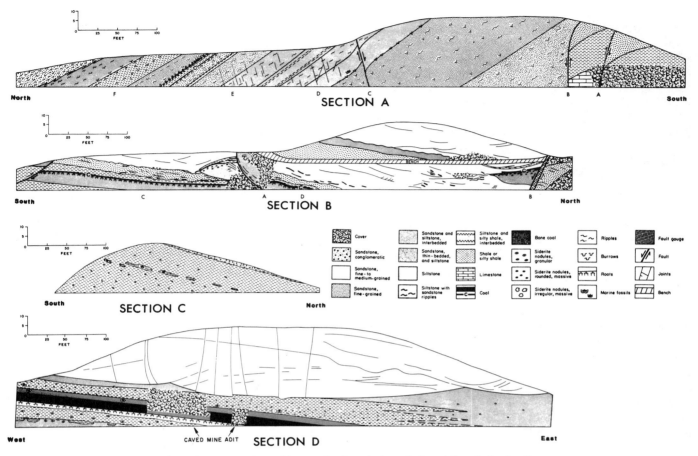

Figure 3. Cross-sections of Stop 1, Sections A through C, and Stop 2, Section D.

a silty shale that contains irregularly shaped siderite nodules (Fig. 3, Section A, Point F) and culminates in a rippled sandstone. The irregular siderite nodules contain carbonized root impressions, and the interbedded rippled sandstone and siltstone part of the sequence is burrowed. The contact between the silty shale and the upper rippled sandstone is gradational.

Geologists for many years have noted that the Caseyville Formation has a variable lithology. However, most natural exposures in cliff faces or stream beds tend to display only the sandy or conglomeratic facies. Oil and gas production from this unit also favors the coarser-grained facies. Cliffs and well cuttings are not the most favorable places to look for trace fossils or for thin zones within a very thick formation that may yield microfossils. Thus, many people have equated the medium-grained to conglomeratic, massive sandstones—the erosion-resistant parts of the Caseyville Formation—with the total lithology of the formation, leading them to think of the formation in general as representing nonmarine deposits. However, the lithologic variations and the occurrence of marine trace fossils and microfossils in this locality establish a different interpretation for parts of the Caseyville Formation, namely, that it is marine in origin.

The part of the Caseyville Formation illustrated in Figure 3, Section A, is interpreted as a shallow-water sand flat or tidal

delta. The several fine- to coarse-grained, relatively thin, detrital sequences, which exist above the burrowed zone that was faulted (Fig. 3, Section A, Point C), contain evidence of shallow-water current activity (thin-bedded, rippled sandstone), animal activity (burrowing), and plant activity (rooting). Thus, these vertical sequences represent less persistent or smaller-scale multiple marine or brackish shallow-water sand flat or tidal delta environments.

Stop 1 B. Lithologic variation in the Caseyville Formation and the influence of lithology on post-depositional structure is documented in the second part of the box cut at Stop 1. The undiagrammed rocks exposed in the east cut contain two striking structural features; a small overturned anticline breached by a thrust fault and a highly fractured sandstone unit bounded by two faults near the north end of the cut. Both features were mapped by Kehn (1977). Their respective counterparts in the west side of the cut are more obscure and are represented as Points A and B in Section B of Figure 3.

The rocks above the coal at the south end of the west cut (Fig. 3, Section B, Point C) are a complex upward-coarsening sandstone sequence that represents two different depositional events. Laterally, the lower fine-grained part of the sandstone thins and terminates along a concave erosional surface. The rocks

Figure 4. Fault in the Caseyville Formation. Two different periods of displacement are indicated by disagreement between stratigraphic displacement and dragfolds.

Figure 5. Horizontal roots preserved as casts in the Caseyville Formation.

above this concave erosional surface are a siltstone that contains siderite nodules in the upper part, a thin coal, and a thin shale that is truncated by a sandstone. This sandstone is the upper part of the sandstone seen at Point C. The fine-grained terrigenous sequence that is capped by the coal was deposited in an abandoned distributary channel. The south side of the channel is the concave erosional surface, whereas the north side of the channel has been displaced by a fault (Fig. 3, Section B, Point A). With later progradation, a less distal part of the distributary system reestablished itself over the older abandoned distal edge of the distributary system. The upper, slightly coarser-grained sandstone cuts out the coal and cuts into the lower finer-grained part of the distributary bar (Fig. 3, Section C, Point C) and represents a more landward part of the distributary system.

In Figure 3, Section B, the sequence of rocks above Point D clearly represents the lateral equivalent of the rocks and environments at Point C. At the north end of the cut (Fig. 3, Section B, Point B), the siltstone-to-coal sequence represents a second fine-grained distributary channel fill. Sixteen feet (5 m) south of this second fine-grained channel fill, a pocket of shale clasts and siderite nodules represents the lag gravel deposited by the prograding distributary system as it reestablished itself over the abandoned distributary channel. Between Points D and B in Section B, Figure 3, the rocks above the bench represent a continued progradation of the deltaic system over the previously deposited material that contained the two fine-grained channel fills.

The small, faulted anticline in the east cut corresponds to Point A, Section B, in the west cut. In the west cut, recent slumping has obscured the view of the fault. The fractured sandstone unit bounded by two faults in the east cut is reduced to a thin zone at Point B in the west cut. When this area was subjected to the tectonic forces that produced the folding and faulting, the fine-grained channel fills were zones of weakness in a predominantly sandstone unit, and faulting occurred in these zones (Fig. 3, Section B, Points A and B). These relationships demonstrate the influence of lithologic variation within an environment of deposition on the subsequent structure of the area.

Points of interest for the paleobotanist are the layers of coal found in the sandstones that represent distributary bar deposits. These coal layers are of two different origins; single coalified logs and pieces of eroded peat mat. The occurrence of leaf scar impressions above coal layers that are split by a thin sand suggests that these coal layers represent logs in which the central pith part was filled with sand. Other layers that are coalified logs have no sand split. Coal layers that represent peat mats have a more equal length-to-width ratio than do the logs. Thus, the rock record is consistent with the occurrence of large pieces of peat mat and single logs found in modern distributary bar sands.

Stop 1 C. The last exposure on the west side of the parkway (Fig. 3, Section C) contains rocks that form an upward-coarsening sequence from clay shale to silty shale. Large siderite nodules located midway up the face of the outcrop are of particular interest. The weathered nodules contain orbiculoid brachiopods, ostracods, broken brachiopod spines, and other fragments that might be broken echinoderm spines. This fossil assemblage represents a brackish-water or perhaps a marine environment. The prominent white layer near the top of the cut is recently weathered clay shale. More than 1 ft (30 cm) of conglomeratic sandstone can be seen at the top of this cut.

The conglomeratic facies of the Caseyville Formation is also exposed among the pines in the northernmost cut on the east side of the parkway. It is overlain by black, silty shales. To date, no fossils that would indicate marine or brackish conditions have been found in the silty shales above the conglomeratic sandstone.

Stop 2

Stop 2 is a small box cut located at mile marker 31.9 on the Western Kentucky Parkway (Fig. 1). On the St. Charles 7½-minute quadrangle, this stop is marked on the parkway as a coal outcrop between Fox and Cane Run. The north side of the cut is illustrated in Figure 3, Section D. Rocks of the Carbondale For-

Figure 6. Slump block in a distributary system within the Caseyville Formation.

mation, including the western Kentucky No. 9 or Springfield Coal, are exposed in this cut. The massive sandstone exposed above the coal, and the sandstones exposed in the large cuts to the east, represent a major distributary system. During construction of the parkway the roadbed intersected a drift in an old underground coal mine. The breaks in the coal bed are the areas where the coal was mined (Figure 3, Section D). Major joints in this cut are controlled by the excavated part of the coal bed.

At the west end of the cut a micritic limestone containing scattered siderite grains occurs within fine-grained clastic sediments. No fossils have been found in this limestone. The limestone is overlain by a fine-grained sequence of rocks that terminates in an underclay and accompanying coal. This is the western Kentucky No. 9, or Springfield Coal. The upper part of this coal contains detrital silt and clay and is termed a bone coal. At the west end of the outcrop the bone coal contains external molds of pelecypods and brachiopods. External and internal molds of brachiopods, ostracods, pelecypods, gastropods, and a coral have been found above the bone coal in a leached, white-colored silty shale layer. In the center of the cut the silty shales exposed at road level above the Springfield Coal contain external molds of brachiopods and pelecypods and segments of crinoid stems. The discovery of fossils in rocks such as these is a matter of fresh exposure, patience, and luck. The rocks above the thick Springfield Coal represent the termination of a substantial swamp by the invasion of marine waters. The fossiliferous silty shales are overlain by siltstones that contain thin beds of sand. This lithology represents the distal part of a deltaic lobe and is best seen below the thick sandstone at the east end of the cut.

As the distributary bar environment prograded over the distal bar, it eroded the majority of the finer-grained distal bar sediments and deposited the thick distributary bar sandstone. At least three shifts in the distributary channel can be documented by the erosional events that are preserved in the distributary bar sandstone. The lower central part of the distributary bar sand-

stone truncates the underlying sediments most deeply and contains a few eroded siderite nodules near its base. The sandstone at the west end of the outcrop cuts into the siltstones that underlie it and also truncates the oldest part of the sandstone in the center of the outcrop. The third period of truncation can be seen in the upper, central part of the outcrop as the third sand unit cuts down to the east through the sands of the two previous units and into the material below them. These crosscutting relations represent the shifting through time of the major axis of sediment dispersion across the distributary bar as the distributary system prograded. The net result of such shifting and progradation of a distributary system is the development of a thick and laterally extensive sand unit. This type of deposition is characteristic of the transition between the upper part of the lower delta plain and the very lower part of the upper delta plain. Similar units are common in the Ashland, Kentucky, area and were called bar/channel sands by Whaley (1969).

Stop 3

Stop 3 is located in the south cut of the southwest loop of the Kentucky 56 exit from the Pennyrile Parkway at the Sebree Toll Plaza (Fig. 1). On the Beech Grove 7½-minute quadrangle, this stop is in the northwest corner of the map just south of the Rough Creek Fault. The rocks exposed in this cut are mostly sandstones and shales plus an unnamed coal in the Caseyville Formation. A spectacular slump block of Pennsylvanian age is preserved. This exposure is on the upthrown side of the Rough Creek Fault. Although the fault is not exposed, drag features associated with the fault system are evident in the rocks exposed in the southeast entrance loop to the parkway and in exposures on the parkway proper.

The lower part of this cut exposes an upward-coarsening sequence capped by a seatrock and coal (top of the lightpole, Fig. 6). Siderite nodules in the lower 5 ft (1.5 m) of this interval contain marine fossils. These rocks represent bayfill deposits. The absence of a moderate or thick sandstone indicates that the culmination of clastic sedimentation in this bay ended before a distributary or crevasse system prograded over the area.

In contrast, the rocks above the coal represent a two-cycle upward-coarsening sequence in which the paleoslumping occurred. The first sequence ends in a fine-grained sandstone that is thickest in the east end of the cut (Fig. 6). These rocks represent the lateral or distal part of a distributary system. Above the fine-grained sandstone, a second sequence of silty shales is capped by a 37-ft (11.2-m) thick sandstone. The rocks above the fine-grained sandstone were deposited when the locus of deposition of the lower distributary switched to another locality or when a younger distributary system prograded over this locality.

The most prominent feature in the exposure is a paleoslump block present in the center of the outcrop (above the lamp, Fig. 6). Above the curved rotational plane of the slump, the steep apparent dip of the sandstone beds is to the east. The sandstone beds in the slump are covered by beds of the 37-ft (11.2-m) thick sandstone, which exhibits a gentle apparent dip to the west. To the right of the lightpole the surface of rotation of the slump block rises from the coal to a position above the fine-grained sandstone. The lower silty shales, the fine-grained sandstone, and the upper silty shales have been displaced by the slump. These features indicate that failure occurred in the lower unit of silty shales above the coal and beneath the fine-grained sandstone (Fig. 6). Several smaller rotational surfaces along which minor movement occurred can be seen to the east of the major slump. About 100 ft (30 m) east of the light pole, a position not visible in Figure 6, the lower sandstone has broken and moved down about 2 ft (0.7 m) along a curved surface that extends through the lower silty shale to the coal.

The gently westward dipping beds of the 37-ft (11.2-m) thick sandstone scoured the top of the slump block and, at the far west end of the exposure, eroded down into the fine-grained sandstone. This relation suggests that the slump occurred at the front of a westward prograding distributary system. Coleman (1976) notes the existence of large slump features in prograding distributary bars in modern deltas.

REFERENCES CITED

Coleman, J. M., 1976, Deltas: Processes of deposition and models for exploration: Champaign, Continuing Education Publishing Company, 102 p.

Crimes, T. P., 1975, The stratigraphic significance of trace fossils, in Frey, R. W., ed., The Study of trace fossils: New York, Springer-Verlag, p. 109–130.

Kehn, T. M., 1977, Geologic map of the Crofton Quadrangle, western Kentucky: U.S. Geological Survey Geologic Quadrangle Map GQ1361, Scale 1:24,000.

Owen, D. D., 1856, Report of the geological survey in Kentucky made during the years 1854 and 1855: Kentucky Geological Survey Ser. 1, v. 1, 416 p.

Palmer, J. E., and Dutcher, Russell R., eds., 1979, Depositional and structural history of the Pennsylvanian System of the Illinois Basin, Part 1: Road log and descriptions of stops (Ninth International Congress of Carboniferous Stratigraphy and Geology, field trip 9): Urbana, Illinois State Geological Survey, 116 p.

Whaley, P. W., 1969, A litho-genetic model for rocks of a lower deltaic plain sequence [Ph.D. thesis]: Louisiana State University, Baton Rouge, Louisiana, 136 p.

Whaley, P. W., Hester, N. C., Williamson, A. D., Beard, J. G., Pryor, W. A., and Potter, P. E., 1979, Depositional environments of Pennsylvanian rocks in western Kentucky: Geological Society of Kentucky, Annual Field Conference, Guidebook: Kentucky Geological Survey, 48 p.

Whaley, P. W., Beard, J. G., Duffy, J. E., and Nelson, W. John, 1980, Structure and depositional environments of some Carboniferous rocks of western Kentucky (American Association of Petroleum Geologist, Eastern Section Guidebook): Evansville, Indiana, Murray State University, 38 p.

Williams, D. A., Williamson, A. D., and Beard, J. G., 1982, Stratigraphic framework of coal-bearing rocks in the western Kentucky coal field: Kentucky Geological Survey, Series XI, Information Circular 8, 201 p.

2

Hydrogeology of Turnhole Spring groundwater basin, Kentucky

James F. Quinlan, National Park Service, Box 8, Mammoth Cave, Kentucky 42259
Ralph O. Ewers, Department of Geology, Eastern Kentucky University, Richmond, Kentucky 40475
Arthur N. Palmer, Department of Earth Sciences, State University of New York College at Oneonta, Oneonta, New York 13820

LOCATION

The Turnhole Spring groundwater basin covers more than 85 mi[2] (220 km[2]) in parts of Edmundson and Hart Counties, Kentucky, approximately 30 miles (50 km) north of Bowling Green, Kentucky and 10 miles (16 km) west of Interstate 65 at Cave City, Kentucky (Fig. 1).

INTRODUCTION

Water levels in underground streams in the Mammoth Cave area can rise and fall as much as 100 feet (30 m) in response to heavy rains. Generally a cave system consists of a dendritic or trellised network of passages in which tributary streams, fed by sinking streams and/or by infiltration through the soil, flow to a main trunk. The flow system of a cave is analogous to that of a surface river which is fed both by tributaries and by seepage from its banks. Mammoth Cave consists of a vertically interconnected, stacked series of dendritic passage networks. The highest passages are the oldest; as Green River successively lowered its valley, a new network of dendritic passages repeatedly developed and conveyed water to springs along its banks. Physiography of the Turnhole Spring basin and surrounding region is shown on Figure 2.

Dye-tracing, cave-mapping, and mapping of water levels in several hundred domestic water wells during base-flow conditions have been used to determine the flow routes of groundwater shown in Figures 3 and 4. Some of the cave streams carry pollutants into the park from Park City, which has no sewage treatment facility, and from the interstate highway interchange at Cave City. The heavy dashed lines in Figure 4 show how, during moderate-flow and flood-flow conditions, some of the water in the major west-flowing cave stream is diverted north from the Turnhole Spring groundwater basin into any of three adjacent groundwater basins (Sand Cave, Echo River, and Pike Spring) and even into Mammoth Cave itself.

The complex evolution of Mammoth Cave has been partially deciphered. The older, upper-level passages conveyed water northward from the Sinkhole Plain to an ancestor of the Echo River Spring shown in Figure 4. The Turnhole Spring groundwater basin subsequently pirated the headwaters of ancestral Echo River (Quinlan and Ewers, 1981). Water from the Sinkhole Plain then went westward to Turnhole Spring. Now water from the Sinkhole Plain flows to Echo River Spring and to the exhibited parts of Mammoth Cave only when floods temporarily divert part of the Turnhole drainage northward into the present-day Echo River groundwater basin. The Turnhole Spring groundwater basin enables illustration of a suite of karst features

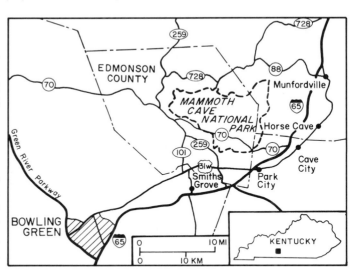

Figure 1. Location of Turnhole Spring, Mammoth Cave area, Edmundson and Hart Counties, Kentucky.

and their interrelationships with one another, the surface, and underground.

For an introduction to the regional geology, detailed discussion of stops and hydrologic concepts, and a guide to Mammoth Cave itself see Quinlan and others (1983) and Quinlan and Ewers (1985). The most complete guide to the cave, however, is by Palmer (1981). We recommend visiting Mammoth Cave the day before, the day after, or on the same day as this 4- to 5-hour trip through Turnhole Spring groundwater basin.

The Historic Tour best exhibits the geologic and historic features of the cave. Organized educational groups visiting the cave are admitted free to one tour if arrangements are made in advance. Check with the National Park Service at Mammoth Cave, KY 42259 (502-758-2251) for the cave tour schedule and for information about motels and campgrounds. Ask for a free copy of *A Visitor's Introduction to Mammoth Cave National Park* by J. F. Quinlan, 1985. It is an excellent supplement to this tour guide, and includes a useful bibliography on the geology, archaeology, biology, and history of the park. Many of the publications cited in it can be ordered from: E.N.P. & M.A., Box 25, Mammoth Cave, KY 42259. Write for a list.

SITE DESCRIPTION

Driving Directions. Exit from I-65 at Park City (Exit #48). Turn south onto Kentucky 255, go 0.6 mile (1.0 km) to T-junction, and turn right (west) onto U.S. 31W. Go 1.1 mile (1.8 km) to the entrance of Park Mammoth Resort and turn right.

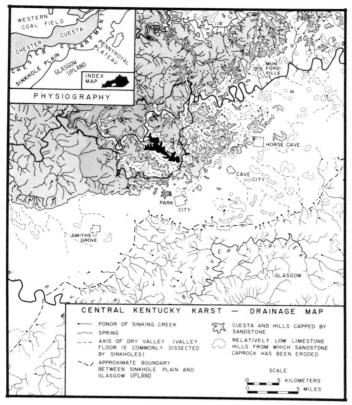

Figure 2. Map showing landforms, surface drainage, and physiographic divisions of the Mammoth Cave Region. Mammoth Cave Ridge is shown in black and the boundaries of Mammoth Cave National Park are shown as long dashed lines.

Proceed to the top of the hill, cross the railroad tracks, and go 0.5 mile (0.8 km) straight ahead to the next set of railroad tracks. Stop. Walk 800 feet (240 m) to the left along the tracks to the overlook.

Stop 1 - Overlook at Park Mammoth Resort

This overlook gives a scenic view of the Sinkhole Plain. You are standing on Big Clifty Sandstone at the top of the Chester Escarpment and at the crest of the Chester Cuesta (Figs. 2 and 3). All rocks seen on this trip are of Mississippian age.

You are at an elevation of about 800 feet (250 m). The west-trending highway nearest you, U.S. 31W, is at an elevation of about 600 feet (180 m). On the south horizon is the Glasgow Upland, at 800 feet (250 m) elevation. It is underlain by the lower half of the St. Louis Limestone. Thin shaley and silty beds within it contain numerous spring-fed, north-flowing allogenic streams which cross the Glasgow Upland and sink at its south margin, near the top of the lower half of the St. Louis. About a hundred sinkholes can be seen from here. The Sinkhole Plain is underlain by the upper half of the St. Louis Limestone and the Ste. Genevieve Limestone. These beds dip north, toward you, at an angle which is slightly steeper than the regional slope of the Glasgow Upland and Sinkhole Plain. Can you recognize the dip slope of this lower cuesta shown in Figure 3? The escarpment face beneath the Big Clifty Sandstone is underlain by the Girkin Limestone. As depicted in Figures 3 and 4, underground streams beneath the Sinkhole Plain flow north and northwest to the Green River, which is behind you. The flow is via Stops 5, 6, and 7.

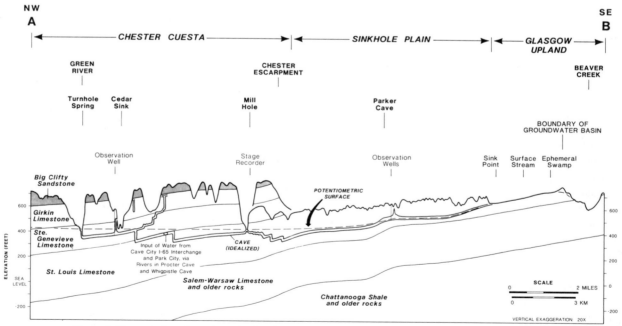

Figure 3. Cross section along the axis of major subsurface drainage in the Turnhole Spring groundwater basin.

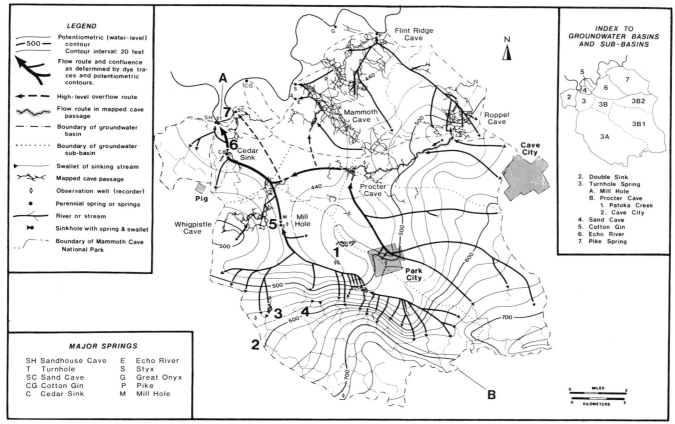

Figure 4. Map summarizing hydrology of Turnhole Spring groundwater basin, the major basin draining into Mammoth Cave National Park. Shown are basin and sub-basin boundaries, the potentiometric surface, sinking streams, underground flow routes, caves, springs, observation wells, and field trip stops.

The big isolated hill 6.5 miles (10.5 km) to the southwest (Pilot Knob) is a relict interstream area that was flanked by streams during the late Tertiary. As you look to the west, parallel to the escarpment face, you are looking along the axis of a monocline where the dip is north and relatively steep, 350 feet/mile (60 m/km). In contrast, the dip over much of the Sinkhole Plain east of here is about 50 to 80 feet/mile (10-15 m/km). The locally narrow width of the Sinkhole Plain is determined by the relative steepness of the dip of the beds most susceptible to karstification. Similarly, the knobs and ridges are relict interstream areas that are best preserved where the beds dip less steeply.

Resume Driving. Retrace route to U.S. 31W and turn left, back to Park City. Go to the traffic light, turn right onto Kentucky 255, and go 2.6 miles (4.2 km) across the Sinkhole Plain to Mt. Vernon Church, on the right. At 0.3 miles (0.5 km) farther, just beyond the base of the hill, start ascending the Glasgow Upland. Continue 1.5 miles (2.4 km), cross the freeway, and immediately turn right onto Oak Grove Church Road. For the next 6.1 miles (9.8 km) you will be traveling on the Glasgow Upland. Note the lack of sinkholes and the presence of surface streams. At 3.4 miles (5.5 km) pass Oak Grove Church and turn left onto the paved road 0.2 miles (0.3 km) beyond it. At 200 feet (60 m) beyond this turn, notice the trickle of the north fork of

Little Sinking Creek. Go 0.9 miles (1.4 km) and cross the south fork of Little Sinking Creek. Note the use of Kentucky riprap for bank stabilization and the width of the floodplain. Continue to the T-junction. Turn right onto U.S. 68-Kentucky 80; go 0.7 mile (1.1 km) to Hays. Turn right onto Kentucky 259. One mile (1.6 km) west-northwest of here are two sinkholes into which 340 tons of whey were dumped in 1970. Water from the well which supplied Smiths Grove was grossly contaminated and unfit to drink for several months. If the whey had been dumped 2 miles (3.2 km) to the northeast, it would have flowed into Mammoth Cave National Park and destroyed much of its unique cave fauna which is blind. Continue 1.7 miles (2.7 km) to the 90-degree turn to the left (north). Park, but stay on the road.

Stop 2 - Cayton Branch

The south fork of Little Sinking Creek is 1,000 feet (300 m) east of here, where the trees end. All of the water seen today becomes groundwater which flows northwest and north to the Green River. After heavy rains, however, Little Sinking Creek overflows its banks and sends as much as 6 feet (2 m) of water across this road and northwest 900 feet (270 m) to the clump of trees on a line bisecting the right angle of the turn. There the

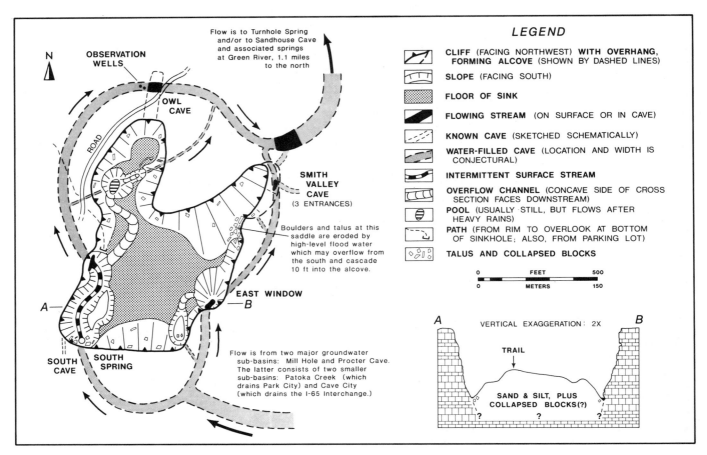

Figure 5. Sketch map hydrology and geomorphology of Cedar Sink.

water sinks into the ground at the base of the hill, flows west, rather than north, and travels 16 miles (25 km) via three caves to the Barren River at Graham Springs, near Bowling Green (Figs. 2 and 3). A temporary lake more than 1,000 feet (300 m) wide will exist here for several days after a heavy rain. We interpret the south fork of Little Sinking Creek to have once flowed west to the Barren River and to have been pirated on the surface, north to the Green River, during Holocene or Pleistocene time. The diversion of the surface stream can be explained in terms of the steeper hydraulic gradient to the north. The headward capture of subsurface drainage by Turnhole Spring (Quinlan and Ewers, 1981) would have provided a steeper, shorter, more efficient route for surface water to flow to base level.

Resume Driving. Continue along the pave road; cross bridge over freeway. Turn right onto paved road 1.1 mile (1.8 km) beyond and bear left at the intersection 250 feet (75 m) from this turn. Go 0.2 miles (0.3 km) and stop at the junction with the dirt road that crosses the stream 100 feet (30 m) on right.

Stop 3 - Swallet of Little Sinking Creek

Do Not Cross Fence. During low flow all water is lost into the base of the nearby bluff or, during very low flow, upstream

from the bridge. During moderate and flood flow there is more water than this swallet can accept, and the stream sinks 1,000 feet (300 m) downstream. Indeed, after heavy rains, a temporary lake 15 feet (5 m) deep and 1,000 feet (300 m) wide is formed.

Resume Driving. Continue 0.2 mile (0.3 km), rejoin Kentucky 259, and turn right. Go 0.7 mile (2.1 km) to junction of Kentucky 1339 at Rocky Hill. Turn right and proceed 2.3 miles (3.7 km). Turn left onto the paved road and stop at the corner.

Stop 4 - Outcrop of Lower St. Louis Limestone

The Lower St. Louis Limestone at this locality consists of shaley and silty beds. The relative impermeability of these beds keeps streams on the surface in the Glasgow Upland. Where these beds are deeper in the subsurface there are no surface streams; sinkholes dominate the landscape.

Resume Driving. Proceed 2.4 miles (3.9 km) to the T-junction with U.S. 31W, turn left (west), and go 1.7 mile (2.7 km). Turn right onto the paved road and travel north, parallel to the course of ancestral Gardner Creek which continues northwestward at Cedar Spring Valley. Go 1.2 miles (1.9 km) north and park at the junction with the gravel road on the left, at the base of the hill.

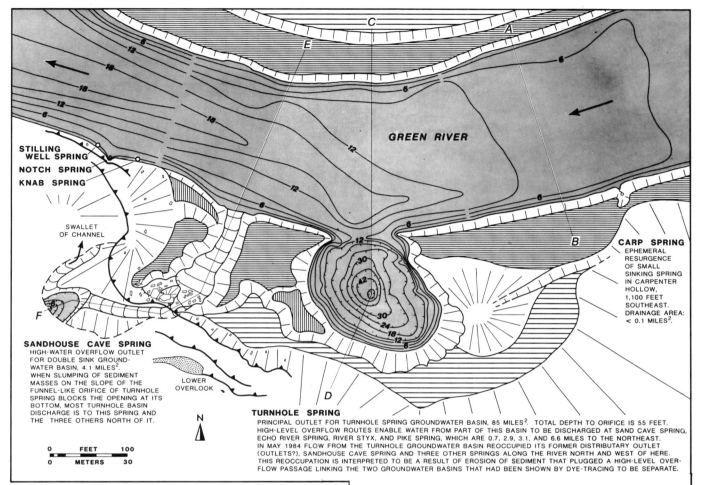

STILLING WELL SPRING

NOTCH SPRING

KNAB SPRING

SWALLET OF CHANNEL

GREEN RIVER

CARP SPRING
EPHEMERAL
RESURGENCE
OF SMALL
SINKING SPRING
IN CARPENTER
HOLLOW,
1,100 FEET
SOUTHEAST.
DRAINAGE AREA:
< 0.1 MILES².

SANDHOUSE CAVE SPRING
HIGH-WATER OVERFLOW OUTLET
FOR DOUBLE SINK GROUND-
WATER BASIN, 4.1 MILES².
WHEN SLUMPING OF SEDIMENT
MASSES ON THE SLOPE OF THE
FUNNEL-LIKE ORIFICE OF TURNHOLE
SPRING BLOCKS THE OPENING AT ITS
BOTTOM, MOST TURNHOLE BASIN
DISCHARGE IS TO THIS SPRING AND
THE THREE OTHERS NORTH OF IT.

LOWER
OVERLOOK

N

TURNHOLE SPRING
PRINCIPAL OUTLET FOR TURNHOLE SPRING GROUNDWATER BASIN, 85 MILES². TOTAL DEPTH TO ORIFICE IS 55 FEET.
HIGH-LEVEL OVERFLOW ROUTES ENABLE WATER FROM PART OF THIS BASIN TO BE DISCHARGED AT SAND CAVE SPRING,
ECHO RIVER SPRING, RIVER STYX, AND PIKE SPRING, WHICH ARE 0.7, 2.9, 3.1, AND 6.6 MILES TO THE NORTHEAST.
IN MAY 1984 FLOW FROM THE TURNHOLE GROUNDWATER BASIN REOCCUPIED ITS FORMER DISTRIBUTARY OUTLET
(OUTLETS?), SANDHOUSE CAVE SPRING AND THREE OTHER SPRINGS ALONG THE RIVER NORTH AND WEST OF HERE.
THIS REOCCUPATION IS INTERPRETED TO BE A RESULT OF EROSION OF SEDIMENT THAT PLUGGED A HIGH-LEVEL OVER-
FLOW PASSAGE LINKING THE TWO GROUNDWATER BASINS THAT HAD BEEN SHOWN BY DYE-TRACING TO BE SEPARATE.

| 0 | FEET | 100 |
| 0 | METERS | 30 |

Figure 6. Sketch map of landforms and bathymetry of the Turnhole Spring area. Cross sections are in Quinlan and others (1983, Fig. 19).

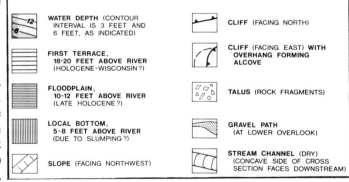

WATER DEPTH (CONTOUR INTERVAL IS 3 FEET AND 6 FEET, AS INDICATED)

FIRST TERRACE, 18-20 FEET ABOVE RIVER (HOLOCENE-WISCONSIN?)

FLOODPLAIN, 10-12 FEET ABOVE RIVER (LATE HOLOCENE?)

LOCAL BOTTOM, 5-8 FEET ABOVE RIVER (DUE TO SLUMPING?)

SLOPE (FACING NORTHWEST)

CLIFF (FACING NORTH)

CLIFF (FACING EAST) WITH OVERHANG FORMING ALCOVE

TALUS (ROCK FRAGMENTS)

GRAVEL PATH (AT LOWER OVERLOOK)

STREAM CHANNEL (DRY) (CONCAVE SIDE OF CROSS SECTION FACES DOWNSTREAM)

Stop 5 - Mill Hole

This sinkhole, on the floor of Cedar Spring Valley, is a spectacular collapse sink with a perennial stream at its bottom. Walk 250 feet (75 m) west, along the gravel road, to the far side of the barn. (Note the Visitor Register 100 feet (30 m) ahead, on the right. The residents, Mr. and Mrs. Don Greene, request that you sign it as you leave.) Turn left, proceed 1,000 feet (300 m) along the dirt road, and turn right. Walk 600 feet (180 m) west, across the field, to the edge of the big sinkhole. Little Sinking Creek (Stop 3) and 16 other sinking streams have been traced to this sinkhole. Notice the north-trending trough in the potentiometric surface near Mill Hole in Figure 4. A trough on a potentiometric surface indicates an axis and direction of maximum flow. Travel times for the same 5-mile (8-km) distance to Mill Hole range from less than 24 hours to as much as 18 days (about 30 to 1,500 feet/hour; 10-500 m/hr).

Water level fluctuations of as much as 100 feet (30 m) have occurred here. Instrumentation for monitoring water temperature, conductivity, velocity, stage, and soil moisture, is installed upstream on the surface and in caves, here, and downstream at Stops 6 and 7. This instrumentation system, when used with data from a rain gauge network, is expected to give new insights into the movement of groundwater and how the aquifer responds to storms.

Mill Hole is named for the floating mill that was operated here during the 19th century. The mill would rise and fall in response to changes in water level of the stream in the sinkhole.

Resume Driving. Return to the highway and proceed up the hill. You will be traveling on a ridge capped by Big Clifty Sandstone. Descend from this ridge into the rolling, streamless floor of Cedar Spring Valley at 1.8 mile (2.9 km). The elevation of the saddles on the valley floor averages 580-620 feet (177-189 m) about 160-200 feet (50-60 m) above the Green River. Continue 1.5 miles (2.4 km) to the T-junction at Cedar Spring. Turn right onto Kentucky 259 and go 1.2 miles (1.9 km) to Pig, a good lunch stop. The axis of a west-trending syncline is near here. Obviously, one of these two is the Pig Trough. Turn right onto Kentucky 422, continue to Mammoth Cave National Park, and note the outcrop of Big Clifty Sandstone, shale, and limestone on the left. The sandstone and shale protect and preserve the caves below. The impermeability of these rocks prevents groundwater from freely percolating into the subsurface and destroying the caves. Most of the passages above base level are remarkably dry, except at the margins of the ridges, where water readily penetrates via the vertical shafts such as those seen in Mammoth Cave.

The sandstone, shale, and uppermost limestone include pyrite that has been shown by Pohl and White (1965) to be the source of sulfur for the gypsum deposits that are so widespread in many local caves. Continue 0.3 mile (0.5 km) and stop at the Cedar Sink parking lot, on the right. Follow the trail to the sink.

Stop 6 - Cedar Sink

This is a larger collapse sinkhole than that seen at the previous stop. It too is located on the floor of Cedar Spring Valley. Take the trail leading to the overlook at the bottom of the sink. Collapse sinkholes form after an initial subsurface collapse occurs, commonly at the intersection of two passages where the roof span is widest and therefore weakest. The collapsed rock may divert a cave stream around one or both sides of the debris pile. Clay and other sediment accumulate on this pile of collapsed rock, thus shielding it from erosion. The stream repeatedly undercuts the cave walls, induces collapse of the cantilever beams thus formed, and widens the passage. Stoping continues and eventually the void reaches the surface.

The hydrology of Cedar Sink is shown in Figure 5. Water in an ephemeral stream can be seen from the overlook and some-

times also at six other places within this sink. As shown in Figure 4, the water in Cedar Sink comes chiefly from two sub-basins with approximately the same area: Mill Hole (which we have seen) and Procter Cave (which includes Park City and the I-65 interchange at Cave City).

Resume Driving. Return to the parking lot and continue north 0.6 miles (1.0 km) to a T-junction. Turn left (west) onto Kentucky 70 and go 0.4 mile (0.6 km) to the parking lot on the right. The road here overlooks Double Sink Valley.

Stop 7 - Turnhole Spring

From the parking lot, hike along the trail to the left and descend to Green River by continuing to the right at the lower of the two overlooks. There are six springs here (Fig. 6), but only two of them, Notch and Stilling Well, are perennial. The largest and most obvious is Turnhole Spring, so-named because steamboats that went to Mammoth Cave before construction of a dam downstream in 1906 had to travel backwards for 6 miles (9.7 km) from the cave in order to turn around here. Prior to the major storm of May 1984, all of the water from Cedar Sink flowed only to Turnhole Spring. Downstream, to the left, is Sandhouse Cave Spring, a high-level overflow discharge point for the two perennial springs that are about 160 feet (50 m) downstream from the large channel leading to the river. Knab Spring flows only during floods. None of the water from Double Sink Valley (Fig. 4) flows to Turnhole Spring; the valley drains only to the four springs west of Turnhole. The 1984 storm reopened a sediment-blocked passage that formerly linked Turnhole Spring and Sandhouse Cave Spring. Also, Turnhole Spring was partly blocked by slumped masses of clay, silt, and sand from its walls. Since 1984, most of the flow from the Turnhole Spring groundwater basin is discharged at Notch Spring. But sometimes the now inactive Turnhole Spring is flatulent; it repeatedly blows its sediment plug and temporarily discharges surges of water.

To get to Mammoth Cave go 3.9 miles (6.3 km) east to the T-junction with a large sign indicating that the cave is to the left; go 2.7 miles (4.3 km).

REFERENCES CITED

Palmer, A. N., 1981, A geological guide to Mammoth Cave National Park. Teaneck, New Jersey, Zephyrus Press, 210 p.

Pohl, E. R., and White, W. B., 1965, Sulfate minerals: Their origin in the Central Kentucky Karst: American Mineralogist, v. 50, p. 1462–1465.

Quinlan, J. F., and Ewers, R. O., 1981, Preliminary speculations on the evolution of groundwater basins in the Mammoth Cave Region, Kentucky: GSA Cincinnati '81 Field Trip Guidebooks. Washington, American Geological Institute, v. 3, p. 496–501.

Quinlan, J. F., and Ewers, R. O., 1985, Ground water flow in limestone terranes:

Strategy rationale and procedure for reliable, efficient monitoring of ground water quality in karst areas: National Symposium and Exposition on Aquifer Restoration and Ground Water Monitoring (5th, Columbus), Proceedings, p. 197–234.

Quinlan, J. F., Ewers, R. O., Ray, J. A., Powell, R. L., and Krothe, N. C., 1983, Ground-water hydrology and geomorphology of the Mammoth Cave Region, Kentucky, and of the Mitchell Plain, Indiana. Field trips in midwestern geology: Indiana Geological Survey, v. 2, p. 1–85.

The Upper Ordovician Fredericktown Section, Nelson County, Kentucky

Martin C. Noger, Kentucky Geological Survey, 311 Breckinridge Hall, University of Kentucky, Lexington, Kentucky 40506

LOCATION

The Fredericktown Section is in Nelson County, Kentucky, in the southwestern quarter of the Maud 7½-minute quadrangle (Peterson, 1972) in a roadcut along the southwestern side of U.S. 150, 4.5 miles (7.2 km) east of the Bluegrass Parkway (Fig. 1).

INTRODUCTION

The Fredericktown Section contains the most complete sequence of Upper Ordovician lithologic units exposed in south-central Kentucky and includes the type section of the Bardstown Member of the Drakes Formation. The dolomitic limestone and mudstone and micrograined limestone lithofacies of the Ashlock and Drake Formations pinch out or intergrade northward to the even-bedded fossiliferous shale and limestone facies of the Bull-fork Formation of north-central Kentucky. Fossil studies indicate that rocks in the Fredericktown Section are of Mid-Cincinnatian (or Maysvillian) and late Cincinnatian (or Richmondian) age.

The Fredericktown Section is an excellent locality for use in academic studies because of the availability of published reports with stratigraphic sections showing lithologies and thicknesses of lithologic units, identification and range of fossils, and depositional environments of lithofacies.

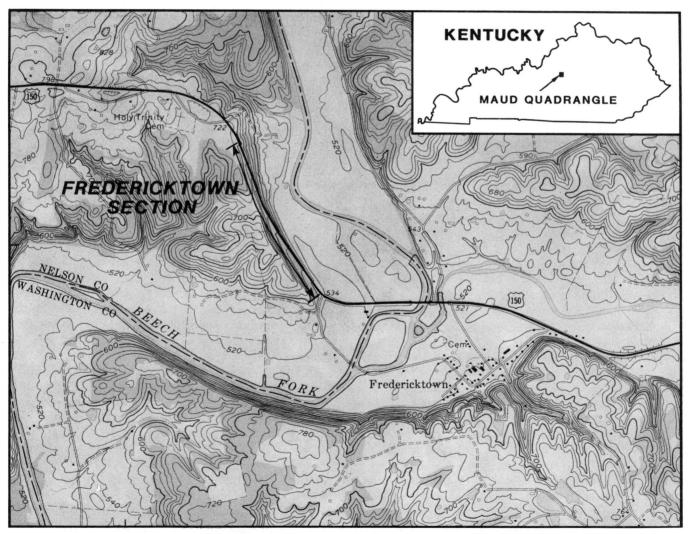

Figure 1. Location map, Fredricktown Section, Nelson County, Kentucky.

SITE DESCRIPTION

The Fredericktown Section exposes 195 feet (59.4 km) of Upper Ordovician strata consisting of the Grant Lake Limestone; the Tate, Gilbert, Terrill, and Reba Members of the Ashlock Formation; and the Rowland, Bardstown, and Saluda Members of the Drakes Formation (Noger and Kepferle, 1985).

The Grant Lake Limestone was defined by Peck (1966, p. B14–B16, B23–B24) from roadcut exposures along Kentucky 1449 in the Maysville East and Orangeburg 7½-minute quadrangles in northeastern Kentucky. It is 36.7 ft (11.2 m) thick in the Fredericktown Section and composed of limestone (75 percent) interbedded and intermixed with shale. The limestone consists of abundant fossils and fossil fragments in a medium- to coarse-grained calcite matrix. Fossils are dominately brachiopods (including strophomenids and *Platystrophia ponderosa*) and branching bryozoans. A bed of crossbedded calcarenite 3.6 ft (1.1 m) thick is present about 2 feet (0.6 m) below the contact with the overlying Terrill Member of the Ashlock Formation (Fig. 2). Shale is greenish-gray, argillaceous, and calcareous, with abundant whole fossils or fossil fragments. The Grant Lake is characterized by thin, nodular beds of limestone lentils separated by shale laminae.

The Tate was named by Foerste (1906, p. 212) for exposures near Tate Creek in the Richmond North 7½-minute quadrangle in Madison County, Kentucky, but no type section was designated. A representative section of the Tate Member of the Ashlock Formation was described in roadcuts along U.S. 27 north of the Dix River in the Lancaster 7½-minute quadrangle in Lincoln County, east-central Kentucky, by Weir and others (1965, p. D25–D27). The Tate Member is 50.6 ft (15.4 m) thick and composed of greenish-gray, fine-grained, argillaceous, silty, dolomitic limestone and dolomitic mudstone; mudcracks and low-amplitude ripples occur sparsely on bedding planes. Megafossils are sparse.

The Gilbert was named by Foerste (1912, p. 18, 23), probably for outcrops near the town of Gilbert in the Lancaster 7½-minute quadrangle in Lincoln County, Kentucky. Weir and others (1965) described a type section from roadcuts along U.S. 27 north of the Dix River in the Lancaster 7½-minute quadrangle, Lincoln County, east-central Kentucky. The Gilbert Member is 13.1 ft (4 m) thick and is composed of limestone and shale. The limestone is dominantly micrograined with interbeds of fine- to medium-grained limestone and shale. Stromatoporids are common near the top of the unit. The basal bed (2.1 ft [0.6 km] thick) is composed of abundant fossil fragments in a fine- to medium-grained calcite matrix, in low-angle crossbeds; brachiopods and bryozoans dominate. The shale is gray, calcareous, and fossiliferous, with mostly brachiopods and gastropods.

The Terrill Member of the Ashlock Formation was named by Weir and others (1965, p. D13, D29). The type section was described from roadcuts along Kentucky 52 in the Moberly 7½-minute quadrangle, Madison County, east-central Kentucky. It is 6.2 ft (1.9 m) thick and is composed of greenish-gray, dolo-

mitic, calcitic shale with interbeds of argillaceous limestone. Fossils are sparse, except for bryozoans in the upper 1 ft (30 cm). The Terrill and overlying Reba Members of the Ashlock Formation were not mapped separately on the Maud 7½-minute quadrangle but were tentatively identified to be present in the Fredericktown Section in later studies by Weir and others (1984).

The Reba Member of the Ashlock Formation was named by Weir and others (1965, p. D13, D28–D29). The type section was described from roadcuts along Kentucky 52 in the Moberly 7½-minute quadrangle, Madison County, east-central Kentucky. The Reba is 5.6 ft (1.7 m) thick and consists mainly of fine-grained to micrograined, nodular-bedded limestone with lenses of argillaceous limestone. Fossils include abundant brachiopods and bryozoans.

The Rowland Member of the Drakes Formation was named by Weir and others (1965, p. D17, D32–D33). The type section was described from roadcuts along U.S. 27 in the Standford 7½-minute quadrangle, Lincoln County, east-central Kentucky. The Rowland is 45 ft (13.7 m) thick and consists of limestone and minor shale interbeds. The limestone is dominately greenish-gray, fine-grained, dolomitic, silty, argillaceous, suncracked, ripple-marked, and sparsely fossiliferous; it contains occasional interbeds of fine- to medium-grained limestone with abundant gastropods, brachiopods, bryozoans, and ostracods.

The Bardstown Member of the Drakes Formation was named by Peterson (1970) and the Fredericktown Section contains the type section. The Bardstown is 29.3 ft (8.9 m) thick and consists of greenish-gray, fine- to medium-grained, silty, argillaceous limestone and medium- to very coarse-grained limestone with abundant fossils. Minor interbeds are composed of greenish-gray, calcareous shale. Fossils include brachiopods, bryozoans, horn corals, and colonial corals. Colonial coral heads occur mostly in two layers near the middle of the unit.

The name "Saluda" was given by Foerste (1902, p. 369) to a unit of dolomite containing minor limestone and shale exposed at Madison in southeastern Indiana. In Kentucky, the Saluda is chiefly a fine-grained dolomite and is a member of the Drakes Formation. The Saluda is 20.6 ft (6.3 m) thick and consists of dolomite and minor shale. The dolomite is greenish-gray, fine-grained, calcareous, poorly fossiliferous, and silty. Shale is greenish-gray. Abundant colonial coral heads occur in the upper 5 ft (1.6 m) of the greenish-gray dolomite.

The position, identification, and range of fossils collected during the Kentucky Geological Survey-U.S. Geological Survey Geologic Mapping Program are shown on Figure 2. Conodont interpretations by Kohut and Sweet (1968) indicate that rocks below the base of the Drakes Formation are of mid-Cincinnatian (or Maysvillian) age. The Drakes is of late Cincinnatian (or Richmondian) age (Ethington and Sweet, 1977).

The Upper Ordovician strata of Kentucky are composed chiefly of limestone, shale, and dolomite. The strata may be divided into lithofacies that have distinctive features by which they may be separated from adjacent units. Lithofacies units generally correspond to formal stratigraphic units used during the Ken-

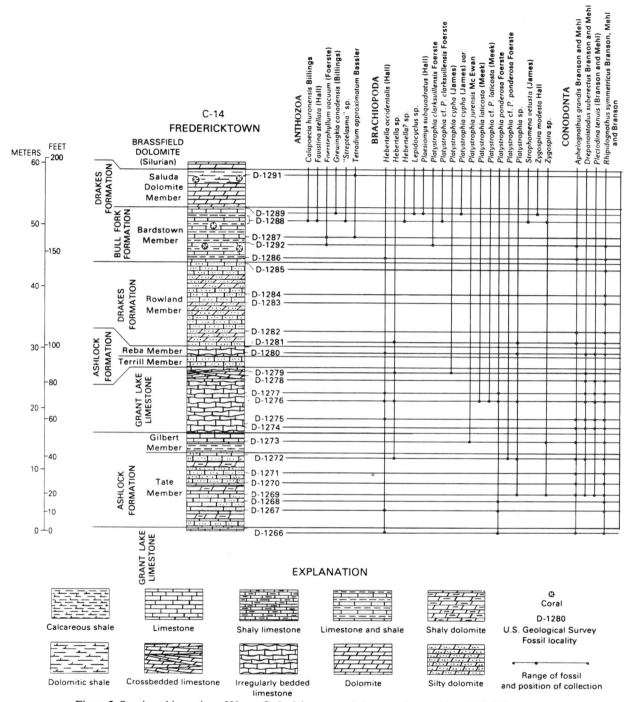

Figure 2. Stratigraphic section of Upper Ordovician exposed along southwest side of U.S. 150, north of Fredericktown, Nelson County, Kentucky. From Weir and others (1984).

tucky Geological Survey-U.S. Geological Survey Geologic Mapping Program (Weir and others, 1984, p. E83, E90).

The Fredericktown Section can be grouped into the following general lithofacies (Fig. 3): (1) nodular-bedded limestone and shale (Grant Lake Limestone); (2) even-bedded fossiliferous shale and limestone (Bardstown Member of Drakes Formation); (3)

micrograined limestone (Reba and Gilbert Members of the Ashlock Formation); (4) dolomite and dolomitic mudstone (Saluda Dolomite Member of the Drakes Formation); and (5) dolomitic limestone and mudstone (Tate and Terrill Members of the Ashlock Formation and Rowland Member of the Drakes Formation). Since the Reba (5.6 ft or 5.1 m thick) is a dominantly

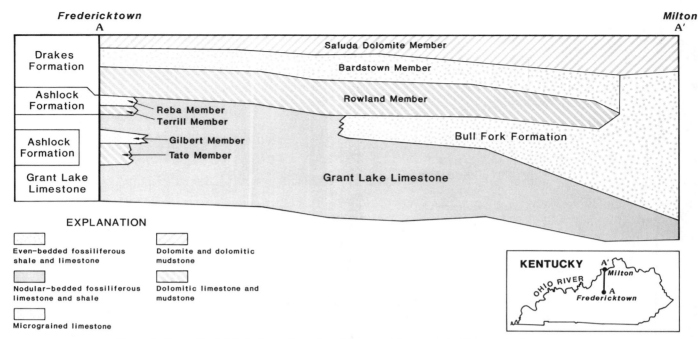

Figure 3. Generalized lithofacies and stratigraphic nomenclature of the Grant Lake Limestone and younger Upper Ordovician formations between Fredericktown and Milton, Kentucky. (Adapted from Section C-C', Plate 7, Weir and others, 1984.)

micrograined to fine-grained limestone, the unit has been placed in the micrograined limestone lithofacies instead of the nodular-bedded fossiliferous limestone and shale lithofacies of other areas.

Lithofacies 5 within the Ashlock Formation interfingers to the north with lithofacies 2 in the Maud Quadrangle (Fig. 3). Lithofacies 5 within the Drakes Formation thins and pinches out near the Ohio River in north-central Kentucky. Lithofacies 3 in the Ashlock Formation north of the Fredericktown Section becomes coarser-grained and is not distinguishable from lithofacies 2. Lithofacies 1 in the Drakes Formation in the Fredericktown Section continues northward to the pinchout of the underlying

lithofacies 5. At this point it becomes unmappable and merges with the Bullfork Formation. Lithofacies 5 is continuous from the Fredericktown Section northward to north-central Kentucky.

From their study of numerous sections of Upper Ordovician rocks exposed in central Kentucky, Weir and others (1984, p. E103) concluded that these lithofacies were deposited on a shelf that sloped gently northward. The northern part of the shelf was mostly covered by fossiliferous limestone and shale, and the southern part by dolomite and dolomitic mudstone. Near the close of the Late Ordovician, the seas withdrew northward and the entire area was covered by dolomite and dolomitic mudstone.

REFERENCES CITED

Ethington, R. L., and Sweet, W. C., 1977, Guide to field excursion 2, Ordovician of Eastern Midcontinent, August 10–13, 1977: Third International Symposium on the Ordovician System, 48 p.

Foerste, A. F., 1902, The Cincinnati anticline in southern Kentucky: American Geologist, v. 30, no. 6, p. 359–369.

—— 1906, The Silurian, Devonian and Irvine formations of east-central Kentucky with an account of their clays and limestones: Kentucky Geological Survey, ser. 3, Bulletin 7, 369 p.

—— 1912, Strophomena and other fossils from Cincinnatian and Mohawkian horizons, chiefly in Ohio, Indiana, and Kentucky: Granville, Ohio, Denison University Scientific Laboratories Bulletin 17, p. 17–173.

Kohut, J. S., and Sweet, W. C., 1968, Upper Maysville and Richmond conodonts from the Cincinnati region of Ohio, Indiana, and Kentucky, [Pt.] 10 of the American Upper Ordovician standard: Journal of Paleontology, v. 42, no. 6, p. 1456–1477.

Noger, M. C., and Kepferle, R. C., 1985, Stratigraphy along and adjacent to the Bluegrass Parkway (Guidebook and Roadlog for Geological Society of Ken-

tucky 1985 Field Excursion): Kentucky Geological Survey, ser. 11, 24 p.

Peck, J. H., 1966, Upper Ordovician formations in the Maysville area, Kentucky: U.S. Geological Survey Bulletin 1244-B, p. B1–B30.

Peterson, W. L., 1970, Bardstown member of the Drakes Formation in central Kentucky, in Cohee, G. V., Bates, R. G., and Wright, W. B., Changes in stratigraphic nomenclature by the U.S. Geological Survey: U.S. Geological Survey Bulletin 1294-A, p. A36–A41.

—— 1972, Geologic map of the Maud Quadrangle, Nelson and Washington Counties, Kentucky: U.S. Geological Survey Geologic Quadrangle Map GQ-1043.

Weir, G. W., Greens, R. C., and Simmons, G. C., 1965, Calloway Creek Limestone and Ashlock and Drakes Formations (Upper Ordovician) in south-central Kentucky: U.S. Geological Survey Bulletin 1224-D, p. D1–D36.

Weir, G. W., Peterson, W. L., and Swadley, W C, 1984, Lithostratigraphy of Upper Ordovician strata exposed in Kentucky: U.S. Geological Survey Professional Paper 1151-E, 121 p.

4

The Jeptha Knob cryptoexplosion structure, Kentucky

C. Ronald Seeger, Department of Geography and Geology, Western Kentucky University, Bowling Green, Kentucky 42101

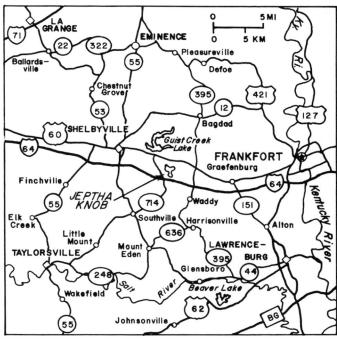

Figure 1. Location of Jeptha Knob.

LOCATION

The Jeptha Knob Structure is on the Shelbyville and Waddy, 7½-minute quadrangles in Shelby County, Kentucky, 5 mi (about 8 km) east of Shelbyville (Figs. 1, 2). The center is at about 38°10′N latitude and 85°07′W longitude. It is a topographic high, extending up to small flat-topped knobs at about 1180 ft (360 m). It can be reached by automobile on routes I-64 or US 60 east from Louisville or west from Lexington. Note that Stop 3 is on a private farm, and permission is required from Mr. and Mrs. Robert Totten, Route 3, Shelbyville, KY, 40065.

INTRODUCTION

The Jeptha Knob Structure lies on the west flank of the Cincinnati Arch; the rocks of the region dip westward, at about 22 ft per mi (4 m per km). The Cincinnati Arch presumably has stood relatively higher than its surroundings since at least Ordovician times, for Ordovician and younger rocks thin over it. Topographically, Jeptha Knob can be considered a monadnock on the surface of the slightly rejuvenated Lexington peneplain of Tertiary times. The rougher, higher, central portion of the structure is wooded. Farmlands surround it and extend into it from the west in the valley of Britton Run, and a few fields occupy the top, flat surfaces of the knobs. The general elevations in the region immediately around the structure vary from about 750 ft (229 m) in the valleys and 850 ft (259 m) on the divides in the southwest to 850 ft (259 m) in the valleys and 920 ft (280 m) on the divides in the east.

However, the main significance of Jeptha Knob is that it is one of a class of geologic structures called "cryptovolcanic" by Bucher (1933), which are now called "cryptoexplosion" and most of which are now generally accepted to represent ancient meteorite or comet impact events. In addition to the reasons presented earlier (Seeger, 1968), which include geologic and geophysical studies as described briefly below, the recent finding of anomalously high iridium in breccias (Seeger and others, 1985) suggests that Jeptha Knob is an impact structure. The rocks of the Jeptha Knob region of Ordovician age are intensively deformed under and surrounding an undeformed cap rock of Silurian age. Breccias, faults, and folds are present, outward from the center, respectively for a total radius of about 8000 ft (2,400 m). Many blocks are substantially uplifted.

The first geologic report on Jeptha Knob was by Linney (1887). He suggested that sinking of the central core of the Knob below the general level of the surrounding countryside protected its Silurian capstone, while beds of the same and greater age were eroded away, thus leaving the downfaulted block finally as an upland mass above the general level of the plain around it. This theory had to be abandoned because of its "maladjustment of certain important structural details of the geology of the area, discovered late in the second decade of the recent century, chiefly the absence of adequate subsidence criteria" (Jillson, 1962). Linney did not describe the remarkable ring of structural features that circle Jeptha Knob.

Bucher (1925) did the next study and mapping of Jeptha Knob. He found that the only reliable stratigraphic section he could employ was a biostratigraphic one. Cressman (1981) compiled this on a chart, along with the later stratigraphic names, which he used in mapping the feature and surrounding region (Cressman, 1975a, 1975b, 1981). Figure 3 is the currently used stratigraphic section of the U.S. Geological Survey (Cressman, 1981). Cressman (1981) also discussed the surface geology.

Seeger (1968) studied Jeptha Knob in the 1960s and concluded that the Jeptha Knob Structure was the result of a violently disruptive process that occurred in its entirety during a very brief period of the early Silurian. The formation of the Jeptha Knob structure was extremely rapid, and all the deformation seems to have been contemporaneous.

To determine the shape of this disturbance in three dimensions, Seeger used geophysical methods, a magnetic and a gravity survey, to supplement the rather meager subsurface information. Such methods would be very likely to reveal a buried structure caused by a volcanic or tectonic event. The magnetic survey showed that there is not likely to be a basement counterpart to the Jeptha Knob Structure, and the gravity work indicated that the

C. R. Seeger

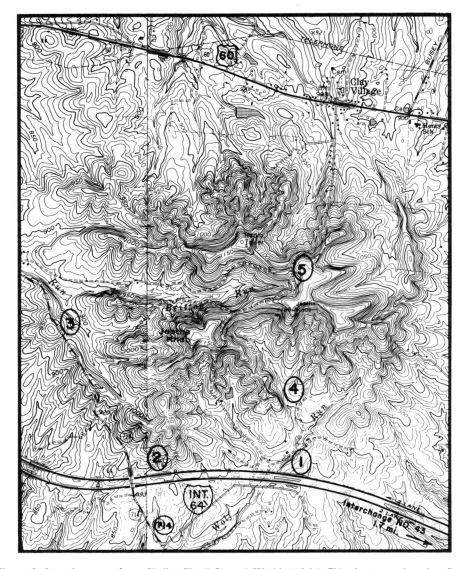

Figure 2. Location map from Shelbyville (left) and Waddy (right) 7½-minute quadrangles. Stops 1–5 indicated.

gravity features observed can be totally explained by the densities of the near surface sedimentary rocks. Furthermore, the drilling information available seemed to show that the disturbance decreases with depth and at rather shallow depth.

The above findings, along with complete absence of any other indication of endogenetic activity, volcanism, or tectonism of any type, forced Seeger to conclude an exogenetic origin for Jeptha Knob. Hypervelocity impact of a meteorite or comet is the most likely exogenetic mechanism, and he therefore modeled the original structure of Jeptha Knob using the parameters determined for such craters (Baldwin, 1963). The model was based on the best known parameter presently available, the extent of the capping undisturbed Silurian rocks. The original crater was found to have the following probable minimum dimensions:

	Feet	*Meters*
Apparent diameter	6600	2012
Apparent depth	920	280
Rim height	290	88
Depth to the limit of major brecciation	1600	488

In summary then, Jeptha Knob has the following features suggestive of an impact origin: (1) approximate circularity, (2) intense brecciation of the country rock, (3) folding beyond the central zone, decreasing rapidly outward (damped), (4) short period of time of formation and contemporaneousness of all structural features, (5) geophysical and drilling evidence that the structure is confined to near surface rocks, (6) distinctive sedi-

SYSTEM	SERIES	GROUP, FORMATION, MEMBER Heavy line to left of column marks units that crop out in structure		THICKNESS, IN FEET	DESCRIPTION
SILURIAN	Middle	Louisville Limestone Waldron Shale Laurel Dolomite Osgood Formation		75	Concealed by soil and chert residuum. Presence inferred from fossils identified in residuum (Foerste, 1931, p. 182) and from thickness of interval
	Lower	Brassfield Formation		18	Finely crystalline calcareous dolomite; contains abundant small vugs; angular fragments of very finely crystalline dolomite present in some beds; basal 3 to 6 ft. in several localities is calcarenite and calcirudite consisting largely of fragments reworked from Upper Ordovician formations
		———————UNCONFORMITY———————			
ORDOVICIAN	Upper	Drakes Formation	Bardstown Member	25–50	Nodular-bedded fossiliferous limestone and shale
			Rowland Member	50	Argillaceous, dolomitic limestone
		Grant Lake Limestone		140	Nodular-bedded fossiliferous limestone and shale
		Calloway Creek Limestone		60	Fossiliferous limestone and minor interbedded shale; 6–8 ft. thick calcarenite at top
	?—	Clays Ferry Formation		300	Interbedded limestone and shale
	Middle	Lexington Limestone'	Sulphur Well Member Perryville Limestone Member Tanglewood Limestone Member Grier Limestone Member Logana Member Curdsville Limestone Member	200	Fossiliferous limestone Calcilutite Calcarenite Fossiliferous limestone Brachiopod coquina, calcisiltite, and shale; 24–56 ft. above base of formation. Calcarenite

Figure 3. Stratigraphic section at Jeptha Knob. Thickness and presence of members based on regional thickness and facies trends (Cressman, 1981).

mentary breccias, (7) general uplift of the folded and crushed zones, decreasing outward, and (8) chaos breccias in the central zone containing units from several horizons including surface beds. No criterion for the recognition of impact structures that could be expected to be present was found to be absent, except that shattercones, which are often (but not always) present, have not been found.

In view of all of the above, Seeger (1968) concluded that the Jeptha Knob Structure is indeed the result of hypervelocity impact of a meteorite or comet. This conclusion seems to be confirmed by the finding of anomalously high iridium (0.108 ± 0.014 ppb, Seeger and others, 1985) in the highest breccias found by Seeger, compared to much lower concentrations in other breccias from Jeptha Knob and the Versailles cryptoexplosion structure (Seeger, 1972) and other rocks, as reported in the literature (Silver and Schultz, 1982).

The model dimensions above assume a simple, cup-shaped crater was formed. It is possible, though, that the structure discussed above is the central peak area of a much larger, complex crater. This would be supported by the finding of breccias in the lower Brassfield Formation 20 mi (32 km) away (Cressman, 1981) or farther (Gauri and others, 1969). This possibility is under investigation.

SITE DESCRIPTION

To visit Jeptha Knob the following approach and stops are

suggested (note location and routes on Figs. 1 and 2). Approaching from the east on I-64, watch for the Waddy–Kentucky 395 interchange. (If you should arrive from the west, you may want to start the field trip in Shelbyville as described below and conclude by visiting Stops 1 and 2.) After you pass the Waddy–Kentucky 395 interchange, the next road observed (passing under I-64) is Buzzard Roost Road. Slow down and look for road cuts between it and the next road, an overpass of Kentucky 714. Here the land rises, and Stops 1 and 2 (see Fig. 2 and sketches in Figs. 4 and 5) are located in the road cuts on the north side of the north lane. As noted by Cressman (1981), better exposures in the road cuts always display more complicated structures than can be observed in nearby fields of the Jeptha Knob area.

At Stop 1 (Fig. 4), deformed rocks of the Rowland Member of Drakes Formation and Grant Lake Limestone (Upper Ordovician) are present. Here faulting and brecciation can be found. The south side of the north lane also shows similar features, although not as well exposed.

After leaving Stop 1 continue over Wolf Run and a road that underpasses I-64 to the next major rise of the land, just before the Kentucky 714 overpass. This is Stop 2 (Fig. 5). Here folded, faulted and brecciated Grant Lake Limestone rocks are present in the north side of the road.

After leaving Stops 1 and 2, proceed under Kentucky 714 to the next interchange (Shelbyville), go north on Kentucky 53 to US 60, and then east (right) on US 60 to Kentucky 714 (see Fig. 1). Turn south (right) on Kentucky 714. As the road rises, good

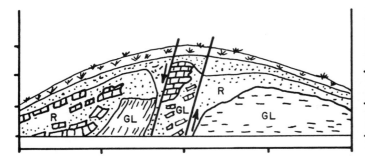

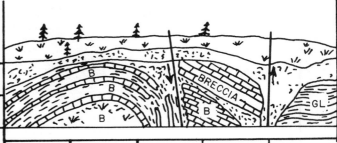

Figure 4. Sketch of deformed rocks of Stop 1 (modified from Jillson, 1962). Tick marks are 10 ft (3 m) vertical, 100 ft (30 m) horizontal, center about 600 ft (180 m) east of Wolf Run on north side of north lane I-64. View in direction N30E. R–Rowland Member, Drakes Formation; GL–Grant Lake Limestone (mostly shale in exposure). Rocks are more covered by soil and vegetation than implied in sketch. Faults shown are major breaks in a zone of deformation.

Figure 5. Sketch of deformed rocks of Stop 2 (modified from Jillson, 1962). Tick marks are 10 ft (3 m) vertical, 100 ft (30 m) horizontal. View covers from about 550 to 1050 ft (168 to 320 m) east of Kentucky 714 overpass on north side of north lane of I-64. View in direction N60W. B–Bardstown Member, Drakes Formation; GL–Grant Lake Limestone. Rocks are more covered by soil and vegetation than implied in the sketch. A stratigraphic separation of over 50 ft (15 m) is implied in the faulting shown here, but, as an indication of the complexity, Cressman (1975b, 1981) mapped Calloway Creek Limestone south of these faults, and Clays Ferry Formation is also involved.

views of the center of Jeptha Knob are to be seen to the left (east) of the road, especially just past the entrance to "the Knobs" farm (Stop 3, Fig. 2). Note also the dipping beds along the road.

Then proceed over I-64 on Kentucky 714 and take the first road to the left. This soon intersects with a farm road that goes back under I-64 and toward the center of Jeptha Knob to the north. Stay to the right, up Wolf Run Valley, and go to the cattle gate of the northernmost farm in this area of Jeptha Knob. This is the farm of Mr. and Mrs. Robert Totten, Route 3, Shelbyville, Kentucky, 40065 (Stop 4, Fig. 2). Ask for permission to visit the property. Look at deformed rocks, faults, folds, and breccias in Wolf Run. Also, the Tottens have collected rocks from all around the property, and excellent examples of the lithologies, including highly fossiliferous beds and breccias, are in the outer walls of the farmhouse and in the barn.

Return via the farm road and Kentucky 714 to US 60 and turn right (east). At Clay Village, on the side road clearly visible on your right, is a road that leads south to the top of Jeptha Knob. The road is rough, but passable, and you can reach the top. You may park and observe the undeformed Silurian caprocks. You may also find breccias below the tops of the knobs in the basal Silurian Brassfield Formation (Cressman, 1981) containing clasts of several typical Ordovician lithologies in a tan- to cream-colored limonitic and dolomitic matrix. These are the rocks that contain the iridium anomaly (Seeger and others, 1985).

REFERENCES CITED

Baldwin, R. B., 1963, The measure of the moon: Chicago, University of Chicago Press, 488 p.

Bucher, W. H., 1925, The geology of Jeptha Knob: Kentucky Geological Survey, ser. 6, v. 21, p. 193–237.

—— 1933, Cryptovolcanic structures in the United States [with discussion]: International Geological Congress, 16th, Washington, D.C., 1933, Report, v. 2, p. 1055–1084.

Cressman, E. R., 1975a, Geologic map of the Shelbyville quadrangle, Shelby County, Kentucky: U.S. Geological Survey Geologic Quadrangle Map GQ-1258.

—— 1975b, Geologic map of the Waddy quadrangle, central Kentucky: U.S. Geological Survey Geologic Quadrangle Map GQ-1255.

—— 1981, Surface geology of the Jeptha Knob cryptoexplosion structure, Shelby County, Kentucky: U.S. Geological Survey Professional Paper 1151-B, 16 p., 1 plate.

Gauri, K. L., Noland, A. V., and Moore, B., 1969, Structurally deformed Late Ordovician to Early Silurian strata in north-central Kentucky and southeastern Indiana: Geological Society of America Bulletin, v. 80, no. 9, p. 1881–1886.

Jillson, W. R., 1931, Structural geologic map of Kentucky: Kentucky Geological Survey, ser. 6.

—— 1962, Geology of a recently discovered faulted area south of Jeptha Knob in Shelby County, Kentucky: Frankfort, Kentucky, Roberts Printing Co., 65 p.

Linney, W. M., 1887, Report on the geology of Shelby County: Frankfort, Kentucky Geological Survey, ser. 2, 16 p.

Seeger, C. R., 1968, Origin of the Jeptha Knob Structure, Kentucky: American Journal of Science, v. 266, no. 8, p. 630–660.

—— 1972, Geophysical investigation of the Versailles, Kentucky, astrobleme: Geological Society of America Bulletin, v. 83, p. 3515–3518.

Seeger, C. R., Asaro, F., Michel, H., Alvarez, W., and Alvarez, L., 1985, Iridium discovery at the Jeptha Knob cryptoexplosion structure, Kentucky: Lunar and Planetary Science 16, p. 757–758.

Silver, L. T., and Schultz, P. H., eds., 1982, Geological implications of impacts of large asteroids and comets on the earth: Geological Society of America, Special Paper 190, 528 p.

Cincinnati region: Ordovician stratigraphy near the southwest corner of Ohio

R. A. Davis, Cincinnati Museum of Natural History, Cincinnati, Ohio 45202

LOCATION

In the Cincinnati region an outstanding and readily accessible exposure of the classic Upper Ordovician section is the Riedlin Road/Mason Road locality, about 5 mi (8 km) south of Cincinnati. The site is a road-cut in Forest Hills, Kenton County, Kentucky (administratively a part of Taylor Mill, Kentucky). Access is provided by I-275 via exit 79 (Fig. 1). From the exit one travels north on Kentucky 16 (Taylor Mill Road) to the first cross-road (Riedlin Road/Mason Road) and turns right (east). From this point the section may be studied by progressing east and south for a distance of some 0.4 mi (0.7 km). The site is located on the Covington, Kentucky-Ohio 7.5-minute quadrangle (Luft, 1971).

INTRODUCTION

The abundantly fossiliferous shales and limestones of the Cincinnati region were deposited during the latter part of the Ordovician Period. Ironically, when the Ordovician was named, in 1879, geologists already had been studying Cincinnati's rocks and fossils for more than four decades. Moreover, the Cincinnati region was well on its way to becoming the type-area for what, in North America, now is called the "Cincinnatian Series." The Riedlin Road/Mason Road locality is a fine exposure of a significant segment of the "type" Cincinnatian.

In the Nineteenth Century the Cincinnatian "layer-cake" of interbedded limestones and shales was divided by geologists into thick, broadly defined lithologic units. Then, during the first half of the Twentieth Century, more detailed studies recognized some eight "formations" and twice that many "members." The quote marks are important. Formations and members today are rock units defined exclusively on lithology. However, in the days before the conventions of stratigraphy had been encoded, it was common practice to use fossil-content as well as lithology to discriminate "formations" and "members." The use of fossils to differentiate rock units in the Cincinnati region was encouraged by the repetitive nature of the sequence of limestones and shales here and by the subtlety of lithologic differences between some adjacent units.

About the time that the definitive compilation of Cincinnatian stratigraphy at mid-century was presented by Caster and others (1955), there began a period of intense study of Cincinnatian rocks that continues to this day. A major stimulus was the joint project of the U.S. Geological Survey and the Kentucky Geological Survey to map the entire Commonwealth of Kentucky geologically (Pojeta, 1979). In any case, there has been an

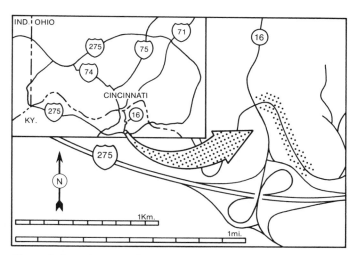

Figure 1. Location map for the Riedlin Road/Mason Road locality (stippled area along road).

avalanche of theses and published papers reporting studies of the Ordovician of Indiana, Kentucky, and Ohio, together and separately. Although the main goal in many of these studies has been to recognize and delineate truly lithologic units, the ultimate objective is an understanding of the paleogeography and paleoenvironments that resulted in those lithologic units and in the nature and distribution of the fossils contained therein. These various attempts to understand the chrono-, bio-, and lithostratigraphy of the region have resulted in a number of differing schemes for subdividing the Cincinnatian of the Ohio-Kentucky-Indiana tristate area. As indicated in Figure 2, unanimity has yet to be reached as to which proposed scheme, if any, best reflects geological reality.

SITE DESCRIPTION

Four lithologic units are exposed at the Riedlin Road/Mason Road locality: the Kope Formation, the Fairview Formation, the Bellevue Limestone, and the Miamitown Shale (Fig. 3). These all are in the lower part of the Cincinnatian Series; in fact, in the "type" Cincinnatian, the Kope-Fairview contact corresponds to the boundary between the Edenian Stage and the Maysvillian Stage. This locality has been studied in detail by Tobin (1982) and has been described and interpreted by Tobin and Pryor (in Meyer and others, 1981).

The Kope Formation exceeds a thickness of 200 ft (60 m). The uppermost 28 ft (8.5 m) are exposed at this locality. The sequence consists of interbedded limestone, siltstone, and shale, with shale comprising almost three-quarters of the total thickness.

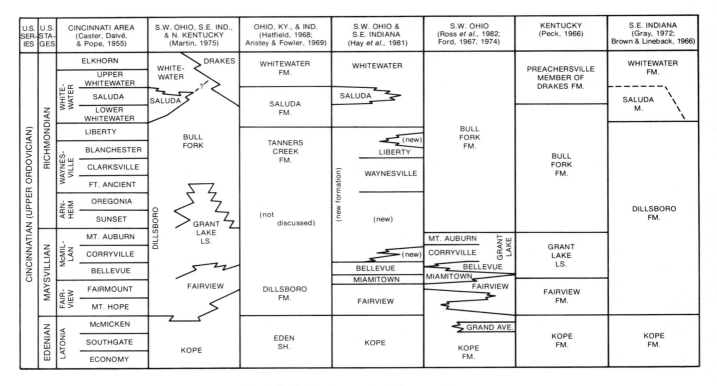

Figure 2. Cincinnatian stratigraphic nomenclature.

Shale beds in the formation here average 7.5 in (19 cm) in thickness. The limestone layers are fairly continuous, but exhibit some pinch-and-swell; they average some 2.5 in (6 cm) in thickness.

The most abundant fossils in the Kope Formation here are the brachiopod *Onniella* and bryozoa. (Superficially these bryozoa appear to include at least *Batostoma, Dekayia, Hallopora,* and *Escharopora.*) Other fossils here include other brachiopods (*Pseudolingula, Sowerbyella, Zygospira*), crinoids (columnals mostly), trilobites (*Cryptolithus, Flexicalymene, Isotelus*—mostly fragments), and trace-fossils. Osgood (in Meyer and others, 1981) reported the trace-fossils *Diplocraterion* and *Trichophycus* in a 2.8-in (7-cm)-thick bed of calcareous siltstone in the Kope Formation at this locality; this layer is about ~16.5 ft (5 m) below the top of the formation.

Overlying the Kope Formation at this locality are 104 ft (31.7 m) of the Fairview Formation. Like the Kope, the Fairview Formation consists of interbedded limestone, siltstone, and shale, but the limestone comprises about half of the total thickness of the Fairview (as contrasted to less than a quarter in the Kope). Shale beds in the Fairview Formation here average 3 in (8 cm) in thickness (less than half the 7.5 in [19-cm] average in the Kope). The limestone layers are fairly continuous, but exhibit some pinch-and-swell; they average 3 in (8 cm) in thickness and are more closely spaced than in the Kope Formation.

Fossils in the Fairview Formation here include bryozoans (*Constellaria, Escharopora, Hallopora*), brachiopods (*Hebertella, Onniella, Platystrophia, Plectorthis, Rafinesquina, Zygospira*), pelecypods (*Ambonychia*), gastropods (*Cyclonema, Loxoplocus*), cephalopods (*?Orthonybyoceras*), crinoids (columnals mostly), trilobites (*Flexicalymene, Isotelus*—fragments mostly), and trace-fossils (*Chondrites, Diplocraterion, Trichophycus,* etc.).

The Bellevue Limestone in this area averages about 20 ft (6 m) in thickness. Here only about 16 ft (5 m) are present. As the name implies, the Bellevue Limestone has much less shale than either the Kope or the Fairview (less than one-fifth, as opposed to three-quarters and one-half, respectively). The shale strata are unfossiliferous and average less than 0.8 in (2 cm) in thickness. Overall, the bedding in the Bellevue is irregular, with much pinch-and-swell. There are wavy, irregular limestone layers that are closely spaced and average 1.6 in (4 cm) in thickness. In addition there are strata that consist of about two-thirds fossils in a shale matrix; most of these fossils are horizontally oriented, flat brachiopods (*Rafinesquina*).

Other fossils in the Bellevue Limestone here include bryozoans (*Hallopora, Monticulopora*), brachiopods (*Hebertella; Platystrophia,* including *P. ponderosa; Zygospira*), pelecypods (*Ambonychia, Caritodens*), gastropods (*Loxoplocus*), crinoids (columnals mostly), trilobites (*Flexicalymene, Isotelus*—fragments mostly), and trace-fossils (for example, borings in bryozoan colonies).

Within the Bellevue Limestone here is a 1.3-ft (0.4-m) thick layer of the Miamitown Shale. The base of this tongue is some 9 ft (2.7 m) above the base of the Bellevue. Slightly more than three-quarters of the thickness of the Miamitown Shale is shale, but limestone nodules are common. Fossils to be encountered

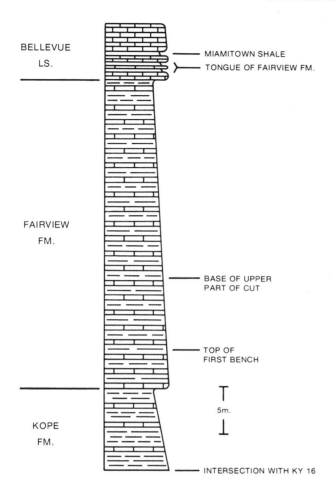

Figure 3. Stratigraphic section at the Riedlin Road/Mason Road locality (measurements after Tobin, 1982).

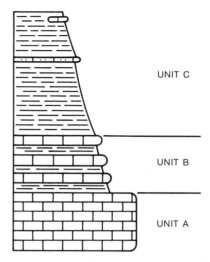

Figure 4. "Ideal" bedding cycle in the Kope and Fairview Formations; thickness varies from about 1.3 ft (0.4 m) to about 13 ft (4 m) (redrawn from Tobin and Pryor [in Meyer 1981]).

include pelecypods (*Ambonychia*) and gastropods (*Loxoplocus*). Note that the Miamitown Shale across the Ohio River commonly lies between the Fairview Formation and the Bellevue Limestone; in other words, the Miamitown transgresses the Fairview-Bellevue contact in the region as a whole.

Tobin (1982) also identified a tongue of the Fairview Formation within the Bellevue Limestone at this locality. As shown in his detailed diagram of the section, the tongue is 3.3 ft (1 m) thick, and its base is somewhat less than 1 m above the base of the Bellevue Limestone (Tobin, 1982, Appendix 3).

Cyclic Bedding. Tobin and Pryor (in Meyer and others, 1981) recognized bedding cycles in the Kope and Fairview Formations at this locality. Their idealized cycle is represented in Figure 4.

The lowermost unit, A, consists of grainstone. This well-sorted, medium- to fine-grained carbonate rock has well-rounded particles of 1.6-mm mean diameter. It is matrix-poor (less than 5%) and has a high degree of fossil-breakage. The beds have an average thickness of 5.9 in (15 cm) and are fairly continuous laterally. Sedimentary structures, such as ripples, channels, and flute and groove marks, are common. The fauna consists mainly

of brachiopods, bryozoans, and crinoids (columnals).

The middle unit, B, consists of interbedded packstones, wackestones, and shale. The two kinds of limestone are moderately to poorly sorted, are coarse-grained (3.1 mm mean), have a moderate to low degree of fossil-breakage, and are matrix-rich (up to about one-half). The limestone beds have an average thickness of 3 in (8 cm) and are less continuous laterally than are the grainstone beds in A. Sedimentary structures are uncommon. The fauna is diverse; it consists largely of bryozoans, brachiopods, pelecypods, gastropods, and crinoids.

The uppermost unit, C, is relatively thick and shale-rich, but it includes thin siltstone beds and silty limestone nodules and beds that may be cross-laminated, rippled, or both. Faunal diversity is low, and the shales are barren of fossils, with even evidence of bioturbation nearly absent. The limestones and siltstones contain highly broken, well-sorted, fine-grained fossil debris.

Tobin and Pryor were fully aware that exposures of these cyclic sequences (Fig. 5) rarely meet the idealized standard (Fig. 4). They concluded that the cycles are highly variable vertically; for example, the middle unit (B) may be absent altogether, or cycles may be separated by strata with no apparent regular bedding order. They concluded that there is great variation from outcrop to outcrop, too, suggesting that the cycles have only local significance.

Paleoenvironmental interpretations. From their study of the sections exposed here and elsewhere in the area Tobin and Pryor (in Meyer and others, 1981) drew several preliminary conclusions about the Ordovician environment in the Cincinnati region, as follows:

Kope, Fairview, and Bellevue sediments were deposited simultaneously in adjacent areas of the sea floor. Kope deposition was on the deepest part of a gently sloping platform, mostly below wave-base. Fairview deposition was in shallower water, but still under generally low-energy conditions. The Bellevue Limestone, on the other hand, is the result of the high-energy environment that existed above ordinary wave-base. Thus, the section at this locality represents a shoaling-upward, offshore-to-shoreface sequence.

The Miamitown Shale does not occur at the same stratigraphic horizon throughout its known areal extent. It may be the record of a particular suite of environmental conditions that existed throughout the region simultaneously; in other words, it may be a "time-line."

The cyclic bedding in the Kope Formation and the Fairview Formation records the alternation of times of carbonate deposition (presumably in clear water) with times of shale deposition from muddy water. The latter might have been due to changes in the nature and amount of materials leaving the source area to the east or to periodic influxes of sediment that had been deposited elsewhere, but was subsequently retransported (for example, by turbidity flows or storms).

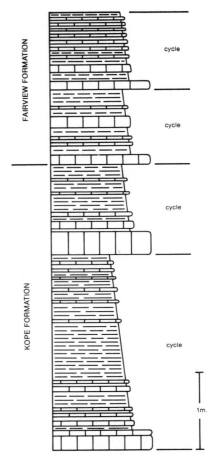

Figure 5. Bedding cycles across the Kope-Fairview contact at the Riedlin Road/Mason Road locality (redrawn from Tobin and Pryor [in Meyer and others, 1981]; their cycles 2, 3, 4, and 5 are represented).

REFERENCES CITED

Anstey, R. L., and Fowler, M. L., 1969, Lithostratigraphy and depositional environment of the Eden Shale (Ordovician) in the tri-state area of Indiana, Kentucky, and Ohio: Journal of Geology, v. 77, p. 668–682.

Brown, G. D., Jr., and Lineback, J. A., 1966, Lithostratigraphy of Cincinnatian Series (Upper Ordovician) in southeastern Indiana: American Association of Petroleum Geologists, Bulletin, v. 50, p. 1018–1023.

Caster, K. E., Dalvé, E. A., and Pope, J. K., 1955, Elementary guide to the fossils and strata of the Ordovician in the vicinity of Cincinnati, Ohio: Cincinnati Museum of Natural History, Cincinnati, Ohio, 47 p. (Date commonly cited incorrectly as 1961, when work was reprinted—see Davis, 1981).

Davis, R. A., ed., 1981, Cincinnati fossils: Cincinnati Museum of Natural History, Cincinnati, Ohio, 58 p.

Ford, J. P., 1967, Cincinnatian geology in southwest Hamilton County, Ohio: American Association of Petroleum Geologists, Bulletin, v. 51, p. 918–936.

——1974, Bedrock geology of the Cincinnati West Quadrangle and part of the Covington Quadrangle, Hamilton County, Ohio: Ohio Division of Geological Survey, Report of Investigations No. 93, 1 sheet.

Gray, H. H., 1972, Lithostratigraphy of the Maquoketa Group (Ordovician) in Indiana: Indiana Geological Survey, Special Report 7, 31 p.

Hatfield, C. B., 1968, Stratigraphy and paleoecology of the Saluda Formation (Cincinnatian) in Indiana, Ohio, and Kentucky: Geological Society of America, Special Paper 95, 34 p.

Hay, H. B., Pope, J. K., and Frey, R. C., 1981, Lithostratigraphy, cyclic sedimentation, and paleoecology of the Cincinnatian Series in southwestern Ohio and southeastern Indiana: in Roberts, T. G. ed., GSA Cincinnati '81 Field trip guidebooks. Volume I: Stratigraphy, sedimentology, American Geological Institute, Falls Church, Virginia, p. 73–86.

Luft, S. J., 1971, Geologic map of part of the Covington Quadrangle, Northern Kentucky: United States Geological Survey, Geologic Quadrangle 955.

Martin, W. D., 1975, The petrology of a composite vertical section of Cincinnatian Series limestones (Upper Ordovician) of southwestern Ohio, southeastern Indiana, and northern Kentucky: Journal of Sedimentary Petrology, v. 45, p. 907–925.

Meyer, D. L., Tobin, R. C., Pryor, W. A., Harrison, W. B., Osgood, R. G., Hinterlong, G. D., Krumpolz, B. J., and Mahan, T. K., 1981, Stratigraphy, sedimentology, and paleoecology of the Cincinnatian Series (Upper Ordovician) in the vicinity of Cincinnati, Ohio: in Roberts, T. G., ed., GSA Cincinnati '81 Field trip guidebooks. Volume I: Stratigraphy, sedimentology American Geological Institute, Falls Church, Virginia, p. 31–71.

Peck, J. H., 1966, Upper Ordovician formations in the Maysville area, Kentucky: United States Geological Survey, Bulletin 1244-B, 30 p.

Pojeta, J., Jr., 1979, The Ordovician paleontology of Kentucky and nearby states—Introduction: United States Geological Survey, Professional Paper 1066-A, 48 p.

Ross, R. J., Jr., and others, 1982, The Ordovician System in the United States: International Union of Geological Sciences, Publication No. 12, 73 p. + 3 sheets.

Tobin, R. C., 1982, A model for cyclic deposition in the Cincinnatian Series of southwestern Ohio, northern Kentucky, and southeastern Indiana [Ph.D thesis]: University of Cincinnati, 483 p.

Middle Ordovician High Bridge Group and Kentucky River fault System in central Kentucky

Gary L. Kuhnhenn *Eastern Kentucky University, Richmond, Kentucky 40475*
Donald C. Haney *Kentucky Geological Survey, Lexington, Kentucky 40506*

LOCATION

The Middle Ordovician High Bridge Group and Kentucky River fault system are exposed along the new part of U.S. 27 just north and south of the Kentucky River near the village of Camp Nelson, Kentucky, in the southwestern quarter of the southwestern section of the Little Hickman 7½-minute quadrangle (Fig. 1). There is ample room to safely park along roadcuts here, but caution is strongly urged because of the common occurrence of heavy traffic. Stop 4 is located on the east side of "Old U.S. 27" along an unnamed road just north of the river and below Stop 1.

INTRODUCTION

The High Bridge Group includes the oldest exposed rocks in Kentucky and represents an excellent example of a sequence of carbonates resulting from tidal-flat deposition on a large, stable cratonic platform (Cressman and Noger, 1976; Kuhnhenn and others, 1981). The High Bridge carbonate sequence is correlated (and very similar) to the well-known Black River carbonate sequence of New York.

The Kentucky River fault system is a major structural feature in Kentucky. This system consists of a narrow band of normal faults and grabens trending northeast across central Kentucky (Black and Haney, 1975) and is a part of the 38th parallel lineament.

SITE DESCRIPTION

High Bridge Group

The Middle Ordovician High Bridge Group is a sequence of carbonate rocks exposed primarily in the bluffs of the gorge of the Kentucky River and its tributaries in central Kentucky. These excellent, but inaccessible outcrops are the result of the combined effects of the location of the Jessamine Dome astride the Cincinnati Arch, vertical displacement along the Kentucky River fault system, and the entrenchment of the Kentucky River. Roadcuts along U.S. 27 (Fig. 1), where it crosses the Kentucky River near Camp Nelson, provide the easiest access to the thickest and most extensive exposure of the High Bridge Group.

The High Bridge Group consists of three formations which are, in ascending order: (1) the Camp Nelson Limestone, (2) the Oregon Formation, and (3) the Tyrone Limestone (Fig. 2). The roadcuts along U.S. 27 combine to provide excellent exposure of

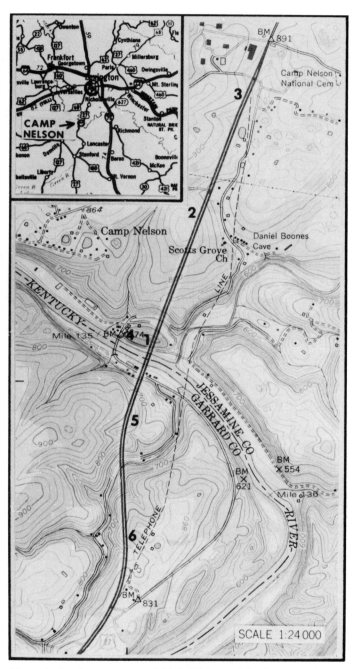

Figure 1. Southwestern part of Little Hickman 7½-minute quadrangle showing Camp Nelson site and described stops (indicated by numbers 1 through 6). Inset in upper left corner shows major highways in Lexington–central Kentucky area.

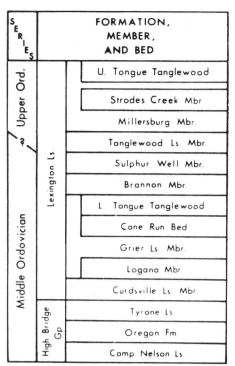

S E R I E S	FORMATION, MEMBER, AND BED		
Upper Ord.	Lexington Ls.	U. Tongue Tanglewood	
		Strodes Creek Mbr.	
		Millersburg Mbr.	
		Tanglewood Ls. Mbr.	
		Sulphur Well Mbr.	
		Brannon Mbr.	
		L. Tongue Tanglewood	
		Cane Run Bed	
		Grier Ls. Mbr.	
		Logana Mbr.	
		Curdsville Ls. Mbr.	
Middle Ordovician	High Bridge Gp.	Tyrone Ls.	
		Oregon Fm.	
		Camp Nelson Ls.	

Figure 2. Stratigraphic section for described stops at Camp Nelson site.

the Tyrone, Oregon, and upper part of the Camp Nelson strati-graphic units for a total of 440 feet (134 m).

The Camp Nelson Limestone is dominantly a thick-bedded, light-gray to buff, dense, micritic limestone that is characterized by dolomite-filled burrows which develop a "honeycomb" appearance on weathered surfaces. The Oregon Formation is a blocky, thick-bedded, buff, finely crystalline dolostone that is often interbedded with micritic limestone. The Tyrone Limestone is primarily a medium, light-gray to white, dense, micritic limestone which commonly contains calcite blebs.

The High Bridge carbonate sequence is generally sparsely fossiliferous, but contains thin layers that are abundantly fossiliferous. Notable fauna include ostracodes, trilobites, brachiopods, bryozoans, pelmatozoans, gastropods, pelecypods, cephalopods, and corals. Of particular interest is the tabulate coral, *Tetradium*, which occurs throughout the High Bridge sequence, but is most abundant in the Tyrone Limestone.

Several bentonites occur in the upper part of the High Bridge sequence. Of notable importance are the locally occurring "Mud Cave" bentonite at the top of the Tyrone Limestone and the "Pencil Cave" bentonite located 14 to 20 ft (4.2 to 6.1 m) below the top of the Tyrone and occurring throughout central Kentucky.

Recent workers (Cressman and Noger, 1976; Kuhnhenn and others, 1981; Horrell, 1981; Lazarsky, 1983) have interpreted the High Bridge carbonate sequence as a product of a large tidal-flat complex located on a stable craton. These workers suggest that a humid tropical climate, similar to that of the Bahama Islands today, existed during deposition of the High Bridge se-

quence. The recent discovery of gypsum crystal molds in the cryptalgalaminated micrites of the Tyrone Limestone suggests the climate may have been more arid.

Stop 1. Stop 1 (Figs. 1, 3) is located adjacent to the north abutment of the new Kentucky River bridge and on the west side of the road. This roadcut consists of the Camp Nelson Limestone and contains a massive, white, unfossiliferous micrite displaying vague color mottling (tan) throughout most of the unit and vague laminations (algal?) at the top. The base of the micrite unit grades downward into thin-bedded, sparsely fossiliferous, micritic limestone of subtidal origin, while the upper contact is abrupt and overlain by similar subtidal deposits. This unique white micrite unit is interpreted as being the product of deposition in a tidal pond or lagoonal portion of the tidal-flat environment where well-oxygenated (and probably hypersaline) waters caused complete oxidation of organic material in the lime muds leading to the nearly white color.

Stop 2. Stop 2 (Figs. 1, 3, 4) is located at the north end of the largest roadcut (on west side and 0.5 mile (0.8 km) north of Stop 1) north of the Kentucky River bridge. The north end of the roadcut consists of the Tyrone Limestone at the top of the outcrop and the underlying Oregon Formation which extends down to road level. Moving south along the roadcut, the upper part of the Camp Nelson Limestone can be seen underlying the Oregon Formation.

The lower part of the Tyrone Limestone is exposed above the upper bench of the roadcut which is marked by a prominent bentonitic layer (thought to be the Pencil Cave bentonite). Directly above the bentonite is a dark, argillaceous, micritic limestone that has abundant, small mud-crack polygons and is devoid of fossils. It is thought this unit may have been influenced by deposition of volcanic ash.

The boundary between the Tyrone Limestone and the Oregon Formation lies below the level of the upper bench of the roadcut and is marked by a thick-bedded dolomicrite displaying cryptalgal laminations, mud-crack polygons and prism cracks, and a prominent dolomitized intrasparrudite layer. The middle part of the Oregon Formation is marked by a thick sequence of wavy, thin-bedded, light-gray, partially dolomitized, micritic limestone with dolomite-filled burrows and is interpreted to be restricted shallow subtidal in origin. A horizon of large, dolomite-filled, vertical burrows can be seen near road level. The lower part of the Oregon Formation is characterized by massive-bedded, blocky, buff dolomicrite displaying cryptalgal laminations, mud-crack polygons of varying thickness, prism cracks of varying depth, and a distinctive color mottling. A prominent bentonitic layer occurs just above the basal contact of the Oregon Formation.

The dolostone units (dolomicrities) characterizing the upper and lower parts of the Oregon Formation are interpreted to have originated in the intertidal and possibly the very shallow subtidal part of a tidal-flat environment based on occurrence of cryptalgal laminations, size and thickness of mud-crack polygons, and depth of prism cracks. The unique color mottling is thought to result

Figure 3. Northward view of large roadcuts along U.S. 27 at the bridge crossing the Kentucky River. Right portion (east side of U.S. 27) of figure (and above car) shows the main fault of the Kentucky River fault system (Stop 5). Roadcut in upper center of figure (indicated by lower arrow) and on west side of U.S. 27 is Stop 1. The upper arrow indicates the location of Stop 2 on west side of highway.

Figure 4. Stop 2 on west side of U.S. 27 north of the Kentucky River bridge. Roadcut shows upper portion of Camp Nelson Limestone, Oregon Formation, and lower portion of Tyrone Limestone. The lower arrow (left side of figure) indicates Camp Nelson–Oregon boundary and upper arrow (right side of figure) shows Oregon–Tyrone boundary.

from the oxidation of organic-rich mud by movement of oxygenated waters along paths of preferred permeability between mud-crack polygons and along prism cracks and burrows.

The basal dolomicrite unit of the Oregon Formation is underlain by a burrowed (dolomite-filled), light-gray to buff, sparsely fossiliferous, micritic limestone of restricted subtidal origin and marks the upper boundary of the Camp Nelson Limestone. The upper part of the Camp Nelson sequence, as seen in this roadcut, contains a prominent unit of dark greenish-gray, argillaceous (bentonitic?), dolomitic limestone displaying abundant, small, mud-crack polygons and apparent "teepee" structures at the upper contact of the unit.

Further observation of the upper part of the Camp Nelson Limestone can be made by walking south along the roadcut. This part of the Camp Nelson sequence is dominantly a dark, burrowed (dolomite-filled), sparsely fossiliferous, dense, micritic limestone of subtidal origin. Various intensely burrowed zones (interpreted as hardgrounds) occur throughout this sequence, as do occasional light-gray micrite units marking a brief return to intertidal deposition.

Stop 3. Stop 3 (Figs. 1, 5) is located 0.5 mile (0.8 km) north of Stop 2 and on the west side of U.S. 27. This isolated roadcut consists entirely of the Tyrone Limestone. The Tyrone is dominantly a light-gray (weathering to white), dense, micritic limestone displaying a variety of cryptalgal laminations, mud-crack polygons, prism cracks, vertical tubiform burrows, and calcite-filled gypsum crystal molds. The micritic limestone is interrupted by thin layers of intrasparrudites and biosparites. Nodular chert replaces the host limestone at various horizons and is often associated with a bentonitic layer. The "Mud Cave" bentonite is located just below the top of the outcrop. A fossiliferous, micritic limestone, containing *Tetradium* colonies (some in place), is found directly above the Mud Cave bentonite, but should still be considered part of the Tyrone Limestone.

The Tyrone Limestone is interpreted as having been deposited in the high intertidal and possibly the low supratidal portions of a tidal-flat environment. This interpretation is based on the abundant cryptalgal laminations, relatively shallow prism cracks, vertical tubiform burrows, generally sparse and restricted fauna, and the occurrence of calcite-filled gypsum crystal molds. The intrasparrudite and biosparite layers represent periodic high-energy events (such as storms), or possibly tidal-channel deposits. While intertidal/supratidal deposits are by far the most common type of rocks in the Tyrone Limestone, the unit also contains wavy, thin-bedded, dolomitic, and sparsely fossiliferous, micritic limestone interpreted as having been deposited in a restricted, very shallow, subtidal environment.

Stop 4. Stop 4 (Figs. 1, 3) is located at the intersection of old U.S. 27 and an unnamed road just north of the old bridge crossing the Kentucky River (below and west of Stop 1). The roadcut on the north side of the unnamed road consists of the lowermost exposure of the Camp Nelson Limestone. The part of the Camp Nelson sequence exposed in this outcrop is dominantly a light- to dark-gray, dense, sparsely fossiliferous, micritic limestone display-

Figure 5. Tyrone Limestone exposed in roadcut (Stop 3) on West side of U.S. 27 north of Stop 2.

ing a variety of cryptalgal laminations, mud-crack polygons, prism cracks, and color mottling reminiscent of the carbonate units seen in the overlying Tyrone Limestone and Oregon Formation (prior to dolomitization). This roadcut also contains a massive, white, unfossiliferous micrite very similar to the white micrite unit described at Stop 1 (stratigraphically higher in the Camp Nelson Limestone). This micrite unit grades downward into burrowed (dolomite-filled), sparsely fossiliferous, dark, micritic limestone of a restricted, shallow subtidal origin, while the upper contact is sharp and overlain by cryptalgalaminated micrite.

Kentucky River Fault System

The Kentucky River fault system is one of the major structural features in Kentucky and consists of a narrow band of normal faults and grabens trending in a generally northeast direction across central Kentucky. The general sense of displacement is down to the southeast with offset being as much as 700 ft (213 m). In addition, there are hundreds of northwest-trending normal faults and grabens, most of which display displacements of less than 100 ft (30 m).

Excellent and easily accessible exposures of the Kentucky River fault system can be examined in a series of roadcuts along a new part of U.S. 27 just south of the Kentucky River. Additional exposures of the fault system can be seen in roadcuts along old U.S. 27 south of the Kentucky River.

Figure 6. Upthrown block of the main fault of the Kentucky River fault system exposed in a roadcut (Stop 5) on west side of U.S. 27 south of the Kentucky River bridge. Note white, micritic limestone bed showing offset along small faults.

In the area just south of the river, the Kentucky River fault system is a graben structure formed by multiple fault blocks downthrown along high-angle normal faults having a maximum displacement of 700 ft (213 m). Other observable structural features include small reverse faults, joints, drag folds, horizontal slickensides, bedding slickensides, and mineralized fractures.

Several sets of joints have been identified in the Camp Nelson area. Interpretation of strike and dip measurements of the joint sets suggest lateral stresses which support other evidence for strike-slip movement along faults related to the Kentucky River fault system (Black and Haney, 1975).

Three fault zones are exposed in the U.S. 27 roadcuts south of the river. The principal faults strike northeastward and dip at angles of 75 - 90°. While the sense of movement along these faults is normal, evidence suggests more than one period of movement. Black and Haney (1975) provide a chronological interpretation of the structural features observed in the Camp Nelson area.

Stop 5. Stop 5 (Figs. 1, 6) is located 0.2 mile (0.3 km) south of the new Kentucky River bridge. The main fault (northwesternmost margin of the graben) of the Kentucky River fault system is well exposed at the south end of the first large roadcut on the east side of U.S. 27. The undifferentiated Grier and Curdsville Limestone Members (Lexington Limestone) on the southeast is downthrown (approximately 300 ft [91 m]) against the Camp Nelson Limestone on the northwest producing a very distorted

fault zone characterized by large angular blocks, drag folds, and fractures. Reverse drag and other associated features (such as slickensides, flexure slip, and small reverse faults) are evident in the Camp Nelson Limestone. On the west side of the highway, a white micritic limestone unit (also observed north of the river) serves as a marker bed which shows normal and reverse movement associated with small block faults.

Another northeast-trending fault is located 1.0 mile (1.7 km) further south on U.S. 27 and represents the southeasternmost margin of the graben. Here the fault has a vertical displacement of 100 ft (30 m) or less, displays drag folds, and is within the Grier Limestone Member.

Stop 6. Stop 6 (Fig. 1) is located on the east side of U.S. 27, 0.4 mile (0.6 km) south of Stop 5. Here a third northeast-trending fault occurs almost in the center of the graben. Again vertical displacement is about 100 ft (30 m) with the undifferentiated Grier and Curdsville Limestone Members in juxtaposition with itself. Small drag folds occur in the southeastern block near the fault.

An excellent example of contorted bedding ("flow rolls") can be examined by walking northward along the roadcut. Note the unique deformational relationship (timing) of the associated chert nodules indicating very early emplacement of the silica.

REFERENCES CITED

Black, D.F.B., and Haney, D. C., 1975, Selected structural features and associated dolostone occurrences in the vicinity of the Kentucky River fault system: Geological Society of Kentucky, 1975 Annual Field Conference, Guidebook, Kentucky Geological Survey, 27 p.

Cressman, E. R., and Noger, M. C., 1976, Tidal-flat carbonate environments in the High Bridge Group (Middle Ordovician) of central Kentucky: Kentucky Geological Survey, Series 10, Report of Investigations 18, 15 p.

Horrell, M. A., 1981, Stratigraphy and depositional environments of the Oregon Formation (Middle Ordovician) of central Kentucky [M.S. thesis]: University of Kentucky, Lexington, 120 p.

Kuhnhenn, G. L., Grabowski, G. J., Jr., and Dever, G. R., Jr., 1981, Paleoenvironmental interpretation of the Middle Ordovician High Bridge Group in central Kentucky, *in* Roberts, T. G., ed., Field Trip Guidebooks, volume 1: stratigraphy and sedimentology (94th Annual Meeting, Geological Society of America, Cincinnati): Falls Church, Virginia, American Geological Institute, p. 1–30.

Lazarsky, J. J., 1983, Petrographic analysis of synsedimentary and diagenetic features of the High Bridge Group (Middle Ordovician) in the subsurface, Fayette County, central Kentucky [M.S. thesis]:, Eastern Kentucky University, Richmond, 110 p.

Devonian and Mississippian black shales of Kentucky

Roy C. Kepferle, Department of Geology, Eastern Kentucky University, Richmond, Kentucky 40475

LOCATION

A series of road cuts exposes black shale of Devonian and Mississippian age along I-64, about 49 mi (79 km) east of Lexington, Kentucky (Fig. 1). The entire sequence can best be visited from separate approaches. The lower part of the sequence (Fig. 1, locality A) should be approached from the west on the eastbound lanes of I-64; the upper part of the sequence (Fig. 1, locality B) should be approached from the east on the westbound lanes. The black-shale section is in the Farmers 7½-minute Quadrangle (McDowell, 1975), and the base of the Ohio Shale at the west end (locality) is at an elevation of 710 ft (216.5 m).

INTRODUCTION

Black shale of Devonian and Mississippian age is found in a sequence of Paleozoic rock that ranges in age from Middle Ordovician to Pennsylvanian (Fig. 2). The black shales are noted for their high content of organic matter, much of which exceeds 8 percent by weight. Thus, the shales are regarded as sources of synthetic fuels, natural gas, and petroleum, as well as potential sources of associated trace elements (U, Th, Mo, V, Co, Ni, and Cu; Leventhal and Kepferle, 1982). The black shales are also sedimentologic examples of deposition in an anaerobic environment. Transition to a dysaerobic environment is characterized by the associated greenish-gray mudstone (Ettensohn and Barron, 1981).

Physiographically, the I-64 exposures lie in the Knobs between the outer Bluegrass and the Cumberland escarpment (Fig. 2). The Knobs are erosional remnants of the escarpment that forms a narrow, horseshoe-shaped belt around the central Bluegrass. The physiographic patterns reflect the combined effects of structure and differential erosion of sedimentary units. The structural uplift and faulting are associated with the Cincinnati Arch, the crest of which crosses the state northeastward from Nashville, Tennessee, to Cincinnati, Ohio. This is the most accessible of the few places where the entire black-shale sequence is exposed on the western flank of the Appalachian Basin. It is a key section for correlation of the Ohio and Sunbury Shales in the Upper Devonian and Lower Mississippian rock sequence to the north and the south; it serves as a reference point for comparison

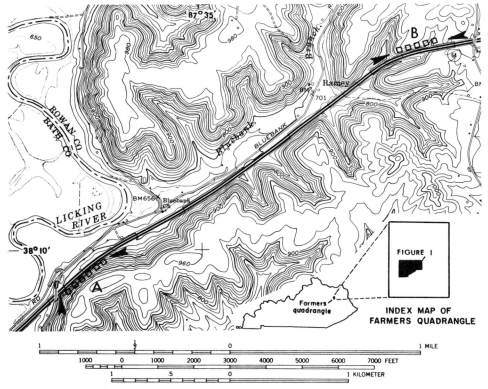

Figure 1. Location: The base of the section (A) is on the south side of I-64, in the Farmers 7½-minute Quadrangle, Rowan County, Kentucky, and commences at the underpass 1.4 mi (2.3 km) east of the Licking River bridge (the Bath-Rowan county line). The upper part of the section (B) is on the north side of I-64 along the westbound lanes, 4.4 mi (7.1 km) west of the Morehead Interchange, and just west the underpass shown at the east edge of the map.

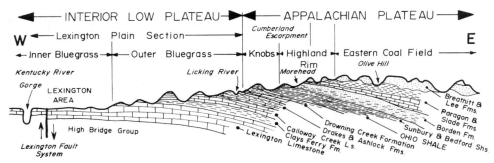

Figure 2. Geologic units and physiographic regions crossed along I-64 between Lexington and Morehead in east-central Kentucky. Field-trip stops A and B are at the approximate boundary between the Interior Low Plateau and the Appalachian Plateau.

with the much thinner Chattanooga Shale across the crest of the Cincinnati Arch to the southwest and with the New Albany Shale in the Illinois Basin to the west. The nature of two systemic boundaries can be examined in detail, as can the bases for rock-stratigraphic, biostratigraphic, and gamma-ray correlations of the sequence.

Other descriptions of this sequence of roadcuts have been included in papers, stratigraphic cross sections, guidebooks, and theses, including Provo and others (1978), Wallace and others (1977), Swager (1978), Kepferle and Roen (1981), and Pryor and others (1981). The Lower Mississippian rocks are described by Chaplin (1980).

SITE DESCRIPTION

Locality A

The base of the section is on the south side of I-64, 1.4 mi (2.3 km) east of the Bath-Rowan county line. The shale sequence and associated rocks, in order of decreasing age, including the Drowning Creek Formation (Silurian; McDowell, 1983), the upper part of the Olentangy Shale, the Huron Member, the Three Lick Bed, and the Cleveland Member of the Ohio Shale (Devonian), the Bedford Shale (Mississippian and Devonian), and part of the Mississippian Sunbury Shale (Figs. 3 and 4). The lower part of the section is more accessible from the road than that part above the Three Lick Bed (the Cleveland Member).

The Drowning Creek Formation of the Crab Orchard Group (McDowell, 1983) is mainly olive-gray argillaceous, dolomitic shale, which is poorly fissile and weathers to thin flakes. Dolosiltite in light olive-gray resistant beds 1.2 to 2.4 in (3 to 6 cm) thick makes up less than 15 percent of the unit and contains *Chondrites*-like branching burrows. The base of the Drowning Creek Formation is not exposed. The top of the Drowning Creek Formation is a planar disconformity and lag zone, which is not readily discernible, owing to the grossly similar argillaceous aspect of the greenish-gray to olive-gray mudstone that comprises the lower 3 ft (0.9 m) of the overlying Devonian shales. Confirmation that the Silurian-Devonian systemic boundary is here at the base of the Olentangy Shale, rather than at the more obvious

contact between gray and black shales of the Huron Member of the Ohio Shale above, comes from the identification of elements of the Devonian conodont *Palmatolepis* sp. in the lag concentrate of coarse quartz grains, phosphorite granules, and other phosphatic fossil fragments at the base of the Olentangy (Anita G. Harris, written communication, 1977).

Close inspection of this basal sequence of greenish-gray mudstone reveals several thin laminae of black shale. Pyrite also locally cements the basal lag concentrate, or "bone bed." Elsewhere, this particular unconformity shows numerous variations in rock types, both in the rocks immediately above the unconformity and in the rocks beneath the unconformity. It should not be viewed as "typical," except for the lag concentrate, and even this characteristic is locally missing.

A more obvious lithologic contrast is marked by the abrupt transition from the greenish-gray and olive-gray mudstone of the Olentangy Shale to the overlying olive-black to brownish-black shale of the basal Huron Member of the Ohio Shale. The base of the black shale is marked by a lag zone less than 0.5 in (1 cm) thick, which contains conodonts. Similar lag zones have been reported from nearby cores at these horizons (Kepferle et al., 1982), as well as at other horizons higher in the section. One of these zones is above the next overlying sequence of alternating black shale and greenish-gray mudstone. These lag zones figure prominently in regional correlation of stratigraphic subunits within the shale sequence.

The more resistant layers of black shale have been found to contain more organic matter than the less resistant layers. Some of the black-shale layers are massive and blocky, as is the layer about 20 ft (6 m) thick, which is at road level at the crest of the rise to the east on I-64. The lower half of this layer was sampled for inclusion in the geochemical benchmark series by the U.S. Geological Survey (Kepferle and others, 1985), and samples are available for distribution to qualifying laboratories for comparative analysis. Massive black-shale layers commonly contain undisturbed calcisilt or quartz-silt laminae and abundant pyrite as disseminated spherules, discontinuous laminae, nodules, and irregular aggregates. Trace fossils are absent or occur in a thin zone at the top of these black-shale layers, where the overlying bed is greenish-gray to olive-gray shale or mudstone. Fissility increases

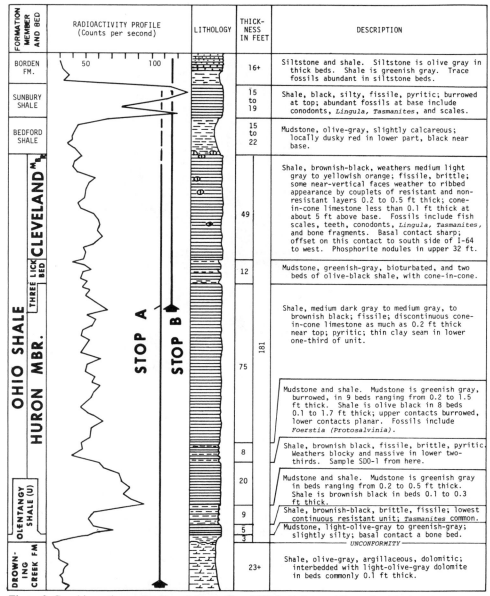

Figure 3. Graphic section and radioactivity profile of Mississippian and Devonian black shales exposed along I-64 in Rowan County, Kentucky. (Modified from Provo and others, 1978, and Kepferle and Roen, 1981, stop 12).

with weathering and with organic matter and sulfide content. Higher in the section the black to brownish-black shale commonly alternates with less black shale in couplets that are generally less than 1 ft (30 cm) thick. On weathering, these couplets impart a ribbed appearance to the exposed slopes. The more resistant ribs may protrude sufficiently to form mini-benches on which light-gray chips of weathered shale accumulate (Fig. 5).

Greenish-gray to olive-gray mudstone contrasts with the black and olive-black shale. Bioturbation has destroyed whatever primary sedimentary lamination might have been present, hence, the rocks are called mudstones rather than shales (according to the classification of Lundegard and Samuels, 1980, for example). This prevalent bioturbation and sparse organic matter attest to

accumulation under more oxygenated conditions than the environment attributed to black-shale accumulation. The nature of the contacts between the greenish-gray to olive-gray mudstones and the black shale provides further evidence of the life-tolerant environment in which the mudstones accumulated. At the base of such mudstone layers, light, mudstone-filled burrows penetrate the upper few centimeters of the underlying black shale and form an irregular contact. The contact between the two, where the black shale lies above greenish-gray mudstone, on the other hand, is generally planar. Burrows filled with black shale are not generally found in the mudstone beds. Studies of some of these trace fossils have been documented by Jordan (1979).

Body fossils, other than the linguloid and orbiculoid brachi-

Figure 4. Photograph of the exposure at stop A along the south side of I-64, Rowan County, Kentucky. (Photo by Paul Edwin Potter).

opods, include fish remains, mainly scales, and in certain layers, bones, conical teeth, and scutes. Woody fragments occur chiefly as long, vitrainlike plant straps, some of which are thick enough to recognize in outcrop. The algae *Foerstia* (= *Protosalvinia*) occurs as a distinct biostratigraphic marker zone near the base of the Huron Member (Schopf and Schwietering, 1970). In this exposure, the *Foerstia* zone approximately coincides with the sequence of greenish-gray mudstone and black shale immediately above the 20-ft (6.1-m) thick, massive, black-shale sequence. The *Foerstia* zone has become a key to understanding the relationship of sedimentation between the Appalachian and Illinois Basins, as well as between the Appalachian and Michigan Basins (Hasenmueller and others, 1983). At Locality A (Fig. 1), the *Foerstia* zone can best be reached along the eastern end of the road cut, as can the next unit of stratigraphic note, the Three Lick Bed.

This composite section (Localities A and B, Fig. 1) is the type locality of the Three Lick Bed (Provo and others, 1978). The Three Lick Bed is characterized by three beds of greenish-gray to olive-gray mudstone separated by two somewhat thicker beds of brownish-black to olive-black shale (Fig. 5). In earlier times, a "lick" was a site where game sought salt by licking the efflorescence from the soil or rock. Such licks were commonly kept bare of vegetation by some combination of salinity, animal activity, and erosion. A small creek more than 1 mi (1.6 km) north of the highway is named Three Lick Branch. Speculation leads us to believe that here the creek was named for the rock unit, which was later named formally for the creek.

Figure 5. Photograph of the exposure of the Three Lick Bed (T) at the west end of stop B on the north side of the westbound lanes of I-64, 2.6 mi (4.2 km) east of stop A and 5.0 mi (8.0 km) west of the Morehead Interchange, Rowan County, Kentucky.

The Three Lick Bed has been recognized in most of the exposures along the western flank of the Appalachian Basin between Chillicothe, Ohio, and Berea, Kentucky, as well as along the crest of the Cincinnati Arch in the Cumberland saddle through Burkesville, Kentucky, to the vicinity of Celina, Tennessee. It has been recognized in cores even farther westward, where it thins and eventually disappears. Eastward and southward, the Three Lick Bed can be recognized on gamma-ray logs by a characteristic three-pronged signature (Provo and others, 1978). This outcrop and several others in the region have been scanned with a scintillometer, and the resulting readings have been used to produce a synthetic gamma-ray log that can be used for correlation between the surface and the subsurface (Ettensohn and others, 1979).

Locality B

The upper part of the section above the Three Lick Bed is best observed on the westbound lanes of I-64. The top of the section first appears just west of the overpass about 4.4 mi (7.1 km) west of the Morehead Interchange (Exit 137) and extends westward past a small ravine on the right. The west end of the exposure is 3.4 mi (5.5 km) from the Licking River bridge at the Bath-Rowan county line. The exposed rocks include a complete sequence of the Farmers Member of the Borden Formation, including the Henley Bed at its base, the underlying Sunbury and Bedford Shales (all Mississippian), and the upper part of the Ohio Shale, including the Cleveland Member and Three Lick Bed (Figs. 3, 5, and 6). Abundant trace fossils can be found in the siltstones of the Borden Formation (Chaplin, 1980).

Fragments of phosphatic brachiopods and conodonts occur in a persistent zone at or near the base of the uppermost black shale in the sequence, the Sunbury Shale. This zone is correlated with the widespread thin basal zone identified northward in the outcrops of the Sunbury Shale by Pepper and others (1954). The brachiopods include *Lingula melie* and *Orbiculoidea herzeri.*

Figure 6. Photographs of the Farmers Member of the Borden Formation and the black Sunbury Shale (S) along the north side of the westbound lanes of I-64, 4.7 mi (7.5 km) west of the Morehead Interchange, Stop B, Rowan County, Kentucky.

Based on faunal evidence found by de Witt (1970), the contact between the Mississippian and Devonian Systems is placed near the base of the Bedford Shale, near a grayish- to dusky-red shale typical of that associated with the Red Bedford delta (Pepper and others, 1954). Most of the Bedford Shale is greenish gray. Phosphorite occurs as scattered nodules in the basal part of the Bedford Shale and in the underlying Cleveland Member of the Ohio Shale.

REFERENCES CITED

Chaplin, J. R., 1980, Stratigraphy, trace fossil associations, and depositional environments in the Borden Formation (Mississippian), northeastern Kentucky; Geological Society of Kentucky Annual Field Conference, Guidebook: Kentucky Geological Survey, 114 p.

de Witt, Wallace, Jr., 1970, Age of the Bedford Shale, Berea Sandstone, and Sunbury Shale in the Appalachian and Michigan basins, Pennsylvania, Ohio, and Michigan: U.S. Geological Survey Bulletin 1294-C, p. G1–G11.

Ettensohn, F. R., and Barron, L. S., 1981, Depositional model for the Devonian-Mississippian black shales of North America: a paleoclimatic-paleogeographic approach, *in* Roberts, T. G., ed., GSA Cincinnati '81 Field Trip Guidebooks, Volume II: Falls Church, American Geological Institute, p. 344–361.

Ettensohn, F. R., Fulton, L. P., and Kepferle, R. C., 1979, Use of scintillometer and gamma-ray logs for correlation and stratigraphy in homogeneous black shales: Summary: Geological Society of America Bulletin, pt. I, v. 90, p. 421–423 (pt. II, v. 90, p. 828–849).

Hasenmueller, N. R., Kepferle, R. C., Matthews, R. D., and Pollock, D., 1983, *Foerstia (Protosalvinia)* in Devonian shales of the Appalachian, Illinois, and Michigan basins, eastern United States, *in* Proceedings, Eastern Oil Shale Symposium, Lexington: Lexington, University of Kentucky, Institute for Mining and Minerals Research Report 83-089, p. 35–58.

Jordan, D. W., 1979, Trace fossils and stratigraphy of Devonian black shale in east-central Kentucky [M.S. thesis]: Cincinnati, University of Cincinnati, 227 p.

——1980, Trace fossils and stratigraphy of Devonian black shale in east-central Kentucky [abs.]: American Association of Petroleum Geologists Bulletin, v. 64, no. 5, p. 729–730.

Kepferle, R. C., and Roen, J. B., 1981, Chattanooga and Ohio Shales of the southern Appalachian basin, *in* Roberts, T. G., ed., GSA Cincinnati '81 Field Trip Guidebooks, Volume II: Falls Church, American Geological Institute, p. 259–330.

Kepferle, R. C., Beard, J. G., and Pollock, J. D., 1982, Lithologic and stratigraphic description of cores of organic-rich shale of Late Devonian and Early Mississippian age in northeastern Kentucky: U.S. Geological Survey Open-File Report 82-219, 75 p.

Kepferle, R. C., de Witt, W., Jr., and Flanagan, F. J., 1985, Ohio Shale (Devonian), SDO-1 from Rowan County, Kentucky: U.S. Geological Survey Open-File Report 85-145, 14 p.

Leventhal, J. S., and Kepferle, R. C., 1982, Geochemistry and geology strategic metals and uranium in Devonian shales of the eastern interior United States, *in* Synthetic fuels from oil shale II (Symposium, Institute of Gas Technology at Nashville, Tennessee, October 26–27, 1981: preprint); Chicago, Institute of Gas Technology, 24 p.

Lundegard, P. D., and Samuels, N. D., 1980, Field classification of fine-grained sedimentary rocks: Journal of Sedimentary Petrology, v. 50, no. 3, p. 781–786.

McDowell, R. C., 1975, Geologic map of the Farmers Quadrangle, east-central Kentucky: U.S. Geological Survey Geologic Quadrangle Map GQ-1236.

——1983, Stratigraphy of the Silurian outcrop belt on the east side of the Cincinnati Arch in Kentucky, with revisions in nomenclature: U.S. Geological Survey Professional Paper 1151-F, 27 p.

Pepper, J. F., de Witt, W., Jr., and Demarest, D. F., 1954, Geology of the Bedford Shale and Berea Sandstone in the Appalachian Basin: U.S. Geological Survey Professional Paper 259, 111 p.

Provo, L. J., Kepferle, R. C., and Potter, P. E., 1978, Division of black Ohio Shale in eastern Kentucky: American Association of Petroleum Geologists Bulletin, v. 62, no. 9, p. 1703–1713.

Pryor, W. A., Maynard, J. B., Potter, P. E., Kepferle, R. C., and Kiefer, J., 1981, Energy resources of Devonian-Mississippian shales of eastern Kentucky; Geological Society of Kentucky Annual Field Conference, Guidebook: Kentucky Geological Survey, 44 p.

Schopf, J. M., and Schwietering, J. F., 1970, The *Foerstia* Zone of the Ohio and Chattanooga Shales: U.S. Geological Survey Bulletin 1294-H, p. H1–H15, 2 pls.

Swager, D. R., 1978, Stratigraphy of the Upper Devonian–Lower Mississippian shale sequence in the eastern Kentucky outcrop belt [M.S. thesis]: Lexington, University of Kentucky, 116 p.

Wallace, L. G., Roen, J. B., and de Witt, W., Jr., 1977, Preliminary stratigraphic cross section showing radioactive zones in the Devonian black shales in the western part of the Appalachian basin: U.S. Geological Survey Oil and Gas Investigations Chart OC-80, 2 sheets.

The Mississippian-Pennsylvanian transition along Interstate 64, northeastern Kentucky

Frank R. Ettensohn, Department of Geology, University of Kentucky, Lexington, Kentucky 40506

LOCATION

A 17 mile segment of I-64 crosses the Mississippian–Lower Pennsylvanian outcrop belt nearly perpendicular to strike in Rowan and Carter counties, northeastern Kentucky (Fig. 1), revealing a series of approximately 30 roadcuts. All cuts are on highway right-of-way on either side of the road. Large-scale relationships and features in some cuts are best viewed from the opposite side of the road.

INTRODUCTION

A series of exposures in northeastern Kentucky provides a premiere opportunity to examine a progression of Middle Mississippian–through–Early Pennsylvanian (Valmeyeran through Morrowan) stratigraphic units and environmental sequences (Fig. 2) in a section nearly perpendicular to strike on the western flank of the Appalachian Basin. Examples of environmental sequences including subaequeous delta (Borden Formation; Chaplin, 1980), shallow-subtidal, intertidal, supratidal and pedogenic carbonates (Newman Limestone; Ferm and others, 1971; Ettensohn, 1979a, 1980, 1981; Dever, 1980a, 1980b; Ettensohn and others, 1984), and shoal-water delta plain (Breathitt Formation; Ferm and others, 1971; Horne and others, 1974; Short, 1979a) are present. However, the rocks also show the unexpected absence of units, abrupt facies changes, deep erosional truncation and prominent disconformities within and between major environmental sequences. These seeming anomalies have led to difficulty and controversy in interpreting the relationships between sequences and the nature of the Mississippian-Pennsylvanian transition. As a result, two controversial models have arisen: the Barrier-Shoreline Model (Ferm and others, 1971; Horne and others, 1974) and a tabular-erosion model (Ettensohn, 1979b, 1980, 1981). The interstate cuts contain the primary evidence for both models. The interstate segment also transects the axis of the Waverly Arch (Fig. 1) which apparently was active during the Carboniferous. Hence, evidence for the effects of synsedimentary tectonism on lithofacies, biofacies, unit distribution and development of disconformities is readily apparent (Dever and others, 1977; Dever, 1980a, 1980b; Ettensohn, 1979a, 1980, 1981; Ettensohn and Dever, 1979a, 1979b, 1979c; Ettensohn and others, 1984).

SITE DESCRIPTION

The series of cuts begins in the east at the junction of I-64 and U.S. 60 in the Grahn 7½-minute Geologic Quadrangle (Englund, 1976), proceeds through the Olive Hill (Englund and Windolph, 1975) and Soldier (Philley and others, 1975a) 7½-

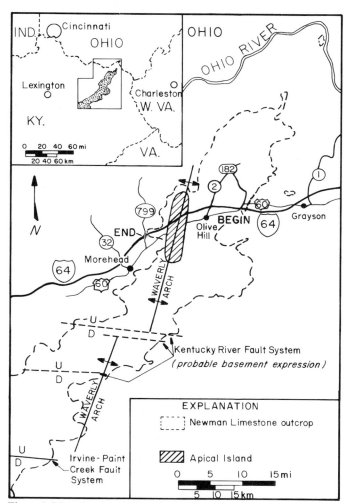

Figure 1. Map showing the location of I-64 where it crosses the outcrop belt of the Newman Limestone. The series of outcrops described herein begins at the intersection of U.S. 60 and I-64 east of Olive Hill and ends at the overpass of Kentucky 799.

minute Quadrangles, and ends at the overpass of Kentucky 799 in the eastern part of the Cranston 7½-minute Quadrangle (Philley and others, 1975b; fig 1). The roadcuts are readily apparent and can be located individually through a locality register (Ettensohn, 1980) or by comparison with a road log and accompanying topographic quadrangle maps (Ettensohn, 1981).

Because of the number of exposures involved, general information about the roadcuts is best provided relative to the Waverly Arch in the tabular-erosion model (Fig. 3B) with reference to geographic features and milepost markers along the highway. The Barrier-Shoreline interpretation of the interstate exposures is shown in Figure 3A.

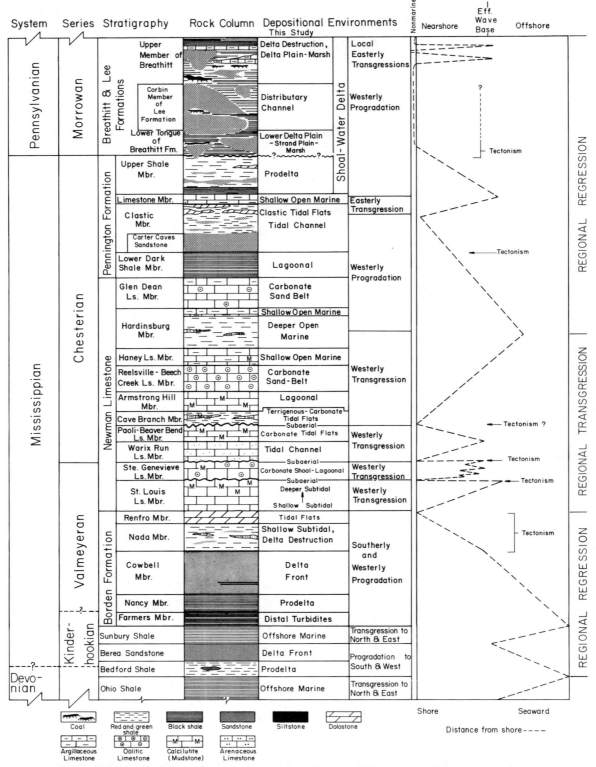

Figure 2. Sequence of Carboniferous units on or near Interstate 64 showing the inferred succession of environmental sequences, environments, and tectonic events on the Waverly Arch (from Ettensohn, 1980).

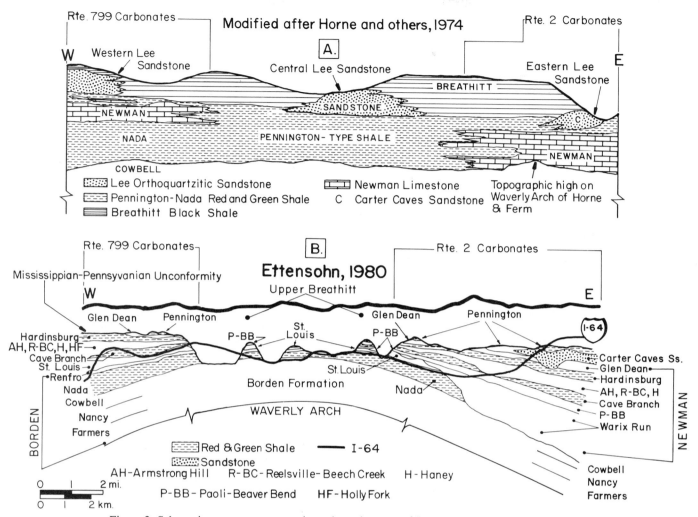

Figure 3. Schematic east-west cross sections along the trace of I-64 showing interpretations of the geology based on the Barrier-Shoreline Model *A* and the tabular-erosion model *B*. The tabular-erosion interpretation is used as the basis for the site description (from Ettensohn, 1981), which is keyed to interstate milepost markers.

The Mississippian carbonates east of Waverly Arch, commonly called the "Route 2 carbonates," thicken in an eastward direction away from the arch (Fig. 3B). Cuts including the "Route 2 carbonates" begin at the east edge with the intersection of U.S. 60 and I-64 (milepost 161) and end at the intersection of Kentucky 2 and I-64 (milepost 156) (Fig. 1). Westward thinning of the "Route 2 carbonates" probably is related to depositional and erosional thinning on the eastern flank of the Waverly Arch. Erosional thinning occurred on intraformational disconformities within the Newman Limestone and on the so-called systemic unconformity separating Mississippian and Pennsylvanian rocks; Pennsylvanian dark shales and orthoquartzites disconformably overlie progressively older Mississippian carbonates and greenish-gray clastics toward the apex of the Waverly Arch (Fig. 3B). The Warix Run Member, which is a massive, highly crossbedded, arenaceous calcarenite is present only on the eastern side of the arch (Fig. 3B) and fills structurally related erosional lows. The lower 15 ft (4.5 m) of a cut on westbound lane at milepost 158.6 is an excellent example of the Warix Run; it exhibits large, high-angle trough crossbeds, channel-fill sequences and a quartzose bar-like body over 200 ft (61.0 m) long. Disconformities above and below the Warix Run (Fig. 2) are better developed near the Waverly Arch and are related to episodes of uplift on the arch. Caliche profiles and subaerial exposure crusts are present on most disconformities, but are thicker and better developed near the arch. An excellent caliche profile exhibiting tepee structures, breccias, crusts, and large carbonate nodules is present atop at Paoli–Beaver Bend Member in the lower 15 ft (4.5 m) (below the lower bench) of the cut at milepost 157.1 (Ettensohn, 1981, Fig. 30). A relatively complete, typical exposure of the "Route 2 carbonates" is present on Kentucky 2 just north of its interchange (Exit 156) with I-64. Crossbedded arenaceous calcarenites of the Warix Run Member define the base of the Newman, which disconformably overlie greenish-gray siltstones and shales of the

Borden Formation (Cowbell Member); at the top, dark Pennsylvanian shales of the Breathitt Formation disconformably overlie green shales (Hardinsburg Member) in the upper Newman. A horizon of exposure crusts occurs on top of Paoli–Beaver Bend calcilutites just below the prominent reentrant formed by the Cave Branch Shale. Approximately 85 ft (25.9 m) of Newman carbonates are present between the major disconformities near the base and top of the cut.

Immediately west of the Kentucky 2 interchange, the Newman carbonates undergo abrupt thinning as the interstate crosses the apical region of the Waverly Arch (Figs. 1, 3). For the next 8 mi (128 km) between the intersection of Kentucky 2 with I-64 (milepost 155.5) and the weigh station on I-64 (milepost 147.5), the highway crosses the former "apical island" area of the Waverly Arch (Ettensohn, 1977, 1980, 1981) where the Newman carbonates thin and pinch out (Fig. 3B). Most of the Newman in this area is present in the form of erosional remnants 20 to 800 ft (6.1 to 243.8 m) in length and 10 to 30 ft (3.0 to 9.1 m) in height (Fig. 3B), which exhibit paleokarst and paleoslumps; distorted bedding and collapsed paleosinks filled with green Mississippian and black Pennsylvanian shales are present in exposures at mileposts 155.5, 154.3, and 153.5. By milepost 153, the interstate crosses the apex of the arch where the greatest thinning occurs. The Newman is no longer present, and black Breathitt shales disconformably overlie red and green shales of the Nada Member of the Borden, a relationship that is best seen at the eastern end of the exposure on the westbound lane at milepost 152.6. Westward from this exposure, remnants of the Newman no more than 100 ft (30.5 m) in length and 10 ft (3.0 m) in height reappear; and a progressive thickening of the Mississippian section occurs below the Mississippian-Pennsylvanian unconformity as the highway "descends" the western flank of the Waverly Arch. This trend is shown at the large exposure at milepost 151, where upper parts of the Borden and the lowermost Newman are present just below the systemic unconformity (Ettensohn, 1981: 239–243). Unfortunately, the unconformity and thin carbonates (3 ft [0.9 m] thick) below are largely hidden by vegetation midway up the exposure, so that all one sees are the green shales and siltstones of the Borden (Nada and Cowbell members) in the lowermost 65 ft (19.8 m) of the cut and the dark shales and sandstones of the Breathitt in the uppermost 80 ft (24.4 m). Above the unconformity in the Breathitt on the westbound lane, a superb example of facies relationship between a sandstone body and a dark shale is exposed (Fig. 3A, center); the relationship is interpreted to represent intertonguing between a beach-barrier and back-barrier shales (Ferm and others, 1971; Horne and others, 1974) or between a splay complex and interdistributary shales (Short, 1979b). (To examine this exposure, climb to the highest bench in the cut along the eastbound lane and look northward.) This bench is the continuation of a strip pit from which the Olive Hill fire clay of Crider (1913) was mined. The gray fire clay is locally present atop the systemic unconformity at the base of the Breathitt and has a distribution that is largely restricted to the axis of the Waverly Arch (Ettensohn, 1981:

242–243). The Breathitt continues to be the dominant unit along the interstate through milepost 148.1. At milepost 149 near the Carter-Rowan county boundary, the Breathitt consists of thick sequences of burrowed siltstones and shales which exhibit cyclic fining-upward sequences and channels (Short, 1979a).

After milepost 148.1, Newman carbonates west of the arch, commonly called the "Route 799 carbonates," reappear and thicken westwardly (Fig. 3B). The "Route 799 carbonates" essentially begin at the weigh station (milepost 147.5) and end at the overpass of Kentucky 799 (milepost 146). The Newman members along this part of the highway, however, are generally thinner and contain a more restricted fauna; a new member, the Holly Fork, which consists of orange-brown, fenestral dolostone, is present only on the west side of the arch. Exposures at the weigh station (milepost 147.5) are typical of the "Route 799 carbonates." The red claystones (Cave Branch Member) that form a prominent reentrant in the middle of the exposure may represent a reworked terra rosa soil that formed on top of an exposure-crust horizon on the St. Louis Member (Ettensohn and others, 1984). In-place brachiopod communities are abundant in the Haney Member on top of the bench. The large exposure at milepost 147 shows a thick sequence of Newman carbonates disconformably overlain by black Breathitt shales. Below the prominent reentrant (Cave Branch Member), the top of the St. Louis exhibits an intraformational caliche and paleokarst horizon up to 7 ft (2.1 m) thick with sink holes up to 12 (3.7 m) deep and 4 ft (1.2 m) wide. The prominent dolostone channel-complex, which occurs above the reentrant but below the bench, represents a tidal-flat complex in the Holly Fork and Armstrong Hill members which formed only on the lee side of the Waverly Arch (Ettensohn, 1977, 1980, 1981). The dolostone channel-fill facies of the Holly Fork thickens westwardly to the point that it becomes a major part of the exposures at mileposts 146.4 and 146.2. At the western end of the exposure at milepost 146.4, a Holly Fork dolostone channel fill nearly 10 ft (3.0 m) thick truncates underlying units to the level of the St. Louis. Well developed exposure crusts and microkarst occur on the disconformity atop the St. Louis. At the eastern end of the same cut, spongiostromatid and girvanellid algal mats (Ettensohn, 1975, 1981) are found in the red nodular calcilutite lenses of the Armstrong Hill Member. In the last Newman exposure (milepost 146.2) the gradation between the upper Borden (lower, yellow-brown dolostones of the Renfro Member) and the lower part of the Newman (St. Louis Member) is apparent. The upper, yellow-brown dolostones of the Holly Fork sit disconformably on the St. Louis; normally intervening Newman units have been eroded away.

REFERENCES CITED

Chaplin, J. R., 1980, Stratigraphy, trace-fossil associations and depositional environments in the Borden Formation (Mississippian), northeastern Kentucky: Geological Society of Kentucky Annual Field Conference: Kentucky Geological Survey Guidebook, 114 p.

Crider, A. F., 1913, The fire clays and fire clay industries of the Olive Hill and Ashland districts of northeastern Kentucky: Kentucky Geological Survey, Series 4, v. 1, pt. 2, p. 589–711.

Dever, G. R., Jr., 1980a, Stratigraphic relations in the lower and middle Newman Limestone (Mississippian), east-central and northeastern Kentucky: Kentucky Geological Survey, Series 11, Thesis Series 1, 49 p.

——1980b, The Newman Limestone—an indicator of Mississippian tectonic activity in northeastern Kentucky, *in* Luther, M. K., ed., Proceedings of the technical sessions, Kentucky Oil and Gas Association, thirty-sixth and thirty-seventh annual meetings, 1972 and 1973: Kentucky Geological Survey, Series 11, Special Publication 2, p. 42–54.

Dever, G. H., Jr., Hoge, H. P., Hester, N. C., and Ettensohn, F. R., 1977, Stratigraphic evidence for late Paleozoic tectonism in northeastern Kentucky: American Association of Petroleum Geologists, Eastern Section Annual Meeting: Kentucky Geological Survey Guidebook, 80 p.

Englund, K. J., 1976, Geologic map of the Grahn quadrangle, Carter County, Kentucky: U.S. Geological Survey, Geologic Quadrangle Map GQ—1262, scale 1:24,000, 1 sheet.

Ettensohn, F. R., 1975, Stratigraphy and paleoenvironmental aspects of Upper Mississippian rocks, east-central Kentucky [Ph.D. thesis]: University of Illinois, Urbana, 320 p.

——1977, Effects of synsedimentary tectonic activity on the upper Newman Limestone and Pennington Formation, *in* Dever, G. R., Jr., Hoge, H. P., Hester, N. C., and Ettensohn, F. R., Stratigraphic evidence for late Paleozoic tectonism in northeastern Kentucky, American Association of Petroleum Geologists Eastern Section Annual Meeting: Kentucky Geological Survey Guidebook, p. 18–29.

——1979a, Generalized description of Carboniferous stratigraphy, structure and depositional environments in east-central Kentucky, *in* Ettensohn, F. R., and Dever, G. R., Jr., eds., Carboniferous geology from the Appalachian Basin to the Illinois Basin through eastern Ohio and Kentucky: University of Kentucky, Lexington, p. 64–77.

——1979b, The Barrier-Shoreline Model in northeastern Kentucky, *in* Ettensohn, F. R., and Dever, G. R., Jr., eds., Carboniferous geology from the Appalachian Basin to the Illinois Basin through eastern Ohio and Kentucky: University of Kentucky, Lexington, p. 124–129.

——1980, An alternative to the Barrier-Shoreline Model for deposition of Mississippian and Pennsylvanian rocks *in* Northeastern Kentucky: Geological Society of America Bulletin, v. 91, no. 3, pt. 1, p. 130–135; pt. II, p. 934–1056.

——1981, Mississippian-Pennsylvanian boundary in northeastern Kentucky, *in* Roberts, T. G., ed., Geological Society of America Cincinnati '81, Field Trip Guidebooks, v. 1—stratigraphy and sedimentology: American Geological Institute, p. 195–257.

Ettensohn, F. R., and Dever, G. R., Jr., 1979a, The Newman Limestone on the east side of the Waverly Arch, *in* Ettensohn, F. R., and Dever, G. R., eds., Carboniferous geology from the Appalachian Basin to the Illinois Basin through eastern Ohio and Kentucky: University of Kentucky, Lexington, p. 96–107.

——1979b, The Waverly Arch apical island, *in* Ettensohn, F. R., and Dever, G. R., Jr., eds., Carboniferous geology from the Appalachian Basin to the Illinois Basin through eastern Ohio and Kentucky: University of Kentucky, Lexington, p. 108–112.

——1979c, The Newman Limestone west of the Waverly Arch, *in* Ettensohn, F. R., and Dever, G. R., Jr., eds., Carboniferous geology from the Appalachian Basin to the Illinois Basin through eastern Ohio and Kentucky: University of Kentucky Lexington, p. 119–124.

Ettensohn, F. R., Dever, G. R., Jr., and Grow, J. S., 1984, Fossil soils and subaerial crusts in the Mississippian of eastern Kentucky, *in* Rast, N., and Hay, H. B., eds., Field Trip guides for Geological Society of America Annual Meeting, Southeastern and North-Central sections: Department of Geology, University of Kentucky and Kentucky Geological Survey, p. 84–105.

Ferm, J. C., Horne, J. C., Swinchatt, J. P., and Whaley, P. W., 1971, Carboniferous depositional environments in northeastern Kentucky, Guidebook, Geological Society of Kentucky, Annual Spring Field Conference: Kentucky Geological Survey, 30 p.

Horne, J. C., Ferm, J. C., and Swinchatt, J. P., 1974, Depositional model for the Mississippian-Pennsylvanian boundary in northeastern Kentucky, *in* Briggs, G., ed., Carboniferous of the southeastern United States: Geological Society of America Special Paper 148, p. 97–114.

Philley, J. C., Hylbert, D. K., and Hoge, H. P., 1975a, Geologic map of the Soldier quadrangle, northeastern Kentucky: U.S. Geological Survey Geologic Quadrangle Map GQ-1233, scale 1:24,000, 1 sheet.

——1975b, Geologic map of the Cranston quadrangle, northeastern Kentucky: U.S. Geological Survey Geologic Quadrangle Map GQ—1212, scale 1:24,000, 1 sheet.

Short, M. R., 1979a, Early Pennsylvanian bay-fill environments, Breathitt Formation, *in* Ettensohn, F. R., and Dever, G. R., Jr., eds., Carboniferous Geology from the Appalachian Basin to the Illinois Basin through eastern Ohio and Kentucky: University of Kentucky, Lexington, p. 115–119.

——1979b, Late Mississippian-Early Pennsylvanian sedimentation on the Waverly Arch apical island, *in* Ettensohn, F. R., and Dever, G. R., Jr., eds., Carboniferous geology from the Appalachian Basin to the Illinois Basin through eastern Ohio and Kentucky: University of Kentucky Lexington, p. 112–115.

Red River Gorge Geological Area (Daniel Boone National Forest) and Natural Bridge State Park, east-central Kentucky

Garland R. Dever, Jr., *Kentucky Geological Survey, University of Kentucky, Lexington, Kentucky 40506*
Lance S. Barron, *Kentucky Center for Energy Research Laboratory, P.O. Box 13015, Lexington, Kentucky 40512*

LOCATION

The Red River Gorge Geological Area and Natural Bridge State Park are located 32.8 and 40 mi (55 and 67 km) southeast of I-64 on Mountain Parkway and Kentucky 11 and 715; Menifee, Powell and Wolfe Counties; Slade and Pomeroyton 7½-minute Quadrangles (Fig. 1).

INTRODUCTION

The concentration of natural arches in the Red River Gorge Geological Area of the Daniel Boone National Forest and adjoining Natural Bridge State Park is unique in the eastern United States (Fig. 2). More than 150 arches have been catalogued by the U.S. Forest Service.

The geological area and state park are in the intricately dissected Pottsville Escarpment that forms the western border of the Cumberland Plateau. Natural arches occur locally along the length of the escarpment in eastern Kentucky, but the greatest number are in the drainage of the Red River where the effects of erosion and mass wasting have created geomorphic conditions,

Figure 1. General location map.

principally narrow sandstone ridges, particularly suitable for the formation of arches. The Pottsville Escarpment is capped by cliff-forming sandstone underlain by less-resistant shale, limestone, and siltstone. Arches principally are in the Corbin Sandstone Member of the Lee Formation (Stops 1, 2, 4, and 5); they also

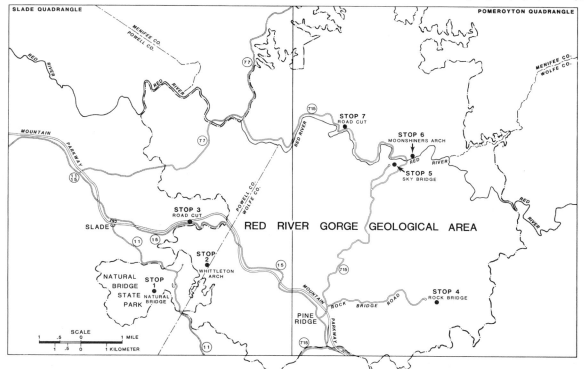

Figure 2. Map of Red River Gorge Geological Area and Natural Bridge State Park showing stops and highways.

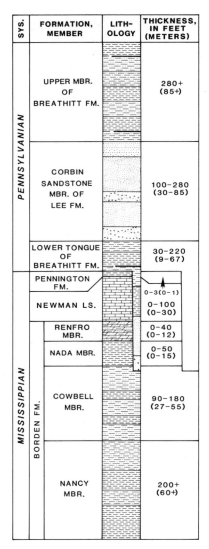

SYS.	FORMATION, MEMBER	LITH-OLOGY	THICKNESS, IN FEET (METERS)
PENNSYLVANIAN	UPPER MBR. OF BREATHITT FM.		280+ (85+)
	CORBIN SANDSTONE MBR. OF LEE FM.		100-280 (30-85)
	LOWER TONGUE OF BREATHITT FM.		30-220 (9-67)
MISSISSIPPIAN	PENNINGTON FM.		0-3(0-1)
	NEWMAN LS.		0-100 (0-30)
	RENFRO MBR.		0-40 (0-12)
	NADA MBR.		0-50 (0-15)
	COWBELL MBR.		90-180 (27-55)
	NANCY MBR.		200+ (60+)

(BORDEN FM.)

Figure 3. Generalized columnar section for Slade and Pomeroyton quadrangles. Adapted from Weir (1974) and Weir and Richards (1974).

occur in the Newman Limestone (Stop 6). With accessibility via roads and hiking trails, the geological area and state park provide an excellent opportunity to observe and study the effects of differential weathering and erosion, mass wasting, and jointing in the development of cliffs, rock shelters, arches, and chimney rocks. Geology of the area was described by McFarlan (1954) and mapped by Weir (1974) and Weir and Richards (1974). Trail descriptions were published by Ruchhoft (1976).

Mississippian and Pennsylvanian sedimentary rocks crop out in the area (Fig. 3). Average dip is about 30 ft per mi (5.6 m per km) east-southeast. The Pennsylvanian-age Breathitt and Lee Formations cap the uplands and underlie upper slopes of the hillsides; the Mississippian-age Borden Formation, Newman Limestone, and, locally, Pennington Formation underlie the lower slopes and valleys. These sedimentary rocks represent prodelta and lower delta-front sediments (lower and middle Borden

Formation), shallow marine platform to supratidal deposits (upper Borden and Newman Limestone), and deltaic-fluvial sediments (Breathitt and Lee Formations) (Dever and others, 1977; Rice, 1984; Smith and others, 1971). The geological area and state park are in the northern part of the Rome Trough, a fault-bounded basin identified in the subsurface. Bounding and internal faults were active intermittently from Cambrian into post-Pennsylvanian time. The Mississippian-Pennsylvanian boundary is a regional unconformity (Stops 3 and 7).

SITE DESCRIPTION

The Red River Gorge Geological Area and Natural Bridge State Park are readily accessible via the Mountain Parkway (Fig. 1). Exit from the Mountain Parkway at the interchange with Kentucky 11 at Slade (Fig. 2); the Slade interchange is 32.8 mi (55 km) southeast of the Mountain Parkway interchange with I-64. Turn onto Kentucky 11 and proceed southeastward 2.6 mi (4.3 km) to the second entrance on the right into Natural Bridge State Park. (First park entrance on the right leads to Hemlock Lodge.) Turn right into the park and travel westward 0.2 mi (0.3 km) to parking area on the west side of the Middle Fork of the Red River. Natural Bridge (Stop 1) is reached by following Park Trail 1 from the north side of the parking area. Newman Limestone crops out locally along the trail and large colluvial sandstone blocks, derived from the Corbin Sandstone Member of the Lee Formation, rest on the slopes. Average walking time, one way, is about 30 minutes.

Stop 1: Natural Bridge

Natural Bridge is a ridge-top arch developed in the upper part of the Corbin Sandstone Member of the Lee Formation. Orientation and configuration of Natural Bridge are controlled mainly by joints (Fig. 4). The length of the arch parallels a nearly vertical joint set with an average strike of 303° (N. 57° W.). An intersecting joint set beneath the arch has an average strike of 270° (due west).

Ridge-top arches form in narrow sandstone ridges on divides. Mass wasting is a major process, in addition to weathering and erosion, in the development of the narrow ridges. Sandstone blocks, formed by the weathering and wedging action of water, ice, and roots penetrating along joints and bedding planes, are detached and move downslope by gravity. Erosional undermining of less-resistant underlying strata contributes to the detachment.

Return to Kentucky 11 and proceed northward 0.3 mi (0.5 km) to the entrance to the Whittleton Branch Camping Area. Turn right into the parking area. Whittleton Arch (Stop 2) is reached by following Forest Service Trail 216 from the northeast end of the camping area up Whittleton Branch. Large colluvial sandstone blocks, derived from the Corbin Sandstone Member of the Lee Formation, rest in the stream bed. At the east end of a prominent limestone outcrop along the south side of the branch,

Figure 4. Intersecting joint sets in southeast end of Natural Bridge. (Courtesy Fred E. Coy, Jr.).

turn right and follow Forest Service Trail 217 to Whittleton Arch. Average walking time, one way, is about 40 minutes.

Stop 2: Whittleton Arch

Whittleton Arch is a valley-head arch developed near the base of the Corbin Sandstone Member of the Lee Formation. Orientation of the arch and its suggested origin are related to a joint set that parallels its length. Average strike is 35° (N. 35° E.). Water from a surface stream flowing across a sandstone ledge penetrated vertical joints in the stream bed on the back (southeast) side of the present arch. The downward-moving water encountered the relatively impermeable shale and siltstone of the lower tongue of the Breathitt Formation underlying the Corbin and issued as springs near the base of the Corbin. Leaching of sandstone cement and erosion resulting from water movement along the joints and through the sandstone (together with probable freeze-thaw action and root penetration) enlarged openings along the joints and started development of a rock shelter (shallow cave) on the front (northwest) side of the sandstone cliff. Intersection of the rock shelter with an enlarged vertical joint formed the arch, which was enlarged by roof falls and erosion. The stream now flows across the sandstone ledge immediately southwest of the arch; part of its present flow enters a vertical joint and issues from the sandstone face.

Return to Kentucky 11 and proceed northwestward 2.3

mi (3.8 km) to the Mountain Parkway interchange at Slade. Enter the eastbound lanes of the Mountain Parkway and travel eastward 1.9 mi (3.2 km) to the roadcut through the Newman Limestone and lower tongue of the Breathitt Formation along the south side of the highway.

Stop 3: Mississippian-Pennsylvanian Boundary

Sandstone and shale of the lower tongue of the Breathitt Formation (Pennsylvanian) rest on a very irregular surface on the Newman Limestone (Mississippian). The irregular surface is related to karst developed in the Newman prior to deposition of Pennsylvanian sediments. Prior karstification is indicated by the presence of deposits of Breathitt-type sandstone and shale in small solution channels in the limestone exposed at the west end and near the middle of the roadcut. The slumped sandstone and shale and deformed bedding of the overlying coal and sandstone in the western part of the roadcut suggest collapse into a sinkhole after deposition of the Pennsylvanian sediments. Original configuration of the sandstone and shale is shown by the undeformed beds in the eastern part of the roadcut.

Continue on the Mountain Parkway and travel southeastward 5.3 mi (8.8 km) to Exit 40, leading to Kentucky 15 and 715. At the end of the exit ramp, turn right onto combined Kentucky 15 and 715 and proceed northward 0.8 mi (1.2 km) to road junction in Pine Ridge. Turn right onto Kentucky 715 and travel northeastward 0.4 mi (0.6 km) to the Red River Gorge Geological Area Information Center. Turn right onto Rock Bridge Road and proceed eastward 3.2 mi (5.3 km) to the Rock Bridge parking area. Rock Bridge (Stop 4) is reached by following Forest Service Trail 207 from the south side of the parking area. Average walking time, one way, is about 30 minutes.

Stop 4: Rock Bridge

Rock Bridge across Swift Camp Creek is a waterfall arch developed in the lower part of the Corbin Sandstone Member of the Lee Formation. The arch is in a local tongue of sandstone, split off the base of the main Corbin body, that intertongues with the lower tongue of the Breathitt Formation (Weir and Richards, 1974). The main body of the Corbin crops out along the upper part of Trail 207. The origin of Rock Bridge outlined by McFarlan (1954) is analogous to the origin of Whittleton Arch. Water from Swift Camp Creek penetrated a vertical joint in the stream bed upstream from a waterfall, moved laterally through the sandstone, and issued from the face of the waterfall. Leaching of sandstone cement and erosion enlarged the pathway through the joint and sandstone, progressively diverting the stream flow beneath the top of the waterfall. With complete diversion, the upper part of the former waterfall was left as a sandstone span (Rock Bridge) across Swift Camp Creek. The position of the waterfall migrated upstream and presently is a short distance upstream from the mouth of Rock Bridge Fork (McFarlan, 1954; Trail Marker 12). Average strike of the joint set paralleling the length of the main span of Rock Bridge is 305° (N. 55° W.). Strikes of joints in the waterfall range from 322° to 287° (N. 38°–73° W.).

Return to the Red River Gorge Geological Area Information Center, turn right onto Kentucky 715, and travel northward 4.5 mi (7.2 km) to the junction of Ridge Drive and Gorge Drive (Kentucky 715). Proceed straight ahead on Ridge Drive and travel northeastward 0.8 mi (1.3 km) to the Sky Bridge parking area. Sky Bridge (Stop 5) is reached by following Forest Service Trail 214 from the east end of the parking area. The trail crosses Sky Bridge and forms a loop leading beneath the arch and ending back at the parking area. Average walking time for the loop is about 20 minutes.

Stop 5: Sky Bridge

Sky Bridge is a ridge-top arch developed on a narrow sandstone ridge in the upper part of the Corbin Sandstone Member of the Lee Formation. It is similar to Natural Bridge, but sculpturing by weathering and erosion is in a more advanced stage. Planar joint surfaces and their role in controlling the orientation and configuration of Sky Bridge consequently are not as readily evident as at Natural Bridge. Orientation of the narrow sandstone ridge and Sky Bridge is controlled by a nearly vertical joint set that is exposed in part of the south face of the arch, in the pier of the small secondary arch, and in cliffs along the north side of the ridge. In Sky Bridge, average strike of the joint set is 70° (N. 70° E.); average strike is 79° (N. 79° E.) in cliffs west of the arch and 63° (N. 63° E.) in cliffs east of the arch. Intersecting joint sets have average strikes of 0° (due north) and 274° (N. 86° W.) and have controlled the formation of sandstone blocks in the lower east end of Sky Bridge. A curved joint, with strike ranging from 54° to 75° (N. 54°–75° E.), is present in the span.

Return to the junction of Ridge Drive and Gorge Drive, turn left onto Gorge Drive (Kentucky 715), and proceed northeastward 1.4 mi (2.3 km) to the concrete bridge across the North Fork of the Red River (Menifee County boundary). Turn right into the parking area on the north side of the bridge. Moonshiners Arch (Stop 6) is reached by walking upstream on the footpath along the river bank. The path crosses springs issuing from dolomite in the upper part of the Borden Formation. Average walking time, one way, is about 5 minutes.

Stop 6: Moonshiners Arch

Moonshiners Arch, a limestone arch developed in the lower part of the Newman Limestone, was formed by local roof collapse in a cave passage exposed by erosion. The cave with its floor in the Newman Limestone formerly was part of an underground system draining southward into the Red River. The underground drainage system is still active, but now exits into the river through passages in the dolomite of the upper Borden Formation (Renfro Member) underlying the Newman. Talus-covered springs issuing from the dolomite are present along the footpath from the parking area; the mouth of an actively draining cave is exposed a short distance upstream from Moonshiners Arch.

Continue on the Gorge Drive (Kentucky 715) and travel westward 2.9 mi (4.8 km) to Bell Falls. Large colluvial sandstone blocks, derived from the Corbin Sandstone Member of the Lee Formation, rest on slopes and in the river bed. Park beside the highway at Bell Falls.

Stop 7: Mississippian-Pennsylvanian Boundary: Bell Falls Channel Fill

The road cut overhanging the highway consists of conglomerate containing abundant chert clasts (derived from Newman Limestone), geodes (derived from Borden Formation), quartz pebbles and sand (source area north and northeast of Kentucky; Rice, 1984), and scattered tree fragments (intrabasin lowland flora). The conglomerate is part of a Pennsylvanian sandstone body, more than 100 ft (30 m) thick, in the lower tongue of the Breathitt Formation (Weir and Richards, 1974). It fills a channel cut into siltstone of the Cowbell Member of the Borden Formation (Mississippian), which is exposed along the highway immediately to the west. The Pennington Formation, Newman Limestone, and upper part (Renfro and Nada Members) of the Borden Formation, all Mississippian units, were removed by erosion prior to deposition of Pennsylvanian sediments (Fig. 3). This channel is part of a southeastward-draining paleovalley, mapped by Rice (1984).

Continue on the Gorge Drive (Kentucky 715) and proceed westward 4.2 mi (7 km) to the junction with Kentucky 77. Turn left onto Kentucky 77 and travel southwestward 5.3 mi (8.8 km) to the junction with combined Kentucky 11 and 15. Turn left onto combined Kentucky 11 and 15 and proceed southeastward 1.5 mi (2.5 km), returning to the Mountain Parkway interchange at Slade.

REFERENCES CITED

Dever, G. R., Jr., Hoge, H. P., Hester, N. C., and Ettensohn, F. R., 1977, Stratigraphic evidence for late Paleozoic tectonism in northeastern Kentucky (American Association of Petroleum Geologists Eastern Section meeting guidebook): Lexington, Kentucky Geological Survey, 80 p.

McFarlan, A. C., 1954, Geology of the Natural Bridge State Park area: Kentucky Geological Survey, ser. 9, Special Publication 4, 31 p.

Rice, C. L., 1984, Sandstone units of the Lee Formation and related strata in eastern Kentucky: U.S. Geological Survey Professional Paper 1151-G, 53 p.

Ruchhoft, R. H., 1976, Kentucky's land of the arches: Cincinnati, Ohio, The Pucelle Press, 132 p.

Smith, G. E., Dever, G. R., Jr., Horne, J. C., Ferm, J. C., and Whaley, P. W., 1971, Depositional environments of eastern Kentucky coals (Geological Society of America Coal Geology Division field conference guidebook): Lexington, Kentucky Geological Survey, 22 p.

Weir, G. W., 1974, Geologic map of the Slade quadrangle, east-central Kentucky: U.S. Geological Survey Geologic Quadrangle Map GQ-1183, scale 1:24,000.

Weir, G. W., and Richards, P. W., 1974, Geologic map of the Pomeroyton quadrangle, east-central Kentucky: U.S. Geological Survey Geologic Quadrangle Map GQ-1184, scale 1:24,000.

"Four Corners": A 3-dimensional panorama of stratigraphy and depositional structures of the Breathitt Formation (Pennsylvanian), Hazard, Kentucky

Donald R. Chesnut, James C. Cobb, and John D. Kiefer, Kentucky Geological Survey, University of Kentucky, Lexington, Kentucky 40506

LOCATION

The "Four Corners" exposure is located at the intersection of Kentucky 80, Kentucky 15, and the Daniel Boone Parkway about 2.7 mi (4.3 km) northward of Hazard in Perry County, Kentucky. The stop is located in Carter coordinate section 14-S-76, on the Hazard North 7½-minute Quadrangle (Fig. 1). The multilane highways offer easy accessibility by vehicle. Buses may park along the shoulder of the Daniel Boone Parkway or Kentucky 80. Multiple benches on these high roadcuts offer access to most of the strata by foot; however, much of the geology is visible from the road. Potential dangers such as traffic, extreme heights, rock falls, and unstable footing should be stressed. Be particularly aware of people below you. Do not enter adits or overhangs of adits.

INTRODUCTION

The intersection of two major highways has created a magnificent series of roadcuts up to 350 feet (107 m) high opening a three-dimensional panorama of the Breathitt Formation (Middle Pennsylvanian) at the Four Corners. The significance of this site lies in the stratigraphic range and the number of depositional environments represented by the rocks exposed here. The Breathitt Formation contains all of the major coals in eastern Kentucky, and a significant number of the mineable coals in the region are present in this section. Named stratigraphic units exposed in these roadcuts are, from oldest to youngest, the Magoffin Member, Haddix coal zone, Hazard (Hazard No. 5A) coal zone, Hazard No. 7 coal, Francis (Hazard No. 8) coal, and the Hindman (Hazard No. 9) coal. The sections contain lithologies representing a variety of depositional environments from marine to terrestrial. Both marine and terrestrial fossils can be found, along with a large variety of sedimentary features including large-scale slump structures, cut-and-fill features, accretion bedding, fining-upward and coarsening-upward sequences, and many other classical features of bay-fill, peat-swamp, and fluvial environments.

This section and the numerous other spectacular outcrops along the Daniel Boone Parkway and Kentucky 80 provide an outstanding field laboratory of clastic environments which can be used in conjunction with the guidebook for the Coal Section field trip of the 1981 Geological Society of America Annual Meeting (Cobb and others, 1981), and the readily available 7½-minute geologic quadrangle maps (e.g., Seiders, 1964) of the area.

The Magoffin Member of the Breathitt Formation represents a marine depositional environment which received terrigenous

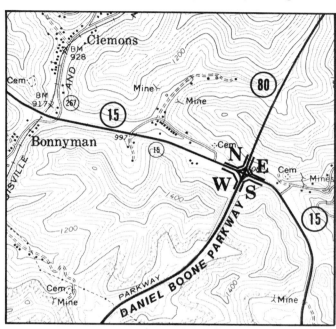

Figure 1. Location map of "Four Corners."

clastic sediments. Upon these deposits the Haddix coal zone peat developed. The Haddix coal is mined in Breathitt County to the north, but in many areas it is absent. Brackish to fresh-water influences were still active during Haddix deposition, as indicated by the fossiliferous black shale between coals of this zone.

Crevasse-splay sediments were deposited over the Haddix peat deposits and built up to a level which supported tree growth. In-place stumps and transported logs can be seen in this zone. A reduction in sediment influx allowed peat accumulation which later became the Hazard coal. A sandstone with accretion beds directly overlying the thickest Hazard coal indicates fluvial processes, particularly point-bar deposition onto the peat swamp. This point-bar-type deposit marks the occurrence of a river channel in the area. The accretion beds are overlain by overbank deposits of siltstone containing abundant plant fossils and in-place stumps. Although sporadic peat accumulation continued (upper part of Hazard coal zone), this sequence is dominated by thick overbank deposits. Compaction developed faster than sediment influx, and marine conditions developed above the overbank deposits. Fossil fragments, calcareous concretions, and burrowing indicate brackish to marine conditions for these deposits, but the grain size still indicates a high terrigenous clastic influx.

After the bay was filled with sediments, peat, which later became the Hazard No. 7 coal, began to accumulate. A succession of overbank deposits covered this extensive peat deposit. The burial of in-place stumps by these overbank deposits preserved a good example of a Pennsylvanian forest in eastern Kentucky. There are, in fact, at least two successive preserved forests separated by a dark claystone and several thin coals midway up the interval. The preserved trees probably represent a mature forest because many reached a diameter of 3.0 ft (91 cm; Jennings, 1981).

A diversion in the major drainage system brought an end to overbank deposition in this area. During this period of emergence, peat, which became the Francis coal, accumulated. A period of more rapid subsidence caused brackish or freshwater to inundate the Francis coal swamp at least once, depositing a fossiliferous claystone parting; however, peat accumulation resumed. The thick sandstone above the Francis coal represents the return of a major distributary. This distributary eroded peat deposits in some places and caused major slumping in other places.

The Francis rider, or Hazard No. 8 rider, may have been deposited under marine influence, resulting in the unusual calcareous concretions in the claystone immediately above the coal. These calcareous concretions can be traced for several miles eastward. Marine fossils have been reported in this interval in cores near here. These sediments are followed by fluvial channel and point-bar deposits.

Peat that formed the Hindman coal accumulated over the thick fluvial deposits during a period of emergence coupled with uniform subsidence. Although a marine zone is present in some areas above this coal, none is evident here; however, marine conditions probably existed here after the peat was formed.

SITE DESCRIPTION

Figure 2 is a diagrammatic sketch showing positions of the coal beds, their stratigraphic nomenclature, and structures and lithologies of enclosing strata.

The Magoffin Member of the Breathitt Formation is exposed only in the lower cuts along Kentucky 15 and the entrance ramp from Kentucky 15 to the westbound Daniel Boone Parkway on the western roadcut. The Magoffin Member is a medium-gray shale containing large calcareous concretions. The Haddix coal zone overlies the Magoffin and is present along the same entrance ramp to the Parkway. The Haddix zone consists of several thin coals separated by siltstone and black fossil shale. A black shale containing the pelecypod, *Naiadites* and occasional shark teeth occurs between two of the Haddix coals. A sandstone containing log impressions, vertical stumps, and cut-and-fill structures extends up to the Hazard coal zone.

The Hazard (Hazard No. 5A) coal zone consists of several coals separated by silty shale, siltstone, and fine-grained sandstone. The two thickest coals in this zone are 1.1 and 4.3 ft (0.33 to 1.3 m) thick. Above the upper coal, which is the thickest, is a fining-upward sequence. Accretion-type beds occur at the top of the sandstone and interfinger with siltstone. This siltstone, which is intensively rooted, contains vertical stumps, *Calamites,*

and siderite nodules. The fining-upward sequence culminates in three thin coals, all less than 0.3 ft (9.1 cm) thick.

The thin coals are overlain by a marine sequence which has been traced into several adjacent counties. Evidence for a marine environment includes calcareous concretions, burrowed zones, and shell fragments.

The Hazard No. 7 coal overlies an argillaceous sandstone with siderite nodules and flaser bedding. The full seam thickness of the Hazard No. 7 coal is 7.1 ft (2.2 m). Recognizing the Hazard No. 7 coal in this area is easy because it is extensively mined and the mine adits, now partially collapsed, are obvious. These collapsed adits give an interesting perspective of roof falls.

Siltstone, fine-grained sandstone, and shale overlie the Hazard No. 7 coal. Ripples, flaser bedding, lenticular bedding, siderite nodules, rooting, and plant fossils are common (Jennings, 1981). Very thin coals also occur. Of particular interest are the more than 20 vertical stumps and trunks found in the interval between the Hazard No. 7 coal and the Francis (Hazard No. 8) coal. The largest of these measures 3.0 ft (91 cm) in diameter, and the average size is more than 1.0 ft (30 cm) in diameter.

The Francis (Hazard No. 8) coal overlies the siltstone and shale above the Hazard No. 7 coal. In the southern corner of this intersection the Francis coal is eroded, in the western corner it is eroded and slumped, and in the northern corner the Francis coal reaches a maximum thickness of 8.6 ft (2.6 m) with up to 3.0 ft (91 cm) of clay parting. The clay parting contains brackish- or fresh-water fossils. The Francis coal is overlain by a thin and discontinuous claystone and a thick, coarse-grained sandstone. This sandstone is the beginning of a fining-upward sequence and is capped by a coal bed, informally known either as the Francis rider or Hazard No. 8 rider. This coal is 0.7 ft (21 cm) thick and is characterized by oddly shaped, grapefruit-size, calcareous nodules. The nodules contain cone-in-cone structures and are enclosed in a dark-gray claystone unit that is eroded at the top. A conglomeratic sandstone that fines upward (fining-upward sequence 3) overlies the claystone. It contains coal spar, pebbles, log impressions, and convoluted bedding.

The Hindman (Hazard No. 9) coal is 3.1 ft (94 cm) thick. It is overlain by a dark-gray shale; a thick, dark-gray, silty shale; siltstone and sandstone interbeds; and a conglomeratic sandstone with log impressions. The shales and siltstones are part of a marine unit known as the Stoney Fork Member. At many locations these shales contain abundant marine fossils; however, none have been reported from this locality. The fossiliferous part of the shale may have been eroded. The Stoney Fork Member is a key stratigraphic marker bed in eastern Kentucky. For details about the coals at this locality, see Hower and others, (1981), especially Tables 8, 9, and 10.

REFERENCES CITED

Cobb, J. C., Chesnut, D. R., Hester, N. D., and Hower, J. C., 1981, Coal and coal-bearing rocks of eastern Kentucky: Guidebook for Geological Society of America Coal Division field trip, November 5–8, 1981, Kentucky Geo-

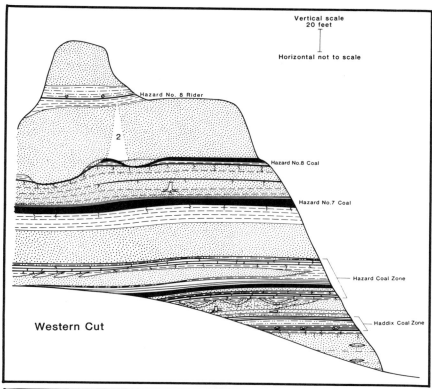

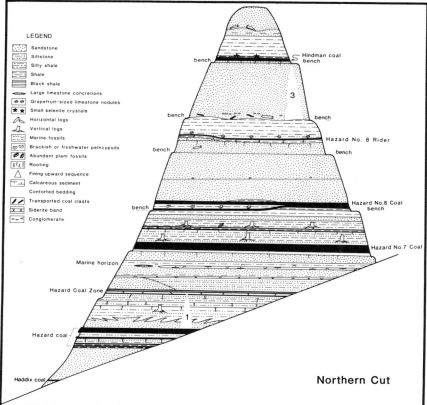

Figure 2. Schematic diagram of rocks in the Breathitt Formation (Middle Pennsylvanian) exposed in two roadcuts at "Four Corners," near Hazard, Kentucky. Numbered triangles represent fining-upward sequences.

logical Survey, University of Kentucky, Lexington: 169 p.

Hower, J. C., Bland, A. E., Fiene, F. L., and Koppenaal, D. W., 1981, Petrology, mineralogy and geochemistry of coals in the central portion of the Eastern Kentucky Coal Field, *in* Cobb, J. C., Chesnut, D. R., Hester, N. D., and Hower, J. C., Coal and coal-bearing rocks of eastern Kentucky: Guidebook for Geological Society of America Coal Division field trip, November 5–8, 1981, Kentucky Geological Survey, University of Kentucky, Lexington: p. 131–146.

Jennings, J. R., 1981, Pennsylvanian plants of eastern Kentucky: Compression fossils from the Breathitt Formation near Hazard, Kentucky, *in* Cobb, J. C., Chesnut, D. R., Hester, N. D., and Hower, J. C., Coal and coal-bearing rocks of eastern Kentucky: Guidebook for Geological Society of America Coal Division field trip, November 5–8, 1981, Kentucky Geological Survey, University of Kentucky, Lexington: p. 147–159.

Seiders, V. M., 1964, Geology of the Hazard North Quadrangle: U.S. Geological Survey Geologic Quadrangle Map GQ 344.

Pennsylvanian-age distributary-mouth bar in the Breathitt Formation of eastern Kentucky

Donald R. Chesnut and James C. Cobb, Kentucky Geological Survey, University of Kentucky, Lexington, Kentucky 40506

LOCATION

A Pennsylvanian-age distributary-mouth bar in the Breathitt Formation can be seen at two localities in Knott County, Kentucky. The northwest locality is on Kentucky 80, 500 ft (152 m) west of the intersection of Kentucky 80 and the Hindman access road ("Hindman Spur") (Fig. 1). This stop is located in Carter coordinate section 16-K-79, in the northwestern corner of the Hindman 7½-minute Quadrangle. The multilane highway offers easy accessibility by vehicle, and buses may park along the

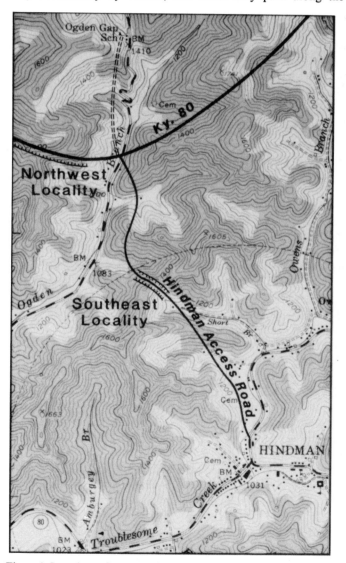

Figure 1. Locations of road cuts in the Hindman 7½-minute Quadrangle where distributary sandstones and bayfill rocks are well exposed.

shoulder of Kentucky 80, between the road cut and the intersection of the Hindman access road.

The southeast locality is a road cut located on the Hindman access road about 0.5 mi (0.8 km) south of the intersection of Kentucky 80 and the access road. This stop is in the northwestern corner of the Hindman 7½-minute Quadrangle in Carter coordinate section 25-K-79. Buses may park in the large flat area on the southwestern side of the road just west of the road cut.

INTRODUCTION

These two road cuts illustrate a sequence of deltaic sedimentation seldom preserved in the Pennsylvanian-age rocks of eastern Kentucky. A coarsening-upward marine "bay" fill grades into overlaying distributary-mouth bar sandstones, which are, in turn, truncated by the distributary and overlain by distributary-channel sandstones. This cycle ended with the formation of peat. A zone of penecontemporaneous slumping is associated with the distributary sandstones and bayfill rocks at the southeast locality.

All the units observed in these road cuts are in the Breathitt Formation (Pennsylvanian). The strata exposed at the northwest locality range from the top of the Magoffin Member through the Hazard coal zone (Fig. 2). The strata exposed at the southeast locality range from the Copland coal zone through the Hazard coal zone and also contain the Magoffin Member (Fig. 3). The Magoffin Member is a very extensive marine member and is very important as a key stratigraphic unit in much of the central Appalachian Basin.

Additional information about the highway geology along this road and others may be found in Cobb, and others, 1981. A highway geological cross section for Kentucky 80 from Hazard, Kentucky, to U.S. 23 (D. R. Chesnut, in preparation) is nearing completion. All 7½-minute Quadrangles for the state are available.

SITE DESCRIPTION

Northwest Locality

The lower part of the section consists of calcareous siltstone that contains siderite layers and burrows. This grades upward into silty, fine-grained sandstones that alternate with layers of fine-grained sandstone. The silty sandstones contain calcareous concretions, and the fine-grained sandstone layers contain convoluted bedding and flow rolls. The next 15 to 30 ft (4.5 to 9 m) of section consists of fine- to medium-grained, flaggy-bedded sandstone. The bedding changes in attitude and thickness, becoming

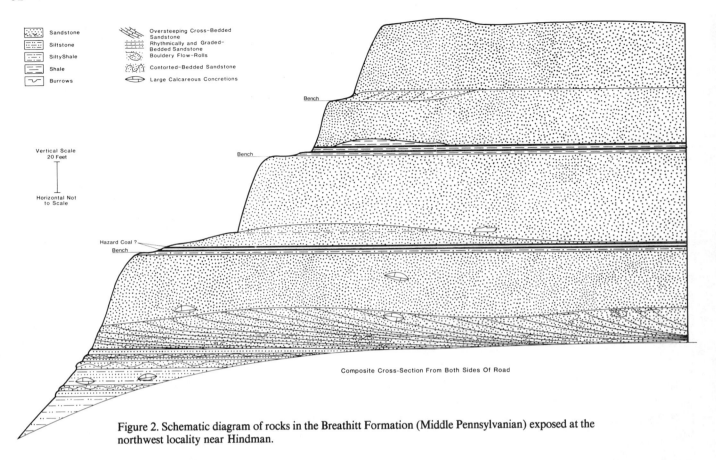

Figure 2. Schematic diagram of rocks in the Breathitt Formation (Middle Pennsylvanian) exposed at the northwest locality near Hindman.

thinner and tangential to the west where the bedding planes are nearly parallel and horizontal, but to the east diverging up dip, attaining maximum dips of 10 to 15 degrees (Fig. 2). The horizontal beds are about 6 in (15 cm) thick, but thicken to 20 in (50 cm) where steeply dipping. These beds are rippled and slightly calcareous and have poorly defined graded bedding. They contain horizontal and vertical trace fossils (Cobb, 1981, p. 35–36). Vertical escape burrows are common; they begin at the bottom and extend to the tops of individual graded beds.

An erosion surface truncates the inclined beds. Overlying this surface is a more massive, medium-grained sandstone. The upper part of this sandstone is rooted and capped by the Hazard coal. Strata overlying the coal include sandstone, shale, and coal.

The section below the lower coal represents a very extensive open marine bay that was progressively filled by increasingly coarser clastic material as deltaic sediments prograded into the bay. The flaggy-bedded, accretionary sands were deposited in floods. These beds were then truncated by the distributary as it eroded its previous deposits. Following abandonment of the distributary systems, vegetation was established and peat accumulated. Overlying these deposits are several sandstone sequences of probably fluvial origin and coals.

Southeast Locality

Strata associated with the Copland coal zone (Fig. 3) include rooted claystone shales, siltstones, and sandstones; in-place stumps; and thin coals. This interval contains abundant plant fossils (flora identified by James Jennings in Cobb and others 1981, p. 37–38) and represents part of a fining-upward sequence (No. 1 on Fig. 3). The next 60 ft (18.2 meters) is clayey, burrowed, calcareous siltstone and silty shale with fossiliferous limestone concretions. At the base is a highly organic shale with a low-diversity, depauperate, restricted marine fauna. A normal Magoffin marine fauna occurs a little higher in the section. This is a coarsening-upward sequence (No. 2 in Fig. 3) known as the Magoffin Member and is the same unit observed at the base of the section at the northwest locality. The next 30 ft (9 m) is disturbed. These beds may be a continuation of the previous coarsening-upward sequence represented by silty sandstones at the northwest locality. The bedding is tilted, generally to the south, and offset by normal faults that are displaced down to the north. The faults continue into the underlying marine siltstones for only a few feet, appearing to become tangential with the bedding of the underlying siltstone. Smaller-scale features observed in the disturbed beds include disharmonic folds, down-dip thickening of beds, boudinage of sideritic layers, and fracturing and mild deformation of calcareous concretions.

An erosional surface occurs at the top of this intensely disturbed section, which is overlain by approximately 50 ft (15 m) of sandstone in a fining-upward sequence (No. 3 in Fig. 3). This

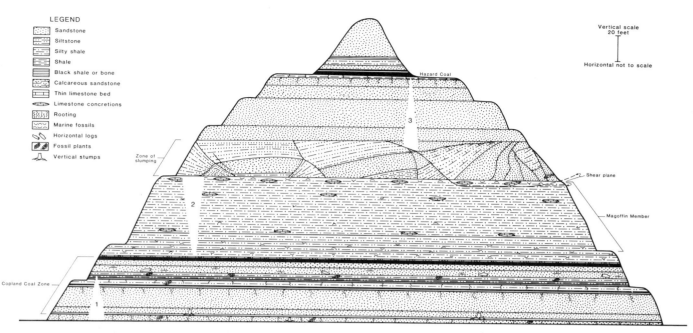

Figure 3. Schematic diagram of rocks in the Breathitt Formation (Middle Pennsylvanian) exposed at the southeast locality near Hindman.

sandstone contains accretion bedding that dips to the north in the upper 14 ft (4.5 m). The primary sedimentary structures in this 15-ft (4.5-m) section, in ascending order, are large-scale trough crossbeds followed by smaller-scale trough crossbeds overlain by ripple-bedded, fine-grained sandstones capped by intensely rooted sandstone, claystone, and coal. The rest of the section is composed of sandstone, clayey siltstone, plant-rich claystone, and canneloid shale.

The in-place stumps, abundant plant fossils, rooting, and fining-upward deposits support a fresh-water origin for the Copland coal zone and associated strata. Subsidence (or eustatic change in sea level) resulted in the transgression of a marine bay over the upper Copland peat, and dysaerobic or anaerobic conditions existed briefly. The overlying 60 ft (18.2 m) of section represents the infilling of this bay with terrigenous clastic deposits. Marine conditions persisted throughout the time of deposition of these finer-grained beds, as indicated by the marine fauna in the upper portion of the sequence. The disturbed zone resulted from

slumping in response to channel erosion. This river channel was likely a part of the distributary system responsible for deposition of the mouth-bar sediments at the northwest locality.

The fluvial-sandstone sequence above the paleoslumps represents the migration of a river across the slumped strata, partially eroding them and then burying them. Peat-forming environments were again established when the river system migrated out of the area.

REFERENCES CITED

Cobb, J. C., Chesnut, D. R., Hester, N. C., and Hower, J. C., 1981, Coal and coal-bearing rocks of eastern Kentucky (Guidebook for Geological Society of America Coal Division field trip, Nov. 5–8): Lexington, Kentucky Geological Survey, University of Kentucky, 169 p.

Chesnut, D. R. (in preparation), Geological highway cross section along Highway 80 from Hazard to Water Gap, Kentucky: Kentucky Geological Survey, open file report.

Burning Springs Anticline, West Virginia

Jonathan K. Filer, West Virginia Geological and Economic Survey, P.O. Box 879, Morgantown, West Virginia 26507-0879

LOCATION

The site is located in northwestern West Virginia, straddling the Wood/Ritchie county line along U.S. 50, 12 mi (20 km) east of the intersection of that highway with Interstate 77 near Parkersburg, West Virginia (Fig. 1). The site is on the southeast corner of the Willow Island, West Virginia, 7½-minute Quadrangle. Exposures are in road cuts on either side of the highway. No permission is necessary for access.

INTRODUCTION

The Burning Springs Anticline is a unique feature of the outer Allegheny Plateau in western West Virginia because of its structural orientation and relief: due N-S trend, dips of up to 70°, and structural relief of over 1600 ft (485 m). Nearby structures generally trend about N30°E, have maximum dips of about 2° or less, and amplitudes of 200 ft (60 m) or less. The identification or repeated beds due to thrust faulting in the Lower and Middle Devonian section of the Sandhill deep well (Wood-351, see Fig. 2) drilled on the crest of the structure (Woodward, 1959) led to the interpretation of the anticline as detached. This resulted in the recognition of the allochthonous nature of the relatively undisturbed plateau between the anticline and the high folds of the inner plateau to the east (Rodgers, 1963; Gwinn, 1964).

The Burning Springs Anticline has been of importance to the development of oil and gas production since the beginning of that industry. The first well drilled specifically for oil in West Virginia was completed in 1860 on the crest of the structure in Wirt County. The importance of the anticlinal structure in localizing oil and gas accumulations was soon recognized by Andrews (1861), based on his study of this area. This is one of the earliest references to the "anticlinal" theory of oil and gas accumulation.

Early development of oil and gas production could take place along the anticline because the vertical movement resulted

in removal of soft, shaly strata of the Permian/Upper Pennsylvanian Dunkard Group, Monongahela Formation, and Upper Conemaugh Group. Early drilling technologies were unable to cope with caving in these soft formations, and for the first three decades of its history the oil and gas industry in West Virginia was essentially restricted to this uplift where relatively competent formations were brought near the surface (White, 1904).

In the early 1980s, the anticline and the area immediately to the east again became the site of the most active oil and gas drilling in West Virginia. Productive zones are fractured Upper Devonian silts and shales, and the fracturing is in large part due to the detached deformation caused by the Allegheny Orogeny that produced the structure (Overbey and Henniger, 1983; Filer, 1984).

SITE INFORMATION

The Burning Springs Anticline is exposed in a series of six road cuts on both sides of a 1.5-mi (2.4-km) stretch of U.S. 50 (see Fig. 2). For discussion purposes the outcrops are labeled 1 through 6 from east to west. Exposed in these outcrops are the Upper Pennsylvanian Conemaugh Group, the Middle Pennsylvanian Allegheny Formation, and the upper part of the Lower Pennsylvanian Pottsville Group (Cardwell and others, 1968; Arkle and Barlow, 1972).

In general, the Pottsville-to-Conemaugh sequence of northwestern Virginia is considered to have been deposited during a gradual transgression from alluvial plain to upper and lower delta plain environments, with minor marine incursions in the Conemaugh (Presley, 1979; Donaldson and Shumaker, 1979). This sequence is marked by rapid lateral and vertical changes in lithology, with the transgression resulting in an upsection decrease in sandstone percentage, sandstone bed thickness, and overall grain size. In addition, coal thickness and continuity increase upsection into the upper delta plain environment. Marine limestones and shales, as well as red and green shales, are present in the Conemaugh Group, but not below.

As shown in Figure 3, a generalized stratigraphic column of the section exposed across the Burning Springs Anticline, these facies changes are apparent in the outcrops. As discussed above, it was the relatively soft shales—Pittsburgh Shale (Conemaugh Group, Fig. 3) and above—that prevented successful early exploration adjacent to the structure.

U.S. 50 crosses the anticline at the highest of several structural culminations along the length of the feature. Thus, exposed in these cuts and along the flanks of the structure is much of the shallow section that produced oil in fields developed on structurally lower parts of the crest of the anticline to the north and south in the 1860s–1870s. The alluvial and deltaic channel sands of the sequence represent the pay zones of one of the first two

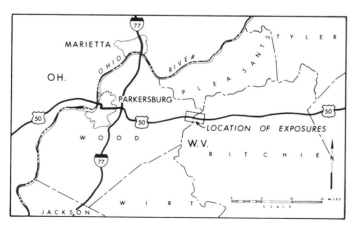

Figure 1. General location map.

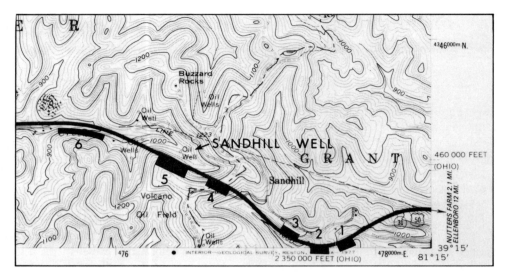

Figure 2. Detailed location map. Numbered black boxes indicate outcrops (stops) referred to in text.

major petroleum provinces of the world. (The other early province was the development of the Upper Devonian Bradford sand, subsequent to the Drake well in northwestern Pennsylvania.) As

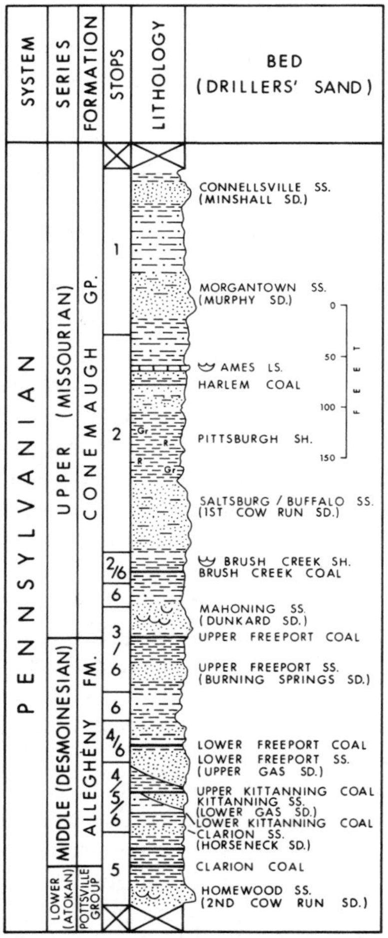

Figure 3. Generalized stratigraphic column of units exposed across the Burning Springs Anticline at U.S. 50.

drilling technology improved, sands in the same zones were found to be productive in somewhat deeper fields to the east and west.

Because of the rapid lateral and vertical variation in alluvial/deltaic sequences of this type, the section exposed here is only generally representative of that encountered by the drill, even in nearby fields. Sands that would be assigned the same stratigraphic and informal "drillers" names vary in thickness, petrology, precise facies, and exact stratigraphic position over short distances and disappear completely in some areas.

The somewhat less steeply dipping east flank of the basically N-S–striking Burning Springs Anticline is better exposed (outcrops 1–4). The high cut (5) just on the west (Wood County) side of the Wood/Ritchie county line is at the crest of the anticline. The west flank is poorly exposed, and the only significant outcrop occurs about 0.4 mile (0.7 km) west of the crest in the gully along the drainage to the south of the highway (6 in Fig. 2).

Figure 3 indicates the portion of the stratigraphic sequence exposed at each cut or outcrop labeled on Figure 2, and Figure 4 is a sketch of the exposures.

Stop 1

The easternmost exposure, 1 (south side of road), is the first at which significant eastward dip from the structure is observable along U.S. 50. The outcrop is divided in the center by a small gully and vegetation. At the eastern end of the exposure, beds dip about 30° to the east and strike almost due N-S. At the west end of the outcrop, dip has reached about 40°, with strike unchanged. Exposed are fine-grained, micaceous sands and shales of the middle of the Conemaugh Group (Fig. 3, 4).

Stop 2

Stop 2 is immediately west and still on the south side of the highway; it also displays Conemaugh strata (Fig. 3). At the east

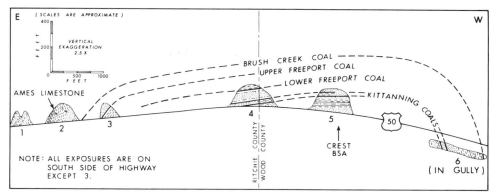

Figure 4. Sketch of exposures across the Burning Springs Anticline along U.S. 50.

end of the outcrop is a thin marine zone, including about 2 ft (61 cm) of poorly exposed, highly fossiliferous Ames Limestone, occurring a few feet above a thin sandstone ledge (Fig. 4). The Ames outcrop may not be readily observable without digging, but pieces of highly fossiliferous limestone may be identified in the float on the outcrop. Fossils include abundant brachiopods, crinoid stem pieces and plates, and some mollusks and gastropods. The prominent, highly weathered red and green shale beneath the Ames is the Pittsburgh Shale, discussed above. Poorly exposed, interbedded, thin, and fine-grained sandstones, siltstones, and gray shales make up the rest of the exposure.

Arkle and Barlow (1972) indicated the presence of the fossiliferous marine Brush Creek Shale, as well as the underlying Brush Creek coal at the western end of this exposure. These zones are apparently now fairly well covered by vegetation, although the coal is observable on the first bench directly across the highway (north side). The dip at this outcrop is about 50°E, the steepest observable in these cuts on the east side of the structure.

Stop 3

The next road cut to the west, on the north side of the highway, straddles the Conemaugh/Allegheny contact. The coal in the center of the cut was identified by Arkle and Barlow (1972) as the Upper Freeport, which marks the top of the Allegheny Formation. The overlying medium-to-coarse-grained massive sandstone, the Mahoning, marks the top of a generally sandier section that lacks marine zones and red shales and that includes the remainder of the Pennsylvanian section to the base of the Pottsville. Trough cross-bed sets are present, especially at the base of the Mahoning ("Dunkard" of drillers). Underlying the Upper Freeport coal is a thick shale zone, which in turn is underlain by a poorly exposed fine-grained sandstone. A sand at about this position, probably somewhat better developed, was the pay zone in the Burning Springs Field several miles (kilometers) to the south along the crest of the anticline. This field was opened in 1860, the first in West Virginia, only a few months after the famous Drake well in Pennsylvania was drilled. The dip of the strata decreases towards the crest and in this exposure is 30°E.

Stop 4

Here the beds are still dipping to the east, but only about 5°. The two relatively thick (2–3 feet (60–90 cm)) coals near the base of these outcrops are the Kittanning coals (Arkle and Barlow, 1972), and the massive sandstone overlying them at the level of the upper bench on both sides is the Lower Freeport or "Upper Gas" sand. The base of this sand occurs about 100 ft (30 m) below the base of the Upper Freeport or "Burning Springs" sand exposed at the western end of Stop 3 (Fig. 3 and 4). Thus, almost no section is missing, in spite of the 2,000 ft (606 m) of cover between two outcrops (Fig. 3).

The channel sand between the two Kittanning coals is the Kittanning Sandstone. This sand pinches out to the west in this exposure, but a second channel sand in the same position develops to the west in the next exposure.

Stop 5

Stop 5 is a few hundred feet (meters) farther west on the south side of the highway; it is on the crest of the Burning Springs Anticline. The top of this high road cut consists of the continuation of beds exposed just to the east (Fig. 4). At the base of the flat-lying exposure is a massive fine-to-coarse-grained sandstone with both tabular and trough cross-bed sets. This is the Homewood Sandstone ("Second Cow Run" of drillers). It is considered to be the uppermost bed of the Pottsville Group and is the oldest unit exposed in the core of the anticline here. Above it are a shaly section and the thin, shaly Clarion coal.

Stop 6

In comparison to the east flank discussed above, the west flank of the structure is very poorly exposed along U.S. 50. The section up to the Brush Creek coal can be seen along a drainage gully, to the south and well below the level of the road. Much of this exposure is heavily vegetated and mantled with soil, but general lithology and bedding can be observed. The Kittanning coals and adjacent sandstones and shales are the best exposed part of the section. The coals can be found at the location of a culvert that crosses under the highway 0.7 mi (1.2 km) west of the Ritchie/Wood county line. The dark gray Brush Creek fossiliferous marine shale and the underlying coal stand out at the western end of the gully. The drainage gully then opens into a small hollow, and no more exposure can be seen. Where good bedding can be observed, dip can be seen to increase to the west to about 70° at the Brush Creek coal.

STRUCTURE

The structural uniqueness of the Burning Springs Anticline has long been recognized, but it was not until after the Sandhill deep well was drilled to basement at 13,331 feet (4,040 m) in 1955 on the crest of the structure and just north of U.S. 50 that the full complexity of the structure was realized. The structure is now considered to have formed at the termination of a detachment at the Silurian Salina level, due to pinch-out of the F_4 Salina salt glide zone (Rodgers, 1963; Gwinn, 1964; Clifford and Collins, 1974). At least three thrusts were identified in highly deformed Lower and Middle Devonian carbonates and clastics penetrated in the Sandhill well (Bayles and others, 1956; Woodward, 1959). These thrusts ramp up from the decollement zone, repeat about 1,600 ft (485 m) of section in the Sandhill well, and die out upwards in overlying, thick, relatively plastic Upper Devonian shales.

Mapping of the Burning Springs structure in this area has shown the surface structure to be "box-shaped," that is, dips of the steep limbs rapidly decrease into a broad, flat, slightly east-dipping crest that fails to reflect the structural complexity at depth (Hennen, 1910; Shockey, 1954). This series of exposures illustrates the basic shape of the anticline at the surface, as shown in Figure 4. The decrease in dip observed between Stops 2 and 3 continues to the west. At Stop 4, beds are still dipping to the east, but only about 5°. As Figure 4 shows, the structure flattens to the west of Stop 3. The broad, relatively flat crest is illustrated by Stops 4 and 5. The projection of the Brush Creek coal across the crest of the structure and into its steeply dipping (70°W) outcrop at Stop 6 (Fig. 4) requires the very steep increase in dip on the west flank, as illustrated.

An alternative interpretation is that thrust faulting, identified deeper in the structure, reaches the surface. The flat-lying crest could then be thrust up and over the west flank of the structure, with the plane of the fault cutting the surface in the covered interval between Stops 5 and 6. Examination of logs of oil and gas wells drilled on the Burning Springs Anticline, both in this area and along the structure to the north and south, has yet to reveal significant thrusting above the Upper Devonian level. This suggests that the major detachment died out upwards as the thick Upper Devonian section deformed in a relatively ductile manner. The overlying competent Mississippian and Pennsylvanian section would have essentially domed up over the Upper Devonian section in response to compression, with only minor faulting evident at the surface.

Recently, the Burning Springs Anticline and areas immediately to the east have been the site of very active oil and gas exploration, with producing zones developed in Upper Devonian shales and thin siltstones. It is apparent that this production is from fractured reservoir rocks and that one of the most significant tectonic events to have affected the area is detached deformation during the Allegheny Orogeny. The anticline is the most obvious structural expression of this deformation, but perhaps more important for Devonian shale oil and gas production has been fracturing due to movement in the trailing Burning Springs Thrust Sheet immediately east of the anticline (Overbey and Henniger, 1983; Filer, 1984). Thus, over 120 years after the beginning of the oil and gas industry in West Virginia, the focus of exploration has returned to the Burning Springs Anticline and vicinity.

REFERENCES CITED

Andrews, E. B., 1861, Rock oil—its geologic relations and distributions: American Journal of Science, Series 2, v. 32, p. 85.

Arkle, T., and Barlow, J. A., 1972, The log of the I. C. White Memorial Symposium field trip: West Virginia Geological and Economic Survey, p. 73.

Bayles, R. E., Henry, W. H., Fettke, C. R., Harris, L. D., Flowers, R. R., and Haught, O. L., 1956, Wood County deep well: West Virginia Geological Survey, Report of Investigations 14, 62 p.

Cardwell, D. H., Erwin, R. B., and Woodward, H. P., 1968, Geologic map of West Virginia: West Virginia Geological and Economic Survey, 2 sheets.

Clifford, M. J., and Collins, H. R., 1974, Structures of southeastern Ohio [abs.]: American Association of Petroleum Geologists Bulletin, v. 58, p. 1891.

Donaldson, A. C., and Shumaker, R. C., 1979, Late Paleozoic molasse of central Appalachians, in Donaldson, A. C., Presley, M. W., and Renton, J. J., eds., Carboniferous coal guidebook: West Virginia Geological and Economic Survey Bulletin B-37-3 (Supplement), p. 1–42.

Filer, J. K., 1984, Geology of Devonian shale oil and gas in Pleasants, Wood, and Ritchie Counties, West Virginia, in Proceedings, Unconventional Gas Recovery Symposium, May 13–15, 1984, Pittsburgh, Pennsylvania: SPE/DOE/GRI 12834, p. 37–46.

Gwinn, V. E., 1964, Thin-skinned tectonics in the plateau and northwestern valley and ridge provinces of the central Appalachians: Geological Society of America Bulletin, v. 75, p. 863–900.

Hennen, R. V., 1910, Oil and gas fields and structural contours, map of Wood, Ritchie and Pleasants Counties: West Virginia Geological and Economic Survey.

Overbey, W. K., and Henniger, B. R., 1983, Development of oil and gas production from Devonian shales and siltstones in Pleasants and Ritchie Counties, West Virginia [abs.]: Fourteenth Annual Appalachian Petroleum Geology Symposium, West Virginia Geological and Economic Survey Circular C-31, p. 49.

Presley, M. W., 1979, Facies and depositional systems of Upper Mississippian and Pennsylvanian strata in the central Appalachians, in Donaldson, A. C., Presley, M. W., and Renton, J. J., eds., Carboniferous coal guidebook: West Virginia Geological and Economic Survey Bulletin B-37-1, p. 1–50.

Rodgers, J., 1963, Mechanics of Appalachians foreland folding in Pennsylvania and West Virginia: American Association of Petroleum Geologists Bulletin, v. 47, p. 1527–1536.

Shockey, P. N., 1954, Some aspects of the surface geology of the northern portion of the Burning Springs Anticline [M.S. thesis]: Morgantown, West Virginia University, 31 p.

White, I. C., 1904, Petroleum and natural gas, precise levels, West Virginia Geological Survey, vol. 1A, 625 p.

Woodward, H. P., 1959, A symposium on the Sandhill deep well, Wood County, West Virginia: West Virginia Geological and Economic Survey, Report of Investigations 18, 182 p.

13 14

Pennsylvanian System stratotype sections, West Virginia

Kenneth J. Englund, *U.S. Geological Survey, Reston, Virginia 22092*
Harold H. Arndt, *U.S. Geological Survey, Denver, Colorado 80225*
Stanley P. Schweinfurth, *U.S. Geological Survey, Reston, Virginia 22092*
William H. Gillespie, *U.S. Geological Survey, Charleston, West Virginia 25314*

INTRODUCTION

The establishment of a standard reference section for the Pennsylvanian System in the Appalachian basin was necessitated by inconsistencies and variations in the application and correlation of the loosely defined Lower, Middle, and Upper Pennsylvanian Series. To provide rock-representative definitions for these units, the Pennsylvanian System stratotype study was undertaken in consultation with interested geologists and paleontologists from State geological surveys, U.S. Geological Survey, academia, and industry. An essential criterion for the designation of the stratotype section was the selection of the most complete and continuous sequence of Pennsylvanian strata that are also conformable with underlying and overlying systems. This requirement was only met in West Virginia where Pennsylvanian rocks

extend from the Mississippian strata of the Pocahontas coal field to the Permian strata of the Dunkard basin, a location that follows the recommendation of the Pennsylvanian Subcommittee of the National Research Committee on Stratigraphy that the best standard section for Pennsylvanian rocks in the Appalachian basin would be in West Virginia (Moore and others, 1944, p. 665).

The Pennsylvanian System is defined by a composite stratotype consisting of 14 outcrop sections extending from the base to the top of the system over a distance of 100 mi (160 km). Fieldguide site 14 includes a description of strata in the Lower Pennsylvanian Series at three successive sites; A, A-B, and B-C in south-central West Virginia (Fig. 1). Field guide site 13 includes descriptions of strata in the Middle and Upper Pennsylvanian Series. Corroborative investigations include: (1) geologic mapping

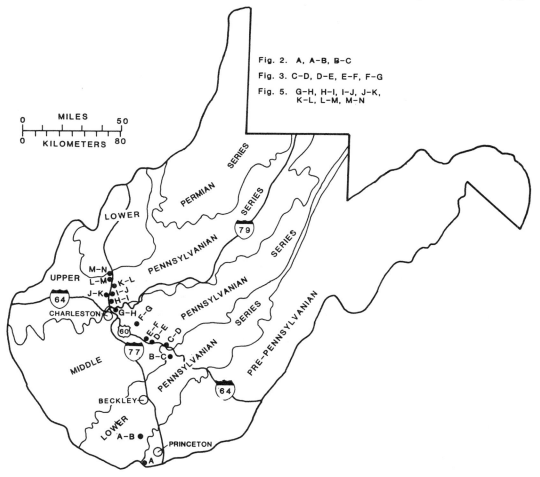

Figure 1. Index map of West Virginia showing locations of Figures 2, 3, and 5.

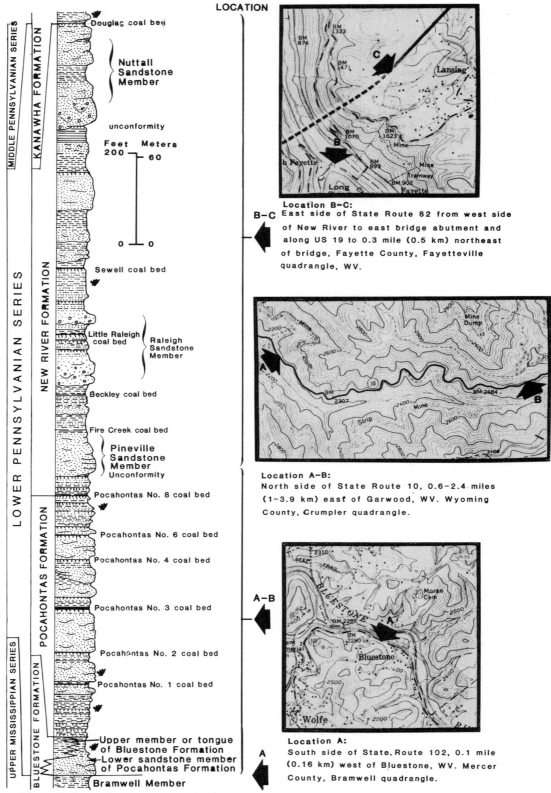

Figure 2. Composite stratotype section for the Lower Pennsylvanian Series, West Virginia, and the localities from which it was constructed.

of areas between sections (Englund, 1968; Stricker, 1981; Meissner, 1981, 1983; Englund and others, 1977a; and Henry, 1984); (2) core drilling (Schweinfurth and others, 1976; and Henry and others, 1981); (3) regional stratigraphic and depositional environment studies (Englund and others, 1977b, 1979; Englund and Henry, 1981); and (4) biostratigraphic analyses (Gillespie and Pfefferkorn, 1979).

SITE 13, LOWER PENNSYLVANIAN SERIES

Location

The Lower Pennsylvanian Series of the Pennsylvanian System is defined by three outcrop sections in south-central West Virginia. Outcrop A is located on West Virginia 102, west of Bluefield, West Virginia and outcrop A-B is located on West Virginia 10 east of Garfield, West Virginia. Both outcrops are west of I-77 between Beckley and Princeton, West Virginia. Outcrop B-C is located on West Virginia 82 and 12 near Lansing, West Virginia, south of U.S. 60 (Figs. 1 and 2).

Locality A

The Mississippian-Pennsylvanian boundary has been defined lithically and biostratigraphically at the conformable contact between the Bluestone and Pocahontas Formations in the vicinity of Locality A (Hennen and Gawthrop, 1915; Reger, 1926; Cooper, 1944) (Fig. 2). This longstanding practice is applicable in the southeasternmost outcrops or trough area of the Appalachian basin where Late Mississippian and Early Pennsylvanian formations attain their maximum thicknesses. To the northwest, this contact relation is modified by the wedging out of the lower sandstone member or tongue of the Pocahontas Formation. There, the systemic boundary extends from the base of the lower sandstone member into the upper part of the Bluestone Formation at approximately the contact between the Bramwell Member and the upper member of the Bluestone. This relationship, observed near the southeastern edge of the Appalachian basin, terminates about 30 mi (48 km) to the northwest where the Pocahontas Formation and upper part of the Bluestone Formation are both truncated by an unconformity at the base of the overlying New River Formation (Lower Pennsylvanian). As a result, the conformable and gradational aspect of beds at the systemic boundary in the southeast is replaced to the northwest by a hiatus that coincides with the widely recognized Mississippi-Pennsylvanian unconformity.

The section exposed at Locality A is on the steeply dipping northwest limb of the Abbs Valley anticline (Fig. 2). It contains the type section of the Bramwell Member of the Bluestone Formation which is the uppermost unit of Late Mississippian age in the Appalachian basin and may be younger than the youngest Late Chesterian of the Upper Mississippi Valley (Gordon and Henry, 1981). Basal beds of the Bramwell consist of about 9 ft

(2.7 m) of black, carbonaceous shale containing abundant ostracodes dominated by *Darwinula,* scattered lamellibranch pelecypods (*Naiadites* and *Anthraconaia?*) and a few inarticulate brachiopods (*Lingula*) indicating a transition from nonmarine to brackish-water deposition. Overlying beds of the member total 86 ft (26.2 m) in thickness and consist of medium-dark-gray calcareous shale and siltstone that coarsen up and grade into very fine to fine-grained ripple-bedded sandstone. These beds are largely prodeltaic and grade upward to coarser distal-bar sediments of a prograding delta (Englund and others, 1979). Abundant and diverse marine invertebrates at several horizons include the following marine faunule with strong Late Chesterian affinities (Gordon and Henry, 1981): *Fenestella* sp., *Polypora?* sp., *Archimedes* sp., *Orbiculiodea* sp. indet., *Orthotetes* aff. *O. kaskaskiensis* (McChesney) n. sp. B, *Diaphragmus* cf. *D. cestriensis* (Worthen), *Ovatia* aff. *O. pileiformis* (McChesney), *Anthracospirifer leidyi* (Norwood and Pratten), *Spiriferellina?* sp., indet., *Composita subquadrata* (Hall), *Eumetria costata* (Hall), *Phestia* sp., *Paleyoldia* sp., indet., *Aviculopecten* sp., *Posidonia?* sp. indet., *Sulcatopinna* sp., *Solenomya* sp., *Sphenotus* sp. indet., *Wilkingia?* sp., *Edmondia* sp., and pelmatozoan fragments. Other marine invertebrates include a trilobite, echinoid spines, several species of gastropods, conodonts, several other taxa of pelecypods, and the coiled nautiloids *Liroceras* sp. and *Peripetroceras* sp.

At Locality A, the overlying lower sandstone member of the Pocahontas Formation consists of slumped, delta-front siltstone and sandstone succeeded by massive, channel-fill sandstone of a distributary lobe (Englund and others, 1979). Pennsylvanian plant fossils, including *Neuropteris pocahontas* D. White and *Mesocalamites* sp., have been collected from the delta-front facies but more commonly occur in the beds overlying the lower sandstone member. The fronds and pinnules of *Neuropteris pocahontas* D. White, its male reproductive bodies, called *Aulacotheca campbelli* (D. White) Halle, and its seed, *Holcospermum maizeretense* Stockmans and Wiliere, occur in profusion about 1.0 mi (1.6 km) to the southwest.

The significance of Locality A is that it defines the systemic boundary in the type area of both the Bluestone and Pocahontas Formations, which represent the youngest and oldest units, respectively, of the Mississippian and Pennsylvanian Systems in the Appalachian basin. Because of its continuity of deposition, this section presents the ideal locality for a point-boundary stratotype for the base of the Pennsylvanian System.

Locality A-B

Locality A-B extends from the Bramwell Member of the Bluestone Formation, through the Pocahontas Formation, and into the lower part of the New River Formation. It is the most completely and continuously exposed section of the Pocahontas Formation in southern West Virginia and is used as the stratotype for the lower part of Lower Pennsylvanian Series. The Bramwell Member and most of the lower sandstone member of the Pocahontas crop out in the stream bed and hillside below West Virginia

10 and the remainder of the section is exposed in cutbanks of the highway (Fig. 2). A tongue of the upper member of the Bluestone, consisting of greenish-gray and grayish-red siltstone, mudstone, shale, and claystone, overlies the distal edge of a distributary lobe of the lower sandstone member of the Pocahontas at this locality. Of extreme biostratigraphic significance is the presence of fronds of *Neuropteris pocahontas*, its male fructification *Aulacotheca campbelli* (D. White) Halle, and its seed *Holcospermum maizeretense*, which confirm the Pennsylvanian age of the upper member of the Bluestone. Overlying strata of the Pocahontas Formation contain several shale beds with abundant, well-preserved compression and impression floras. The flora about 50 ft (15.2 km) below the Pocahontas No. 1 coal bed includes *N. pocahontas, Lepidodendron dichotomum,* and *Lyginopteris hoeninghausii.* These taxa are also present in the roof shale of the Pocahontas No. 1 coal bed. Several thin, carbonaceous shale beds, some of which contain coaly laminae and are underlain by rooted underclay, also occur in the lower part of the Pocahontas. One of these carbonaceous shale beds about 100 ft (30.5 m) below the Pocahontas No. 1 coal bed contains molds of ostracodes and sparse *Lingula,* suggesting an intermittent, marginal-marine influence during deposition of these strata.

Other plant-bearing beds are present throughout most of the overlying Pocahontas Formation. A prolific and diverse flora occurs in a local channel filled with light-gray silty shale just below the Pocahontas No. 8 coal bed. The flora from this bed consists of *Neuropteris pocahontas* var. *inaequalis, Sphenopteris pottsvillea, M. eremopteroides, Alethopteris evansi, A. davreuxi, Aneimites pottsvillensis, Sphenopteris preslesensis,* several other species of *Sphenopteris,* and other taxa. This assemblage is typical of the middle and upper parts of the Pocahontas Formation.

Locality B-C

The entire New River Formation from the top of the Pocahontas Formation (immediately above river level) to the base of the Kanawha Formation crops out along West Virginia 82 and in excavations of the New River Gorge bridge abutment (Fig. 2). This section is in the type area for the New River Formation and presents the most continuously exposed sequence of beds for a stratotype of the upper part of the Lower Pennsylvanian Series. Outcrop data were supplemented by information from core drilling in the Lower Pennsylvanian Series at this locality (Englund and others, 1979, p. 18). The Pocahontas Formation, which is present mostly in the subsurface, is 80 ft (24.3 m) thick and wedges out about 6 mi (9.7 km) to the northwest. It is disconformably overlain by the Pineville Sandstone Member at the base of the New River Formation. The New River Formation is 910 ft (277.4 km) thick and characteristically includes several beds of quartzose conglomeratic sandstone that crop out in precipitous cliffs or resistant ledges along the gorge walls. Intervening beds consist of shale, siltstone, sandstone, coal, and underclay of a back-barrier facies including bay-fill sediments that are flaser bedded and burrowed. The coal beds in this sequence are typically widespread but thin. The Nuttall Sandstone Member, the uppermost member of the New River Formation, consists of quartzose sandstone of a barrier-bar complex. Fossil plants and fresh- or brackish-water invertebrates occur in several beds. Shale about 25 ft (7.6 m) below the Sewell coal bed contains a good florule which includes: *Alethopteris decurrens, Neuropteris pocahontas* var. *inaequalis, N. pocahontas* var. *pentias, Sphenopteris pottsvillea, Sphenopteris preslesensis, Stigmaria ficoides, Calamites carinatus, Neuropteris heterophylla,* and *Cordaites sp.*

The contact between the New River and the overlying Kanawha Formation is exposed on the east side of U.S. 19 at the base of a thin coal bed that was correlated with the Lower Douglas coal bed in McDowell County, West Virginia, by Hennen and Teets (1919, p. 276). This contact also forms the boundary between the Lower and Middle Pennsylvanian Series and is marked by a lithic change from a sequence dominated by the quartzose, conglomeratic sandstone of the New River Formation to a largely bay-fill, back-barrier sequence of shale, siltstone, sandstone, coal and underclay in the lower part of the Kanawha Formation. Differences in the floras have been noted across this boundary. The flora occurring just above the coal is dominated by *Sphenopteris schatzlarensis* and includes a few *Neuropteris gigantea* and *Alethopteris lonchitica* pinnules and *Cordaites sp.*

SITE 14, MIDDLE AND UPPER PENNSYLVANIAN SERIES

Location

The Middle and Upper Pennsylvanian series of the Pennsylvanian System stratotype are defined by 11 outcrop sections in central West Virginia. Outcrops are located in road cuts and strip mines adjacent to U.S. 60 and I-77, east and north of Charleston, West Virginia (Figs. 1, 3, and 5).

Locality C-D

The lower part of the Middle Pennsylvanian Series, exposed at locality C-D, extends from the Lower-Middle Pennsylvanian Series boundary at the top of the New River Formation upward through the lower part of the Kanawha Formation to the top of the Gilbert coal bed which is about 160 feet (48.8 m) above the base of the Kanawha Formation (Fig. 3). The Douglas coal bed, which delineates the boundary between the Lower and Middle Pennsylvanian Series at Locality C, is comprised of three coal benches ranging from 3-6 in (7-15 cm) in thickness separated by 6-23 in (15-58 cm) of gray shale and underclay. The strata between the Douglas and Gilbert coal beds are mostly bay-fill sediments in which medium- to medium-dark-gray shale and siltstone predominate. This interval also includes comparatively thin and lenticular beds of sandstone and coal about 90 ft (27.4 m) above the base of the Kanawha Formation. A 32-ft (9.8 m) thick distributary-channel sandstone near the top of the interval is overlain locally by rooted levee deposits composed of interlami-

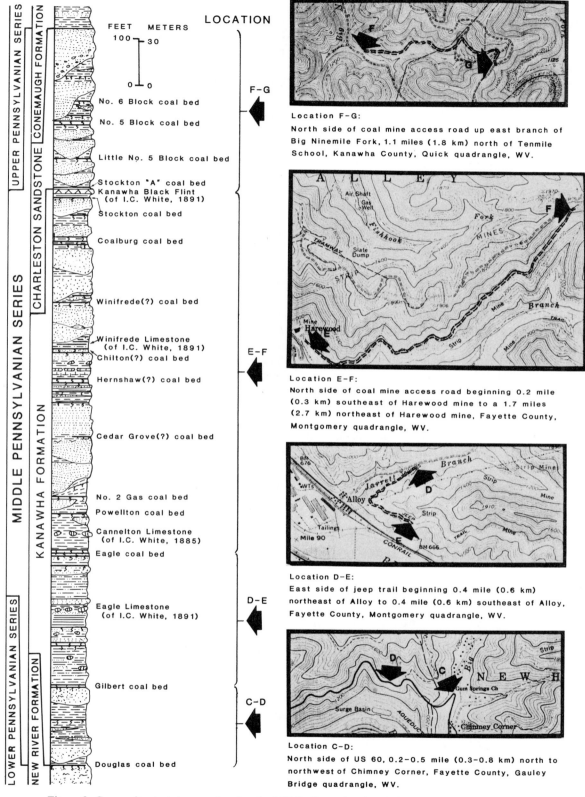

Figure 3. Composite stratotype section for the Middle Pennsylvanian Series, West Virginia, and the localities from which it was constructed.

nated shale, siltstone, and fine-grained sandstone. The Gilbert coal bed is typically multiple-bedded and at this locality includes three coal splits that range from 3-4 in (7-10 cm) in thickness. Fossil plant impressions are common in shale beds of the lower part of the Kanawha Formation and include: matted *Cordaites* leaves, *Annularia radiata, Calamites* spp., *Sphenopteris schlatzlarensis,* and *Neuropteris gigantea.*

Locality D-E

The lower part of the Kanawha Formation crops out east of Alloy, West Virginia along the south side of Jarrett Branch and in roadcuts ascending the mountain southeast of Jarrett Branch (Fig. 3). Locality D-E includes strata from the Gilbert coal bed through the Eagle coal bed in the middle to upper part of the section near Alloy. The Gilbert coal bed is 10-13 in (15-30 cm) thick and is overlain by a bay-fill sequence similar to that at Locality C-D. Dark-gray shale in the lower part of the sequence contains scattered brackish-water pelecypods which are most abundant about 9 ft (2.7 m) above the Gilbert coal. Strata at Locality D-E total about 236 ft (72 m) in thickness and consist chiefly of medium-dark-gray shale with scattered siderite layers and concretions, comparatively thin crevasse-splay sandstone beds, and numerous thin and discontinuous coal beds, typical of a lower delta plain environment of deposition. Burrows are common and some of the shale contains root penetrations. The Eagle Limestone of White (1891) consists of 21 in (50.5 cm) of calcareous sandy siltstone with abundant marine fossils. A Morrowan faunule was identified from the Eagle Limestone by Henry and Gordon (1979). About 2 ft (0.6 m) of dark-gray shale, which includes limestone concretions up to 8 in (20 cm) in thickness, overlies the fossiliferous siltstone. Strata between the Eagle Limestone and the Eagle coal bed consist of small-scale crossbedded silty shale, siltstone, and very fine-grained sandstone, indicating a more rapid influx of detrital sediments into an interdistributary bay.

The Eagle coal is generally multiple-bedded and consists of five benches ranging from 8-39 in (20-99 cm) thick. Plant fossils collected from the roof shale of the Eagle coal bed include: *Stigmaria ficoides, Calamites suckowi, Neuropteris heterophylla, Alethopteris lonchitica, Mariopteris muricata,* and *Trigonocarpus.*

Locality E-F

Locality E-F (Fig. 3) includes 734 ft (224 m) of strata between the Eagle coal bed and the top of the Kanawha Black Flint of White (1891) in one of the most accessible and best exposed sections of the middle part of the Middle Pennsylvanian Series in central West Virginia. The strata consist principally of terrestrial sandstone, siltstone, shale, underclay, and coal, but also contain several marine incursions represented by argillaceous limestone, ellipsoidal limestone and siderite concretions, and siliceous shale and chert. Thick distributary channel sandstone predominates in the section above the No. 2 Gas coal bed.

A flint-clay parting near the base of the Hernshaw(?) coal bed is widely distributed in the Appalachian basin. It is grayish-black, breaks with a conchoidal fracture, and occurs in lenses as much as 2 in (5 cm) thick. The distribution of the flint clay bed support a volcanic origin; sanidine has been identified in a correlative bed in Kentucky (Seiders, 1965).

The upper part of the section at Locality E-F shows a transition from the thick bay-fill deposits of the lower delta plain to similar but thinner bay-fill deposits represented by shale units that commonly contain marine and brackish-water invertebrate fossils. The dark-gray calcareous shale overlying the Hernshaw(?) coal bed bears molds and casts of chonetide brachiopods and other marine invertebrates of late Morrowan age (Henry and Gordon, 1979). Crevasse-splay sandstone and bay-fill shale separate the Hernshaw(?) coal bed from the overlying Chilton(?) coal bed. The Winifrede Limestone of Hennen (1914), overlying the Chilton(?) coal bed, consists largely of calcareous shale that contains a diverse marine fauna, including brachiopods, gastropods, and pelecypods of probable Atokan age (Henry and Gordon, 1979). The shale underlying the Coalburg coal contains abundant plant impressions including *Lepidodendron* twigs, *Calamites suckowi, Pecopteris miltoni,* and *Sphenopteris* (al. *Diplothmema) cheathami.* Shale beds above and below the Lower Stockton coal bear a diverse plant assemblage that includes *Neuropteris scheuchzeri, N. semireticulata, N. rarinervis, Sphenophyllum cuneifolium, Sphenopteris* sp., and *Pecopteris* sp. among others. The Kanawha Black Flint of White (1891) is the most distinctive bed in the Middle Pennsylvanian Series. It is comprised of 6.7 ft (2 m) of black chert that grades to highly siliceous silty shale at its base. The lower part of the chert and the underlying siliceous shale contain marine and brackish-water invertebrates.

Locality F-G

The Charleston Sandstone at Locality F-G (Fig. 3) consists of 367 ft (111.9 m) of strata between the top of the Kanawha Black Flint and the Middle-Upper Pennsylvanian Series boundary at the base of the Conemaugh Formation. About 70 percent of this sequence is fine- to coarse-grained partly conglomeratic, feldspathic sandstone that occurs in thick coalescing beds. Intervening beds consist of siltstone, shale, underclay, coal, and flint clay. Deposition occurred in an upper delta plain environment that was characterized by multiple-shifting channels extending across broad flood basin peat swamps. Erosion and channel migration resulted in partial or complete truncation of the underlying coal bed. Correlative strata 1.5 mi (2.4 km) to the southeast contain the following florule: *Sphenophyllum majus, Pecopteris arborescens, P. unita, P. hemitelioides, Neuropteris ovata, N. rarinervis, Cordaites principalis,* and *Artisia.* The Middle-Upper Pennsylvanian Series boundary is at the base of a dark-greenish-gray shale at the gradational contact between the Charleston Sandstone and Conemaugh Formation.

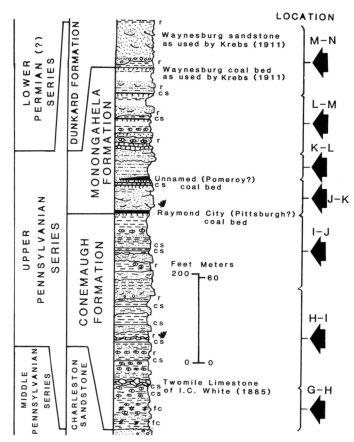

Figure 4. Composite stratotype section for the Upper Pennsylvanian Series, West Virginia. cs, carbonaceous shale; fc, flint clay; r, grayish-red.

UPPER PENNSYLVANIAN SERIES

Locality G-H

The base of the Upper Pennsylvanian Series is at the contact between the Charleston Sandstone and the overlying Conemaugh Formation at Locality G-H (Fig. 4). The massively bedded sandstone in the lower 82 ft (25 m) of this roadcut on I-79 is the uppermost part of the Charleston Sandstone. It grades upward into a persistent grayish-red mudstone that is typical of the Conemaugh Formation in central West Virginia. About 108 ft (33 m) of strata from the base of the Conemaugh Formation to the base of the Twomile Limestone of White (1885) has been proposed as the lowermost segment of the stratotype for the Upper Pennsylvanian Series in this section (Englund and others, 1979; p. 44). These strata consist predominantly of terrestrial deposits of grayish-red, greenish-gray, and variegated mudstone that include two widely recognized flint-clay beds (Fig. 4).

Locality H-I

Locality H-I includes about 200 ft (65 m) of strata in the middle part of the Conemaugh Formation from the base of the Twomile Limestone of White (1885) at the northwest end of the roadcut up through a massive sandstone at the southeastern end of the cut (Fig. 5). The Twomile Limestone is a lacustrine deposit consisting of largely bioturbated, carbonate mudstone about 6.5 ft (2 m) thick. It contains the pulmonate gastropod *Anthracopupa*, fresh water ostracodes (notably *Carbonita*), *Spirorbis,* rare conchostracons, and debris of fish or amphibian origin. Dessication features are common and several beds of laminar, possibly stromatolitic boundstone are present near the top of the limestone. A grayish-red shale about 99 ft (30 m) above the Twomile Limestone contains the following plant fossils: *Calamites* sp., *Sphenophyllum* cf. *thoni, Pecopteris hemitelioides, Neuropteris ovata, Pseudomariopteris ribeyroni,* and *Cordaites principalis.*

Locality I-J

The uppermost 172 ft (52 m) of the Conemaugh Formation are exposed in a roadcut at Locality I-J. These rocks consist largely of variegated grayish-red and greenish-gray, partly calcareous sandstone and shale that are typical of the Upper Series in central West Virginia. A fine-grained, locally conglomeratic fluviatile sandstone in the road ditch below the base of this section correlates with the massive sandstone at the top of the section at Locality H-I. A few thin beds of carbonaceous shale or clay in overlying strata were also correlated with nearby sections.

Locality J-K

The section at Locality J-K (Fig. 5) represents the lowermost part of the Monongahela Formation from the Raymond City coal bed to the unnamed (Pomeroy ?) coal bed. The correlation of the Raymond City coal with the Pittsburgh coal bed (Hotchkiss, 1880) is corroborated by both microspore and compression floras. Macerations of this coal reveal that *Thymospora thiessenii,* the so called "Pittsburgh spore," constitutes about 93 to 95 percent of the microspores. Although this form species ranges above and below the Pittsburgh coal bed in the northern part of the central Appalachian basin, the traditional view is that its epibole is in the Pittsburgh coal bed proper and that it can be used to differentiate the Pittsburgh from other coal beds (Kosanke, 1943). The megaflora in strata overlying the Raymond City coal is dominated by *Neuropteris scheuchzeri, Danaeites emersonii, Calamites, Pecopteris arborescens,* and *P. unita.* The unnamed coal bed about 71 ft (23 m) above the Raymond City coal bed is tentatively correlated with the Pomeroy coal bed of Ohio.

Locality K-L

Variegated shale and sandstone beds of the Monongahela Formation from the unnamed (Pomeroy?) coal bed to a thin arenaceous limestone, about 59 ft (18 m) higher, are exposed in an access road on the east side of I-77 at Locality K-L (Fig. 5).

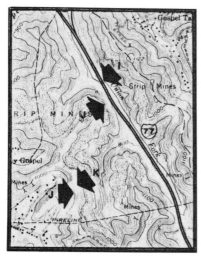

Location J–K:

Highwall of abandoned strip mine and road to landfill, about 4.7 miles (7.6 km) north of Charleston, WV, Kanawha County, Pocatalico quadrangle.

Location I–J:

Both sides of I-77 about 0.6 mile (1.0 km) south of Tupper Creek Road, about 5.6 miles (9.0 km) north of Charleston, WV, Kanawha County, Pocatalico quadrangle.

Location H–I:

West side of I-77 just south of the Twomile Creek Road interchange; about 2.8 miles (4.5 km) north of Charleston, WV, Kanawha County, Pocatalico quadrangle.

Location G–H:

Northwest side of I-79 0.3 mile (0.5 km) northeast of I-77 interchange, 2.5 miles (4.0 km) east of Charleston, WV. Kanawha County, Big Chimney quadrangle.

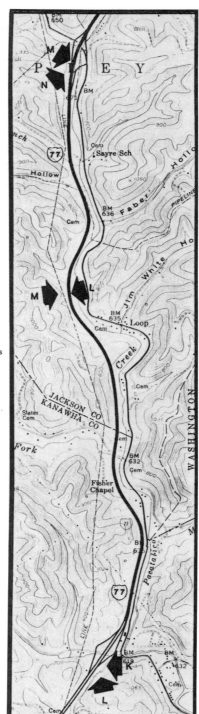

Location M–N:

West side I-77, in southwest quadrant of Goldtown Road interchange, 15.1 miles (24.3 km) north of Charleston, WV, Jackson County, Sissonville quadrangle.

Location L–M:

West side of I-77 13.7 miles (22.0 km) north of Charleston, WV, Jackson County, Sissonville quadrangle.

Location K–L:

East side of road near southeast quadrant of I-77/Haines Creek Road interchange; 11.7 miles (18.8 km) north of Charleston, WV, Kanawha County, Sissonville quadrangle.

Figure 5. Localities of exposures that make up the composite stratotype section for the Upper Pennsylvanian series shown in Fig. 4.

The most significant compressional flora in this sequence occurs about 58 ft (17 m) below the unnamed (Pomeroy?) coal bed in this roadcut. The fossils are well preserved and consist of *Neuropteris scheuchzeri, N. ovata, Pecopteris arborescens, P. candollena, P. hemitelioides, P. miltoni, P. polymorpha, P. unita, Danaeites emersonii, Sphenophyllum oblongifolium, Asterophyllites equisetiformis, Calamites* sp., *Cordaites palmaeformis, Pseudomariopteris ribeyroni, Sphenopteris* sp., and *Odontopteris brardii.* This diverse assemblage, particularly with the profusion of species of the genus *Pecopteris,* is typical of strata above the Pittsburgh coal bed in northern West Virginia and southwestern Pennsylvania.

Locality L-M

Locality L-M (Fig. 5) contains 138 ft (42 m) of limestone and variegated shale, claystone, mudstone, and sandstone in the upper-middle part of the Monongahela Formation. An unnamed lacustrine limestone at the base of this section attains a maximum thickness of 20.5 ft (6.3 m) at the south end of the roadcut. It was also identified at the top of the section at locality K-L and contains similar features and fossils as those described in the Twomile Limestone at Locality H-I. Overlying beds consisting

largely of grayish-red, very calcareous, slickensided claystone and mudstone extend upward to a thin carbonaceous shale.

Site M-N

The uppermost sequence of the Monongahela Formation, including the top of the Upper Pennsylvanian Series is exposed at Locality M-N (Fig. 5). About 58 ft (17.5 m) of mostly greenish-gray and grayish-red claystone and mudstone and a few beds of fine-grained sandstone extend up to the Waynesburg coal zone. This zone, consisting of coaly carbonaceous shale and claystone was correlated with the Waynesburg coal bed of southwestern Pennsylvania and the sandstone above the coal bed was correlated with the Waynesburg Sandstone by Krebs, 1911. Overlying strata near Liberty, West Virginia, contain a flora marked by the first stratigraphic appearance of *Callipteris conferta,* which historically has been used to identify the base of the Permian System. This flora also contains several species from older units (Gillespie and Pfefferkorn, 1979, p. 94). For these reasons placement of the Pennsylvanian-Permian boundary at the top of the Waynesburg Sandstone as used by Krebs (1911) is followed in the present report. When other criteria are established for a boundary stratotype for the base of the Permian, the top of the Upper Pennsylvanian Series can be adjusted accordingly.

REFERENCES CITED

Cooper, B. N., 1944, Geology and mineral resources of the Burkes Garden quadrangle, Virginia: Virginia Geological Survey Bulletin 60, 229 p.

Englund, K. J., 1968, Geologic map of the Bramwell quadrangle, West Virginia-Virginia: U.S. Geological Survey Geologic Quadrangle Map GQ-745, scale 1:24,000.

Englund, K. J., Arndt, H. H., and Henry, T. W., (eds.), 1979, Proposed Pennsylvanian System Stratotype, Virginia and West Virginia (Ninth International Congress of Carboniferous Stratigraphy and Geology meeting Guidebook Field Trip 1): Washington, D.C., American Geological Institute, 136 p.

Englund, K. J., Arndt, H. H., Henry, T. W., Meissner, C. R., Jr., Windolph, J. F., Jr., and Warlow, R. C., 1977a, Geologic map of the New River Gorge area, Fayetteville, Raleigh, and Summers Counties, West Virginia: U.S. Geological Survey Open-File Report OF-77-76, Map A.

Englund, K. J., and Henry, T. W., 1981, Mississippian-Pennsylvanian boundary in the central part of the Appalachian Basin (Part I: southwestern Virginia-southern West Virginia), *in* Roberts, T. G., ed., (Geological Society of America annual meeting guidebooks, v. 1, Stratigraphy, sedimentology, field trip 4): Washington, D.C., American Geological Institute, p. 153–194.

Englund, K. J., Windolph, J. F., Jr., Warlow, R. C., Henry, T. W., Meissner, C. R., Jr., and Arndt, H. H., 1977b, Stratigraphic section of the New River Gorge area, Fayette, Raleigh, and Summers Counties, West Virginia: U.S. Geological Survey Open-File Report OF-77-76, Map C.

Gillespie, W. H., and Pfefferkorn, H. W., 1979, Distribution of commonly occurring plant megafossils in the proposed Pennsylvanian System stratotype, *in* Englund, K. J., and others, eds., Proposed Pennsylvanian System Stratotype, Virginia and West Virginia (Ninth International Congress of Carboniferous Stratigraphy and Geology meeting guidebook field trip 1): Washington, D.C., American Geological Institute, p. 86–96.

Gordon, Mackenzie, Jr., and Henry, T. W., 1981, Late Mississippian and Early Pennsylvanian invertebrate faunas, east-central Appalachians—A preliminary report, *in* Roberts, T. G., ed., (Geological Society of America annual meeting guidebooks, v. 1, Stratigraphy, sedimentology, field trip 4): Washington, D.C., American Geological Institute, p. 165–171.

Hennen, R. V., 1914, General section, Kanawha series, Kanawha County; *in* Krebs, C. E., and Teets, D. D., Jr., Kanawha County: West Virginia Geological Survey [County Report], p. xxvi–xxviii.

Hennen, R. V., and Gawthrop, R. M., 1915, Wyoming and McDowell Counties: West Virginia Geological Survey [County Report], 783 p.

Hennen, R. V., and Teets, D. D., Jr., 1919, Fayette County: West Virginia Geological Survey [County Report], 1002 p.

Henry, T. W., 1984, Geologic map of the Mammoth quadrangle, Kanawha and Clay Counties, West Virginia: U.S. Geological Survey Geologic Quadrangle Map GQ-1576, scale 1:24,000.

Henry, T. W., Englund, K. J., Johnson, P. L., Mory, P. C., and Windolph, J. F., Jr., 1981, Description and correlation of core from five deep drill holes in Carboniferous rocks along the New River Gorge, West Virginia: U.S. Geological Survey Open-File Report OF-81-1339, 88 p.

Henry, T. W., and Gordon, Mackenzie, Jr., 1979, Late Devonian through Early Permian(?) faunas in area of proposed Pennsylvanian System stratotype: *in* Englund, K. J., and others, eds., Proposed Pennsylvanian System Stratotype, Virginia and West Virginia (Ninth International Congress of Carboniferous Stratigraphy and Geology meeting guidebook field trip 1): Washington, D.C., American Geological Institute, p. 97–103.

Hotchkiss, Jedediah, 1880, The Coal fields of West Virginia and Virginia in the Great Ohio, or Trans-Appalachian, Coal Basin, illustrated by Maps and Geological Sections: The Virginias, v. 1, no. 2, p. 18–21 [supplemental map between p. 24–25].

Kosanke, R. M., 1943, The characteristic plant microfossils of the Pittsburgh and Pomeroy coals of Ohio: American Midland Naturalist, v. 29, no. 1, p. 119–132.

Krebs, C. E., 1911, Jackson, Mason, and Putnam Counties: West Virginia Geological Survey [County Report], 387 p.

Meissner, C. R., Jr., 1981, Geologic map of the Shady Spring quadrangle, Raleigh and Summers Counties, West Virginia: U.S. Geological Survey Geologic Quadrangle Map GQ-1546, scale 1:24,000.

——— 1983, Geologic map of the Ripley quadrangle, Jackson County, West

Virginia: U.S. Geological Survey Geologic Quadrangle Map GQ-1569, scale 1:24,000.

Moore, R. C. (chairman), and others, 1944, Correlation of Pennsylvanian formations of North America: Geological Society of America Bulletin, v. 55, p. 657–706.

Reger, D. B., 1926, Mercer, Monroe, and Summers Counties: West Virginia Geological Survey [County Report], 963 p.

Schweinfurth, S. P., Arndt, H. H., and Englund, K. J., 1976, Description of core from three U.S. Geological Survey core holes in Carboniferous rocks of West Virginia: U.S. Geological Survey Open-File Report 76-159, 61 p.

Seiders, V. M., 1965, Volcanic origin of flint clay in the Fire Clay coal bed, Breathitt Formation, eastern Kentucky: *in* Geological Survey Research 1965: U.S. Geological Survey Professional Paper 525D, p. D52–D54.

Stricker, G. D., 1981, Geologic map of the Crumpler quadrangle, West Virginia: U.S. Geological Survey Geologic Quadrangle Map GQ-1547, scale 1:24,000.

White, I. C., 1885, Resume of the work of the U.S. Geological Survey in the Great Kanawha Valley during the summer of 1884: The Virginias, v. 6, p. 7–16.

——1891, Stratigraphy of the bituminous coal field of Pennsylvania, Ohio, and West Virginia: U.S. Geological Survey Bulletin 65, 212 p.

Greenland Gap, Grant County, West Virginia

Katharine Lee Avary, West Virginia Geological and Economic Survey, P.O. Box 879, Morgantown, West Virginia 26507

LOCATION

Greenland Gap is located in Grant County, West Virginia (Fig. 1), and are on the Greenland Gap and Medley 7½-minute Quadrangles (Fig. 2). The exposures are along West Virginia 42 and 93, west of Scherr, and along Grant County Rt. 1 (Greenland Road) and Grant County Rt. 3/3 (Greenland Gap Road), east of Scherr.

INTRODUCTION

Greenland Gap is a spectacular water gap through the westernmost major anticline (Wills Mountain anticline) in this part of the Valley and Ridge province. The Allegheny (topographic and structural) Front is just west of Wills Mountain and Devonian- through Pennsylvanian-aged rocks are exposed as one ascends the front. The Greenland Gap area is unique in West Virginia; it is the only place where a major water gap is east of a state or federal highway that crosses the front. Thus, almost 10,000 ft (3,030 m) of rocks ranging in age from Early Pennsylvanian to Late Silurian can be seen at the various stops, along with changes in structural style, as one drives eastward from the Appalachian (High Fold) plateau province, down the front to the Valley and Ridge province.

Stops 1–10 are located along West Virginia Rts. 42 and 93, west of Scherr. Stop 11 is located along Greenland Road and Stops 12 and 13 are located along Greenland Gap Road.

Although exposure is not continuous along West Virginia 42 and 93 or along Greenland Road and Greenland Gap Road, enough rocks are exposed to allow the visitor to get an idea of the overall stratigraphic sequence (Table 1). Visitors are encouraged to read the references before visiting, so that they may have a better understanding of the regional setting of these units. For information on the regional geology of the Greenland Gap area, see Cardwell and others, (1968), Reger and Tucker (1924), Clark (1967), Clark (1976), and Perry (1978).

SITE DESCRIPTION

Stop 1

A few feet (meters) of Pottsville Formation, dipping gently to the northwest, are exposed here. The Pottsville is mostly medium- to coarse-grained sandstone with some conglomeratic beds. The rocks are fractured and have slumped slightly. Some iron-staining is apparent on the weathered surfaces. An unconformity separates the Pottsville from the underlying Mauch Chunk Formation. The gentle dips observable here can be compared with the much steeper dips that can be observed eastward down the Allegheny Front. The Pottsville is interpreted as being

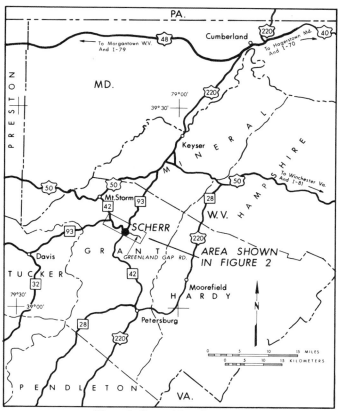

Figure 1. General location map.

of fluvial origin, with the source area to the east. The conglomerates seen here are not present further west (Presley, 1977).

Stops 2a, 2b, 3

The Mauch Chunk is visible at these three stops, as dark reddish-gray, fine-grained sandstone, with some cross-bedding, interbedded with red and green shales and siltstones. Scattered iron nodules can be found at Stop 2b. The Mauch Chunk is considered to be of fluviodeltaic origin (Dennison and Wheeler, 1975). The red shales of the Mauch Chunk are extremely prone to sliding; a landslide in the spring of 1984 removed the entire eastbound lane of West Virginia 42/93 at Stop 3. Scars from the slide should remain visible for many years.

Stop 4

The Greenbrier Limestone, which underlies the Mauch Chunk, cannot be seen along West Virginia 42/93, but is exposed in an abandoned quarry just east of the highway. The Greenbrier is interpreted as a shallow marine carbonate deposit, which is con-

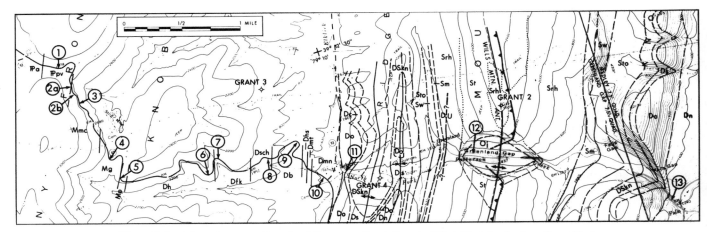

Figure 2. Location of stops in Greenland Gap area, Grant County, West Virginia. Geology from Clark (1976), Dennison (1970), and Reger and Tucker (1924). Pa-Allegheny; Ppv-Pottsville; Mmc-Mauch Chunk; Mg-Greenbrier; Mp-Pocono; Dh-Hampshire; Dfk-Foreknobs; Dsch-Scherr; Db-Brallier; Dhs-Harrell; Dmt-Mahantango; Dmn-Marcellus–Needmore; Do-Oriskany; Ds-Shriver; Dc-Corriganville; DSkn-Keyser–New Creek; Sto-Tonoloway; Sw-Wills Creek; Sm-Mifflintown; Srh-Rose Hill; St-Tuscarora; Oj-Juniata.

siderably thinner here than in the type area in southeastern West Virginia. Approximately 100 ft (30 m) of Greenbrier are exposed in the old quarry. The open fields and relatively flat terrain visible from the highway are characteristic of areas underlain by the Greenbrier.

Stop 5

The basal Mississippian, the Pocono Formation, can be seen where the power line crosses the highway. Only a few feet (meters) of weathered, reddish-brown sandstones interbedded with shale crop out, but in this area the Pocono is usually about 300 ft (90.9 m) thick. These shallow marine clastics overlie the Hampshire Formation and represent the waning phase of Catskill delta sedimentation.

Stop 6

About 1,900 ft (575 m) of Hampshire Formation are fairly well-exposed here. The Hampshire consists of grayish-red to brownish-red, micaceous, fine- to very fine-grained sandstone, interbedded with grayish-red to brownish-gray siltstones, shales, and mudstones. The Hampshire is of nonmarine fluviodeltaic origin.

Stop 7

The Pound Sandstone Member of the Foreknobs Formation is exposed here. This is the type section of the Foreknobs and Scherr Formations and of the Greenland Gap Group (Dennison, 1970). The Pound marks the Fammenian/Frasnian European series boundary (McGhee and Dennison, 1976) and represents offshore barrier sands. A *Cyrtospirifer-Camarotoechia* community (McGhee, 1976) is found in the Pound. The overlying Red Lick Member (poorly exposed here between the Pound and the Hampshire) was deposited in a lagoonal, near-shore environment and contains the characteristic *Leptodesma-Tylothyris* community fauna. The Pound is underlain by the Blizzard Member, which was deposited in a broad, flat, sublittoral marine zone, inhabited by the *Atrypa-Cypricardella* community. The Briery Gap Sandstone, which underlies the Blizzard, is environmentally similar to the Pound and marked a time of shallower-water, offshore bar deposition, with the *Cyrtospirifer-Camarotoechia* community present. The lowest member of the Foreknobs Formation, the Mallow Member, was deposited in a deeper-water offshore environment. The *Ambocoelia-Chonetes* community is characteristic. Overall, the Foreknobs represents a general shallowing-upward sequence. Fossils are scattered, so careful searching may be needed to observe these communities. The western equivalents of both the Hampshire and Foreknobs Formations are important reservoirs for oil and gas across north-central West Virginia.

Stop 8

The Scherr Formation, the lower portion of the Greenland Gap Group, can be examined here. The Scherr is finer grained and thinner bedded than the overlying Foreknobs Formation. Siltstones and shales and only scattered, fine-grained sandstones comprise the Scherr, in contrast to the more abundant sandstones of the Foreknobs. The Scherr is only sparsely fossiliferous and is thought to represent deeper-water deposition than the Foreknobs.

Stop 9

A small, though fairly typical, exposure of the Brallier Formation consists of thin, olive-gray siltstones, interbedded with shales. The Brallier represents deeper-water deposition than the overlying Scherr and can be distinguished from the Scherr by the

Table 1. Stratigraphic Summary, Greenland Gap, Grant Co., WV

Age	Group Formation	Lithology	Thick-ness (ft)*	Significant References	Seen At Stop(s)
Pennsylvanian	Pottsville Formation	sandstone, conglomerate	350-450 in area— only few feet at stop 1	Pressley, 1977	1
Mississippian	Mauch Chunk Formation	red and green sandstone, siltstone and shale	~600	Pressley, 1977, Dennison and Wheeler, 1975	2a, b, 3
Mississippian	Greenbrier Limestone	limestone, gray	~100	Yielding, 1984	4
Mississippian	Pocono Formation	yellowish-brown sandstone, siltstone, and shale	~300	Dally, 1956	5
Devonian	Hampshire Formation	red sandstone, siltstone and shale	1890	Dennison, 1970	6
Devonian	Greenland Gap Group		2325	Dennison, 1970, McGhee and Dennison, 1976	
Devonian	Foreknobs Formation	sandstone, siltstone and shale	1321		7
Devonian	Scherr Formation	siltstone, fine sandstone, and shale	1004	Dennison, 1970	8
Devonian	Brallier Formation	siltstone and shale	1323	Dennison, 1970, Avary and Dennison, 1980	9
Devonian	Harrell Shale	shale	240	Hasson, 1972	
Devonian	Mahantango Formation	shale and siltstone	123	Jolley, 1983	
Devonian	Marcellus Shale	shale	862	Hasson and Dennison, 1977	10
Devonian	Needmore Shale	shale	~200	Dennison, 1961, Newton, 1979	
Devonian	Oriskany Sandstone	sandstone, fossiliferous	120	Diecchio, et al, 1984	11, 13
Devonian	Helderberg Group Shriver Chert Mandata Shale Corriganville Limestone New Creek Limestone Keyser Limestone	limestone, nodular cherty limestone, argillaceous limestone and shale	~450	Head, 1974, Travis, 1971	11, 13
Silurian	Tonoloway Limestone	limestone, laminated, argillaceous	364	Smosna, et al, 1977	11
Silurian	Wills Creek Formation	calcareous shale and limestone	280	Smosna and Patchen, 1978	
Silurian	Mifflintown Formation	shale, limestone and sandstone	~250'	Patchen, 1968, 1973, Patchen and Smosna, 1975	
Silurian	Rose Hill Formation	shale and sandstone	~440	Smosna and Patchen, 1978	
Silurian	Tuscarora Sandstone	quartz sandstone	245	Smosna and Patchen, 1978	12
Ordovician	Juniata Formation	red sandstone, shale and siltstone	?	Diecchio, 1984	12

*Thickness from: Chen, 1981, Clark, 1976, Dennison, 1970, Reger and Tucker, 1924, and Woodward, 1943

increase in shale content and decrease in sandstone content. The Brallier is considered to be of turbidite origin (Avary and Dennison, 1980) as characterized by siltstone beds with abrupt lower boundaries and gradational upper boundaries. These alternating siltstones and shales represent the Bouma D and E sequences, associated with outer fan turbidite deposits. Compare the dip of these rocks with the more gentle dip of the Pottsville and Mauch Chunk observed near the crest of the front.

Stop 10

A few feet (meters) of dark gray, fissile Marcellus Shale are exposed at this stop. The other Devonian shales (Harrell and Mahantango) between the Marcellus and the Brallier, and the Needmore Shale which underlies the Marcellus, are not well-exposed. The units are not as resistant to weathering as the Upper Devonian coarser clastics that form the Foreknobs topography. Also, the shales have been highly deformed and exhibit plastic deformation.

Stop 11

Just east of Scherr, Greenland Road and the North Fork of Patterson Creek cut through Elk Lick Run anticline (Walker Ridge) and Elk Lick Run syncline. These are minor parasitic folds on the northwest flank of the larger amplitude fold, Wills Mountain anticline (Clark, 1976). Oriskany, Helderberg, and Tonoloway are exposed here. The Oriskany is a shallow marine, fine- to coarse-grained quartz sandstone, mostly silica-cemented, fossiliferous, and massively bedded. The Oriskany is underlain by the Shriver Chert, which weathers to a distinctive cherty residium. The Shriver is best seen in an abandoned quarry near the axis of the Elk Lick Run syncline on the north side of the road. The Tonoloway can be seen in an abandoned quarry on the north side of Greenland Road, east of the Shriver quarry.

Stop 12

The actual gap through New Creek Mountain (Wills Mountain anticline) was formed by the North Fork of Patterson Creek. The Tuscarora Sandstone is at the crest of the asymmetrical anticline, which is slightly overturned to the northwest. A "high-angle, longitudinal reverse fault" (Clark, 1976) is visible (especially with binoculars) in the Tuscarora, at the crest of the structure. The Tuscarora is a highly fractured but extremely resistant quartz arenite. The Tuscarora boulder field is better developed on the northeast side of the gap, since this south-facing slope is more subject to mechanical weathering. The Juniata is difficult to see under the Tuscarora talus, but some may be seen in the creek bed. The formations between the Tuscarora and the Tonoloway are poorly exposed on either limb of Wills Mountain anticline.

Stop 13

The waterfalls at the eastern end of Greenland Gap are formed by the Oriskany Sandstone. The Helderberg and Tonoloway Limestones can be examined along the road and also along the stream. These exposures can be compared with the exposure in the Elk Lick Run anticline and syncline to the northwest (Stop 11). Note the gentler dips and lack of the parasitic minor folds on the southeast limb, in contrast to the steeper dips and presence of the Elk Lick Run anticline and syncline on the northwest flank of Wills Mountain anticline. The Needmore, Marcellus, and Mahantango are exposed along the road east of the Oriskany outcrops. Note also the presence of siltstones in the Mahantango in contrast to the shales west of the gap at the base of the Allegheny Front.

The Oriskany is interpreted as shallow marine sandstone with a distinctive large shell fauna of mostly brachiopods. When weathered, the Oriskany is quite friable, due to leaching of calcite cement. Where it is silica cemented, the Oriskany is not as friable. The Oriskany is an important gas reservoir in a good portion of West Virginia and has been the prime target for eastern overthrust-belt exploration. About 10 mi (16.7 km) south of Greenland Gap, the Jordan Run field discovery well encountered significant volumes of gas in the Oriskany. Fractures seem to be important in enhancing a reservoir. Observe the fractures in the Oriskany and compare them with the fractures seen on the other side of the gap, in a structural setting probably somewhat comparable to that found at 8,000–10,000 ft (2,424–3,030-m) depths in the productive area.

The Helderberg rocks are quite cherty and hence frequently have a knobby appearance when weathered. They are also fairly fossiliferous. The Tonoloway, characterized by thinly laminated beds, represents a carbonate tidal flat environment.

REFERENCES CITED

Avary, K. L., and Dennison, J. M., 1980, Back Creek Siltstone Member of the Devonian Brallier Formation in Virginia and West Virginia: Southeastern Geology, v. 21, p. 121–153.

Cardwell, D. H., Erwin, R. B., and Woodward, H. P., eds., 1968, Geologic map of West Virginia: West Virginia Geological and Economic Survey, 2 sheets, scale 1:250,000.

Chen, Ping-fan, 1981, Lower Paleozoic stratigraphy, tectonics, paleogeography, and oil/gas possibilities in the central Appalachians (West Virginia and adjacent states), Part 2, measured sections: West Virginia Geological and Economic Survey Report of Investigations RI-26-2, 300 p.

Clark, G. M., 1967, Structural geomorphology of a portion of the Wills Mountain anticlinorium, Mineral and Grant Counties, West Virginia [Ph.D. thesis]: State College, Pennsylvania State University, 165 p.

Clark, K. A., 1976, Bedrock geology and karst development in Grant and Mineral Counties, West Virginia [M.S. thesis]: Toledo, Ohio, University of Toledo, 100 p.

Dally, J. L., 1956, Stratigraphy and paleontology of Pocono Group in West Virginia [Ph.D. thesis]: New York, Columbia University, 248 p.

Dennison, J. M., 1961, Stratigraphy of Onesquethaw Stage of Devonian in West Virginia and bordering states: West Virginia Geological and Economic Survey Bulletin 22, 87 p., 8 pls.

——1970, Stratigraphic divisions of Upper Devonian Greenland Gap Group ("Chemung Formation") along Allegheny Front in West Virginia, Maryland, and Highland County, Virginia: Southeastern Geology, v. 12, p. 53–82.

Dennison, J. M., and Wheeler, W. H., 1975, Stratigraphy of Precambrian through Cretaceous strata of probable fluvial origin in southeastern United States and their potential as uranium host rocks: Southeastern Geology, Special Publication no. 5, 210 p.

Diecchio, R. J., 1985, Post-Martinsburg Ordovician stratigraphy; Virginia and West Virginia: Virginia Division of Mineral Resources Publication 57, 77 p.

Diecchio, R. J., Jones, S. E., and Dennison, J. M., 1984, Oriskany Sandstone—regional stratigraphic relationships and production trends: West Virginia Geological and Economic Survey, Map WV-17, 9 plates.

Hasson, K. O., 1972, Stratigraphy and structure of the Harrell Shale and part of the Mahantango Formation between Bedford, Pennsylvania and Scherr, West Virginia, in Dennison, J. M., ed., Guidebook, 37th annual field conference of Pennsylvania geologists, p. 81–89.

Hasson, K. O., and Dennison, J. M., 1977, Devonian Harrell and Millboro Shales in parts of Pennsylvania, Maryland, West Virginia, and Virginia, in Schott, G. L., Overbey, W. K., Jr., Hunt, A. E., and Komar, C. A., eds., Proceedings, First Eastern Gas Shales Symposium, October 17–19, 1977, Morgantown Energy Research Center, Publication MERC/SP-77/5, p. 634–637.

Head, J. M., 1974, Correlation and paleogeography of Helderberg Group (Lower Devonian) of central Appalachians: American Association of Petroleum Geologists Bulletin, v. 58, p. 247–259.

Jolley, R. M., 1983, The Clearville Siltstone Member of the Middle Devonian

Mahantango Formation in parts of Pennsylvania, Maryland, West Virginia, and Virginia [M.S. thesis]: Chapel Hill, University of North Carolina, 192 p.

McGhee, G. R., Jr., 1976, Late Devonian benthic marine communities of the central Appalachian Allegheny Front: Lethia, v. 9, p. 111–136.

McGhee, G. R., Jr., and Dennison, J. M., 1976, The Red Lick Member, a new subdivision of the Foreknobs Formation (Upper Devonian) in Virginia, West Virginia, and Maryland: Southeastern Geology, v. 18, p. 49–57.

Newton, C. R., 1979, Biofacies pattern in the Needmore Shale: paleoenvironmental and paleobathymetric implications, in Avary, K. L., ed., Devonian clastics in Maryland and West Virginia, field trip guidebook for Eastern Section, American Association of Petroleum Geologists annual meeting, Morgantown, West Virginia, Oct. 1–4, 1979, p. 77–82.

Patchen, D. G., 1968, Keefer Sandstone gas development and potential in West Virginia: West Virginia Geological and Economic Survey Circular C-17, 20 p.

——1973, Stratigraphy and petrography of the Upper Silurian Williamsport Sandstone, West Virginia: Proceedings of the West Virginia Academy of Science, v. 45, p. 250–265.

Patchen, D. G., and Smosna, R. A., 1975, Stratigraphy and petrology of Middle Silurian McKenzie Formation in West Virginia: American Association of Petroleum Geologists Bulletin, v. 59, p. 2266–2287.

Perry, W. J., 1978, The Wills Mountain anticline: a study in complex folding and faulting in eastern West Virginia: West Virginia Geological and Economic Survey Report of Investigations RI-32, 34 p.

Presley, Mark W., 1977, A depositional systems analysis of the Upper Mauch Chunk and Pottsville Groups in northern West Virginia [Ph.D. thesis]: Morgantown, West Virginia University, 157 p.

Reger, D. B., and Tucker, R. C., 1924, Mineral and Grant Counties: West Virginia Geological Survey County Report and Maps, 866 p., 4 maps.

Smosna, R. A., and Patchen, D. G., 1978, Silurian evolution of the central Appalachian Basin: American Association of Petroleum Geologists Bulletin, v. 62, p. 2308–2328.

Smosna, R. A., Patchen, D. G., Warshauer, S. M., and Perry, W. J., Jr., 1977, Relationships between depositional environments, Tonoloway Limestone, and distribution of evaporites in the Salina Formation, West Virginia, in Fisher, J. H., ed., Reefs and evaporites—concepts and depositional models, Studies in Geology no. 5, American Association of Petroleum Geologists, p. 125–143.

Travis, J. W., 1971, Paleoenvironmental study of the Helderberg Group of West Virginia and Virginia [Ph.D. thesis]: East Lansing, Michigan State University, 150 p.

Woodward, H. P., 1943, Devonian system of West Virginia: West Virginia Geological Survey, v. 15, 655 p.

Yeilding, C. A., 1984, Stratigraphy and sedimentary tectonics of the Upper Mississippian Greenbrier Group in eastern West Virginia [M.S. thesis]: Chapel Hill, University of North Carolina, 117 p.

The Taconic sequence in the northern Shenandoah Valley, Virginia

Lynn S. Fichter, Department of Geology and Geography, James Madison University, Harrisonburg, Virginia 22807
Richard J. Diecchio, Department of Geology, George Mason University, Fairfax, Virginia 22030

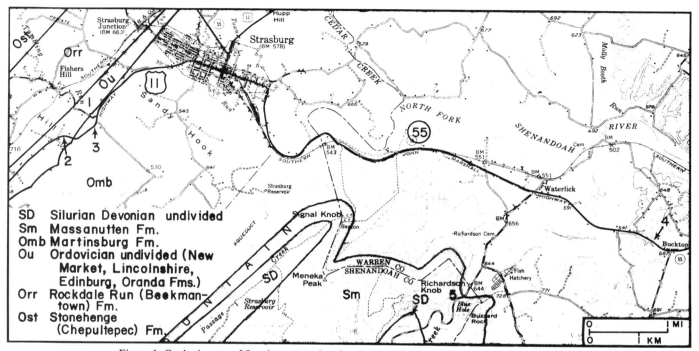

SD Silurian Devonian undivided
Sm Massanutten Fm.
Omb Martinsburg Fm.
Ou Ordovician undivided (New Market, Lincolnshire, Edinburg, Oranda Fms.)
Orr Rockdale Run (Beekmantown) Fm.
Ost Stonehenge (Chepultepec) Fm.

Figure 1. Geologic map of Strasburg area showing stop locations. Strasburg 15-minute quadrangle.

LOCATION

The Taconic sequence in the northern Shenandoah Valley is located on the Strasburg and Toms Brook, 7½-minute Quadrangles, or the Strasburg, 15-minute Quadrangle. In addition, geologic maps are available for both 7½-minute quadrangles (Rader and Biggs, 1976). All stops are readily accessible on paved heavy duty and medium duty roads and are numbered on the index map (Fig. 1) in stratigraphic order. All stops are on public land or right-of-way and require no special permission for access.

INTRODUCTION

The rocks discussed for the northern Shenandoah Valley are a complete sequence of formations representing the transitions from: (1) the stable, carbonate-dominated, Proto-Atlantic continental margin of the early Ordovician (Beekmantown, New Market, and Lincolnshire Formations); (2) through a transition zone representing rapid subsidence (Edinburg, Oranda Formations); to the (3) rapidly filling flysch basin of the Taconic Orogeny (Martinsburg Formation); to the (4) post-orogenic shallow shelf and quartzitic fluvial deposits which prograded across the filled and stabilizing basin of the Early Silurian (Massanutten Formation). This sequence illustrates the marked changes in lith-

ology, processes of sedimentation, and depositional systems which occurred when the stable margin of northern Virginia was involved in the continent-island arc collision of the Taconic Orogeny.

This Strasburg cluster stop is interesting by itself, but a greater regional significance can be gained by comparing it with the Germany Valley section at the Allegheny Front (Fig. 2; and Diecchio, this volume). The sequence of formations from Strasburg, Virginia to Germany Valley, West Virginia show marked differences that are not just in the names; the sedimentology, lithologies, and depositional environments are markedly different. Diecchio (1980a, 1985) has proposed that two basins existed in this part of the geosyncline separated by a low arch (Figs. 3 and 4), and the greater significance of this Strasburg cluster stop is its comparison with the Germany Valley section. The arch itself probably existed just east of Brocks Gap, Virginia (not shown in Fig. 1) (Virginia 259, 5.0 mi [8 km] west of Broadway, Virginia). Brocks Gap is a complex and spectacular exposure cut by the North Fork of the Shenandoah River through Little North Mountain (Rader and Perry, 1976; Sherwood and others, 1977); formations range from the Middle Ordovician Reedsville (Martinsburg) through to the Lower Devonian Oriskany at Chimney Rock (far west end of the ap). The well exposed Middle Ordovi-

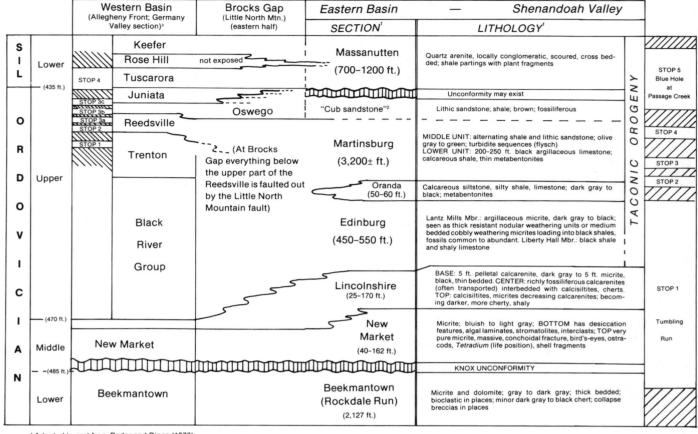

Figure 2. Ordovician and Lower Silurian stratigraphy and correlation diagram for the northern Shenandoah Valley (Strasburg cluster stop), Brocks Gap (Little North Mountain) and Germany Valley. Germany Valley stratigraphy is described in Diecchio (this volume).

[1] Adapted in part from Rader and Biggs (1976)
[2] "Cub sandstone" (Thornton, 1953)
[3] Stops in the fieldguide by Diecchio (this volume) Correlations from Rader 1976, Patchen, et al (in press)

cian to Lower Silurian rocks in the east half of Brocks Gap are a sharp contrast to the Strasburg and Germany Valley sections, sharing some characteristics with both, but also being uniquely distinct. Further regional differences in stratigraphy and paleogeography can be seen in southwestern Virginia (Markello and others, 1979; Read, 1980).

Evidence for the larger historic, plate tectonic, and sedimentary tectonic framework in which all this took place is diverse (see Thomas, 1977; Diecchio, 1980a, 1985; Rader and Henika, 1978), but, in essence, the Taconic Orogeny occurred when an island arc chain converged on the east coast from the southeast. The earliest signs of the orogeny are in the ash falls preserved in the Lower–Middle Ordovician carbonates, e.g., Stop 2. During mountain building the Shenandoah Valley was in a protected reentrant (Thomas, 1977, p. 1266) and suffered only minor flexure folding but it was this which created the eastern basin preserved in the Shenandoah Valley (Strasburg cluster stop), the western basin preserved at Germany Valley (Allegheny Front), and the arch (preserved at Brocks Gap) in between.

The eastern basin was much deeper than the western, especially to the south, and although the Taconic island arc was off the east coast, paleocurrent data indicates the Martinsburg flysch did not come from the east but from the southwest down a long, rapidly subsiding basin parallel to the coast (McBride, 1962). The clastics of the western basin exposed at Germany Valley have a different history. Although the Reedsville correlates with the upper Martinsburg, the Reedsville is markedly different in sedimentology and depositional environments (Diecchio, 1980a, 1985; Kreisa, 1981; note, Kreisa's Martinsburg is the same as our Reedsville). Furthermore, the Juniata Formation which is so well developed at the Allegheny Front is completely absent by facies change in the northern Shenandoah Valley. At Brocks Gap, the Oswego sandstone—the facies equivalent of the Juniata—comprises almost all of the section from the top of the Reedsville to the base of the Tuscarora; the only sign of the Juniata at Brocks Gap is a few red zones within the coarse sands identified as Oswego.

Thus what we see in the Ordovician and Early Silurian of

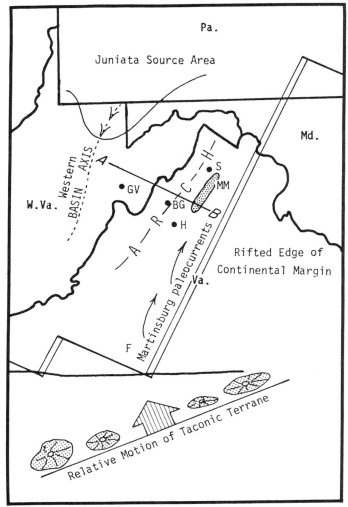

Figure 3. Paleogeographic reconstruction of the Late Ordovician in Virginia, West Virginia, and Pennsylvania. Position of the rifted edge of the continental margin (which opened in the late Precambrian) and relative direction of the convergence of the Taconic terrane in the Ordovician (from Thomas, 1977). At the time of this map, the Taconic terrane had already collided with south-central Virginia, and flexure folding to the north had created the eastern and western basins separated by an arch (paleogeography from Diecchio, 1980a, 1985). Line *A–B* is the cross section in Figure 4; *BG,* Brocks Gap; *F,* Fincastle conglomerate; *GV,* Germany Valley; *H,* Harrisonburg; *MM,* Massanutten Mountain; *S,* Strasburg.

Virginia, West Virginia, and surrounding areas is a very complex history. And within all this is the Strasburg cluster stop which is interesting with regard to sedimentology, paleogeography, sedimentary tectonics, and plate tectonics.

The Strasburg cluster stops are included within the west flank and northern end of the Massanutten synclinorium (see Fig. 1). Massanutten Mountain, held up by the Lower Silurian Massanutten sandstone, is the most prominent topographic feature in the Shenandoah Valley. The steep outside flanks of the mountain

are underlain by the Upper Ordovician Martinsburg Formation. The topographically much lower Shenandoah Valley is underlain by Cambro-Ordovician carbonates. Within the Massanutten synclinorium is Fort Valley, readily accessible by continuing south on Virginia 656 from Stop 5; Silurian and Lower and Middle Devonian rocks above the Massanutten Formation are present.

Seven formations seen at five localities are part of the Strasburg cluster stop. Figure 2 summarizes the stratigraphy and provides basic descriptions as well as the stop positions in the stratigraphic column. Six of the formations are well exposed, distinctive, and easily recognized. The Martinsburg, however, which underlies most of the map area, is poorly exposed, and varies markedly from bottom to top. Rader and Biggs (1976) divided the Martinsburg into three sections—a lower black shale and limestone, a middle flysch sequence, and an upper sandstone; Thornton (1953) informally named the upper section the "Cub sandstone" after its best exposure at Cub Run, near Catherines Furnace, and we will use his terminology here.

In addition to the localities discussed in this guide, much other geology is in the northern Shenandoah Valley area, including a nearly complete sequence up through the Acadian Orogeny (Fichter and Poche', 1979; Rader and Biggs, 1976).

SITE DESCRIPTION

Stops are arranged in stratigraphic order bottom to top.

Stop 1. Tumbling Run Section

This is a classic section (Cooper and Cooper, 1946) where the Beekmantown, New Market, Lincolnshire, and Edinburg Formations are exposed. The exposure is a long, low roadcut on the north side of Virginia 601, 0.3 mi (0.5 km) west of the Virginia 601 and U.S. 11 junction; Virginia 601 is about 1.5 mi (2.4 km) south of Strasburg. A small bridge crosses Tumbling Run at the New Market Formation and is indicated on Figure 5, which summarizes most of the exposure. The Beekmantown Formation (not shown in Fig. 5) is exposed west of the bridge and is largely an interbedded micritic and dolomitic supertidal deposit. The Knox unconformity between the Beekmantown and New Market, which represents the sea level drop at the Sauk/-Tippecanoe sequence boundary, is not exposed here. An alternate to Stop 1 is a roadcut located at the U.S. 11/I-81 northbound exit ramp. The best exposure is along I-81; the ramp exposure is masked by a crust.

This stop contains the transition from the carbonate-dominated supertidal deposits of the stable continental margin (Beekmantown Formation), through intertidal/subtidal environments (New Market Formation) to shelf/reef deposits (Lincolnshire Formation) to a basin under increasing clastic influence (Edinburg Formation). Bentonites derived from the Taconic island arc source are in the Edinburg; these are the earliest evidence of Taconic orogenic activity preserved in these strata.

Also of interest at this stop are the extensive, well developed tufa deposits in the run found above and below the bridge.

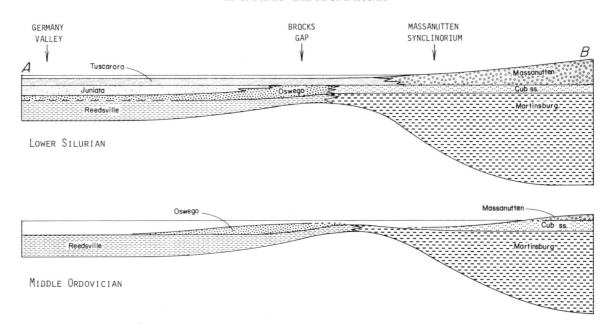

Figure 4. Generalized east-west cross section through the Middle Ordovician and Lower Silurian of northern Virginia and West Virginia showing the development of sedimentary facies from the eastern Taconic forearc basin (Strasburg cluster stop) across the arch of Brocks Gap to the Central Appalachian Basin (exposed in the Germany Valley section) in West Virginia.

Stop 2. Oranda Formation

The Oranda Formation is exposed along the west side of U.S. 11 0.35 mi (0.6 km) south of Virginia 601 (Fig. 1). "The Oranda is composed of gray to brownish-gray calcareous silt-stone; black, silty shale; dark-gray calcareous mudstone; cobbly weathering, gray, fossiliferous and argillaceous limestone; and tan to brown metabentonites . . . one of the thicker metabentonites has a highly fossiliferous silicified zone;" thickness ranges from 50–60 ft (15–18 m) (Rader and Biggs, 1976). This formation is a transition from the carbonate-dominated Edinburg Formation below to the clastic-dominated Martinsburg flysch above. The abundant metabentonites indicate the Taconic arc island was now close. The collision had already taken place 10 to 15 million years earlier in the southwest (Abingdon and Wytheville, Virginia) as indicated by the Paperville Shale and Knobs Formation (Rader, 1982; Markello and others, 1979), and the Fincastle Conglomerate (Bartholomew and others, 1982).

Stop 3

This is the lower unit of the Martinsburg Formation (Rader and Biggs, 1976); it is exposed on the west side of U.S. 11 0.1 mi (0.2 km) south of its junction with Virginia 601 (i.e., Tumbling Run section). U.S. 11 is a divided highway here and additional exposures of the lower unit of the Martinsburg can be found along the north bound lane. It consists of 200–250 ft (61.0–76.2 m) of argillaceous limestone, calcareous shale, and meta-bentonites.

Compared to the carbonate-dominated Tumbling Run sec-

tion, the Martinsburg Formation shows the full influence of the Taconic orogenic source area. The distal portion of the clastic wedge with its rhythmically thin-bedded, black, anoxic, deep water deposits still shows the influence of two sources, carbonate debris from the western platform, and clastics from the Taconic sourceland, but it will soon give way upsection to the thickest and more typical unit of this formation at Stop 4.

Stop 4

This is the middle unit of the Martinsburg Formation (Rader and Biggs, 1976); it is a low, small, obscure exposure on the north side of Virginia 55, a few hundred yards west of its junction with Virginia 626. Despite the tiny nature of this exposure it is quite adequate to illustrate the characteristic graywacke flysch of this formation with distinctive Bouma sequences of a mid-fan type.

Stop 5. Blue Hole Section: "Cub sandstone" of the Martinsburg, and the Massanutten Formation

Blue Hole is located at a water gap along Passage Creek on Virginia 678 leading south into Fort Valley within the Massanutten synclinorium; the upper unit of the Martinsburg ("Cub" of Thornton, 1953) and the Massanutten are exposed on the west side of the road, 1.9 mi (3.1 km) south of the Virginia 55 and 678 junction at Waterlick, Virginia. The Massanutten sandstone apparently lies directly on top of the Martinsburg (Fig. 2) although the contact is not exposed.

Note that compared to the Germany Valley section (Fig. 2)

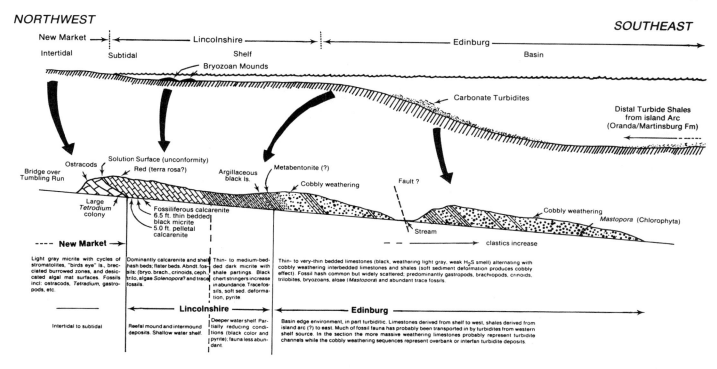

Figure 5. Summary of the geology present in the Tumbling Run section.

the Oswego and Juniata Formations are not here; for a long time these missing formations were interpreted as a major unconformity, which would make the Massanutten area part of the Taconic uplift. However, recent interpretations by the authors indicate that the Martinsburg Formation shallows up from the basin turbidites of Stop 4 to a shelf ("Cub sandstone") just before deposition of the Massanutten Formation. It is still possible that an unconformity exists between the Martinsburg ("Cub sandstone") and Massanutten Formations. There is no biostratigraphic control, but it now seems the Oswego and Juniata were never deposited in the eastern basin of Massanutten Mountain (Diecchio, 1980a, 1985). Thus the "Cub sandstone"–Massanutten Formation transition is one of the more interesting events in the area.

The "Cub sandstone" at Blue Hole is not well exposed and is badly weathered; individuals interested in the Cub should see it at Catherines Furnace (see above) where it is still badly weathered, but well exposed.

The Massanutten Formation is a conglomeratic quartz arenite with scouring and abundant large scale cross bedding. The sudden appearance of so much quartz is still a problem, but it has been suggested (Diecchio, 1980b) that the source of the abundant quartz can be attributed to sedimentation from an eastern continental source area. The lower portion of the Massanutten Formation has been interpreted as a fluvial system (braided stream; Pratt, 1979) with paleocurrents generally from the southwest (Roberts and Kite, 1978). At the Passage Creek section shale partings present between some of the sandstones have produced fragments of some of the oldest known vascular plants (Pratt et al., 1978).

The transition from the lithic wackes of the upper Martinsburg ("Cub sandstone") to the quartz arenites of the Massasnutten represents the cessation of the influence of the Taconic source area.

REFERENCES CITED

Bartholomew, M. J., Schultz, A. P., Henika, W. S., and Gathright, T. M., III, 1982, Geology of the Blue Ridge and Valley and Ridge at the junction of the central and southern Appalachians, *in* Lyttle, P. T. ed., Central Appalachian Geology, NE-SE Geological Society of America 1982 Field Trip Guidebooks, p. 121–170.

Cooper, B. N., and Cooper, G. A., 1946, Lower Middle Ordovician stratigraphy of the Shenandoah Valley, Virginia: Bulletin Geological Society of America, v. 57, p. 35–113.

Diecchio, Richard J., 1980a, Post-Martinsburg Stratigraphy, Central Appalachian Basin [Ph.D. thesis]: Chapel Hill, University of North Carolina, 196 p.

—— 1980b, Stratigraphic and petrologic evidence for partial closure of the Proto-Atlantic during the Taconic orogeny: Geologic Society of America National Meeting Abstracts, v. 12, No. 7, p. 413.

—— 1985, in press, Post-Martinsburg Ordovician stratigraphy of the Virginias: Virginia Division of Mineral Resources.

Fichter, Lynn S., and Poche', David J., 1979, Sedimentary Paleoenvironments and Paleozoic Evolution of the Appalachian Geosyncline in the Northern Shenandoah Valley, *in* Exline, Joseph D., ed., Guidebook for Field Trips in

Virginia: National Association of Geology Teachers—Eastern Section, April 1979: p. 25–42.

Kreisa, R. D., 1981, Storm-generated sedimentary structures in subtidal marine facies with examples from the Middle and Upper Ordovician of southwestern Virginia: Journal of Sedimentary Petrology, v. 51, p. 823–848.

Markello, J. R., Tillman, C. G., and Read, J. F., 1979, Lithofacies and Biostratigraphy of Cambrian and Ordovician Platform and Basin Facies Carbonates and Clastics, Southwestern Virginia, *in* Glover, Lynn, and Read, J. Fred, eds., Guides to Field Trips 1–3 for Southeastern Section Meeting Geological Society of America, Blacksburg, Virginia: p. 43–109.

McBride, E. F., 1962, Flysch and associated beds of the Martinsburg Formation (Ordovician), Central Appalachians: Journal of Sedimentary Petrology, v. 32, p. 39–91.

Patchen, D. G., Avary, K. L., and Erwin, R. B., in press, Correlation of stratigraphic units of North America–Appalachian Basin: American Association Petroleum Geologists.

Pratt, L. M., Phillips, T. L., and Dennison, J. M., 1978, Evidence of non-vascular land plants from the early Silurian (Llandoverian) of Virginia, U.S.A.: Review of Paleobotany and Palynology, v. 25, p. 121–149.

Pratt, Lisa M., 1979, The Stratigraphic Framework and Depositional Environment of the Lower to Middle Silurian Massanutten Sandstone [M.S. thesis]: Chapel Hill, University of North Carolina, 76 p.

Rader, Eugene K., and Biggs, Thomas H., 1976, Geology of the Strasburg and Toms Brook Quadrangles, Virginia: Virginia Division of Mineral Resources, Report of Investigations 45.

Rader, E. K., and Perry, W. J., Jr., 1976, Reinterpretation of the geology of Brocks Gap, Rockingham County, Virginia: Virginia Minerals, v. 22, p. 37–45.

Rader, E. K., and Henika, W. S., 1978, Ordovician shelf-to-basin transition, Shenandoah Valley, Virginia, *in* Contributions to Virginia Geology—III: Virginia Division of Mineral Resources Publication 7, p 51–65.

Rader, Eugene K., 1982, Valley and Ridge Stratigraphic Correlations, Virginia: Virginia Division of Mineral Resources, Publication 37, (chart).

Read, J. F., 1980, Carbonate ramp-to-basin transitions and foreland basin evolution: American Association of Petroleum Geologists Bulletin, v. 64, p. 1575–1612.

Roberts, W. P., and Kite, J. S., 1978, Syntectonic deposition of Lower to Middle Silurian sandstones, central Shenandoah Valley, Virginia: Virginia Minerals, v. 24, p. 1–5.

Sherwood, W. C., Campbell, F. H., Kearns, L. E., Rader, E. K., and Perry, W. J., Jr., 1977, Geology of Little North Mountain and the Central Shenandoah Valley: Ninth Annual Virginia Geology Field Conference, James Madison University, Harrisonburg, Virginia.

Thomas, William A., 1977, Evolution of Appalachian-Ouachita salients and recesses from reentrants and promontories in the continental margin: American Journal of Science, v. 277, p. 1233–1278.

Thornton, C. P., 1953, The geology of the Mount Jackson quadrangle, Virginia [Ph.D. thesis]: New Haven, Yale University, 215 p.

Devonian to Mississippian section, Elkins, West Virginia

J. S. McColloch and J. F. Schwietering, *West Virginia Geological and Economic Survey, Mogantown, West Virginia 26507-0879*

LOCATION

The site is a series of twenty road cuts along the four-lane section of U.S. 33 between Canfield and Bowden in northern Randolph County, West Virginia. It is depicted on the Elkins and Bowden 7½-minute Quadrangles (Fig. 1). The seven-mile-long section begins 2.1 miles (3.4 km) east of the intersection of U.S. 33, 250, and 219 in Elkins, West Virginia, and is accessible by car and bus.

INTRODUCTION

Rock exposures along U.S. 33 provide an opportunity to study the geology of 4,200 ft (1280 m) of strata which include: (1) the Upper Devonian Scherr, Foreknobs, and Hampshire Formations (Scherr and Foreknobs Formations comprise the Greenland Gap Group, equivalent to the Chemung Formation as described by Dennison [1970, 1971]); (2) the Lower Mississippian Pocono Formation; (3) the Middle Mississippian Greenbrier Group; and (4) the Upper Mississippian Mauch Chunk Group (Figs. 2, 3). Upper Devonian strata exposed at the site are part of the Catskill deltaic complex, and have been correlated with drillers' sands encountered in wells of the nearby Cassity Field (Schwietering, 1979; Lewis, 1983). Limestones of the Greenbrier Group quarried in the Elkins area are mainly used for aggregate. However, the limestone quarry and mine in the Greenbrier Group at the east end of the section is inoperative. A highly radioactive sandstone bed occurs in the Pocono Formation, but it is not ore grade.

SITE DESCRIPTION

Geological Setting

The section is located on the southeast limb of the Elkins Valley anticline, with dips ranging between 17 and 4 degrees. These rocks were folded and faulted during the Pennsylvanian and Permian Periods. A small thrust fault near the western end of the site (Fig. 2) is an example of an uplimb thrust fault (Berger and others, 1979). This fault, which has a stratigraphic throw of 200 feet (61 m), shortened beds as the Elkins Valley anticline grew. It may be related to a décollement which ramps up under the Elkins Valley anticline (Gwinn, 1964). Several small reverse faults are present in the section, and mineralized fractures are common.

Location of Geological Features

Geological features are located approximately by mileage, and specifically by highway station numbers (such as 100+00) which are stamped in the edge of the pavement at 100-ft (30.5 m) intervals (Fig. 2). The numbers increase from west to east (Figs. 1, 2). The following provides the mileage (and kilometers) for discussion of geological features and directions.

0.0 mile (km). Intersection of U.S. 33, 250, and 219.
2.1 miles (3.4 km). Section begins at west end of four-lane section of highway at station number 92+00.

Exposures of the Scherr Formation from 93+00 to 170+00. These olive-gray, gray, and brown marine shales, siltstones, and sandstones were deposited as turbidites in delta front and prodelta environments, and display sedimentary features such as graded bedding, sharp basal contacts with either sharp or gradational upper contacts, sole marks, flute casts, groove marks, ball-and-pillow structures, ripple marks, and hummocky cross-stratification. Some units are fossiliferous.

2.2 mi (3.5 km). Thick siltstone exposed in core of small anticline at 97+25 is at or near the base of the Scherr Formation. Small-scale ball-and-pillow structures. *Ambocoelia* (brachiopod) and other fossils present.
2.3 mi (3.7 km). Discontinuous bedding, sole marks, and fossils in siltstone beds 17.5 ft (5.3 m) above ditch at 102+66. Hummocky cross-stratification in shale and siltstone unit 13 ft (4.0 m) above ditch at 103+75. Hydroplastic deformation features. Ripple marks on float. Zone of thick, very-fine-grained sandstone at 105+00. Crinoid stems and brachiopods in siltstone beds interbedded with shale 21.6 ft (6.6 m) above ditch at 105+65.
2.4 mi (3.9 km). Two relatively clean, oil-stained sandstone beds have sharp upper and lower contacts, and are separated by 10 ft (3.0 m) of silty shale from 108+32 to 108+65. Thin drag marks on float. At 110+40, ball-and-pillow structures in siltstone unit 22 ft (6.7 m) above ditch. Brachiopods and cephalopods in siltstone unit 38.3 ft (11.7 m) above ditch. Reddish-brown shale beds between 110+40 and 112+50 may represent time markers (suggested by Dennison, 1970, p. 66).
2.5 mi (4.0 km). Clay mylonite zone at 113+00 cuts across bedding. Zone almost parallel to thrust fault at 130+00. Sole marks on some siltstone beds of the siltstone/shale unit from 113+25 to 113+82. Coquinite zones in massive siltstones at 114+25.
2.6 mi (4.2 km). Rocks in ditch leading into borrow pit at 118+50 contain corals, bryozoans, and brachiopods. Zones of coquinite in basal part of siltstone beds. Ripple marks on upper surface of sandstone bed at 119+35. Oil-stained sandstone unit from 119+50 to 119+75 has small crinoid stems in base. Groove

Figure 1 (this and facing page). Maps showing location of site in Randolph County, West Virginia. Site includes 20 road cuts along U.S. 33 from Canfield to Bowden. Geological Contacts and highway station numbers along U.S. 33 are shown.

casts on float. Swales between hummocks in hummocky cross-stratification filled with organic mud between 121+15 and 121+50.

2.8 mi (4.5 km). Short, narrow groove marks on bases of some siltstone beds at 126+50. Thrust fault, marked by zone of fault breccia, has approximately 200 ft (61 m) of stratigraphic throw at 130+00. Kink band on upthrown side of fault at 130+62.

2.9 mi (4.7 km). Abundant flute casts, drag marks, and some cross-bedding in siltstone unit from 133+65 to 133+88.

3.0 mi (4.8 km). Brachiopods and coalified wood fragments at 139+00. Siltstone beds from 140+00 to 147+00, correlated to the Elkins Sandstone by Reger (1931), contain ball-and-pillow struc-

tures (which Reger misinterpreted as fossil trees). Hummocky cross-stratification with coquinites at base of beds. Ripple marks on float.

3.2 mi (5.1 km). Sandstone, which laterally truncates against dark-gray shale at 147+85, contains approximately 15% fossil hydrocarbon (Heald, 1980).

3.5 mi (5.6 km). Sandstones and siltstones in ditch from 162+30 to 163+50 have load structures and contain abundant crinoid stems.

Exposures of the Foreknobs Formation from 170+00 to 275+90. The Foreknobs Formation consists of predominantly

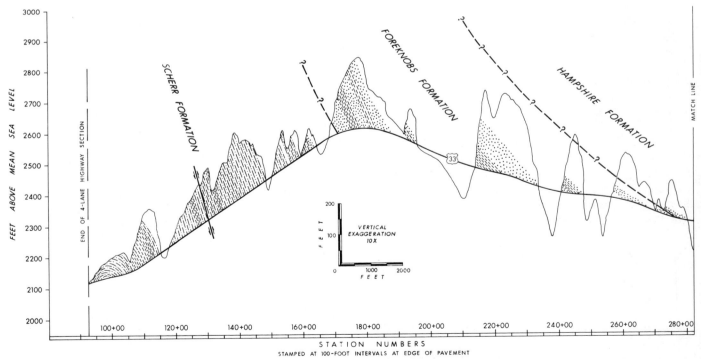

Figure 2 (this and facing page). Cross section showing geology along U.S. 33 from Canfield to Bowden, Randolph County, West Virginia.

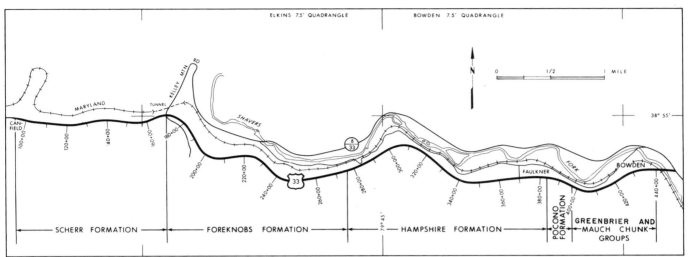

marine gray sandstones and dark-gray shales deposited in fuvio-deltaic and nearshore environments, such as tidal flats, barrier islands, and lagoons (Lewis, 1983). Sedimentary features include flaser bedding, rip-up clasts, clay balls, pseudoanticlines, sole marks, hydroplastic deformation features, ripple marks, slump features, and channel cuts. Woody fragments are found in many units. Some units contain marine fossils.

3.6 mi (5.8 km). Intersection of U.S. 33 with Kelley Mountain Road. Contact of the Scherr and Foreknobs Formations at base of the lower of two sandstone-filled channels at 170+00 (McColloch and Schwietering, 1985). These sandstones may represent channel fill in abandoned tidal channels. Large-scale rip-up clasts and plant debris. Elliott (1982) found *Cooperidiscus alleganius* (Clarke) (edrioasteroid) near base of Foreknobs Formation.

3.7 mi (6.0 km). Flaser bedding, convolute bedding, load structures, gas-heave structures, and pseudonodules from 175+37 to 177+79. Bradford Zone, a drillers' sand (Schwietering, 1979) from 178+12 to 182+50. Spheroidal weathering of shale unit.

3.9 mi (6.3 km). Reddish-brown sandstone underlain by red beds and overlain by marine sandstones and shales at 182+50.

4.5 mi (7.2 km). Ball-and-pillow structures, load casts, scour-and-fill features, and graded bedding in sandstones and shales at

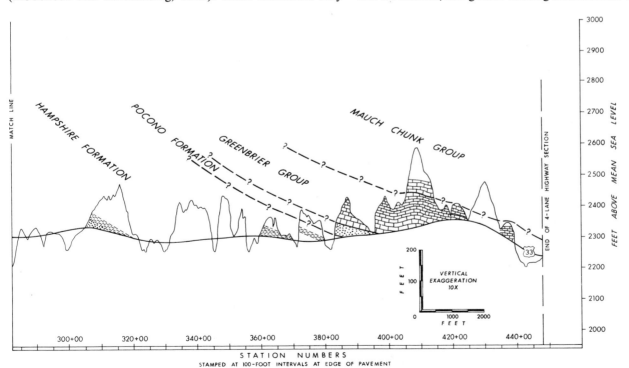

215+00 to 218+00. Shale rip-up clasts in gray sandstone. Clay-filled channel in sandstone unit at 218+50. Hydroplastic deformation. Abundant plant debris in sandstones. Pyrite occurs in organic muds.

4.6 mi (7.4 km). Zone of clay balls, pyrite, and plant material from 220+50 to 221+50. Large-scale load features at base of sandstone at 225+00. Vertical burrows in sandstone beds.

4.7 mi (7.6 km). At 226+50 channel in massive sandstone filled with very dark-gray shale at base and red shale at top. Ball-and-pillow structures, load structures, and conglomeratic lenses in sandstone unit from 227+50 to 233+50. Pseudoanticlines in red shale bed. Channel in sandstone is filled with interbedded shale and sandstone on north side of highway.

5.0 mi (8.0 km). Spectacular cut-and-fill features in sandstones and red shales from 241+00 to 247+00. Cross-bedding, scour-and-fill structures, abundant plant fossils, and compressed coalified logs in sandstone unit 241+30. *Stigmaria* in black shale unit at 244+00.

5.3 mi (8.5 km). Beginning of massive red beds (Hampshire Formation) in cut benches above highway at 256+70.

5.4 mi (8.7 km). Large-scale cross-bedding from 260+50 to 269+30. Slump feature at 262+00.

Exposures of the Hampshire Formation from 275+90 to 384+00. Most of the red shales, gray sandstones, and yellow-brown siltstones of the Hampshire Formation were deposited in nonmarine environments. A few layers were deposited in a brackish or marine environment. Much of the Hampshire Formation is covered or poorly exposed. These rocks contain plant fossils, linguloid brachiopods, and fish scales. The oldest known seeds have been found in the Hampshire Formation in these road cuts (Gillespie and others, 1981).

5.7 mi (9.1 km). Contact of Foreknobs and Hampshire Formations at road level at 275+90 (McColloch and Schwietering, 1985).

6.9 mi (11.1 km). Carbonaceous shale with plant fossils at 339+00 to 343+00.

7.3 mi (11.7 km). Small reverse fault with two feet of displacement on north side of highway at 361+00. Abundant plant fossils, including *Rhacophyton,* linguloid brachiopods, fish scales, and vertebrate fossils from 362+30 to 365+00.

7.5 mi (12.1 km). Faulkner exit at 370+00.

7.7 mi (12.4 km). Black shale channel fill at 383+00 has middle Famennian (Fa2c, Fa2d) plant fossils (Gillespie, personal communication). Disconformable contact of Hampshire Formation with overlying Pocono Formation at 384+00.

Exposures of the Pocono Formation from 384+00 to 397+50. Predominantly marine, coarse-grained, brown sandstones and conglomerates, and dark-gray shales of the Pocono Formation represent the destructive phase of the Catskill deltaic complex (Dally, 1956; Dennison, 1971). A disconformity exists between the Hampshire and Pocono Formations at this location,

based on fossil evidence. Another disconformity occurs between the Pocono Formation and the Greenbrier Group. In this exposure bedding is parallel at the contact, but regional subsurface data indicate that an angular unconformity of considerable magnitude is present (Lewis, 1983).

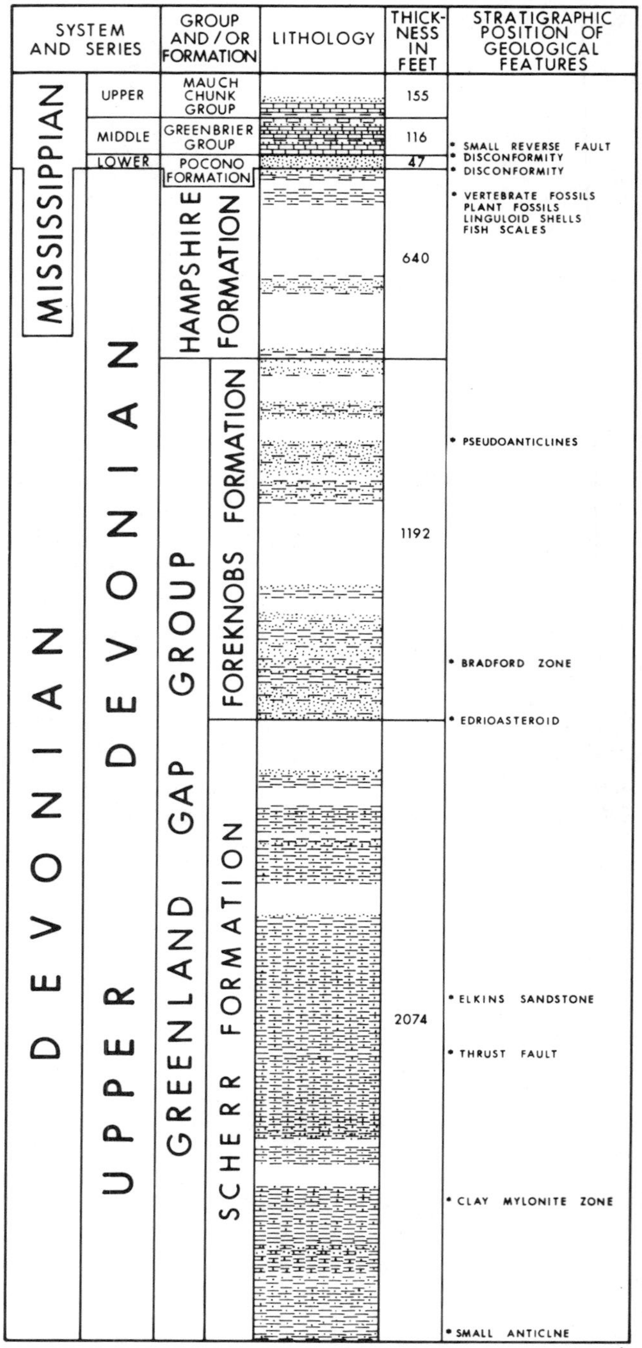

Figure 3. Generalized geological column showing Upper Devonian to Upper Mississippian rocks exposed along U.S. 33 from Canfield to Bowden, Randolph County, West Virginia.

7.8 mi (12.5 km). Highly radioactive brown, conglomeratic sandstone bed in Pocono Formation contains 0.94 pounds (.4 kg) U_3O_8 per ton (Kirstein, personal communication). Abundant Mississippian brachiopods.

Exposures of the Greenbrier and Mauch Chunk Groups from 390+00 to 439+20. Carbonates of the Greenbrier Group were deposited in supratidal and turbulent shallow-water shelf environments (Leonard, 1968). Terrigenous clastics in the Greenbrier Group resulted from the periodic influx of sediments from the north, northeast, and east (Yeilding, 1984). Deposition of the Greenbrier Group resulted from a shift of the depositional axis toward the southeast (Rittenhouse, 1949). According to Yeilding (1984, p. 66) "the Greenbrier Group in Randolph County represents two major transgressive-regressive cycles of sedimentation." The Greenbrier Group is gradationally overlain by the Mauch Chunk Group.

Basal Mauch Chunk limestones were deposited in supratidal and shallow-water shelf environments. Basal Mauch Chunk sandstones and shales were derived from land areas north and northeast of the shallow marine environment, and were deposited in prograding transitional environments.

7.9 mi (12.7 km). Inoperative quarry in limestones of Greenbrier Group from 390+00 to 403+50. Reverse faults with approximately two feet (0.6 m) of displacement visible in the highwall. Disconformable contact between Pocono Formation and Greenbrier Group at 397+50.

8.0 mi (12.9 km). Entries into room-and-pillar limestone mine at 403+50.

8.1 mi (13.0 km). Three red sandstone and shale units (Taggard Formation). Gradational contact of Greenbrier and Mauch Chunk Groups between third and fourth benches of road cut. Pink calcite fracture fillings, slickensided fault planes, and solution features in limestone from 404+00 to 411+50.

8.2 mi (13.2 km). Fining upward sequence of gray sandstone to red shale and thin sandstone beds to gray shale from 411+50 to 414+00.

8.7 mi (14.0 km). Pink calcite fracture fillings and slickensided fault planes in massive limestone unit.

8.8 mi (14.2 km). East end of bridge over Shavers Fork. End of section.

REFERENCES CITED

Berger, P. S., Perry, W. J., Jr., and Wheeler, R. L., 1979, Three-stage model of brittle deformation in Central Appalachians: Southeastern Geology, v. 20, p. 59–67.

Dally, J. L., 1956, The stratigraphy and paleontology of the Pocono Formation in West Virginia [Ph.D. thesis]: Columbia University, New York, New York, 248 p.

Dennison, J. M., 1970, Stratigraphic divisions of Upper Devonian Greenland Gap Group ("Chemung Formation") along Allegheny Front in West Virginia, Maryland, and Highland County, Virginia: Southeastern Geology, v. 12, p. 53–82.

—— 1971, Petroleum related to Middle and Upper Devonian deltaic facies in Central Appalachians: American Association of Petroleum Geologists Bulletin, v. 55, p. 1179–1193.

Elliott, D. K., 1982, An edrioasteroid from the Foreknobs Formation (Upper Devonian) of West Virginia: Proceedings of the West Virginia Academy of Science, v. 53, p. 69–73.

Gillespie, W. H., Rothwell, G. W., and Scheckler, S. E., 1981, The earliest seeds: Nature, v. 293, p. 462–464.

Gwinn, V. E., 1964, Thin-skinned tectonics in the Plateau and Northwestern Valley and Ridge Provinces of the Central Appalachians: Geological Society of America Bulletin, v. 75, p. 863–900.

Heald, M. T., 1980, Petrographic study of Devonian shales and siltstones at road-cut on Route 33 east of Elkins, WV: U.S. Department of Energy, Morgantown Energy Technology Center, Morgantown, West Virginia, UGR File #278, 20 p.

Leonard, A. D., 1968, The petrology and stratigraphy of Upper Mississippian Greenbrier Limestone of eastern West Virginia [Ph.D. dissertation]: West Virginia University, Morgantown, West Virginia, 219 p.

Lewis, J. S., 1983, Reservoir rocks of the Catskill Delta in northern West Virginia: Stratigraphic basin analysis emphasizing deposystems [M.S. thesis]: West Virginia University, Morgantown, West Virginia, 148 p.

McColloch, J. S., and Schwietering, J. F., 1985, Geology along U.S. Route 33, Canfield to Bowden, Randolph County, West Virginia: West Virginia Geological and Economic Survey, Publication MAP-WV19.

Reger, D. B., 1931, Randolph County: West Virginia Geological Survey (County Report), 989 p.

Rittenhouse, G., 1949, Petrology and paleogeography of the Greenbrier Formation: American Association of Petroleum Geologists Bulletin, v. 33, p. 1704–1730.

Schwietering, J. F., 1979, Cross section *in* Avary, K. L., ed., 1979, Devonian clastics in West Virginia and Maryland, field trip guide, October 3–5, 1979: conducted for Eastern Section, American Association of Petroleum Geologists Meeting, October 1–4, 1979, Lakeview Inn, Morgantown, West Virginia, 100 p. [Available from the Appalachian Geological Society, P.O. Box 2605, Charleston, WV 25329.]

Yeilding, C. A., 1984, Stratigraphy and sedimentary tectonics of the Upper Mississippian Greenbrier Group in eastern West Virginia [M.S. thesis]: University of North Carolina, Chapel Hill, North Carolina, 117 p.

Taconian clastic sequence and general geology in the vicinity of the Allegheny Front in Pendleton County, West Virginia

Richard J. Diecchio, Department of Geology, George Mason University, Fairfax, Virginia 22030

LOCATION

Outcrops of the Taconic clastic sequence and the general geology in the vicinity of the Allegheny Front are exposed at Judy Gap on U.S. 33, Pendleton County; Circleville and Spruce Knob, 7½-minute quadrangles about 8 miles (13 km) west-northwest of Franklin, West Virginia (Fig. 1). West Virginia 28 and 33 provide access for any kind of vehicle including busses. All other travel in the area will be on dirt roads or narrow paved roads that provide access only by passenger cars and vans.

Figure 1. Location map of the Judy Gap section of the Taconian clastic sequence, Pendleton County, West Virginia.

INTRODUCTION

In the area around the Allegheny Front in Pendleton County (Fig. 2) each of the following can be observed: an almost complete exposure of the Taconian (Queenston) clastic sequence on North Fork Mountain; the contrast in structural styles on either side of the Allegheny Front as seen along the Spruce Knob Road; and a number of panoramic views of structurally and stratigraphically controlled geomorphology of the High Plateau Province and the Valley and Ridge Province on either side of the Allegheny Front (Fig. 3).

Germany Valley is developed over Middle Ordovician limestones that occur in the core of the Wills Mountain anticline (Fig. 2). These limestones are the oldest rocks exposed anywhere in this area. The Wills Mountain anticline (alternately referred to as the Hightown or Bluegrass anticline in Virginia) forms the western margin of the Nittany anticlinorium and extends from Bedford County, Pennsylvania, to Alleghany County, Virginia, a distance of 185 mi (298 km) (Perry, 1978). The Wills Mountain anticline is situated at the western edge of the Valley and Ridge Province, just east of the Allegheny structural front.

Exposed on the ridges surrounding Germany Valley (North Fork Mountain and River Knobs) and on the flanks of the Wills Mountain anticline are strata from the Upper Ordovician Reedsville Formation through the Lower Silurian Tuscarora Sandstone (Fig. 2). These strata make up the Taconian, or Queenston, clastic sequence (Fig. 4).

Stratigraphy

A generalized stratigraphic column of the strata exposed within the area is given in Figure 4. Only Taconian stratigraphy will be discussed in detail in this guide. The general geology and stratigraphy of Pendleton County are presented by Tilton and others (1927) and by the Marietta Geological Society (1974). The reader is also referred to the volumes on the Ordovician, Silurian, and Devonian Systems of West Virginia (Woodward, 1941, 1943, 1951).

Taconian Clastic Sequence. The Taconian (Queenston) sequence is a coarsening- and shallowing-upward progradational sequence whose deposition accompanied the Taconic Orogeny. Detailed measured sections of the Taconian sequence are presented by Chen (1981) for the Martinsburg Formation (Reedsville Formation) and Oswego Sandstone, and by Diecchio (1985) for the upper Reedsville Formation to the top of the Tuscarora Sandstone.

The Taconian clastics are underlain by Middle Ordovician limestones. The New Market Limestone (Chazyan, or upper Middle Ordovician) is the oldest unit exposed in Germany Valley and is a high-calcium quarry stone. The New Market (On of Fig. 2) is overlain by bioclastic limestones of the Black River Group (Obr). The Black River Group is overlain by the Trenton Group (Ot), limestones with shale interbeds and bentonites, the earliest detritus from the Taconic sourceland. The upper formation of the Trenton Group is the Dolly Ridge Formation (described at Stop 1). Additional information on the Ordovician limestones can be obtained from Kay (1956) and Perry (1972).

The Reedsville Formation (Or), also referred to as the Martinsburg Formation (Woodward, 1951; Chen, 1981), is the lowest primarily clastic unit exposed within the field area. It is in excess of 1,000 ft (over 300 m) thick and is predominantly shale. Lower beds of the Reedsville are very calcareous shale interbedded with calcarenite (Stop 2). The middle of the Reedsville contains many fewer limestone interbeds and some siltstone interbeds (Stop 3a). The lower and middle parts of the Reedsville contain a normal marine fauna. The uppermost part of the Reedsville (Stop 3b) contains bioturbated terrigenous mudstone, with interbedded shale, siltstone, and sandstone, and contains the well-known *Orthorhynchula* assemblage biozone, a shallow water fauna that was studied in detail by Bretsky (1969, 1970).

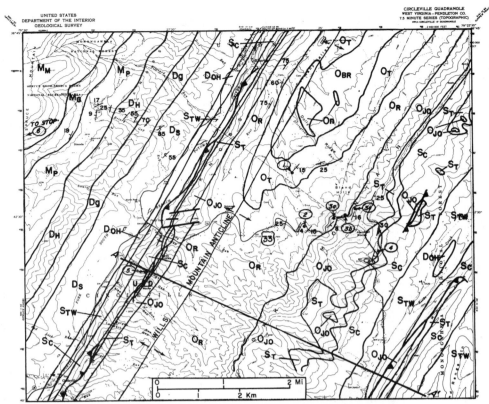

Figure 2. Geologic map of the Wills Mountain anticline and the Allegheny Front in Pendleton County. Circled numbers refer to field stops. *A-A-'* is line of cross section in Figure 5. Geology in part from Perry (1972, 1978).

The Reedsville provides an excellent example of the upward increase in amount and grain size of clastic detritus so characteristic of a clastic wedge. The lithologic and faunal change that occurs in the upper part of the Reedsville (*Orthorhynchula* biozone) is thought to represent a eustatic sea-level drop associated with the Late Ordovician glaciation (Diecchio, 1985).

The Oswego Sandstone conformably overlies the Reedsville Formation (Stop 3b). The Oswego is approximately 180 ft (55 m) thick and represents the continuation of upward coarsening that occurs in the Reedsville.

The Juniata Formation conformably overlies the Oswego Sandstone. The Juniata contains about 740 ft (225 m) of cyclically interbedded red sandstone and mudstone (Stops 3c and 4) that were deposited during the Late Ordovician low stand of sea level. The Juniata represents the top of the regressive Taconian clastic wedge. The Juniata is overlain by the Tuscarora orthoquartzite, probably a nearshore marine sand that prograded across the basin in the Early Silurian. Paleocurrent measurements indicate that the sandy beds in the Oswego, Juniata, and Tuscarora were transported from the eastern basin margin westward into the basin (Diecchio, 1985; Yeakel, 1962).

The Taconian sequence described here is radically different than equivalent strata in the eastern part of the Valley and Ridge. The reader is referred to Diecchio (1985) and Fichter and Diecchio (this volume) for a description of the eastern facies and for a discussion of lateral variations in the Taconian sequence from the Allegheny Front to the Shenandoah Valley.

Structural Geology

A detailed treatment of the structural geology of the Wills Mountain anticline is given by Perry (1975, 1978). Dennison and Naegele (1963) provide information on the Allegheny structural front. The Allegheny Front is the boundary between the severely folded and faulted surficial strata of the Valley and Ridge and the gently folded strata of the High Plateau. East of the Allegheny Front the strata are commonly found to dip steeply on the limbs of many of the anticlines, many of which are asymmetrical, with more steeply dipping to overturned western limbs. The western limbs are also preferentially faulted with respect to the eastern limbs, as is the case at Stop 5. West of the Allegheny Front the strata dip much less steeply, usually less than 30 degrees, and surface faulting is rare.

The Wills Mountain anticline is a good example of the style of folding typical of the Valley and Ridge. The anticline is developed above a major thrust fault that ramps westward from the Cambrian Rome Shale decollement zone to the Reedsville decollement zone (Fig. 5). The Wills Mountain anticline plunges to the northeast and southwest. Germany Valley is developed in a culmination along the crest of the anticlinal axis, where the lower part of the section (Ordovician limestones) is exposed.

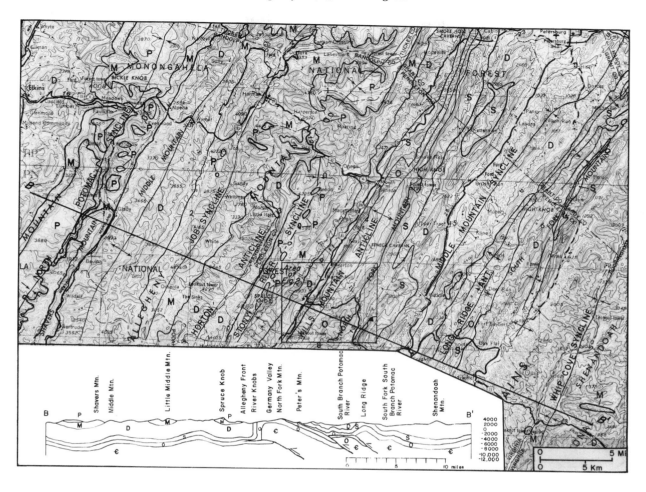

Figure 3. Geologic map and cross section of the area visible from observtion tower on Spruce Knob. Modified from Cardwell and others (1968).

SITE DESCRIPTION

Stop 1. Drive about 2.5 mi (4.1 km) east of Judy Gap on U.S. 33 and turn left on to Bland Hills Road. The Dolly Ridge Formation (upper Trenton Group, Ot) is most accessible along Bland Hills Road on Dolly Ridge, about 1.6 mi (2.7 km) north of U.S. 33. The Dolly Ridge Formation is a dark gray, dense, fine grained, medium to thinly bedded limestone that weathers yellowish brown and contains olive gray shale beds and bentonites. These limestones represent the waning stages of an expensive carbonate bank (Rodgers, 1968) that was situated on the eastern margin of North America during the Cambrian and Ordovician prior to Taconian clastic sedimentation. The onset of these clastic sediments is evident here as shale beds in the Dolly Ridge Formation. The bentonites are not identifiable here.

Stop 2. The lower, calcareous part of the Reedsville Formation (Or) can be observed along U.S. 33 where it begins to climb the western slope of North Fork Mountain, about 0.8 mi (1.3 km) east of the junction with Bland Hills Road. Here the Reedsville consists of medium gray to grayish olive, calcareous shale that weathers to a light gray limestone-like surface. It is interbedded with laminae of very thin beds of medium gray,

bioclastic calcarenite that weathers moderate yellow brown and rare thin interbeds of medium gray, calcareous siltstone that weathers light olive gray. Fossils in this part of the section include the brachiopods *Rafinesquina, Sowerbyella,* and *Zygospira;* the bryozoans *Prasopora* and *Hallopora;* the gastropod *Sinuites;* the trilobite *Cryptolithus;* the cephalopod *Orthoceras;* crinoid stalks and columnals; and the graptolites *Diplograptus* and *Climacograptus* (Woodward, 1951). The bioclastic layers are probably storm rip-up deposits as described by Kreisa (1981). This outcrop represents the transition from the carbonate-dominated regime below to the clastic-dominated lower part of the Taconian clastic sequence.

Stop 3. The transition from the middle shaly part of the Reedsville to the upper mudstones of the Reedsville (*Orthorhynchula* zone) to the Oswego Sandstone to the red beds of the Juniata Formation can be observed in a series of three closely spaced outcrops along U.S. 33 just west of the crest of North Fork Mountain.

Stop 3a is at the Germany Valley overlook along U.S. 33, about 0.7 mi (1.1 km) east of Stop 2. Exposed here is the shaly middle part of the Reedsville Formation that is composed of light

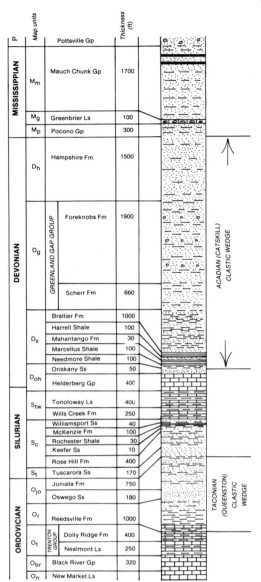

Figure 4. Generalized stratigraphic column of the strata exposed within the area of Figure 3.

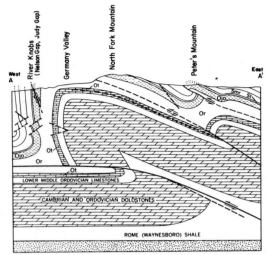

Figure 5. Structural cross section of the Wills Mountain anticline in Pendleton County. After Perry (1978). Not to scale. See Figure 2 for location of section.

olive-gray shale, with thin interbeds of medium gray, calcareous siltstone that weathers grayish orange, rare thin interbeds of medium gray, bioclastic calcarenite that weathers grayish orange, and rare medium to thin interbeds of yellow brown, fine-grained sandstone that weathers light olive gray. The percentage of sandstone increases up-section, and the percentage of calcarenite decreases upward, typical of a clastic wedge.

At the overlook (Figs. 2 and 3) you can observe Germany Valley (Ordovician limestones at the culmination of the Wills Mountain anticline); the ridge crest of North Fork Mountain (Tuscarora Sandstone on the east limb of Wills Mountain anticline) to the east; and River Knobs (Tuscarora Sandstone on the west limb of Wills Mountain anticline), which also form the first set of "razorback" hills or ridges to the west. At the northeast end of Germany Valley the Tuscarora ridge forms a spectacular ex-

ample of closure at the nose of a plunging anticline. Also to the west, beyond the River Knobs, are the Fore Knobs that are developed along the Upper Devonian Greenland Gap Group/ Hampshire Formation contact east of the Allegheny Front. West of the Fore Knobs is Spruce Mountain, atop of which is Spruce Knob, formed on a resistant bed of Pennsylvanian Pottsville Sandstone.

Stop 3b is the next outcrop uphill (east) along U.S. 33, about 0.2 mi (0.3 km) from Stop 3a. Exposed is the upper Reedsville Formation and the transition to the Oswego Sandstone (mapped with the Juniata as Ojo). The Reedsville is a bioturbated, medium gray, fossiliferous mudstone that weathers light olive gray. The upper Reedsville contains a completely different fauna (*Orthorhynchula* assemblage biozone) than in its lower beds, including the brachiopods *Orthorhynchula* and *Lingula;* the bivalves *Ambonychia, Ischyrodonta, Modiolopsis,* and *Tancrediopsis;* the trilobite *Isotelus;* and phosphatized remains of the gastropod *Plectonotus.* The *Orthorhynchula* zone is a shallow water fauna, studied in detail by Bretsky (1969, 1970). Small phosphate nodules and *Lingula* are common in the uppermost part of the Reedsville. The abrupt lithologic and faunal change that occurs in the upper part of the Reedsville (*Orthorhynchula* zone) is thought to represent a eustatic sea-level drop associated with the Late Ordovician glaciation (Diecchio, 1985). The phosphate and *Lingula* zone probably represents continued shallowing to shoaling or brackish water conditions.

The Oswego Sandstone, which overlies the Reedsville Formation in gradational contact, is a light brownish gray, medium- to fine-grained, medium bedded, cross-bedded sublitharenite, with interbedded silt-shale. The lower contact of the Oswego is arbitrarily chosen at the base of the lowest cross-bedded sandstone above the *Orthorhynchula* zone. The Oswego represents the culmination of the upward-coarsening cycle that started at the base of the Reedsville. The immature composition of the Oswego reflects its provenance in an orogenic sourceland to the east.

Stop 3c is the next exposure uphill along U.S. 33, about 0.3 mi (0.5 km) east of stop 3b, and across from the picnic area. This outcrop contains the contact between the Oswego and the Juniata Formations. This contact is chosen at the base of the lowest red mudstone, but is uncertain here because of the large covered interval that occurs between the top of the outcrop at 3b and the base of the outcrop at 3c. The Oswego makes up about the lower 10 ft (3 m) of the exposure at Stop 3c.

The overlying red beds of the Juniata Formation clearly exhibit their cyclical nature at this exposure. A typical cycle consists of a lower red, cross-bedded, fine-grained sandstone (sublitharenite); a middle red, sub-lithic wacke with vertical burrows; and an upper red bioturbated mudstone. The Juniata represents deposition in nonmarine to marginal marine conditions, probably a delta plain that existed during the glacio-eustatic low stand of sea level.

Stop 4. The Juniata Formation (Oj) and its contact with the overlying Tuscarora Sandstone (St) are well exposed along U.S. 33 on the east slope of North Fork Mountain, about 1.5 mi (2.5 km) east of Stop 3c. U.S. 33 cuts through the Tuscarora three times, and this stop is the middle of these three exposures. The top of the Juniata is very similar to its base, implying that the same conditions existed until the end of the Ordovician. The contact with the overlying Tuscarora Sandstone is gradational. This contact is arbitrarily regarded as the Ordovician-Silurian boundary.

The Tuscarora Sandstone is a very fine-grained to very coarse-grained quartzarenite. Vertical *Skolithos* burrows occur near the base, overlain by a thinly bedded to massive cross-bedded interval that is in turn overlain by a burrowed interval. This probably represents nearshore marine conditions, implying that sea level rose at the beginning of the Silurian. About two-thirds of the way up through the Tuscarora section is an interval of thinly bedded sandstone interbedded with dark gray shale, containing *Arthrophycus* (annulated bedding plane trails) at the base of the sandstone beds. This interval may represent lagoonal, or possibly marine conditions. The upper part of the Tuscarora, like the lower part, contains cross-bedded sandstone and *Skolithos,* probably indicating a return to nearshore marine conditions. The Tuscarora Sandstone, on the eastern limb of the Wills Mountain anticline, is about 170 ft (52 m) thick and dips about 25 degrees to the east.

The Tuscarora Sandstone is overlain by the Rose Hill Formation, a sequence of interbedded hematitic sandstone and olive-gray marine shale. Because a covered interval occurs at the top of the Tuscarora exposure, the contact with the Rose Hill is not observable. In fact, much of the remainder of the Silurian is poorly exposed in the field area.

Stop 5. Turn around and head west on U.S. 33 to its junction with West Virginia 28. The vertical western limb of the Wills Mountain anticline is clearly evident in Nelson Gap, the first gap through the River Knobs south of Judy Gap, or the first gap to the east of West Virginia 28 if going south from the junction with U.S. 33. As shown in Figure 5, the asymmetrical Wills

Mountain anticline is overturned to the west with backfaults occurring on the western limb. A down-to-the-east contraction fault (rotated uplimb thrust) is responsible for a repetition of the Tuscarora Ridge and is visible here as a double spine of Tuscarora Sandstone. As an alternate to this stop, the same features are exposed, but not as well, in Judy Gap, on U.S. 33 just east of its junction with West Virginia 28.

There is a good exposure of strata from the top of the Upper Silurian Williamsport Sandstone to the middle of the Tonoloway Limestone in a quarry on the southeast side of the intersection where West Virginia 28 meets U.S. 33 at Judy Gap. A measured section in this quarry is provided by Woodward (1941, p. 237); however, Woodward is incorrect in that the upper Tonoloway Limestone (Silurian) and basal Keyser Limestone of the Helderberg Group (Silurian and Devonian) are not exposed in the quarry (W. J. Perry, Jr., personal communication). The Silurian Williamsport Sandstone through Keyser Limestone (Silurian and Devonian) is exposed in Lambert Gap, south of Judy Gap and Nelson Gap.

Stop 6. Traveling north from Judy Gap along West Virginia 28 and U.S. 33, the first road to the left (west) will take you to Spruce Knob. Just south of this intersection, there is a good exposure of black shale, probably the Middle Devonian Marcellus Shale, on the west side of West Virginia 28 and U.S. 33. Part of the Spruce Knob road along Briery Gap Run passes through the Devonian (Acadian) clastic sequence, or the Catskill "delta", a coarsening- and shallowing-upward progradational sequence accompanying the Acadian Orogeny. McGhee and Dennison (1976) provide a measured section of the Devonian clastics exposed along Briery Gap Run. Additional information on the Devonian shales along the Allegheny Front can be found in Dennison and Naegele (1963). Fichter (this volume) provides a discussion of the Devonian clastic sequence in eastern Pendleton County.

The Brallier Formation (shale and siltstone) and the Greenland Gap Group (refer to Fig. 2) are structurally disrupted and dip steeply or are overturned to the west. In the lower part of the Hampshire red beds, bedding flattens to a dip of about 25 degrees to the west, and thus marks the Allegheny structural front in this area. From here to the top of Spruce Knob, beds of the upper Hampshire Formation and the Mississippian strata have similar dip attitudes. The structural front coincides with a topographic feature known as the Fore Knobs developed along the Greenland Gap Group/Hampshire contact. The very subtle geomorphic change that occurs at the Fore Knobs is observable on the northwest part of the Circleville topographic quadrangle. East of the Fore Knobs the terrain has a steeper slope than west of the Fore Knobs. The steeper slope is developed over the steeply dipping strata east of the front. The gentler slopes are developed over the more gently dipping strata west of the front.

At the first intersection, turn left and proceed through the gently dipping Mississippian strata that are poorly exposed along the Spruce Knob road. Near the top of Spruce Mountain, the red soil visible in the road bed is residuum from the red beds of the

Upper Mississippian Mauch Chunk Sandstone (Mm). On Spruce Mountain, turn right and proceed to the overlook on Spruce Knob—the highest point in West Virginia (4,861 ft [1,473 m] above sea level)—where slabs of Pennsylvanian Pottsville quartzitic sandstone and conglomerate mark the axis of the Stony River syncline. Notice the lopsided or "flagged" spruce trees: their branches are prevented from growing toward the west by the harsh and constant westerly wind.

On a clear day a number of geomorphic features of the Plateau and the Valley and Ridge can be seen from the Spruce Knob observation tower. Using Figure 3 as a guide, you can identify some structural and stratigraphic landforms in the area. Looking toward the west, in the plateau you can see Cunningham Knob (N85W), Yokum Knob (N70W), and Pharis Knob (N58W) on the south end of Little Middle Mountain. These are all developed on sandstones of the Mississippian Mauch Chunk Group (Mm) on the eastern limb of the Job syncline. West of these knobs is Middle Mountain, which is developed along the Greenland Gap Group (Dg)/Hampshire (Dh) contact on the western limb of the Middle Mountain anticline.

Looking toward the north (N15E) you can see a line of knobs formed by the Pocono Sandstone. These knobs are on strike with Spruce Knob along the axis of the Stony River syncline.

Toward the east you can look back over the Allegheny Front across much of the Valley and Ridge. North Fork Mountain, Tuscarora Sandstone on the eastern limb of the Wills Mountain anticline, is the closest prominent ridge, along which Harmon Rocks (due east) form a noticeable landmark. The Tuscarora "razorbacks" of River Knobs, the western limb of the Wills Mountain anticline, are barely visible through the trees looking toward Harmon Rocks. The Fore Knobs are obscured by the trees. The farthest prominent ridge that is noticeable along the skyline on a clear day is Shenandoah Mountain, Devonian Hampshire Formation (Dh) along the axis of the Whip Cove syncline. Shenandoah Mountain forms the Virginia/West Virginia state line. On a very clear day, the Blue Ridge can be seen farther to the east, beyond Shenandoah Mountain.

The path to the south of the observation tower leads to an overlook from which you may observe the Greenbrier "bench" to the south. This feature, made prominent by agricultural clearing over the Mississippian Greenbrier Limestone, includes some magnificent karst features.

REFERENCES CITED

Bretsky, P. W., 1969, Central Appalachian Late Ordovician communities: Geological Society of America Bulletin, v. 80, p. 193–212.

——, 1970, Upper Ordovician ecology of the Central Appalachians: Peabody Museum of Natural History, Yale University, Bulletin 34, 150 p.

Cardwell, D. H., Erwin, R. B., and Woodward, H. P., 1968, Geologic map of West Virginia: West Virginia Geological and Economic Survey.

Chen, Ping-fan, 1981, Lower Paleozoic stratigraphy, tectonics, paleogeography and oil/gas possibilities in the Central Appalachians (West Virginia and adjacent states): Part 2, measured sections: West Virginia Geological and Economic Survey, Report of Investigation 26-2, 300 p.

Dennison, J. M., and Naegele, O. D., 1963, Structure of Devonian strata along Allegheny Front from Corriganville, Maryland to Spruce Knob, West Virginia: West Virginia Geological and Economic Survey Bulletin 25, 42 p.

Diecchio, R. J., 1985, Post-Martinsburg Ordovician stratigraphy of Virginia and West Virginia: Virginia Division of Mineral Resources Publication no. 57, 77 p.

Kay, Marshall, 1956, Ordovician limestones in the western anticlines of the Appalachians in West Virginia and Virginia northeast of the New River: Geological Society of America Bulletin, v. 67, p. 55–106.

Kreisa, R. D., 1981, Storm-generated sedimentary structures in subtidal marine facies with examples from the Middle and Upper Ordovician of southwestern Virginia: Journal of Sedimentary Petrology, v. 51, p. 823–848.

McGhee, G. R., Jr., and Dennison, J. M., 1976, The Red Lick Member, a new subdivision of the Foreknobs Formation (Upper Devonian) in Virginia, West Virginia, and Maryland: Southeastern Geology, v. 18, p. 49–57.

Marietta Geological Society, 1974, Some geological features in Pendleton County, West Virginia, and Highland County, Virginia: 25th Annual Ohio Inter-collegiate Field Conference Guidebook, Marietta, Ohio.

Perry, W. J., Jr., 1972, The Trenton Group of Nittany anticlinorium, eastern West Virginia: West Virginia Geological and Economic Survey Circular 13, 30 p.

——, 1975, Tectonics of the western Valley and Ridge foldbelt, Pendleton County, West Virginia—a summary report: Journal of Research of the U.S. Geological Survey, v. 3, p. 583–588.

——, 1978, The Wills Mountain anticline: a study in complex folding and faulting in eastern West Virginia: West Virginia Geological and Economic Survey, Report of Investigation no. 32, 29 p.

Rodgers, John, 1968, The eastern edge of the North American continent during the Cambrian and Early Ordovician, in Zen, E-An, et al., eds., Studies of Appalachian Geology: Northern and Maritime: New York, Interscience Publishers, p. 141–149.

Tilton, J. L., Prouty, W. F., and Price, P. H., 1927, Pendleton County: West Virginia Geological Survey, 384 p.

Woodward, H. P., 1941, Silurian System of West Virginia: West Virginia Geological Survey, v. 14, 326 p.

——, 1943, Devonian System of West Virginia: West Virginia Geological Survey, v. 15, 655 p.

——, 1951, Ordovician System of West Virginia: West Virginia Geological Survey, v. 21, 627 p.

Yeakel, L. S., 1962, Tuscarora, Juniata, and Bald Eagle paleocurrents and paleogeography in the Central Appalachians: Geological Society of America Bulletin, v. 73, p. 1515–1540.

ACKNOWLEDGMENTS

The author is grateful to Nick Byrnes, Jeanne Casey, and Kris Dennen for their assistance in the field.

The Catskill clastic wedge (Acadian Orogeny) in eastern West Virginia

Lynn S. Fichter, Department of Geology, James Madison University, Harrisonburg, Virginia 22807

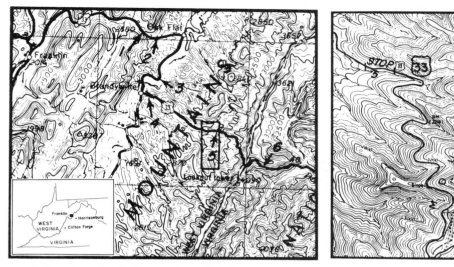

Figure 1. Location of Oak Flat to Shenandoah Mountain cluster stops in Pendleton County, West Virginia found on the Brandywine and Fort Seybert 7½-minute quadrangles. The left hand topographic map showing all six stops along U.S. 33 is an enlarged portion of the 2 degree Charlottesville quadrangle. Right hand map is the Brandywine 7½-minute topographic quadrangle showing location at Stop 5.

LOCATION

Exposures of the Catskill clastic wedge are located in Pendleton County, West Virginia, and include parts of the Brandywine and Fort Seybert 7½-minute quadrangles (Fig. 1). The left hand topographic map in Figure 1 shows all six stops along U.S. 33; it is an enlarged portion of the 2 degree Charlottesville quadrangle. The right hand map is a part of the 7½-minute Brandywine quadrangle showing location of Stop 5.

INTRODUCTION

The rocks of the Catskill clastic wedge begin (Stop 1) with Late Silurian and Early Devonian carbonates and quartz arenites deposited between the Taconic and Acadian Orogenies and continue through a complete sequence of formations illustrating the Acadian Orogeny in eastern West Virginia (Stops 2 through 6). The most interesting aspects of these rocks are their sedimentology and depositional systems. Much can be seen and learned from studying and comparing the formations present. The level of fascination and controversy increases greatly, however, in light of the diverse interpretations made of the same age rocks in other parts of the Catskill clastic wedge. The following papers are recommended as background reading: Friedman and Johnson (1966), Allen and Friend (1968), McCave (1968), Dennison (1970a, 1971), Meckel (1970), Walker (1971, 1972), and Walker and Harmes (1971).

The most prominent topographic feature in the area is She-nandoah Mountain, which trends NE to SW and reaches elevations of over 3,700 ft (950 m); its crest forms the Virginia-West Virginia state line. The rocks discussed at these stops are exposed between Shenandoah Mountain, on the east, and Long Ridge, the next mountain to the west. The stops are arranged in stratigraphic order so that traveling east on U.S. 33 one travels up section through the Late Silurian-Early Devonian pre-orogenic carbonates and quartz arenites to the Mid to Late Devonian Acadian clastic wedge. Thus, as one travels east and climbs in elevation from the Potomac River Valley to the crest of Shenandoah Mountain (Stops 2 through 6) one climbs stratigraphically from deep water to shallow water deposits. Metaphorically, it is as if climbing up through the stratigraphy stop by stop is to climb not only up the side of Shenandoah Mountain but also to climb up out of that ancient Devonian basin, across the submarine fan, up the slope, across the shelf, and onto the shoreline. An interesting exercise while viewing these stops is to try simultaneously to reconstruct the Middle to Late Devonian paleogeography and basin structure and to imagine climbing up out of that basin. Alternatively, from the crest of Shenandoah Mountain, we might attempt to imagine standing on the ancient alluvial plain and, while looking down into the present day Potomac River valley, observe the clastic wedge prograding into the Devonian sea.

SITE DESCRIPTION

Thirteen formations are present in this cluster stop. All the

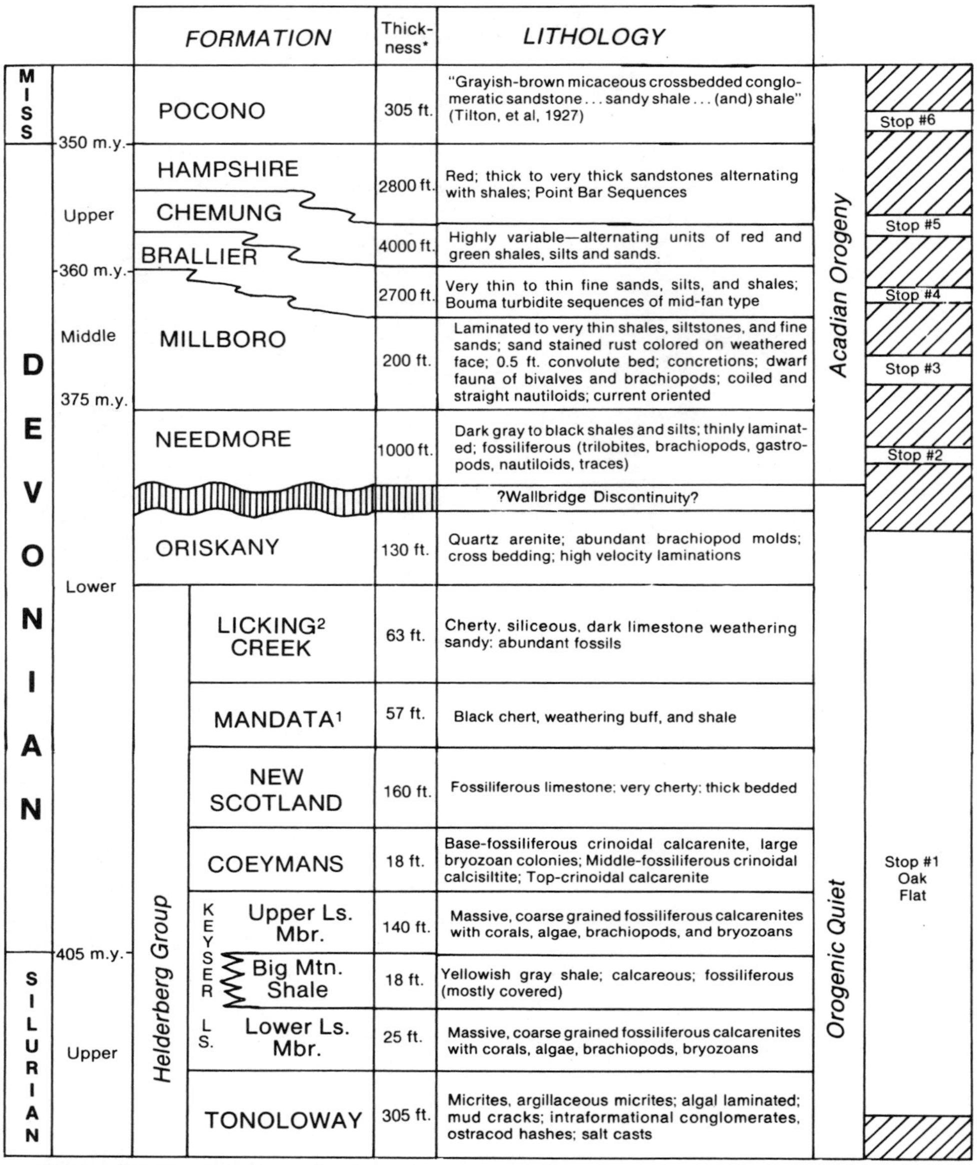

		FORMATION	Thick-ness*	LITHOLOGY		
M I S S	350 m.y.	POCONO	305 ft.	"Grayish-brown micaceous crossbedded conglomeratic sandstone . . . sandy shale . . . (and) shale" (Tilton, et al, 1927)		Stop #6
D E V O N I A N	Upper	HAMPSHIRE	2800 ft.	Red; thick to very thick sandstones alternating with shales; Point Bar Sequences	Acadian Orogeny	Stop #5
		CHEMUNG				
	360 m.y.	BRALLIER	4000 ft.	Highly variable—alternating units of red and green shales, silts and sands.		Stop #4
			2700 ft.	Very thin to thin fine sands, silts, and shales; Bouma turbidite sequences of mid-fan type		Stop #3
	Middle	MILLBORO	200 ft.	Laminated to very thin shales, siltstones, and fine sands; sand stained rust colored on weathered face; 0.5 ft. convolute bed; concretions; dwarf fauna of bivalves and brachiopods; coiled and straight nautiloids; current oriented		
	375 m.y.	NEEDMORE	1000 ft.	Dark gray to black shales and silts; thinly laminated; fossiliferous (trilobites, brachiopods, gastropods, nautiloids, traces)		Stop #2
		?Wallbridge Discontinuity?				
	Lower	ORISKANY	130 ft.	Quartz arenite; abundant brachiopod molds; cross bedding; high velocity laminations		
		Helderberg Group — LICKING[2] CREEK	63 ft.	Cherty, siliceous, dark limestone weathering sandy; abundant fossils		
		MANDATA[1]	57 ft.	Black chert, weathering buff, and shale		
		NEW SCOTLAND	160 ft.	Fossiliferous limestone; very cherty; thick bedded		
		COEYMANS	18 ft.	Base-fossiliferous crinoidal calcarenite, large bryozoan colonies; Middle-fossiliferous crinoidal calcisiltite; Top-crinoidal calcarenite	Orogenic Quiet	Stop #1 Oak Flat
		KEYSER LS. — Upper Ls. Mbr.	140 ft.	Massive, coarse grained fossiliferous calcarenites with corals, algae, brachiopods, and bryozoans		
S I L U R I A N	405 m.y.	Big Mtn. Shale	18 ft.	Yellowish gray shale; calcareous; fossiliferous (mostly covered)		
	Upper	Lower Ls. Mbr.	25 ft.	Massive, coarse grained fossiliferous calcarenites with corals, algae, brachiopods, bryozoans		
		TONOLOWAY	305 ft.	Micrites, argillaceous micrites; algal laminated; mud cracks; intraformational conglomerates, ostracod hashes; salt casts		

*Thickness for the Oak Flat section from Chen (1981) except for Manadata and Licking Creek, based on Woodward (1943); other thicknesses from Tilton, et al (1927).

[1]Port Jervis of Woodward (1943).

[2]Port Ewen of Woodward (1943); also referred to as Shriver Chert by some authors.

Figure 2. Summary stratigraphy of the Late Silurian to Early Mississippian formations seen in the Oak Flat to Shenandoah Mountain cluster stops. Note that Woodward (1943) and Chen (1981) do not agree on all thicknesses for formations in the Oak Flat section.

stops are road cuts along a 13 mi (22 km) stretch of U.S. 33, an east-west primary highway, and are on public land or right-of-way requiring no special permission for access. From the east the stops can be reached by driving 20 mi (33 km) west of Harrisonburg, Virginia on U.S. 33. From the west the first destination should be Franklin, West Virginia at the junction of U.S. 33 and 220. Figure 2 summarizes the stratigraphy and provides formation thicknesses, basic descriptions, and the stop positions in the stratigraphic column.

Stop 1—Oak Flat Section

Stop 1 is 6.7 mi (11.2 km) east of the U.S. 220 and 33 junction in Franklin, W.Va. (1.1 miles [1.8 km] west of Oak Flat) on the south side of U.S. 33, across the highway from a small roadside park. The exposure is a large but discontinuous 0.3 mi (0.5 km) long road cut.

Seven steeply dipping Upper Silurian and Lower Devonian formations from the Tonoloway to the Oriskany are exposed.

This stop is an excellent example of almost the full Helderberg Group and the overlying Oriskany (and down the road a little, the Needmore).

The formations in the Oak Flat Section have been measured and described by Woodward (1943) and Chen (1981). Note, however, that Chen and Woodward do not agree on all thicknesses, and Chen's measured section stops at the New Scotland Formation. Brief descriptions of the formations are in Figure 2. The following summary of the formations and their environmental and paleogeographic interpretations are based on Dennison (1970a), Dennison and Head (1975), Head (1969), Smosna and Patchen (1978), and Smosna and others (1977).

At the east end of the cut the Tonoloway Formation is exposed in a generally broad, gentle anticline extending a couple of hundred yards along the highway. The (algal?) laminated micrites appear uniformly monotonous at first, but on closer study many other features can be found, including ostracods (*Leperditia*) and intraformational conglomerates. Near the formation top, clay content increases and the rocks weather out in thin, fissile plates with occasional mud cracks and salt casts. All of these features indicate a tidal flat environment with occasional restricted circulation and high evaporation. (The Salina salt beds to the west are a facies of the Tonoloway.) The Keyser is divided into a lower and upper member separated by the Big Mountain Shale. Both the Lower and Upper Keyser (Byers Island Member and Jersey Shale Member respectively of Head, 1969) are largely structureless, coarse-grained, fossiliferous calcarenites. The abundant well-washed fossils imply a shallow, clear-water, high energy environment such as a carbonate barrier or large areas of fossil debris in a shallow subtidal region. With the Keyser Formation a major transgression began to advance northeastward into the Silurian Central Appalachian Basin easing the restricted circulation that prevailed during the Tonoloway.

The Big Mountain Shale, although exposed, may be hard to find, especially with summer vegetation. The shale is brown, fissile, and locally calcareous and fossiliferous. The Big Mountain is interpreted as a clastic influx transported north and northeastward by longshore currents during a minor regressive phase in the basin. The source was low and to the southeast in the Clifton Forge, Virginia area.

The Coeymans (of Woodford, 1943; New Creek of some others) is well exposed above the Upper Keyser in the large, prominent road cut. The base of the formation can be recognized by a bed with numerous, large bryozoan colonies. Although the Coeymans is very fossiliferous, it contains more calcisiltite and is darker than the Upper Keyser, and is interpreted as a deeper subtidal environment than the Keyser. It represents a continuation of the major transgression that began with the Lower Keyser.

The New Scotland (of Woodward, 1943; Corriganville of some others) is a thick-bedded fossiliferous limestone with numerous chert beds that are easily visible from the road. The Mandata (Port Ewen of Woodward, 1943) and Licking Creek (Port Jarvis of Woodward, 1943; Shriver Chert of some others) are poorly exposed, badly weathered, and hidden in the trees. The New Scotland, Mandata, and Licking Creek, like the Coeymans, are deeper subtidal, closer to the basinal axis, and were deposited during the maximum stages of the transgression.

The quartz arenitic Oriskany contains, in places, abundant molds of brachiopods. Near the top of the formation is a very thick bed with laminations parallel to bedding. Above that are large-scale planar cross beds, usually seen because brachiopod molds are aligned along the foresets. Systematic studies of the depositional environments of the Oriskany have not been done, but it is generally interpreted as a beach and near shore sand. Unconformities have been recognized both below and above (the Wallbridge discontinuity) the Oriskany at various places in the Central Appalachians, but no direct evidence for their presence exists at this stop. Overall, however, the Oriskany represents a time of shallowing and even complete withdrawal of the Devonian sea from the craton at the boundary between the Tippecanoe and Kaskaskia sequences, just prior to the Acadian Orogeny.

Stop 2—Needmore Formation

Stop 2 is 0.5 mi (0.8 km) east of Stop 1 and 2.3 mi (3.8 km) west of Stop 3. It is a high, vertical road cut on the south side of U.S. 33. The rocks here are nearly horizontal, but large folds can be seen along the highway between Stops 1 and 2, which bring the Needmore into the same steeply dipping attitude as the Oak Flat section.

The Needmore is made up of dark gray to black, laminated shales and silts. Fossils are scattered through the section but tend to concentrate in richly fossiliferous zones of gastropods, cephalopods, brachiopods, and trilobites. Bioturbated beds are also common. These characteristics indicate a quiet (?shelf) environment with generally good circulation and oxygen levels.

The Needmore is an abrupt change from the underlying Oriskany. The simplest interpretation is that the Oriskany to Needmore sequence was the result of a transgressive sea onto the craton. The Millboro Formation overlying the Needmore suggests a different hypothesis, however. At Stop 3 the fossils in the Millboro are dwarfed, implying lower oxygen levels. To the north of Pendleton County there is a facies change and the Millboro name is dropped and replaced by the (older to younger) Marcellus, Mahantango, and Harrell Formations, but at Stop 3 the Marcellus is black and highly organic implying anoxic conditions. This trend to anoxic conditions from the Needmore to the Millboro and Marcellus implies a rapidly subsiding basin rather than a shelf in a simple transgressing sea. Support for the rapidly subsiding basin interpretation is in the overlying formations (Stops 3, 4, and 5). Therefore the Needmore most likely marks the first clastic influx from the Acadian mountains rising to the east, into a rapidly subsiding foreland basin to the west.

Stop 3—Millboro Formation

Stop 3 is 2.3 mi (3.8 km) east of Stop 2 and 2.0 mi (3.3 km) west of Stop 4; 1.5 mi (2.5 km) west of the U.S. 33 and West

Virginia 23 junction at Brandywine, West Virginia. The road cut is about 300 ft (77 m) long on both sides of the highway. Here, the Millboro dips a few degrees to the south.

The Millboro consists of dark gray to black, laminated to very thin bedded shales, silstones, and fine grained sandstones. Small-scale cross bedding is sometimes seen in the rust-weathering sands. In the east cut a 6 in (15 cm) thick zone is convolute bedded, produced by fluid escape and loss of grain-to-grain contact causing flowage. On some bedding planes the fauna are abundant, but consist mostly of dwarf brachiopods and clams, which are current oriented. Occasional straight and coiled nautiloids can also be found but these are generally larger than the brachiopods and clams.

The Millboro reflects a deeper, more anoxic environment (the dwarf fauna) with periods of higher energy deposition (cross-bedded sands and current oriented fossils) than the Needmore at Stop 2. These two interpretations seem paradoxical since deeper environments are usually quieter. One major source of energy in a deep water environment is a turbidity current. Support for the idea that these deposits in the Millboro are the first, thin, distal deposits of a submarine fan prograding into the foreland basin can be found in the overlying Brallier Formation at Stop 4.

Stop 4—Brallier Formation

Stop 4 is 2.0 mi (3.3 km) east of stop 3, 0.5 mi (0.8 km) east of the U.S. 33 and West Virginia 23 junction at Brandywine, and 4.1 mi (6.8 km) west of Stop 5. The cut is on the north side of the highway at a salt storage shed.

These rocks represent Bouma turbidite sequences of a mid-fan type (T_{CDE} dominates, but other sequences including T_{ABCDE} are present). These sequences are clearly visible and easily studied in the lower part of the cut. The upper portion of the cut is more weathered and illustrates the typical weathering pattern of the Brallier, which consists of jutting sand beds alternating with weathered shales. Various pieces of sandstone float have a variety of trace fossils and current marks such as shallow flutes and grooves.

By the time the Brallier Formation appears in the section the former deepwater nearly anoxic basin of the Millboro Formation has begun to rapidly fill in with large submarine fans prograding from the east.

Stop 5—Chemung-Hampshire Transition

Stop 5 is 4.1 mi (6.8 km) east of Stop 4 and 2.6 mi (4.3 km) west of the crest of Shenandoah Mountain. The long road cut begins (traveling east) at a gentle "S" curve in the highway (see Figs. 1 and 3) and extends up the hill for several hundred yards (meters); Figure 3 is a detail map showing the location of the strip log in Figure 4.

Whereas other formations in this region tend to be distinctive enough so that they can be easily recognized even at 55 MPH (80 km/hr), the Chemung varies markedly from place to place in

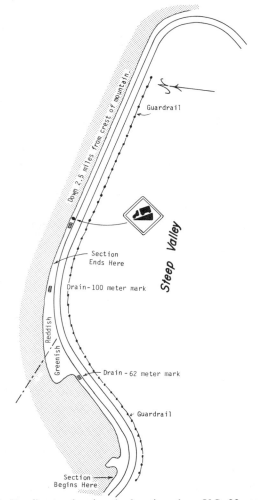

Figure 3. Detail map showing the location along U.S. 33 at Shenandoah Mountain of the strip log in Figure 4. Note that the exposed section extends up and down the highway beyond the strip log.

color, texture, and structures. For example, the underlying Brallier is recognized by its characteristic mid-fan type Bouma sequences. The Hampshire above is characterized by its distinctive red color, alternating sandstones and shales, and absence of marine fossils. In between the Brallier and the Hampshire, the Chemung consists of gray, brownish-gray, green, and red shales, siltstones and sandstones (becoming thicker upsection), occasional quartz pebbles or conglomerates, and scattered marine fossils and drifted plant fragments. Furthermore, contacts of the Chemung with the underlying Brallier and overlying Hampshire are not lithologically distinct. Butts (1939: 322) placed the Brallier-Chemung contact "at the zone or horizon at which large fossils, such as *Leiorhyncus, Productella, Spirifer,* and *Leptodesma* appear . . . the appearance of Chemung fossils is not accompanied by any marked change in lithology." For the Chemung-Hampshire contact Butts states (1939: 333) "the Hampshire is bounded below everywhere by the fossiliferous green and gray . . . shale and sandstone of the Chemung." Dennison (1970b) subdivided the Chemung (Dennison's [1970a]

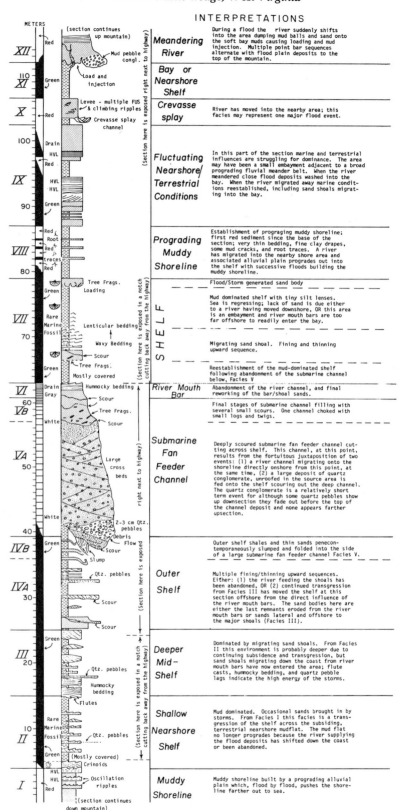

Figure 4. Strip log summarizing a portion of the Chemung-Hampshire section exposed at Stop 5, with suggested interpretations. For purposes of this field guide the Hampshire begins with the establishment of point bar sequences at Facies XII.

Greenland Gap Formation) into two formations, one of which contains five members, but his subdivisions are not used here.

Stop 5 is at the top of the Chemung and includes the base of the Hampshire. For purposes of this field guide the base of the Hampshire is defined as the first appearance of distinctive point bar sequences. Some of these fining upward point bar sequences are close to the ideal model with scoured bases, mud-pebble lag gravels, and large-scale trough cross beds grading up into small-scale trough cross beds. Many variations of the point bar sequence are present, however, including those variations illustrated by Allen (1965: p. 235), and reproduced in Reading (1978: pp. 53–55). On the strip log (Fig. 4) point bar sequences come in at Facies XII, but continue for the next 2.5 mi (3.8 km) to the top of Shenandoah Mountain. The exposed section also continues down the mountain below Facies I on the strip log and, while driving from Stop 4 to Stop 5, the poorly defined transition from Brallier to Chemung can be detected in scattered outcrops.

Allen and Friend (1968: p. 58) stated that the Hampshire (their Catskill) "was deposited in a vast coastal plain of alluviation" with the characteristics of a meandering river (point bar sequences). Thus, the Chemung is environmentally transitional between the submarine fan of the Brallier below and the alluvial plain of the Hampshire, above. Beyond that, specific environmen-tal interpretations of the Chemung are as widely different as is its color and lithology (see reading list above).

The strip log in Figure 4 summarizes some detailed environmental interpretations of the Chemung-Hampshire transition along U.S. 33 at Shenandoah Mountain; suggestions, disagreements, and alternate interpretations are encouraged.

Stop 6—Pocono Formation

Stop 6 is 3.0 mi (5.0 km) east of Stop 5 and 0.4 mi (0.6 km) east of the crest of Shenandoah Mountain. The road cut exposes Hampshire at the crest of the mountain, then, going eastward (and down the east side of the mountain), climbs up section into the Pocono before going back down section again into the Hampshire.

Little or no recent work has been done on the Pocono in this region. The generalized description in Figure 2 is from Tilton and others (1927). Dennison (1971) reports that along the Allegheny Front west of this stop the Pocono contains marine fossils, whereas to the north it "appears totally to represent subaerial delta deposits, (p. 1186)." Although a deltaic interpretation for these formations is brought into question by Allen and Friend (1968), Walker (1970, 1972), and Walker and Harmes (1971), the Pocono does represent the waning effects of the Acadian Orogeny as sediment supply decreased and sediment winnowing increased.

REFERENCES CITED

Allen, J.R.L., 1965, Fining-upwards cycles in alluvial successions: Geological Journal, v. 4, p. 229–246.

Allen, J.R.L., and Friend, P. F., 1968, Deposition of the Catskill Facies, Appalachian Region, *in* Klein, G. D., ed., Late Paleozoic and Mesozoic Continental Sedimentation, northeastern North America: Geological Society of America Special Paper 106, p. 21–74.

Butts, Charles, 1939, Geology of the Appalachian Valley in Virginia: Commonwealth of Virginia, Virginia Geological Survey, University of Virginia, 568 p.

Chen, Ping-Fan, 1981, Late Paleozoic stratigraphy, tectonics, paleogeography, and oil gas possibilities in the Central Appalachians (West Virginia and adjacent states): Part 2. Measured Sections: West Virginia Geological and Economic Survey Report of Investigations, No. R1-26-2, p. 147–148.

Dennison, J. M., 1970a, Silurian stratigraphy and sedimentary tectonics of southern West Virginia and adjacent Virginia, *in* Silurian Stratigraphy Central Appalachian Basin: Appalachian Geological Society Field Conference, Roanoke, Virginia, p. 2–33.

——— 1970b, Stratigraphic divisions of Upper Devonian Greenland Gap Group ("Chemung Formation") along Allegheny Front in West Virginia, Maryland and Highland County, Virginia: Southeastern Geology, v. 12, p. 53–82.

——— 1971, Petroleum related to Middle and Upper Devonian Facies in Central Appalachians: American Association of Petroleum Geologists Bulletin, v. 55, p. 1179–1193.

Dennison, J. M., and Head, J. W., 1975, Sea level variations interpreted from the Appalachian Basin Silurian and Devonian: American Journal of Science, v. 275, p. 1089–1120.

Friedman, G. M., and Johnson, K. G., 1966, The Devonian Catskill deltaic complex in New York, type example of a "tectonic deltaic complex," *in* Shirley, M. L., ed., Deltas in Their Geologic Framework: Houston Geological Survey, p. 171–188.

Head, J. W., III, 1969, The Keyser Limestone at New Creek, West Virginia: An illustration of Appalachian Early Devonian depositional basin evolution, *in* Donaldson, A. C., and Erwin, R., eds., Some Appalachian Coals and Carbonates: Models of Ancient Shallow-Water Deposition (Geological Society of America preconvention field trip, November, 1969): Geology Department, West Virginia University, and West Virginia Geological and Economic Survey, p. 323–355.

Meckel, L. D., 1970, Paleozoic alluvial deposition in the Central Appalachians: A summary, *in* Fisher, G. W., Pettijohn, F. J., Reed, J. C., and Weaver, K. N., Studies of Appalachian Geology, Central and Southern: New York, Wiley-Interscience Publishers, p. 49–67.

McCave, I. N., 1968, Shallow and marginal marine sediments associated with the Catskill Complex, *in* Klein, G. D., ed., Late Paleozoic and Mesozoic Continental Sedimentation, northeastern North America: Geological Society of America Special Paper 106, p. 75–107.

Reading, H. G., 1978, Sedimentary Environments and Facies: Blackwell Scientific Publications, New York, 657 p.

Smosna, R. A., Patchen, D. G., Washauer, S. M., and Perry, W. J., Jr., 1977, Relationships between depositional environments, Tonoloway Limestone, and distribution of evaporites in the Salina Formation, West Virginia, *in* Fisher, J. H., ed., Reefs and Evaporites—concepts and depositional models: American Association of Petroleum Geologists Studies in Geology 5, p. 125–143.

Smosna, Richard, and Patchen, Douglas, 1978, Silurian evolution of the Central Appalachian Basin: American Association of Petroleum Geologists Bulletin, v. 62, p. 2308–2328.

Tilton, J. L., Prouty, W. F., Price, P. H., 1927, Pendleton County, West Virginia Economic and Geological Survey, p. 134–185.

Walker, R. G., 1971, Nondeltaic depositional environments in the Catskill Clastic Wedge (Upper Devonian) of Central Pennsylvania: Geological Society of America Bulletin, v. 82, p. 1305–1326.

——— 1972, Upper Devonian marine–non-marine transition, Southern Pennsylvania: Pennsylvania Geological Survey, Fourth Series, General Geology Report 62, 25 p.

Walker, R. G., and Harmes, J. C., 1971, The "Catskill Delta": a prograding muddy shoreline in Central Pennsylvania: The Journal of Geology, v. 79(4), p. 387–399.

Woodward, H. P., 1943, Devonian System of West Virginia: Volume XV, West Virginia Geological Survey, p. 212–213.

Trimble Knob basalt diatreme and associated dikes, Highland County, Virginia

Eugene K. Rader, Thomas M. Gathright II, John D. Marr, Jr., Virginia Division of Mineral Resources, Charlottesville, Virginia 22903

LOCATION

The Trimble Knob basalt diatreme and associated dikes are located adjacent to U.S. 250 and 220, near Monterey and Highland, Highland County Virginia (Fig. 1).

INTRODUCTION

The igneous bodies in Highland County are unique in the Appalachians in terms of composition, trends, form, and age. Dikes with an andesitic composition are the most common types. Basalt, olivine basalt, trachyte, and diatreme breccia are also present in the area. Mica (biotite) pyroxenite and peridotite have been reported to the north in Pendleton County, West Virginia (Garner, 1956). Trends of the dikes are north 30° to 40° east and about north 45° west. In addition to typical dike forms, elliptical shaped pipes are common. There appears to be no direct relationship between mineralogical composition and trend or form of the igneous bodies. The radiometric ages of the igneous bodies are Eocene (Fullagar and Bottino, 1969).

Highland County is in the western part of the Valley and Ridge physiographic province approximately 50 mi (80.4 km) north of the southern end of the Central Appalachians (Fig. 1). Southwestward trending, regionally extensive doubly plunging, asymmetric folds are the characteristic structures of the area. In

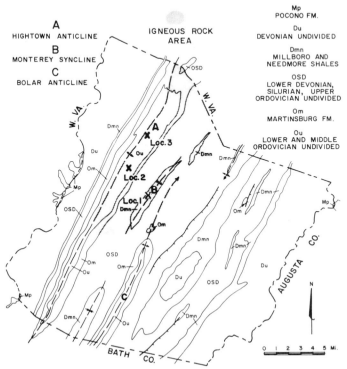

Figure 2. Generalized geologic map of Highland County, Virginia.

western Highland County only the Hightown and Bolar anticlines and Monterey syncline have intrusive igneous rocks. Paleozoic rocks in Highland County (Fig. 2) range from Lower Ordovician dolomites (Beekmantown Formation) exposed in the anticlines to Mississippian sandstones (Pocono Formation) in the westernmost syncline, which locally defines the Alleghany structural front. The igneous bodies intrude rocks as young as Middle Devonian age (Millboro Shale) in Highland County and Upper Devonian age (Brallier Formation) in adjacent Pendleton County, West Virginia.

Highland County is on the axis of a major Appalachian gravity low. The gravity low also crosses the center of the Thermal Springs district about 35 mi (56 km) southwest of the center of the igneous area and there reaches a local minimum of –80 milligals (Fig. 3). Dennison and Johnson (1971) suggest that the gravity low marks an area of maximum uplift due to isostatic rebound.

To the east in Augusta and southern Rockingham counties is an area of alkalic intrusions (Fig. 1). These intrusions have been dated by Zartman and others (1967) at 150 Ma (Jurassic). Included in the alkalic intrusions are nepheline syenite, teschenite, and picrite (Johnson, Milton, and Dennison, 1971). Mole Hill,

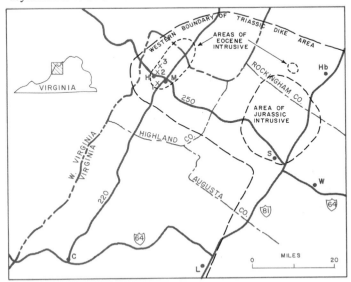

Figure 1. Index map showing access to Highland County igneous area, and regional distribution of Eocene, Jurassic and Triassic igneous rocks (modified from Dennison and Johnson, 1971). *M*–Monterey; *H*–Hightown; *Hb*–Harrisonburg; *L*–Lexington; *S*–Staunton; *C*–Covington; *W*–Waynesboro.

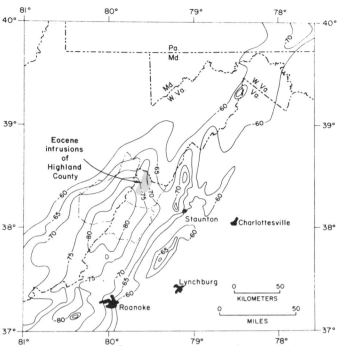

Figure 3. Simple Bouguer gravity map of westcentral Virginia.

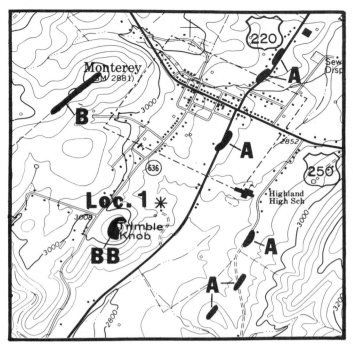

Figure 4. Locality map for Trimble Knob (Locality 1), 0.7 mi (1 km) southwest of junction of U.S. 220 and 250 in Monterey, Virginia. From Monterey 7½-minute quadrangle. Access to property may be obtained from Mrs. Frank O. Bird at the house marked with an asterisk. Black areas are igneous bodies: *BB*–basalt diatreme; *B*–basalt; *A*–andesite.

an olivine basalt plug, also in southern Rockingham County, (Fig. 1) has a radiometric age of 47 ± 1 Ma (Eocene; Wampler, 1975).

Surrounding the Eocene and Jurassic intrusive areas and also in the eastern Valley and Ridge, Blue Ridge, and Piedmont region are diabase intrusions (Fig. 1). Radiometric dates have not been reported for the diabase intrusions in Virginia; however, by analogy with other similar east coast igneous bodies, a Late Triassic or Early Jurassic age is assumed for a portion of the dikes. Two distinct trends are recognized in central Virginia: 1) north 30°-50° west; and 2) north 10° west to north 10° east.

Available information suggests that the 38th parallel fracture zone may extend from Kentucky and West Virginia into the Highland, Augusta and Rockingham county areas. Dennison and Johnson (1971) suggest that igneous activity along the 38° fracture zone has been sporadic from late Precambrian to Eocene, the center of activity migrating westward with respect to time.

Prior to 1884 igneous dikes were known to occur in Highland County (Rogers, 1884, Plate VII, Geologic Section No. 8). Rogers' cross section was reproduced by Campbell and Campbell (1885) in their discussion of the rocks of Highland County. This appears to be the first published description of the igneous rocks in the Valley and Ridge province. In 1890 Darton and Diller published detailed descriptions of several of the igneous rocks from the central Appalachians. Darton (1899) described the igneous rocks as basalt necks and basalt and granite felsophyre dikes. Three necks were shown on Darton's map (1899): (1) Pyramid Hill (Now Trimble Knob); (2) Sounding Knob, 4 mi (6.4 km) south of Trimble Knob; and (3) Strait Creek, 3 mi (4.8 km) northeast of Trimble Knob. Also, several dikes were

reported in Monterey and Bluegrass valleys. Watson and Cline (1913) described the granite-felsophyre as a light to dove gray, fine-grained, porphyritic rock with a groundmass composed of sodic plagioclase, orthoclase, quartz, biotite, and magnetite with phenocrysts of albite, orthoclase, biotite, and augite. Dennis (1934) reviewed all known igneous exposures in the Valley and Ridge. Rader and Griffin (1960), Kettran (1970) and Johnson,

Figure 5. Trimble Knob, view looking west, relief 125 ft (38 m) above valley floor: locality 1.

Milton, and Dennison (1971) have studied the igneous bodies and their reports contain petrographic descriptions. Fullagar and Bottino (1969) have obtained radiometric ages for the andesite dikes in Bluegrass Valley (Locality 3).

SITE DESCRIPTION

Three localities, illustrating the variety of intrusive rocks in the Highland County igneous complex, were selected for this cluster stop.

Trimble Knob (Locality 1; Figs. 4, 5), 0.7 mi (1.1 km) southwest of Monterey is an excellent example of a basalt diatreme. Exposures on the crest and uppermost slopes of the hill are relatively unweathered, xenolith-free basalt. The dark gray basalt has a porphyritic sub-ophitic texture with a microlitic groundmass of plagioclase, pyroxene, and ilmenite. Phenocrysts are olivine, augite, and hypersthene. Most of the augite phenocrysts are zoned but appear to be relatively unaltered. Hypersthene phenocrysts show resorption. Minor flow structure around larger phenocrysts is present near the center of the pipe.

Xenoliths of shale, limestone, and sandstone occur in the outer one-third of the diatreme. The xenolith content varies between 50 and 90 percent of the breccia mass. Flat, angular fragments of black shale ¼-in to 36 in (.6 to 90 cm) in maximum dimension are the most abundant clast type. These clasts were derived from the enclosing Millboro Shale (Middle Devonian). Rounded limestone and sandstone clasts derived from the underlying Helderberg Group (Lower Devonian) are less common. These clasts range from ½-in to 8 in (1.3 to 20 cm). Recognizable fossils (crinoids and brachiopods) are present in many of the white limestone clasts which appear to have been calcined. At other localities in Highland County clasts as old as Upper Ordovician have been identified in the breccia. The matrix is basalt. Analcime was identified by x-ray diffraction from a sample near the outer edge of the intrusion.

Localities 2 and 3 in Bluegrass Valley (Fig. 6) are good exposures of a compound basalt/andesite dike, basalt dike, andesite diatreme, and several andesite dikes. A compound dike exposed along Virginia 637 (Locality 2) is basaslt at the west end and andesite toward the east (Darton, 1899; Dennis, 1934; Johnson, Milton, and Dennison, 1971). This dike, approximately 2,500 ft (762 m) long intrudes rocks from Lower Ordovician (Beekmantown Formation) to Middle Ordovician (Dolly Ridge Formation) age. Along the road dark greenish gray basalt with large (up to ½-in [12.7 cm]), filled myrolitic cavities is exposed. In thin section the rock consists of microcrystalline plagioclase, hypersthene, augite, and magnetite. Myrolitic cavity fillings consist of analcime, calcite, and chalcedony.

The eastern approximately 1/3 of the dike is of andesitic composition. The light gray andesite is porphyritic with a microcrystalline groundmass composed of subhedral plagioclase (oligoclase?) and finely divided magnetite. Phenocrysts are subhedral plagioclase, hornblende, and biotite. The zone of transition from basaltic composition to andesitic is very poorly exposed.

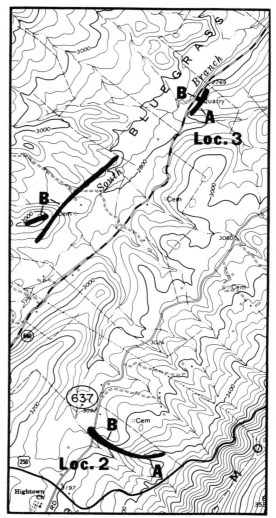

Figure 6. Localities 2 and 3, Monterey 7½-minute quadrangle, Virginia. Black areas are igneous bodies: *B*–basalt; *A*–andesite.

An andesite diatreme, a basalt dike, and several andesite dikes are exposed in the inactive quarry on Virginia 640 in Bluegrass Valley (Fig. 7; Locality 3). The enclosing rock is dolomite and limestone of the Beekmantown Formation and dips steeply southeast. Three thin zones of andesite breccia discordant to subparallel to bedding are exposed in the northern half of the quarry. The light-greenish gray breccia has an andesite matrix with clasts of dolomite, black shale, and obsidian. Plagioclase, probably sanidine, hornblende, and biotite comprise the matrix.

Three porphyritic andesite dikes are exposed. The two dikes exposed in the quarry face are 3 to 5 ft (.9 to 1.5 m) wide. A dike about 20 ft (6.1 m) wide is exposed at the top of the highwall. Carbonate xenoliths comprise up to 50 percent of the two small dikes. Xenoliths are restricted to a 1 to 2 ft (.3 to .6 m) thick zone at the margins of the larger dike. These corroded pebbles were derived from the wall rock by gas-fluidization and corrasion (Johnson, Milton, and Dennison, 1971, p. 42). The groundmass is microcrystalline plagioclase, hornblende, and magnetite with phenocrysts of plagioclase (andesine to oligoclase), hornblende, and biotite.

Near the center of the quarry a dark greenish gray basalt dike occurs (Fig. 7). This northwest trending basalt dike intersects and is normal to the andesite dikes. The basalt is composed

Figure 7. Basalt (*B*) and andesite (*A*) dikes in Virginia Department of Highways and Transportation quarry; locality 3. Note branching of the basalt dike; width approximately 1 m.

of plagioclase (labradorite), pyroxene, olivine altered to serpentine, and magnetite (Rader and Griffin, 1960). Figure 7 shows a branching of this dike in the quarry face.

Prior to 1969 the igneous bodies in Highland County were assumed to be of Late Triassic or Early Jurassic age. Fullagar and Bottino (1969) obtained ages of 37 ± 3 (Rb-Sr) and 47.1 ± 1.3 (K-Ar) Ma. They considered the K–Ar derived age to be the more reliable. An age of 47 Ma corresponds to the Middle Eocene. These igneous bodies are evidence of the youngest known igneous activity in the eastern United States (Dennison and Johnson, 1971). At Locality 3 a northwest trending basaltic dike intrudes and offsets two northeast trending andesite dikes. Therefore, the basalt is somewhat younger than 47 Ma.

Loevlie and Opdyke (1974) determined the paleomagnetic pole direction for several felsite (andesite) intrusions in Highland County. The pole direction agrees with early Tertiary poles from the western United States and with the radiometric ages of Fullagar and Bottino (1969).

Fullagar and Bottino (1969) report an initial Sr^{87}/Sr^{86} ratio of 0.7037 ± 0.0006 for the andesite dikes. The relatively low Sr^{87}/Sr^{86} initial ratio is indicative of magma derived from the lowermost crust or uppermost mantle. "Magmas produced by anatexis of sialic crust would be expected to have significantly higher initial ratios" (Fullagar and Bottino, 1971, p. 59).

REFERENCES CITED

Butts, Charles, 1940, Geology of the Appalachian Valley in Virginia: Virginia Geological Survey Bulletin 52, Pt. 1, 568 p.

Campbell, H. D. and Campbell, J. L., 1885, William B. Rogers' Geology of the Virginias; a review: American Journal of Science, series 3, v. 30, p. 357–374.

Darton, N. H., and Diller, J. S., 1890, On the occurrence of basalt dikes in Upper Paleozoic series in the central Appalachian Virginias: American Journal of Science, Series 3, v. 39, p. 269–271.

Darton, N. H., 1899, Monterey Folio: U.S. Geological Survey Geologic Atlas No. 61.

Dennis, W. C., 1934, Igneous rocks of the Valley of Virginia: [M.S. thesis], University of Virginia, Charlottesville, 76 p.

Dennison, J. M., and Johnson, R. W., 1971, Tertiary intrusions and associated phenomena near the thirty-eighth parallel fracture zone in Virginia and West Virginia: Geological Society of America Bulletin, v. 82, p. 501–508.

Fullagar, P. D., and Bottino, M. L., 1969, Tertiary felsite intrusions in the Valley and Ridge provinces, Virginia: Geological Society of America Bulletin, v. 80, p. 1853–1858.

——, 1971, Radiometric ages of igneous rocks in the Valley and Ridge province of Virginia, *in* Johnson, R. W., Jr., Milton, C., and Dennison, J. M., Field trip to the igneous rocks of Augusta, Rockingham, Highland, and Bath counties, Virginia: Virginia Division of Mineral Resources, Information Circular 16, Appendix IV, p. 57–63.

Garner, T. E., 1956, The igneous rocks of Pendleton County, West Virginia: West Virginia Geological Survey Report Investigation 12, 31 p.

Hunt, S. T., 1975, Mineralogy, chemistry, and petrogenesis of some hypabyssal

intrusions, Highland County, Virginia [M.S. thesis]: Virginia Polytechnic Institute and State University, Blacksburg, 72 p.

Johnson, R. W., Jr., Milton, Charles, and Dennison, J. M., 1971, Field trip to the igneous rocks of Augusta, Rockingham, Highland, and Bath counties, Virginia: Virginia Division of Mineral Resources, Information Circular 16, 68 p.

Kettran, L. P., Jr., 1970, Relationship of igneous intrusions to geologic structures in Highland County, Virginia [M.S. thesis]: Virginia Polytechnic Institute and State University, Blacksburg, 46 p.

Loevlie, R., and Opdyke, N. D., 1974, Rock magnetism and paleomagnetism of some intrusions from Virginia: Journal of Geophysical Research, v. 79, no. 2, p. 343–349.

Rader, E. K., and Griffin, V. S., 1960, A petrographic study of some dikes in a quarry in Bluegrass Valley, Highland County, Virginia (Abstract): Virginia Journal of Science, v. 11, p. 213.

Rogers, W. B., 1884, A reprint of annual reports and other papers on the geology of the Virginias: D. Appleton and Co., New York, 832 p.

Wampler, J. M., 1975, Potassium-argon determination of Triassic and Eocene igneous activity in Rockingham County, Virginia: Geological Society of America Abstracts with Programs, v. 6, p. 547.

Watson, T. L., and Cline, J. H., 1913, Petrology of a series of igneous dikes in Central Western Virginia: Geological Society of America Bulletin, v. 24, p. 301–334.

Zartman, R. E., Brock, M. R., Heyl, A. V., and Thomas, H. H., 1967, K-Ar and Rb-Sr ages of some alkalic intrusive rocks from central and eastern United States: American Journal of Science, v. 265, p. 848–870.

The Browns Mountain anticlinorium, West Virginia

Byron R. Kulander, *Department of Geological Sciences, Wright State University, Dayton, Ohio 45435*
Stuart L. Dean, *Department of Geology, University of Toledo, Toledo, Ohio 43606*
Peter Lessing, *West Virginia Geological Survey, Morgantown, West Virginia 26507*

LOCATION

The Browns Mountain section is located on the Marlinton and Minnehaha Springs 7½-minute quadrangles, along West Virginia 39 between Huntersville and Minnehaha Springs, Pocahontas County, West Virginia (Fig. 1). All stops along the section are adjacent to, or a short walk from, the main road and are easily accessible by bus or automobile.

INTRODUCTION

The Browns Mountain anticlinorium is a major structure within the Allegheny Plateau (Price, 1929; Gwinn, 1964; Kulander and Dean 1972, 1978, 1986). The structural style resembles that of the Valley and Ridge province and is characterizead by tight folds, fore- and back-limb thrust faults, strike-slip faults, and local passive slip and flow structures, including a small shale diapir. Upper Ordovician rocks commonly crop out along thrusts and in anticlinal cores, attesting to the high structural amplitude (10,000 ft [3,047 m]) of the anticlinorium. The cross-section stops provide an opportunity to investigate a structural style similar to that encountered in hydrocarbon exploration in other folded plateau anticlinoria. Rocks ranging in age from Late Ordovician to Middle Devonian are exposed along the described section (Fig. 2). In addition, many individual structures can be examined and photographed directly from West Virginia 39.

Local Tectonics

The complex structures within the Browns Mountain anticlinorium developed over two major decollement horizons. The uppermost lies within Upper Ordovician Martinsburg Shales. The underlying decollement is in thin-bedded carbonates and shales of the Early Cambrian Waynesboro Formation. These detachment zones, stippled in Figure 3, bound two major litho-tectonic units, or plates. The lower plate, soled in the Waynesboro, contains massive, competent, Middle Cambrian through Middle Ordovician carbonates. In contrast, the overlying plate, seated in Martinsburg Shale, contains a repetitive sequence of thinner-bedded shales, sandstones and carbonates.

Different mechanical properties of these two plates have contributed to the development of contrasting inter-plate structural styles. For example, the low ductility contrast and competent nature of Lower Waynesboro plate carbonates led to shortening by long-wavelength folding over imbricate step-up thrusts, soled in the Waynesboro Formation (Fig. 3). On a regional scale, the imbricated, lower-plate structures, roofed in Martinsburg Shale, form a classic duplex as described by Boyer and Elliott (1982). The lower plate rocks are not exposed within the anticlinorium.

The markedly different structural style of the upper plate is characterizead by short wavelength, asymmetrical to overturned folds, and closely-spaced fore- and back-limb thrust faults, rooted in Martinsburg and younger Shales (Fig. 4). Structures mapped

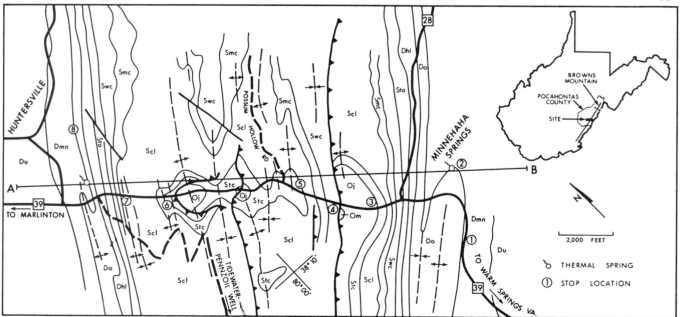

Figure 1. Location map showing geological boundaries of the Browns Mountain section with numbered stops circled.

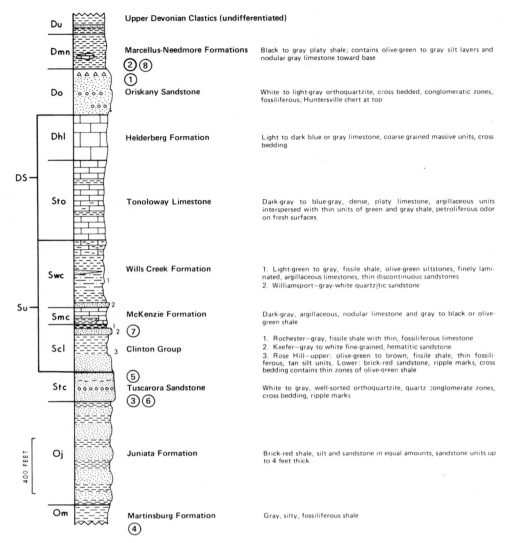

Figure 2. Stratigraphic column of Browns Mountain area with numbered stops circled.

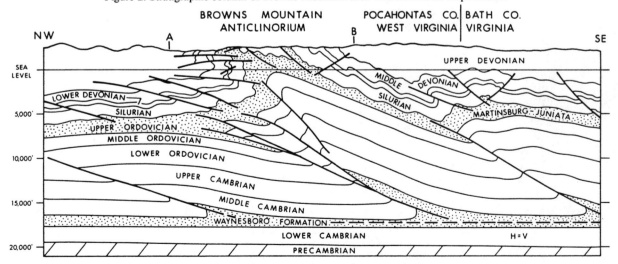

Figure 3. Generalized structural section of Browns Mountain anticlinorium. Detachment zones are stippled. Horizontal scale equals vertical scale.

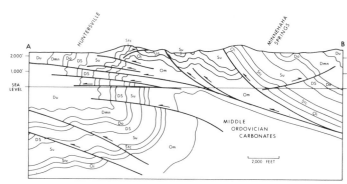

Figure 4. Detailed structural section A-B (see Figures 1 and 3 for location).

in the upper plate of Browns Mountain are controlled primarily by several competent lithotectonic units, including the Juniata Formation, Tuscarora Sandstone, and Helderberg Limestone/Oriskany Sandstone. These units seldom exceed aggregate thicknesses of 700 ft (213 m) and are surrounded by extensive sequences of ductile shales, siltstones, and thin-bedded carbonate units.

A chronology of structural development within the Browns Mountain anticlinorium is suggested by contrasting plate structure. For example, the Martinsburg decollement is arched by underlying step-up faults soled in the Waynesboro, and upperplate structures are rotated by this folding (for example the folded back-thrust in figures 1 and 4). These observations imply that deformation above the Martinsburg preceded significant lower-plate shortening at any given locality. In addition, the tight folds and closely-spaced faulting of upper plate rocks within Browns Mountain (and the entire overthrust belt) are necessary to permit upper-plate shortening to match or exceed lower-plate shortening. Deformation of both plates has produced a total structural amplitude within the Browns Mountain anticlinorium that approaches 10,000 ft (3,047 m).

Folds

Three anticlines are prominent along the West Virginia 39 section (Fig. 1). The eastern fold just west of Minnehaha Springs shows the largest amplitude, and Martinsburg Shale crops out in a small exposure within its core. More resistant rocks from the Oriskany to the Juniata can be viewed along West Virginia 39 and 28, across the eastern flank of the anticline. The east flank has been thrust over the adjoining syncline and, consequently, Tuscarora and Clinton rocks are truncated on this western anticlinal flank.

The central anticlinal complex, traversed by Possum Hollow Road, is marked by numerous small structures. Several Tuscarora folds are asymmetrical to the west. In addition, the Juniata Formation crops out along a small thrust fault along West Virginia 39, approximately 1,800 ft (548 m) west of Possum Hollow Road.

The western anticline exposes Juniata sandstones and shales within its axial section. A small subsidiary anticline is situated immediately west of the major structure. Both of these structures,

known locally as the "big and little arch," are held up by Tuscarora Sandstone, and can be viewed from West Virginia 39. The "big arch" is cut by a folded back-thrust evident on low-altitude aerial photographs.

Numerous minor folds can be observed along West Virginia 39 and in Knapp Creek. These structures are best-developed in ductile shales and thin-bedded Silurian siltstone and limestone units.

Faults

The major thrust fault, cropping out immediately west of Minnehaha Springs, has a mapped length of twenty miles (Cardwell and others, 1968) with a displacement within the cross section of 1,500 ft (457 m). Further to the south, this fault displaces Juniata rocks onto Middle Devonian black shales. Several unmapped small back thrusts show displacements of several feet and cut the Tuscarora Sandstone along West Virginia 39. Direct indications of the folded back thrust at the "big arch" and thrust fault approximately 1,800 ft (549 m) west of Possum Hollow (Fig. 1) are not readily visible from the road, but can be seen on low-altitude aerial photographs.

Browns Mountain Deep Well

The Tidewater-Pennzoil #1USA 357 well (Permit # Pocahontas 21), a major Appalachian overthrust test, was drilled several miles south of Huntersville. The well was spudded in Lower Silurian strata, drilled 11,379 ft (3,468 m) to a subsea depth of 7,905 ft (2,395 m) and bottomed in the Middle Ordovician Chazy Group. A number of thrust faults, substantiated by strata duplication, corroborate the structural style shown on the structure sections and map. Specifically, at a drilling depth of 410 ft (125 m), Tuscarora was thrust over Clinton; at a drilling depth of 450 ft (137 m), Tuscarora was thrust over Wills Creek; at a drilling depth of 2,560 ft (780 m), Martinsburg was thrust over Clinton (overturned); at a drilling depth of 3,320 ft (1,011 m), McKenzie was thrust over Wills Creek (overturned); and at a drilling depth of 10,440 ft (3,182 m), a thrust fault repeated Middle Ordovician units. In fact, the Middle Ordovician Black River Group was encountered three times between drilling depths of 8,470 and 10,440 ft (2,581 and 3,182 m). These lowermost faults can be interpreted as splays from a primary, Waynesboro-soled step-up. The overlying faults are most likely rooted in Martinsburg Shale and resemble those cropping out along West Virginia 39.

STOP DESCRIPTION

The geologic traverse of the Brown Mountain anticlinorium begins on the southeast limb of the structure and ends on the northwest limb. Stop locations are shown on Figure 1 and stratigraphic position and relationship are shown on Figure 2.

Stop. 1 Oriskany Sandstone Exposure

Immediately south of Minnehaha Springs, along the

southwest trend of West Virginia 39, a dip slope of Devonian Oriskany Sandstone can be observed. Bedding is well-defined by fossil molds. Joint sets approximately parallel and perpendicular to the highway are not fold-related; rather they fit a regional pre-fold joint pattern (Lessing and others, 1984).

Stop 2. Thermal Spring

A structurally-controlled thermal spring at Camp Minnehaha, in Minnehaha Springs, discharges from the Oriskany at the Marcellus-Needmore Shale contact, along the axial trace of a northeast-plunging anticline. Here, the shale serves as an effective seal, restricting flow to the highly-fractured sandstone. A minimum circulation depth of 1,475 ft (449 m) has been calculated for water recharged from Oriskany outcrop areas up-plunge (southwest) of the spring. Water temperature is constant at 20.4°C, and 97% of the water is over 30 years old based on tritium content (Lessing and others, 1984). Deep circulation is restricted by a postulated back-thrust (Fig. 4), whereas the anticlinal plunge controls subsurface flow direction. Permission to enter this site must be obtained from camp personnel.

Stop 3. Tuscarora-Juniata Contact

A sharp contact between Ordovician Juniata red sandstones and shales and the Silurian Tuscarora Sandstone can be viewed along West Virginia 39. Note local Tuscarora duplication by fault wedging. In addition, some Tuscarora bedding planes show well-developed bioturbation. An almost complete section of Juniata is exposed in a continuous road cut to the west. Excellent bedding wedging can be seen in low-dipping siltstone and sandstone beds.

Stop 4. Martinsburg Outcrop

A small expanse of gray to brown fossiliferous Ordovician Martinsburg shale directly underlies stream deposits along the fault trace in Knapp Creek. Deformed fossils and cleavage attest to a high degree of plastic shortening within this outcrop. The exposure is difficult to find during summer and is not visible from the road.

Stop 5. Shale Diapir and Adjacent, Near-Recumbent, Tuscarora Folds

A shale diapir cuts through rigid Tuscarora Sandstone along West Virginia 39. Here, density and viscosity contrasts, coupled with tectonic loading by the now-eroded hanging-wall rocks of the major fault to the east, have presumably produced the diapir and near-recumbent folds 100 ft (30 m) to the east. The folds are asymmetrical to the east and may represent an early phase of back thrust development. Again, note early-formed extension joints cutting Tuscarora rocks within the folds. The detachment flooring the diapir and adjacent flexures lies at or just above road

level. Note horizontal Tuscarora below the detachment level in the creek bed.

Stop 6. Little Arch Anticline

The "little arch" is a subsidiary structure to the larger anticline ("big arch") immediately to the east. The flexure is defined by resistant Tuscarora Sandstone with Juniata rocks exposed in the core.

Stop 7. Clinton Iron Ore

A 1 ft (30 cm) thick bed of oolitic hematite, contained within the Keefer Formation of the Silurian Clinton Group, crops out in the west flank of a small westwardly asymmetrical anticline along West Virginia 39. Note S- and Z-shaped kink zones on west and east flanks, respectively, that reveal the proper shear sense for this flexure.

Stop 8. Huntersville Chert—Marcellus–Needmore Contact

The contact between the Huntersville Chert and overlying black shales is exposed in an outcrop at the Ambassadors for Christ campground at the eastern end of the main facility. These cherts show typical irregular-bedded nature. Zones within the black shale illustrate the complex folding characteristic of this ductile unit. Permission to study this outcrop can be obtained from the permanent resident of the campground adjacent to the exposure.

REFERENCES CITED

Boyer, S. B., and Elliott, D., 1982, Thrust systems: American Association of Petroleum Geologists Bulletin, v. 66, no. 9, p. 1196–1230.

Cardwell, D. H., Erwin, R. B., and Woodward, H. P., 1968, Geologic map of West Virginia: 1:250,000 scale, West Virginia Geological and Economic Survey.

Gwinn, V. E., 1964, Thin-skinned tectonics in the Plateau and northwestern Valley and Ridge provinces of the central Appalachians: Geological Society of America Bulletin, v. 75, p. 863–900.

Kulander, B. R., and Dean, S. L., 1972, Gravity and structures across Browns Mountain, Wills Mountain, and Warm Springs anticlines—gravity study of the folded plateau, West Virginia, Virginia, and Maryland, *in* Lessing, P., Hayhurst, R. I., Barlow, J. A. and others, editors, Appalachian structures—origin, evolution, and possible potential for new exploration frontiers—a seminar: West Virginia University and West Virginia Geological and Economic Survey, p. 141–180.

——1978, Gravity, magnetics, and structure—Allegheny Plateau/western Valley and Ridge in West Virginia and adjacent states: West Virginia Geological and Economic Survey, Report of Investigations 27, 91 p.

Kulander, B. R., and Dean, S. L., 1986, Structure and tectonics of the central and southern Appalachian Valley and Ridge provinces, West Virginia and Virginia: American Association of Petroleum Geologists Bulletin (in press).

Lessing, P., Hobba, W. A., Jr., Dean, S. L., and Kulander, B. R., 1984, Relations between geology and warm springs in West Virginia using airborne radar: U.S. Geological Survey, Professional Paper (in press).

Price, P. H., 1929, County report—Pocahontas County: West Virginia Geological and Economic Survey, 531 p.

Stratigraphic and structural features of Fincastle Valley and Eagle Rock Gorge, Botetourt County, Virginia

Eugene K. Rader and Thomas M. Gathright II, Virginia Division of Mineral Resources, Charlottesville, Virginia 22903

LOCATION

Stratigraphic and structural features of Fincastle Valley and Eagle Rock Gorge are exposed along U.S. 220 between Fincastle and Eagle Rock, Botetourt County, Virginia (Fig. 1).

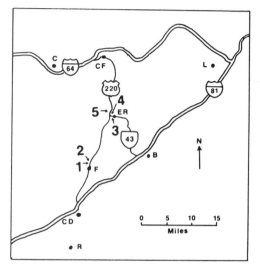

Figure 1. Generalized map showing access to the Fincastle–Eagle Rock area. *C,* Covington; *CF,* Clifton Forge; *L,* Lexington; *ER,* Eagle Rock; *F,* Fincastle; *B,* Buchanan; *CD,* Cloverdale; *R,* Roanoke. Numbers indicate locations of stops.

INTRODUCTION

Five localities that illustrate stratigraphic features of the Taconic flysch and molasse sequences and structural features of the eastern Valley and Ridge physiographic province are described in this "cluster stop" (Fig. 1). The southern most area (Localities 1 and 2) represents evidence for the onset of the Taconic flysch sequence (Middle Ordovician). At Eagle Rock (Localities 3 and 4) the eastern Taconic molasse sequence is exposed. Both of these areas were described and interpreted by Woodward (1936). The decollement at the base of the Pulaski–North Mountain thrust system is exposed at Locality 5.

SITE DESCRIPTION

Fincastle Conglomerate Member of the Martinsburg Formation: Localities 1 and 2

The conglomerate beds north of Fincastle were first described in a manuscript by Woodward (1936, p. 135). Later, Stow and Bierer (1937) briefly described these rocks as the Athens conglomerate near Fincastle. Subsequently, the Fincastle

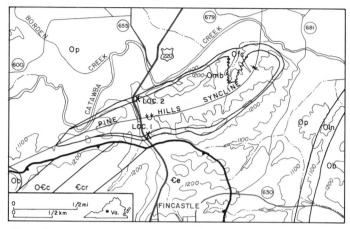

Figure 2. Geologic map of the type area of the Fincastle Conglomerate Member (modified from McGuire, 1970 and Karpa, 1975; Oriskany and Salisbury 7.5 minute quadrangles). Map symbols: *Omb,* Martinsburg Formation; *Ofc,* Fincastle Conglomerate Member; *Op,* Paperville Shale; *Oln,* Lincolnshire and New Market Limestone; *Ob,* Beekmantown Formation; *OЄc,* Conococheague Formation; *Єcr,* Copper Ridge Formation; *Єe,* Elbrook Formation.

has been treated as a member of various formations by different workers. We consider it as a member of the Martinsburg Formation for reasons discussed later. Roadcuts and outcrops described or referred to by Woodward (1936); Stow and Bierer (1937); Butts (1940); Decker (1952); Kellberg and Grant (1956); and Cooper (1960) have been destroyed by recent highway construction. Because no previous investigator designated a type section, except by inference, and their exposures have been destroyed, the writers herein designate Locality 1 (cut 1 of Bartholomew and others, 1982) as the principal reference section for the Fincastle Conglomerate Member of the Martinsburg Formation. Sections described by Karpa (1975) are designated as reference sections. The known surface extent of the Fincastle Conglomerate Member is shown on Figure 2.

Three lithologies are dominant in the Fincastle: (1) conglomerate, (2) sandstone, and (3) shale. Figure 3 shows the stratigraphic distribution of the three basic rock types at these localities.

Two basic types of conglomerate are present in the Fincastle. The most striking type consists of beds of poorly sorted, clast-supported, pebble to boulder conglomerate. Clasts are limestone, dolomite, quartzite, sandstone, chert, vein quartz, granite gneiss, quartz pebble conglomerate, greenstone, and shale. Clast diameter ranges from one-half inch to seventeen inches. Clasts of quartzite and other resistant materials greater than three inches are well rounded. The limestone, dolomite, sandstone, and siltstone clasts are rounded to subrounded. Intermediate size quartz-

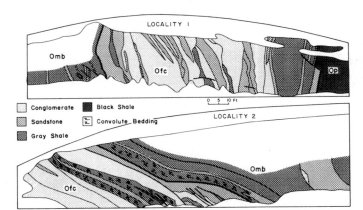

Figure 3. Sketches of exposures at Localities 1 and 2 (modified from Bartholomew and others, 1982, p. 139). *Omb,* Martinsburg Formation; *Ofc,* Fincastle Conglomerate Member; *Op,* Paperville Shale.

ite and vein quartz clasts are well rounded but the limestone and dolomite clasts are subrounded to tabular. In the less than 0.5-in (1.3-cm) class the quartzite and vein quartz clasts are subrounded and contrast with the subangular limestone, dolomite, and chert clasts. Approximately 90 percent of the clasts are limestone, green quartzite, and vein quartz (Kellberg and Grant, 1956).

In contrast to the clast supported conglomerate are beds of conglomerate in which the clasts appear to be "floating" in the litharenite matrix. This finer grained conglomerate is composed of clasts of quartzite, vein quartz, limestone, and chert up to one inch in size. Vein quartz and black chert comprise about 80 percent of the clasts (Kellberg and Grant, 1956). The quartz, chert, and quartzite clasts are subrounded to well rounded and contrast somewhat with the subangular to subrounded limestone and dolomite clasts.

Most of the clasts in both conglomerates are fractured. Fractures that are restricted to the clasts are filled with quartz. Pervasive fractures are filled with calcite (Kellberg and Grant, 1956).

Matrix framework grains in both conglomerates are quartz (monocrystalline and polycrystalline), limestone, and dolomite with lesser amounts of chlorite, sericite, and clay. The cement is calcite. Limonite and goethite are common, giving the rock a brown color.

The clasts were derived from rocks deposited in the eastern part of the Appalachian basin and uplifted basement igneous and metamorphic units. Clasts from the following formations have been identified: Lincolnshire and New Market limestones; Beekmantown, Chepultepec, Conococheague, Elbrook, and Rome formations; Chilhowee Group, Catoctin Formation, and basement meta-igneous units (Woodward, 1936; Cooper, 1960; McGuire, 1970; Karpa, 1975).

Scour channels are common and many have tabular shale clasts in the basal portion. The conglomerates fine upward to sandstone.

Sandstone in the Fincastle Member is medium- to very-coarse-grained litharenite. Monocrystalline and polycrystalline

quartz, limestone, and chert comprise the framework grains. The matrix is composed of chlorite, sericite, and clay with minor limonite and goethite (Kellberg and Grant, 1956). Calcite cement is a minor component. Cross stratification is rare in the massive beds.

Interbedded with the sandstones and conglomerates are gray sandy shales and siltstones. These gray, fine-grained clastic rocks contrast sharply with the underlying fissile, black shale of the Paperville Shale. Convolute bedding is common in the shale at Locality 2 (Fig. 3).

Discussion

These coarse clastic deposits have been interpreted as river channel deposits (Woodward, 1936); as resedimented stream or nearshore deposits (Kellberg and Grant, 1956); as broad, shallow marine channels (Karpa, 1975); and as a submarine fan complex (Bartholomew and others, 1982). Preserved primary sedimentological features and associated stratigraphic units tend to support a submarine fan complex type of deposit. The underlying Paperville Shale is a fissile, black, graptolitic shale with black, aphanic limestone conglomerate beds (Lowry and others, 1972), interpreted as a basinal deposit (Read, 1980). Rocks overlying the Fincastle are base-truncated Bouma cycles interpreted to represent a basin fill sequence. All stratigraphic contacts are conformable. The rocks are interpreted as a shallowing-upward sequence beginning with basinal black shale, to debris fans, to turbidites, to sandstone.

The five main facies of the radial fan model proposed by Walker (1978) are represented in the Fincastle Conglomerate Member. At Locality 1 the lower sandstone-shale units represent the inner lower fan to outer mid-fan position. The conglomerates, pebbly sandstones, and massive sandstones that make up the bulk of the member represent deposition in mid-fan to upper fan channels. The thin associated silty shales probably are levee deposits. The convolute bedded silty shale (Locality 2) represents slumps developed on the upper fan.

Assignment of the Fincastle Conglomerate Member to the Martinsburg Formation is based on practical mappability. The contact between the Paperville Shale and the base of the Fincastle is the most readily mapped contact; the rocks are interpreted to represent the local onset of the coarse phase of the Taconic flysch sequence, thus the Fincastle is considered to be a member of the Martinsburg Formation.

SITE DESCRIPTION:

Eagle Rock Gorge: Localities 3 and 4

Recent highway construction in Eagle Rock gorge (Localities 3 and 4) has exposed an exceptional sequence of polydeformed Ordovician through Lower Devonian strata (Figs. 4 and 5). The Ordovician to Middle Devonian imbricated rock sequence in and east of the gorge is a footwall-derived horse block

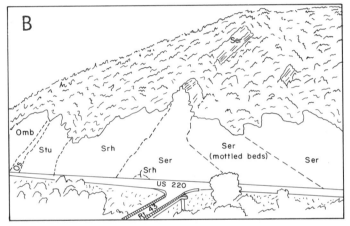

Figure 4. *(A)* View of the southwestern face of Eagle Rock gorge and the road cut exposures of the Silurian quartz arenites and "red beds" of the eastern structural block. *(B)* Diagram of the formational boundaries in the photograph above. *Omb,* Martinsburg Formation; *Os,* Oswego Formation; *Stu,* Tuscarora Formation; *Srh,* Rose Hill Formation; *Ser,* Eagle Rock Sandstone.

implaced at the leading edge of the Pulaski–North Mountain thrust system (Bartholomew and others, 1982; Rader and Gathright, 1984). The horse block lies above a décollement developed in the Lower and Middle Devonian limestones and shales beneath the Fincastle Valley area and beneath the Cambro–Ordovician-age dolomites of the main body of the thrust system. The new exposures along U.S. 220 (Locality 3) make the gorge the area of best exposure of any one of the many horse blocks in the fensters or along the leading edge of the Pulaski--North Mountain thrust system. These exposures (Locality 3) are of a fault duplicated and unusually thick sequence of Silurian quartzarenites. Similar exposures on Virginia 43 (Locality 4) on the northeast side of the gorge display the same rock sequence at a slightly higher structural level. Both the anomalous stratigraphy and complex structure have elicited much interest and several interpretations from geologists that have studied the area in the last 50 years.

Three Silurian formations are of particular interest: the Tus-

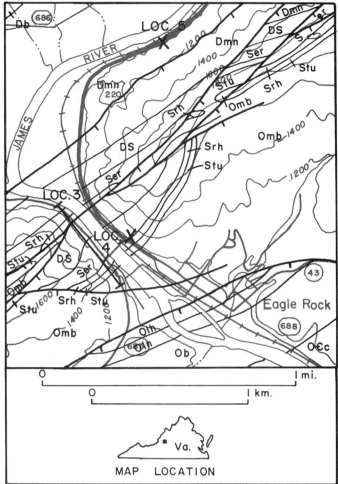

Figure 5. Geologic map of the Eagle Rock area, modified from McGuire (1970). Map symbols: *Db,* Brallier Formation; *Dmn,* Millboro and Needmore Shales; *DS,* Lower Devonian–Upper Silurian undivided; *Ser,* Eagle Rock Sandstone; *Srh,* Rose Hill Formation; *Stu,* Tuscarora Formation; *Omb,* Martinsburg Formation; *Oln,* Lincolnshire and New Market Limestones; *Ob,* Beekmantown Formation; *OCc,* Conococheague Formation.

carora Formation, a light gray to white, medium-grained, thin- to thick-bedded, cross-bedded, locally conglomeratic quartzarenite (140 ft [42.7 m]); the Rose Hill Formation, a sequence of dark dusky red, hematite cemented quartzarenites with interbedded light gray quartzarenites and dark dusky red or dark greenish gray shales (83 + ft [25 +m]), with ball-and-pillow structures (Locality 4); and the Eagle Rock Sandstone of Lampiris (1975), a quartzarenite unit like the Tuscarora near the base but becoming finer-grained, thinner bedded and calcareous at the top, with a zone of pale dusky red or green mottled sandstones and thin shales interbedded with light gray quartzarenites between the lower and upper quartzarenites (412 ft) [125.6 m]).

At Locality 3 these formations are best exposed in the eastern of two, folded, west facing, fault separated sequences (Fig. 4) that also includes the Upper Martinsburg and Oswego clastics at the base and the Tonoloway Formation and calcareous sandstone and limestone of Upper Silurian and Lower Devonian age at the top. A few feet (m) of olive-gray shale of the Needmore (?) Formation lies below the fault separating the two sequences. The western sequence includes a remnant of the Tuscarora, Rose Hill and Eagle Rock sandstones westward to the end of road cut exposures. Both sequences and the separating fault are folded into a west facing syncline/anticline pair visible at road level and above.

Structures visible along U.S. 220 (Locality 3) that have been useful in developing a structural history for the Eagle Rock area include small displacement thrust faults in the quartz-arenites, small scale kink folds in the Tonoloway limestones, and two cleavage orientations of which one is consistent with stresses developed during thrusting and a second cross-cutting cleavage that is axial-plane parallel to the anticline–syncline pair (Bartholomew and others, 1982).

Discussion

The first recorded structural interpretation was by Woodward (1936). Although he did not recognize the direction of movement on the fault in the gorge or that the Silurian rocks were part of a detached block, he did recognize the necessity for early faulting and later folding during emplacement of the Pulaski–North Mountain thrust sheet. Structure sections developed from detailed mapping of the Eagle Rock area (Fig. 5) by McGuire (1970) and by Bartholomew and others (1982) suggest that the western sequence was thrust over the eastern sequence

while the rocks were sub-horizontal, and that the folds that warp the thrust fault and both sequences developed later. Folded imbricate thrust faults within similar Silurian stratigraphic sequences on Sugar Loaf and North mountains to the east and northeast (Bick, 1972) occur in the footwall block of the now eroded Pulaski–North Mountain thrust system. These imbricated sequences, had they been detached and rotated by the Pulaski–North Mountain thrusting, could duplicate the structure at Eagle Rock.

SITE DESCRIPTION

Pulaski-North Mountain Décollement: Locality 5

The décollement at the base of the Pulaski–North Mountain thrust system is exposed along State Highway 43, 1.5 mi (2.4 km) north of the highway bridge in Eagle Rock gorge (Fig. 5; Locality 5). Isoclinal folds and high angle faults, some of which may be rotated northwest dipping thrusts, are visible in strongly cleaved rocks of the Millboro Shale. The deformed rocks are thought to lie in the immediate footwall of the Pulaski–North Mountain thrust system.

Along this structural front the Pulaski–North Mountain thrust system appears to be ramping up from a Lower to Middle Devonian décollement level to a level above the Lower Mississippian Price Formation. Southwest of Eagle Rock in Montgomery County the thrust system lies above a bedding plane decollement in the Lower Mississippian Maccrady Shale that overlies the Price Formation.

REFERENCES CITED

Bartholomew, M. J., Schultz, A. P., Henika, W. S., and Gathright, T. M., II, 1982, Geology of the Blue Ridge and Valley and Ridge at the junction of the central and southern Appalachians, in Lyttle, P. T., ed., Central Appalachian Geology: NE-SE Geological Society of America, Field Trip Guidebooks, p. 121–170.

Bartholomew, M. J., 1981, Geology of the Roanoke quadrangle, Virginia: Virginia Division of Mineral Resources Publication 34, 23 p.

Bick, K. F., 1973, Complexities of overthrust faults in central Virginia: American Journal of Science, Cooper Vol. 273-A, p. 343–352.

Butts, C., 1940, Geology of the Appalachian Valley in Virginia: Virginia Geological Survey Bulletin 52, pt. 1, 568 p.

Cooper, B. N., 1960, The geology of the region between Roanoke and Winchester in the Appalachian Valley of Virginia: Johns Hopkins University Studies in Geology 18, 84 p.

Decker, C. E., 1952, Stratigraphic significance of graptolites of Athens shale: American Association of Petroleum Geologists Bulletin, v. 36, p. 1–145.

Gathright, T. M., II, and Rader, E. K., 1981, Field guide to selected Paleozoic rocks, Valley-Ridge Province, Virginia: Virginia Minerals, Vol. 27, No. 3, p. 18–23.

Karpa, J. B., III, 1975, The Middle Ordovician Fincastle Conglomerate north of Roanoke, Virginia, and its implications for Blue Ridge tectonism: [M.S. thesis], Virginia Polytechnic Institute and State University, Blacksburg, Virginia, 164 p.

Kellberg, J. M., and Grant, L. F., 1956, Coarse conglomerates of the Middle Ordovician in the southern Appalachian Valley: Geological Society of America Bulletin 67, p. 697–716.

Lampiris, N., 1975, Stratigraphy of the clastic Silurian rocks of central western Virginia and adjacent West Virginia [Ph.D. thesis]: Blacksburg, Virginia Polytechnic Institute, 237 p.

Lowry, W. D., McDowell, R. C., and Tillman, C. G., 1972, The Diamond Hill and Fincastle conglomerates—evidence of great structural relief between the Blue Ridge anticlinorium and Salem synclinorium in Middle Ordovician time: Geological Society of America Abstracts with Programs, v. 2, no. 4, p. 88.

McGuire, O. S., 1970, Geology of the Eagle Rock, Strom, Oriskany, and Salisburg quadrangles, Virginia: Virginia Division of Mineral Resources Report of Investigations 24, 39 p.

Rader, E. K., and Gathright, T. M., II, 1984, Road Log in Stratigraphy and structure in the thermal springs area of the western anticlines: Sixteenth Annual Virginia Geologic Field Conference Guidebook, p. 23–55.

Read, J. F., 1980, Carbonate ramp-to-basin transitions and foreland basin evolution, Middle Ordovician, Virginia Appalachians: American Association of Petroleum Geologist Bulletin, v. 64, p. 1575–1612.

Stow, M. H., and Bierer, J. C., 1937, Some significance of an Athens conglomerate near Fincastle, Virginia (abs.): Virginia Academy of Science Proceedings, 1936–1937, p. 71.

Walker, R. G., 1978, Deep-water sandstone facies and ancient submarine fans: Models for exploration for stratigraphic traps: American Association of Petroleum Geologists, v. 62, no. 6, p. 932–966.

Woodward, H. P., 1936, Geology and Mineral resources of the Natural Bridge Region, Virginia: Virginia Division of Mineral Resources Open File Report 12, 350 p.

Batoff Creek section of Pennsylvanian-Mississippian strata, Raleigh County, West Virginia

Gayle H. McColloch, Jr., West Virginia Geological and Economic Survey, P.O. Box 879, Morgantown, West Virginia 26507-0879

LOCATION

Located along West Virginia 41 between Beckley and Prince, Raleigh County, West Virginia (Fig. 1), the site is partially in the New River Gorge National River area on the north side of Batoff Creek.

INTRODUCTION

The Batoff Creek section (Fig. 2) is the best exposure of Upper Mississippian–through–Lower Pennsylvanian strata in the broad, relatively flat area surrounding the New River, locally known as the Fayette Plateau. The section is located in the area where most units of the New River Formation were described and includes two economically important coal beds: the Fire Creek and Sewell coals.

The Pocahontas Formation exposed in the section is typical of the unit in the Fayette Plateau area. The unit is approximately 450 ft (136 m) thick, which is thinner than its 700-ft (212-m) thickness in the type locality 40 mi (66.7 km) to the southwest at Pocahontas, Virginia. The No. 3 and No. 6 Pocahontas coals, although too thin and impure to mine in the section, are economically important nearby. Most of the Mississippian Bluestone Formation is also exposed here.

The section displays three transitional environments that developed during late Mississippian and early Pennsylvanian times. (1) The lower part of the interval, from below the Princeton Sandstone to the base of the Pocahontas Formation, was deposited in a coastal setting. (2) A delta system was built from the east into the basin at the base of the Pocahontas, overriding the coastal sediments of the Bluestone Formation. (3) A trans-

gressive event occurred, separating deltaic sediments of the underlying Pocahontas Formation from the coastal sediments of the New River Formation.

Besides coal, two other important commodities are represented in the section. The quartzose sandstones of the New River Group are used in the region for aggregate, and the carbon-rich black shales of the Pride Shale Member of the Bluestone Formation have been suggested as a source of expanded aggregate elsewhere in the region (Dewey Kirstein, West Virginia Geological and Economic Survey, personal communication).

The site (Figs. 1 and 3) is in the Appalachian Plateau physiographic province. The Fayette Plateau is an upland formed by several sandstones of the New River Formation. The New River and its major tributaries, such as Piney Creek (near the lower

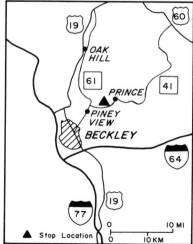

Figure 1. General location map Batoff Creek Section, Raleigh County, West Virginia.

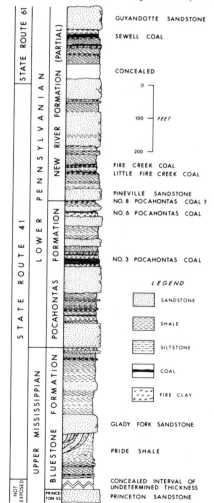

Figure 2. Columnar section of rocks exposed along West Virginia 41 and 61 between Beckley and Prince, West Virginia.

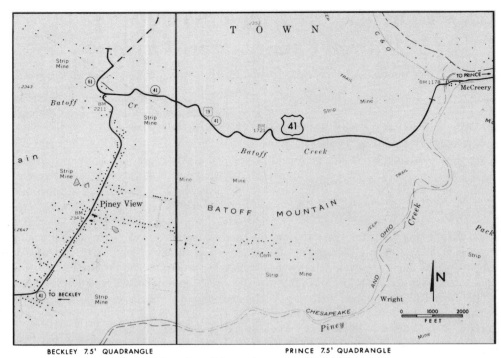

Figure 3. Location of the section north of Batoff Mountain.

portion of the section), are rejuvenated streams that have formed deep gorges. Locally, the course of the New River, before rejuvenation, appears to have been influenced by the presence of the Mann Mountain anticline. The river flows in a band of entrenched meanders that closely follow the west flank of the fold (Fig. 4) between Grandview (south of the site) and Gauley Bridge (shown near top of map in Fig. 4).

The regional structure is generally monoclinal. The dip direction is to the northwest and ranges from as little as 100 ft to as much as 400 ft per mile (18.75 to 75 m per km) (1° to 4°). This variation alternates between areas of relatively low dip (5 mi [8.3 km] or more wide) and high dip (1 to 2 mi [1.6 to 3.3 km] wide). Superimposed on this regional dip are a number of small folds with maximum structural relief of 200 to 300 ft (60 to 90 m) and one large fold, the Mann Mountain anticline, that displays over 1,000 ft (303 m) of structural relief. Kulander and Dean (1980) have suggested that this fold is detached at the lower part of the Devonian.

SITE DESCRIPTION

Of the 1,500 ft (455 m) of section exposed along Batoff Creek, the lower 1,400 ft (424 m) is accessible along West Virginia 41 (Prince 7½-minute Quadrangle) (Fig. 3). This road has moderate-to-light traffic; parking on the south side is adequate for smaller vehicles, but too small for buses. The upper 100 ft (30 m) of the section is, unfortunately, on private, unreclaimed surface mine property northwest of West Virginia 61 (Beckley 7½-minute quadrangle) and is not currently well exposed.

Mississippian Bluestone Formation. The base of the section (near McCreery) is in the Bluestone Formation above the top of the Princeton Sandstone. Gamma ray and density logs from a nearby gas well (Raleigh County 619) show the Princeton to be a clean sandstone. The grayish-red and grayish-green siltstone and

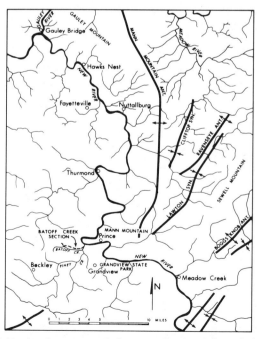

Figure 4. Regional structure, drainage, and cultural features including those mentioned in the text.

Figure 5. Contemporaneous slump block at bottom of Glady Fork Sandstone Member on West Virginia 41.

Figure 6. *Neuropteris pocahontas.*

shales of the basal Bluestone Formation are part of the Pride Shale Member. These lithologies are characteristic of the lower portion of the Pride Shale in the area (Englund, 1979). The upper part of the Pride Shale Member is characterized by gray shales and siltstones that are typical of the unit. Sparse pelecypods and inarticulate brachiopods can be found in this unit.

The top of the Pride Shale Member occurs at the base of a large contemporaneous slump block (Fig. 5). The slumping caused chevron folds to form in the top of the Pride Shale. The gray sandstones, siltstones, and shales of the slump block are assigned to the Glady Fork Sandstone Member. A sharp contact at the top of the slump block separates it from the rest of the Glady Fork Sandstone Member. Above this unit, the remainder of the Bluestone Formation is sandstone, siltstone, argillaceous limestone, underclay, carbonaceous shale, and coal. This interval includes a number of fossil horizons containing a fresh-water and brackish-water fauna of ostracodes, pelecypods, and inarticulate brachiopods. The Bluestone Formation rocks appear to have been deposited in lagoons, tidal flats, marshes, and bars shoreward of a Princeton Sandstone barrier.

The Mississippian-Pennsylvanian boundary is generally placed at the first occurrence of *Neuropteris pocahontas* (Fig. 6), although none have been found in the section. The boundary is conformable here, as it is along all its outcrop in southeastern West Virginia. This boundary occurs somewhere in the sandstone at the base of the Pocahontas Formation (Fig. 2). Elsewhere it can occur in the upper part of the Bluestone Formation.

Pennsylvanian Pocahontas Formation. This formation consists of a wedge of clastic rocks that thins from a maximum of about 700 ft (212 m) near the center of the basin (southwest of the section) to about 450 ft (136 m) at this location. The sediments are argillaceous sandstone, siltstone, shale, coal, and under-

clay. This unit, which extends from the basal Pocahontas Sandstone to the Pineville Sandstone, is marked by gray and brown units that contrast with the red and green rocks of the Bluestone Formation.

The Pocahontas Formation is the lowest unit to contain commercially minable coals in West Virginia. However, no coals of minable thickness occur in this section. The most important Pocahontas Formation coals in the region, the No. 3 and No. 6 Pocahontas seams, are low-to-medium-volatile coals averaging less than 1 percent sulfur, less than 6 percent ash, and about 14,500 Btu per pound.

Maps of large coal mines show seam discontinuities of dendritic, carbonaceous shale-filled channels with sinuous tributaries. The channel trends do not seem to show any preferred orientation.

Well-preserved, abundant plant fossils are found in the Pocahontas Formation. Invertebrate fossils are sparse, although pelecypods, *Lingula,* and other inarticulate brachiopods can be found at various locations and parts of the section.

Rocks of the lower part of the Pocahontas Formation were deposited on the front of a delta system that overrode the coastal sediments of the Bluestone Formation. The upper part of the Pocahontas Formation was deposited farther back on the delta plain.

Pennsylvanian New River Formation. The base of this formation is traditionally placed at the base of the No. 8 Pocahontas coal (Hennen and Gawthrop, 1915). Basin-wide, in out-crop areas, this contact is conformable to the northwest until it passes below drainage where it is reported as an erosional contact at the base of the Pineville Sandstone (Englund, 1979). To the northwest, Miller (1974) and Englund and others (1977, 1979) have reported that rock units disappear until the entire Pocahontas and much of the underlying Mauch Chunk are gone. This erosional contact is reported to form the Mississippian-Pennsylvanian boundary in the subsurface.

The New River Formation is composed of argillaceous and clean sandstones, siltstones, shales, conglomerates, underclays, and coals. The clean-washed, mature, cross-bedded sandstones and conglomerates of the New River Formation distinguish it from the Pocahontas Formation. Quartzarenites contrast with the argillaceous subgreywacke sandstones of the Pocahontas and the overlying Kanawha Formations. These New River sandstone units, including the Raleigh, Guyandotte, and Nuttall, form the plateau area that surrounds the section. The thick gray shales of the formation are nearly as prominent as the sandstones in road-cuts and excavations.

Both the Fire Creek and Sewell coals are minable at this location. West of the section the Beckley coal reaches minable thickness. Normally the interval between the Beckley and Fire Creek seams is about 100 ft (30 m). Low-to-medium-volatile coals of the New River Formation average less than 1 percent sulfur, less than 6 percent ash, and around 14,400 Btu per pound.

Well preserved, abundant plant fossils are found in the New River Formation. Invertebrate fossils of the New River Formation are sparse and consist of pelecypods, *Lingula,* and other inarticulate brachiopods. Marine fossils have been reported in a few locations near the base of the New River Formation many miles (km) southwest of this section (Henry and Gordon, 1979).

Maps of coal mines that cover large areas show seam discontinuities of dendritic, carbonaceous, shale-filled channels with sinuous tributaries. Unlike similar channels in the Pocahontas, these show a preferred orientation trending to the northwest. Some of these same maps also show small, short washouts filled with sandstone.

The New River Formation is dominated by thick sequences of sandstones that are quartzose to the northwest and argillaceous to the southeast. These sandstones appear as barrier islands or bars. The coals were deposited in back-barrier swamps or coastal-plain swamps on the mainland. The prominent gray shales were deposited in quiet lagoons and bays.

REFERENCES CITED

Englund, K. J., 1979, Mississippian System and lower series of the Pennsylvanian System in the proposed Pennsylvanian System stratotype area, *in* Englund, K. J., Arndt, H. H., and Henry, T. W. (eds.)., Proposed Pennsylvanian stratotype, Virginia and West Virginia: Guidebook for IX-ICC Field Trip no. 1, American Geological Institute Selected Guidebook series no. 1, p. 69–72.

Englund, K. J., Arndt, H. H., Gillespie, W. H., Henry, T. W., and Pfeffercorn, H. W., 1977, A field guide to proposed Pennsylvanian System stratotype, West Virginia: American Association of Petroleum Geologists/Society of Economic Paleontologists and Minerologists Convention 1977 pre-meeting field trip guide, 80 p.

Englund, K. J., Cecil, C., and Stricker, G. D., 1979, Depositional environments of the Mississippian System and lower Pennsylvanian series in proposed Pennsylvanian stratotype area, *in* Englund, K. J., Arndt, H. H., and Henry, T. W. (eds.), Proposed Pennsylvanian stratotype, Virginia and West Virginia: Guidebook for IX-ICC Field Trip no. 1, American Geological Institute Selected Guidebook Series no. 1, p. 113–114.

Hennen, R. V., and Gawthrop, R. M., 1915, Wyoming and McDowell Counties: West Virginia Geological and Economic Survey (County Report), 783 p.

Henry, T. W., and Gordon, M., 1979, Late Devonian through early Permian(?) invertebrate faunas in proposed Pennsylvanian system stratotype area, *in* Englund, K. J., Arndt, H. H., and Henry, T. W. (eds.), Proposed Pennsylvanian stratotype, Virginia and West Virginia: Guidebook for IX-ICC Field Trip no. 1, American Geological Institute Selected Guidebook series no. 1, p. 97–103.

Kulander, B. R., and Dean, S. L., 1980, Basement structure relationships to stress history, fracture domains and décollement tectonics in southern West Virginia [abs.]: Program and abstracts of the eleventh annual Appalachian Petroleum Geology Symposium, West Virginia Geological and Economic Survey, Circular no. C-16, p. 6.

Miller, M. S., 1974, Stratigraphy and coal beds of upper Mississippian and lower Pennsylvanian rocks in southwestern Virginia: Virginia Division of Mineral Resources Bulletin 84, 211 p.

Appalachian Valley and Ridge to Appalachian Plateau transition zone in southwestern Virginia and eastern West Virginia: Structure and sedimentology

Robert C. Whisonant, Department of Geology, Radford University, Radford, Virginia 24142
Arthur P. Schultz*, Virginia Division of Mineral Resources, Charlottesville, Virginia 22903

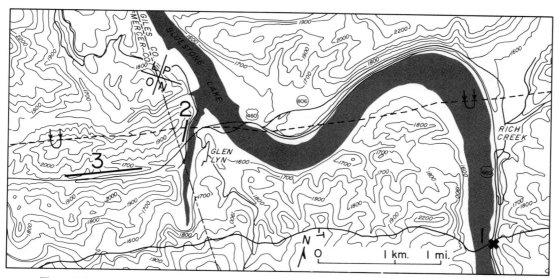

Figure 1. Location map to the Appalachian Valley and Ridge-Plateau Transition Zone in southwestern Virginia showing Stops *1, 2,* and *3.* Trace of St. Clair Thrust (*solid line with tick on hanging wall*) and axis of Glen Lyn Syncline (*dashed line with overturned syncline symbol*) are shown. Corners of 7½-minute quadrangles are: *N,* Narrows; *O,* Oakvale; *L,* Lerona; and *P,* Petertown.

LOCATION

The Appalachian Valley and Ridge to Appalachian Plateau transition is exposed in a series of road cuts located in Giles County, Virginia, and Mercer County, West Virginia, and include portions of the Narrows, Oakvale, Lerona, and Peterstown 7½-minute Quadrangles (Fig. 1). The exposures may be reached via U.S. 460 approximately 15 mi (24 km) east of Princeton, West Virginia, or 2 mi (3 km) north of Narrows, Virginia.

INTRODUCTION

The St. Clair–Clinchport Thrust exposed at Stop 1 is the northwesternmost of several major southeast-dipping thrusts in the Valley and Ridge and the Appalachian Plateau Provinces (Figs. 2, 3; Milici, 1973; Harris and Milici, 1977). Associated with the St. Clair Thrust is the Glen Lyn Syncline, an important sub-thrust fold clearly exposed at Stop 2 that marks the boundary between the folded rocks of the Valley and Ridge and the nearly flat-lying strata of the Plateau (Fig. 3; McDowell, 1982). Stops 1 and 2, therefore, show two of the most significant regional structural features in this part of the Southern Appalachians.

Stop 3 exposes the middle portion of the Middle Mississippian Bluefield Formation. The facies within this sequence reflect an important change in Appalachian Paleozoic sedimentation. The strata at Stop 3 are part of the transition from the final stages of carbonate deposition in this part of the Appalachians to the initial phases of Alleghanian orogenic terrigenous deposition (Whisonant and Scolaro, 1980). A great number and variety of primary sedimentary features are present at Stop 3 and offer opportunities for detailed paleoenvironmental analysis of a mixed carbonate-siliciclastic sequence.

SITE DESCRIPTION

Stop 1. St. Clair Thrust

Stop 1 is found along U.S. 460 approximately 1 mi (1.6 km) south of Rich Creek, Virginia (Fig. 1). The stop must be approached from the northbound lane of U.S. 460. Park at the gravel pull-off on the east side of the highway about 300 yds (274 m) north of the exposure. Walk, single file, back to the exposure. *NOTE:* Extreme caution must be used at all times to

*Present address: U.S. Geological Survey, Reston, Virginia 22070

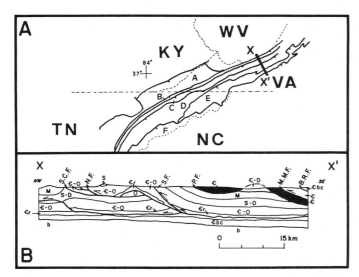

Figure 2. *a*, tectonic Index Map (Harris and Milici, 1977) and *b*, regional cross section *X-X'* (Milici, 1973). Thrust sheets are: *A*, Cumberland; *B*, St. Clair–Clinchport; *C*, Narrows–Copper Creek; *D*, Saltville; *E*, Pulaski–Max Meadows; *F*, Blue Ridge. Cross section symbols: *S.C.F.*, St. Clair Fault; *N.F.*, Narrows Fault; *S.F.*, Saltville Fault; *P.F.*, Pulaski Fault; *M.M.F.*, Max Meadows Fault; *B.R.F.*, Blue Ridge Fault; *M*, Mississippian; *S-D*; Silurian through Devonian; *€-O*, Cambrian through Ordovician; *€r*, Cambrian Rome Formation; *€bc*, Cambrian basal clastics; *b*, basement rocks. Saltville and Blue Ridge faults are covered with Coastal Plain sediments in Alabama; Pulaski and Blue Ridge faults continue into the Central Appalachians; St. Clair, Saltville and Narrows faults terminate at Southern–Central Appalachian boundary.

keep *off* this very fast, busy, and dangerous highway. For larger groups, we suggest a flagman on the side of the road to the south, to warn oncoming drivers.

Exposed at Stop 1 is the St. Clair Thrust, one of several major thrust faults of the Southern Appalachians (Fig. 2a, b). The fault is considered part of a regional structural boundary, the Appalachian structural front (Price, 1931) or the Alleghanian front (Rodgers, 1964). At Stop 1 (Figs. 1–4) dolomites of the Ordovician Beekmantown Formation (upper Knox Group) are thrust onto Devonian shales, siltstones and sandstones, a stratigraphic throw of 5,000–6,000 feet (1,524–1,828 m) (Butts, 1933). Recent structural models (Perry et al., 1979; Milici, 1973) attribute up to 18 mi (29 km) of horizontal displacement on the St. Clair Thrust in the study area. The fault (Fig. 4) is unusually well exposed at Stop 1 and dips 30°–35° to the southeast. The fault contact is sharp and deformation in the hanging wall dolomites is fairly inconspicuous, consisting of very minor, small-scale faulting and well-developed jointing. A thin (1–3 in [2.5–7.6 cm]) zone of brecciated dolomite and carbonate gouge occurs along the contact surface. Deformation in the footwall Devonian shales is more pronounced and consists of mesoscopic folding and faulting and cleavage development near the thrust contact. The cleavage and bedding in the Devonian shale dip approximately parallel to the fault contact. Bedding is overturned to the southeast and is part of the southeast overturned limb of the Glen Lyn Syncline.

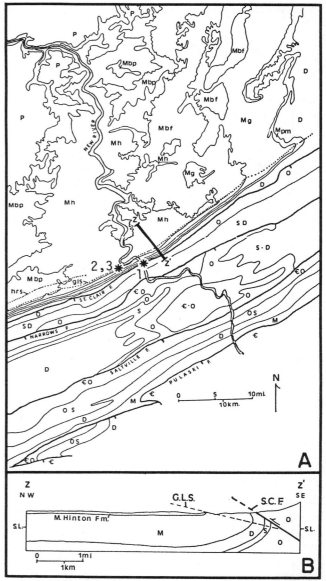

Figure 3. *a*, Geologic Map (Butts, 1933; Cardwell and others, 1968) and *b*, cross section *Z-Z'* (McDowell, 1982). Symbols same as in Figure 2 with the addition of *Mpm*, Mississippian Price and Maccrady Formations; *Mg*, Mississippian Greenbrier Formation; *Mbf*, Mississippian Bluefield Formation; *Mh*, Mississippian Hinton Formation; *Mbp*, Mississippian Bluestone and Princeton formations; *P*, Pennsylvanian. Cross section symbols: *G.L.S.*, Glen Lyn Syncline with axis (dashed line); *S.C.F.*, St. Clair Fault. Note contrast in the outcrop pattern across the structural front where western boundary is defined by the trace of the axis of the Glen Lyn Syncline (*dotted line* with *gls*). Axis of the Hurricane Ridge Syncline (*dotted line* with *hrs*) is just to the northwest.

Stop 2. Glen Lyn Syncline

Stop 2 is located near Glen Lyn, Virginia, and is approached by turning north off U.S. 460 onto Virginia 648 near the west end of the East River Bridge (Fig. 1). Park on the open area to the right just after the turn and walk north along Virginia 648 about 600 ft (183 m).

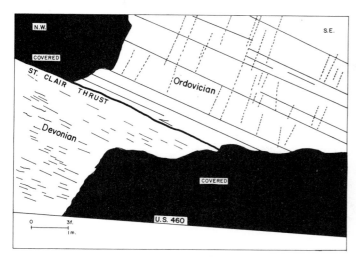

Figure 4. Outcrop sketch from photograph of the St. Clair Thrust fault (Stop 1) exposed along the northbound lane of U.S. 460. Trace of bedding and cleavage is shown in the Devonian (footwall) shales; bedding and joints (*dashed lines*) are shown in the Ordovician dolomites of the hanging wall.

The roadcut at Stop 2 exposes the Glen Lyn Syncline (McDowell, 1982). This is a regional-scale structure associated with the Appalachian structural front and the St. Clair Thrust. Earlier workers (Campbell, 1896; Reger, 1926; Cooper, 1961, 1971) refer to the Hurricane Ridge Syncline as the major sub-thrust structure whose axis defines the boundary between the Valley and Ridge structural province and the Appalachian Plateau structural province. Recent work (McDowell, 1982) has shown that the Hurricane Ridge Syncline is actually three separate structures of differing form, trend, and tectonic significance. The synclinal axis seen at Stop 2 is that of the Glen Lyn Syncline (Fig. 5). Its axial trace (Fig. 3a, b) parallels the trace of the St. Clair Fault and its southeast limb is overturned to the southeast. The axis of the Glen Lyn Syncline separates the thrust-faulted and regionally folded rocks typical of the Valley and Ridge structural province from relatively flat-lying, undeformed rocks of the Appalachian Plateau Province. This change is evident by comparing outcrop patterns across the structural front (Fig. 3a, b).

The rocks at Stop 2 (Fig. 5) are sandstones, siltstones, and shales of the Upper Mississippian Hinton Formation. Several minor structures and sedimentary features, including a normal fault with associated drag folding, coal with underclay, and upside-down mudcracks, can be seen in the rocks near the syncli- nal axis.

Stop 3. Sedimentological Features in the Bluefield Formation

Stop 3 is located along U.S. 460 about 0.5 mi (1 km) west of the Virginia–West Virginia state line (Fig. 1). Ample parking is available along the wide shoulder of U.S. 460.

Approximately 200 ft (61 m) of strata located in the

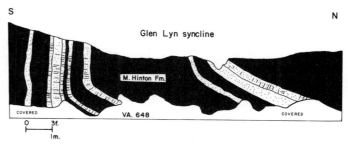

Figure 5. Axial part of the Glen Lyn Syncline exposed at Stop 2. Rocks of the Valley and Ridge are to the south, rocks of the Appalachian Plateau are to the north. *Shaded areas* are shales and siltstones; *dot pattern* indicates siltstones and sandstones; closely-spaced joints in the sandstones and siltstones are shown as *short solid lines*.

middle portion of the Bluefield Formation (Mississippian–Middle Chester) are well-exposed at this stop, except for the basal 20 ft (6.1 m) which are partially covered (Fig. 6). The section lies above predominantly carbonate units (Greenbrier Limestone and lower Bluefield) and below primarily terrigenous material (upper Blue-field and Hinton-Princeton-Bluestone Formations). Therefore, the middle Bluefield, a complex of interbedded limestone and silici-clastic layers, records the Late Paleozoic change in sedimentary regime from carbonate to terrigenous conditions in this region of the Southern Appalachians (Whisonant and Scolaro, 1980).

The strata at Stop 3 (Fig. 1) are in the overturned, southeast limb of the Glen Lyn Syncline and strike north 70 degrees east and dip 60 degrees southeast. At Stop 3, large areas of the under-sides of beds as well as numerous cross-sectional views of layering are well exposed, revealing a great variety of easily observable primary sedimentary structures. Fossils are abundant in many parts of the section. Most of the discussion below is taken from Whisonant and Scolaro (1980) who described the sedimentologi-cal features of this exposure in great detail; Humphreville (1981) analyzed the paleoecology of the local fauna.

The sedimentological features at Stop 3 are best observed and understood in terms of an analysis of the various depositional environments represented (shown in lateral relationship in Figure 7). The lowest strata are poorly exposed, but can be seen near the end of the concrete drainage ditch at the west end of the cut. These layers consist of approximately 9 ft (2.7 m) of medium- to thick-bedded ooidal and skeletal grainstones-packstones displaying uneven erosional bases (Fig. 6). Fragments of pelmatozoans and bryozoans are especially abundant. Strata of this type (designated Facies 1 in Figure 6) are interpreted as having formed in medium- to high-energy ooid shoals and bars above wave base (Whisonant and Scolaro, 1980).

Above the lowest Facies 1 strata are about 30 ft (9.1 m) of very thin-bedded to laminated calcareous shales and argillaceous wackestones-mudstones. The beding generally displays horizontal parallel laminations but some ripple bedding is also present. Gi-gantic blastoid calices (up to 2.4 in [6.1 cm] in height) and delicately preserved *Archimedes* colonies (some with fronds still attached) may be seen. These attached fenestrates are not found

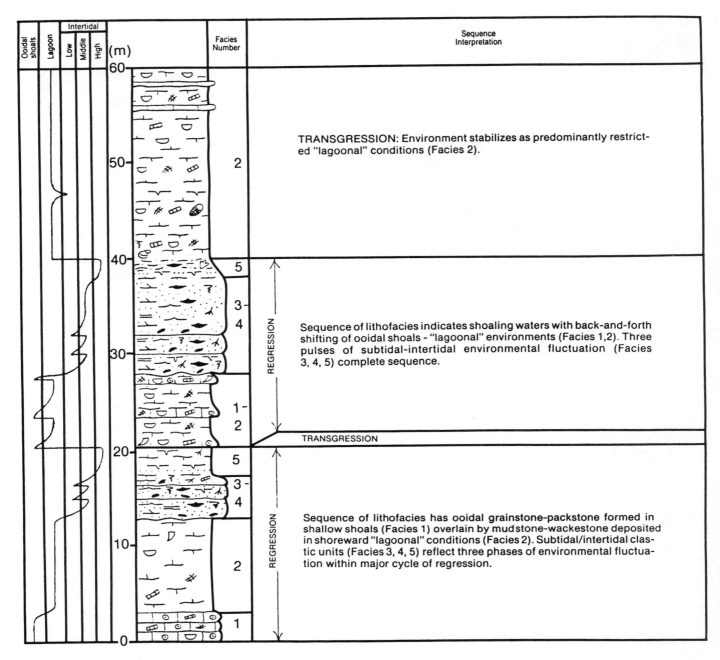

Figure 6. Section at Stop 3 showing fluctuating shore-related environments superimposed on two major cycles of regression.

in growth position but apparently fell over and came to rest on the soft muddy substrate with minimum damage to the colony. Brachiopods and solitary corals are common in this interval. Higher in the section (especially above the upper sandstone, Fig. 6), brachiopods are generally the principal fossils to be found. Pavements of well-preserved productid brachiopods, which Humphreville (1981) interpreted as storm generated, are common in the higher beds. These dark-colored mudstones-wackestones (Facies 2, Fig. 6) are thought to have originated in calm, protected waters ("lagoonal" or back-barrier) shoreward of the ooidal environments (Whisonant and Scolaro, 1980).

Above the lower muddy beds is the first of two major quartz sandstone units (Fig. 6). The arenaceous layers are thick- to very thin-bedded, display channeled bases, and consist of fine- to very fine-grained silty micaceous sandstones containing 15% to 25% matrix and minor calcite cement. Two distinct lithofacies are represented within the sandstones (Fig. 6). Facies 3 beds are medium- to thick-bedded with parallel laminations common, both horizontal and gently inclined. Cross-bedding is present, most typically as small-scale trough cosets, but some large scale trough sets also occur. Bimodal-bipolar current patterns typify the cross-bedded strata. Soft-pebble conglomerates, transported large plant fragments, and abundant sole marks, including groove casts, drag and bounce marks, and brush marks also occur. Other primary structures include ripple marks (both straight-crested and linguoid types with some mud-filling of the ripple troughs), parting lineation, current crescents, load casts, and some burrows and trails. In the lower sandstone, small dune forms and an exceptionally clear set of rhomboidal rill marks (formed by beach run-off) are found in beds of this facies. The sediments of Facies 3 are interpreted as shallow subtidal to lower intertidal in origin (Whisonant and Scolaro, 1980). The basal units of the two major sand cycles have scoured contacts and abundant sole marks and soft-pebble conglomerates, suggesting initial deposition in tidal channels. These environments change upward into subtidal to lower intertidal sand-dominated flats.

Facies 4 sediments (Fig. 6) are found in both lower and upper sandstones. These strata are thin- to very thin-bedded silty sandstones. Parallel laminations, both horizontal and gently inclined, are common. Cross-stratification is present (again, bimodal-bipolar patterns are typical) and consists of cross-laminated trough foresets, some of which show a herringbone aspect. Facies 4 beds resemble the sandstones of Facies 3 but are distinguished by the following criteria: (a) a far greater abundance of ripple forms; (b) flaser beds, including ripple troughs filled with mud or finely-macerated plant fragments; (c) double-crested and flat-topped ripple marks; and (d) an abundance of both vertical burrows and horizontal trails, but with burrows generally predominating. In addition, desiccation cracks (seen as very well-developed curled platelets near the top of the lower sandstone), load casts (most typically squamiform), sole marks (groove casts),

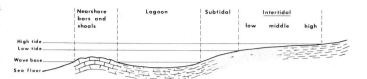

Figure 7. Schematic diagram of inferred depositional environments represented in the middle Bluefield Formation.

parting lineation, current crescents, and strongly current-oriented ("drifted-in") plant fragments are abundant.

In the upper sandstone, one occurrence each of a pseudonodule (hydroplastic failure features found in water-saturated intertidal sediments [Klein, 1977]) and runzel (wrinkle) marks may be seen. The runzel marks are especially interesting features that superficially resemble squamiform load casts but apparently result from shear stresses transmitted to muddy sediment through thin sheets of water (less than 0.5-in 1.3 cm] deep) by wind blowing over tidal flat environments (Reineck, 1969). The sediments of Facies 4 probably formed in a lower to middle intertidal setting (Whisonant and Scolaro, 1980).

Immediately above each of the sandstones are very thin-bedded to laminated calcareous shales and siltstones that constitute a separate facies (Facies 5, Fig. 6); a few thin sandstones and limestones are included here also. Primary sedimentary structures in these layers are the least well-developed (particularly above the lower sandstone) of the three clastic facies. The most diagnostic structures of this facies are lenticular to wavy bedding (as defined by Reineck and Singh, 1980) and couplets of fine rhythmites (pinstripe "tidal" bedding of Reineck, 1967; Wunderlich, 1970). Abundant mudcracks, a general predominance of horizontal trails over vertical burrows, flaser bedding, and ripple marks are typical of this facies. Groove casts and macerated plant fragments occur in these strata.

Humphreville (1981) found tetrapod tracks in this facies above the higher sandstone; these tracks are older than any terrestrial vertebrate fossils yet found in the Southern Appalachians. Sediments of Facies 5 are thought to have formed in a mud-dominated middle to higher intertidal environment (Whisonant and Scolaro, 1980).

The facies analysis developed above and depicted in Figure 6 shows that two major regressions controlled the formation of this section of the Bluefield. Smaller-scale facies shifts are evident within the two shoaling-upward cycles. The small scale changes probably reflect local variations in sediment supply, wave and current energy, or other basin conditions. The two principal regressive sequences must result from more regional factors, such as changes in rates of uplift or subsidence. The possibility of eustatic sea-level changes associated with Carboniferous Gondwana glacial activity as a cause of the cyclical sedimentation in this portion of the Bluefield should be considered also.

REFERENCES CITED

Butts, Charles, 1933, Geologic map of the Appalachian Valley in Virginia with explanatory text: Virginia Geological Survey Bulletin 42, 568 p.

Campbell, M. R., 1896, Description of the Pocahontas sheet (Va.–W.Va.): U.S. Geological Survey Geologic Atlas: Folio 26.

Cardwell, D. H., Erwin, R. B., and Woodward, H. P., 1968, Geologic Map of West Virginia: West Virginia Geological and Economic Survey.

Cooper, B. N., 1961, Grand Appalachian Field Excursion: Geologic guidebook No. 1, The Virginia Polytechnic Institute Engineering Experiment Series, 187 p.

—— 1971, Appalachian structural and topographic front between Narrows and Beckley, Virginia and West Virginia: Guidebook to Appalachian tectonics and sulfide mineralization of southwestern Virginia, Guidebook No. 5, Virginia Polytechnic Institute and State University, Field Trip No. 3: p. 87–142.

Harris, L. D., and Milici, R. C., 1977, Characteristics of thin-skinned style of deformation in the Southern Appalachians, and potential hydrocarbon traps: U.S. Geological Survey Professional Paper 1018, 40 p.

Humphreville, R. G., 1981, Stratigraphy and paleoecology of the Upper Mississippian Bluefield Formation [M.S. thesis]: Virginia Polytechnic Institute and State University, Blacksburg, 365 p.

Klein, G. D., 1977, Clastic tidal facies: Continuing Education Publishing Company, Champaign, Illinois, 149 p.

McDowell, R. C., 1982, The Hurricane Ridge, Glen Lyn, and Caldwell Synclines of southeastern West Virginia and southwestern Virginia: a reinterpretation: Southeastern Geology, v. 23, p. 83–88.

Milici, R. C., 1973, Tectonics of the Southern Appalachian Cumberland Plateau and adjacent parts of the Valley and Ridge—a review: Geological Society of America Abstracts with Programs, v. 5, p. 197.

Perry, W. J., Jr., Harris, A. G., and Harris, L. D., 1979, Conodont-based reinterpretation of Bane Dome—a structural re-evaluation of Alleghany Frontal zone: American Association of Petroleum Geologists Bulletin, v. 63, p. 647–675.

Price, P. H., 1931, The Appalachian structural front: Journal of Geology, v. 39, p. 24–44.

Reger, D. B., 1926, Mercer, Monroe, and Summers Counties: County Report, West Virginia Geological and Economic Survey, 963 p.

Reineck, H. W., 1967, Layered sediments of tidal flats, beaches, and shelf bottoms of the North Sea, *in* Lauff, G. H., ed., Estuaries: American Association for the Advancement of Science Publication 83, p. 191–206.

—— 1969, Die Enstehung von Runzelmarken: Natur and Museum, v. 99, p. 386–388.

Reineck, H. W., and Singh, I. B., 1980, Depositional sedimentary environments: Springer-Verlag, New York, 549 p.

Rodgers, John, 1964, Basement and no-basement hypotheses in the Jura and the Appalachian Valley and Ridge, *in* Lowry, W. D., ed., Tectonics of the Southern Appalachians: Virginia Polytechnic Institute Department of Geological Sciences Memoir 1, p. 71–80.

Whisonant, R. C., and Scolaro, R. J., 1980, Tide-dominated coastal environments in the Mississippian Bluefield Formation of eastern West Virginia, *in* Tanner, W. F., ed., Shorelines, past and present: Proceedings, fifth symposium on coastal sedimentology, Tallahassee, Department of Geology, Florida State University: p. 593–637.

Wunderlich, Friedrich, 1970, Genesis and development of the "Nellenköpfenschichten" (Lower Emsian, Rheinian Devonian) at locus typicus in comparison with modern coastal environment of the German Bay: Journal of Sedimentary Petrology, v. 40, p. 102–130.

Max Meadows tectonic breccia at Pepper, Virginia

Arthur P. Schultz*, Virginia Division of Mineral Resources, Charlottesville, Virginia 22903

LOCATION

Outcrops of extensive Max Meadows tectonic breccia are along the bluffs overlooking New River at Pepper, Virginia (Fig. 1). A few feet (m) northeast of the New River bridge on Virginia 114 turn north into a dirt road. Park in the wide area just onto the dirt road and do not block the driveway to the private house. Do not drive to the outcrops on the dirt road which is a railroad right of way. Walk northeast about 1,000 ft (300 m) to the railroad tracks along the dirt road.

INTRODUCTION

The regionally extensive and spectacular Max Meadows tectonic breccia (Cooper and Haff, 1940; Cooper, 1946, 1970; Schultz, 1983) are unique not only when compared with fault breccias on other thrusts in the Appalachians (Rodgers, 1970) but also with breccias in other thrust belts of the world (Gretener, 1977). The excellent outcrops at Pepper, Virginia, are both historically and scientifically important.

The breccias here have been studied since the late 1800's by geologists, who through the years have proposed several theories on their genesis. Hundreds of geologists have visited the outcrops and have made their own interpretations. Recent work (Bartholomew and Lowry, 1979; Schultz, 1983) has interpreted the breccias in the regional tectonic framework of the Pulaski thrust sheet, a major fault in the Southern Appalachian Valley and Ridge Province.

The Pepper, Virginia locality has exposures of the Pulaski fault surface, extensive outcrops of Max Meadows breccia and outcrops with folded, faulted and cleaved dolomite. The section of the Elbrook Dolomite (Cambrian) is one of the best exposures in the Valley and Ridge of Virginia.

The Pulaski thrust is one of several major southeast-dipping Alleghanian thrusts of the Southern and Central Appalachians (Fig. 2). It has been traced along strike approximately 310 mi (500 km; Cooper, 1970; Rodgers, 1970) from near Staunton, Virginia southward into Tennessee where it is overridden by rocks of the Blue Ridge thrust sheet. Estimates of the displacement of the thrust near the study area range from 9.3 mi (15 km; Cooper, 1970; Lowry, 1971) to 31.3 mi (50 km; Bartholomew, 1979). Near Pulaski, Virginia, Cambrian rocks are thrust over rocks of Mississippian age (Figs. 2, 3).

The Pulaski thrust sheet in the study area has undergone a two-stage deformation sequence during the Alleghanian orogeny. The first stage of deformation involved decollement and ramp thrusting with final emplacement of the Pulaski thrust sheet on a Mississippian level decollement (Saltville thrust-sheet rocks). The second stage involved regional scale, open folding, and thrust

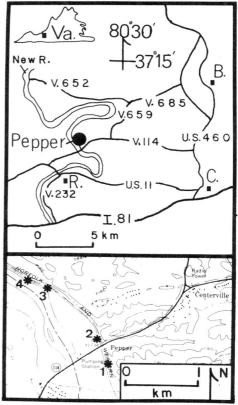

Figure 1. Highway map showing Interstate (*I*), United States (*U.S.*) and Virginia (*V*) highways in the vicinity of the Max Meadows breccia field stop at Pepper, Virginia (black circle symbol on New River). Black squares are major towns: *B*, Blacksburg; *C*, Christiansburg; *R*, Radford. Topographic map (bottom) of Pepper, Virginia shows location of Localities *1–4* described in the text (see Fig. 4). Map is part of Radford North Quadrangle.

faulting of both the Pulaski and Saltville thrust sheets. Conodont color alteration indices (Epstein and others, 1977) indicate that rocks of the Pulaski thrust sheet were subjected to temperatures in the range of 300°–400°C as the result of either initial burial or tectonic burial.

In the study area (Figs. 2–4) the Pulaski sheet consists of two distinctively different lithotectonic units. Rocks of the lower part of the sheet are chiefly laminated and thinly-bedded dolomites, argillaceous dolomites, and calcareous mudstones, alternating with massive dolomites and limestones of the Rome Formation and the lower part of the Elbrook Formation. The upper part of the Pulaski sheet consists of massive dolomites and thin- to thick-bedded limestones of the middle and upper part of the Elbrook Formation and the Conococheague Formation.

Broken formations in the lower portions of the Pulaski thrust sheet, near Pulaski, Virginia, contain some of the most strongly deformed sedimentary rocks in the foreland fold and

*Present address: U.S. Geological Survey, M.S. 955, Reston, Virginia 22092.

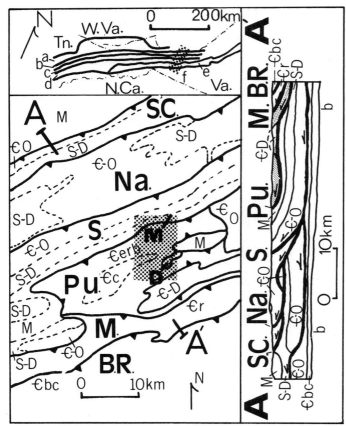

Figure 2. Regional geologic maps and cross section. (*1*), State and tectonic index map (Harris and Milici, 1977) showing the major thrust sheets of the Southern Appalachian Valley and Ridge Province: *a,* Cumberland; *b,* Clinchport–St. Clair; *c,* Narrows–Copper Creek; *d,* Saltville; *e,* Pulaski–Max Meadows; *f,* Blue Ridge. Shaded area is figure 2-2. (*2*) Regional geologic map (Butts, 1933) shows major southeast dipping thrust faults and distribution of Paleozoic rocks near the study area. Shaded area is figure 3. Bold letters: *S.C.,* Saint Clair thrust sheet; *Na.,* Narrows thrust sheet; *S.,* Saltville thrust sheet; *Pu.,* Pulaski thrust sheet; *M.,* Max Meadows thrust sheet; *B.R.,* Blue Ridge thrust sheet. Stratigraphic units are: *M,* Mississippian; *S-D,* Silurian through Devonian; *C-O,* Cambrian through Ordovician; *Cc,* Cambrian Conococheague Formation; *Cerb,* Cambrian Elkbrook and Rome formations and Max Meadows breccia; *Cr,* Cambrian Rome Formation; *Cbc,* Cambrian basal clastics. (*3*), Structure section along *A–A'.* Symbols same as geologic map except for *b,* basement rocks. (Figure modified from Lowry [1979]).

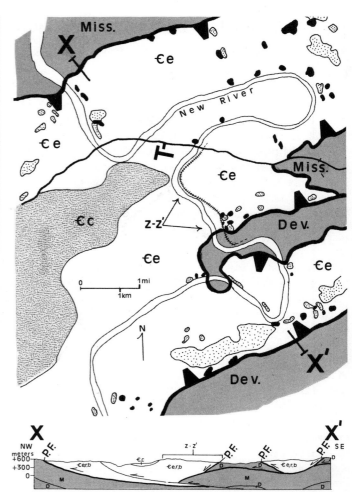

Figure 3. Geologic map (Schultz, 1983) of Pepper, Virginia, area showing rocks of the Saltville thrust sheet (light gray shading): *Miss.,* Mississippian; *Dev.,* Devonian; overlain by rocks of the Pulaski thrust sheet: *Cc,* Cambrian Conococheague Formation; *Ce,* Cambrian Elbrook Formation; *dot pattern,* Cambrian Rome Formation; *black shading,* Max Meadows breccia. Note position of section *Z–Z'* (fig. 4). Cross section *X–X'* (Schultz, 1983) shows: *Z–Z',* outcrop section (fig. 4) at Pepper, Virginia in relation to local geologic setting; *P.F.,* Pulaski fault; *Miss.,* Mississippian; *Dev.,* Devonian; *C,* Cambrian Conococheague, Elbrook and Rome formations and Max Meadows breccia.

thrust belt of the Appalachians (Schultz, 1983, 1984). These broken formations consist of lesser amounts of folded and faulted mudstones and carbonates of the Rome Formation and the Max Meadows breccia in a complexly folded and faulted terrain of carbonates of the Elbrook Formation (Fig. 3). Deformation in the broken formations ranges from grain-scale cataclasis to pervasive macroscopic faulting. The broken formations of the Pulaski thrust sheet are unique in their thickness (980 to 1,640 ft) [300–500 m]), degree of rock disruption, and their association with the extensive and spectacular Max Meadows tectonic breccia (Cooper and Haff, 1940; Cooper, 1970; Rodgers, 1970; Schultz, 1983).

The tectonic breccias were first interpreted (Campbell, 1894) as marine sedimentary conglomerates indicative of early Paleozoic unconformities. Subsequent interpretations (Campbell, 1925) include a sedimentary conglomerate of recent origin and a contact-spring calcareous tufa deposit. The tectonic nature of the breccias was first recognized by Cooper (1946, 1970; see also Cooper and Haff, 1940) who showed that they were generated by cataclastic deformation (brittle fracture) associated with the emplacement of the Pulaski and Max Meadows faults. Cooper (1970) defined two "facies" of breccia, autobreccia and tectonic breccia, which are two end-member types of breccia in a continuous series of cataclastically deformed dolomite. Autobreccia consists of highly fractured, thin-bedded dolomite with literally thousands of intersecting extension fractures. Although the bedding has been extended in several directions, the "clasts" have not been rotated and the original sedimentary layering is preserved. In contrast, tectonic breccia consists of clasts that have undergone rigid body rotation and size reduction within a continuously deforming matrix of finely-ground dolomite.

From 1950 to the present, many workers have studied the Max Meadows tectonic breccias as part of detailed mapping in the Valley and Ridge of Virginia and Tennessee. This work has shown that the breccias are found on the Pulaski thrust sheet along its entire strike length, some 310 mi (500 km). Most recent (Bartholomew and Lowry, 1979; Schultz, 1983, 1984) detailed 7½-minute quadrangle mapping (Fig. 3) has shown that the breccia is an integral part of the highly deformed rocks near the base of the Pulaski thrust sheet. Tectonic breccia occurs well above the basal Pulaski fault surface (up to 990 ft [300 m] above) but decreases in abundance away from this contact, where breccia zones are less common and thinner. Tectonic breccias occur either as sill-like bodies (bedding-parallel) that are folded by macro- and mesoscopic structures or as dike-like bodies which truncate bedding (Fig. 4). Sills and dikes of breccia range from less than an inch to several hundreds of feet (6 cm to tens of m) thick. Map patterns of the breccia (Fig. 3) are irregular shaped, isolated masses of various size.

Recent tectonic models (Schultz, 1983, 1984) for breccia formation agree well with those of Cooper (1970). These models suggest that the Max Meadows tectonic breccias were generated by pervasive fracturing and continued cataclastic deformation, i.e., rigid body rotation and comminution of dolomite clasts in a continuously deforming matrix of crushed dolomite. This deformation occurred along the basal thrust surface and along numerous faults in the broken formations at the base of the Pulaski thrust sheet during the major Alleghanian thrusting episode. During subsequent and continued Alleghanian deformation, breccia was mobilized, perhaps under elevated fluid pressures (Schultz, 1983) to form dikes and sills of irregular shapes.

SITE DESCRIPTION

The outcrops can be seen by walking along the gravel road that parallels the tracks. Locality 1 is southeast of Virginia 114; Localities 2, 3, and 4 are northwest of the highway. It is suggested that you follow the outcrops in numerical sequence. Do not walk on the railroad tracks. This is Norfolk and Southern Railway property, so please do not climb fences along the cliffs, or disturb any railway equipment. Keep ears and eyes open for trains at all times.

Walking from Locality 1 to Locality 4 (Fig. 4), a typical vertical section through the lower part of the Pulaski thrust sheet can be seen. The 4 Localities are on the northwest limb of a major fold in the thrust sheet. Localities 1 and 2 show typical relations in the broken formations at the base of the Pulaski thrust sheet. Between Localities 2 and 3, the transition from broken formations to the upper less deformed parts of the Pulaski thrust

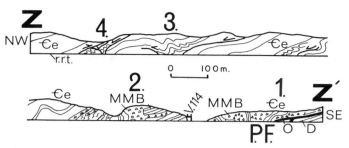

Figure 4. Outcrop section *Z–Z'* (Schultz, 1983) at Pepper, Virginia (note Virginia 114 bridge which crosses New River at Pepper): *MMB*, Max Meadows breccia; *Ce*, Cambrian Elbrook Formation; *O*, Ordovician Knox Group(?) dolomite; *D*, Devonian; *P.F.,* Pulaski fault; *r.r.t.,* Railroad track. Numbers *1–4* refer to Localities described in the text.

sheet occurs. The structurally higher rocks of Localities 3 and 4 are less deformed and folding and thrusting can be related to movement on decollements below the thin zones of deformation. Walking northwest beyond Locality 4 reveals relatively undeformed rocks of the higher parts of the Pulaski thrust sheet and one can walk to the axis of the large scale fold approximately one-half mile northwest of Locality 4 on the railroad tracks. The entire section of the Cambrian Elbrook Formation can be seen from Locality 1 to the fold axis beyond Locality 4.

Locality 1. The Pulaski fault surface is well exposed at this Locality (Fig. 4). Thick masses of Max Meadows breccia are present above the fault. The breccia (Fig. 4) consists of poorly sorted, angular to subrounded clasts of massive dolomites, laminated dolomites, and finely laminated greenish-gray calcareous mudstones in a fine- to very fine-grained matrix of crushed dolomite. Clasts range from less than one inch to more than 3 feet (cm to >1 m) long. The breccias are massive to crudely layered and are well to poorly indurated. Breccia is finest grained along the fault contact. Note that at this Locality (Fig. 4) a small footwall-derived horse block of Ordovician Knox (?) Dolomite is present between the tectonic breccias of the hanging wall of the Pulaski thrust and the Devonian Millboro Shale of the footwall. The block is internally deformed and much of the original bedding has been obliterated during thrusting. The Devonian shales are strongly cleaved and faulted. Above the fault surface, large blocks of dolomite are engulfed in breccia. These blocks are fractured, veined, and faulted. This is one of the few places that the actual fault surface is exposed. The fault dips to the northwest, indicating it was folded after emplacement.

Locality 2. An excellent exposure of the Max Meadows breccia can be observed at Locality 2 (Fig. 4). There, a large but typical dike of breccia (Fig. 5) has intruded massive, gently-dipping dolomites of the Elbrook Formation. This outcrop (fig. 4) was first described by Campbell (1894) and shows the typical occurrence of breccia found along the entire Pulaski thrust sheet.

At the southeast margin of the large breccia dike, thick bedded, finely laminated dolomites of the Elbrook Formation are sharply truncated. Wall rock deformation is limited to less than 4 in (10 cm) from the breccia-dolomite contact and consists of a zone of closely spaced fractures parallel to the contact. The

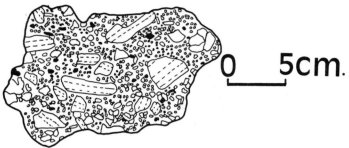

Figure 5. Sketch from a slab of Max Meadows breccia collected at outcrop location 2 (*Z–Z'*). Clasts are laminated dolomite (dashed lines) and massive dolomite (unshaded) and gray-green mudstones (black shading).

western margin of the dike is more complicated by folding and faulting. In some places the contact is gradational, with breccia grading into folded dolomites. Elsewhere a sharp contact is present. In the central part of the dike, large blocks of highly deformed dolomite are surrounded by breccia. These mega-clasts are internally fractured and show brecciated margins. A small cave in the breccia is located on the lower southeast side of the main dike where a thick breccia sill occurs. The breccia sill (Fig. 4) is approximately parallel to the massive dolomites above and contains breccia similar to that of the adjacent dike. Thin-bedded dolomites in outcrops just to the northwest of the dike of breccia are complexly folded, faulted, and fractured. This type of deformation (autobreccia of Cooper, 1970) is thought to represent the initial state of Max Meadows breccia formation. Folds are disharmonic and some show rotation and disruption by fracturing and small scale faulting. These shaly dolomites are typical of the lower part of the Cambrian Elbrook Formation, and the majority of clasts in the Max Meadow breccia are derived from these rocks.

Locality 3. An excellent exposure of decollement associated folds and cleavage occurs at Locality 3 (Fig. 4). Northwest facing, asymmetric folds have developed in response to bed parallel thrusting along a fault just below the folded sequence. Rocks here are thin to thick bedded dolomites of the middle portion of the Cambrian Elbrook Formation. Particular attention should be paid to the exceptionally well developed axial plane and fanning cleavage in the more shaly dolomite beds. This cleavage is a spaced, stylolitic type related to pressure solution during fold tightening. A comprehensive discussion of this cleavage can be found in Schweitzer (1984). The northwest asymmetry of these folds on a generally northwest dipping sequence of rocks (Fig. 4) indicates that they have been rotated following their formation.

Locality 4. The structurally highest zone of mesoscopic deformation in the Pulaski thrust sheet in the study area (Figs. 3, 4) can be seen at Locality 4 (Fig. 4). This zone of folding, faulting and cleavage is bounded above and below by decollements. The upper bounding decollement is exceptionally well exposed. This sharp contact separates all mesoscopic deformation of the lower part of the Pulaski thrust from relatively undeformed rocks higher in the Pulaski thrust sheet. Some of the best developed cleavage in

dolomites in the Appalachian Valley and Ridge Province can be seen associated with the numerous small folds. Near the eastern end of the deformed zone, cleavage is so well developed it is often mistaken for bedding. In places, this cleavage is warped by small faults indicating a complex history of thrusting, folding, cleavage formation, and continued faulting and folding which modified earlier structures.

REFERENCES CITED

Bartholomew, M. J., 1979, Thrusting component of shortening and a model for thrust fault development of the central/southern Appalachian junction: (abstract), Geological Society of America Abstracts with Programs, v. 11, p. 384–385.

Bartholomew, M. J., and Lowry, W. D., 1979, Geology of the Blacksburg quadrangle, Virginia: Virginia Division of Mineral Resources Publication 14, text and 1:24,000 scale map.

Butts, Charles, 1933, Geologic map of the Appalachian Valley of Virginia with explanatory text: Virginia Geological Survey Bulletin 42.

Campbell, M. R., 1894, Paleozoic overlaps in Montgomery and Pulaski Counties, Virginia: Geological Society of America Bulletin, v. 5, p. 171–190.

——, 1925, The Valley coal fields of Virginia: Virginia Geological Survey Bulletin 25, 322 p.

Cooper, B. N., 1946, Metamorphism along the "Pulaski" fault in the Appalachian valley of Virginia: American Journal of Science, v. 244, p. 95–104.

——, 1970, The Max Meadows Breccias: A reply, *in* Fisher, G. W., Pettijohn, F. J., Reed, J. C., Jr., and Weaver, K. N., eds., Studies of Appalachian Geology: Central and Southern: Wiley-Interscience, New York, p. 179–191.

Cooper, B. N., and Haff, J. C., 1940, Max Meadows fault breccia: Journal of Geology, v. 48, p. 945–947.

Epstein, A. G., Epstein, J. B., and Harris, L. D., 1977, Conodont color alteration—an index to organic metamorphism: U.S. Geological Survey Professional Paper 995, 27 p.

Gretener, P. E., 1977, On the character of thrust faults with particular reference to the basal tongues: Bulletin of Canadian Petroleum Geology, v. 25, p. 110–122.

Harris, L. D., and Milici, R. C., 1977, Characteristics of thin-skinned style of deformation in the Southern Appalachians and potential hydrocarbon traps: U.S. Geological Survey Professional Paper 1018, 40 p.

Lowry, W. D., 1971, Appalachian overthrust belt, Montgomery County, southwest Virginia: Field Trip No. 4, *in* Lowry, W. D., ed., Guidebook to Appalachian Tectonics and Sulfide Mineralization of Southwestern Virginia: Virginia Polytechnic Institute and State University, Department of Geological Sciences Guidebook No. 5, 178 p.

——, 1979, Nature of thrusting along the Allegheny front near Pearisburg and of overthrusting in the Blacksburg-Radford area of Virginia: Virginia Polytechnic Institute and State University, Department of Geological Sciences Guidebook No. 8, 66 p.

Rodgers, J., 1970, The Pulaski fault, and the extent of Cambrian evaporites in the Central and Southern Appalachians, *in* Fisher, G. W., and others, eds., Studies of Appalachian Geology: Central and Southern: Wiley-Interscience, New York, p. 175–178.

Schultz, A. P., 1983, Broken-formations of the Pulaski thrust sheet near Pulaski, Virginia [Ph.D. thesis]: Virginia Polytechnic Institute and State University, Blacksburg, 112 p.

——, 1984, Broken formations of the Pulaski thrust sheet near Pulaski, Virginia: Virginia Polytechnic Institute and State University, Department of Geological Sciences, Memoir Series (in press).

Schweitzer, J., 1984, Cleavage development in dolomite, Elbrook Formation, Pulaski thrust sheet, southwest Virginia [M.S. thesis]: Virginia Polytechnic Institute and State University, Blacksburg, 111 p.

Chestnut Ridge fenster: Illustration of a thin-skinned deformation of the Pine Mountain block, Lee County, Virginia*

Robert C. Milici, *Virginia Division of Mineral Resources, Charlottesville, Virginia 22903*

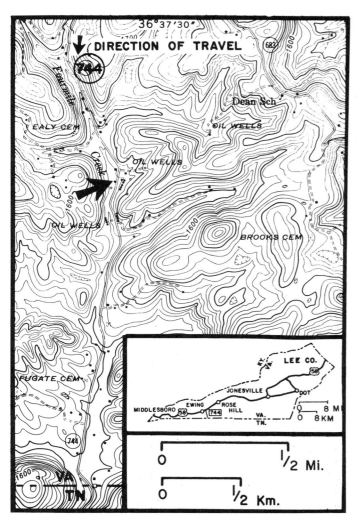

Figure 1. Location of Chestnut Ridge fenster area, in northeastern quarter of Coleman Gap 7½-minute Quadrangle, Lee County, Virginia.

LOCATION

The Chestnut Ridge fenster area is located southeast of Ewing, Lee County, Virginia, in parts of the Coleman Gap, Ewing and Back Valley 7½-minute Quadrangles (Fig. 1). From Ewing, Virginia, proceed northeast 1.7 mi (2.7 km) on U.S. 58, turn right on Lee County road 744; proceed 1.1 mi (1.8 km) to road fork, take right fork. (Route has been on rocks of the Knox Group; outcrops of the Maynardville Formation are exposed on left side of road 0.7 mi [1.1 km] south of road fork.) The fenster is exposed in the road and adjacent fields 1.0 mi (1.6 km) south of the road fork 0.3 mi (0.5 km) beyond the Maynardville Formation exposure.

INTRODUCTION

The Chestnut Ridge fenster is the southwesternmost of several fensters exposed along the axial region of the Powell Valley anticline (Fig. 1). Butts (1927), upon discovery of the fensters, proposed that the Pine Mountain fault, which heretofore was thought to dip steeply toward basement (Wentworth, 1921), was in reality a low-angle thrust, extending southwestward from Pine Mountain to the fenster area. Remarkably, although he had a minimum of surface data, Butts constructed a cross-section to near sea level that differs little from those produced today with the aid of subsurface control.

Later, Rich (1934), utilizing Butts' data, formulated the general characteristics of thin-skinned deformation.

SITE DESCRIPTION*

At this locality the Pine Mountain fault is exposed on a side road leading to the west (Figs. 1, 2). The Pine Mountain fault brings the Maynardville Formation (Upper Cambrian) in contact with the Rose Hill Formation (Silurian) in the autochthonous plate. West of the intersection on the north side of the valley, the Pine Mountain fault is arched upward from road level about 80 ft (24 m), thus carrying the Maynardville from road level to tree line on the hill. The Rose Hill is generally in pasture land, whereas the rocky Maynardville is in woodland, making easy the demarcation of the Pine Mountain fault along the edge of the Chestnut Ridge fenster.

Surface and subsurface studies by Miller and Fuller (1954) and Miller and Brosgé (1954) throughout most of the fenster area of southwest Virginia clearly show that the Pine Mountain fault is arched on the order of 5,000 ft (1,500 m) beneath the Powell Valley anticline. Drilling by Shell Oil Company, south of the Chestnut Ridge fenster (Fig. 3), determined that this arching was not the result of the folding process operating within the sedimentary prism; rather, arching resulted from the subsurface duplication of more than 5,600 ft (1,700 m) of beds by the unexposed Bales thrust fault (Harris, 1967). The source of the duplicated beds is within the autochthonous plate beneath the Pine Moun-

---

*Adapted from Harris and Milici, 1977.

*This description was originally prepared by the late Leonard D. Harris.

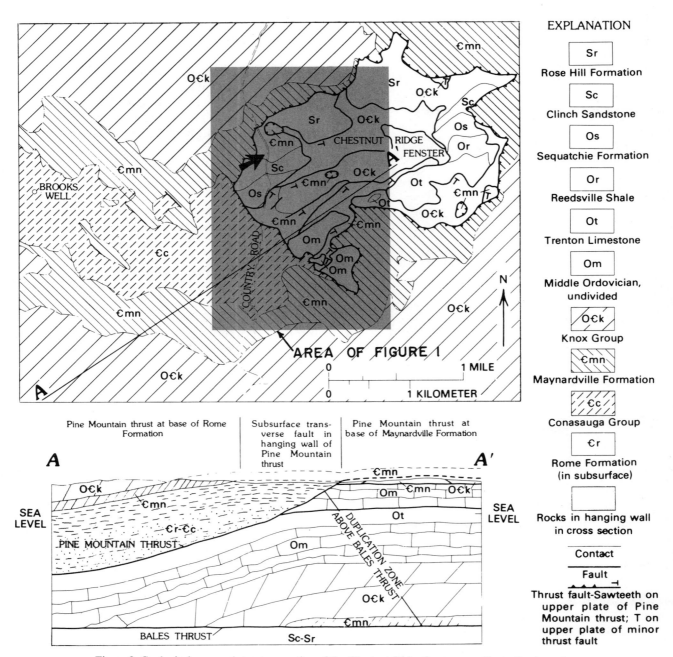

Figure 2. Geological map and structure section of the Chestnut Ridge fenster area, illustrating how the Pine Mountain thrust fault in the hanging wall initially changed stratigraphic position by a transverse fault from the near base of the Rome Formation up to the base of the Maynardville Formation. The transverse crosscut zone was later deformed by massive duplication above the subsurface Bales thrust. Hanging-wall rocks shaded and patterned in cross section and patterned on map.

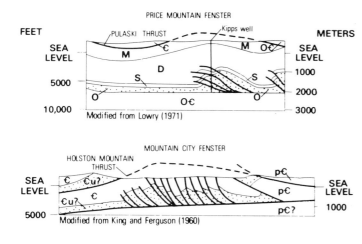

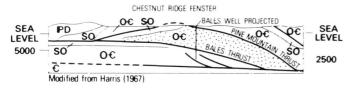

EXPLANATION

Areas of subsurface duplication are stippled

IPD, Pennsylvanian to Devonian rocks
M,   Mississippian rocks
D,   Devonian rocks
S,   Silurian rocks
SO,   Silurian and Ordovician rocks
O,   Ordovician rocks
OЄ,   Ordovician and Cambrian rocks
Є,   Cambrian rocks
Єu?,   Unicoi Formation
pЄ,   Precambrian rocks
pЄ?,   Precambrian(?) rocks

Figure 3. Diagrams showing essentially vertical uplift accompanying tectonic thickening in the subsurface, which is thought to be a mechanism responsible for the "folding" of thrust faults. Subsurface thickening occurs either by repetition through many minor thrusts or by massive duplication above a single thrust fault.

tain thrust, which is contrary to the generally accepted concept that rocks of that plate are passive elements in thin-skinned deformation (Rich, 1934). Oil exploration in the area has been confined to the Trenton in the duplicated slab, and the autochthonous plate below the slab has not been explored.

From this locality to the southwest edge of the fenster, the road traverses the truncated edges of the Rose Hill Formation, Clinch Sandstone, and Sequatchie Formation, all within the autochthonous plate. Exposed in a small quarry on the west side of the road is a fault slice of the Maynardville Formation, which on the road is in fault contact with a small slice of the Trenton Limestone. Detailed mapping of the Chestnut Ridge fenster (Miller and Fuller, 1954; Harris and others, 1962) shows that a series of fault slices of the Maynardville Formation and Knox Group exposed in the fenster intervenes between the older Cambrian rocks of the Pine Mountain thrust sheet and the younger formations in the autochthonous plate (Fig. 2). This series of slices resembles the broken-formation zone that overlies the Pennsylvanian décollement at Dunlap (Harris and Milici, 1977).

We suggest that these slices may be abandoned parts of the Knox derived from the broken-formation zone beneath the crosscut in the allochthonous plate.

This exposure of the Conasauga Group marks the southwest edge of the Chestnut Ridge fenster. Note on the map (Fig. 2) the sharp stratigraphic change, from east to west, of the Pine Mountain fault—from the base of the Maynardville Formation over the fenster to the Conasauga at the southwest edge. Drilling west of the fenster shows that the Pine Mountain thrust changes stratigraphic position along strike to the southwest from the base of the Maynardville to near the base of the Rome Formation along a steeply dipping transverse fault (Harris, 1970). The sharp contact of fenster rocks and the Conasauga may be the trace of the transverse fault, which enabled the Pine Mountain fault to form along strike at different stratigraphic positions.

REFERENCES CITED

Butts, Charles, 1927, Fensters in the Cumberland overthrust block in southwestern Virginia: Virginia Geological Survey Bulletin, 28, 12 p.

Harris, L. D., 1967, Geology of the L. S. Bales well, Lee County, Virginia—a Cambrian and Ordovician test, *in* Proceedings of the Technical Session, Kentucky Oil and Gas Association, 29th Annual Meeting, June 3–4: Kentucky Geological Survey, ser. 10, Special Publication 14, p. 50–55.

——— 1970, The Valley and Ridge and Appalachian Plateau, structure and tectonics; details of thin-skinned tectonics in parts of Valley and Ridge and Cumberland plateau provinces of the southern Appalachians, *in* G. W. Fisher et al., eds., Studies of Appalachian geology, central and southern: Interscience, New York, p. 161–173.

Harris, L. D., and Milici, R. C., 1977, Characteristics of thin-skinned style of deformation in the southern Appalachians and potential hydrocarbon traps: U.S. Geological Survey, Professional Paper 1018, 40 p.

Harris, L. D., Stephens, J. G., and Miller, R. L., 1962, Geology of the Coleman Gap quadrangle, Tennessee-Virginia: U.S. Geological Survey Geological Quadrangle Map GQ-188.

Miller, R. L., and Brosgé, W. P., 1954, Geology and oil resources of the Jonesville district, Lee County, Virginia: U.S. Geological Survey Bulletin 990, 240 p.

Miller, R. L., and Fuller, J. O., 1954, Geology and oil resources of the Rose Hill district—the fenster area of the Cumberland overthrust block—Lee County, Virginia: Virginia Geological Survey Bulletin 71, 383 p.

Rich, J. L., 1934, Mechanics of low-angle overthrust faulting as illustrated by Cumberland thrust block, Virginia, Kentucky, and Tennessee: American Association Petroleum Geologist Bulletin, v. 18, no. 12, p. 1584–1596.

Wentworth, C. K., 1921, Russell Fork fault, *in* The geology and coal resources of Dickerson County, Virginia: Virginia Geological Survey Bulletin 21, p. 53–67.

Mississippian facies of the Newman Ridge area, Hancock County, Tennessee

Kenneth O. Hasson, East Tennessee State University, Johnson City, Tennessee 37614

LOCATION

Excellent exposures of Mississippian strata occur in road cuts on Indian and Newman Ridges, Sneedville and Howard Quarter 7½-minute quadrangles (Hancock and Claiborne counties) (Fig. 1). All outcrops are on public roads accessible by car; the Newman Ridge area can be reached via Tennessee 33 either from the south or east. Parking can be a problem at Stops 1 and 3 because the shoulders of the road are narrow (refer to Figs. 1 and 2 for stop locations).

INTRODUCTION

Facies changes and regional relationships in the Mississippian rocks of east Tennessee are demonstrated in this group of stops. Observable in a compact area are replacement of Fort Payne Chert by Maccrady Formation, in a manner similar to a textbook diagram; pinchout of the Fort Payne; transition between green Grainger Formation and red Maccrady Formation; transition between Newman Limestone and clastic Pennington Formation; and partial exposures of Chattanooga Shale and Sneedville Limestone. Also, this is the type area of the Newman Limestone, in which one can find also the distal margin of the red Taggard Formation of the Greenbrier Group of southern West Virginia. This red bed is the principal marker bed within the Newman.

The stratigraphic units exposed in the Newman Ridge area are the Devonian-Mississippian Chattanooga Shale, the Early Mississippian Grainger Formation, the Maccrady Formation, Fort Payne Chert, Newman Limestone and Pennington Formation.

Chattanooga Shale. The Chattanooga Shale is partly exposed on the steep northwest face of Newman Ridge at Localities 2 and 3 (Fig. 2). The formation is about 500 ft (152 m) thick, as calculated from plane table measurements, and consists of very dark gray to grayish black, thinly laminated shale. It overlies unconformably the Silurian Sneedville Limestone, which is exposed in the valley of Blackwater Creek on the northwest side of Newman Ridge, and is conformably overlain by the Grainger Formation. The upper part of the Chattanooga is exposed also at Locality 1 (Fig. 2).

Grainger Formation. The Grainger Formation consists of a basal siltstone, a middle silty shale, and an upper siltstone in the Sneedville area. Thickness is between 200 and 400 ft (60.8 and 121.6 m). The Grainger shales are typically olive color, and the siltstones are in distinct beds up to about 1 ft (0.3 m) thick (Hasson, 1972).

Maccrady Formation. The Maccrady Formation is typi-

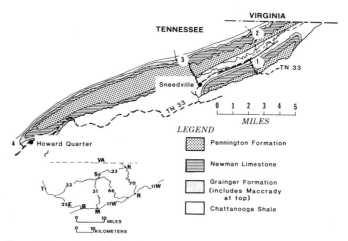

Figure 1. Location of field trip area. On the inset map, which shows the road access to Sneedville area, the symbols are: K=Kyles Ford; S=Sneedville; T=Tazewell; B=Bean Station; M=Mooresburg; R=Rogersville. The large map is a geologic map of the Newman Ridge area, Hancock County, Tennessee. The numbers refer to stops described in text. Refer to Figure 2 for road details. (Map taken from Rodgers, 1953.)

cally a red to maroon shale, siltstone, and sandstone sequence. On the northwest side of Newman Ridge it is between 113 and 123 ft (34.3 to 37.3 m) thick. The uppermost part of the formation is a silt-free clay shale that has an abundant fauna (*Dictyoclostus, Breviphillipsia*). The shale was probably once red but now has only a faint red or pinkish color, the result of deep leaching. Its basal contact is transitional with the underlying Grainger Formation; this is particularly well exposed at Localities 2 and 3 (Fig. 2). The Maccrady is overlain with minimal transition by the Newman Limestone. The Maccrady is replaced by the Fort Payne Chert to the southwest along Newman Ridge (Fig. 3). Detailed measured sections of the Maccrady are given by Kuczynski and Hicks, 1979, and by Hasson, 1972.

Fort Payne Chert. The Fort Payne Chert occupies the interval between the Grainger Formation and Newman Limestone at the southwest terminus of Newman Ridge (Locality 4). Although the Fort Payne has never been officially recognized in this area, this is the identical sequence described by Butts (1940) and by Hasson (1972) at Cumberland Gap. At Locality 3 the Fort Payne occurs as thin, distinct beds that superficially resemble siltstone interbedded with the Maccrady clay shale. In thin section the chert is chalcedonic to fibrous, silty, with spherical and rod-shaped glauconite grains. The chert is replaced northeastward by Maccrady red beds and is totally absent at Locality 2 (Fig. 2).

Newman Limestone. Newman Ridge is the type area of the Newman Limestone. On Newman Ridge the formation is approximately 400 ft (121 m) thick (Localities 2 and 3, Fig. 2); it

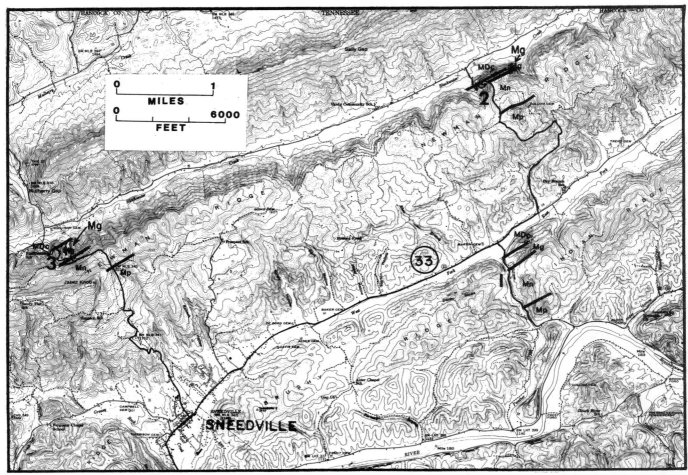

Figure 2. Part of the Sneedville 7½-minute Quadrangle. This map is reduced from original scale; refer to bar scale for actual distances. Geologic contacts taken from Greene, 1959.

is about 922 ft (281 m) thick at Locality 1 (Beckman, 1985). At Localities 2 and 3 the Newman contains a fair amount of oolite and calcarenite; at Locality 1 it is dominantly calcilutite. This most likely represents increased water depth down the carbonate ramp (deep water would be to the east or southeast). Chert is a very minor component. In the approximate middle of the Newman section, a tongue of red or reddish calcilutite occurs that is thought to be the distal end of the Taggard Formation of the Greenbrier Group of southern West Virginia (Moody and Woods, 1977; Hasson, 1983; Yeilding, 1984). This red bed is present at Localities 1, 2, and 3 (Fig. 2) and in the road bank on the southeast side of Newman Ridge on the dirt road leading to Locality 2.

The upper contact between the Newman and the Pennington is gradational and is easily seen at Localities 1, 2, and 3 (Fig. 2). At these places the limestone is overlain by dark gray, olive-weathering shale. The shale is succeeded by limestone and shale before sandstone becomes the dominant lithology. At this writing the upper boundary of the Newman is not fixed.

Pennington Formation. The Pennington Formation is a heterogeneous, principally clastic unit of red and green shale, varicolored sandstone, minor limestone, and some conglomeratic sandstone. Only a few hundred feet (meters) are preserved on Newman Ridge.

SITE DESCRIPTION

Stop 1

Stop 1 is located in cuts along the northeast side of Tennessee 33 through Indian Ridge at 83°09'30"W, 36°33'N on the Sneedville 7½-minute Quadrangle.

Beginning at and extending in a southeast direction from the bridge over the East Fork of Panther Creek are exposures of the upper part of the Chattanooga Shale, Grainger Formation, Maccrady Formation, an almost-complete Newman section, and the Pennington (Fig. 2). The section is particularly long because of the low dip. Vehicles will have to be moved at least once. Parking is available near the base and top of the Newman Limestone.

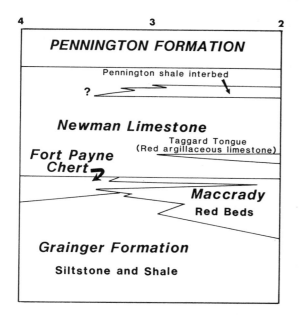

Figure 3. Interpretive facies diagram showing the stratigraphic relationships exposed in the field trip area and discussed by Hasson, 1983. Details are discussed in text; numbers refer to localities described in text. The diagram is not to scale. Refer to Figure 1 for locations.

The upper part of the Grainger Formation here is sandstone and siltstone, which is overlain directly by the Maccrady clay shale. At Stop 2 you will see typical red Maccrady sandstone and siltstone at this same position. Question: "Is the non-red sandstone Maccrady or Grainger?" About 5 ft (1.5 m) stratigraphically above the top of the sandstone, in the Maccrady clay shale, is a covered bed of glauconite and red shale. This glauconite bed is present at the same stratigraphic position at Stop 2. The clay shale is well exposed a few feet up section in the bed of a logging road. The shale is deeply leached and contains fossils that are very fragile. About 11 ft (3.3 m) stratigraphically below the Newman in the road bank at the level of the drainage ditch are four siltstone beds each about 2 in (5 cm) thick. These beds are actually cherty siltstone and represent the ultimate pinchout of the Fort Payne Chert. **PLEASE LEAVE THE ROCKS THERE FOR OTHERS.** The Newman is exposed almost continuously for about 2,000 ft (600 m) along the road. In the road bank opposite the field, 500 or so ft (152 m) above the base of the section, are red calcilutite beds of the Taggard Formation. The transitional contact of alternate shale and limestone between the Newman and Pennington is well exposed in the curve in the road at the top of the section.

Stop 2

Stop 2 is at 83°09′40″W, 36°34′48″N, located on the steep northwest face of Newman Ridge, Sneedville 7½-minute quadrangle. The exposure is accessible via a dirt road from the southeast. From Stop 1 drive toward Sneedville and turn right on the county road at the intersection where Tennessee 33 turns south to Sneedville (Fig. 2). Go about 0.5 mi (0.8 km) until you see a barn on the right and mail boxes and a small creek on the left side of the road. Turn left onto this dirt road and follow it to Stop 2. This road crosses Newman Ridge. On your right will be outcrops of Grainger and Maccrady Formations and a few limestone beds. Stop if you want. Continue across the syncline axis. The site of the collapsed building on your left is at the approximate top of the Newman Limestone. The Newman is exposed in the road bank and in the roadbed; the shaly lower beds of the Pennington Formation are exposed in the road (not shown on the topographic map) that leads to Collins Cemetery. The cemetery is in the shaly part of the Pennington.

Proceed down section (northwest) through the Newman Limestone, which here is 394 ft (120 m) thick (Beckman, 1985). The Taggard Formation red beds are exposed in the roadbed 239 ft (73 m) below the top of the Newman. The base of the Newman is in the drainage ditch on the northwest side of the road; basal beds of the Newman Limestone at this locality are dominantly oosparite, in contrast with Stop 1. The oosparite is 121 ft (37 m) thick (Beckman, 1985). The clay shale seen at Stop 1 is exposed in the road bed below the oolite. The glauconite bed from Stop 1 is present at the same stratigraphic position and is exposed in the roadbed. This part of the section is best seen after a heavy rain when the road is washed free of dust. No chert, as seen at Stop 1, has ever been found in this interval. Continue down section.

Exposed below the clay shale are red Maccrady sandstones, siltstones, and shales. *Zoophycus* is common, and a few brachiopods have been found. A few thin beds of almost pure hematite occur also. The transitional nature of the contact between the Grainger and Maccrady Formations is particularly well exposed here. The contact is placed where the shale is dominantly red. The tripartite character of the Grainger is also well exposed, and the upper siltstone is quite thin; 24 mi (40 km) west, this siltstone changes to a shale, as in the exposure at Cumberland Gap (Hasson, 1972).

Continuing downhill, the Chattanooga-Grainger contact is exposed in the road bank on the right. The contact consists of massive siltstones of the Grainger Formation overlying black Chattanooga Shale. As the road descends Newman Ridge into the valley, some Chattanooga Shale is exposed.

Follow this road across the valley to its end, turn left, and continue to the end of the road. Limestone outcrops along the road and in the fields on the left are the Silurian Sneedville Limestone. At the end of the road turn left up the mountain to Stop 3. If it is wet or rainy, don't drive past the Grainger-Maccrady contact because the road is very steep and slippery. Instead, retrace the route back to Sneedville, turn right at College Street (traffic light), and go on over Newman Ridge from the southeast to Stop 3.

Stop 3

Stop 3 is located approximately 2.5 mi (4 km) north of

Sneedville at 83°14′18″W, 36°33′12″N (Sneedville quadrangle). The principal feature of interest at this stop is the interbedding of the Maccrady and Fort Payne Chert. **Parking will be a problem—BE CAREFUL.** Going up the mountain from Stop 2, you will go through a cut in the Chattanooga on the right. The Chattanooga-Grainger contact is exposed at the curve, as is the rest of the Grainger. You can park here and walk through the lower part of the section. Continue up section (uphill).

The Maccrady-Grainger transition is exposed on the next curve, and there is a little room to park. Here, red and green shale and siltstone alternate; *Zoophycus* is particularly abundant in the red beds. In the cut on the right, thin (2-in; 5-cm) chert beds are interbedded with the Maccrady. The chert, a tongue of Fort Payne, is slightly silty and a translucent green. The interval of interbedded chert and shale is 8 ft (2.4 m) thick. On the left is the clay shale seen at Stops 1 and 2. Here, there is no mistaking the reddish cast. The shale contains abundant *Dictyoclostus.* The top of the Maccrady is the red bed at the beginning of the next curve. The base of the Newman is not exposed along the road. Continue uphill.

The section of road on which the Newman crops out is very narrow, with no shoulders. To see the Newman, it is best to walk through it, but it is a long, uphill walk because the road parallels strike. Of principal interest is the presence of the red Taggard Formation, which crops out a few hundred feet (meters) above the curve.

At the top of the section, a cut on the right exposes the dark gray shale of the Pennington Formation. A few feet (m) of limestone overlie the shale; the limestone is succeeded by sandstone, which is exposed at the base of Jabez Knob (Fig. 2).

Proceed down the mountain toward Sneedville. You may encounter broken road. This is a textbook example of an active creep that constantly displaces the road. Hummocky topography, ponding, and tilted trees are all quite evident. You cross the slide again on the next curve.

If coming from Sneedville, follow the above text in reverse order.

Stop 4

Stop 4 is located on the Howard Quarter 7½-minute Geologic Quadrangle (Harris and Mixon, 1970) at 83°24′52″W, 36°28′40″N. Follow Tennessee 33 south from Sneedville. At the edge of town on the right is a borrow pit that exposes beds of highly contorted Chattanooga Shale. In a short distance you cross the Hunter Valley fault, which has brought rocks of the Cambrian Rome Formation over the Chattanooga Shale. If not overgrown, the fault zone can be seen. Continue on Tennessee 33. About 2 mi (3.2 km) south of the Hancock-Claiborne county line turn right onto the dirt road. Proceed about 0.25 mi (0.4 km) to Stop 4, which is at the bridge.

Stop 4 is at the southwest-plunging end of the Newman syncline. At this point you are looking northeast along the axis of the syncline as it plunges toward you. At road level are about 20 ft (6 m) of massive chert (Fort Payne Chert), overlain by Newman Limestone. Note particularly that red beds are absent. The Maccrady has been replaced by Fort Payne Chert. The Grainger Formation is not exposed at road level. To see the top of the Grainger–Fort Payne–Newman, follow the tributary to Sycamore Creek on the northwest side of the syncline, where this part of the section is well exposed.

REFERENCES CITED

Beckman, C. S., 1985, Petrology and depositional environments of the Newman Limestone in Hancock and Claiborne counties, Tennessee [M.S. thesis]: Chapel Hill, University of North Carolina, 93 p.

Butts, C., 1940, Geology of the Appalachian Valley in Virginia: Virginia Geological Survey Bulletin 52, pt. I, 568 p.

Greene, A. V., 1959, Geology of Newman Ridge and Brushy–Indian Ridge between Sneedville, Hancock County, Tennessee, and Blackwater, Lee County, Virginia [M.S. thesis]: Knoxville, University of Tennessee, 55 p.

Harris, L. D., and Mixon, R. B., 1970, Geologic map of the Howard Quarter quadrangle, northeastern Tennessee: U.S. Geological Survey Geologic Quadrangle Map GQ-842.

Hasson, K. O., 1972, Lithostratigraphy of the Grainger Formation (Mississippian) in northeast Tennessee [Ph.D. thesis]: Knoxville, University of Tennessee, 142 p.

—— 1983, The Carboniferous succession, basin tectonics and sea level fluctuations, eastern Tennessee, USA [abs.]: X Congreso Internacional de Estratigrafica y Geologias del Carbonifero, Madrid, p. 241.

Kuczynski, J. C., and Hicks, M. C., 1979, A study of the Mississippian red beds of Hancock County, Tennessee [B.S. thesis]: Johnson City, East Tennessee State University, 23 p.

Moody, J. R., and Woods, L. D., 1977, Type and standard reference section of the Newman Limestone, northeast Tennessee [B.S. thesis]: Johnson City, East Tennessee State University, 19 p.

Rodgers, J., 1953, Geologic map of east Tennessee with explanatory text: Tennessee Division of Geology Bulletin 58, pt. II, 168 p.

Yeilding, C. A., 1984, Stratigraphy and sedimentary tectonics of the upper Mississippian Greenbrier Group in eastern West Virginia [M.S. thesis]: Chapel Hill, University of North Carolina, 117 p.

Thorn Hill: A classic Paleozoic stratigraphic section in Tennessee

Don W. Byerly, K. R. Walker, W. W. Diehl, M. Ghazizadeh, R. E. Johnson, C. T. Lutz, A. K. Schoner, W. A. Simmons, J.C.B. Simonson, L. J. Weber, and J. E. Wedekind, *Department of Geological Sciences, University of Tennessee, Knoxville, Tennessee 37916*

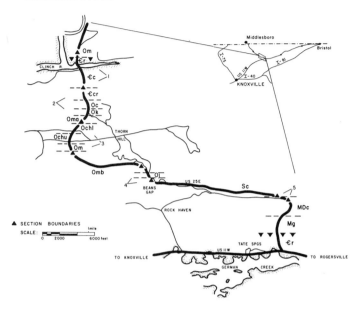

Figure 1. Location. The Thorn Hill Section occurs along 8.8 mi (13 km) of U.S. 25-E from its intersection with U.S. 11-W, 44 mi (70 km) north of Knoxville, Tennessee.

INTRODUCTION

The "Thorn Hill Section" is an amazingly complete and well exposed section of rocks that has served at least four generations of geologists as a reference for early Cambrian through early Carboniferous stratigraphy of the Southern Appalachians. Eastward from War Ridge along 8.8 mi (14 km) of U.S. 25-E, across Clinch Mountain to Poor Valley Ridge, Grainger County, Tennessee, virtually every unit between the Rome and the Grainger formations in the Southern Appalachians is exposed in the 10,912 ft (3,326 m) thick section (Fig. 1). The section is bounded by two major thrust faults—the Copper Creek fault and the Saltville fault; the Rome Formation is the hanging wall for both faults. Most of the section is embraced by the Avondale, 7½-minute topographic quadrangle with the remainder on the Dutch Valley, 7½-minute Quadrangle.

The stratigraphic descriptions given here are synopses from Walker (1985). The stratigraphy here represents a westward thinning edge of a thick wedge of Paleozoic sedimentary rocks whose history is recorded in three stratigraphic sequences bounded by unconformities of regional extent (Fig. 2). Sequence I embraces Cambrian through Early Ordovician, Sequence II the middle Ordovician through Early Devonian, and Sequence III Late Devonian through Early Mississippian.

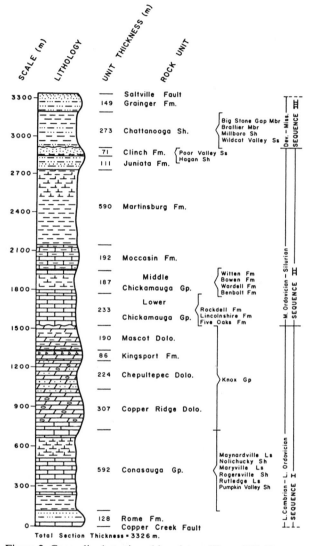

Figure 2. Generalized stratigraphic column. Thorn Hill, Tennessee.

Sequence I

The first sequence in the Thorn Hill Section includes the Rome Formation, the Conasauga Group, and the Knox Group. Outcrops of units in this sequence begin at the northwest end of the section, near the bridge across Indian Creek (Figs. 3 and 4). The rocks change upward from terrigenous clastic rocks derived from western sources (Rome Formation) to carbonate rocks (Knox Group) representing a westward transgressing sea. As the sea transgressed, the Rome (Lower-Middle Cambrian) was deposited in a shallow tidal environment west of a carbonate bank (Shady Dolomite) developing along the margin of the subsiding Appalachian Basin. Continued gradual subsidence throughout the

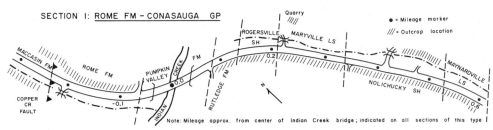

Figure 3. Section No. 1, Rome Formation/Conasauga Group. Approximate distances in miles eastward from the center of the Indian Creek bridge.

Cambrian is manifest by the overlying Conasauga Group. Facies of the Conasauga at this locality represent a north-south transitional zone between a carbonate shelf to the east, and an irregularly subsiding but deeper western basin where terrigenous mud derived from the west accumulated. By latest Cambrian and early Ordovician time much of this carbonate platform had become subaerially exposed instigating the development of the paleokarstic disconformity separating the first two depositional sequences. (Walker and Byerly, 1985).

Rome Formation. The upper 420 ft (128 m) of the Rome exposed west of Indian Creek mark the base of the Thorn Hill section (Fig. 3). The lowest part of the Rome at War Ridge is a thin mylonite along the Copper Creek fault, which rests upon the maroon and fossiliferous Moccasin Formation.

The red beds with subordinant hues of green seen in this section are very typical of Rome exposures throughout the basin. The green is largely due to glauconite and the red is iron-oxide resulting from weathering in a tropical to sub-tropical environment in both the sediment source area and in the depositional environment. Body fossils are scarce in the Rome, but exposures are notable for well preserved sedimentary structures and trace fossils that allow paleoenvironmental interpretation.

The fossils and sedimentary structures of the Rome Formation suggest deposition in a supratidal to shallow subtidal environment. Supratidal conditions can be recognized by the presence of halite hoppers, desiccation cracks, algally laminated dolostone, and vugs or birdseyes that may be pseudomorphs after evaporites.

Mudflats in an intertidal environment can be inferred from red beds of laminated shale and siltstone, and fine-grained, laminated to thin-bedded sandstone containing desiccation cracks, ripple marks, and signs of bioturbation. Greenish gray, brownish gray, and grayish orange, thin- to thick-bedded, cross-stratified sandstone, lacking mud and containing few biogenic structures can be interpreted as sand flats in an intertidal or shallow subtidal environment.

Conasauga Group. The Rome grades upward into the Conasauga Group. The Conasauga is 1,942 ft (592 m) thick here and consists of an alternation of carbonate (mostly limestone) and siliciclastic (mostly shale) rocks that can be divided into six formations, each a facies in a "central phase" of the Conasauga, which becomes mostly carbonate to the east and becomes dominantly shale to the west.

Pumpkin Valley Shale. The Pumpkin Valley Shale, approximately 358 ft (109 m) thick, is not exposed along the highway, but can be seen along side roads. The Pumpkin Valley is variegated in color and therefore difficult to distinguish from the shale in the underlying Rome. The base of the Pumpkin Valley is customarily placed at the top of the highest sandstone in the Rome. Also, the Pumpkin Valley does not contain evidence of subaerial exposure like the Rome. This suggests a subtidal environment for the Pumpkin Valley.

Rutledge Limestone. The Rutledge Limestone, 177 ft (54 m) thick, consists dominantly of internally thick-laminated, to thin-bedded, externally massive, fine- to medium-grained, mottled limestone and dolostone that are regularly interbedded. The "ribboned" appearance of the rock is due to thick laminae or thin beds of fine-grained limestone which protrude between more deeply weathered, thinner, more dolomitic and clayey laminae. Trilobite fragments, though rare, constitute the fauna. The depositional environment was subtidal.

Rogersville Shale. The Rogersville Shale, 89 ft (27 m) thick, which overlies the Rutledge, consists of three parts. Lower and upper shale members are separated by the Craig Limestone Member. The shale members consist of grayish green, fissile, laminated, silty clay shale with very subordinate thin intercalations of fine-grained, mottled limestone. Small trilobites and inarticulate brachiopods are present within the shales. The Craig Limestone Member is 26 ft (8 m) thick and consists of fine-grained, clayey dolostone, with minor beds of oolite and intraclastic packstone.

Maryville Limestone. The Maryville Limestone, above the Rogersville, is well exposed in the road cuts and very well exposed in the quarry northeast of the highway where it is quarried for "Imperial Black marble." A progradational facies and a trans-

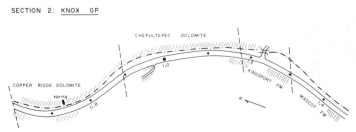

Figure 4. Section No. 2. Knox Group. Approximate distances in miles eastward from the center of the Indian Creek bridge.

gressive facies alternate several times within the 554 ft (169 m) of the Maryville. The three progradational sequences are characterized by mottled and ribboned limestone with high micrite content (30 to 50 volume percent) and thin interlaminae of clayey dolostone. Other lithologic aspects associated with the progradational facies include oolitic grainstone, intraclastic packstone, and oncolitic grainstone and packstone. Trilobite debris and echinoderm grains are ubiquitous.

The three transgressive facies alternating with the progradational facies are characterized by terrigenous quartz silt and interbeds of silty clay shale. The dominant rock type of this facies is quartzose silty pelsparite. Bedding ranges from slightly irregular to nodular, with abundant dolomicrite associated with stylolite swarms separating limestone laminae. The upper transgressive facies of the Maryville grades into the overlying Nolichucky Shale.

Nolichucky Shale. The Nolichucky, 587 ft (179 m) thick, is predominantly laminated dark brown, quartz silty shale in medium to thick external beds. Most of the shale, where fresh, has a high carbonate mud content. Three limestone lithotypes occur as interbeds. These are ooid grainstone to packstone, intraclastic grainstone to packstone (intraformational conglomerate), and thick-bedded, mottled packstone to mudstone (the latter dominant in the upper Nolichucky). The contact between the Nolichucky and the overlying Maynardville Limestone is conformable and is at the base of the first thick, ribboned limestone above a preponderance of shale.

Maynardville Limestone. The basal 89 ft (27 m) of the Maynardville is the Low Hollow Member, a mottled, ribboned to nodular limestone that is mostly fine-grained, but which contains occasional beds of intraclastic conglomerate. Dark gray shale intercalations are also common. Trilobite debris is abundant and stromatolitic lamination is prevalent in the upper part. The upper 89 ft (27 m) of the Maynardville, the Chances Branch Dolomite Member, are more dolomitic, finer-grained, and contain more stromatolitic laminae than the Low Hollow. The contact with the overlying Copper Ridge Dolomite of the Knox Group is placed at the base of the first thick and more massive dolostone, which here is immediately above a one meter thick, partly dolomitized limestone.

Knox Group. The lower part of the Knox (Copper Ridge Dolomite and Chepultepec Dolomite), 1,742 ft (531 m) thick, represents carbonate and terrigenous clastic sedimentation during the close of Sequence I (Fig. 4). The Copper Ridge and the Chepultepec, consist of three facies: 1) crystalline dolostone, 2) arenaceous dolostone, and 3) calcitic dolostone. The lowermost facies constitutes the lower 902 ft (275 m) of the Copper Ridge Dolomite and consists of laminated and non-laminated dolostone largely devoid of silt and coarse-grained terrigenous clastics. Dolomitized ooids, oncoids, peloids, intraclasts, and domal and columnar stromatiolites are ubiquitous in this interval.

Facies 2, an arenaceous dolostone with shaly interbeds and quartzose sandstone, occupies the uppermost 105 ft (32 m) of the Copper Ridge and the lower 328 ft (100 m) of the Chepultepec.

Dolostone intraclasts, desiccation cracks, quartz-filled channels, and cross-laminated quartz sandstone are other characteristics of this facies. Facies 3, occupying the uppermost 407 ft (124 m) of the Chepultepec, is characterized by an increase in limy dolostone and shale intercalations.

Thickness of the Copper Ridge alone is 1,007 ft (307 m) and the thickness of the Chepultepec is 735 ft (224 m). The Chepultepec extends upward from the base of a dolomitic quartz sandstone to include a thin-bedded, chert-rich, dolostone overlain by the sandy, cherty, dolostone of the basal Kingsport Formation.

The Kingsport and Mascot Formations comprising the upper Knox Group are Early Ordovician in age and have a combined thickness of 906 ft (276 m). Both units are of particular interest because they contain economically important Pb-Zn deposits in several districts in eastern Tennessee. Here the Kingsport is 282 ft (86 m) thick, dominantly made up of medium- to coarse-grained dolostone with minor amounts of limestone and fine-grained dolostone. Horizons of chert nodules are abundant in the lower part of the unit (formerly Longview Dolomite), and a 30 ft (9 m) thick brecciated and mineralized zone is conspicuous near the top.

The Mascot Formation ranging in thickness from 246 to 666 ft (75 to 203 m) due to the unconformity at its top, is 623 ft (190 m) thick here. It is composed of thick beds of fine-grained dolostone with thin interbeds of shale and chert. The chert occurs as nodules and beds. The shale beds are typically 2 to 5 in (5 to 10 cm) thick and impart a greenish gray color to the unit.

Petrographic studies by Churnet and others, 1982, suggest a depositional model consisting of a supratidal flat, shallowing to the northwest (Mascot), flanked by a shallow marine environment with islands (Kingsport).

Sequence II

The termination of Sequence I was associated with regression of the sea and karstification of the exposed Knox. As Sequence II developed, the major provenance for terrigenous clastic sediment shifted from a western source to a more easterly source. Sequence II deposition began in Middle Ordovician and is comprised of two sub-sequences. The lower sub-sequence, comprising about one half of Sequence II, consists of sediments influenced by the development of a foredeep toward the southeast. Within the basin, graptolitic shale and turbidites, derived primarily from eastern sources, dominate. Patch reefs and skeletal sand banks developed west of the foredeep separating the deeper basin from a shallow water carbonate shelf that extended westward toward the Nashville Dome. The lowermost three-fourths of the Middle Ordovician rocks in this section (lower and middle Chickamauga Group) represent deposition on this shelf.

Ultimately, during the Middle Ordovician, the foredeep filled and shallow peritidal environments prograded northwestward to the Thorn Hill area, with more open marine conditions prevailing to the west and northwest. This reversal in bathymetry is manifest in the Moccasin Formation.

The upper sub-sequence—the Martinsburg, Juniata, and Clinch formations—commenced with development of an intrashelf basin that was filled by the sediment of the Martinsburg.

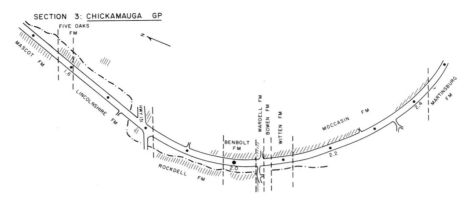

Figure 5. Section No. 3. Chickamauga Group (Five-Oaks through Moccasin Formations). Approximate distances in miles eastward from the center of the Indian Creek bridge.

Sequence II ends with deposition of the Juniata and Clinch on top of this basin fill as a set of shallow, near-shore, terrigenous facies.

Chickamauga Group. Five Oaks Formation. This basal unit, which is 62 ft (19 m) thick, unconformably overlies the dolostone of the Knox Group (Fig. 5). The lower part consists of dolomitic, fenestral lime mudstone with subordinate peloidal-skeletal wackestone and packstone and represents a tidal flat. The upper lithotype consists of peloidal-echinoderm grainstone, which represents a shallow subtidal environment.

Lincolnshire Formation. The Lincolnshire, overlying the Five Oaks, is 308 ft (94 m) thick and is composed of skeletal wackestone, packstone, and grainstone with a diverse marine fauna dominated by echinoderms and bryozoans. Several genera of calcareous algae are present with *Girvanella* commonly occurring as coatings on skeletal grains. The depositional environment for the Lincolnshire was a shallow marine platform with sufficient wave energy to work the lime sand.

Rockdell Formation. The Rockdell, 394 ft (120 m) thick, consists of echinoderm-bryozoan packstone and grainstone with intermixed lenses of skeletal wackestone containing in situ fenestrate and encrusting bryozoans and sponges. The unit can be interpreted as a shallow water sand bank complex deposited seaward of the Lincolnshire. The bryozoan-sponge wackestone lenses represent growth of lime-mud mounds in response to biotic activity and the hydrodynamics of the local setting. Birdseyes, diagenetic peloids, and the red color of the micrite in some mounds suggest periods of subaerial exposure.

Benbolt Formation. The Benbolt, 246 ft (75 m) thick, is a cobbly-weathering, highly fossiliferous wackestone and packstone with intercalated shale. Included in the diverse biota are echinoderms, bryozoans, brachiopods, and sponges. The upper Benbolt is primarily made up of sparsely fossiliferous calcareous shale. The Benbolt was deposited on a subtidal, sometimes turbid-water, shelf.

Wardell Formation. The Wardell is 66 ft (20 m) thick. It includes a lower unit of interbedded fossiliferous limestone and shale containing abundant brachiopods (*Sowerbyella* sp.), ramose bryozoans, and calcareous algae, and an upper unit consisting of coral-algal packstone/boundstone high in fine-grained siliciclastic content. The unit is capped by a medium-bedded fossiliferous packstone. A nearshore shelf environment is inferred for the Wardell.

Bowen Formation. The Bowen, 76 ft (23 m) thick, is predominantly brownish red and tan, finely laminated, calcareous mudstone containing interlaminar birdseyes, vertical burrows, and desiccation cracks. High intertidal to supratidal conditions are inferred for deposition of the Bowen.

Witten Formation. The Witten is 226 ft (69 m) thick and consists of flaggy bedded biomicrite and dismicrite with interbedded shale. Fossils including gastropods, trilobites, brachiopods, and ostrocodes are locally abundant and horizontal burrows (*Planolites*) are found within undulatory and rippled beds. The Witten probably represents a return to shallow subtidal conditions following the deposition of the Bowen. The Witten is distinguished from the overlying Moccasin by the absence of reddish coloration, vertical burrows, and mudcracks.

Moccasin Formation. The Moccasin, 630 ft (192 m) thick, it represents deposition in a mixed carbonate-siliciclastic tidal-flat environment. The lower 151 ft (46 m) of the Moccasin consists of maroon, argillaceous, desiccation-cracked, vertically burrowed, peloidal, fenestral mudstone and wackestone with local concentrations of arthropods and intraclasts. This part represents several tidal flat sub-environments such as tidal channels and tidal ponds. The next 112 ft (34 m) of Mocassin is blue-gray, bioturbated, peloidal, fossiliferous wackestone and packstone with a diverse biota including brachiopods, bryozoans, corals (*Tetradium*), echinoderms, arthropods, molluscs, sponges, calcareous algae, and cephalopods, suggesting a return to subtidal conditions.

The upper 367 ft (112 m) of the Moccasin is similar to the lowermost interval of the formation and represents a return to a tidal flat complex of ponds, channels, and supratidal conditions. Two distinctive bentonite beds occur near the top of the Moccasin. Both are very thin-bedded, very fine-grained, "chippy" weathering, and pistachio green in color. One occurs about 61 ft (18.5 m) and the other about 10 ft (3 m) from the top of the Moccasin. The contact with the overlying Martinsburg is conformable and is drawn above the last bed of dominantly maroon, intraclastic, sparsely fossiliferous, silty, argillaceous mudstone below the tan calcareous shale and grayish fossiliferous packstone of the lower Martinsburg.

Martinsburg Formation. Only the lower and upper Martinsburg appear on strip maps; however, excellent exposures of all facies occur between mile 2.4 and mile 4.2. The lower "shale facies" of the Martinsburg, 1,214 ft (370 m) thick, is about 80% shale and 20% limestone. It consists predominantly of thinly laminated, graptolitic, medium to dark gray shale that accumulated within a deeper water environment than the underlying Moccasin. The intercalated limestone in the lowermost Martinsburg is similar to that comprising a 328 ft (100 m) thick middle "carbonate facies" (about 50% limestone). Many of the limestone beds form fining-upward sequences with coarse fossil debris grading into micrite. Some limestone beds, however, are in situ accumulations of fossils with a muddy matrix (packstone and wackestone), and are size graded with larger fossils near the base. Fossils within the limestone are mainly brachiopods and trilobites.

The uppermost 394 ft (120 m) of the Martinsburg are a "mixed carbonate-siliciclastic facies" with characteristics indicative of shallow water deposition. These sediments grade upward into the peritidal facies of the overlying Juniata. The limestone of this facies (about 50% of the thickness) is fossiliferous packstone, most of which forms fining upward sequences containing considerable amounts of terrigenous material. Some beds, especially high in the facies, contain as much as 50% quartz silt and sand. The fining upward limestones commonly have a basal fossil lag that grades upward into fossiliferous packstone and finally into bioturbated shale. Limestones with in situ biotas are wackestone to packstone (thus, with more micrite than current-influenced beds). The diverse biota of the upper Martinsburg includes brachiopods, bryozoa, pelecypods, and gastropods, echinoderms, and trilobites.

The upper 16 ft (5 m) of the Martinsburg consists of less fossiliferous, gray, fine siltstone in which terrigenous quartz may reach 70% by volume. Thin skeletal-rich laminae of brachiopods and trilobite debris are scattered within this siltstone. The uppermost meter of the Martinsburg contains red siltstone clasts that are indistinguishable from the sediment of the overlying Juniata, yet the Martinsburg-Juniata contact is knife-sharp where the color changes from gray to red.

Juniata Formation. The uppermost Ordovician Juniata, 364 ft (111 m) thick, can be divided into four facies. In ascending order these are 1) a basal unit of siltstone and silty, fine-grained sandstone with occasional limestone laminae, 2) a fine- to medium-grained sandstone, 3) a shaly unit, and 4) a shale to silty shale and siltstone with minor sandstone (Fig. 6). The lower three facies contain desiccation cracks, ripple cross-laminations, flaser bedding, vertical burrows, and soft-sediment deformation features, all of which suggest tidal flat deposition. The upper 131 ft (40 m) comprising facies 4 lack physical as well as biogenic evidence for a tidal flat deposition. A deeper water, subtidal environment has been inferred for this part of the Juniata.

Clinch Formation. The Clinch (Lower Silurian) is the ridge-forming unit of Clinch Mountain, a conspicuous topograph-

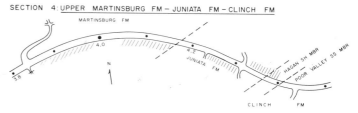

Figure 6. Section No. 4. Martinsburg Formation/Juniata Formation/ Clinch Formation. Approximate distances in miles eastward from the center of the Indian Creek bridge.

ic feature in northeastern Tennessee and southwestern Virginia. The highway traverses the Clinch, mostly along strike, from mile 4.4 to mile 6.7. This exposure of the Clinch, perhaps the best for the unit, is divisible into a lower Hagan Shale Member about 49 ft (15 m) thick and an upper Poor Valley Ridge Member about 184 ft (56 m) thick (Schoner, 1985).

The Hagan Shale Member, consisting of clayshale interbedded with siltstone and thin- to medium-bedded quartzose sandstone that becomes more prominent in the upper part of the unit, was deposited under shallow marine conditions. The increase in sand in the upper Hagan signifies an increase in the energy of water motion in the environment, perhaps during storms. Marine trace fossils present in the Hagan. Evidence of subaerial exposure has not been observed in either the Hagan Shale or the Poor Valley Ridge Members here. Quartzose sand of the upper Hagan and the bulk of the Poor Valley Ridge Member is probably reworked sediment from uplifted Chilhowee (Cambrian) or Ocoee (Precambrian) rocks to the southeast.

Four facies in the Poor Valley Ridge Member suggest barrier/bar prograding sequences in a shallow marine setting. The facies represent subtidal migration of rip current and tidal inlets cutting a barrier/bar system. Facies 1 shows medium-to-large-scale trough cross beds up to 3 ft (1 m) thick in fine- to coarse-grained sandstone with clay-drape laminae containing *Arthrophycus*. Facies 2 consists of medium- to coarse-grained sandstone. Facies 3 consists of massive, densely bioturbated, medium- to coarse-grained sandstone beds up to 10 ft (3 m) thick. *Diplocraterion, Arencolites,* and *Skolithos* are most abundant. Facies 4 consists of thin to medium horizontally laminated beds of very fine- to medium-grained sandstone with common *Arthrophycus* and *Paleophycus.*

Wildcat Valley Sandstone. Following regression of the sea after deposition of the Clinch, a thin sandstone unit, the Wildcat Valley Sandstone, was deposited. The Wildcat Valley, although not well exposed in this section, is considered to be unconformable with both the overlying Chattanooga Shale and the underlying Clinch. The Wildcat Valley represents an Early Devonian transgression and regression. The unit is 15 ft (4.5 m) thick and characterized by medium- to coarse-grained, usually friable sandstone that weathers yellowish orange and contains the brachiopod, *Costispirifer.*

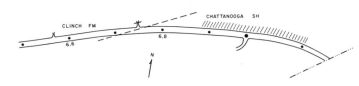

Figure 7. Section No. 5. Clinch Formation/Chattanooga Shale. Approximate distances in miles from the center of the Indian Creek bridge.

Sequence III

Chattanooga Shale. The Chattanooga Shale-Wildcat Valley Sandstone unconformable contact occurs at approximately mile 6.7 (Fig. 7). Here, the Chattanooga Shale is divisible into three members, a lower black shale, the Millboro Member; distinctly bedded siltstone and silty shale, the Brailler Member; and an upper black to gray shale, the Big Stone Gap Member. The Millboro Member (423 ft [129 m] thick) although not well exposed here, is a grayish black to very dark gray, thinly laminated shale containing inarticulate brachiopods, conodonts, plant matter, and *Tasmanites.*

The silty Brailler Member, 150 ft (46 m) thick, represents a minor regression that resulted from a lowering of sea level and/or progradation of a Devonian deltaic sequence. The siltstone may be distal turbidites of such a progradation. Most of the 300 ft (91 m) of the Big Stone Gap Member are covered, but the black to gray shale boundary and the transition with the Grainger are observable. The Devonian-Mississippian boundary is believed to be within the middle of the Big Stone Gap Member.

Grainger Formation. The 489 ft (149 m) of the Grainger exposed at this site represent nearly a complete section. The lower 135 ft (41 m) are characterized by thick- to very thick-bedded gray siltstones that weather to a distinctive light olive gray or grayish red, with bright yellowish orange joint surfaces.

The middle 302 ft (92 m) are predominantly thickly laminated, silty to very silty shales with interbeds of siltstone. The uppermost 53 ft (16 m) are thin- to very thick-bedded, very fine- to coarse-grained sandstone. The upper sandstone is commonly feldspathic with quartz pebble conglomerate and conglomeratic sandstone interbeds.

A distinctive glauconite zone, almost 8 ft (2.5 m) thick, occurs 36 feet (11 m) below the contact between the middle silty shale and the upper sandstone. This zone, consisting of interbedded glauconitic shale and silty glauconitic fine-grained limestone, is located at approximately mile 8.5. The zone is variegated, but hues of green and purple are especially notable.

The trace of the Saltville fault crosses the highway at mile 8.6.

REFERENCES CITED

Churnet, H. G., Misra, K. C., and Walker, K. R., 1982, Deposition and dolomitization of upper Knox carbonate sediments, Copper Ridge district, East Tennessee: Geological Society of America Bulletin, v. 93, p. 76–96.

King, P. B., and others, 1944, Geology and Manganese deposits of northeastern Tennessee: Tennessee Division of Geology Bulletin 52, 204 p.

Walker, K. R., ed., 1985. The geological history of the Thorn Hill Paleozoic section (Cambrian–Mississippian), eastern Tennessee: University of Tennessee Department of Geological Sciences Studies in Geology 10, 128 p.

Saltville fault at Sharp Gap, Knoxville, Tennessee

Robert D. Hatcher, Jr., Department of Geology, University of South Carolina, Columbia, South Carolina 29208

LOCATION

Sharp Gap is located on the north side of Frontage Road parallel to I-275 in Sharp Gap, which passes through Sharp Ridge, Knoxville, Tennessee (Fig. 1). It also includes exposure on the south side of Sharp Gap at road level off the side of I-275, Knox County, Tennessee, Knoxville and Fountain City 7½-minute Quadrangles, Tennessee (Cattermole, 1958, 1966).

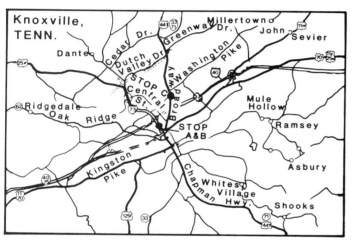

Figure 1. Location of Sharp Gap and Saltville Fault exposures.

INTRODUCTION

The Appalachian Valley and Ridge consists of a classic foreland fold and thrust belt. At the latitude of Knoxville the Valley and Ridge consists of a dominantly thrust-faulted terrane with 10 major thrusts across the belt. One of the largest of these thrusts is the Saltville fault which extends from Virginia to Georgia and has a maximum stratigraphic throw of 12,000 ft (3658 m) with Lower Cambrian rocks thrust over the Mississippian in parts of Tennessee and Virginia. Its maximum horizontal displacement is on the order of several tens of miles. The Sharp Gap stop affords an excellent opportunity to observe both a fresh exposure of this major thrust, as well as some of the complexity which occurs along it.

Rocks of the Middle Ordovician Chickamauga Group, the upper part of the Knox Group (Lower Ordovician), Middle Cambrian Conasauga Group, and the Lower Cambrian Rome Formation are exposed at Sharp Gap. The purposes of this stop are to illustrate an excellent exposure of one of the largest thrusts in the Valley and Ridge, the Saltville fault, and to examine exposures of the Rome Formation where it is freshly exposed and readily accessible in fault contact along with a footwall ramp with Knox Group dolomites and limestones. Complications in both large-scale and mesoscopic structural geometry occur at Sharp

Gap. As a result, several interesting features may be observed along the fault contact which are otherwise not readily exposed at other localities.

SITE DESCRIPTION

Stop A. Interstate 275 Road Level

Please pull all vehicles completely off the pavement. Exposed at road level on the southwest side of Sharp Gap on Interstate 275 is a very fresh suite of rocks of the Lower Cambrian Rome Formation thrust over Lower Ordovician Knox Group (Mascot) dolomites and limestones. The Rome Formation at this locality consists of interlayered variegated shales and sandstones with some layers of carbonate. The sandstone layers range up to 1.5 ft (40 cm) thick and the few dark blue-gray carbonate layers range up to 2 ft (60 cm) thick. Shale zones range up to 3 ft (91 cm) and all consist of variegated shales and sandstones colored from red to gray and green to brown. Most of the carbonates are gray and weather to a lighter tan color.

Abundant primary sedimentary structures are present in individual beds in the Rome Formation which indicate deposition occurred under intertidal conditions. Mud cracks, ripple marks, rain imprints and other features indicating shallow water to littoral conditions of deposition prevail. These are best exposed about 40 ft (15 m) above the fault and carbonate sequences.

Rocks of the Knox Group consist of massive layers of dolomite and limestone ranging in thickness from 6 to 8 in (15-20 cm) up to 3 ft (1 m). Layers have a predominantly micritic texture and contain abundant stylolites.

The Saltville fault contact is a very sharp break consisting of a brittle zone which contains abundant thin layers of well-cemented gouge. The fault zone may consist of either Rome Formation carbonates or shales on the Knox carbonates. A zone of triboplastic cataclasite is present which initially appears mylonitic. This cataclasite may have been the product of grinding and pulverization along the fault, perhaps under hydrofracturing conditions or under conditions which produced very fine gouge which flowed to produce a pseudomylonite. The fault must have been reactivated several times because of the crack-seal nature of the brittle deformation, and the footwall rocks continued to be broken.

Stop B. Saltville Fault on the North Side of Sharp Gap

The exposure of the Saltville fault on the northeast side of Sharp Gap is, on first impression, relatively simple. Variegated shales and sandstones with some interbedded carbonate of the Rome Formation are in fault contact with Mascot Dolomite of the Knox Group. However, as the cut is traversed from southeast

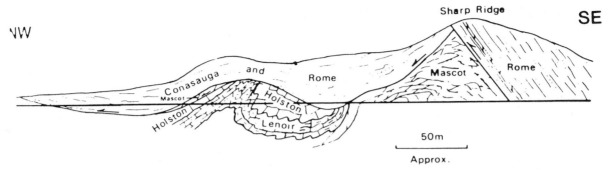

Figure 2. Cross-section showing inferred relationships of the Saltville fault on the frontage road on the northeast side of I-275 in Knoxville, Tennessee.

to northwest, the structural complexity becomes apparent (Fig. 2). The Mascot is highly deformed and Conasauga shale appears to be in sharp fault contact on top of the Mascot again at the northwest end of the first cut. Walking down the gutter beside the road into the valley to the northeast side of the road, then across the valley and into the trees next to the roadfill, exposures of Holston "Marble" may be observed. The Holston here is typical of the reddish to white clastic biosparite of this belt of Middle Ordovician rocks. Some 30 ft (10 m) northwest of this first exposure of Holston are exposures of fossiliferous shaly limestone belonging to the Lenoir Limestone, which lies in proper stratigraphic order beneath the Holston. A few yards (meters) further northwest lie exposures of Mascot Dolomite again. A fault contact must be present between the two exposures since there is observable disharmony between layering in the more highly deformed Lenoir and the less deformed (more competent) Mascot. The Mascot may actually be traced into the forest above the Holston where we first observe it. There is also a block of Holston Marble on the contact at its northwesternmost exposure as well. Climbing back up to road level and walking a few yards (meters) to the west, there is another sharp contact in which the Conasauga shale is above Mascot Dolomite.

The structure of the hanging wall at Sharp Gap consists of a low-angle sediment of the Saltville thrust which has been broken and thrust over by the back limb, bringing the Rome Formation over the top and forming the more steeply-dipping section at the southeast end of the cut (Fig. 2). The footwall structure consists of what appears to be a much more complex structure, yet it yields to interpretation as a syncline of Middle Ordovician rocks resting in a highly deformed but largely upright condition atop the Mascot Dolomite. The Mascot at this location has been interpreted by Roeder and others (1978, Fig. 6) as being locally overturned and the major structure an isoclinal syncline. This interpretation is really unnecesary. Isoclinal folding on a large scale has never been demonstrated elsewhere in the Valley and Ridge of Tennessee, although it has been demonstrated in a very different tectonic sense near Blacksburg, Virginia (Bartholomew and Lowry, 1979).

Stop C.

An additional, excellent exposure of the Saltville fault also occurs about 3 mi (5 km) to the northeast at the intersection of Dutch Valley Road and Broadway (U.S. 441) (Fig. 1). This cut is on the northeast side of U.S. 441, and the Rome Formation is in sharp contact upon upper Knox Group carbonate rocks. The fault dips steeply toward the southeast but must flatten quickly at depth, since the dip of the hanging-wall rocks also decreases to the southeast.

Dip on the principal segment of the Saltville fault at Sharp Gap (Stop A) is about 45 degrees to the southeast. As one traces the fault to the northeast, particularly on the northeast side of the Gap (Stop B), the variation in dip and the apparent folding of the fault becomes quite evident. This folding must be real, since the hanging wall has been broken through again by the main segment of Saltville fault. The exposure at Dutch Valley Road and Broadway (Stop C) is actually the northwestern segment of the Saltville fault to the northwest of its exposure at Sharp Gap (Cattermole, 1958, 1966).

REFERENCES CITED

Bartholomew, M. J., and Lowry, W. D., 1979, Geology of the Blacksburg Quadrangle, Virginia: Virginia Division of Mineral Resources, Scale 1/24,000.
Cattermole, J. M., 1958, Geologic Map of the Knoxville Quadrangle, Tennessee: U.S.G.S. Map GQ-147.

Cattermole, J. M., 1966, Geologic Map of the Fountain City Quadrangle, Knox County, Tennessee: U.S.G.S. Map GQ-513.
Roeder, D. H., Yust, W. W., and Little, R. L., 1978, Folding in the Valley and Ridge Province of Tennessee: American Journal of Science, v. 278, pp. 477–496.

Middle Ordovician shelf-edge skeletal sand-bank, organic build-ups, and quartz-sand-waves of the Holston and Chapman Ridge Formations near Knoxville, Tennessee

Kenneth R. Walker, Department of Geological Science, University of Tennessee, Knoxville, Tennessee 37996

LOCATION

This set of three outcrops is located south of the city of Knoxville, on the east side of Alcoa Highway (U.S. 129) one mi (1.6 km) south of the James E. Karnes Bridge over the Tennessee River (Fort Loudon Lake). The northernmost of the three outcrops is directly across Alcoa Highway from the Naval Reserve Station. The locality is in the southern part of the Knoxville 7½-minute Quadrangle (Fig. 1). The site is easily accessible, although Alcoa Highway is heavily traveled, and you are cautioned to cross the highway with great care. Parking is available on the east side of the highway just south of the Naval Reserve Station.

INTRODUCTION

The rocks at this locality represent the carbonate shelf-edge environments which restricted circulation and in part created the lagoon-like environment of the Lenoir Formation (visible, for example, at the Midway Road Exit from I-40 about 10 mi (16 km) east of Knoxville). Also demonstrated at this locality are the down-slope, deeper-water carbonate build-ups of the upper Holston Formation. The organic build-ups are not as well developed here as at some localities, but the other shelf-edge facies are better shown here. The Holston Formation thickens and thins markedly along strike in this outcrop belt, but is near its thickest development at this locality.

This locality is significant for several reasons: 1) The shelf margin facies exposed here contain some of the earliest *platform edge* reef masses in the geological record; 2) the organic build-ups here are the earliest red-mud/encrusting-organism, down-slope masses (thus, similar to the "Walsortian" type) in the geological record; 3) various rock types here demonstrate both current- and noncurrent-induced sedimentary structures (e.g., cross lamination); and 4) samples from this locality demonstrate the great abundance and variety of early, marine cements in rocks of Middle Ordovician age.

The reader is referred to Walker and others, (1983), Ruppel and Walker (1982, 1984), Walker and others, (1980), Benedict and Walker (1978), and Shanmugam and Walker (1978) for more complete treatment of the entire Middle Ordovician sequence of Tennessee. A fuller treatment of the reefy parts of the Holston is in Walker and Ferrigno, 1973.

One of the common characteristics of all parts of the Holston at every locality is the great abundance of pseudofibrous and radiaxial calcite cements. These types of cements have been interpreted by other workers as representing marine cements of very

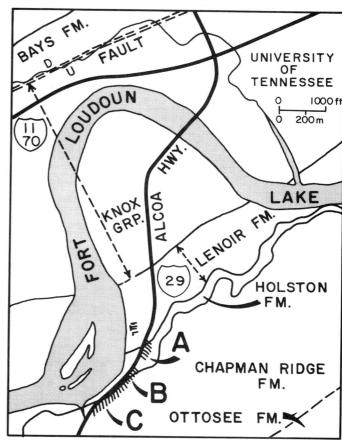

Figure 1. Location of outcrops A, B, and C of the Middle Ordovician Holston Formation along U.S. 129 (Alcoa Highway) south of Knoxville, Tennessee. North is toward the top of the map, and the distance from Fort Loudoun Lake to Outcrop A is one mile.

early diagenetic origin. Many grainstone samples from the Alcoa Highway locality also demonstrate the contemporaneity of other marine cements and syntaxial calcite overgrowths on echinoderm grains indicating that the latter are also of marine origin.

SITE DESCRIPTION

The outcrops here can be divided readily into three parts, here called parts A, B, and C (Fig. 1). Outcrop A, which is the northernmost outcrop, consists of rocks of the lower Holston Formation. Bryozoan/pelmatozoan organic build-ups (muddy boundstones) are rare in this part of the formation, and the lithologies are mostly pelmatozoan/ramose-bryozoan skeletal sands (grainstones). The most striking feature of these lithologies is the

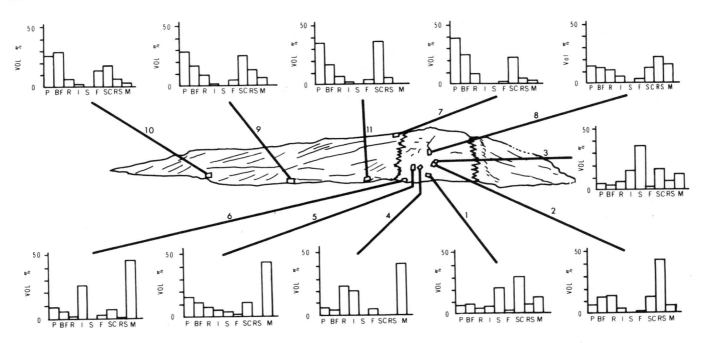

Figure 2. Drawing of small reef body (about 100 ft [30 m] across) near the southern end of outcrop B, showing the composition of samples from its various parts. Samples 1, 3-6, and 8 are from the core facies; 2, 7, and 9-11 are from the flank facies. Components are: P = pelmatozoan debris, BF = bryozoan fragments, R = ramose bryozoa, I = incrusting bryozoa *in situ*, S = stromatactis structures, F = other fossils, SC = mosaic calcite cement, RS = radiaxial calcite cement, M = fine-grained matrix. Note complete absence of algae.

large scale, debris-slope and avalanche type cross beds. The general dip at this outcrop is 20° to 25° to the southeast, and apparent dip is less. Thus, all of the steeply dipping bed-sets here are cross beds. Corrected dips of these exceed 30°, and the sets are roughly triangular in cross section. Occasional sets of bidirectional cross laminae indicate the activity of tidal ebb and flood currents. On the north end of the outcrop, grainstones of the Holston are interbedded with lagoonal wackestones of the Lenoir Formation.

The biota of Outcrop A is difficult to collect because of the massive nature of the rocks; our knowledge is restricted to that derived from thin section study and the conodont work of Bergstrom and Carnes, 1976. The megafauna consists of pelmatozoan debris and less abundant ramose bryozoans (dominantly species of *Batostoma* and *Helopora*). The bryozoan fauna is probably as diverse as that higher in the Holston, but is much less well studied. Algae are present although their abundance decreases upward. The dominant forms are *Girvanella* (lumps), *Solenopora, Contexta,* and *Vermiporella* (Moore, 1977). Algal and fungal borings are common in skeletal grains. The lower Holston lies within the upper *Pygodus anserinus* Conodont Zone according to Bergstrom and Carnes (1976).

These rocks represent a shallow water, skeletal sand-bank environment, based on the abundant current and debris-slope cross bedding, and the presence of algae. Toward the southeast, shales of deep water origin are coeval with these rocks (Shanmu-gam and Walker, 1980), so these skeletal sand-banks had a shelf marginal position.

Outcrop B of this locality (Fig. 1) consists in the lower part of rocks similar to those of Outcrop A, but the upper 66 ft (20 m) are made up of pelmatozoan/bryozoan organic build-ups (Walker and Ferrigno, 1973). These consist of anastomosing, bryozoan boundstone masses of dark red to maroon, lime mudstone which represent reef core deposits. The masses are separated from each other by cross beds of pink to white, pelmatozoan/bryozoan grainstone which are interpreted as reef flank deposits. Reef core facies only becomes common near the southern end of Outcrop B, in the upper quarter of the Holston. In most outcrops of the Holston, this subdivision of the formation is usual, with the lower two-thirds to three-quarters dominated by skeletal sand-bank deposits, and the upper quarter to third containing abundant build-ups. Although there are numerous boundstone areas in this outcrop, one small reef mass at the extreme southern end of Outcrop B is particularly instructive. Figure 2 shows the shape of this mass and its associated flanking beds, and the composition of samples from the reef. The core facies is characterized by its high content of red lime mud and abundance of bryozoa. The most abundant of the bryozoa are: the bifoliate ramose *Stictopora,* the subcylindrical ramose *Bythopora,* and the incrusters *Amplexopora, Mesotrypa, Constellaria,* and *Hemiphragma.* The flank beds consist of pelmatozoan grains with subordinate ramose bryozoa (chiefly *Stictopora* and *Bythopora*).

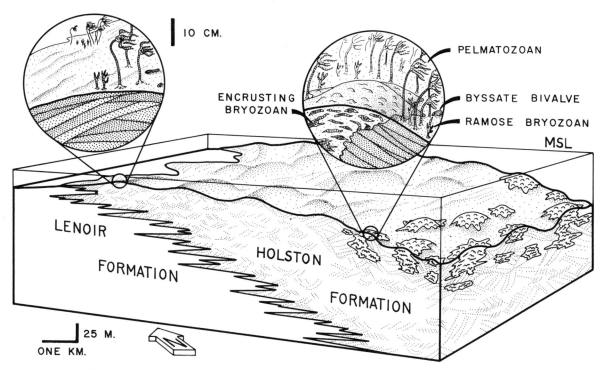

Figure 3. Postulated environments of deposition for the lower and upper Holston Formation.

Incrusting bryozoa are virtually absent from the flank beds. The cross beds in the upper (reefy) part of the Holston are not of current origin, but formed as skeletal debris-slopes.

The most significant aspect of the upper Holston is the complete absence of algae. In addition, although the upper Holston was deposited near a shelf edge, current-induced sedimentary structures are virtually absent. For these reasons, the upper Holston is interpreted as a deposit of a down-slope environment below algal compensation depth and below effective wave base at a depth of about 330 ft (100 m). Figure 3 shows a depositional model for the skeletal sand-banks and reefs of the Holston.

The top of the Holston is often characterized, as it is here, by the presence of a thin, gray, nodular, shaley limestone containing phosphatic and manganiferous nodules. This unit represents a period of very slow deposition and a pronounced change in environmental conditions. The reappearance of *Girvanella* (algal) coatings on some nodules indicates a shallowing of water at this locality. This unit also marks the first appearance of abundant fine-grained, terrigenous clastic material in the shelf-edge sequence. Although a fall in sea level may be indicated by this unit, its regional significance is problematical.

Outcrop C of this locality begins at the top of the nodular bed and continues southward to Woodson Drive (about 1,650 ft) (500 m). The outcrop consists of variably quartzose skeletal sands of the Chapman Ridge Formation. The lower few meters of this formation contain little quartz silt or sand, but consist of pelmatozoan/bryozoan skeletal sand superficially similar to the grainstones of the Holston. The Chapman Ridge lithology differs

significantly from that of the Holston, however, in its characteristic thin to medium, originally horizontally bedding, and its darker maroon color. This lowermost part of the Chapman Ridge Formation may also represent unusually slow deposition, for large numbers of orthoconic cephalopod shells are sometimes concentrated along bed surfaces. This lower quartz-poor lithology grades upward into rocks containing more quartz silt and fine sand which make up the bulk of the Chapman Ridge. Current-induced sedimentary structures are particularly common in all of these rocks. "Herringbone" (bidirectional) cross beds indicate the action of tidal currents, and ripple-drift cross lamination the activity of wave-induced currents. Gently dipping sets of low-angle cross beds become common a few tens of feet (meters) above the base of the formation. The current cross bedding of the Chapman Ridge is distinctly different in character from the debris-slope cross bedding of the underlying Holston. Tracks and trails left by invertebrates are abundant on many bedding plains; some surfaces are crowded with trilobite resting tracks (usually preserved as casts on the bottoms of overlying beds). Many beds show oscillation ripple marks. Small-scale festoon cross bed-sets occur only rarely in the Chapman Ridge. The fauna is dominated by pelmatozoan grains, bifoliate ramose bryozoa (particularly *Pachydictya* and *Stictopora*), and trilobite debris. Although not very abundant, a relatively diverse algal flora was present in the Chapman Ridge environment.

The presence of skeletonized algae and abundant wave- and current-induced sedimentary structures indicate deposition above algal compensation depth and effective wave base in a few to a few tens of ft (m) of water. The environment of deposition of the Chapman Ridge is interpreted as an offshore area of subtidal sand waves.

The Chapman Ridge sand body grades southeastward and northeastward into deeper water shales (the Blockhouse and Sevier Formations), and northwestward into shallow water shelf carbonates. Because most earlier workers looked directly southeastward for the source region for the terrigenous sands and silts of this unit, they were puzzled by the absence in that direction of coeval coarse clastics in the adjacent fault blocks. A regional viewpoint clears up that puzzle. The terrigenous sands of the Chapman Ridge can be traced southwestward along strike to the Athens, Tennessee, area where they can be correlated with a thick mass of very coarse, terrigenous sands toward the southeast across strike in the area of Mount Vernon, Tennessee. This relationship is well shown in the facies maps of this sequence published by Walker and others (1983).

REFERENCES CITED

Benedict, G. L., III, and Walker, K. R., 1978, Paleobathymetric analysis in Paleozoic sequences and its geodynamic significance: American Journal of Science, v. 278, p. 579–607.

Bergstrom, S. M., and Carnes, J. B., 1976, Conodont biostratigraphy and paleoecology of the Holston Formation (Middle Ordovician) and associated strata in Eastern Tennessee, in Conodont Paleoecology, Barnes, C. R. (ed.), p. 27–58, Geological Association of Canada, Special Publication 15.

Moore, N. K., 1977, Distribution of the benthic algal flora in Middle Ordovician carbonate environments of the Southern Appalachians, in Ruppel, S. C., and Walker, K. R., The ecostratigraphy of the Middle Ordovician of the Southern Appalachians (Kentucky, Tennessee, and Virginia) U.S.A., University of Tennessee, Studies in Geology, No. 1, p. 18–29.

Ruppel, S. C., and Walker, K. R., 1982, Sedimentology and distinction of carbonate buildups: Middle Ordovician, east Tennessee: Journal of Sedimentary Petrology, v. 52, p. 1055–1071.

—— 1984, Petrology and depositional history of a Middle Ordovician carbonate platform: Chickamauga Group, northeastern Tennessee: Geological Society of America Bulletin, v. 95, p. 568–583.

Shanmugan, G., and Walker, K. R., 1978, Tectonic significance of distal turbidites in the Middle Ordovician Blockhouse and lower Sevier formations in east Tennessee: American Journal of Science, v. 278, p. 551–578.

Walker, K. R., and Ferrigno, K. F., 1973, Major Middle Ordovician reef tract in east Tennessee: American Journal of Science, v. 273-A (Cooper Volume), p. 294–325.

Walker, K. R., Broadhead, T. W., and Keller, F. B., 1980, Middle Ordovician carbonate shelf to deep water basin deposition in the Southern Appalachians: Guidebook for Field Trip 10, Geological Society of America Annual Meeting, University of Tennessee, Studies in Geology, No. 4.

Walker, K. R., Shanmugan, G., and Ruppel, S. C., 1983, A model for carbonate to terrigenous clastic sequences: Geological Society of America Bulletin, v. 94, p. 700–712.

Cumberland Plateau décollement zone at Dunlap, Tennessee

Robert L. Wilson, *The University of Tennessee at Chattanooga, Chattanooga, Tennessee 37403*
Steven F. Wojtal, *Oberlin College, Oberlin, Ohio 44074*

LOCATION

The Cumberland Plateau décollement zone is exposed on Tennessee 8, 2.3 mi (3.8 km) northwest of the junction of the Tennessee 8 and U.S. 127 north of Dunlap, Tennessee (Fig. 1), Savage Point, 7½-minute quadrangle. Deformed rocks, exposed along the northeast side of the highway, in almost continuous outcrop for more than 1.3 mi (2.2 km), denotes the décollement zone. Wide shoulders along the roadway will accommodate groups of up to 50 persons.

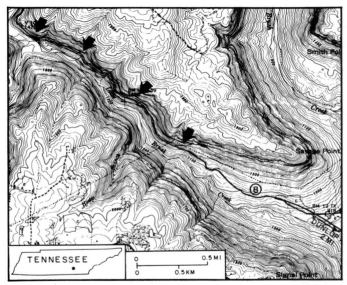

Figure 1. Part of the Savage Point quadrangle (104-NW) showing the location of the Cumberland Plateau décollement zone along Tennessee 8.

INTRODUCTION

The Cumberland Plateau thrust, first described by Stearns (1954), marks the western limit of Alleghanian deformation in East Tennessee. Cumberland Plateau thrust sheet, is one of several large structural features in this part of the Appalachians (Fig. 2). The thrust along the leading edge of this sheet has been traced as far southwest as Spencer, Tennessee (in Van Buren County). The Emory River cross fault marks the northeastern edge of the sheet, and the frontal thrust joins the Emory River cross fault southwest of Wartburg, Tennessee (in Morgan County). From Harriman, Tennessee, southwest to Chattanooga, Tennessee, the Rockwood thrust marks the trailing edge of the Cumberland Plateau sheet. In the Harriman and Crossville areas, the Cumberland Plateau sheet underlies a broad area of high, rugged topography nearly 15 mi (25 km) wide across strike. Farther southwest, the Cumberland Plateau is split by the Sequatchie Valley. This valley, over 1600 ft (487 m) deep for most of its length, has also breached the Cum-

berland Plateau sheet and divides it into two parts. The trailing edge of the western part of the Cumberland Plateau sheet, the Cumberland Plateau thrust, is exposed along the west wall of the Sequatchie Valley. The leading edge of the eastern part of the Cumberland Plateau sheet, the Sequatchie Valley thrust, crops out along the west side of the valley floor (Wilson and Stearns, 1958; Milici, 1963; 1970). The two faults join at the north end of the Sequatchie Valley, indicating that the Cumberland Plateau thrust is an upper glide horizon associated with the steeply dipping Sequatchie Valley thrust (Milici, 1963; 1970; Fig. 3). The southward extent of the trailing thrust of the western part of the sheet is not certain, but small outliers of older rocks thrust on younger rocks near the Alabama-Tennessee state line support Wilson and Stearns' (1958) contention that it extends into Alabama. The leading thrust of the eastern part of this large sheet, the Sequatchie Valley thrust, on the other hand, extends several tens of miles to the southwest (Rodgers, 1970).

Undeformed footwall strata below and to the west of the Cumberland Plateau thrust and to the northeast of the Emory River fault dip uniformly 1°E, away from the axis of the Nashville dome 45 mi (75 km) to the west. Strata within the Cumberland Plateau sheet are not extensively deformed. Broad domains of flat-lying strata in the hanging wall are separated by large Rich-type anticlines formed by duplicated strata above the step-shaped thrust surface. For example, the Sequatchie Valley anticline, the largest structure in the Cumberland Plateau sheet, was formed by duplication of the thick Cambro-Ordovician Knox carbonates above the Sequatchie Valley thrust (Rich, 1934). Likewise, smaller, doubly plunging anticlines in the Crab Orchard Mountains (east of Crossville) lie above frontal and lateral ramps in the footwall of the Cumberland Plateau thrust fault (Wilson and Stearns, 1958; Milici, 1963).

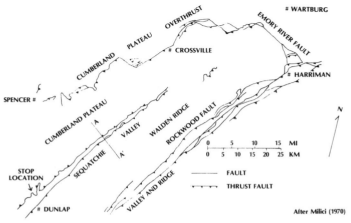

Figure 2. Major structural features of the Cumberland Plateau in East Tennessee.

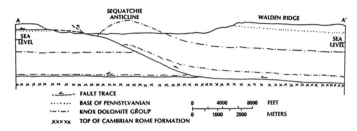

Figure 3. Geologic cross section of the Sequatchie Valley–Cumberland Plateau thrust and the Sequatchie Anticline.

Ordovician limestones from the Knox Group are thrust on Mississippian rocks along the Sequatchie Valley thrust, indicating that the displacement along this thrust is at least 1.8 mi (3 km) (Fig. 3). Displacement on this thrust must decrease up its dip, either by interbed slip within hanging wall units or by distributing slip onto some now-eroded fault branch. Harris and Milici (1977) state that displacement on the upper glide horizon, where Pennsylvanian units are thrust on slightly younger Pennsylvanian units, is not more than 0.6 mi (1 km) and probably about 1,000 ft (305 m).

Deformation at the Dunlap exposure of the Cumberland Plateau thrust was, for the most part, accomplished by minor faults. These minor faults occur in very regular arrays (Fig. 4 and 5), indicating that they are part of this foreland sheet's ordered response to the forces of westward transport. Features seen here are typical of foreland fold-and-thrust-belt thrusts (Harris and Milici, 1977; Wojtal, 1982; 1985). Two factors make the Dunlap exposure unique. First, the deformed rocks exposed at Dunlap are part of a thrust sheet of regional extent, yet they have not traveled a great distance. Harris and Milici (1977) argued that displacement on this part of the Cumberland Plateau thrust is less than 0.6 mi (1 km). The décollement zone rocks at Dunlap provide insight into the history of the early stages of a typical foreland thrust. Second, rocks at Dunlap have never been over a major footwall ramp. The décollement zone rocks at Dunlap

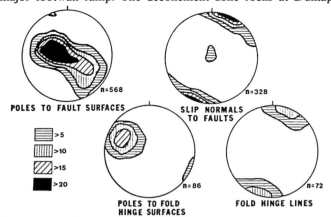

Figure 4. Lower-hemisphere equal-area net projections of structural data from the Cumberland Plateau décollement zone. Contours in multiples of sigma, the calculated standard deviation of a random distribution of the same size (Kamb, 1959). Values of 2 to 3 sigma indicate point concentrations significantly different from those in random distributions.

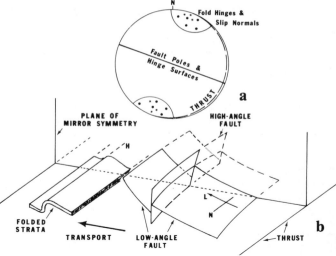

Figure 5. (a) Lower hemisphere stereographic projection summarizing the geometry of deformation elements from the Cumberland Plateau décollement zone. Poles to fold hinge surfaces and poles to fault surfaces define a great circle that is perpendicular to the thrust and is aligned parallel to transport. Slip normals to faults and fold hinge lines cluster are generally aligned normal to the great circle. (b) Sketch showing the general geometry of structures in the Cumberland Plateau décollement zone. Note transport direction, plane of mirror symmetry parallel to great circle defined in (a), and low-angle fault cut by high-angle fault. H = fold hinge line. L = slip direction on low-angle fault. N = slip normal to low-angle fault.

developed while a hanging-wall flat composed of Pennsylvanian sediments rode over a footwall flat composed of slightly younger to laterally equivalent Pennsylvanian strata. That the deformation in this décollement zone is similar to that observed elsewhere suggests that the "transport fabric" of low- and high-angle faults in foreland sheets develops early in the sheet's existence and is not dependent upon travel over a major footwall ramp.

In addition to their regular geometry, the minor faults locally developed in a recognizable sequence. Low-angle faults precede high-angle faults at all points. Low-angle faults at different points in the décollement zone may not be contemporaneous. Low-angle faults in duplexes, which are common here (and along other décollement zones, according to Dahlstrom, 1970), probably developed sequentially. Harris and Milici (1977) argued that duplexes seen here were formed by hanging-wall imbrication and that this imbrication led to the "abandonment" of fault-bound slivers of hanging-wall rocks along thrusts and to the thinning of hanging-wall units during transport. Wojtal (1982), on the other hand, argued that local fault interactions indicate that footwall imbrication was more common at this exposure. Likewise, the décollement zone probably developed during a time period of sizable length. Stratigraphic evidence from other mountain chains shows that thrust sheets move with mean velocities of 3.98^{-10} ft/sec to 3.98^{-9} ft/sec (10^{-10} m/s to 10^{-9} m/s) Eliott, 1976). Assuming that these estimates are applicable to thrust sheets in the Appalachians and that the Cumberland Plateau sheet moved

0.6 mi (1 km) to the west-northwest, the deformation in the décollement zone developed during a time period of approximately 500,000 years in length.

SITE DESCRIPTION

The best exposure of the Cumberland Plateau thrust is found along the trailing edge of the western part of the Cumberland Plateau thrust sheet, on Tennessee Highway 8 west of Dunlap, Tennessee. Deformed rocks, up to 300 ft (91 m) from the thrust surface, outcrop in nearly continuous exposures along the north side of the highway. As expected for rocks near the western limit of deformation, the deformation is not penetrative but occurs by sliding on discrete minor faults and by outcrop scale folding. Harris and Milici (1977) first described the structures exposed at this location in their study of the characteristics of foreland thrust faults. They assigned deformed rocks near this thrust to two distinct zones based upon the local character of the minor faults: a lower broken zone and an upper fractured zone. Their lower broken zone is characterized by beds cut by low-angle "splay" faults with reverse offsets and high-angle "shear and rotational normal" faults that typically have normal offsets of bedding. Their upper fractured zone is characterized by predominantly normal faulting; normal faults in this upper part of the deformed hanging wall are either shear and rotational normal faults or "gravity normal" faults. Plate 3 in Harris and Milici (1977) is an interpretative sketch section of the décollement zone seen along Tennessee Highway 8. Note that all parts of their sketch section are not parallel: some are nearly normal to the thrust's strike while others are nearly parallel to it. A similar sketch section of a part of the décollement nearly 1 mi (about 1.6 km) long was presented by Bartholomew and others (1980).

Wojtal's (1982) detailed analysis of minor structures from the 330-ft (100-m) thick deformed zone above the Cumberland Plateau thrust (called the décollement zone in the remainder of this text) elaborated on several features noted by Harris and Milici (1977). Wojtal (1982) assigned minor faults throughout the décollement zone to three groups, based upon the dihedral angle between bedding and the fault. Some minor faults parallel either primary or secondary bedding. Bedding-parallel faults are characterized by surface markings such as slickensides or fibrous mineral coatings. They usually cannot be traced more than a few feet at these exposures, although one bedding-parallel fault follows a locally persistent coal horizon and can be traced several hundred feet. Sliding lineations on bedding-parallel faults are usually aligned west-northwest, parallel to the inferred transport direction of the Cumberland Plateau sheet.

Faults that cut bedding at angles less than 45° are common throughout the decollement zone (Fig. 5). Low-angle faults typically strike parallel to the underlying thrust. Lineations on low-angle faults are usually oriented down the dip of the minor faults, and reverse offsets of bedding are most common on low-angle faults. Harris and Milici (1977) called these low-angle faults splays; they are contraction faults that result in shortening of bedding (Norris; 1958). Near the thrust, east-dipping contraction

Figure 6. Photograph of a part of the intensely deformed rock near the base of the Cumberland Plateau décollement zone. Note low-angle faults cut and offset by high-angle faults. Scale is 12 in. (30.5 cm).

faults are more common than those that dip west. The east-dipping low-angle faults commonly occur in small duplexes. Farther from the thrust, opposed dip complexes are common (Fig. 6).

Faults that cut bedding at angles ≥45° are also common throughout the décollement zone. There are at least two kinds of high-angle faults; the character of an individual high-angle fault is largely dependent upon the position of that individual fault relative to the thrust. Large, regularly spaced high-angle faults are common near the thrust (within 30 ft; 9 m) (see Fig. 6). These faults typically strike parallel to the underlying thrust and dip west. These regularly spaced high-angle faults cut both bedding and low-angle faults; both features are offset down-to-the-west, resulting in a normal series of vertical separations. Lineations on these regularly spaced high-angle faults usually have rakes ≤45°, and in some cases, lineations are nearly parallel to the minor fault's strike. Strike-parallel lineations and the offsets of bedding and low-angle faults across these faults indicate that most are oblique-slip faults. Offsets on these faults decrease up the dip of minor faults, and these oblique-slip faults end in drape folds of overlying sedimentary strata. Harris and Milici (1977) called these high-angle faults "shear and rotational normal faults."

High-angle faults greater than 30 ft (9 m) from the thrust plane are subtly different from those nearer the thrust plane (Fig. 7). Like the high-angle faults near the thrust plane, those faults

Figure 7. Photograph of high-angle faults underlying the Whitwell Shale at the west end of the Dunlap exposure of the Cumberland Plateau thrust.

farther from the thrust plane generally strike parallel to the underlying thrust plane and dip steeply to the west. They also cut both bedding and low-angle faults, and the sense of vertical separation of both is generally "normal." Lineations on high-angle faults more than 30 ft (9 m) from the thrust plane are usually down dip, though, and there is no evidence for any component of movement parallel to the strike of the fault. East-dipping high-angle faults are more common at this location. The high-angle faults found farther from the thrust plane are the extensional faults that result in extension parallel to bedding (Norris, 1958); Harris and Milici (1977) called them "gravity normal" faults.

In addition to mesoscopic minor faults, other deformation features are visible. Within 33 ft (10 m) of the thrust surface, the rock between minor faults is cut by numerous microfaults. Microfaults typically have offsets less than 0.1 in. (1 mm), but they are locally quite numerous. The rock is also cut by numerous mineral-filled veins. More than 33 ft (10 m) from the thrust surface, microfaults generally are restricted to rock immediately adjacent to minor faults, and they commonly are not present. Mineral-filled veins are common throughout the décollement

Figure 8. Photograph of small, angular, inclined horizontal fold, asymmetric to west. Scale is 12 in. (30.5 cm).

zone, although they are less numerous in rock farther from the thrust surface. Inclined horizontal folds (Ragan, 1985), which are overturned to the west, are common in the strata between minor faults throughout the décollement zone (Fig. 8). Folds in sandstone units have roughly parallel geometries, whereas those in the more shaly horizons often depart from parallel geometry. Even where folding is noncylindrical and disharmonic, fold hinge surfaces strike parallel to the underlying thrust and dip steeply east. Interlimb angles are typically 90°, but may be as tight as 30°.

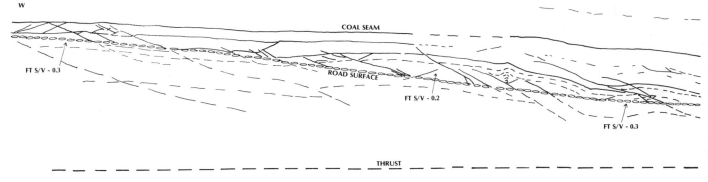

W

COAL SEAM

FT S/V - 0.3

ROAD SURFACE

FT S/V - 0.2

FT S/V - 0.3

THRUST

Figure 9. 1:600 scale downplunge projection of Cumberland Plateau décollement zone. Plane of section is parallel to great circle defined in Figure 5 (a). Cross section constructed by computer compilation of information from 1:50 scale photographs of exposure.

Most folds are intimately related to faults. Anticlines are often cored by contraction faults, and synclines pass upward into contraction faults. In the remaining cases, folds are related to steps in minor fault surfaces or are drag folds associated with movement on minor faults.

Stereographic projections show that deformation elements from the décollement zone are consistently oriented and geometrically related (Fig. 4). Minor faults strikes are typically subparallel to the thrust's strike, so poles to minor faults define a great circle that is normal to the fault and contains the direction of sheet movement. Slip normals, which are the poles to the great circles that contain both the pole to a sliding surface and the direction of the sliding lineation (Price, 1967) for minor faults, are also well oriented; they either lie on the great circle defined above or are oriented normal to it. Poles to fold hinge surfaces are centered in a point maximum that also lies on the great circle; fold hinge lines are typically normal to the great circle. The overall deformation has a distinct monoclinic symmetry, with a plane of mirror symmetry oriented normal to the thrust surface and parallel to the direction of sheet movement (Fig. 5).

Harris and Milici (1977) argued that the deformation observed in the decollement zone is part of a laterally extensive layer along the base of the thrust sheet formed during its movement. Wojtal (1982) provided further support for their interpretation, but suggested that deformation style changes more gradually with increasing distance from the thrust surface than is implied by Harris and Milici's (1977) two-part division. Figure 9, a large-scale downplunge projection of the decollement zone, shows that structures in the decollement zone are part of a laterally extensive deformed layer. This is, perhaps, most clearly seen in strata immediately adjacent to the thrust, where both large-offset low-angle contraction faults and regularly spaced, oblique-slip high-angle faults are common. The style and intensity of deformation in this 33-ft (10-m) thick layer are markedly consistent along a line parallel to transport nearly 660 ft (201 m) long. Above this 33-ft (10-m) thick layer is a 100–135-ft (30–41-m) thick layer of rock cut by low-angle faults with rare high-angle faults. Deformation style is generally uniform along the length of this second layer. At different locations in the cross section, one can observe a similar transition in deformation style from the base of the layer, where large-offset low-angle (contraction) faults are

Figure 10. Photograph of laterally continuous coal seam described in text. Thickness of coal seam is structurally controlled. Note deformation beneath coal seam, while overlying sediments are relatively undisturbed.

relatively common, to the top, where small-offset low-angle (contraction) faults interact with a laterally persistent, 4–8-in. (10–20-cm) thick coal seam.

This coal seam, which pinches out to the east but persists continuously to the west, can be examined in continuous outcrop along the road for more than 0.5 mi (1 km). Nowhere along the entire exposure of the seam do any major structural features cut through the coal, despite the fact the style of deformation above the coal is markedly different from that below the coal. The structural behavior of material near the coal horizon underscores the interplay of the stratigraphy with deformation in the decollement zone (Fig. 10). By cataloging the offsets on low-angle faults beneath the coal seam, Wojtal (1982) showed that the seam is the locus of an important bedding-parallel fault. An aggregate of more than 30 ft (9 m) of top-to-the-west layer parallel slip was accommodated by the coal seam. The morphology of the seam is consistent with the interpretation that it was the locus of slip. The rooted seat earth immediately beneath the coal is highly fractured, and many of the fractures have indications of sliding such as scratches and striations. Locally along the exposure, the coal layer suddenly thickens into forms that have lens- or pod-shaped cross sections. The three-dimensional shape of these forms cannot be directly observed, but features associated with the forms sug-

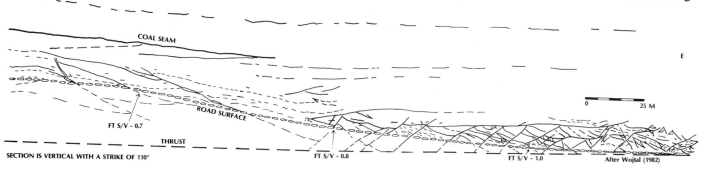

COAL SEAM

ROAD SURFACE

FT S/V - 0.7

THRUST

SECTION IS VERTICAL WITH A STRIKE OF 110°

FT S/V - 0.8

FT S/V - 1.0

After Wojtal (1982)

0   25 M

Figure 9. (*Continued*).

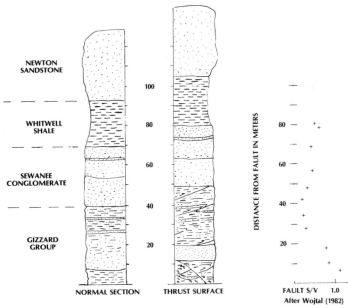

NEWTON
SANDSTONE

WHITWELL
SHALE

SEWANEE
CONGLOMERATE

GIZZARD
GROUP

DISTANCE FROM FAULT IN METERS

NORMAL SECTION   THRUST SURFACE

FAULT S/V   1.0

After Wojtal (1982)

Figure 11. Comparison of deformed and undeformed stratigraphic sections for décollement zone at Dunlap. Fault surface area per unit volume measurements are shown in relation to distance from the fault.

cited are consistent with top-to-the-west displacement of strata above the coal seam.

Deformation in the strata above the coal horizon is restricted to the occurrence of regularly spaced high-angle faults that extend strata along a line more northerly than the sheet's transport direction (Fig. 7). Harris and Milici (1977) argued that localized occurrences of high-angle faults such as these were due to alternating periods of extension and compression parallel to transport during emplacement of the sheet. Wojtal (1982) noted that open folds with wavelengths of about 1,000 ft (305 m), amplitudes of 5 to 10 ft (1.5 to 3 m), and axes with bearings of N40E are apparent from stereographic projections of poles to bedding from the upper part of the décollement zone. He suggested that the layer-parallel extension above the coal horizon was due to "outer-arc" extension around these folds.

The gradual overall change in structural style across the décollement zone suggests that deformation intensity decreases more smoothly with distance from the thrust than is implied by Harris and Milici's (1977) two-fold division of the décollement zone. Figure 11, which gives fault surface area per unit volume (measured using methods outlined in Underwood, 1970) for minor faults with an observed length greater than 3 ft (0.9 m), provides further support for this inference. Values of fault surface area per unit volume are greatest (on the order of 0.3 sq ft/cu ft [1.0 m^2/m^3]) in the 33-ft (10-m) thick layer near the thrust. Fault surface area per unit volume values decrease irregularly with increasing distance from the thrust; the fault surface area per unit volume in the upper part of the décollement zone is usually about 0.06 to 0.15 sq ft/cu ft (0.2 to 0.5 m^2/m^3).

gest that they persist for some distance normal to the plane of the outcrop (Wojtal, 1982). Fractures in the coal within the lenses have sigmoidal traces that look like stylized "Z folds" when looking north. These fractures strike roughly parallel to the thrust's strike and are often marked by down-dip lineations. All features

REFERENCES CITED

Bartholomew, M. J., Milici, R. C., and Schultz, A. P., 1980, Geologic structure and hydrocarbon potential along the Saltville and Pulaski thrusts in Southwestern Virginia and Northern Tennessee: Virginia Division of Mineral Resources, Publication no. 23, Part A, Sheet 1.

Dahlstrom, C.D.A., 1970, Structural geology in the eastern margin of the Canadian Rocky Mountains: Bulletin Canadian Petroleum Geology, v. 18, no. 3, p. 332–406.

Elliott, D., 1976, The energy balance and deformation mechanisms of thrust sheets: Royal Society of London Philosophical Transactions, Series A, v. 283, p. 289–312.

Harris, L. D., and Milici, R. C., 1977, Characteristics of thin-skinned style of deformation in the Southern Appalachians, and potential hydrocarbon traps: U.S. Geological Survey Professional Paper 1018, 40 p.

Kamb, B. C., 1959, Ice petrofabric observations from Blue glacier, Washington, in relation to theory and experiment: Journal of Geophysical Research, v. 64, no. 11, p. 1889–1909.

Milici, R. C., 1963, Low-angle overthrust faulting, as illustrated by the Cumberland Plateau–Sequatchie Valley fault system: American Journal of Science, v. 261, p. 815–825.

——1970, The Allegheny structural front in Tennessee and its regional tectonic implications: American Journal of Science, v. 268, p. 127–141.

Norris, D. K., 1958, Structural conditions in Canadian coal mines: Geological

Survey of Canada Bulletin, v. 44, 54 p.

Price, R. A., 1967, The tectonic significance of mesoscopic subfabrics in the southern Rocky Mountains of Alberta and British Columbia: Canadian Journal of Earth Sciences, v. 4, p. 39–70.

Ragan, D. M., 1985, Structural geology: An introduction to geometric techniques (3rd edition): New York, John Wiley and Sons, 393 p.

Rich, J. L., 1934, Mechanics of low-angle overthrust faulting as illustrated by Cumberland thrust block, Virginia, Kentucky, and Tennessee: American Association of Petroleum Geologists Bulletin, v. 18, no. 12, p. 1634–1654.

Rodgers, J., 1970, The tectonics of the Appalachians: New York, Wiley-Interscience Publishers, 271 p.

Stearns, R. G., 1954, The Cumberland Plateau overthrust and geology of the Crab Orchard Mountains area, Tennessee: Tennessee Division of Geology Bulletin 60, 47 p.

Underwood, E. E., 1970, Quantitative stereology: Reading, Addison-Wesley Publishing Company, 274 p.

Wilson, C. W., Jr., and Stearns, R. G., 1958, Structure of the Cumberland Plateau, Tennessee: Geological Society of America Bulletin, v. 69, no. 10, p. 1283–1296.

Wojtal, S. F., 1982, Finite deformation in thrust sheets and their material properties [Ph.D. thesis]: Baltimore, Johns Hopkins University, 303 p.

——1985, Deformation within foreland thrust sheets by populations of minor faults: Journal of Structural Geology (in press).

Whiteoak Mountain synclinorium, Bradley County, Tennessee

Robert L. Wilson, Department of Geology, University of Tennessee at Chattanooga, Chattanooga, Tennessee 37401

LOCATION

Part of the Whiteoak Mountain synclinorium is exposed near the Hamilton-Bradley County line along I-75 about 21 mi (33.8 km) east of downtown Chattanooga (Fig. 1), Snow Hill, 7½-minute quadrangle. Wide shoulders along the roadway permit vehicular parking, although prior permission for stopping should be obtained from the Tennessee Highway Patrol. This stop displays superb exposures of Valley and Ridge faults and folds.

INTRODUCTION

The western half of the Valley and Ridge is dominated structurally by a series of thrust faults. In almost all cases, especially in the northeastern parts of Tennessee, the Rome Formation of Cambrian age is thrust over younger strata. The faults of this belt can be grouped into three major families of faults (Rodgers, 1953), namely, the Kingston, the Clinchport, and Saltville.

The relationship of the Whiteoak Mountain synclinorium to the various structural features of the Hamilton-Bradley County areas can be seen in Figure 2. Hamilton County to the west consists mainly of a faulted anticlinorium which lies between the Walden Ridge syncline to the west and the Whiteoak Mountain synclinorium to the east. Both these synclines preserve rocks of Pennsylvanian age. In contrast, the central part of Hamilton County contains extensively faulted strata of Cambrian and Or-

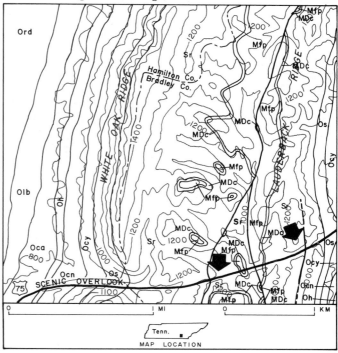

Figure 1. Part of the Snow Hill 7½-minute geologic quadrangle (after Wilson, 1983). Western arrow: location of Figure 4; eastern arrow: location of Figures 5-8. Ord = Ridley Limestone; Olb = Lebanon Limestone; Oca = Carters Limestone; Oh = Hermitage Formation; Ocn = Cannon Formation; Ocy = Catheys Formation; Os = Sequatchie Formation; Sr = Rockwood Formation; MDc = Chattanooga Shale; Mfp = Fort Payne Formation.

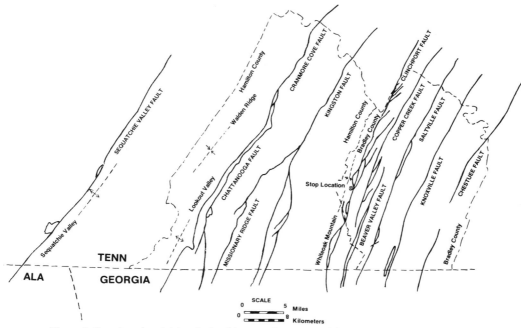

Figure 2. Stop location (x) in relationship to major structural features of southeast Tennessee.

Figure 3. Gently dipping beds of the Sequatchie Formation on the western limb of the Whiteoak Mountain synclinorium.

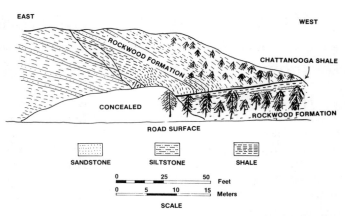

Figure 4. Sketch of a part of the Whiteoak Mountain synclinorium faulted over the west limb. Chattanooga Shale acts as glide plane.

dovician age. The western limb of the Whiteoak Mountain synclinorium is marked by the gently dipping beds near the Bradley County line (Fig. 1; Wilson, 1983). The eastern limb is steep and marks the footwall of the Clinchport Fault. Eastward from the Clinchport Fault there are a series of thrust faults: Copper Creek, Beaver Valley, Saltville, Knoxville, and Chestuee (Fig. 2). The footwall rocks of all of these consist primarily of Cambro-Ordovician Knox and Chickamauga carbonate rocks, while the hangwall sequences are normally the less competent Lower and Middle Cambrian shales of the Rome and Conasauga.

The Whiteoak Mountain synclinorium is the name applied to the series of synclinal structures along the eastern edge of the Kingston Fault block. The Kingston Fault originates in northeast Alabama and extends entirely across the northwest corner of Georgia into Tennessee where it continues past Kingston and dies out on the edge of the Powell Valley anticline.

The deepest part of the Whiteoak Mountain synclinorium includes Grindstone Mountain, which is located just east of Ooltewah, Tennessee. Grindstone Mountain is capped by the Warren Point sandstone of Lower Pennsylvanian age (Wilson, in press) and repesents the only preservation of Pennsylvanian strata in the Valley and Ridge province. This part of the Kingston Fault block contains the longest continuous section of strata exposed anywhere in Tennessee. Roughly some 10,000 ft (3,050 m) of rocks, from the shales of the Conasauga Formation of Middle Cambrian age to the sandstones of Lower Pennsylvanian age, are included.

The part of the Whiteoak synclinorium exposed along I-75 in eastern Hamilton County consists of a western limb which dips gently to the east (about 15°) and an eastern limb which is sharply folded and faulted. The eastern limb comprises the footwall of the Clinchport fault, and provides an excellent opportunity to see regional structural features.

Along the toe of the Clinchport fault are structures which are generally more complex than in the toes of the other major thrusts in the Valley and Ridge province. "Horses" or fault slices of Knox carbonates along the trace of the Clinchport Fault (Milici, 1979) are not uncommon.

The Clinchport Fault extends southwestward into northwest Georgia and dies out in the Armuchee anticline just north of Rome (Butts and Gildersleeve, 1948). To the northeast the Clinchport fault passes into southwest Virginia where it merges with the Hunter Valley Fault and joins the Russell Fork Fault at the northeast corner of the Pine Mountain overthrust (Rodgers, 1953).

SITE DESCRIPTION

Excellent exposures can be seen in both the east and westbound lanes of I-75 as it passes through the sharply upturned eastern limb of the Whiteoak Mountain synclinorium (Fig. 1).

Approaching the structure from the west along the eastbound lane of I-75 as the roadway climbs the western slope of Whiteoak Mountain, the first exposures are those of the Catheys Formation of Late Ordovician age which is overlain by the Sequatchie Formation of Late Ordovician age. The distinctive white band opposite the scenic overlook is the result of calcite precipitation along joints and fractures in the beds of the upper Sequatchie Formation (Fig. 3). Just beyond the scenic overlook is a completely exposed section of the Rockwood Formation of Early Silurian age. Thick sandstones mark the base of this nearly 600-ft (182.9 m) thick unit which contains several thin bands of hematite. Numerous trenches and pits throughout the area are evidence that iron ore was extracted for about fifty years beginning in 1886 (Wilson, 1981).

Normally, the Chattanooga Shale of Late Devonian age overlies the Rockwood Formation in this area, but here there is a small thrust fault which has used the Chattanooga Shale as a glide plane, and only the lower few inches of the Chattanooga Shale remain. Figure 4 shows a sketch of this fault in which the Rockwood Formation is thrust over the Chattanooga Shale. A close inspection of the fault surface reveals that the Chattanooga has been reduced to only a few inches (cm) of highly deformed shale. Directly beneath the black shale lies a coarse-grained quartz sandstone which represents the Dowelltown member of the Chattanooga Shale. In the next exposure on the right or south side,

Figure 5. North side of the interstate shows nearly vertical beds of the Rockwood Formation on the eastern limb of the Whiteoak Mountain synclinorium. Note the prominent fold. Highway guardrails are set on 76 in (1.9 m) centers.

Figure 6. North side of highway shows near-vertical beds of the Rockwood Formation, along with minor faulting.

Figure 7. Westward-dipping fault near the east end of the I-75 exposure.

between mile marks 16 and 17, the hanging wall of this thrust is characterized by numerous minor structures including splay thrusts, antithetic splays, and gravity normal faults (Milici, 1978).

Continuing eastward along I-75, steeply-dipping to overturned beds of the Rockwood Formation are exposed along both sides of the interstate. The faults in this east limb have small displacements, and the majority appear to be east-dipping thrusts caused by shear in the footwall after the strata were rotated to vertical. Milici (1978) states that these faults are clearly extensional and if the stratification is returned to horizontal they appear to be normal and, perhaps, were formed in the same manner as the faults in the hanging wall of the Cumberland Plateau decollement at Dunlap, Tennessee.

On the north side of I-75, adjacent to the westbound lane,

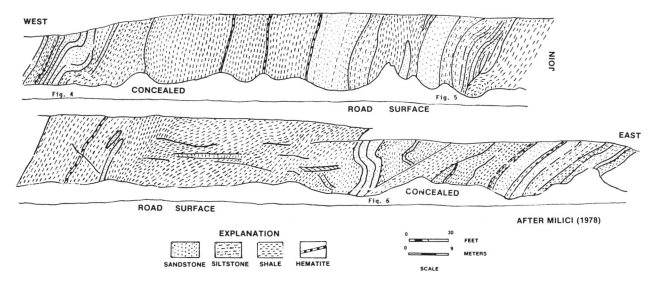

Figure 8. Sketch of the Rockwood Formation along the north side of I-75 at Lauderback Ridge. Location of Figures 5, 6 and 7 are indicated.

there are a series of Rockwood sandstones, shales, and hematite bands in nearly vertical position (Figs. 5 and 6). A westward-dipping fault is contractional as shown by the bed repetition (Fig. 7). The entire sequence here is highly deformed. Splays, thrusts, and recumbent folding are only a few of the more obvious features which can be observed. Folding, faulting, and repetition of beds has resulted in the Rockwood Formation being some 20

percent thicker at this location than at exposures found on the western limb of the synclinorium.

Figure 8 is a sketch of the north side of the I-75 cut through Lauderback Ridge where the broken and folded east limb of the Whiteoak synclinorium is continuously exposed. Indicated on the sketch are the locations of the photographs in Figures 5, 6, and 7.

REFERENCES CITED

Butts, Charles, and Gildersleeve, Benjamin, 1948, Geology and mineral resources of the Paleozoic area in northwest Georgia: Georgia Geological Survey Bulletin 54, 176 p.

Milici, R. E., 1978, The Whiteoak Mountain Synclinorium along Interstate 75, at Green Gap, Tennessee Trip 1, Stop 2, *in* Hatcher, R. D., Jr., Merschat, C. E., Milici, R. C., and Weiner, L. S., Field Trip 1, A Structural Transect in the Southern Appalachians, Tennessee and North Carolina: Tennessee Division of Geology Report of Investigations, no. 37, p. 5–52.

——1979, Structure of Hamilton County, Tennessee, *in* Geology of Hamilton County, Tennessee: Tennessee Division of Geology Bulletin 79, p. 39–45.

Rodgers, John, 1953, Geologic map of east Tennessee with exlanatory text: Tennessee Division of Geology Bulletin 58, pt. 2, 168 p.

Wilson, R. L., 1981, Guide to the geology along the interstate Highways in Tennessee: Tennessee Division of Geology Report of Investigations, no. 39, 79 p.

——1983, Geologic map and mineral resources summary of the Snow Hill quadrangle, Tennessee: Tennessee Division of Geology Geologic Map 112-NE, scale 1:24,000.

——in press, Geologic map of the Ooltewah quadrangle, Tennessee: Tennessee Division of Geology, Geologic Map 112-SE, scale 1:24,000.

33

Geology of Lookout Mountain, Georgia

Howard Ross Cramer, Emory University, Atlanta, Georgia 30322

LOCATION

Lookout Mountain is in Dade and Walker Counties, Georgia, in a relatively rural part of the state. However, it is near interstate access roads and lends itself well to field trips. Six stops, designed to show the major stratigraphic units of the region, are, save for one, on public land readily accessible by automobile. The Mountain cradles Cloudland Canyon State Park, which includes camping facilities. Most of the stops are on the Durham 7½-minute quadrangle; one is on the adjacent Trenton quadrangle, to the west. Figure 1 shows the locality and accessibility of Lookout Mountain.

INTRODUCTION

Lookout Mountain contains exposures of Carboniferous rocks which have resulted from shoreline progradation over carbonate banks. Shelf, delta front, barrier bar, lagoon, delta plain, and coal-bearing deposits are admirably illustrated (Thomas and Cramer, 1979). Rocks from the Upper Devonian Chattanooga Shale to the Lower Pennsylvanian Rockcastle Sandstone are present. The entire stratigraphic sequence on Lookout Mountain is reviewed in each of the following: Allen and Lester (1954), Butts and Gildersleeve (1948), Cressler (1964), Croft (1964), Sullivan (1942), and Thomas and Cramer (1979). Figure 2 illustrates the section; the terminology used is that of Thomas and Cramer (1979), Culbertson (1963), and Crawford (in press).

Devonian rocks of northwestern Georgia, including those on Lookout Mountain, are discussed in Oliver and others (1967). The Chattanooga Shale conformably overlies the Silurian Red Mountain Formation; its incompetency makes precise thickness measurements meaningless, but it is only a few tens of ft (m) thick. The Chattanooga is black and fissile, and usually weathers to gray and rarely to red. A lagoon deposit (Conant and

Swanson, 1961), it contains inarticulate brachiopods elsewhere in the state, and Hass (1956), on the basis of conodonts, considers it to be Late Devonian in age.

Mississippian rocks of northwestern Georgia, including those on Lookout Mountain, are discussed in DeWitt and Wenger (1979) and McLemore (1971).

The Maury Shale conformably overlies the Chattanooga Shale, but structural contortion makes the discrimination of the

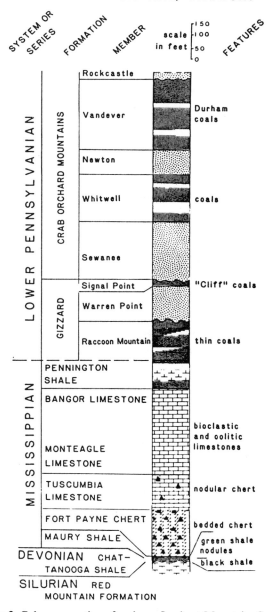

Figure 2. Columnar section of rocks on Lookout Mountain, Georgia.

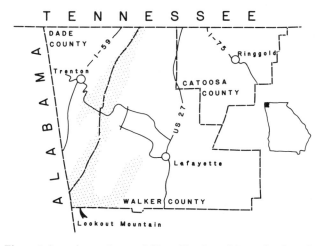

Figure 1. Location and accessibility of Lookout Mountain, Georgia.

two shales sometimes difficult. The Maury is less than five ft (1.5 m) thick and consists of partly silty to sandy, green and gray shale; it is commonly glauconitic and characteristically contains phosphate nodules (Wheeler, 1955). It is Kinderhookian and possibly Osagean, on the basis of conodonts (Hass, 1956).

The Fort Payne Chert conformably overlies the Maury and is about 200 ft (51 m) thick. Much of the formation is composed of bedded chert and cherty dolostone, and its age, based upon abundant fossils, is Osagean. The chert and dolomitization notwithstanding, the Fort Payne is an example of shallow marine shelf deposit (Thomas and Cramer, 1979).

The Tuscumbia Limestone gradationally overlies the Fort Payne, and ranges from 115 to 213 ft (35 to 65 m) thick. It is composed largely of bioclastic limestone with abundant chert nodules; beds of lime mudstone and finely crystalline dolostone are also present. Fossils show it to be Meramecian. The Tuscumbia also is an open marine shelf limestone (Thomas and Cramer, 1979).

The Monteagle Limestone conformably overlies the Tuscumbia, and on Lookout Mountain, where the Hartselle Sandstone is not present, the Bangor Limestone rests upon the Monteagle. The Monteagle and Bangor are lithologically similar and cannot be readily distinguished; both are from carbonate-bank deposition. Collectively they range from 443 to 900 ft (135 to 274 m) thick, are mainly oolitic and bioclastic, and unlike the Tuscumbia with which they might otherwise be confused, contain no chert. The Monteagle and Bangor are Meramecian and Chesterian (Thomas and Cramer, 1979).

The Mississippian and Pennsylvanian boundary is not readily distinguished. Two shale-dominated formations lie athwart the boundary and because they are so similar (resulting from continually changing vertical and horizontal near-shore and littoral sedimentary regimes) they are commonly mapped as one unit. These are the Late Mississippian Pennington Formation and the Early Pennsylvanian Raccoon Mountain Member of the Gizzard Formation. Milici (1974) provides a good description of these rocks and their sedimentary-tectonic setting from nearby Tennessee, as do Thomas and Cramer (1979).

The Pennington Formation conformably overlies the Bangor Limestone and is characterized by maroon and green shale with interbedded limestone and mudstone which grade upward into gray shale with interbedded siltstone, sandstone, siderite, and thin coal seams. These are prodelta and delta clastic rocks. Chesterian aged fossils are known from the Pennington (Thomas and Cramer, 1979).

The Raccoon Mountain Member of the Gizzard Formation gradationally overlies the Pennington and, being composed of gray shale with interbedded sandstone and siderite and coal, is difficult to distinguish from it. Marine invertebrates and plants from the Raccoon Mountain Member show it to be Early Pennsylvanian (Crawford, in press). These rocks are those expected from a near-shore delta setting. The thickness of the two formations varies from 213 to 426 ft (65 to 130 m); the Pennington is generally less than 200 ft (60 m) thick.

Pennsylvanian rocks of Georgia, including those on Lookout Mountain, are discussed in Crawford (in press), Culbertson (1963), Johnson (1946), McCallie (1904), Schlee (1963), Stearns and Mitchum (1962), Troxell (1946), Wanless (1946, 1975), and Wilson and Churnet (1981). Pennsylvanian rocks, because of their economic potential, have been investigated intensively, and because their sedimentary environmental origin (near shore, littoral, and possibly fluvial) has resulted in marvelously complex sedimentation, they have been subjected to a great variety of interpretations, correlations, and terminologies. These problems have not yet been resolved. Largely because of chronology, the interpretations and classifications of Crawford (in press) have been adopted for this guidebook.

The Warren Point Sandstone Member of the Gizzard Formation disconformably overlies the Raccoon Mountain Member. It is about 150 ft (45 m) thick and is composed largely of coarse-grained, cross bedded, conglomeratic sandstone; a little coal is present. No fossils are known, but it is interbedded between Early Pennsylvanian-aged rocks. Chen and Goodell (1964) describe the petrology of this unit, as does Schlee (1963) (therein called the Sewanee Sandstone, however). This is possibly a barrier-bar deposit (Stearns and Mitchum, 1962).

The Signal Point Shale Member of the Gizzard Formation is about 35 to 40 ft (11 to 13 m) thick, is very persistent, and is composed of gray shale with discontinuous sandstone and coal beds. No fossils are reported. Its origin is unclear, but because of the discontinuous coal, it is likely a lagoon deposit.

The Crab Orchard Mountains Formation is the uppermost Paleozoic unit in the Lookout Mountain area. It is composed of several interlayered shale and sandstone members, each of which is characterized by more lateral persistence than the members of the underlying Gizzard Formation, and are therefore with one exception likely to be lower delta plain deposits (Thomas and Cramer, 1979).

The Sewanee Sandstone Member of the Crab Orchard Mountains Formation unconformably overlies the Signal Point Shale member of the Gizzard Formation and is composed of about 250 ft (76 m) of coarse-grained, conglomeratic, cross-bedded sandstone; it contains a discontinuous coal seam near its base. No marine fossils are known. It is a barrier-bar deposit resting upon a coal-bearing lagoon deposit.

The Whitwell Shale Member of the Crab Orchard Mountains Formation is 200 ft (61 m) thick and conformably overlies the Sewanee Sandstone; it is composed predominantly of gray shale with a few interbedded coals and thin sandstone beds. Marine invertebrates and plants show the Whitwell to be Early Pennsylvanian in age (Crawford, in press).

The Newton Sandstone Member of the Crab Orchard Mountains Formation conformably overlies the Whitwell Member and is a coarse-grained, conglomeratic, cross-bedded, bench-forming sandstone about 110 ft (33 m) thick. No fossils are reported but it underlies and overlies Lower Pennsylvanian rocks.

The Vandever Shale Member of the Crab Orchard Moun-

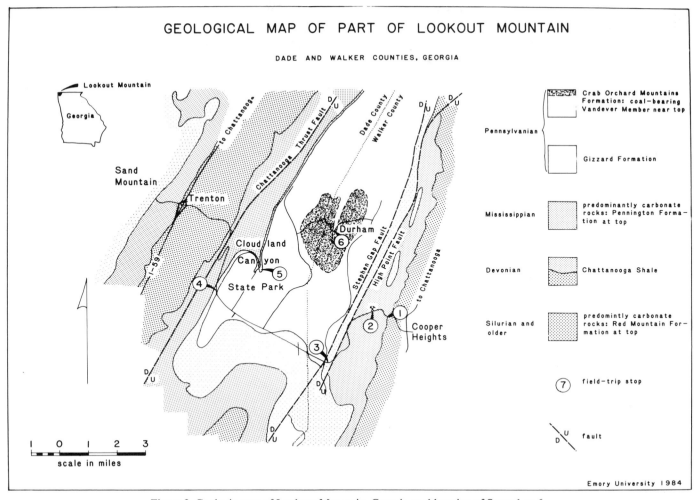

Figure 3. Geologic map of Lookout Mountain, Georgia, and location of Stops 1 to 6.

tains Formation is about 300 ft (91 m) thick and is largely gray shale with interbedded sandstone and several well-developed, persistent coal seams, the Durham coals. Marine fossils and plants show it to be Early Pennsylvanian in age (Crawford, in press).

The Rockcastle Sandstone Member of the Crab Orchard Mountains Formation is the youngest rock unit exposed on the mountain. It unconformably overlies the Vandever Shale and caps the high hills near Durham. Like the Newton, it is coarse-grained, conglomeratic, and cross-bedded. Only about 30 ft (9 m) remain. No fossils are known.

Structurally, Lookout Mountain is a broad, gently undulating syncline, with the youngest rocks, the Rockcastle Sandstone Member, in the topographically highest and structurally lowest position. The mountain is the remnant of the upper plate of a regional decollement, the sliding planes of which were probably the Chattanooga Shale and shale in the Pennington Formation. The root of the decollement has been covered by younger thrusting from the east. On the eastern edge of the mountain, two steeply dipping reverse faults have been mapped with throws of several hundred ft (m). The westernmost one is the Steven

Gap fault and the eastern one is the High Point fault (Crawford, 1983). These are ramps on the décollement.

SITE DESCRIPTIONS

Six stops illustrating the highlights of the stratigraphy are arranged in stratigraphic order. All but one are more or less accessible from public roads and are on public land or in long-abandoned mines and quarries of uncertain ownership. All of the stops are on the Durham 7½-minute quadrangle except for number 4 which is on the Trenton quadrangle. Figure 3 shows a generalized geological map of Lookout Mountain and the location of the six stops.

Stop 1. Silurian Red Mountain Formation, Devonian Chattanooga Shale, and Mississippian Maury and Fort Payne Formations. 0.9 mi (1.5 km) west of Cooper Heights crossroad on the Lookout Mountain Scenic Highway. The nearest safe parking is about 1400 ft (305 m) farther west where the berm is wide, near the road junction. Here, in a road cut on the north side of the road, the three shale formations form a steep slope below the bench-forming Fort Payne Chert. Talus and vegetation generally obscure the contacts of the shale formations, but the 6-in (15 cm) beds of the chert are very pronounced.

Stop 2. *Mississippian Tuscumbia and Monteagle Limestones.* A quarry on the north side of the Lookout Mountain Scenic Highway about 3,600 ft (1,100 m) west of stop 1. There is safe parking on the berm directly across the road from the quarry. The exposures at the bottom of the quarry vary from time to time depending upon the water and dumping, but the exposures in the walls of the quarry and adjacent hill slopes are excellent. About 50 ft (15 m) of the chert-nodule-bearing, massively bedded Tuscumbia Limestone forms the lower part of the quarry wall and is at road level. Conformably above the Tuscumbia about 50 ft (15 m) of the Monteagle Limestone is exposed in the quarry wall, and there is much more on the adjacent hillslopes. Here the oolitic, massively bedded, cross-bedded nature of the Monteagle is easily seen. Fossils, largely corals, are to be found in the exposures on the hillslopes.

Between stops 2 and 3, where the highway turns southward in its ascent of the mountain, an abrupt change of dip is apparent; here the otherwise gently, westward, into-the-mountain dipping limestones are more nearly vertical. The road has crossed the High Point fault and has turned southward, parallel to its trace. The steep dips are from the drag on the lower plate of the ramp. Farther along the road, to the south and west, the rocks assume a more nearly regional dip of about 6 or 8 degrees westward.

Stop 3. *Coal exposure in the Raccoon Mountain Member of the Gizzard Formation, and contact with the Warren Point Member of the Gizzard.* Stop 3 marks Steven Gap, where the Lookout Mountain Scenic Highway passes the lip of the plateau and turns abruptly westward. Parking is convenient along the berm of the highway a few hundred ft (meters) farther west. The road cut has exposed a bituminous-coal seam a few in (cm) thick. The coal is interlayered with coaly shale and shaley coal, and all are overlain by the Warren Point Sandstone. Stratigraphically below this exposure, in road cuts down the mountain, various sand bodies interlayered in the shales of the Raccoon Mountain Member are admirably exposed, and sedimentary structures, such as cross-bedding, cut and fill, channeling, sole marks, etc., may be seen.

At the crossroad a few hundred ft (meters) west of Steven Gap are exposure of the cross-bedded Warren Point Sandstone Member; these are steeply dipping beds, having been turned up as drag on the lower plate of the Steven Gap fault.

Stop 4. *Warren Point Sandstone Member of the Gizzard Formation.* This stop is on the Trenton 7½-minute quadrangle, a few hundred ft (m) west of the Durham quadrangle where the road is on western-facing edge of the mountain, overlooking Trenton. Safe parking can be obtained along the berm of the road opposite the cut.

Here a road cut has exposed several tens of ft (m) of the Warren Point Sandstone; cross-bedding, channeling, cut and fill, ripped up clasts, and many other features anticipated in an offshore-bar deposit are to be seen. The coal-bearing shale below it in the underlying Raccoon Mountain Member, is irregularly exposed because of the vagaries of weathering, vegetation, and slumping.

Stop 5. *Contact of the Gizzard and Crab Orchard Mountains Formations.* Cloudland Canyon State Park. Abundant parking and camping facilities in the park. The park includes Sitton Gulch which was cut by obsequent drainage into the western edge of the plateau. Griffin and Atkins (1983) provide a popular account of the geology in the canyon.

The lip of the canyon is composed of the coarse-grained, conglomeratic, cross-bedded Sewanee Sandstone, and one can see the thin Signal Point Shale and the Warren Point Sandstone in the canyon walls below. Vegetation obscures the lower formations from the top of the canyon, but these, and many sedimentation features, can be seen on the nature trails in the park.

Wilson and Churnet (1981) interpret the rocks at the lip of the canyon as Warren Point Sandstone, much thicker than elsewhere, and as evidence for the presence of what they call the Cloudland Canyon sedimentary basin. Clearly, more stratigraphic work remains.

Stop 6. *Durham Coal Basin.* Parking is adequate along the narrow berms of seldom-used roads and in long-abandoned strip mines. Ownership of the land is unclear and unsettled because of the presence of still-valuable bituminous coal. Crawford (in press), Johnson (1946), McCallie (1904), and Troxell (1946) provide good descriptions of the coals and coal-bearing rocks.

In the valley upstream from where the road crosses the creek in a hairpin turn, is the site of the railhead and loading trestle from the now-abandoned underground slope mines. The residue from the dumping is being actively cannibalized by the Highway Department, and is diminished annually. A concrete tunnel, built in the early part of this century, was designed to allow the creek to continue to flow even though the valley was being filled with mine tailing. The tunnel and trestle-foundations are still to be seen, as are remnants of the old townsite on the hill slope to be west.

The railhead in the valley was in the structurally low area of the coal basin as well as in the topographically low area. The formations dip gently toward the railhead from all directions; this allowed the slope mines to work upward, away from the valley, facilitating drainage and allowing gravity to take the coal from the mine face to the railhead.

Two major coals (the Durham seams) crop out as haloes around the outside of the hill (called Round Top) and can be identified on the Durham quadrangle from the contours which show strip mining. Currently, mining is sporadic and unpredictable, so that exposures of the coal are purely fortuitous. The coals, where seen, are generally less than 2.0 ft (0.5 m) thick but are of good metallurgical quality. Fossil plants are abundant in the spoil piles of the mines. A very accessible fossil-collecting area is in the mines which cross the road about 100 ft (33 m) short of one mile (1.6 km) east of Durham.

REFERENCES CITED

Allen, A. T., and Lester, J. G., 1954, Contributions to the paleontology of northwest Georgia: Georgia Geological Survey Bulletin 62, 166 p.

Butts, Charles, and Gildersleeve, Benjamin, 1948, Geology and mineral resources of the Paleozoic area in northwest Georgia: Georgia Geological Survey Bulletin 54, 176 p.

Chen, C. S., and Gooddell, H. G., 1964, The petrology of the Lower Pennsylvanian Sewanee Sandstone, Lookout Mountain, Alabama and Georgia: Journal of Sedimentary Petrology, v. 34, p. 46–72.

Conant, L. C., and Swanson, V. E., 1961, Chattanooga Shale and related rocks of central Tennessee and nearby areas: United States Geological Survey Professional Paper 357, 91 p.

Crawford, T. J., 1983, High Point and Steven Gap faults, Lookout Mountain, Georgia [abstract]: Georgia Journal of Science, v. 41, p. 22.

——in press, Pennsylvanian geology of northwestern Georgia: Georgia Geological Survey.

Cressler, C. W., 1964, Geology and ground-water resources of Walker County, Georgia: Georgia Geological Survey Information Circular 29, 15 p.

Croft, M. G., 1964, Geology and ground-water resources of Dade County, Georgia: Georgia Geological Survey Information Circular 26, 17 p.

Culbertson, W. C., 1963, Pennsylvanian nomenclature in northwest Georgia: United States Geological Survey Professional Paper 450E, p. E 51-E57.

DeWitt, Wallace, Jr., and Wenger, L. M., 1979, The Appalachian basin region, *in* Introduction and regional analyses of the Mississippian System in the United States: United States Geological Survey Professional Paper 1010, p. 13–48.

Griffin, M. M., and Atkins, R. L., 1983, Geologic guide to Cloudland Canyon State Park: Georgia Geological Survey Geologic Guide 7, 35 p.

Hass, W. H., 1956, Age and correlation of the Chattanooga Shale and the Maury Formation: United States Geological Survey Professional Paper 286, 47 p.

Johnson, V. H., 1946, [Map of] Coal deposits on Sand and Lookout Mountains, Dade and Walker Counties, Georgia: United States Geological Survey Preliminary Map, scale 1:48,000.

McCallie, S. W., 1904, A preliminary report on the coal deposits of Georgia: Georgia Geological Survey Bulletin 12, 121 p.

McLemore, W. H., 1971, The geology and geochemistry of the Mississippian System in northwest Georgia and southeast Tennessee [Ph.D. thesis]: University of Georgia, Athens, 296 p.

Milici, R. C., 1974, Stratigraphy and depositional environments of Upper Mississippian and Lower Pennsylvanian rocks in the southern Cumberland Plateau of Tennessee, *in* Garrett Briggs (ed.), Carboniferous of the southeastern United States: Geological Society of America Special Paper 148, p. 115–133.

Oliver, W. A., Jr., DeWitt, Wallace, Jr., Dennison, J. M., Hoskins, D. M. and Huddle, J. W., 1967, Devonian of the Appalachian basin, United States, *in* International symposium on the Devonian System: Calgary, Alberta Society of Petroleum Geologists, vol. 1, p. 1001–1040.

Schlee, J. S., 1963, Early Pennsylvanian currents in the southern Appalachian Mountains: Geological Society of America Bulletin, v. 74, p. 1439–1451.

Stearns, R. G., and Mitchum, R. M., Jr., 1962, Pennsylvanian rocks of the southern Appalachians, *in* Pennsylvanian System in the United States: Tulsa, American Association of Petroleum Geologists, p. 74–96.

Sullivan, J. W., 1942, The geology of the Sand-Lookout Mountain area, northwest Georgia: Georgia Geological Survey Information Circular 15, 68 p.

Thomas, W. A., and Cramer, H. R., 1979, The Mississippian and Pennsylvanian (Carboniferous) Systems in the United States—Georgia: United States Geological Survey Professional Paper 1110 H, p. H1-H37.

Troxell, J. R., 1946, Exploration of Lookout Mountain and Sand Mountain coal deposits, Dade and Walker Counties, Georgia: United States Bureau of Mines Report of Investigations 3936, 10 p.

Wanless, H. R., 1946, Pennsylvanian geology of part of the southern Appalachian coal field: Geological Society of America Memoir 13, 162 p.

——1975, Appalachian region, *in* E. D. McKee and F. J. Crosby (coordinators), Introduction and regional analyses of the Pennsylvanian System, part 1 *of* Paleotectonic investigations of the Pennsylvanian System in the United States: United States Geological Survey Professional Paper 853, part 1, p. 17–62.

Wheeler, G. E., 1955, Occurrence, possible origin, and geological significance of the phosphate concretions in the Maury Shale: Georgia Academy of Science Bulletin, v. 13, p. 22–27.

Wilson, R. L., and Churnet, H. G., 1981, Carboniferous coal basins of the southeastern United States—a case study on Lookout Mountain, Georgia [abstract]: Geological Society of America Abstracts with Programs, v. 13, p. 582.

Ringgold Gap: Progradational sequences in the Ordovician and Silurian of northwest Georgia

Andrew K. Rindsberg, *Department of Geology, Auburn University, Auburn, Alabama 36849*
Timothy M. Chowns, *Department of Geology, West Georgia College, Carrollton, Georgia 30117*

LOCATION

North end of Taylor Ridge, on south side of I-75, southeast of Chattanooga, Tennessee, 10 mi (16 km) southeast from junction of I-24 and I-75. Catoosa County, Georgia, Ringgold 7½-minute quadrangle. Accessible from I-75 or via frontage road eastward from the western of two Ringgold exits (Georgia 151, Fig. 1). The cut may be conveniently examined away from traffic on the first or second terrace.

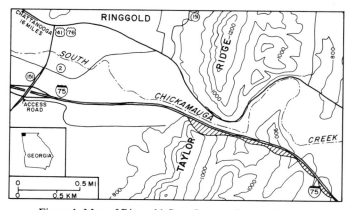

Figure 1. Map of Ringgold Gap, Catoosa County, Georgia.

INTRODUCTION

The interstate highway cuts at Ringgold Gap afford the best exposures in Georgia of Middle Ordovician to Lower Silurian rocks (Catheys, Sequatchie, and Red Mountain formations). The main cut (Fig. 2) is about 2,600 ft (800 m) long and more than 200 ft (60 m) high, with four terraces allowing nearly complete access. The transition from pre-Taconic carbonates to Taconic clastics is well illustrated, in six progradational sequences. Sedimentary structures are particularly well exposed, both in the cliffs and in abundant talus. Ordovician units represent tidal-flat and shallow-marine carbonate and clastic facies. Silurian facies are wholly clastic, including shallow-water turbidites, grading upward through hummocky-bedded strata to littoral units. Ringgold Gap has been intensively studied, including work on stratigraphy (Chowns, 1972b, c), trace and body fossils (Frey and Chowns, 1972; Rindsberg, 1983), palynomorphs (Goldstein, 1970; Colbath, 1983), and paleomagnetism (Morrison, 1983; Morrison and Ellwood, 1983). Chowns (1972a, p. 97–100) provides a measured section. Ringgold Gap is thus a key source of data for studies of regional stratigraphy (e.g., Chowns and McKinney, 1980).

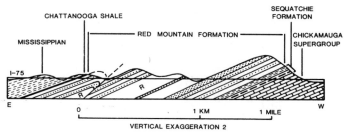

Figure 2. Diagrammatic cross-section of Ringgold road cuts, Georgia (after Chowns, 1972a, p. 76.).

The main cut at Ringgold Gap consists of four progradational sequences in a progressively more clastic milieu. The sequences may be related in part to Saharan glaciation during the Late Ordovician, although this is conjectural. The influx of clastics is possibly a function of the unroofing history of the Taconic Mountains during and after the Late Ordovician orogenic event.

Clastic influx is dramatically recorded at Ringgold Gap. The 930 ft (284 m) of Red Mountain clastics represents about the same amount of time (8 to 10 m.y.) as 222 ft (68 m) of Sequatchie carbonates and clastics. The basal 3 ft (1m) of the Red Mountain Formation is a lag sandstone winnowed from such material. Abundant quartz, muscovite, and clay, presumably derived from Late Precambrian clastics and low-grade metamorphics, account for the relatively great thickness of the Red Mountain Formation. The resulting progradational sequences of the Red Mountain Formation, passing upward through turbiditic, hummocky-bedded, and littoral facies, are strikingly similar to sequences described from the Mesa Verde Group and other Mesozoic shelf deposits of the U.S. Western Interior (cf. Hubert and others, 1972; Hamblin and Walker, 1979; Walker, 1984).

SITE DESCRIPTION

Roadcuts on I-75 expose an almost complete stratigraphic section of Upper Ordovician to Mississippian rocks (Fig. 2). The main cut, which is most often visited, exposes the upper Catheys, the entire Sequatchie, and the lower Red Mountain formations (Fig. 3). Only the main cut is described here, with emphasis on depositional environments as deduced from lithology, physical sedimentary structures, trace fossils, and body fossils. Facies descriptions below are drawn largely from Chowns (1972a, b, c), Chowns and McKinney (1980), and Rindsberg (1983).

Catheys Formation (Middle to Late Ordovician: Shermanian to Maysvillian). The upper 188 ft (57 m), consisting

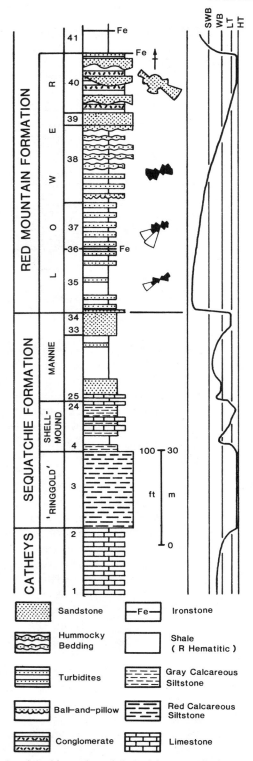

Figure 3. Stratigraphic section of Ordovician and Silurian rocks in the main cut at Ringgold Gap, Georgia (modified after Chowns, 1972c, Fig. 4; bed numbers from Chowns, 1972a, p. 97–100). Width of column suggests the relative percentage of ledge forming strata in the section. Equal area current roses show: stipple-crossbedding; black-tool and gutter marks; blank-flutes. HT - high tide; LT - low tide; WB - wave base; SWB - storm wave base.

of alternating, thin-bedded calcarenite and calcareous shale, are exposed at the west end of the outcrop (Fig. 3, units 1, 2). Fossils are diverse and abundant; calcarenites are composed largely of brachiopod and bryozoan fragments, whereas shales contain a sparse assemblage of bivalves and orthoconic nautiloids. These sediments were evidently deposited on a shallow subtidal shelf where clastics dominated at fine grain sizes and skeletal debris at coarse. The high-energy calcarenites show hummocky bedding and are interpreted as storm beds.

Sequatchie Formation (Late Ordovician: Maysvillian? to Richmondian). At Ringgold Gap, the Sequatchie Formation is divided into an unnamed member, the Shellmound Member, and the Mannie Shale Member. Exposure is complete (Fig. 3).

The basal unnamed member (informal "Ringgold Member" of Rindsberg [1983]) is an 80 ft (24 m)-thick unit of red, green, and gray calcareous siltstone, sandstone, and shale (Fig. 3, unit 3). The lower contact appears to be conformable but abrupt. Physical sedimentary structures consist of birdseyes, desiccation cracks, and mudchip intraclasts. Bioturbation is slight but ubiquitous. Vertical to oblique dwelling burrows (*Skolithos, Arenicolites, Rhizocorallium*) and meniscate feeding burrows indicate that conditions were upper intertidal to supratidal. Body fossils are nearly absent, as are palynomorphs (Goldstein, 1970; Colbath, 1983), rendering the age of this member uncertain.

The distinctive red color of the "Ringgold" Member seems to be an early diagenetic (synsedimentary) feature related to depositional environment. (1) The red color is essentially concordant with bedding and restricted by lithology. (2) Local examples of discordant color are usually associated with burrows, occasionally with extension fractures. (3) The red color is due to minute hematite blebs disseminated in the matrix (Manley and others, 1975). As grain size decreases, iron content rises (up to about 5 percent Fe^{+3}) and red color intensifies. (4) Both red and gray rocks of similar lithology contain similar percentages of total Fe, although the oxidation state differs. (5) According to Manley and others (1975), both red and gray rocks contain chlorite in the matrix, but in the gray rocks the chlorite is iron-rich, and in the red, iron-poor. This suggests that in the red beds iron has been expelled from the chlorite and oxidized to hematite. Thus the primary iron-bearing mineral in these rocks was iron-rich chlorite, and the original color gray; red hematite developed after deposition. However, the fact that the red color is concordant with the depositional interface and strongly influenced by bioturbation suggests that it may have developed very early. The occurrence of red beds specifically in desiccated strata indicates exposure as the probable cause of oxidation and staining.

The base of the overlying Shellmound Member is marked by an abrupt transition from red, slightly bioturbated siltstone to variegated, grayish, wholly bioturbated siltstone (units 4 to 7). This 10-ft (3 m)-thick, variegated siltstone contains a bright green, iron-bearing clay, possibly Fe-chlorite or chamosite, particularly as replacements in round intraclasts, burrow fills, and inarticulate brachiopods. Slow, authigenic deposition is indicated. Shallow-water inarticulate brachiopods (*Pseudolingula*) and bivalves (*Modiolopsis*) occur in life position. Overlying the variegated beds is a very thin (1 in or 2 cm) lag of coarse-grained sandstone including phosphatic granules.

Above the sandstone lag is 5.7 ft (1.7 m) of dark green-gray shale which, despite its thinness, can be traced westward for at least 18 mi (30 km) (unit 8). The shale contains a few very thin to thin, ripplemarked beds of sandstone and siltstone containing bryozoan debris. The shale itself contains no calcareous body fossils, but includes the chitinous remains of eurypterids (*Megalograptus*). The overlying unit is conformable and also contains *Megalograptus*.

The middle Shellmound Member (units 9 to 19; 17 ft or 5 m thick) consists of interbedded bryozoan calcarenite and calcareous siltstone, with a normal marine assemblage of body fossils, especially bryozoans (?*Dekayia*, ?*Hallopora*) and brachiopods (*Leptaena, Hebertella, Strophomena*). The bryozoans attain large size (domal to branched, up to 7 in [15 cm] wide), and some are in life position at the bases of calcarenites. The massive forms contain unbranched borings (*Trypanites*). The calcarenites are broadly lenticular; some have loadcasted bases. In the basal 17 ft (5 m), bioturbation is slight to absent in both calcarenite and siltstone facies. Evidently, the lower Shellmound consists of subtidal, inner-shelf deposits.

In the top 20 ft (6 m) of the Shellmound Member (units 20 to 24), evidence of further shallowing is seen in the siltstone interbeds. Their reddish color, occasional flaser bedding, firmground trace fossils (*Rhizocorallium*), and restricted body-fossil assemblage (*Modiolopsis, Pseudolingula*) point toward intertidal flats as the environment of deposition. The associated calcarenites are lenticular and may represent tidal channels.

The Mannie-Shellmound contact is a thin, black, phosphatic-chloritic hardground, indicating a second hiatus within rocks of Richmondian age. The hardground, which is burrowed and has attached bryozoan bases, is also present along strike at Collegedale and Green Gaps, respectively 7 mi (12 km) and 15 mi (25 km) north in White Oak Mountain, Tennessee. The 8.8 ft (2.7 m) of calcareous shale and brachiopod calcarenite (units 25 to 27) that immediately overlies the hardground has a rich assemblage of brachiopods (especially *Lepidocyclus manniensis*), as well as bivalves, gastropods, and bryozoans. Considering the diverse fauna and alternation of fine- and coarse-grained beds deposited in shallow, normal-marine water, the lowest 8.8 ft (2.7 m) of the Mannie must be the facies equivalent to the calcarenite-siltstone sequence of the Shellmound Member.

The middle part of the Mannie Member, consisting of green-gray to red-gray noncalcareous shales, is thickest at Ringgold Gap (units 27 to 32; 59 ft or 18 m thick). Sedimentary structures are easily discerned only in relatively coarse beds of sandy mudstone and sandstone. Bioturbation ranges from slight to complete, including *Chondrites* and the network burrow *Trichophycus*, recognizable in outcrop by its vertical spreite. Body fossils are small, sparse, and inconspicuous, including the inarticulate *Lingulella*, bivalve *Nuculites*, and gastropod *Liospira*. Originally calcareous shells are preserved as delicate molds. The Mannie was probably

an offshore, outer-shelf mud deposited below the wave base. The greater availability of mud during the late Richmondian can be attributed to the increasing tempo of erosion on the newly arisen Taconic Mountains.

The upper Mannie Member (units 33 to 34; 22 ft or 7 m thick) records a progradational event like that of the "Ringgold" and upper Shellmound members. Sandstones are increasingly thick, shelly, and bioturbated, with *Pseudolingula* and *Modiolopsis* in life position. Trace fossils are monospecific in any one bed, but diverse in the unit as a whole; vertical burrows of suspension-feeders are especially common (*Diplocraterion, Monocraterion, Skolithos*). The environment of deposition was probably lower intertidal to upper subtidal.

Red Mountain Formation (Early Silurian: Llandoverian). Only the lowest (units 35 to 40; 268 ft or 82 m of the total 930 ft or 284 m) of three progradational sequences in the Red Mountain will be discussed. The lowest sequence has the most diverse facies and can be seen at the main Ringgold Gap cut. It is also traceable regionally throughout the southern Appalachians (Chowns and McKinney, 1980).

The Ordovician-Silurian contact is now placed between units 34 and 35 of Chowns (1972b), on the basis of recent fossil collections (Rindsberg, 1983; see Fig. 3). The basal bed of unit 35 is a thick (2 to 3 ft; 0.6 to 1.0 m) sandstone of unique mineralogy for the section; it contains both coarse muscovite and reworked phosphatized fossils. Coarse muscovite is common in local Silurian, but not Ordovician rocks; phosphatic fossil debris is common in the Mannie, but occurs only in this basal sandstone lag of the Red Mountain. The latest Ordovician is missing. The overlying 266 ft (81 m) exposed at the main cut is a shallowing-upward clastic sequence, from turbidites (111 ft or 34 m) to storm beds (81 ft or 25 m) to littoral beds (73 ft or 22 m) (Fig. 3).

At the base, this sequence consists of graded shales with thin bioturbated sandstones and siltstones (units 35 to 37). Higher in the section, the percentage and thickness of sandstones increase, with graded beds showing Bouma sequences (mainly plane lamination overlain by ripple laminae), and diverse sole marks. These include tool marks, gutter marks, and occasional flutes indicating transport from the northwest, as well as ball-and-pillow structures formed by loading. Trace and body fossils are diverse, especially in shales and soles of sandstone beds. The trace-fossil assemblage includes both dwelling and feeding burrows, especially *Planolites, Palaeophycus, Chondrites,* and *Dictyodora.* Body fossils include brachiopods (?*Dalmanella, Leptaena*), gastropods, crinoids, *Tentaculites,* and orthoconic cephalopods. The mode of preservation (aragonitic fossils preserved as compressed molds, calcitic fossils as uncompressed molds) indicates that aragonite dissolved early, before compaction of shales. The depth of water in which these turbidites were deposited cannot have been great, because lithology and fossil assemblages grade upward with no break into those of the overlying hummocky beds. If subsidence and late compaction are discounted, then the initial water depth was equivalent to the thickness of the sequence, i.e., about 260 ft (80 m).

The hummocky beds comprise 81 ft (25 m) of shale and sandstone, with percentage and thickness of sandstones increasing upward (Fig. 3, unit 38). Silty shales and turbidite beds with gradational upper contacts are gradually replaced by fissile clay shales and coarser sandstone beds with hummocky bedding and sharp upper contacts. The trace-fossil assemblage is similar to that of the turbidites, but includes trackways and trails on the upper surfaces of sandstones as well as scoured burrows on soles. Body fossils are also similar to those in underlying beds, but are increasingly fragmented upward in the section. These hummocky-bedded strata are believed to have been deposited and scoured by storm waves, fine-grained sand and silt having been winnowed out and deposited as turbidites like those seen lower in the section. Turbidites were deposited below storm wave base, and hummocky beds between storm and normal wave bases.

Hummocky sandstones and shales grade upward into 81 ft (25 m) of very thick-bedded, dusky red hematitic sandstones, and then into interbedded, fine-to-coarse, bimodally crossbedded sandstone, shale-pebble conglomerate, and shale (Fig. 3, units 39 to 40). Body fossils are fragmented; trace fossils are absent. According to Colbath (1985, personal communication), the uppermost beds in this sequence contain possible nonmarine phytoplankton. Sedimentary structures and taphonomy indicate a littoral environment of deposition, probably mostly subtidal but perhaps partly intertidal. These sandstones cap the dip slope of Taylor Ridge, and mark the east end of the main cut in Ringgold Gap.

Across South Chickamauga Creek, these beds are capped by a thin Clinton-type ironstone made of comminuted and hematite-replaced fossil hash, which was formerly tried for iron ore. This in turn is overlain by a further shallowing-upward sequence of shales and turbidites, which is largely hidden in the small valley east of the main cut. The ironstone was therefore formed during a period of low-clastic influx and reworking which marked the end of shoreline progradation and the beginning of renewed submergence.

REFERENCES CITED

Chowns, T. M., compiler, 1972a, Sedimentary environments in the Paleozoic rocks of northwest Georgia: Georgia Geological Society, Guidebook, no. 11, 100 p.

—— 1972b, Depositional environments in the Upper Ordovician of northwest Georgia and southeast Tennessee: Georgia Geological Society, Guidebook, no. 11, p. 3–12.

—— 1972c, Molasse sedimentation in the Silurian rocks of northwest Georgia: Georgia Geological Society, Guidebook, no. 11, p. 13–23.

Chowns, T. M., and McKinney, F. K., 1980, Depositional facies in Middle-Upper Ordovician and Silurian rocks of Alabama and Georgia: in Frey, R. W., ed., Excursions in Southeastern geology [Geological Society of America, Annual Meeting (Atlanta, 1980)]: Falls Church, Virginia, American Geological Institute, v. 2, p. 323–348.

Colbath, G. K., 1983, Paleoecology and palynomorphs from the Upper Ordovician-Lower Silurian of the southern Appalachians [Ph.D. thesis]: Eugene, Oregon, University of Oregon, 331 p.

Frey, R. W., and Chowns, T. M., 1972, Trace fossils from the Ringgold road cut (Ordovician and Silurian), Georgia: Georgia Geological Society, Guidebook, no. 11, p. 25–55.

Goldstein, R. F., 1970, Comparison of Silurian chitinozoans from Florida well samples with those from the Red Mountain Formation in Alabama and Georgia [M.S. thesis]: Tallahassee, Florida, Florida State University, 90 p.

Hamblin, A. P., and Walker, R. G., 1979, Storm dominated shallow marine deposits: The Fernie-Kootenay (Jurassic) transition, southern Rocky Mountains: Canadian Journal of Earth Sciences, v. 16, p. 1673–1690.

Hubert, J. F., Butera, J. G., and Rice, R. F., 1972, Sedimentology of Upper Cretaceous Cody-Parkman delta, southwestern Powder River Basin, Wyoming: Geological Society of America Bulletin, v. 83, p. 1649–1670.

Manley, F. H., Ogren, D. E., and Webb, L. C., 1975, Mottled Upper Ordovician carbonates in northwest Georgia: Journal of Sedimentary Petrology, v. 45, p. 615–617.

Morrison, J., 1983, Paleomagnetism of the Silurian and Ordovician Red Mountain, Catheys and Sequatchie Formations from the Valley and Ridge Province, northwest Georgia [M.S. thesis]: Athens, University of Georgia, 64 p. [not seen].

Morrison, J., and Ellwood, B. B., 1983, Paleomagnetism of Silurian-Ordovician sediments from the Valley and Ridge Province, northwest Georgia [abstract]: EOS American Geophysical Union, Transactions, v. 64, p. 216.

Rindsberg, A. K., 1983, Ichnology and paleoecology of the Sequatchie and Red Mountain Formations (Ordovician-Silurian), Georgia-Tennessee [M.S. thesis]: Athens, University of Georgia, 381 p.

Walker, R. G., 1984, Shelf and shallow marine sands, in Walker, R. G., ed., Facies models (second edition): Geological Association of Canada, Geoscience Canada Reprint Series, no. 1, p. 141–170.

Early Cambrian clastics in northern Cleburne and Calhoun Counties, Alabama

Denny N. Bearce, Department of Geology, University of Alabama at Birmingham, Birmingham, Alabama 35294

LOCATION

Northeast side of Cleburne County Road 70, 1.5 mi (2.4 km) west of Borden Springs and 6.25 mi (10.1 km) east of Piedmont, NW¼ Section 5, T.13S., R.11E., Cleburne County, Borden Springs 7½-minute quadrangle, Alabama (Fig. 1).

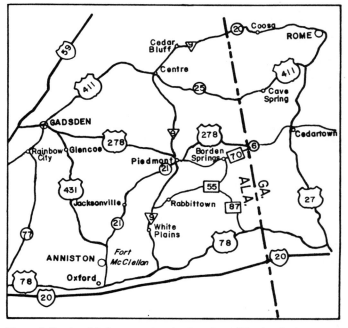

Figure 1. Regional index map showing location of Borden Springs area.

INTRODUCTION

Exposures of the Early Cambrian Chilhowee Group and Shady Formation are present in northern Cleburne and Calhoun Counties 6.5 to 9 mi (10 to 14 km) east of the town of Piedmont, Alabama. These strata comprise a series of far travelled, imbricated nappes, faulted onto younger Paleozoic rocks of the Valley and Ridge province and rooted in a sole fault that underlies the Talladega slate belt, the western border of which, in Alabama, is the Talladega fault (Figs. 2 and 3). The Talladega slate belt is the westernmost belt of the Piedmont metamorphic province in Alabama and western Georgia, and its rocks record the complex sedimentary and tectonic history of a part of the North American continental margin from late Precambrian or Early Cambrian to Devonian time. Exposures within this series of nappes are significant in that a telescoped but detailed record of the eastward transition in sedimentary environments of a part of the North American Early Cambrian shelf and slope is revealed within a cross-strike distance of less than 2.5 mi (4 km). These Cambrian strata also reveal progressive increase in strain and metamorphic recrystallization eastward, from open flexural slip folds with sporadic solution cleavage at the western edge of the nappes to isoclinal, multiply deformed flow folds and a pervasive phyllitic foliation east of the Talladega fault (Bearce, 1982).

Radiometric age dates from the Talladega slate belt indicate a single metamorphic event of Early to Middle Devonian (Acadian) age. The isoclinal folds with phyllitic axial planar foliation developed at this time. Refolding and solution cleavage development within the Talladega belt, and folding and solution cleavage

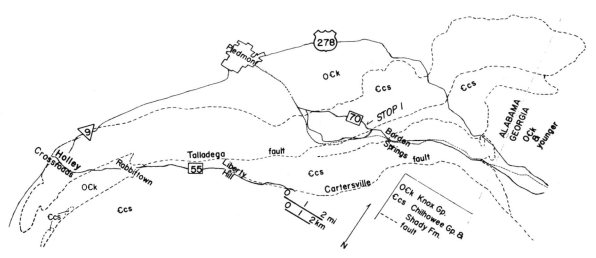

Figure 2. Generalized geology of nappes bordering the Talladega belt in northern Cleburne County, Alabama. Note location of Stop 1 on County Road 70.

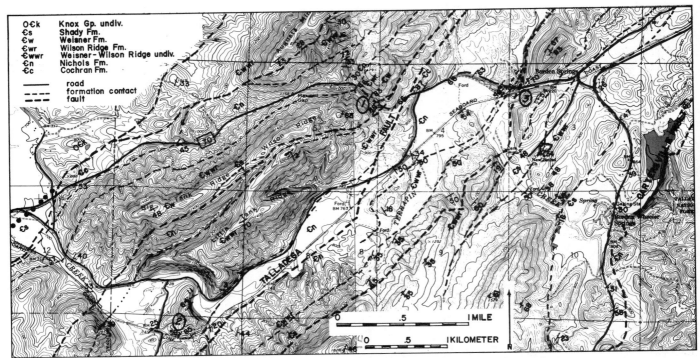

Figure 3. Geology in the vicinity of Borden Springs, showing locations of Stops 1–3.

development within the nappes west of the Talladega fault are probably all Carboniferous (Alleghanian) events (Tull, 1984).

SITE DESCRIPTION

A cluster of three sites has been selected to demonstrate the eastward changes in sedimentary character and structural style within the nappes. Site locations are shown with circled numbers on Figure 3.

Site 1

Site 1 is located along the north side of Cleburne County Road 70 approximately 1.2 mi (2 km) west of the small community of Borden Springs, Alabama. This site is an almost continuous exposure extending approximately 1,640 ft (500 m) and including more than 1,150 ft (350 m) of section of the Chilhowee Group and basal Shady Formation (Fig. 4). The Chilhowee Group consists of sandstones and siltstones deposited in various marine environments. Chilhowee lithologies were designated as the Weisner Formation by early workers in Alabama and Georgia, although their resemblance with the Chilhowee Group of eastern Tennessee has long been recognized (Butts, 1926). Mack (1980) recommends restricting the term Weisner to the uppermost part of this clastic section, which he divides into four formations; in ascending order, Cochran, Nichols, Wilson Ridge, and Weisner. The continuous exposure here contains part of the Nichols, all of the Wilson Ridge and Weisner Formations, and a small amount of carbonate and chert residuum from the basal part of the Shady Formation.

The Nichols Formation is exposed at the northwest end of the section (Fig. 4) and consists of greenish-gray to black micaceous shale and mudstone with thin interbeds and laminations of very fine-grained sandstone. Worm trails and burrows are abundant. Small wave and current ripples are present, although much of the lithology contains persistent planar laminations. Approximately 460 ft (140 m) of Nichols is exposed in the continuous section. Isolated exposures of lower parts of the Nichols occur along County Road 70 westward from the continuous section for about 1,300 ft (400 m) upslope to Maxwell Gap (Fig. 3). Mack (1980) interprets the depositional environment of the Nichols as off-shore marine shelf. East of the Talladega fault the Nichols appears to represent deeper water deposition, possibly a slope environment. The Talladega fault crosses County Road 70 approximately 330 ft (100 m) west (upslope) from the intersection of an unnamed tar road on the south side of the highway approximately 1,300 ft (400 m) east of the east end of the continuous section. At this location, the Nichols slate in the hanging wall of the Talladega fault contains thin, parallel laminations of very fine sand, in part graded, and also contains an interval of pebbly quartzite incorporating lithic clasts and graded beds.

The Wilson Ridge Formation overlies the Nichols Formation with conformable, gradational contact. In the continuous section dominant lithologies of the Wilson Ridge weather maroon. The formation consists mainly of interbedded, fine-grained feldspathic sandstone and ripple-marked siltstone. Rippled beds are cross-laminated. Load casts of fine sand, approximately 0.5 in (1 cm) in cross-sectional width and depth, project from the bases of many sandstone beds into underlying siltstone. Mud cracks and *Skolithus* burrows are common. Sandstone beds tend

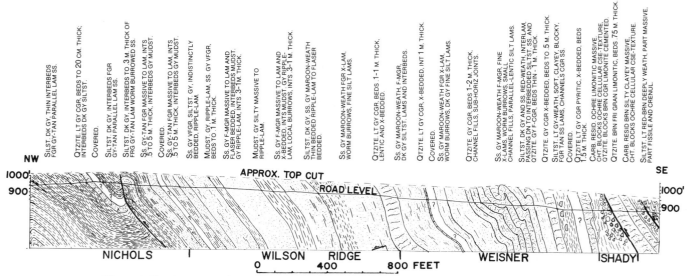

Figure 4. Continuous section exposed at Stop 1 on northeast side of County Road 70, NE ¼ sec. 5, T.13S, R.11E, Borden Springs 7½-minute Quadrangle.

to be lenticular. One coarse-grained, light gray quartzite interval, approximately 66 ft (20 m) thick, is present at the midpoint of the continuous section and extends from the highway southwestward up the nose and onto the crest of Wilson Ridge (Fig. 3). Mack (1980) interprets the environment of deposition of the Wilson Ridge as nearshore, high energy, with much of the deposition accomplished or affected by tidal processes. The Wilson Ridge Formation is approximately 230 ft (70 m) thick at the continuous section location (Fig. 4).

The Weisner Formation, as defined by Mack (1980), overlies the Wilson Ridge Formation conformably but with sharp contact. Although intervals of maroon weathering Wilson Ridge–type lithology are present in the Weisner, the dominant lithology is coarse-grained, cross-bedded orthoquartzite containing channel fills. The lowest interval of the quartzite, forming the base of the Weisner, is approximately 112 ft (34 m) thick and is located 656 ft (200 m) west of the east end of the continuous section. Conglomeratic quartzite, in beds up to 5 ft (1.5 m) thick, with rounded quartz pebbles up to 1.5 in (4 cm) in diameter, is present in the upper part of the Weisner. Siltstone is a minor component of the Weisner, and is generally greenish-gray, laminated in part and rippled. One fine sandy siltstone interval near the top of the Weisner is light tan to light gray, finely cross-laminated, and contains apparent channel deposits of coarse sandstone up to 40 in (1 m) thick that terminate abruptly laterally. *Skolithus* burrows are common throughout the Weisner.

The Weisner Formation, as defined by Mack (1980), comprises approximately one third of the continuous section and is about 460 ft (140 m) thick at this location (Fig. 4).

Mack (1980) interprets the Weisner depositional setting primarily as beach or barrier island with siltstone-mudstone intervals possibly reflecting tidal flats. In the writer's opinion this

interpretation is valid locally and for selected quartzite and mudstone intervals. The Weisner and Wilson Ridge contain different proportions of the same lithologies in the continuous section. The Wilson Ridge is generally finer grained. Both formations display abundant examples of structures suggesting rapid sedimentation, such as load casts, flame structures, and soft sediment folds. Both contain channel fills up to a few ft (m) wide in cross section. A more regional characterization of the environment these formations and the underlying Nichols Formation represent seems to be a prograding shallow marine delta.

The Shady Formation is poorly exposed at the southeastern end of the continuous section. Mack (1980) interprets the Weisner-Shady interval as reflecting marine transgression. The Weisner-Shady contact is therefore probably conformable, although this is not clear in the exposure. Weathering has reduced the Shady to a mixture of yellowish-brown silty carbonate residuum and yellowish-brown, coarse textured, cellular chert exposed in a zone a few ft (m) wide. Faulting has juxtaposed a slice of uppermost Weisner quartzite, overlain by a similar thickness of Shady carbonate residuum and chert, on the east side of this zone. The slice is bordered on its east side by a breccia of chert boulders a few meters thick. A thrust fault juxtaposes sheared, deeply weathered Chilhowee (possibly Nichols Formation) siltstone against the east side of the breccia (Fig. 4).

This section displays a variety of secondary structural features. Average strike is north-northeast. Dips range from 45 degrees southeast to sub-vertical. No large folds are present. Minor folding in the form of a tight anticline-syncline pair with wavelengths less than 16 ft (5 m) is present immediately above the basal Weisner quartzite interval. Higher in the Weisner section a set of smaller, more open drag folds is present.

Aside from the thrust faults at the southeast end of the continuous section mentioned above, no major thrust faults are present within the continuous section. A high-angle thrust fault of probable minor but indeterminable displacement is visible in the Nichols Formation immediately east of a covered zone approximately 160 ft (50 m) from the northwest end of the continuous section. A contemporaneous fault, high angle at the time of formation, is visible in thick bedded sandstone near the top of the Wilson Ridge Formation.

A cleavage, approximately axial planar to folds of several ft (m) wavelength in the exposures of Nichols Formation extending upslope along County Road 70 from the northwest end of the continuous section to Maxwell Gap (Fig. 3), is especially well developed in these fine-grained rocks, and is weakly and sporadically developed to the east in siltstone intervals of the continuous section. Thin sections reveal this cleavage to be essentially close spaced jointing accompanied by dissolution with no apparent reorientation of elongate quartz and detrital mica grains. There is, however, secondary fine white mica oriented in part parallel to cleavage planes.

Folds and cleavage developed in Nichols siltstone in exposures northwest of the continuous section show no obvious evidence of refolding. This can also be said for strata of the continuous section and of the discontinuous exposures of the next thrust block for 330 ft (100 m) along County Road 70 southeast of the continuous section. The Nichols Formation in the hanging wall of the Talladega fault at the intersection of County Road 70 and a blacktop road 1,300 ft (400 m) southeast of the continuous section has a different structural aspect. Here greenish-gray siltstone possesses a slaty cleavage with sub-phyllitic luster. The cleavage shows extensive recrystallization of fine white mica, as well as alignment of quartz grains in thin section. Evidence of multiple deformation is present in that the cleavage, which is axial planar to small isoclinal folds, has been crenulated. The isoclinal folds have been refolded as well.

Site 2

At Site 2, contrasts in structural style northwest and south-

east of the Talladega fault can be studied at two locations within short driving distance of Site 1. The site can be reached by turning south onto the blacktop road that joins County Road 70 at the base of the slope in the vicinity of the Talladega fault (Fig. 3). Follow this road approximately 2.5 mi (4 km), crossing one bridge over Terrapin Creek, and turn left onto an unpaved road. Follow this road approximately 0.4 mi (0.7 km) to a railroad crossing. Park and walk east along the railroad 1,300 ft (400 m) to a large cut in which folds with amplitudes of several ft (meters) and wavelengths of several ft (meters) to several tens of ft (meters) are present in the Wilson Ridge Formation. A well-developed axial planar solution cleavage is present. Folds are northeast trending, gently plunging, and overturned to the northwest. These folds are on the southeast limb of a large northeast trending, gently plunging syncline that occupies most of the width of the thrust block northwest of the Talladega fault and southeast of the continuous section at Site 1. Again, only one folding phase is apparent.

Site 3 is on County Road 70, (Fig. 3) 0.7 mi (1.2 km) east of the Talladega fault and immediately west of the Borden Springs community. A blacktop road joins County Road 70 on the north side 330 ft (100 m) west of the site. At this location phyllites and quartzites of the Weisner and/or Wilson Ridge Formation(s) display small isoclinal flow folds with axial planar foliation that have been refolded into tight folds with wavelengths and amplitudes of a few to several ft (meters) and with axial planar and refracted cleavages that are non-penetrative. Additional folding examples can be seen in cuts along the railroad that runs adjacent and parallel to County Road 70 at this location.

Quartzite beds at Site 3 are either massive or graded, a characteristic common to coarse clastics of the nappes southeast of the Talladega fault. These strata appear to have been deposited by gravity flow and turbidity currents in deeper water (lack of bottom current sedimentary structures) than nappe strata northwest of the Talladega fault. The two sequences, juxtaposed by the Talladega fault, may represent a part of the North American Early Cambrian eastern shelf and slope.

REFERENCES CITED

Bearce, D. N., 1982, Lower Cambrian metasediments of the Appalachian Valley and Ridge province, Alabama; possible relationship with adjacent rocks of the Talladega metamorphic belt: Geological Society of America Special Paper 191, p. 35–45.

Butts, C., 1926, The Paleozoic rocks: Geological Survey of Alabama Special Report 14, p. 40–223.

Mack, G. H., 1980, Stratigraphy and depositional environments of the Chilhowee Group (Cambrian) in Georgia and Alabama: American Journal of Science, v. 280, p. 497–517.

Tull, J. F., 1984, Polyphase late Paleozoic deformation in the southeastern foreland and northwestern Piedmont of the Alabama Appalachians: Journal of Structural Geology, v. 6, p. 223–234.

Middle and Upper Ordovician stratigraphy of the southernmost Appalachians

Thornton L. Neathery, *Geological Survey of Alabama, P.O. Box O, University Station, Tuscaloosa, Alabama 35486*
James A. Drahovzal, *ARCO Exploration and Technology Company, Plano, Texas 75075*

LOCATION

Facies changes in Middle and Upper Ordovician stratig-raphy can be seen in a series of exposures on I-59 and Alabama 77 in the vicinity of Gadsen, Alabama (Fig. 1).

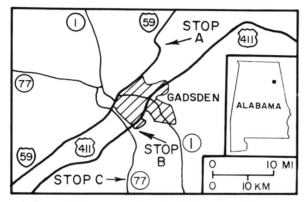

Figure 1. General highway map showing the location of Stops *A, B,* and *C* in Middle and Upper Ordovician strata in the vicinity of Gadsden, Alabama.

INTRODUCTION

Middle and Upper Ordovician stratigraphy of the Alabama part of the southernmost Appalachians can be divided into two distinct sedimentary facies: an eastern, dominantly clastic facies; and a western, dominantly carbonate facies (Drahovzal and Neathery, 1971). The line separating the two facies in Alabama coincides with the Helena-Rome fault system (Fig. 2). Interme-diate or transitional facies between the eastern and western facies are poorly exposed, having been faulted out by large-scale re-gional imbricate thrusting (Neathery and Thomas, 1982). How-ever, partial exposures of the transitional facies can be found on a few scattered fault blocks in northwest Georgia and on Dirtseller Mountain, Cherokee County and Dunaway–Hensley Mountains in Etowah County, Alabama (Fig. 2). Lithologies common to both Middle Ordovician facies are (1) a basal conglomerate; (2) a very fine-grained, gray limestone in the lower part of the se-quences; and (3) a series of thin bentonites near the middle of the sequences. Red calcareous siltstone is common to both facies of the Upper Ordovician. The three stops described herein provide an opportunity to study and evaluate the two contrasting sedi-mentary facies within short traveling distances.

Western Facies

The western facies of the Middle Ordovician (Fig. 2) consists of three mappable lithostratigraphic units: the basal Attalla Chert

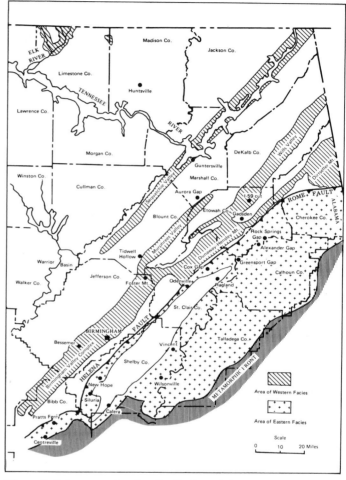

Figure 2. Index map showing distribution of the eastern and western facies of the Middle and Upper Ordovician rocks in northeast Alabama.

Conglomerate Member of the Stones River Formation, an un-named upper carbonate member of the Stones River Formation, and the Nashville Formation; together these units form the Chickamauga Group. The Upper Ordovician consists of three units: the Inman, Leipers, and Sequatchie Formations (Table 1). A general lithologic description of each unit can be found in Wilson (1949), Milici and Smith (1969), and Drahovzal and Neathery (1971).

Units of the western facies of the Middle Ordovician are exposed along the flanks of eroded and faulted anticlinal valleys of the fold and thrust belt and in limited outcrops in the Elk River drainage basin of north Alabama (McMasters, 1965a, 1965b; Fig. 2). The stratigraphic succession of the western facies, which can be recognized as far south as Gadsden, Etowah County, Alabama, is very similar to that of the Stones River and Nashville Groups of central Tennessee and northeast Georgia. Southwest of Gadsden individual units of the Middle Ordovician interval begin to change facies and thin in total thickness. Lithofacies and depositional environments of the western carbonate sequence of the Alabama Middle Ordovician have been discussed by Ward (1983) who recognized three major environments: supratidal, intertidal, and subtidal. Representatives of these environments can be found at the I-59 (Stop A) section although they may be more prominent at other localities. In general, the lithologic succession is one of regression from subtidal conditions at the base of the sequence to alternating supratidal and intertidal conditions near the top. Near the middle of the carbonate sequence are a series of bentonite beds correlative with regionally extensive bentonite beds of central Tennessee (Fox and Grant, 1944). The most distinctive bentonite bed is the biotite-rich T-3 bed (Drahovzal and Neathery, 1971). Above the T-3 bentonite, a general transgressive trend from intertidal to subtidal depositional conditions is observed. The depositional pattern of the western facies coincides generally with a northeast-trending shelf edge similar to units described by Walker (1974) in Tennessee.

TABLE 1. CORRELATION CHART FOR THE
MIDDLE AND UPPER ORDOVICIAN ROCKS
OF ALABAMA

	1	2	3
	WILSON (1949)	WESTERN FACIES	EASTERN FACIES
Richmond	Sequatchie	Sequatchie	?
Maysville	Leipers	(Leipers)	Sequatchie Formation
Eden	Inman	(Inman)	?
Nashville Group — Catheys		Nashville Formation	?
Nashville Group — Bigby-Cannon			
Nashville Group — Hermitage			
Nashville Group — Carters		T-4 T-3 Bentonite	Colvin Mt. Sandstone
Stones River Group — Lebanon			Greensport Formation
Stones River Group — Ridley		Stones River Formation	Little Oak-Athens Shale / Lenoir Limestone
Stones River Group — Pierce			
Stones River Group — Murfreesboro			
Stones River Group — Wells Creek		Attalla Chert Conglomerate Member	
Knox Dolomite		Longview Chepultepec	Newala Longview Chepultepec

In the western facies of the Upper Ordovician, the Inman is characteristically a distinctive maroon and gray mottled argillaceous limestone which thins to the west until it is absent. The overlying Leipers also thins westward, and, where the Inman is missing, lies unconformably upon the lithologically indistinguishable carbonates of the Nashville Formation. The Sequatchie Formation can be subdivided into three recognizable members: the Mannie Shale, the Fernvale Limestone, and an upper sandstone unit. Sediment distribution reflects regional upwarping associated with the Nashville Dome during or shortly after deposition of the Sequatchie. Locally, parts of the Upper Ordovician have been removed by post-Sequatchie–pre-Silurian erosion as evidenced by the occurrence of incised paleochannel deposits in the Sequatchie Formation (McMasters, 1965a, 1965b). Distribution of lithologies suggests a series of transgressions and regressions of a westward prograding deltaic complex with possible longshore current redistribution (Thompson, 1971).

Eastern Facies

The eastern facies (Fig. 2) of the Middle Ordovician consists of five mappable lithostratigraphic units: the Lenoir Limestone, Athens Shale, Little Oak Limestone, Greensport Formation, and Colvin Mountain Sandstone. The Upper Ordovician is represented by the Sequatchie Formation (Table 1). The Greensport, Colvin Mountain, and Sequatchie Formations constitute a red-bed sequence.

Stratigraphic succession of the eastern Middle Ordovician units is one of gradational facies relationships, both vertically and areally. Crucial to regional correlations are the bentonite beds, especially the T-3 bed which is recognized in the Colvin Mountain Sandstone. Lithostratigraphic units of the eastern facies occur primarily along the northwest escarpment of the northeast-trending ridges that mark the frontal edge of southeast dipping Late Paleozoic imbricate thrust sheets. Little is known of the southeastward limits of the Middle and Upper Ordovician. Only at a few isolated localities, such as at Wilsonville (Fig. 2), are one or two of the units exposed. Lithofacies and depositional environments of some of the eastern facies lithologies have been discussed by Jenkins (1984). The lower part of the red-bed sequence is probably equivalent to the Moccasin-Bays sequence of Georgia and Tennessee (Chowns and McKinney, 1980).

The Sequatchie Formation of the eastern facies is probably older than the western Sequatchie (Table 1). Because of structural complications the eastern facies Sequatchie cannot be physically traced to the northeast and its detailed relationship to the Martinsburg, Juniata, and Sequatchie of eastern Tennessee and northwest Georgia is uncertain. The Sequatchie together with the lower red-bed sequence is interpreted as the distal part of a regional clastic wedge associated with an orogenic uplift to the northeast (Thomas, 1977; Chowns and McKinney, 1980).

STOP A. INTERSTATE 59 SECTION, WESTERN CARBONATE FACIES

Location: I-59 road cut through Big Ridge, approximately

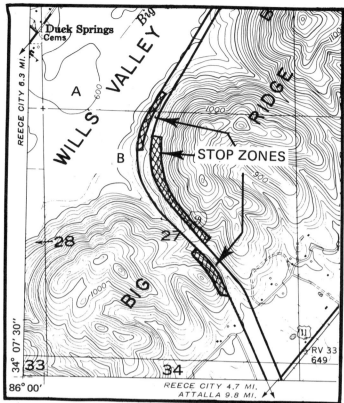

Figure 3. Location map of Stop A: *A,* low rounded hills; *B,* limestone outcrops.

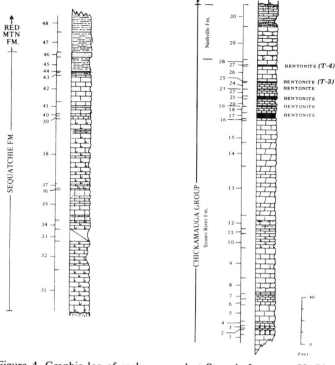

Figure 4. Graphic log of rocks exposed at Stop A, Interstate 59, Big Ridge, Etowah County, Alabama.

10 mi (16.7 km) north of Gadsden, in the N½ sec. 27, T. 10 S., R. 6 E., Keener 7½-minute Quadrangle, Etowah County, Alabama (Figs. 1 and 3).

The I-59 section through Big Ridge is one of several spectacular road cuts in Alabama. Exposed here is more than 1,050 ft (320 m) of rock ranging from the basal beds of the Middle Ordovician, Chickamauga Group, to the basal beds of the Mississippian, Fort Payne Chert. A graphic log of the Ordovician rock units at Big Ridge is shown in Figure 4 (corresponding lithologic descriptions of the various rock units may be found in Drahovzal and Neathery [1971]). Of particular interest is the lower 276 ft (84.1 m) of the section consisting of medium- to dark-gray limestone which represents the Middle Ordovician Chickamauga Group and the upper 217 ft (66.1 m) of maroon and gray mottled limestone and olive shale of the Upper Ordovician Sequatchie Formation. The petrography of the carbonates has been discussed by Wilson (1971a, 1971b) and Ward (1983); McKinney (1971) discussed the bryozoan fauna from the Chickamauga Group in Wills Valley.

The Chickamauga Group has been subdivided into the Stones River and the Nashville Formations (Drahovzal and Neathery, 1971). All the limestone equivalents of the Stones River Group (Tennessee), excluding the Murphreesboro, are exposed as a series of ledges or in excavated cuts from the northwest base of Big Ridge to about 20 ft (6 m) above the bentonite zone exposed on the northbound lane. Murphreesboro equivalents

occur at the base of Big Ridge and can be seen in the stream in the valley floor adjacent to the toe of the ridge. The low rounded hills (Fig. 3, *A*) several hundred yds (m) immediately to the northwest of Big Ridge are developed on the Longview–Newala Limestone of Early Ordovician age. The exposed rocks consist of dolomite clasts in a carbonate matrix. This interval marks the Middle Ordovician unconformity in this area. The topographic expression and characteristic lithology can be traced for approximately 40 mi (64 km) along strike in Wills Valley.

The first distinctive limestone lithology which forms the lower part of the I-59 stratigraphic section crops out just over the road shoulder on the northwest side of the southbound lane (Fig. 3, *B*). These limestones resemble the Ridley Limestone of the Tennessee section. Exposed along the southeast side of the southbound lane is the most complete section of Lebanon-like limestone exposed in Alabama. Above the Lebanon are limestones equivalent to the Carters Limestone. The contact occurs about halfway up the large drainage ditch which connects the south- and northbound lanes. Typical Carters lithology can be seen at the uppermost part of the drainage ditch and in the lower part of the road cut on the southeast side of the northbound lane.

Exposed near the top of the Carters are a number of bentonites and bentonitic shales. The abundant biotite flakes in the bentonite layer, (unit 25, Fig. 4), identifies it as equivalent to the T-3 bentonite of the Tennessee section. Because of the nearly ubiquitous regional occurrence of the biotite in this one bentonite unit, the Stones River–Nashville boundary of Tennessee can be traced southwestward into Alabama.

Above the Stones River Formation are two limestone lithologies indicative of the Nashville Formation. About 4 ft (1.2 m) above the T-4 bentonite (unit 27, Fig. 4) are a series of nodular, irregularly-bedded limestones which are equivalent to the Hermitage Limestones of the Tennessee section. Above the Hermitage are thin-bedded and silty laminated limestones of the Cannon Limestone of the Tennessee section.

Overlying the Cannon along the step-back ledge cut in the side of Big Ridge are the poorly preserved units of the Sequatchie Formation. The more argillaceous units have weathered to siltstone and silty clay. The ledges which stand in relief are principally lavender to medium-gray fossiliferous limestones indicative of the Fernvale member. At the top of the Sequatchie interval is 16 ft (4.9 m) of thin-bedded fossiliferous olive-gray shale—the Mannie Shale(?) (unit 45, Fig. 4)—capped by a 4-ft (1.2 m) bed of dark-brown ferruginous fine-grained sandstone (unit 46, Fig. 4).

Overlying the Ordovician section at the I-59 cut are clastic and carbonate units of the Red Mountain Formation of Silurian age. The remaining part of the I-59 section includes the Devonian Chattanooga Shale and the Mississippian Maury Formation and Fort Payne Chert.

STOP B. HENSLEY MOUNTAIN SECTION: TRANSITIONAL ZONE—INCLUDING ATTALLA CHERT CONGLOMERATE AND POST–KNOX UNCONFORMITY

Location: Section along valley face cutting ridge just northeast of Hensley Mountain at the home of Dr. and Mrs. Frank A. Finney, near Rainbow City, in NE¼ SW¼ SW¼ sec. 28, T. 12 S., R. 6 E., Dunaway Mountain 7½-minute Quadrangle, Etowah County, Alabama, approximately 3.5 mi (5.6 km) south of Gadsden and 1.0 mi (1.6 km) north of junction U.S. 411 and Alabama 77. *Permission to visit outcrop required* (Figs. 1 and 5).

Exposures on Dunaway and Hensley Mountains provide a partial view of the transition zone between the western carbonate and eastern clastic facies as well as spectacular exposures of the regional Middle Ordovician unconformity. The contact between the Attalla Chert Conglomerate and the eroded surface of the Longview Limestone of Early Ordovician age exposed here is undoubtedly one of the finest examples of the Lower–Middle Ordovician unconformity exposed anywhere in Alabama. Approximately 70 ft (21 m) of Attalla Chert Conglomerate overlies about 45 ft (13.7 m) of exposed karstic dolomitic Longview Limestone.

The Longview Limestone is essentially a light-gray medium-grained, saccharoidal dolomitic limestone with poorly developed bedding. Rounded to ellipsoidal chert nodules up to 0.2- by 0.6-ft (6 by 18 cm) diameter occur in the rock; however, the majority of the nodules average 0.1- by 0.2-ft (3 by 6 cm) diameter. Several 0.3- to 0.6-ft (9 to 18 cm) thick, light- to medium-gray chert beds can be seen in the upper part of the exposure. Most of the bedded chert is fractured; apparently it was

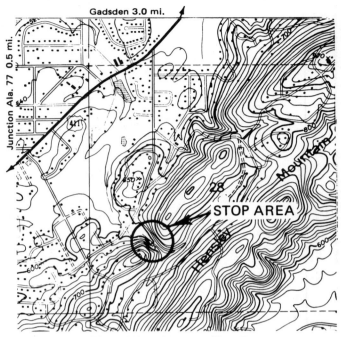

Figure 5. Location map of Stop B.

subjected to tension prior to the deposition of the Attalla. Note that the thin near vertical tension fractures have been filled by fine-grained subangular to subrounded chert fragments and fine sand.

The interface between the Longview Limestone and the Attalla Chert Conglomerate is sharp and irregular. Maximum relief on the Longview surface is approximately 43 ft (13 m) as indicated by the Attalla-filled solution tube adjacent to the barbecue pit. Close examination of the material in the overlying beds, in the tension fractures, and in solution tubes shows no dolomite is mixed with the Attalla. The absence of dolomite in the Attalla suggests that the Longview was deeply weathered prior to the Attalla deposition. The structures do not appear to be collapse features as seen elsewhere along the unconformity.

Note the present-day solution development at the contact between the Attalla Chert Conglomerate and the Longview Limestone. The Longview is being dissolved and frost action spalls off the chert clasts to open the irregular interface that forms the irregular tubes and pockets. The thick mosaic network of large ribs (or ridges) on the basal surface of the Attalla is interpreted as representing impressions of erosional depressions developed in the underlying dolomite prior to deposition of the Attalla. In the upper end of the glen, large blocks of Attalla have been lowered as a result of the active dissolution of the dolomite.

The Attalla Chert Conglomerate is composed of subangular to subrounded chert boulders, cobbles and pebbles cemented together in a matrix of fine chert particles, quartz sand, and clay. Although the larger particles in most of the Attalla usually measure less than 0.3 ft (9 cm) in diameter, clasts up to 3 ft (91 cm) in diameter at Hensley Mountain occur in the lower part of the interval. Bedding and cross bedding are faint. About 40 ft

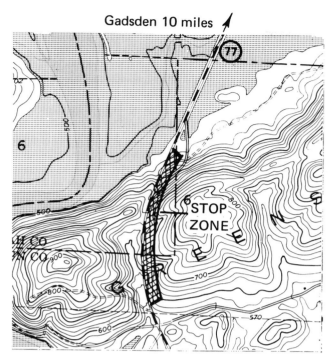

Figure 6. Location map of Stop C.

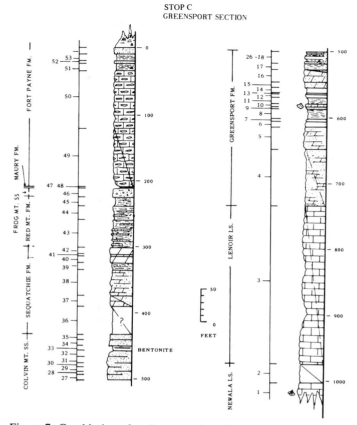

Figure 7. Graphic log of rocks exposed at Stop C, Greensport Gap, Alabama 77, Etowah County, Alabama.

(12 m) above the base of the Attalla, the average clast size suddenly increases slightly suggesting a renewal of source material flooding the depositional basin.

In the lower part of the Attalla, red claystone and shale are common. At the termination of the large chert-filled solution tube adjacent to the barbecue pit, a dense red shale can be seen separating the dolomite from the chert. The shale is deep red to maroon, slightly silty, brittle and blocky. A light-red to pink claystone is mixed with the chert clasts within 0.3 ft (9 cm) of the Attalla-Longview interface. This claystone may represent *terra rossa* from the post-Knox weathering surfaces. The chert clasts in the lower part of the Attalla are slightly more angular than clasts higher in the section. Chert clasts which fill the solution tubes and channels in the Longview Limestone are distinctly more angular than those several feet (meters) above the interface.

Overlying the Attalla is an irregularly occurring medium-gray carbonate unit of unknown thickness which grades upward into a sequence of poorly exposed red, maroon, and dusky-yellow-brown shales, siltstones, and sandstones that represent the distal edge of the Middle and Upper Ordovician Greensport, Colvin Mountain and Sequatchie formations of the eastern red-bed facies. Overlying the Sequatchie units are the similar appearing dark-red and dusky-yellow-brown sandstones and siltstones of the Silurian Red Mountain Formation. Because this outcrop area is in a suburban subdivision, additional exposures are rare.

STOP C. GREENSPORT GAP SECTION: EASTERN FACIES

Location: Greensport Gap cut on Alabama 77, through

Greens Creek Mountain, near the center of sec. 6, T. 14 S., R. 6 E., Ohatchee 7½-minute Quadrangle, Etowah County, Alabama. Approximately 11 mi (18 km) south of Gadsden (Figs. 1 and 6).

The highway cut at Greensport Gap offers a rare opportunity to study a complete section of the northern part of the eastern red-bed facies of the Middle and Upper Ordovician. Exposed are rock units ranging from Early Ordovician to Late Mississippian in age. Other sections can be viewed at Rock Springs Gap and Alexander Gap (Fig. 2; Drahovzal and Neathery, 1971).

The measured section (Fig. 7) begins at the high water level of the river pool at the north side of the cut. At low water, numerous pinnacles of light-gray fine-grained limestone are exposed, with gastropods and maclurites visible on the weathered surfaces (units 1 and 2, Fig. 7).

Typical units of dark-gray Lenoir Limestone crop out along the northwest escarpment of Greens Creek Mountain and are exposed both along the gravel road and in the woods downslope of the road at the north end of the highway cut. The Lenoir Limestone at Greensport Gap is poorly exposed because of colluvial cover derived from the overlying Greensport Formation. In a low bluff in the woods off the gravel road on the north side of the ridge, light-gray micrite intraclasts occur in a matrix of dark-gray Lenoir limestone. On other low ledges, silty zones form the typical anastomosing pattern characteristic of the Lenoir Limestone.

Overlying the Lenoir Limestone is slightly more than 230 ft (70 m) of red clastic rock comprising the Greensport Formation. The Greensport is sparingly fossiliferous but some fossils were collected from unit 9 (Fig. 7) of the measured section. The lowest part of the Greensport is calcareous, whereas the upper part is non-calcareous. The division between the calcareous and non-calcareous facies occurs across a slight unconformity which is visible in the east wall of the cut and marked by vegetative growth on a shelf of debris (between units 16 and 17, Fig. 7). The bedding of the Greensport Formation is monotonously regular and even.

The Colvin Mountain Sandstone is about 60 ft (18 m) thick and includes two sandy bentonite beds in the upper part of the interval (units 33 and 34, Fig. 7). The uppermost bentonite bed contains traces of biotite and is believed to be the T-3 bento-

nite which is the marker in the upper part of the Stones River Formation to the west. The lower beds of the Colvin Mountain Sandstone, adjacent to the highway, have been sharply upturned or rolled. A well-developed fracture cleavage averaging S. 85°E., 77°SW, has broken up the sandstone. Slickensides are common on the bedding planes of the sandstone beds. Above the Colvin Mountain Sandstone is a partially covered interval of 41 ft (12.5 m). The rocks exposed here are mapped as the eastern facies of the Sequatchie Formation. The upper Sequatchie units in Greensport Gap are even-bedded dolomite and calcareous siltstones. The overlying Red Mountain Formation (Silurian) is dominantly coarse sandstone and the basal contact is sharply defined between units 43 and 42 (Fig. 7).

The remaining part of the Greensport section includes the Frog Mountain Formation of Devonian age and the Maury and Fort Payne Formations of Mississippian age. Overlying the Fort Payne Formation to the southeast are black shales of the Floyd Shale of Mississippian age.

REFERENCES CITED

Chowns, T. M., and McKinney, F. K., 1980, Depositional facies in Middle–Upper Ordovician and Silurian rocks of Alabama and Georgia, *in* Frey, R. W., ed., Excursions in Southeastern Geology, v. 2: American Geological Institute, p. 323–348.

Drahovzal, J. A., and Neathery, T. L., 1971, Middle and Upper Ordovician stratigraphy of the Alabama Appalachians, *in* Drahovzal, J. A., and Neathery, T. L., eds., The Middle and Upper Ordovician of the Alabama Appalachians, Alabama Geological Society Guidebook, 9th Annual Field Trip, p. 1–62.

Fox, P. P., and Grant, L. F., 1944, Ordovician bentonites in Tennessee and adjacent states: Journal of Geology, v. 52, p. 319–332.

Jenkins, C. M., 1984, Depositional environments of the Middle Ordovician Greensport Formation and Colvin Mountain Sandstone in Calhoun, Etowah, and St. Clair Counties, Alabama [M.S. thesis]: University of Alabama, University, 156 p.

McKinney, F. K., 1971, Bryozoan stratigraphy of the lower Chickamauga Group, Wills Valley, Alabama, *in* Drahovzal, J. A., and Neathery, T. L., eds., The Middle and Upper Ordovician of the Alabama Appalachians, Alabama Geological Society Guidebook, 9th Annual Field Trip, p. 101–114.

McMasters, W. M., 1965a, Geology of the Elkmont quadrangle, Alabama-Tennessee: U.S. Geological Survey Miscellaneous Geologic Investigation Map I-419, scale 1:24,000, with text.

—— 1965b, Geology of the Salem quadrangle, Alabama-Tennessee: U.S. Geological Survey Miscellaneous Investigation Map I-420, scale 1:24,000, with text.

Milici, R. C., and Smith, T. W., 1969, Stratigraphy of the Chickamauga Supergroup in its type area: Tennessee Department of Conservation, Division of Geology Report of Investigation 24, 35 p.

Neathery, T. L., and Thomas, W. A., 1982, Geodynamics transect of the Appalachian Orogen in Alabama, *in* Rast, N., and Delang, F. M., eds., Profiles of orogenic belts: American Geophysical Union Geodynamics Series, Washington, D.C., v. 10, p. 301–307.

Thomas, W. A., 1977, Evolution of Appalachian-Ouachita salients and recesses from reentrants and promontories in the Continental margin: American Journal of Science, v. 277, p. 1233–1278.

Thompson, A. M., 1971, Clastic-carbonate facies relationships and paleoenvironments in Upper Ordovician rocks of northeast Alabama, *in* Drahovzal, J. A., and Neathery, T. L., eds., The Middle and Upper Ordovician of the Alabama Appalachians, Alabama Geological Society Guidebook, 9th Annual Field Trip, p. 63–78.

Walker, K. R., 1974, Community patterns: Middle Ordovician of Tennessee, *in* Ziegler, A. M., et al., Principles of benthic community analysis: Comparative Sedimentation Laboratory, University of Miami, Sedimenta IV, p. 91–95.

Ward, W. I., 1983, Lithofacies and depositional environments of a portion of the Stones River Formation, Etowah and DeKalb Counties, Northeast Alabama [M.S. thesis]: University of Alabama, University, 255 p.

Wilson, C. W., 1949, Pre-Chattanooga stratigraphy in central Tennessee: Tennessee Department of Conservation, Division of Geology Bulletin 56, 407 p.

Wilson, A. O., 1971a, Petrology of part of the Black River age strata of the Chickamauga Limestone, Etowah and DeKalb Counties, Alabama, *in* Drahovzal, J. A., and Neathery, T. L., eds., The Middle and Upper Ordovician of the Alabama Appalachians, Alabama Geological Society Guidebook, 9th Annual Field Trip, p. 79–100.

—— 1971b, Petrogenesis of parts of Black River age strata of the Chickamauga Limestone, Etowah and DeKalb Counties, Alabama [Ph.D. thesis]: University of North Carolina, Chapel Hill, 112 p.

The Coosa deformed belt, thin imbricate thrust slices in the Appalachian fold-thrust belt in Alabama

William A. Thomas, Department of Geology, University of Alabama, University, Alabama 35486
James A. Drahovzal, ARCO Exploration and Technology Company, Plano, Texas 75075

LOCATION

A deep road cut on Alabama 77 (sec. 29, T.14S., R.6E.) near Ohatchee, Calhoun County, Alabama (Fig. 1), provides an exceptionally good view of imbricate thrust faults in the frontal tier of the Coosa deformed belt.

INTRODUCTION

The Coosa deformed belt includes three distinct parallel tiers of thin imbricate thrust slices of middle Paleozoic rocks. Complexity of structure increases across the belt from the frontal tier on the northwest to the interior tier on the southeast (Thomas and Drahovzal, 1974a). Differences in stratigraphic sequences between some adjacent thrust slices suggest juxtaposition of rocks from palinspastic sites of differing pre-thrust sedimentary and structural history (Thomas and Drahovzal, 1974a); however, at this locality the stratigraphy is consistent from slice to slice.

The frontal tier of the Coosa deformed belt overrides the Eden thrust sheet, and the Eden fault (which crosses Alabama

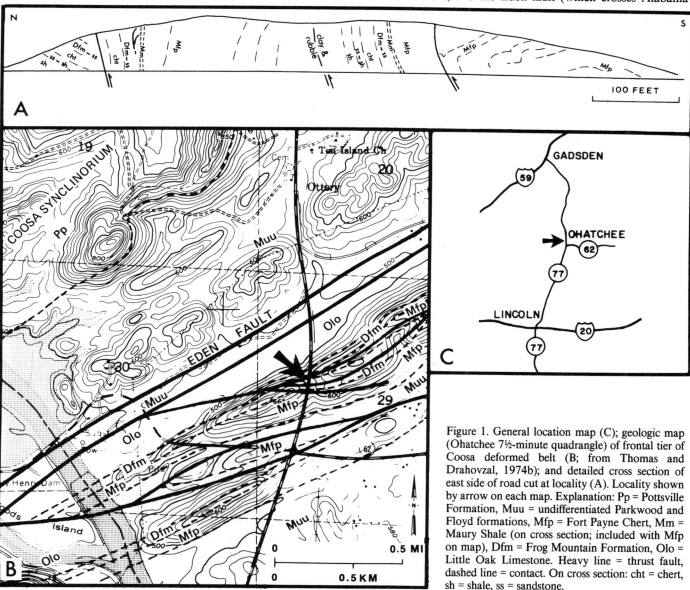

Figure 1. General location map (C); geologic map (Ohatchee 7½-minute quadrangle) of frontal tier of Coosa deformed belt (B; from Thomas and Drahovzal, 1974b); and detailed cross section of east side of road cut at locality (A). Locality shown by arrow on each map. Explanation: Pp = Pottsville Formation, Muu = undifferentiated Parkwood and Floyd formations, Mfp = Fort Payne Chert, Mm = Maury Shale (on cross section; included with Mfp on map), Dfm = Frog Mountain Formation, Olo = Little Oak Limestone. Heavy line = thrust fault, dashed line = contact. On cross section: cht = chert, sh = shale, ss = sandstone.

77 0.4 mi [0.6 km] north of this road cut) cuts the southeast limb of the Coosa synclinorium (Fig. 1). On the southeast, along the Pell City thrust fault (which crosses Alabama 77 9.9 mi [15.9 km] south of this road cut), rocks of the Cambrian-Ordovician Knox Group (exposed in road cuts) are emplaced over younger Paleozoic rocks in the interior tier of the Coosa deformed belt (Fig. 2).

Structures of the Appalachian fold-thrust belt may be divided into three domains (Thomas, 1982). The northwestern domain of the fold-thrust belt is typified by broad, flat-bottomed synclines and narrow, asymmetric anticlines above a shallow basement. The central domain includes thrust ramps and associated folds (including the Coosa synclinorium) that have higher relief and overlie a deeper basement than the structures of the northwestern domain. The Eden fault and Coosa deformed belt constitute the leading part of the southeastern domain, which is characterized by large-scale, low-angle, multiple-level thrust sheets (including the Pell City thrust sheet) (Fig. 2). Both the northwestern and central domains have single-level large-scale thrust sheets above a décollement near the base of the Paleozoic sedimentary sequence (Thomas, 1982; Neathery and Thomas, 1983). Although most of the large-scale stacked thrust sheets in the southeastern domain reflect the regional décollement near the base of the Paleozoic sedimentary sequence, the thrust sheets of the Coosa deformed belt are detached at higher stratigraphic levels ranging from within the Upper Cambrian part of the Knox Group (locally) to within the lower part of the Mississippian.

The three-tiered structural profile of the Coosa deformed belt persists along strike northeastward approximately 10 mi (16 km) to the Anniston cross-strike structural discontinuity and southwestward 40 mi (64 km) to the Harpersville cross-strike structural discontinuity (Fig. 2). At both cross-strike structural discontinuities, the Coosa deformed belt and adjacent structures exhibit abrupt along-strike changes in trend and structural style (Thomas and Drahovzal, 1974a).

SITE DESCRIPTION

The road cut exposes rocks and structures within the frontal tier of thrust faults in the Coosa deformed belt. The stratigraphic section contains the Devonian Frog Mountain Formation and the Mississippian Maury Shale and Fort Payne Chert. Within the cut, the Devonian-Mississippian succession is repeated by three local faults (Fig. 1A), a degree of imbricate repetition unusual along the frontal tier.

The Frog Mountain Formation includes three subdivisions: the upper 12 ft (3.7 m) of sandstone; the middle 14 ft (4.3 m) of chert; and a lower unit (minimum 45 ft or 13.7 m) that is mainly claystone but contains a few interbeds of sandstone and chert (Fig. 1A). The Maury Shale consists of less than 2 ft (0.6 m) of purple-weathering, yellowish-green clay shale. The Fort Payne Chert locally is approximately 150 ft (45 m) thick. Ordovician limestone, stratigraphically below the Frog Mountain, is exposed in the valley northwest of the cut (Fig. 1B).

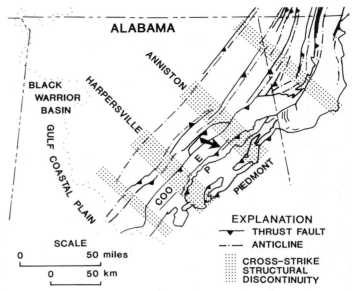

Figure 2. Map of regional structural setting of Coosa deformed belt (locality shown by arrow). Abbreviations: E = Eden fault; P = Pell City fault; COO = Coosa synclinorium.

The Ordovician limestone forms the base of the frontal thrust sheet of the Coosa deformed belt and is thrust northwestward over Mississippian Floyd-Parkwood sandstone and mudstone in the hanging wall of the Eden fault (termed Coosa fault by Thomas and Drahovzal, 1974a) along the southeast limb of the Coosa synclinorium. On the horizon, far to the northwest, are Pennsylvanian rocks in the up-plunge northeast end of the trough of the synclinorium (Fig. 1B). Devonian-Mississippian rocks in the imbricate slices of the frontal tier of the Coosa deformed belt dip southeastward beneath Mississippian Floyd-Parkwood mudstones and sandstones that comprise outcrop areas between the tiers of thrust faults in the Coosa deformed belt (Fig. 1B).

REFERENCES CITED

Neathery, T. L., and Thomas, W. A., 1983, Geodynamics transect of the Appalachian orogen in Alabama, in Rast, N., and Delany, F. M., eds., Profiles of orogenic belts: American Geophysical Union, Geodynamics Series, v. 10, p. 301–307.

Thomas, W. A., 1982, Stratigraphy and structure of the Appalachian fold and thrust belt in Alabama, in Thomas, W. A., and Neathery, T. L., eds., Appalachian thrust belt in Alabama: Tectonics and sedimentation (Field Trip Guidebook, Geological Society of America 1982 Annual Meeting, New Orleans, Louisiana): Tuscaloosa, Alabama Geological Society, p. 55–66.

Thomas, W. A., and Drahovzal, J. A., 1974a, Geology of the Coosa deformed belt: Alabama Geological Society, 12th Annual Field Trip, Guidebook, p. 45–75.

Thomas, W. A., and Drahovzal, J. A., 1974b, A field guide to the Coosa deformed belt: Alabama Geological Society, 12th Annual Field Trip, Guidebook, p. 1–43.

Murphrees Valley anticline, a southeast-verging anticline in the Appalachian fold-thrust belt in Alabama

William A. Thomas, *Department of Geology, University of Alabama, University, Alabama 35486*
Thornton L. Neathery, *Geological Survey of Alabama, Tuscaloosa, Alabama 35486*

LOCATION

A traverse of the Murphrees Valley anticline is provided by three stops along and near U.S. 231 in the vicinity of Oneonta, Blount County, Alabama (Fig. 1).

INTRODUCTION

The Murphrees Valley anticline is a long, narrow, asymmetric fold that, unlike most folds in the Appalachian fold-thrust belt, has a northwest-dipping axial plane. Beds on the southeast limb have a nearly vertical to steep overturned northwest dip, whereas beds on the northwest limb dip less than 30° northwest. The steep southeast limb of the Murphrees Valley anticline is broken by the Straight Mountain fault (Fig. 1). By 1890, A. M. Gibson (1893) had recognized the southeast-directed geometry of the anticline and concluded that the strata on the southeast had been *"thrust under"* the Murphrees Valley structure.

Vertical beds on the southeast limb of the Murphrees Valley anticline bend through a sharp hinge into the broad, flat-bottomed Blount Mountain syncline. The long, gently dipping northwest limb extends northwestward into the broad, flat-bottomed Coalburg syncline (Fig. 2). The narrow, asymmetric Murphrees Valley anticline and the adjacent broad, flat-bottomed synclines are characteristic of the northwestern domain of the Appalachian fold-thrust belt in Alabama (Thomas, 1982; Neathery and Thomas, 1983).

The structural profile of the Murphrees Valley anticline persists along strike approximately 45 mi (72 km) (Fig. 2). Northeastward 25 mi (40 km) from this locality, the anticline plunges northeastward and ends at a point that defines the Anniston cross-strike structural discontinuity (Fig. 2). Southwestward, the Murphrees Valley anticline continues into the northwestern anti-

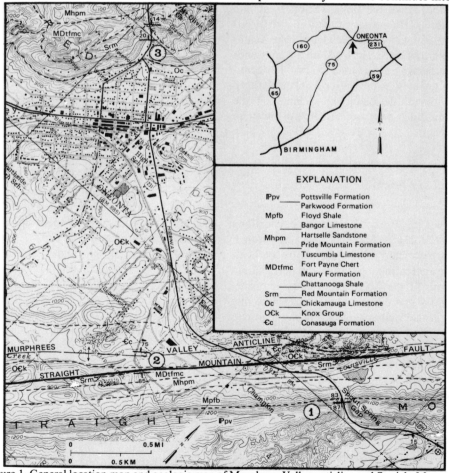

Figure 1. General location map and geologic map of Murphrees Valley anticline and Straight Mountain fault (Oneonta 7½-minute Quadrangle), showing locations of Stops 1, 2, and 3.

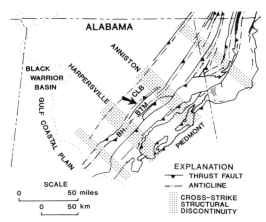

Figure 2. Map of regional structural setting of Murphrees Valley anticline (locality shown by arrow). Abbreviations: BTM = Blount Mountain syncline, CLB = Coalburg syncline, BH = Birmingham anticlinorium.

cline of the Birmingham anticlinorium. However, at the juncture of the Murphrees Valley anticline with the Birmingham anticlinorium approximately 20 mi (32 km) southwest of this locality, an abrupt along-strike change in structural profile defines the trace of the Harpersville cross-strike structural discontinuity (Fig. 2). There, southwestward along strike, the dip of the northwest limb changes abruptly from low-angle northwest to steep overturned southeast, and the sense of fold vergence changes from southeastward to northwestward. The Straight Mountain fault ends southwestward at the Harpersville cross-strike structural discontinuity, and the Blount Mountain syncline has local, relatively steep northeast plunge. The broad, flat-bottomed syncline on the northeast contrasts with a much narrower structure along strike on the southwest.

SITE DESCRIPTION

Stop 1. Southeast Limb of Murphrees Valley Anticline at Spout Spring Gap

Stop 1 (sec. 33, T.12S., R.2E.) is in a road cut along the southwest side of U.S. 231 in Spout Spring Gap through Straight Mountain approximately 2 mi (3 km) southeast of Oneonta (Fig. 1). The approximately vertical dip here characterizes the southeast limb of the Murphrees Valley anticline. Sandstone beds in the lower part of the Pennsylvanian Pottsville Formation are exposed in the southeastern part of the road cut, and mudstone and sandstone of the Mississippian-Pennsylvanian(?) Parkwood Formation are exposed in the northwestern part of the cut. Straight Mountain is formed on the lower Pottsville sandstones. Pre-Parkwood Mississippian formations underlie the valley to the northwest, and the low hills beyond the valley are formed on Cambrian rocks across the Straight Mountain fault (Fig. 1). An abrupt change in dip to the flat-bottomed trough of the Blount Mountain syncline is indicated by subhorizontal Pottsville beds exposed along U.S. 231 approximately 0.6 mi (1 km) east of this road cut (Fig. 1).

The upper part of the Mississippian System in the Blount

Mountain syncline includes units of a northeastward-prograding clastic facies (Parkwood Formation) and a regionally extensive carbonate facies (Bangor Limestone). The Mississippian clastic facies is thicker and more extensive in the Blount Mountain syncline than in the Coalburg syncline northwest of the Murphrees Valley anticline (Thomas, 1972). The stratigraphic differences suggest tectonic movement during latest Mississippian deposition.

Stop 2. Straight Mountain Fault Southeast of Oneonta

Stop 2 (sec. 5, T.13S., R.2E.) illustrates mapping of the Straight Mountain fault at outcrops adjacent to an unnumbered paved road and in road ditches approximately 1 mi (1.6 km) southeast of Oneonta (Fig. 1). The location of the fault trace can be determined approximately here, although the rocks are poorly exposed. On the northwest, in the hanging wall, carbonate rocks of the Cambrian Conasauga Formation are exposed in the wooded valley northeast of the road (at the northwest end of a northwestward curve in the road). Deeply weathered Mississippian Fort Payne Chert in the footwall is poorly exposed in low road cuts on the hilltop southeast of the valley. About halfway down the northwestward slope (northwest of the Fort Payne outcrops) black Chattanooga Shale may be dug from the road ditch (south side of road). Along much of the length of the Straight Mountain fault, as at this stop, the oldest beds in the footwall are the Devonian Chattanooga Shale; however, locally part of the Silurian Red Mountain Formation is preserved in the footwall.

Stop 3. Northwest Limb of Murphrees Valley Anticline on Red Mountain

Stop 3 (sec. 30–31, T.12S., R.2E.) displays gently dipping beds on the northwest limb of the Murphrees Valley anticline in road cuts on both sides of U.S. 231 across the top of Red Mountain in the northwestern part of the city of Oneonta (Fig. 1). A stratigraphic sequence from Middle Ordovician Chickamauga Limestone to Mississippian Fort Payne Chert is exposed. The beds dip 14° to 20° northwest, typical of the northwest limb of the anticline.

REFERENCES CITED

Gibson, A. M., 1893, Report on the geological structure of Murphree's Valley, and its minerals and other materials of economic value: Alabama Geological Survey [Special Report 4], 132 p.

Neathery, T. L., and Thomas, W. A., 1983, Geodynamics transect of the Appalachian orogen in Alabama, *in* Rast, N., and Delany, F. M., eds., Profiles of orogenic belts: American Geophysical Union, Geodynamics Series, v. 10, p. 301–307.

Thomas, W. A., 1972, Mississippian stratigraphy of Alabama: Alabama Geological Survey Monograph 12, 121 p.

Thomas, W. A., 1982, Stratigraphy and structure of the Appalachian fold and thrust belt in Alabama, *in* Thomas, W. A., and Neathery, T. L., eds., Appalachian thrust belt in Alabama: Tectonics and sedimentation (Field Trip Guidebook, Geological Society of America 1982 Annual Meeting, New Orleans, Louisiana): Tuscaloosa, Alabama Geological Society, p. 55–66.

Sequatchie anticline, the northwesternmost structure of the Appalachian fold-thrust belt in Alabama

William A. Thomas, *Department of Geology, University of Alabama, University, Alabama 35486*

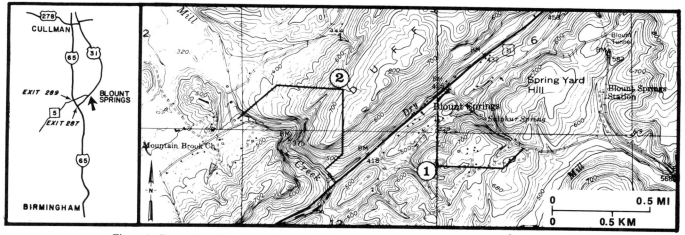

Figure 1. General location map and map of Stops 1 and 2 (Blount Springs 7½-minute Quadrangle).

LOCATION

Two stops at Blount Springs, Blount County, Alabama (Fig. 1), provide traverses of parts of the Sequatchie anticline on the Blount Springs culmination. Blount Springs is on U.S. 31 approximately 4 mi (6 km) east of I-65. Interchange 287 on I-65 has *only* a northbound exit and a southbound entrance connecting with U.S. 31. Interchange 289 (north of Interchange 287 at U.S. 31) on I-65 has exits and entrances both northbound and southbound, and provides access to Blount Springs via Blount County Road 5.

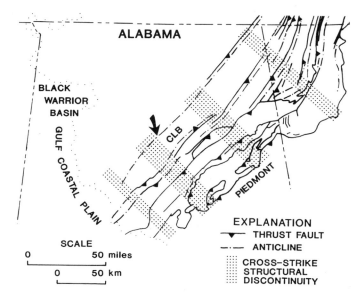

Figure 2. Map of regional structural setting of Sequatchie anticline (locality shown by arrow). Abbreviation: CLB = Coalburg syncline.

INTRODUCTION

The Sequatchie anticline is the northwesternmost structure of the Appalachian fold-thrust belt in Alabama (Fig. 2). Wells drilled to Precambrian rocks indicate that depth to basement is approximately 8,400 ft (2,550 m) beneath the Sequatchie anticline (Neathery and Copeland, 1983). One well on the southeast limb of the anticline documents no duplication within the stratigraphic succession and suggests that the structure is underlain by a regional décollement near the base of the Paleozoic cover sequence. The asymmetric, northwest-verging Sequatchie anticline is associated with the blind northwestward termination of the regional décollement (Thomas, 1985). Although no thrust fault is exposed at this locality, a thrust fault emerges on the northwest limb approximately 50 mi (80 km) northeastward along strike, and the anticline and thrust fault extend northeastward to east-central Tennessee.

The steep northwest limb of the Sequatchie anticline is bordered on the northwest by the Black Warrior foreland basin (Fig. 2), one of the foreland basins along the margin of the North American craton adjacent to the Appalachian-Ouachita orogen. The gently dipping southeast limb of the Sequatchie anticline extends into the broad, flat-bottomed Coalburg syncline (Fig. 2). Narrow, asymmetric anticlines and broad, flat-bottomed synclines characterize the northwestern domain of the Appalachian fold-thrust belt in Alabama (Thomas, 1982). The structures of the northwestern domain contrast with the higher relief thrust ramps and deep synclinoria of the central domain, and with the multiple-level, low-angle thrust sheets of the southeastern domain.

The Blount Springs inlier of Silurian through Mississippian rocks is the most southwesterly of several small culminations

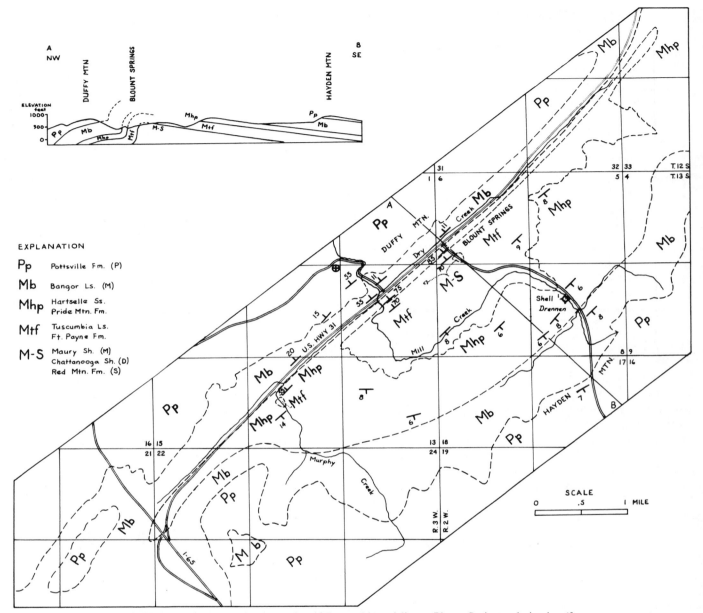

Figure 3. Geologic map and cross section of Sequatchie anticline at Blount Springs culmination (from Thomas and Bearce, 1969).

along the Sequatchie axis. The anticlinal crest can be traced approximately 35 mi (55 km) southwest from Blount Springs; down plunge, toward the southwest, structural relief gradually decreases.

The stratigraphic section exposed at Blount Springs is typical of this part of the Alabama Appalachians (Thomas, 1982). A regional pre–Late Devonian unconformity truncates greenish-gray mudstones of the Silurian Red Mountain Formation. Above the unconformity, the Upper Devonian Chattanooga Shale is approximately 40 ft (12 m) thick, and the overlying Lower Mississippian Maury Shale is less than 3 ft (1 m) thick. The Mississippian System is dominated by carbonate rocks: the Fort Payne

Chert and Tuscumbia (cherty) Limestone in the lower part (425 ft or 130 m thick), and the Bangor Limestone (690 ft or 210 m thick) in the upper part (Thomas, 1972). Between the Tuscumbia and Bangor Limestones, the Pride Mountain Formation (shale, sandstone) and Hartselle Sandstone constitute a clastic tongue (215 ft or 65 m thick) that extends far northeastward from the lower part of a regional clastic facies on the southwest (Thomas, 1972). The Bangor Limestone is overlain by massive sandstones of the lower part of the Pennsylvanian Pottsville Formation.

Blount Springs is named for sulfur-water springs, which result from water percolating through the iron sulfide–bearing Chattanooga black shale at the crest of the Sequatchie anticline.

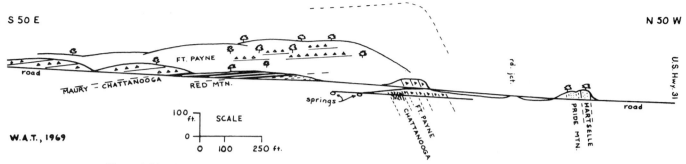

Figure 4. Sketch cross section of crest of Sequatchie anticline along Blount County Road 7 across Spring Yard Hill at Blount Springs (Stop 1 traverse).

The springs provided the attraction for a resort and bottling works as described on a historical marker: "Blount Springs, famous health resort, 1843–1914. Here fashionable ladies and gentlemen of the South vacationed with their families."

SITE DESCRIPTION

Stop 1. Sequatchie Anticline at Spring Yard Hill

The traverse at Stop 1 (sec. 6–7, T.13S., R.2W.) extends 0.6 mi (1 km) southeast along Blount County Road 7 from U.S. Highway 31 in the village of Blount Springs and crosses the hinge of the asymmetric Sequatchie anticline on Spring Yard Hill (Figs. 1, 3, 4). Dip on the southeast limb averages less than 10° southeast. At the hinge, the gentle southeast dip bends abruptly to steep northwest dip. Dip along the northwest limb ranges from 85° to 45° northwest. The dip domains of the opposite limbs are displayed in road cuts along Blount County Road 7; however, the hinge is not exposed. Steeply dipping, deformed Chattanooga Shale on the northwest limb is exposed in the creek downstream from the springs. On the southeast limb, approximately 1.1 mi (1.8 km) southeast of the crest of the anticline, the Shell Oil Company No. 1 Drennen well (Fig. 3) penetrated a complete sequence of beds from Mississippian Pride Mountain Formation to the lower part of the Cambrian-Ordovician Knox Group. The well bottomed in Knox at a total depth of 4,350 ft (1,326 m).

Stop 2. Sequatchie Anticline at Mill Creek

The traverse at Stop 2 (sec. 1–12, T.13S., R.3W.) extends 0.6 mi (1 km) northwest along Blount County Road 5 from

U.S. 31 in the village of Blount Springs and crosses the northwest limb of the Sequatchie anticline along the Mill Creek water gap through Duffy Mountain (Figs. 1, 3, 5). The steeply dipping northwest limb of the anticline is complicated by an abrupt structural terrace. Just northwest of the axial trace, beds dip approximately 85° northwest. Farther northwest, dip decreases abruptly to less than 5° across a structural terrace about 1,000 ft (300 m) wide. Low dips are exposed in the Bangor Limestone along Blount County Road 5 and along Mill Creek, and in the Pottsville Formation at the crest of Duffy Mountain.

Northwest of the terrace, dip steepens to about 55° northwest. The northwestward steepening of dip is exposed in the succession of beds in the upper Bangor Limestone and lower Pottsville Formation (sandstone) along Blount County Road 5 and Mill Creek. The fold is completely exposed in the lower Pottsville sandstone, which forms the crest and northwest dip slope of Duffy Mountain.

Northwest of the Sequatchie anticline, dip abruptly flattens; farther northwest, the beds are nearly horizontal. The structural boundary of the fold-thrust belt with the Black Warrior foreland basin and North American craton to the northwest is defined by the northwest limb of the Sequatchie anticline and is exposed in the lower Pottsville along Blount County Road 5 and Mill Creek.

The structural terrace on the northwest limb of the Sequatchie anticline may indicate a small subsurface splay thrust, a step on a subsurface ramp, or a duplicated slice in the subsurface. Subsurface data indicate no décollement above the Cambrian-Ordovician Knox Group. Whether or not older structures, possibly including basement faults, controlled location of the Sequatchie anticline is unknown.

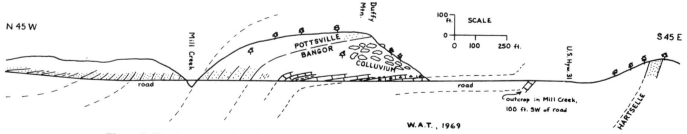

Figure 5. Sketch cross section of northwest limb of Sequatchie anticline along Blount County Road 5 and Mill Creek at Blount Springs (Stop 2 traverse).

REFERENCES CITED

Neathery, T. L., and Copeland, C. W., 1983, New information on the basement and lower Paleozoic stratigraphy of north Alabama: Alabama Geological Survey Open File Report, 28 p.

Thomas, W. A., 1972, Mississippian stratigraphy of Alabama: Alabama Geological Survey Monograph 12, 121 p.

Thomas, W. A., 1982, Stratigraphy and structure of the Appalachian fold and thrust belt in Alabama, *in* Thomas, W. A., and Neathery, T. L., eds., Appalachian thrust belt in Alabama: Tectonics and sedimentation (Field Trip Guidebook, Geological Society of America 1982 Annual Meeting, New Orleans, Louisiana): Tuscaloosa, Alabama Geological Society, p. 55–66.

Thomas, W. A., 1985, Northern Alabama sections, *in* Woodward, N. B., ed., Valley and Ridge thrust belt: Balanced structural sections, Pennsylvania to Alabama (Appalachian Basin Industrial Associates): University of Tennessee Department of Geological Sciences Studies in Geology 12, p. 54–61.

Thomas, W. A., and Bearce, D. N., 1969, Sequatchie anticline in north-central Alabama: Alabama Geological Society, 7th Annual Field Trip, Guidebook, p. 26–43.

Flood-tidal deltas and related back barrier systems: Bremen sandstone, "Pottsville" Formation, Black Warrior Basin, Alabama

Christopher A. Haas and Robert A. Gastaldo, Department of Geology, Auburn University, Auburn, Alabama 36849

LOCATION

Two outcrops of the Bremen Sandstone are located on or near River Highway, 2.4 mi (3.9 km) north of Alabama 69, in Sec.1,T.13S,R.6W, and Sec.6,T.13S,R.5W, Walker County, Alabama, on the Lewis Smith Lake Dam 7½-minute Quadrangle.

INTRODUCTION

The Bremen Sandstone exposed in northwestern Alabama is part of the base of the middle Early Pennsylvanian "Pottsville" Formation located within the Black Warrior Basin (Gillespie and Rheams, 1985; Lyons and others, 1985). At Lewis Smith Lake Dam (Fig. 1), Bremen Sandstone is well exposed and is directly underlain by a black shale unit and the Black Creek Coal seam. Bremen Sandstone is a mineralogically mature, cross-bedded and rippled quartz sandstone (Rheams and Benson, 1982). This sandstone was interpreted by Rheams and Benson (1982) and other researchers to represent a "barrier island" depositional environment similar to the base of the "Pottsville" Formation. After detailed examination of the lithologic characteristics, sedimentary structures, and stratigraphic relationships, the Bremen Sandstone at the Smith Lake Dam site is interpreted as a flood-tidal delta sequence deposited upon black lagoonal shales and a back-barrier coal (the Black Creek Coal).

The section below the Black Creek Coal (outcrop A) is interpreted to represent back-barrier systems similar to rock units described in other parts of the Black Warrior Basin (Hobday, 1974; Horne and Ferm, 1978; Cleaves, 1981; Horsey, 1981; Rheams and Benson, 1982). Silty shale and fossiliferous shale facies represent back-barrier lagoonal, quiet-water deposition, based on their high clay content, laminar sedimentary structures, and preserved organic material. Sandstone facies represent back-barrier washover deposits based on the lithology, sedimentary structures, and stratigraphic relations to other lithofacies. Sandstone facies, composed primarily of medium- to fine-grained quartz, displaying loading structures into underlying shale facies, indicate high energy conditions and quick deposition. Medium- to large-scale rippled beds, micro cross-stratification to large-scale cross-bedding, and thin low angle beds are common features characteristic of washover structures and related deposits. Anomalous low angle beds reported by Cleaves (1981) as placers are interpreted as being derived from washover origin based on their stratigraphic relationship with the overlying Black Creek Coal seam, a back-barrier peat accumulation. However, bedding orientation is problematic if these are primary washover structures, because beds strike N65°E and dip to the north at 3° to 5°, which indicates a seaward direction to the southwest.

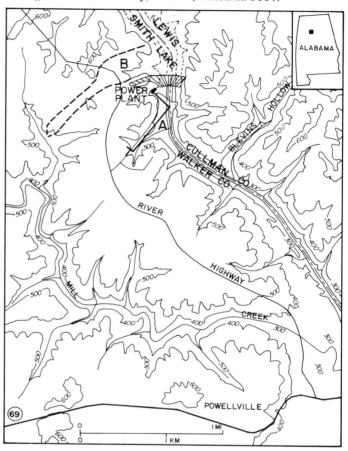

Figure 1. Location map showing outcrops *A* and *B*.

The section above the Black Creek Coal at outcrop B is interpreted, in agreement with Shadrovi (1984), to represent multiple flood-tidal delta systems that were deposited upon and cut through muddy organic back-barrier deposits. We base this on the lithology, sedimentary structures, and paleocurrent data represented by the Bremen Sandstone at this site. The Bremen Sandstone is a well sorted, sub-rounded to rounded quartz sandstone, which supports a high-energy environment of deposition. The S-shaped and planar large-scale, stacked cross-bed sets also indicate a high energy environment with cyclic or flood-tidal conditions. Rippled beds between cross-beds and cross-bed sets indicate lower energy periods during which water level was higher at the beginning of the ebb tide; i.e., ripples formed until the flow reversed into the flood direction. Ripple paleocurrent directions have a fan-like orientation that represents flow around the flood delta, seaward during ebb tide. Also, mud clasts within rippled beds and their absence in cross-bed sets indicate flow was in the ebb direction during the formation of ripples. Plant detritus present in mud clasts supports the contention that clasts were derived from a local back-barrier source aiding in the confirmation of the model for a back-barrier flood tidal depositional system. The lack of bioturbation and indigenous fauna or flora indicates evidence for shifting sands, supporting a tidally influenced system. Thus, a flood-tidal delta depositional system is represented in the exposure of the Bremen Sandstone at the Lewis Smith Lake Dam.

SITE DESCRIPTION

Two outcrops are included at this locality. Outcrop A is located along the west side of the Alabama Power Company private access road leading to the Plant site adjacent to the dam. Outcrop B is located on each side of River Highway at the top of Lewis Smith Lake Dam (Fig. 1). Beginning at the base of outcrop A, the sequence is exposed sporadically for approximately 100 ft (30 m) to the outcropping of the Black Creek Coal seam. The section continues within a box cut excavated by Alabama Power for dam fill between 1957 and 1962. The Black Creek Coal is exposed at low water level in a small bay west of the dam. Above the coal and best exposed in the box cut is approximately 98 ft (30 m) of the Bremen Sandstone.

Outcrop A

Sandstones, silty shales, and thin coal seams are exposed adjacent to the power generating plant. The base of the section consists of 21 ft (6.4 m) of horizontally bedded, light grey mudstone that grades into the overlying silty sandstone. A reddish brown, fine-grained, silty sandstone coarsens upward into rippled sandstone. This sandstone facies is 2.3 ft (70 cm) thick and shows loading structures into the underlying mudstone. Directly overlying is 11.5 ft (3.5 m) of light grey, fine-grained, thin- to medium-bedded, well cemented quartz sandstone. Medium thickness beds are nearly horizontal and contain medium- to large-scale ripples near the top of the sand facies. This unit is directly overlain by a 4.3 ft (1.3 m) thick, light grey, rooted mudstone capped by a thin coal seam 1 to 2 in (2 to 5 cm) thick. This coal layer is overlain by 1.3 ft (40 cm) of thinly interbedded, rippled sandstone and black shale. Fragmentary transported plant axes and possible horizontal burrows are present near the top of this facies. Overlying this facies is 20.5 ft (6.2 m) of fine-grained, rippled, thinly- to medium-bedded quartz sandstone with clay drapes and scattered shale layers. Micro cross-stratification is present within the rippled beds and local small scale cross-bedding is present in thicker sandstone layers. Directly overlying is 4.9 ft (1.5 m) of light grey, fine- to medium-grained, rippled, quartz sandstone. Clay drapes are present over ripples and shale clasts are found within sandstone beds. A light grey shale 4.3 ft (1.3 m) thick with interbeds of rippled silty sandstone overlie the preceding sandstone facies. This is capped by a 1.3 ft (40 cm) thick light brownish grey, fine- to medium-grained, thinly-bedded, rippled, quartz sandstone. Abundant horizontal burrows are present as well as branch and stem plant remains within clay drapes over ripples.

Figure 2. A, Bremen Sandstone exposure in box cut showing stacked cross-bed sets. B, Enlargement of two cross-bed sets. Lower set illustrating S-shaped cross-beds; upper set illustrating planar cross-beds. Scale 1 foot.

The next 1.6 ft (50 cm) of the section is covered.

The supradjacent facies consists of 21 ft (6.4 m) of medium-bedded quartz sandstone, multi-directional cross-bedded quartz sandstone, and low angle, thinly-bedded quartz sandstone with high concentrations of carbonaceous material, magnetite(?), and mica that have been reported previously as heavy mineral placers (Cleaves, 1981). Low angle beds are striking N 065 E and dip to the north at 3° to 5°. These deposits are capped by 1.6 ft (50 cm) of medium-bedded, quartz sandstone. Directly overlying is 1.3 ft (40 cm) of dark grey shale that is topped by a 1 ft (30 cm) massively bedded quartz sandstone sheet. Light grey shale and extensively rooted underclay-paleosol 3.6 ft (1.1 m) overlies the sheet sandstone and is directly beneath the Black Creek Coal seam. The coal seam is 20–24 in. (50–60 cm) thick and is in gradational contact with an overlying black shale that contains an *in situ* plant assemblage dominated by Pteridosperm (seed fern) stems and branches. This black shale grades into grey silty shales with occasional flaser bedding and finally into rippled sandstone 1.3 ft (40 cm) up from coal seam.

Outcrop B

The section continues at the top of the dam where the Black Creek Coal seam is exposed. The black fossiliferous shale unit above the coal is very thin or missing and is overlain by approximately 98 ft (30 m) of the Bremen Sandstone (Fig. 2). This quartz arenite facies is light grey, fine- to medium-grained, cross-bedded and rippled. Approximately 23 stacked cross-bed sets are observable. Each cross-bed set is approximately 3 ft (1 m) thick composed of planar and S-shaped beds with rippled layers between each set or bed. Cross-bedding indicates an average paleo-current direction to the southwest (S10°W, $n = 23$; Fig. 3). The cross-bed foresets throughout the Bremen Sandstone section all have nearly the same angle of dip (21°, $n = 20$) and ripples may be present on the upper surface of foresets (Fig. 2). Shale clasts containing fragmented plant detritus are abundant within troughs of rippled beds but are absent in foreset beds. Ripples on the top of cross-bed sets are mostly symmetrical and indicate a paleocurrent direction of west-northwest to east-southeast, with the greatest concentration of readings in a northwest to southeast pattern (Fig. 3).

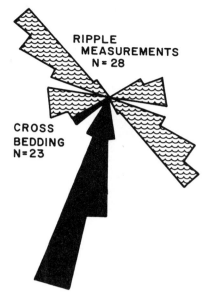

Figure 3. Rose diagram showing paleocurrent direction of ripple structures (*patterned*) plotted against cross-bedding (*solid*) from the Bremen Sandstone.

REFERENCES CITED

Cleaves, A. W., 1981, Resource evaluation of the lower Pennsylvanian (Pottsville) depositional systems of the western Warrior coal field, Alabama and Mississippi: Mississippi Mineral Resources Institute, Final Technical Report 81-1, 125 p.

Gillespie, W. H., and Rheams, L. J., 1985, Plant megafossils from the Carboniferous of Alabama, USA, *in* Escobedo, J. L., Grandos, L. F., Meléndez, B., Pignatelli, R., Rey, R., and Wagner, R. H., eds. Dixiéme Congrés International de Stratigraphie de Géologie du Carbonifére, v. 2: Madrid, Instituto Geologica y Minero de Espana, p. 191–202.

Hobday, D. K., 1974, Beach barrier-island facies in the Upper Carboniferous of northern Alabama, *in* Briggs, G., ed., Carboniferous of the southeastern United States: Geological Society of America, Special Paper 148, p. 209–224.

Horne, J. C., and Ferm, J. C., 1978, The northern Alabama area, *in* Ferm, J. C., Horne, J. C., and Milici, R. C., eds., Carboniferous Environments in the Appalachian Region, Field Guide: Carolina Coal Group, University of South Carolina, p. 730–751.

Horsey, C. A., 1981, Depositional environments of the Pennsylvanian Pottsville Formation in the Black Warrior Basin of Alabama: Journal of Sedimentary Petrology, v. 51, p. 799–814.

Lyons, P. C., Meissner, C. R., Jr., Barwood, H. L., and Adinolfi, F. G., 1985, North American and European megafloral correlations with the upper part of the Pottsville Formation in the Warrior Coal Field, Alabama, USA, *in* Escobedo, J. L., Granados, L. F., Meléndez, B., Pignatelli, R., Rey, R., and Wagner, R. H., eds., Dixiéme Congrés International de Stratigraphie el de Géologie du Carbonifére, Madrid, v. 2: Madrid, Instituto Geologico y Minero de Espana, p. 203–245.

Rheams, L. J., and Benson, J. D., 1982, Depositional setting of the Pottsville Formation in the Black Warrior Basin: Alabama Geological Society, 19th Annual Field Trip, Guidebook, November 5–6, 94 p.

Shadroui, J. M., 1984, Depositional environments of the Pennsylvanian Bremen Sandstone Member of the Pottsville Formation, Alabama, *in* Ettensoh, F., ed., Appalachian Basin Industrial Associates Program–Fall Meeting, University of Kentucky, vol., 7, 107–129.

ACKNOWLEDGMENTS

The senior author would like to thank Ms. Joanne Shadroui for interactive discussions on the Bremen Sandstone. This research has been funded by NSF EAR8407833 awarded to RAG.

Pottsville Formation in Alabama

Lawrence J. Rheams, *Geological Survey of Alabama, P.O. Box O, University Station, Tuscaloosa, Alabama 35486*
D. Joe Benson, *Department of Geology, The University of Alabama, University, Alabama 35486*

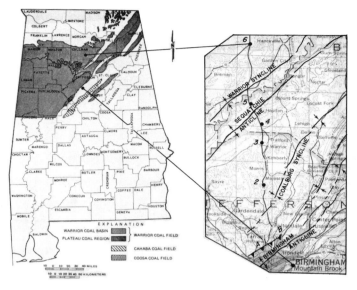

Figure 1. General location map of the Warrior Coal Field in Alabama, and the field guide stops in the Pennsylvanian Pottsville Formation.

LOCATION

Six exposures of the Early Pennsylvanian Pottsville Formation in the Warrior coal field of Alabama can be seen along I-65 starting just north of Birmingham, Alabama and ending about 36 mi (60 km) farther to the north (Fig. 1). The stops illustrate the stratigraphy, structure, and depositional environments of a part of the Pottsville interval in the Warrior coal field.

INTRODUCTION

The Pennsylvanian Pottsville Formation, containing economically important coal seams in Alabama, is a southwestward thickening wedge of clastic sediments occurring in three coal fields, the Coosa, Cahaba, and Warrior (Fig. 1). The Coosa and Cahaba fields lie within the Appalachian Valley and Ridge province as northeast–southwest trending folded, faulted, and truncated synclinoria. The Warrior field is situated northwest of the Valley and Ridge province and extends southwestward across the Appalachian Plateau to the Black Warrior Basin.

In Alabama the Pottsville Formation is best exposed in the Warrior coal basin, the southernmost of a series of Pennsylvanian basins of the Appalachian Plateau. This basin is bounded on the southeast by the northeast–southwest trending Appalachian Valley and Ridge province. To the northeast, a small transitional area called the Plateau coal region connects the Warrior basin with smaller basins in southeastern Tennessee. To the south and west,

the Pottsville is overlain by Cretaceous and Tertiary deposits. However, information from oil and gas wells indicates that the Pottsville clastics extend southwestward in the subsurface and terminate against the buried Ouachita-Appalachian structural front (Metzger, 1965). Erosion across the southern flank of the Nashville Dome has removed the Pottsville across most of extreme northern Alabama.

Structurally the eastern part of the Warrior coal basin consists of a series of northeast–southwest trending folds and northwest-striking normal faults. Along the eastern edge of the basin the beds are highly folded to overturned against the Valley and Ridge province (Birmingham Anticlinorium and the Opposum Valley thrust fault). Toward the northwest the anticlines and synclines are more gentle (Fig. 1).

The Pottsville in the Warrior basin consists of over 6,000 ft (1,820 m) of sandstone, siltstone, shale, and coal that overlies clastics of the Mississippian-Pennsylvanian Parkwood Formation or carbonates of the Mississippian Bangor Limestone (Metzger, 1965). The base of the Pottsville is placed at the base of a thick orthoquartzitic sandstone termed the Boyles Sandstone Member. However, the base of the Pennsylvanian System in Alabama is within the underlying Parkwood Formation (Wanless, 1975).

Stratigraphic subdivision of the Pottsville in the Warrior basin is difficult because of complex facies changes among clastic sedimentary units. McCalley (1900) used the more continuous coal beds and sandstone units to subdivide the Pottsville into two general units; a lower, predominantly sandstone-siltstone, sequence with thin discontinuous coal seams and an upper shale-sandstone sequence containing thicker, more continuous coal seams. This upper sequence has been further subdivided into seven coal groups (McCalley, 1900; Culbertson, 1964).

The Pottsville Formation in the Warrior field was deposited in fluvial to marginal marine environments (Metzger, 1965). The lower part of the Pottsville has been generally interpreted as a barrier island and lagoonal sequence; and the upper part of the Pottsville as a high constructive deltaic complex.

SITE DESCRIPTION

A series of stops has been designed to illustrate the stratigraphy and depositional environments of the Pottsville Formation in Alabama using outcrops along or adjacent to U.S. 31 or I-65 north of Birmingham, Alabama.

Stop 1. Boyles Sandstone Member

Birmingham North 7½-minute Quadrangle, Jefferson County, Alabama (A—SE¼NW¼ section 6, T17S., R2W and B—NE¼SE¼ Section 15,T17S,R3W). This stop includes two

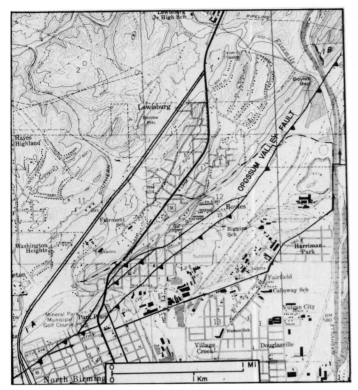

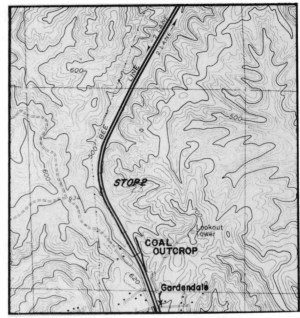

Figure 3. Stop 2, Morris Section, Gardendale 7½-minute Quadrangle, Jefferson County, Alabama (from SW¼SE¼ section 23 to NW¼SE¼ section 14, T15S, R3W).

Figure 2. Stop 1, Boyles Sandstone Member, Birmingham North 7½-minute Quadrangle, Jefferson County, Alabama (A-SE¼NW¼ section 6, T17S, R2W and B-NE¼SE¼ section 15, T17S, R3W).

nearby exposures of the Boyles Sandstone (also called the Millstone Grit), the lowermost member of the Pottsville Formation in the Warrior Coal Field. Site 1A is the I-65 cut through Sand Mountain north of Birmingham. Site 1B, the type locality of the Boyles (Butts, 1910), is located along a secondary road some 5 mi (8.3 km) northeast of Site 1A. Site 1A exposes approximately 200 ft (61 m) of the Boyles Sandstone, whereas the exposure at Site 1B includes 40 ft (12 m) of the underlying Parkwood Formation as well as 300 ft (91 m) of Lower Pottsville above the Boyles. This overlying Lower Pottsville sequence includes shale, siltstone, sandstone, and four thin coal seams of the Tidmore coal horizon. Macerals from these coals indicate a middle Westphalian A or middle New River age (Dr. William H. Gillespie, personal communication).

Both localities lie just west of the Opossum Valley thrust fault on the northwest limb of the Birmingham Anticlinorium. The Opossum Valley fault is a northeast striking thrust fault with a displacement of 7,000 ft (2,120 m) or more, which brings rocks of Cambrian age into juxtaposition with younger Paleozoic rocks (Kidd, 1979). As is frequently the case adjacent to the Opossum Valley fault, strata at both Sites 1A and 1B are overturned and dip at approximately 50° to the southeast.

The Boyles Sandstone Member is characterized by several thick quartzarenitic sandstones that are usually recrystallized and conglomeratic in their lower part. The sandstones are interbedded

with varying amounts of shale, siltstone, and at some localities thin, discontinuous coals. At both Sites 1A and 1B the Boyles consists of two distinct sandstone units, though a third sandstone unit has been reported to the southwest in Tuscaloosa County (Culbertson, 1964). To the west and southwest, in the subsurface of the Black Warrior Basin, several distinct sandstones occur within an interval of several hundred feet (meters) in the lower Pottsville. Three of these sandstones are areally mappable and are commonly termed the Benton, Robinson, and Chandler Sandstones by the oil and gas industry (Kidd, personal communication in Rheams and Benson, 1982).

The Boyles Sandstone Member ranges in thickness from 200 to 700+ ft (61 to 212+ m) in the Warrior Coal Field (Culbertson, 1964). The section exposed at Sites 1A and 1B is unusually thin but the Boyles contains numerous slickensides and both the underlying Parkwood Formation and the overlying Lower Pottsville units exhibit indications of faulting. This suggests that part of the Lower Pottsville may be faulted out at this location and that the exposed thickness of the Boyles Member is not a true thickness.

Stop 2. Morris Section

Gardendale 7½-minute Quadrangle, Jefferson County, Alabama (from SW¼SE¼ sec. 23 to NW¼SE¼ sec. 14, T15S, R3W). This stop consists of a series of exposures along U.S. 31 near Morris, Alabama. Ehrlich (1964) and Metzger (1965) have described these exposures. Graham (1976), using Metzger's stratigraphic subdivision, used sandstone from this locality for his provenance study (samples B-6, 7).

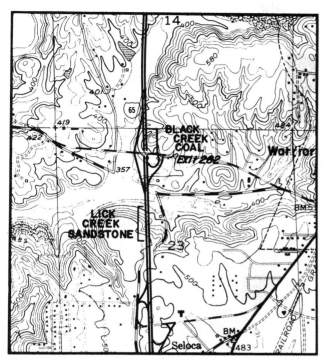

Figure 4. Stop 3, Warrior Section, Warrior 7½-minute Quadrangle, Jefferson County, Alabama (Section 23, T14S, R3W).

Coal exposed at the top of the hill (east side of U.S. 31) is considered to be the Gillespy seam (McCalley, 1900). The Gillespy is the lowest seam of the Pratt Coal Group. The channel-type sandstone across the highway (west side) was deposited at about the same horizon or just above the coal. However, due to the possible faulting in the area, the relationship between the coal and the sandstone has not been ascertained.

The sandstone exposed below the Gillespy seam is a fine-grained, thin- to medium-bedded, 70- to 80-ft-thick (21 to 24 m) delta front and distributary mouth bar deposit. Data obtained from numerous drill records and surface exposures demonstrate that this sandstone has a broad areal extent: It is informally called the Gardendale Sandstone Member of the Pottsville Formation. The lower part of the sequence was reported to contain marine fossils (Ehrlich, 1964). This exposure is typical of the delta front sequences that commonly occur between coal groups in the Warrior coal basin in Alabama.

Stop 3. Warrior Section

Warrior 7½-minute Quadrangle, Jefferson County, Alabama (Section 23, T14S, R3W). This section, previously described by Horne and others (1976) and Rheams and Benson (1982), provides an excellent view of lower delta plain deposits within the Black Creek coal group, the lowest coal group in the upper part of the Pottsville. To the east and south this interval should pass into transitional and upper delta plain deposits. Northward, this interval passes into barrier island deposits. Two

coals are exposed at this stop. The lower coal is the Black Creek coal. Some 28 ft (8.5 m) above the Black Creek are two coal splits of the Jefferson coal seam.

The lower part of the exposed section consists of organic shale that grades upward into fine-grained siltstone, a thick underclay, and then the Black Creek Coal. This sequence represents filling of an open interdistributary bay to the point where vegetation was established. Once established, this vegetation accumulated in a marsh environment parallel to the distributary channel.

With continued subsidence, the coal swamp was gradually submerged and interdistributary bay deposits spread across the area. Sediment deposited in this interdistributary bay was fine-grained silt and clay suspension load material.

This depositional phase was interrupted by an influx of coarser-grained sediment derived from a crevasse in the levee of an adjacent distributary channel, as indicated by the upward coarsing sequence from shale to sideritic siltstone and finally sandstone. Sedimentary structures include current and climbing ripples and horizontal lamination. In the exposure along I-65 northeast of the interchange, the crevasse splay sequence is cut by a poorly defined channel sequence. This channel sequence fines upward from coarse-grained sideritic pebble lags at the base into fine-grained sandstone. The channel deposit shows well-defined lateral accretion bedding (epsilon bedding), which is the product of the south to north migration of the channel across the surface of the splay. These channels are common features of splays representing pathways for sediment transport as the splay builds bayward. On the basis of exposures in strip mines in the vicinity, Horne and others (1976, p. 22) estimated this splay to have been 3 to 5 mi (5 to 8.3 km) wide.

The splay sequence is capped by two thin splits of the Jefferson coal seam that resulted from the re-establishment of vegetation on top of the splay. With continued subsidence the coal-producing vegetation was once again killed off. The upper part of the exposure consists of an upward coarsening bay sequence.

Outcrops along the southbound lane of I-65, south of the interchange, are exposures of a sequence that is stratigraphically higher than exposures north of the interchange. This sequence illustrates a typical lower delta plain distributary channel sequence and also provides an excellent exposure of the Lick Creek Sandstone Member of the Pottsville Formation (Butts, 1910).

The lower part of this exposure consists of an upward coarsening shale-siltstone sequence. The lithologies are heavily burrowed, sideritic, and show fine horizontal lamination grading to lenticular bedding in the upper part of the interval. The strata shows a prominent 4° to 6° depositional dip to the north after the regional dip has been removed, indicating northerly progradation (Horne and others, 1976, p. 26). These features are all characteristic of prodelta or open bay sequences.

These prodelta or bay fill deposits are capped by a sequence of distributary mouth bar sands. Horne and others (1976) divided these sands into two depositional sequences. The lower sequence is composed of fine-grained sandstone and siltstone. The deposits

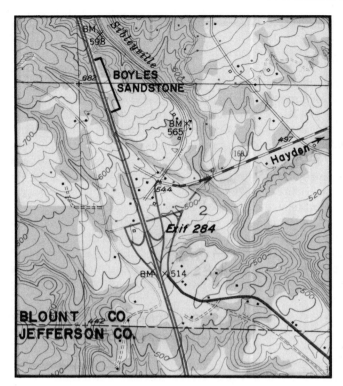

Figure 5. Stop 4, Hayden Section, Warrior 7½-minute Quadrangle, Blount County, Alabama (NE¼NE¼ section 2, T14S, R3W).

are moderately burrowed, flaser bedded, and show occasional convolute lamination. They have been interpreted as distal distributary mouth bar deposits (Horne and others, 1976). The distal bar deposits are overlain by fine- to medium-grained sandstone that is trough crossbedded and contains numerous internal truncation surfaces and occasional convolute lamination. This sequence is interpreted to represent distributary mouth bar crest and back bar deposits.

Truncating the distributary mouth bar sandstone is a coarse-grained distributary channel sandstone. The sandstone contains abundant gravel lag deposits at its base and fines upward. It is trough crossbedded with paleocurrent measurements indicating a north to northwest transport direction (Horne and others, 1976, p. 28).

Both the distributary mouth bar and distributary channel sandstones are litharenites to sublitharenites that contain abundant metamorphic and sedimentary rock fragments. These sands were sampled by Graham and others (1976) in their study of provenance of Pottsville sandstones in the Warrier basin. These sandstones are compositionally distinct from the clean quartzarenites of the barrier island deposits of the lower part of the Pottsville.

Stop 4. Hayden Section

Warrior 7½-minute Quadrangle, Blount County, Alabama (NE¼NE¼ Section 2, T14S, R3W). Barrier island sandstone

units dominate the lower part of the Pottsville in the Warrior coal basin (Horne and others, 1976). The barrier island sandstone deposits are quartzarenites to high level litharenites that contrast sharply with the litharenites and low level sublitharenites comprising most of the delta plain sandstone deposits of the upper part of the Pottsville. This mineralogic maturity is a reflection of the high energy and high level of reworking of the barrier island deposits. The labile constituents that characterize the delta plain deposits are rapidly destroyed in the high energy barrier environment, concentrating the more stable constituents to produce a mineralogically mature sediment.

The Hayden outcrop extends for almost 0.5 mi (0.8 km) along the east side of I-65. Exposed are parts of two barrier island sequences flanked by open shelf and lagoonal deposits. The lowermost lithology, exposed at the extreme northern end of the outcrop consists of fine- to medium-grained sandstone that is ripple laminated and bioturbated. This lithology is interpreted to represent open marine shelf deposition.

The ripple laminated sandstone is sharply overlain by a medium- to coarse-grained quartzarenite, which fines upward. The unit is characterized by well-developed tabular and trough crossbedding. Bed sets range from 1- to 3-ft (30 to 91 cm) in height. Paleocurrent analysis indicates two distinct transport directions; N10°E and N60°W. This bimodal crossbedding is characteristic of barrier deposits in the Warrior basin with one mode representing tidal channel and delta deposition and the other representing longshore transport occurring in the upper shoreface (Hobday, 1974). Here the northwesterly mode, which consists primarily of tabular crossbeds, represents tidal channel deposition, whereas the northeasterly mode, which consists primarily of trough crossbeds, represents longshore transport. The northwesterly orientation of the tidal channels suggests the barrier system had a roughly northeast–southwest orientation.

The lower sandstone sequence is overlain by a 4-ft (1.2 m) interval of interbedded sandstone and shale. The lower 2 ft (61 cm) of this interval consists of finely laminated, silty shale containing thin sandstone flasers. The lithology becomes lighter colored and more arenaceous upward and is sharply overlain by a 1-ft (30 cm) bed of fine- to medium-grained sandstone. The lower contact of this sandstone is an erosional disconformity that is slightly undulatory. The sandstone is rooted and shows no evidence of stratification. It is overlain by a 1-ft (30 cm) clay-shale that grades upward into a heavily rooted underclay overlain by an irregular bed of very organic shale or coal smut.

This sandstone-shale interval represents lagoonal deposition adjacent to the barrier system. The shales and clays are typical lagoonal deposits and the sandstone is a washover deposit derived from the barrier during storms.

The lagoonal sequence is truncated by a second highly crossbedded sandstone. The contact is an erosional disconformity and is undulatory. The sequence shows a distinct upward decrease in grain size. Trough crossbedding is more common in this interval than in the lower sandstone, but paleocurrent measurements show similar transport directions. The upper sandstone

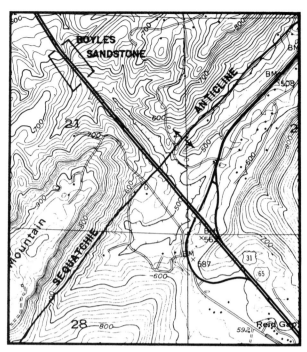

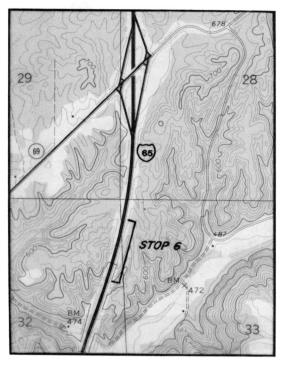

Figure 6. Stop 5, Reid Gap Section, Blount Springs 7½-minute Quadrangle, Blount County, Alabama (NW¼NE¼ section 21, T13S, R3W).

Figure 7. Stop 6, Marriott Creek, Hanceville 7½-minute Quadrangle, Cullman County, Alabama (NW¼NW¼ section 33, T11S, R3W).

represents a second barrier deposit that overrode the underlying lagoonal and barrier deposits.

Stop 5. Reid Gap Section

Blount Springs 7½-minute Quadrangle, Blount County, Alabama (NW¼NE¼ section 21, T13S, R3W). This stop is an exposure of the Boyles Sandstone Member of the Pottsville Formation on the northwest limb of the Sequatchie anticline. At this locality the Boyles is approximately 650 ft (198 m) thick as compared to only 200 ft (61 m) of thickness at Boyles Gap (Stop 1). The thickness of the quartzarenitic sandstone is probably a truer representation of the actual thickness of the Boyles Sandstone in the northern part of the Warrior basin than that at Boyles Gap. The lower part of the sandstone appears to be recrystallized, with some scattered pebbles and pebbly zones. Several primary sedimentary structures, such as crossbeds and channels, are easily viewed along the cut.

Dip on this sandstone is 60 to 70 degrees northwest (notice the decrease in dip through the cut to the northwest). This exposure shows a good view of the northwest limb of the narrow northeast-trending Sequatchie anticline, which is asymmetrical to the northwest and plunges a few degrees to the southwest. This anticline becomes more symmetrical in the direction of the plunge and appears to die out in southwest Jefferson County. To the northeast, the northwestern limb is cut longitudinally by a thrust fault.

Stop 6. Marriott Creek Section

Hanceville 7½-minute Quadrangle, Cullman County, Alabama (NW¼NW¼, Section 33, T11S, R3W). This section, previously described by Horne and others (1976) and Rheams and Benson (1982), lies stratigraphically above the quartzarenites of the lower part of the Pottsville and just below the Black Creek coal group. Lithologically the section exposes a number of upward coarsening cycles that grade from shale at the base to siltstone or fine-grained sandstone at the top.

The shales that make up the lower parts of each of the cycles are dark-colored and highly organic. They are thinly laminated, but highly bioturbated. They become silty upward and contain some lenticular bedding in their upper parts. Siltstone becomes more common upward within the cycles, appearing first as flasers, then thin lenses, and finally as thin beds. These siltstones are bioturbated and sideritic. Symmetrical wave ripples are very common in the siltstones and occasional convolute lamination is present. The siltstones coarsen upward and some of the sequences are capped by fine-grained sandstone. The sandstone is lithic, commonly sideritic, and contains well-defined wave ripples and horizontal lamination. While lithologic contacts within the cycles are gradational, individual cycles are abruptly overlain by shales of the overlying sequence.

These cycles represent prodelta or open marine bay deposition that occurred basinward of a northerly or northwesterly prograding delta front. Each individual cycle represents one progradational event ranging from distal shale at the base to more proximal siltstone and sandstone at the top. Deposition was probably entirely subaqueous, as evidenced by the lack of coal at the top of the cycles. Further updip on the delta platform, interdistributary bay cycles are commonly capped by coal or highly siteritic sandstone reflecting infilling above sea level. In more distal settings such as this subsidence is much greater and sediment accumulation less rapid so that cycles rarely build near sea level.

REFERENCES CITED

Butts, Charles, 1910, Description of the Birmingham quadrangle, Alabama, U.S. Geological Survey Geologic Atlas, Folio 175, 24 p.

Culbertson, W. C., 1964, Geology and coal resources of the coal-bearing rocks of Alabama: U.S. Geological Survey Bulletin 1182-B, 79 p.

Ehrlich, R. L., 1964, Field trip guidebook to the Pottsville Formation in Blount and Jefferson Counties, Alabama: Alabama Geological Society, 1st Annual Field Trip Guidebook, 12 p.

Graham, S. A., Ingersoll, R. V., and Dickinson, W. R., 1976, Common provenance for lithic grains in Carboniferous sandstones from Ouachita Mountains and Black Warrior basin: Journal of Sedimentary Petrology, v. 46, no. 3, p. 620–632.

Hobday, D. K., 1974, Beach and barrier island facies in the upper Carboniferous of northern Alabama, in Briggs, Garrett, ed., Carboniferous of the Southeastern United States: Geological Society of America Special Paper 148, p. 209–224.

Horne, J. C., Ferm, J. C., Hobday, D. K., Saxena, R. S., and Robinson, L. D., eds., 1976, A field guide to carboniferous littoral deposits in the Warrior Basin: American Association of Petroleum Geologists Guidebook, New Orleans, Louisiana, 80 p.

Kidd, J. T., 1979, Areal geology of Jefferson County, Alabama: Alabama Geological Survey Atlas Series 15, 89 p.

McCalley, H., 1900, Report on the Warrior coal basin: Alabama Geological Survey Special Report 10, 327 p.

Metzger, W. J., 1965, Pennsylvanian stratigraphy of the Warrior basin, Alabama: Alabama Geological Survey Circular 30, 80 p.

Rheams, L. J., and Benson, D. J., 1982, Depositional setting of the Pottsville Formation in the Black Warrior Basin: Alabama Geological Society, 19th Annual Field Trip Guidebook, 94 p.

Wanless, H. R., 1975, Appalachian region, in McKee, E. D., and Crosby, E. J., coordinators, Introduction and regional analyses of the Pennsylvanian System, Pt. 1 of Paleotectonic investigations of the Pennsylvanian System in the United States: U.S. Geological Survey Professional Paper 853, pt. 1, p. 17–62.

Birmingham anticlinorium in the Appalachian fold-thrust belt, basement fault system, synsedimentary structure, and thrust ramp

William A. Thomas, Department of Geology, University of Alabama, University, Alabama 35486
Denny N. Bearce, Department of Geology, University of Alabama at Birmingham, Birmingham, Alabama 35294

INTRODUCTION

At Birmingham, in the central part of the Appalachian fold-thrust belt in Alabama, the Birmingham anticlinorium is associated with a large-scale ramp on the regional décollement and includes the Opossum Valley and Jones Valley thrust sheets (Figs. 1, 2, and 3). The southeast limb of the asymmetric anticlinorium in the hanging wall of the Jones Valley fault dips gently southeast into the Cahaba synclinorium (Locality 42); the northwest limb in the footwall of the Opossum Valley fault is overturned and dips steeply southeast (Locality 43) (Fig. 3). The youngest rocks in the anticlinorium are Early Pennsylvanian, indicating thrusting during the Alleghanian orogeny. The Middle Cambrian Conasauga Formation is the oldest unit in the hanging walls of the Opossum Valley and Jones Valley thrust faults, thus indicating the stratigraphic level of the regional décollement (Fig. 3).

The Appalachian fold-thrust belt in Alabama may be divided into three domains (Thomas, 1982). The Birmingham anticlinorium is the most northwesterly structure in the central domain, which is characterized by large-scale thrust ramps and associated folds having structural relief of more than 15,000 ft (4,500 m). In contrast, the northwestern domain is characterized by broad, flat-bottomed synclines and narrow, asymmetric anticlines having relief of less than 10,000 ft (3,000 m). The southeastern domain is characterized by broad, low-angle, multiple-level thrust sheets. Amplitude of structures within the central domain, as indicated by dip angles on fold limbs and by preserved thickness of Paleozoic rocks in the synclines, defines a minimum depth to basement of approximately 23,000 ft (7,000 m). In contrast, in the northwestern domain, depth to basement is approximately 8,200 to 10,000 ft (2,500 to 3,000 m) as indicated by wells drilled to Precambrian basement rocks and by the preserved stratigraphic thicknesses in the synclines (Kidd and Neathery, 1976; Neathery and Copeland, 1983; Thomas, 1985). Depth to basement beneath the central domain thus is indicated to be approximately 13,000 ft (4,000 m) greater than that beneath the northwestern domain, and the domain boundary is interpreted to be defined by a system of steep down-to-southeast basement faults beneath the large décollement ramp along the Birmingham anticlinorium (Fig. 3). The basement fault system is confirmed by seismic profiles.

Along the Birmingham anticlinorium, variations in stratigraphic thickness and facies, unconformities, local sources of clastic sediment, synsedimentary tectonic faults, and sedimentary slump faults indicate episodic synsedimentary structural movement beginning at least as early as Middle Ordovician and con-

tinuing into the Pennsylvanian. Regionally, the Early and Middle Cambrian Rome and Conasauga Formations aggregate several hundred feet thicker in thrust sheets southeast of the anticlinorium than in the subsurface to the northwest (Kidd and Neathery, 1976), suggesting synsedimentary structural movement beginning in Early Cambrian. The evidence of synsedimentary structural movement suggests that the basement fault system beneath the Birmingham anticlinorium was initiated by down-to-southeast displacement as early as Early Cambrian and was episodically reactivated throughout the Paleozoic (Thomas, 1986). Most of the episodic synsedimentary movement from Middle Ordovician through Early Mississippian evidently was expressed as gentle southeastward tilting of strata now in the southeast limb of the anticlinorium. Tilting was accompanied by movement on at least one down-to-southeast normal fault. Regional differences in Mississippian-Pennsylvanian stratigraphy indicate an accelerated rate of subsidence southeast of the anticlinorium. Paleoslopes produced by the subsidence resulted in sedimentary slumping. The inferred configuration of the synsedimentary structure suggests drape folding and minor faulting in the sedimentary cover sequence above the episodically reactivated basement fault system (Thomas, 1986). Finally, the present structural configuration of the Birmingham anticlinorium suggests that the shape and location of the large-scale Alleghanian ramp were controlled by the older basement fault system (Fig. 3). As a result of the position of the ramp, the beds that were deposited at the crest of the drape fold over the episodically active basement fault system were transported above the ramp, and the crest of the earlier synsedimentary structure was transported as the crest of the thin-skinned ramp anticline.

The structural profile of the Birmingham anticlinorium as exhibited at Birmingham persists along strike approximately 30 mi (48 km) from the Harpersville cross-strike structural discontinuity on the northeast to the Bessemer cross-strike structural discontinuity on the southwest (Fig. 2). At the Harpersville cross-strike structural discontinuity, the northwest-verging ramp anticline in the hanging wall of the Opossum Valley fault changes northeastward along strike into the southeast-verging Murphrees Valley anticline. The along-strike change is expressed where the Opossum Valley fault ends, and dip of the northwest limb of the anticlinorium changes abruptly from steep overturned southeast to less than 30° northwest (northwest limb of Murphrees Valley anticline). The trailing part of the Opossum Valley thrust sheet abruptly plunges northeastward into the broad, flat-bottomed Blount Mountain syncline, which is bordered on the northwest by the steep southeast limb of the Murphrees Valley anticline and the Straight Mountain fault (Fig. 2). Southwestward at the Bessemer

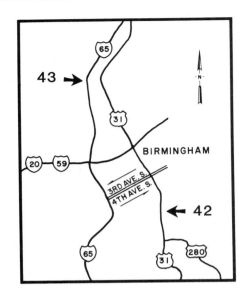

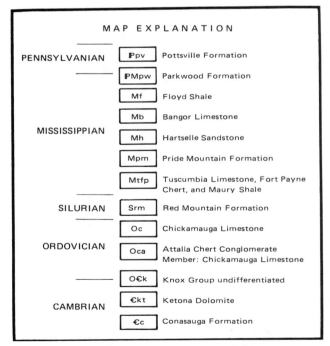

MAP EXPLANATION

PENNSYLVANIAN	Ppv	Pottsville Formation
	PMpw	Parkwood Formation
	Mf	Floyd Shale
	Mb	Bangor Limestone
MISSISSIPPIAN	Mh	Hartselle Sandstone
	Mpm	Pride Mountain Formation
	Mtfp	Tuscumbia Limestone, Fort Payne Chert, and Maury Shale
SILURIAN	Srm	Red Mountain Formation
	Oc	Chickamauga Limestone
ORDOVICIAN	Oca	Attalla Chert Conglomerate Member: Chickamauga Limestone
	OЄk	Knox Group undifferentiated
CAMBRIAN	Єkt	Ketona Dolomite
	Єc	Conasauga Formation

Figure 1. Geologic map of Birmingham anticlinorium at Birmingham (from Kidd and Shannon, 1977) and map of access routes to Localities 42 and 43.

cross-strike structural discontinuity, the Opossum Valley fault ends, and the anticlinorium continues as a single-crested structure aligned with the Jones Valley fault.

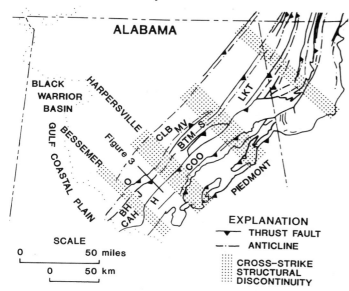

Figure 2. Map of regional structural setting of Birmingham anticlinorium (localities are along line of cross section of Fig. 3). Abbreviations: H = Helena fault, J = Jones Valley fault, O = Opossum Valley fault, S = Straight Mountain fault, BTM = Blount Mountain syncline, CAH = Cahaba synclinorium, CLB = Coalburg syncline, COO = Coosa synclinorium, LKT = Lookout syncline, BH = Birmingham anticlinorium, MV = Murphrees Valley anticline.

The topography at Birmingham is typical of the Appalachian Valley and Ridge. The cut through Red Mountain (Locality 42) for the Red Mountain Expressway (U.S. 280 and 31) exposes Paleozoic strata from Cambrian to Mississippian age on the southeast limb of the anticlinorium (Figs. 1, 4, and 5). Southeast of Red Mountain, Shades Mountain is formed on Pennsylvanian sandstones on the southeast limb of the anticlinorium. The Cambrian Conasauga Formation, the oldest formation in the hanging wall of the Jones Valley thrust sheet, underlies the broad valley northwest of Red Mountain in Birmingham (Figs. 1, 3). Flint Ridge, a mid-valley ridge, is formed on the Cambrian Copper Ridge Formation of the Knox Group at the top of the Opossum Valley thrust sheet; and farther northwest, a broad valley is formed on the Conasauga Formation at the base of the thrust sheet (Fig. 3). Still farther northwest, a low ridge (Sand Mountain) is formed on Pennsylvanian sandstones along the overturned northwest limb of the anticlinorium (Locality 43) (Figs. 1, 3).

LOCALITY 42. SOUTHEAST LIMB OF BIRMINGHAM ANTICLINORIUM

Location

Beds from Cambrian to Mississippian age on the southeast limb of the Birmingham anticlinorium are exposed in the cut through Red Mountain for the Red Mountain Expressway (U.S. 280 and 31; also named Elton B. Stephens Expressway).

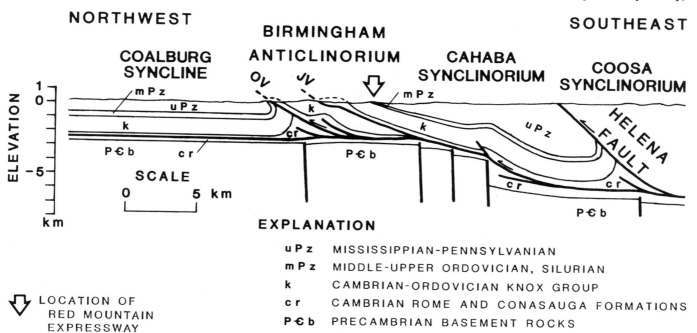

Figure 3. Structural cross section of Birmingham anticlinorium. Location of cross section shown on Figure 2. Abbreviations: OV = Opossum Valley fault, JV = Jones Valley fault. Locality 42 is at Red Mountain Expressway cut on southeast limb; Locality 43 is northwest of Opossum Valley fault on northwest limb.

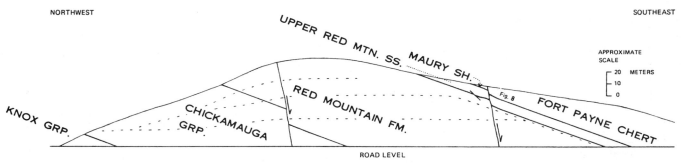

Figure 4. Sketch cross section of northeast face of Red Mountain Expressway cut (Locality 42). Light dashed lines show bench levels on cut face; the walkway from the Red Mountain Museum is on the lowest bench above road level.

Walking access to the cut (sec. 6, T.18S., R.2W.) is through the Red Mountain Museum at 1421 22nd Street South, Birmingham, Alabama (Figs. 1, 6). To reach the Museum from the Red Mountain Expressway northbound, take the Highland Avenue-Arlington Avenue Exit and follow signs from the exit ramp (right on 22nd Street South, right on 23rd Street South, right on Arlington Avenue, left on 22nd Street South). From the Red Mountain Expressway southbound, drive south past the Museum and the deep cut through Red Mountain, take the 21st Avenue South Exit, turn left on 21st Avenue South and go under Expressway, turn left onto entrance ramp to northbound Expressway, and follow Expressway 0.5 mi (0.8 km) north to the Highland Avenue-Arlington Avenue Exit.

Outcrops of Mississippian and Pennsylvanian rocks southeast of Red Mountain may be reached from the Red Mountain Museum by the following route (Figs. 1, 6). From the museum, drive north on 22nd Street South 0.2 mi (0.3 km), northwest (turn left) on Arlington Avenue 0.1 mi (<0.2 km), north (turn right) on Highland Avenue 0.1 mi (<0.2 km), south (turn right) on entrance ramp to Red Mountain Expressway (U.S. 280 and 31); follow Expressway southbound through the Red Mountain cut (no parking along Expressway) 0.5 mi (0.8 km); and either take 21st Avenue South Exit for parking access to Hartselle Sandstone or continue on U.S. 280 east to outcrops of Parkwood and Pottsville Formations.

Synsedimentary slump faults in the Hartselle Sandstone are exposed along the Expressway in a cut 0.3 mi (0.5 km) south of the 21st Avenue South Exit, but parking is not permitted on the Expressway. For parking access to the Hartselle cut (illustrated in Fig. 9) exit to 21st Avenue South; turn east (left) on 21st Avenue South and drive under Expressway overpass; turn south (right, just beyond Expressway) on 20th Place South; continue 0.1 mi (<0.2 km) to 22nd Court South at the crest of Sandstone Ridge; and park on 22nd Court South. From 20th Place South, one may look across the Expressway to the west cut, and may climb over the guard rail and down the slope to the base of the east cut on the Expressway. To return to U.S. 280, retrace route to 21st Avenue South; turn west (left) on 21st Avenue South and continue under Expressway overpass; just beyond the Expressway, turn south (left) on Woodcrest Place;

follow Woodcrest Place and 19th Street 0.6 mi (1 km); turn east (left) on Robins Drive; drive 0.2 mi (0.3 km) to intersection and go straight to rejoin U.S. 280 east.

Farther southeast (approximately 1.6 mi [2.6 km] southeast of the 21st Avenue South Exit), along U.S. 280, the lower part of the Parkwood Formation is exposed in road cuts. A section of the Parkwood Formation is displayed in road cuts along the next 1.5 mi (2.4 km) southeastward along U.S. 280, and at the top of Shades Mountain (approximately 3.1 mi [5 km] southeast of the 21st Avenue South Exit), the base of the Pottsville Formation is exposed. Parking is permitted along this part of U.S. 280.

Site Description

The Red Mountain Expressway (U.S. 280 and 31) cut through Red Mountain is designated as a national site of geologic interest by the American Geological Institute. A paved walkway on a cut bench 60 ft (20 m) above road level on the northeast side of the cut provides easy access to the cut from the Red Mountain Museum, an agency of the City of Birmingham (Fig. 4). The stratigraphic succession exposed in the cut includes strata from the Upper Cambrian Copper Ridge Formation to the Mississippian Fort Payne Chert (Fig. 5) and exhibits features that indicate episodic synsedimentary structural movement during much of the Paleozoic. Stratigraphically higher units can be seen in smaller cuts to the southeast along U.S. 280.

The southeast limb of the Birmingham anticlinorium (the northwest limb of the Cahaba synclinorium) dips approximately 20° southeast. Steep faults and local folds interrupt the southeast limb of the anticlinorium (Simpson, 1965). Most of the faults are normal and are down-to-southeast. Displacement is as much as 425 ft (130 m), but most of the faults have throw of only a few feet. Two down-to-southeast normal faults exposed in the cut have displacement of a few feet (Fig. 4).

The oldest strata exposed in the cut are dolostone and chert of the Upper Cambrian Copper Ridge Formation of the Cambrian-Ordovician Knox Group. These beds can be seen approximately 100 ft (30 m) northwest of the northwest end of the

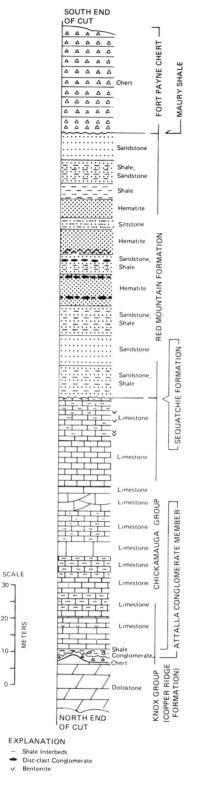

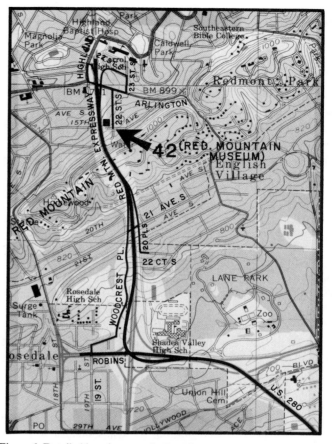

Figure 6. Detailed location map for Locality 42 (Birmingham North and Birmingham South 7½-minute Quadrangles).

paved walkway (Fig. 4); however, the exposure is poor because of vegetation. An apparent fault zone in the lowest exposed dolostone is a few inches wide and contains a gouge-like fill of chert fragments in silty clay of uncertain age. The fault can be traced upward about 10 ft (3 m) from ground level to the base of a massive chert bed near the top of the Copper Ridge. Nature of movement on the fault is unclear.

The Upper Cambrian Copper Ridge Formation is overlain disconformably by Middle Ordovician strata of the Chickamauga Group. On the northeast side of the cut, the unconformity has more than 5 ft (1.5 m) of relief filled with a clay and limestone matrix-supported breccia of gray mottled chert (Attalla Chert Conglomerate Member, the basal member of the Stones River Formation of the Chickamauga Group, Drahovzal and Neathery, 1971). Differences in thickness of the lower part of the Chickamauga indicate more than 30 ft (9 m) of paleotopographic relief between the northeast and southwest sides of the cut. The Lower Ordovician part of the Knox (Chepultepec, Longview, and Newala Formations), regionally extensive elsewhere in Alabama and the southern Appalachians in general, is unconformably absent along the Birmingham anticlinorium and northeastward along the southeast limb of Lookout syncline (Butts, 1926). The local, anomalous, stratigraphically deep truncation of the

Figure 5. Stratigraphic section in the Red Mountain Expressway cut (Locality 42) (from measured section by W. A. Thomas, D. N. Bearce, and J. A. Drahovzal, p. 231–240, *in* Drahovzal and Neathery, eds., 1971).

NORTHWEST   SOUTHEAST

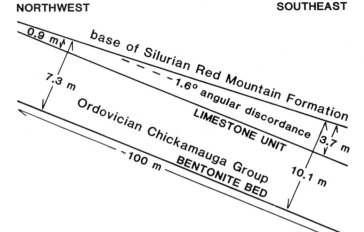

Figure 7. Diagrammatic cross section of angular unconformity between Ordovician and Silurian rocks on southeast limb of Birmingham anticlinorium (measured on northeast face of Red Mountain Expressway cut).

Knox Group suggests differential relative uplift and erosion of more than 2,000 ft (600 m) of section in a linear area along the anticlinorium before Middle Ordovician.

At Birmingham, the Middle Ordovician section above the post-Knox unconformity is dominated by shallow-marine limestones of the Chickamauga Group (Fig. 5). On the southeast limb of the Birmingham anticlinorium, the Chickamauga is 247 ft (75.3 m) thick. Regionally, both to the southeast and northwest, the equivalent section is more than twice as thick. Regional thickness distribution suggests paleodips of approximately 75 ft/mi (15 m/km) away from the crest of a synsedimentary structure coincident with the present southeast limb of the anticlinorium.

Locally, on the southwest side of the cut, the Chickamauga Group is overlain by a thin interval (less than 3 ft; 1 m) of limestone and siltstone of the Late Ordovician Sequatchie Formation (section measured by W. A. Thomas, D. N. Bearce, and J. A. Drahovzal, p. 231–240, *in* Drahovzal and Neathery, eds., 1971). Elsewhere in the cut the Sequatchie is unconformably absent.

Regionally, Ordovician rocks are truncated by a pre-Silurian unconformity that is reflected by the absence of the Sequatchie Formation in most of the cut. Detailed measurements made possible by the complete exposure in this deep cut illustrate truncation of beds up dip in the upper part of the Ordovician section. The truncation indicates an angular discordance of approximately 1.6° between Ordovician beds and the Silurian Red Mountain Formation (Fig. 7). The direction of angular discordance coincides with present dip direction of the southeast limb of the anticlinorium and demonstrates an incremental increase in paleodip from Middle Ordovician to Early Silurian.

The Silurian Red Mountain Formation contains units of coarse-grained, massive, cross-bedded sandstone consisting of

variable amounts of quartz sand, bioclasts, and hematite ooids and matrix. These beach and intertidal sandstones contain conglomerates of discoid limestone clasts that were reworked from thin lenses and interbeds within the formation (Bearce, 1973). Distribution of the intraformational conglomerates and paleocurrent directions suggest a shoal coincident with the crest of the anticlinorium during the Silurian (Bearce, 1973). A thick-bedded coarse-grained sandstone (previously identified as Devonian Frog Mountain Formation, but later shown to be Silurian, Ferrill, 1984) at the top of the Red Mountain Formation locally contains hematite that was probably reworked from lower hematite-bearing sandstones. Although no measure of paleoslope is available from these data, some synsedimentary relief during the Silurian is indicated.

A down-to-southeast normal fault exposed near the southeast end of the cut displaces the sandstone unit (25 ft or 7.6 m thick) at the top of the Red Mountain Formation, as well as underlying strata (Fig. 4). In the lower beds, the fault appears as a brittle, post-sedimentary fault that intersects bedding at an angle of 60° and has approximately 10 ft (3 m) of stratigraphic separation. In contrast, near the fault, the sandstone at the top of the Red Mountain exhibits soft-sediment deformation, slump faults, and local thickening on the downthrown fault block (Fig. 8). These details indicate approximately 3 ft (0.9 m) of synsedimentary fault movement during Silurian and an additional 7 ft (2.1 m) of post-Silurian movement. The normal fault extends upward into the lowest beds of the Mississippian Fort Payne Chert, but approximately 6 ft (1.8 m) above the base of the Fort Payne, chert beds are draped monoclinally across the fault.

The Mississippian Fort Payne Chert is the youngest formation exposed in the Red Mountain Expressway cut. Below the Fort Payne, 7 in (18 cm) of light green-gray and dark purple clay shale constitutes the Lower Mississippian Maury Shale and rests unconformably on the Red Mountain Formation. The Upper Devonian Chattanooga Shale, present elsewhere in the Birmingham area and farther north, is unconformably absent in the Red Mountain Expressway cut.

The southeast slope of Red Mountain is a dip slope on the Fort Payne Chert. The narrow strike-parallel valley southeast of Red Mountain is on the outcrop trace of the Mississippian Tuscumbia Limestone and Pride Mountain Formation which is dominated by mudstone and locally contains thin sandstone beds.

Sandstone Ridge, the low ridge next southeast from Red Mountain, is formed on the Mississippian Hartselle Sandstone, a quartzose sandstone deposited in a barrier-island and shelf-bar system overlying the Pride Mountain mudstones (Thomas and Mack, 1982). In the cut for the Red Mountain Expressway through Sandstone Ridge, the lower part of the Hartselle Sandstone is displaced by synsedimentary rotational downdip slump (listric normal) faults (Fig. 9). The synsedimentary faults indicate slumping and rotational downdip sliding on paleoslopes in the direction of present structural dip on the southeast limb of the anticlinorium (Thomas, 1968). The listric faults apparently flat-

Figure 8. Normal fault, associated with slump structures and local downthrown thickening in sandstone unit at top of Red Mountain Formation, on southeast limb of Birmingham anticlinorium near southeast end of Red Mountain Expressway cut. Explanation: FP = Fort Payne Chert; dashed line = top of sandstone unit; U-RM = sandstone unit at top of Red Mountain Formation; dotted line = base of sandstone unit; RM = Red Mountain Formation below upper sandstone.

ten downward into mudstones of the Pride Mountain Formation and do not extend down into older rocks. Thus, the slump faults in the Hartselle differ in origin from the normal fault that was active during deposition of the upper part of the Red Mountain Formation.

Mississippian and Pennsylvanian rocks crop out southeast of Red Mountain on the southeast limb of the Birmingham anticlinorium and in the Cahaba synclinorium. Shades Mountain, locally visible to the south from Red Mountain, is formed on massive sandstone and quartz-pebble conglomerate that define the base of the Pennsylvanian Pottsville Formation. Between the Hartselle Sandstone on Sandstone Ridge and the Pottsville Formation on Shades Mountain, the succession includes a southwestward-thinning carbonate facies (Mississippian Bangor Limestone) and a northeastward-prograding clastic facies (Mississippian Floyd Shale and Mississippian-Pennsylvanian Parkwood Formation sandstone and mudstone) (Thomas, 1972). In contrast to the region to the northwest, southeast of the anticlinorium, the Mississippian section is thicker, the carbonate facies comprises pro-

portionally less of the section, and the clastic facies extends farther northeast. The greater thickness and greater northeastward progradation of the clastic facies suggest synsedimentary differential subsidence southeast of the anticlinorium (Thomas, 1974). Similarly, the Pennsylvanian is significantly thicker southeast of the anticlinorium, and the succession of coal beds that is regionally recognizable northwest of the anticlinorium is in detail unlike that in the Cahaba synclinorium (Butts, 1926).

LOCALITY 43. NORTHWEST LIMB OF BIRMINGHAM ANTICLINORIUM

Location

Overturned beds on the northwest limb of the Birmingham anticlinorium are exposed in a cut through Sand Mountain on I-65 (sec. 15, T.17S., R.3W.) (Fig. 10). No parking is permitted along I-65; for access to the cut use Exit 263 from I-65. Exit 263 northbound joins 19th Street North, and Exit 263 southbound

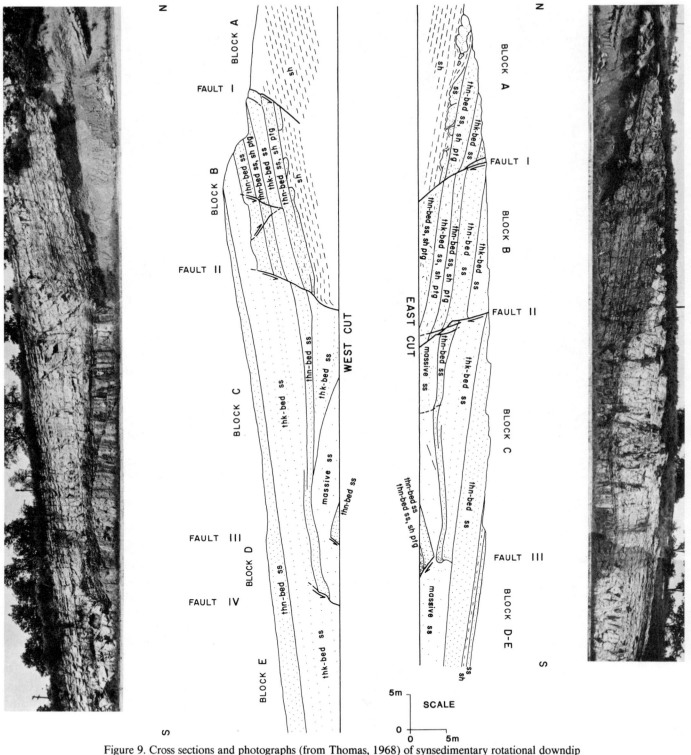

Figure 9. Cross sections and photographs (from Thomas, 1968) of synsedimentary rotational downdip slump faults in Hartselle Sandstone on southeast limb of Birmingham anticlinorium on opposite sides of cut for Red Mountain Expressway through Sandstone Ridge southeast of Red Mountain (vertical cuts are approximately 150 ft [50 m] apart). Explanation: ss = sandstone; sh = clay shale; thn-bed = thin bedded; thk-bed = thick bedded, average more than 1 ft (0.3 m) thick; massive = unit in single bed; ptg = partings, thin interbeds of clay shale within sandstone; areas not patterned are covered. Parts of these cuts are now covered with concrete.

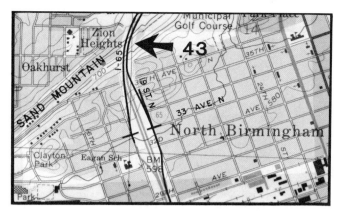

Figure 10. Detailed location map for Locality 43 (Birmingham North 7½-minute Quadrangle).

intersects 33rd Avenue North one block west of 19th Street North. Park on 19th Street North just north of 33rd Avenue North, and walk approximately 0.25 mi (0.4 km) north along northbound entrance ramp of Interstate 65 to the cut.

For a view across the Birmingham anticlinorium, return to I-65 southbound. The valley is formed on Middle Cambrian Conasauga Formation and Ketona Dolomite in the Opossum Valley thrust sheet.

Continue south on I-65 1.9 mi (3 km) to interchange at top of Flint Ridge (I-65, I-59, and I-20); continue south on I-65. Flint Ridge is formed on very poorly exposed Upper Cambrian Copper Ridge Formation (chert) of the Knox Group in the Opossum Valley thrust sheet (Fig. 1). The valley in view to the south is formed on Middle Cambrian Conasauga Formation and Ketona Dolomite in the Jones Valley thrust sheet. The trace of the Jones Valley fault extends along the southeast base of Flint Ridge. Farther south, on the horizon, is Red Mountain.

Site Description

Silurian to Pennsylvanian rocks on the northwest limb of the Birmingham anticlinorium are overturned and dip 65° to 85° southeast (Fig. 11). Along part of the northwest limb, the Opossum Valley thrust fault displaces Cambrian rocks northwestward against overturned Pennsylvanian rocks, but at this locality beds

as old as Silurian are present in the footwall (Fig. 1). The fault trace extends along the southeastern base of the ridge at this locality, and the flat valley on the southeast is formed on the Middle Cambrian Conasauga Formation in the Opossum Valley thrust sheet (Figs. 1, 3).

The thick quartzose sandstone and quartz-pebble conglomerate in the northern part of the cut through Sand Mountain belong to the lower part of the Pennsylvanian Pottsville Formation (Fig. 11). Cross-beds and channel-filling conglomerates demonstrate that the lower Pottsville beds face northwest. Drag folds in shale interbeds also are consistent with an overturned section. Interbedded sandstone and mudstone in the southern part of the cut are stratigraphically below the Pottsville and belong to the Mississippian-Pennsylvanian Parkwood Formation. In low cuts southeast of Sand Mountain, poorly exposed Silurian Red Mountain Formation and Mississippian Fort Payne Chert are part of the overturned sequence (Fig. 11). Between the Fort Payne and Parkwood are quartzose sandstone and gray mudstone that probably constitute an unusually thin, but complete, Mississippian section of Pride Mountain Formation (mudstone), Hartselle Sandstone, and Floyd Shale. Presence of these stratigraphic units at this locality is consistent with the regional distribution of Mississippian facies. A southwest-thinning tongue of Bangor Limestone (predicted from regional facies distribution) might underlie the valley between Hartselle and Parkwood outcrops. Although available data do not preclude possible unexposed faults or tectonic thinning, the anomalously thin Mississippian section here suggests synsedimentary structural movement during Mississippian deposition.

REFERENCES CITED

Bearce, D. N., 1973, Origin of conglomerates in Silurian Red Mountain Formation of central Alabama; their paleogeographic and tectonic significance: American Association of Petroleum Geologists Bulletin, v. 57, p. 688–701.

Butts, C., 1926, The Paleozoic rocks, *in* Geology of Alabama: Alabama Geological Survey Special Report 14, p. 41–230.

Drahovzal, J. A., and Neathery, T. L., 1971, Middle and Upper Ordovician stratigraphy of the Alabama Appalachians: Alabama Geological Society, 9th Annual Field Trip, Guidebook, p. 1–62.

Drahovzal, J. A., and Neathery, T. L., eds., 1971, The Middle and Upper Ordovician of the Alabama Appalachians: Alabama Geological Society, 9th Annual Field Trip, Guidebook, 240 p.

Ferrill, B. A., 1984, Frog Mountain Formation, southwestern Appalachian fold

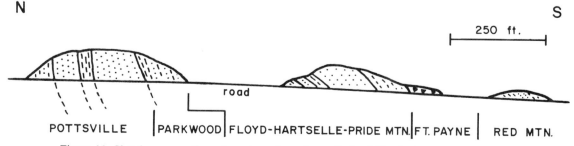

Figure 11. Sketch cross section of overturned northwest limb of Birmingham anticlinorium at Locality 43.

and thrust belt, Alabama [M.S. thesis]: Tuscaloosa, University of Alabama, 178 p.

Kidd, J. T., and Neathery, T. L., 1976, Correlation between Cambrian rocks of the southern Appalachian geosyncline and the interior low plateaus: Geology, v. 4, p. 767–769.

Kidd, J. T., and Shannon, S. W., 1977, Preliminary areal geologic maps of the Valley and Ridge province, Jefferson County, Alabama: Alabama Geological Survey Atlas Series 10, 41 p.

Neathery, T. L., and Copeland, C. W., 1983, New information on the basement and lower Paleozoic stratigraphy of north Alabama: Alabama Geological Survey Open File Report, 28 p.

Simpson, T. A., 1965, Geologic and hydrologic studies in the Birmingham red-iron-ore district, Alabama: U.S. Geological Survey Professional Paper 473-C, C47 p.

Thomas, W. A., 1968, Contemporaneous normal faults on flanks of Birmingham anticlinorium, central Alabama: American Association of Petroleum Geologists Bulletin, v. 52, p. 2123–2136.

——— 1972, Mississippian stratigraphy of Alabama: Alabama Geological Survey Monograph 12, 121 p.

——— 1974, Converging clastic wedges in the Mississippian of Alabama: Geological Society of America Special Paper 148, p. 187–207.

——— 1982, Stratigraphy and structure of the Appalachian fold and thrust belt in Alabama, *in* Thomas, W. A., and Neathery, T. L., eds., Appalachian thrust belt in Alabama: Tectonics and sedimentation (Field Trip Guidebook, Geological Society of America 1982 Annual Meeting, New Orleans, Louisiana): Tuscaloosa, Alabama Geological Society, p. 55–66.

——— 1985, Northern Alabama sections, *in* Woodward, N. B., ed., Valley and Ridge thrust belt: Balanced structural sections, Pennsylvania to Alabama (Appalachian Basin Industrial Associates): University of Tennessee Department of Geological Sciences Studies in Geology 12, p. 54–61.

——— 1986, A Paleozoic synsedimentary structure in the Appalachian fold-thrust belt in Alabama: Virginia Polytechnic Institute Department of Geological Sciences Memoir 3, (in press).

Thomas, W. A., and Mack, G. H., 1982, Paleogeographic relationship of a Mississippian barrier-island and shelf-bar system (Hartselle Sandstone) in Alabama to the Appalachian-Ouachita orogenic belt: Geological Society of America Bulletin, v. 93, p. 6–19.

Harpers Ferry water gap

Douglas G. Patchen and Katharine Lee Avary, *West Virginia Geological and Economic Survey, P.O. Box 879, Morgantown, West Virginia 26507*

LOCATION

Harpers Ferry is located at the confluence of the Potomac and Shenandoah Rivers where the Potomac has cut an impressive water gap through the Blue Ridge. Three states—West Virginia, Virginia, and Maryland—meet at this point, which is located on the Harpers Ferry 7½-minute Quadrangle. Access to Harpers Ferry from the east or west is by way of U.S. 340 (Fig. 1). Recommended stops are along U.S. 340 south of the Potomac River, Sandy Hook Road north of the Potomac, and the B&O Railroad tracks and within Harpers Ferry National Historical Park (Fig. 2).

INTRODUCTION

Harpers Ferry is located at the boundary of two major physiographic and structural provinces: the Great Valley and Blue Ridge physiographic provinces and the Folded Valley and Ridge and detached Blue Ridge Anticlinorium structural provinces.

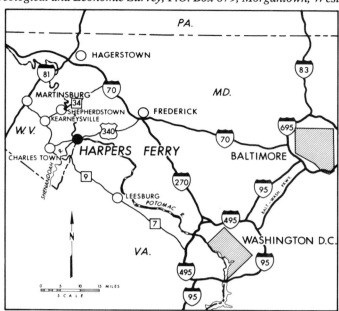

Figure 1. General location map.

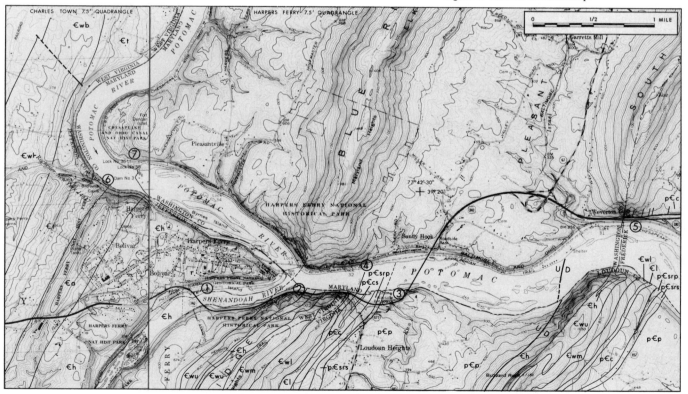

Figure 2. Location of suggested stops (circled numbers) on the Harpers Ferry and adjacent Charles Town 7½-minute Quadrangle maps. Surface geology from Nickelsen (1956). pCp-Pedlar; pCsrs=Swift Run sericite phyllite; pCsrp=Swift Run phyllite; pCc-Catoctin; pCcs=Catoctin tuffaceous sediments; Cl=Loudoun; Cwl=lower Weverton; Cwm=middle Weverton; Cwu=upper Weverton; Ch=Harpers; Ca=Antietam; Ct=Tomstown.

Table 1. Stratigraphic Summary, Harpers Ferry Water Gap

Age	Formation Member	Reference	Local Thickness	Description (after Nickelsen, 1956)
Early Cambrian	Tomstown Fm. upper dolomitic, siliceous limestone lower dolomite	Stose, 1906	1045'	white to dark gray, vuggy sacchroidal dolomite, and white to medium gray dolomitic or siliceous limestone. Scattered low-grade residual iron ore.
	Antietam Fm.	Keith, 1892	400-500'	white, well-sorted, well-rounded, fine to very-fine grained quartzite, and darker, silty phyllite and poorly-sorted sericitic quartzite. *Olenellus.*
	Harpers Fm.	Keith, 1892	800-2000' (est.)	light brown-gray to light green-gray phyllite, with fine-grained, calcareous quartzites at top. *Scolithus.*
	Weverton Fm. Dismal Hollow Mbr. Oregon Hollow Mbr. Loft Mountain Mbr.	Keith, 1892 Nunan, 1979 (members)	350-480'	hard, coarse-grained, purplish quartzites, with phyllite. 3 informal members: more phyllite at base, cross-beds common in middle, conglomerates at top. Sorting decreases upwards.
	Loudoun Fm. upper conglomerate lower phyllite	Keith, 1892	55-110'	red-purple to dark blue-gray phyllite and green to purple-black conglomerate.
Precambrian	Catoctin Fm.	Geiger and Keith, 1891	500-1000'	altered volcanic tuffs and flows, now greenstone and metavolcanics, mostly gray, green, and purple metabasalt.
	Swift Run Fm. upper phyllite lower quartzite	Jonas and Stose, 1939	0-200'	altered volcanic tuffs and flows, now metagraywacke, meta-arkose, and phyllite.
	Pedlar Fm.	Bloomer and Werner, 1955	?	dark gray to gray-green, massive, coarse-grained, hypersthene granodiorite, with dikes of metabasalt or greenstone. Garnets, pyroxene, hornblende and quartz common.

Water gaps through Blue Ridge–Elk Ridge to the west and Short Hill–South Mountain to the east expose the repeated beds of the western limb of the Blue Ridge Anticlinorium. These rocks include the earliest Paleozoic sediments exposed, as well as the youngest Precambrian rocks known, in this area. Harpers Ferry also is an area of historic interest, and one where the geology and topography played an important role in that history. Finally, it is an area of scenic beauty that can be enjoyed in any of the four seasons.

After spending a short time in the water gap you will soon realize that the geology at Harpers Ferry is more than the outcrops, water gaps, rivers, rapids, and floods. It encompasses the town itself and is reflected in the native building stone, carved steps and walkways, historic industries, and sites of historic military fortifications.

Several local landmarks are within the Harpers Formation (Table 1). The hand-carved steps that serve as an entrance to "Public Way" in the village of Harpers Ferry were carved in this formation. A short walk up these steps will take a tourist (or geologist) past Robert Harper's house, which was built between

1775 and 1782 with Harpers Phyllite quarried on the hill, and St. Johns Episcopal Church, built in 1852 with this same rock. The nearby St. Peters Roman Catholic Church (1830–1896) was constructed using Weverton Quartzite (Table 1). At the top of the path lies Jefferson Rock, a balanced boulder of Harpers Phyllite, reinforced by sandstone pillars in 1860. Thomas Jefferson visited here in 1783 and wrote of this view:

On your right comes up the Shenandoah, having ranged along the foot of the mountain a hundred miles to seek a vent. On your left approaches the Potomac, in quest of a passage also. In the moment of their junction, they rush together against the mountain, rend it asunder and pass off to the sea . . . this scene is worth a voyage across the Atlantic.

STRATIGRAPHY

The rocks exposed in the gorge at Harpers Ferry are the oldest in this vicinity and range in age from Precambrian to Early Cambrian. Some controversy still exists as to the age of the Chilhowee Group clastics, but most geologists now agree that the Pedlar, Swift Run, and Catoctin Formations are Precambrian and that the Chilhowee clastics—the Loudoun, Weverton, Harpers, and Antietam Formations—although nonfossiliferous and slightly metamorphosed, are the oldest Paleozoic rocks. Brief descriptions of the rocks exposed in the area are summarized in Table 1, condensed from Nickelsen (1956).

STRUCTURAL GEOLOGY

All of the Precambrian and Lower Cambrian rocks exposed in the Harpers Ferry area have been intensely deformed, and most have been metamorphosed. The small folds that can be observed in Blue Ridge–Elk Mountain and South Mountain–Short Hill are local structures developed on the west limb of the large Blue Ridge or South Mountain Anticlinorium. Also, the Harpers Formation, in its type locality, is on the lower limb of a large recumbent anticline that involves the Blue Ridge to the east. Minor folds, cleavage, and thrust faults have been developed on this lower limb; the upper limb has been eroded away.

The Blue Ridge Anticlinorium is asymmetrical. The slightly overturned, steeper western limb is represented in this area by Blue Ridge–Elk Mountain and South Mountain–Short Hill, and a more gently dipping, right-side-up eastern limb is represented by Catoctin Mountain. Numerous smaller folds that have been developed on the western limb of the anticlinorium have been the subject of controversy for nearly a century. Early workers (Geiger and Keith, 1891; Keith 1894) interpreted Elk Mountain–Blue Ridge and South Mountain–Short Hill to be synclines, and Keith (1894) added an overthrust beneath Blue Ridge. Cloos (1951) interpreted Elk Ridge to be an asymmetric, overturned anticline with a second anticline, but no overthrust, to the west. Nickelsen (1956) interpreted both ridges to be an overturned anticline-syncline combination. He also added a normal fault on the west side of Short Hill that repeats the folds in the western limb of the Blue Ridge Anticlinorium.

Sketches included here are from Chen (in Burford et al., 1964) and present a more complex structural interpretation through Blue Ridge–Elk Mountain. At least three major anticlines, several minor folds, and as many as nine thrust faults are interpreted in Harpers Ferry gorge. An older photo by Woodward (1949, pl. XIX A, p. 75) clearly illustrates one of these large, overturned anticlines.

Recent structural interpretations of seismic profiles through the Blue Ridge and Piedmont have suggested that the entire Blue Ridge is detached and has been thrust westward over lower Paleozoic clastics and carbonates. Thus, basement rocks exposed in Harpers Ferry gorge are correlative with rocks 4–5 mi (6.7–8.3 km) below the surface of the Blue Ridge.

STOP DESCRIPTIONS

Seven stops (Fig. 2) are described in the Harpers Ferry area: (1) the type section of the Harpers Formation in the National Park; (2) the section through the Blue Ridge along U.S. 340 south of the Potomac; (3) the exposure of Pedlar Formation at the Potomac Wayside rest area at the south end of the U.S. 340 bridge to Maryland; (4) the section through Elk Ridge north of the Potomac along Sandy Hook Road; (5) the type section of the Weverton Formation 4 mi (6.4 km) east along U.S. 340; (6) exposures of the Tomstown and Antietam near John Brown's Cave; and (7) the abandoned iron ore deposit in the Chesapeake and Ohio Canal National Historical Park.

Harpers Ferry Historical Park (Stop 1)

The type section of the Harpers Formation consists of several exposures scattered over a 0.6-mi (1-km) interval along the park road from the Armory Worker's House, below Jefferson Rock, westward to U.S. 340 (Fig. 2, No. 1). Intense deformation has obliterated bedding and other sedimentary features resulting in a highly cleaved, dark greenish-gray or brown phyllite, often cut by irregular quartz veins. In several exposures, however, thin (2–4-in, 5–10-cm) iron-stained beds can be observed cutting across cleavage. One of these is approximately 1,000 ft (303 m) west of the Armory Worker's House across from the bridge to Virginus Island (Fig. 3). Here an iron-stained bed defines a small anticline. Cleavage on the east limb is nearly horizontal to slightly east, but on the western limb cleavage nearly parallels the steep westward-dipping beds. At the western edge of the exposure, folded cleavage is prominent.

The next exposure to the west is across from the ruins of the Savory and Company pulp and paper mill that operated here in the late 1800s. Bedding is nearly vertical with two recumbent folds developed. Cleavage dips gently to the east, nearly parallel to the axial plane of the folds. Bedding is still vertical another 630 ft (191 m) west where a thrust fault can be observed on the western edge of the exposure. Two iron-stained beds are present below the thrust, defining bedding.

The westernmost exposure is near the junction of the park

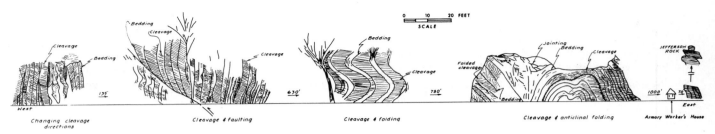

Figure 3. Section along Harpers Ferry National Park road north of the Shenandoah River (from Chen, in Burford and others, 1964).

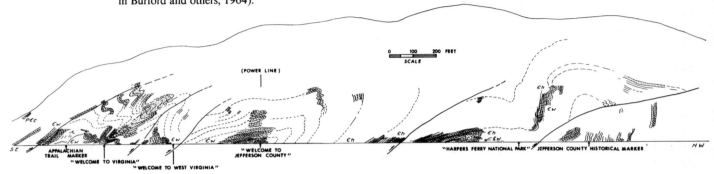

Figure 4. Structural section through Elk Ridge, south side of Potomac River along U.S. 340 (from Chen, in Burford and others, 1964).

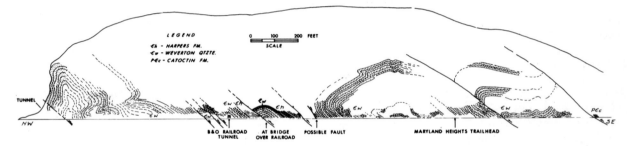

Figure 5. Structural section through Elk Ridge, north side of Potomac River along Sandy Hook Road (after Chen, in Burford and others, 1964).

road and U.S. 340. Bedding is again vertical, but cleavage changes directions, dipping both east and west in alternate beds.

Although bedding is very hard to see in these exposures and cleavage is obvious, the strike of the beds can readily be observed by looking at the trend of the rapids in the Shenandoah River to the south.

The Blue Ridge (Stop 2: Use Caution!)

Be extremely careful at this location. The shoulders along U.S. 340 are quite narrow, and *traffic is fast and heavy*. The same section can be observed north of the Potomac through Elk Ridge, along old U.S. 340 that is much less traveled. You may wish to consider that as an alternate stop.

Nickelsen (1956) mapped the structure through the Blue Ridge as an overturned syncline in the Harpers Formation paired with an overturned anticline in the upper Weverton. Chen (in Burford and others, 1964) observed at least three major anticlines and several minor folds in a highly disturbed zone and interpreted

at least five thrust faults. However, from late spring to early fall this section is covered with heavy vegetation, unlike the scenes shown in old photos taken by Woodward (1949) when the U.S. 340 bypass was under construction. The following brief description of what can be seen late in the year along this section is tied to road signs along U.S. 340. (These signs cannot be considered "permanent," but should last quite a few years.)

Starting from the east, the Appalachian Trail ascends from the road near the Catoctin-Loudoun contact (Fig. 4). Bedding here dips steeply to the east approximately parallel to the first of several thrust faults. A few hundred feet (meters) farther west, the "Enter Loudoun County" and "Welcome to Virginia" signs are opposite a cliff with a highly disturbed zone in the Weverton Formation below this fault. A second fault is present near the "Welcome to Wild, Wonderful West Virginia" sign. West of this fault is the largest anticline in the gorge, below the power line at the "Welcome to Jefferson County" sign.

Still farther west, approximately midway through the gorge, the Harpers Formation is poorly exposed. However, a marked change in dip in the relatively small exposures at road level was interpreted by Chen as the position of a fourth thrust fault. Below this fault, two anticlines and a syncline were sketched by Chen above the fifth fault. This final fault intersects the road near the National Park Service sign "Harpers Ferry National Park Exit 1 mile." Below this fault, bedding is nearly vertical at the Jefferson County historical marker sign.

Potomac Wayside Rest Area (Stop 3)

This rest area is along the U.S. 340 bypass at the south end of the Potomac River bridge between Maryland and Virginia. Across the road the Precambrian Pedlar Formation is exposed in a steep northwest-facing cliff. A few hundred feet (meters) farther west, an outcrop of the Pedlar is accessible along the north side of U.S. 340 at the intersection with Virginia 671. Other beds can be observed along Piney Run as it descends 60 ft (18 m) from road level to the Potomac.

Although this is one of the best outcrops of this garnet-rich granodiorite in the area, it is easily the most dangerous. *Use extreme caution* along both sides of U.S. 340.

Elk Ridge (Stop 4)

The section through Elk Ridge is readily accessible by Sandy Hook Road (old U.S. 340) between Sandy Hook and the B&O bridge over the Potomac to Harpers Ferry. Although both the road and shoulders are narrow, traffic is relatively light and less dangerous than at the two previous stops. The trailhead to Maryland Heights is near the east end of the exposure; this trail can be taken to two good outcrops of Weverton.

Chen's sketch (Fig. 5) includes three major anticlines, several smaller folds, and eight thrust faults. In general, his sketches north and south of the Potomac are in good agreement with regard to the position of major folds, highly faulted zones, and outcrops of the Harpers Formation near the middle of the water gap. From east to west the following sequence of beds and structures can be observed. The topographic slope on the east is interpreted as the first thrust fault, with Weverton beds below both vertical and horizontal. Sharp folds at road level form several wedges, whereas the upper beds are nearly horizontal until folded to nearly vertical to the west. Overall, the structure is the first of three large anticlines, overturned to the west. The western limb is cut by a second thrust fault just east of the Maryland Heights trailhead.

Relatively good exposures over the next 850 ft (257.6 m) are all Weverton beds folded into a large anticline, overturned to the west. Beds in the center of the fold are deformed into several minor folds, with many quartz veins. The structure is faulted on the west. A short, but steep, 5–10 minute walk up the trail to Maryland Heights will allow you to more closely examine this structure. At a point beneath the power lines overlooking the Potomac to the south, an excellent section of Weverton is exposed. The quartzite beds are white to light gray with reddish-purple laminae and are highly cross-bedded. The exposure is on strike with rapids below in the Potomac and the large fold in Blue Ridge. However, the dip on these beds is to the west, whereas bedding in the rapids below definitely dips to the east. Also, above and to the east of this outcrop, Weverton beds become nearly horizontal. Thus, this exposure on the western, overturned limb of the anticline is strategically located where the key changes in dip can be observed.

If you continue along the trail to Maryland Heights you can observe a very small outcrop of Harpers Formation phyllite midway between the outcrop described above and the famous overlook of Harpers Ferry. This overlook is near the Harpers-Weverton contact, which is nearly vertical and can be estimated visually along the change in slope on the west side of Blue Ridge to the south and Elk Ridge to the north. The thick beds of Weverton are very conglomeratic at this exposure, especially near the top of the formation.

Descending along the trail and continuing west along the road, beds in the Harpers and Weverton are exposed in several small folds cut by faults. The total length of this interval is approximately 400 ft (121.2 m), continuing 100 ft (30 m) west of the B&O tunnel through the ridge. Chen interprets five thrust faults in this short interval.

The western side of Elk Ridge is essentially one large, overturned anticline with numerous highly contorted beds on both flanks. The Harpers Ferry overthrust of Keith (1894) and Stose and Stose (1946) was envisioned to be present beneath this structure. Chen indicates a minor thrust may be present, but does not believe it to be of the scale suggested by Keith (1894).

Weverton Type Section (Stop 5)

Nunan (1979) measured and described the type section of the Weverton Formation along U.S. 340 through South Mountain (Fig. 2, No. 5), less than 3 mi (4.8 km) east of the Weverton exposures through Elk Ridge along old U.S. 340. The Weverton is folded into a large anticline overturned to the west, and beds measured by Nunan are on the east limb of the fold. Deformation is apparent, but less than in equivalent beds in Elk Ridge. Nunan (1979) subdivided the formation into three members, all of which are readily exposed. Approximately 1,000 ft (303 m) east of this exposure a small outcrop of Catoctin can be observed along U.S. 340 near the crest of the hill.

John Brown's Cave (Stop 6)

The Antietam and Tomstown Formations are exposed along the B&O Railroad tracks northwest of Harpers Ferry near John Brown's Cave (Fig. 2, No. 6). Access is by way of the Bakerton road, less than 0.5 mi (0.8 km) west of the ridge of Antietam Sandstone west of Harpers Ferry. After turning north on the road to Bakerton, proceed 1.4 mi (2.3 km) to a *dangerous* one-lane B&O underpass. Stop, sound your horn, and proceed under the railroad. Park along the roadway to the left.

The Tomstown Formation is exposed along the railroad tracks west of the underpass and is highly deformed. Beds here are in the eastern limb of a large syncline centered in the overlying Waynesboro Formation to the west. Another excellent Tomstown exposure is to the east in a vertical cut south of the tracks. John Brown's Cave (gated) is in this Tomstown outcrop. The Antietam Formation is exposed about 0.25 mi (0.4 km) east of the underpass, also along the B&O tracks. The Antietam is overturned, with beds dipping 50 degrees to the southeast. Cleavage dips approximately 20 degrees in that direction (Burford and others, 1964). A stream to the east separates the Antietam from exposures of the Harpers Formation. The Harpers is exposed nearly continuously for 1.5 mi (2.4 km) from here to the railroad station in Harpers Ferry. If you choose to visit these exposures, *use extreme caution!* Train traffic is heavy along these tracks from the west toward Harpers Ferry, and their approach is hidden from view by the steep bluffs along the gentle curves in the tracks.

The "Maryland Bank" (Stop 7)

Across the Potomac from John Brown's Cave are the ruins of an old iron mine, the "Maryland Bank" deposit (Fig. 2, No. 7). Ore taken from the Tomstown was shipped to the Antietam Ironworks, 7 mi (11 km) upstream. Access is by foot walking northwest along the old C&O Canal towpath in the National Park. The old mine is located in a steep, southwest-facing bank west of an abandoned building above lock number 36; the building and lock have been preserved by the National Park Service.

REFERENCES CITED

Bloomer, R. O., and Werner, H. J., 1955, Geology of the Blue Ridge region in central Virginia: Geological Society of America Bulletin, v. 66, p. 579–606.

Burford, A. E., Chen, P-f., Donaldson, A. C., Erwin, R. B., Dean, S., Perkins, R., and Arkle, T., Jr., 1964, The Great Valley in West Virginia: Guidebook, Appalachian and Pittsburgh Geological Societies, Oct. 16–17, 62 p.

Cloos, E., 1951, The physical features of Washington County: State of Maryland, Dept. of Geology, Mines, and Water Resources, 333 p.

Geiger, H. R., and Keith, A., 1891, The structure of the Blue Ridge near Harpers Ferry: Geological Society of America Bulletin, v. 2, p. 156–164.

Jonas, A. I., and Stose, G. W., 1939, Age relations of the pre-Cambrian rocks in the Catoctin Mountain–Blue Ridge and Mount Rogers anticlinoria in Virginia: American Journal of Science, v. 237, p. 573–593.

Keith, A., 1892, The geologic structure of the Blue Ridge in Maryland and Virginia: American Geologist, v. 10, p. 362–368.

——1894, Harpers Ferry folio: U.S. Geological Survey Geologic Atlas, folio 10.

Nickelsen, R. P., 1956, Geology of the Blue Ridge near Harpers Ferry, West Virginia: Geological Society of America Bulletin, v. 67, p. 239–270.

Nunan, W. E., 1979, Stratigraphy of the Weverton Formation, northern Blue Ridge anticlinorium [Ph.D. thesis]: Chapel Hill, University of North Carolina, 215 p.

Stose, A. J., and Stose, G. W., 1946, Geology of Carroll and Frederick: State of Maryland, Dept. of Geology, Mines, and Water Resources, p. 11–128.

Stose, G. W., 1906, The sedimentary rocks of South Mountain, Pennsylvania: Journal of Geology, v. 14, p. 201–220.

Woodward, H. P., 1949, Cambrian system of West Virginia: West Virginia Geological Survey, v. 20, 317 p.

Extrusional environments of part of the Catoctin Formation

M. T. Lukert, *Edinboro University of Pennsylvania, Edinboro, Pennsylvania 16444*
Gautam Mitra, *University of Rochester, Rochester, New York 14627*

LOCATION

This site is located in Shenandoah National Park on the Appalachian Trail near the small parking area at mile 39.1 on the Skyline Drive (Fig. 1). It is accessibly by any vehicle including bus. Some agility is required in the "rock-scrambling" necessary to view part of the exposure. NOTE: Collecting is not allowed in Shenandoah National Park without special permission.

INTRODUCTION

The metavolcanics of the late Precambrian (?) Catoctin Formation and apparent correlatives are almost continuously exposed along a belt extending from Pennsylvania southwestward to North Carolina. In Maryland and Virginia, the Catoctin gives topographic expression to the limbs of the breached Catoctin–Blue Ridge Anticlinorium, with the northwestern limb forming the Blue Ridge. The rocks of the Catoctin Formation in the anticlinorium are typically aphanitic, dark green to dark bluish gray, massive, metamorphosed basalts. Interbedded with the metabasalts, however, are thin sedimentary units, phyllitic tuff, and massive rhyolitic tuff. The common occurrence of columnar jointing, the scoriaceous nature of many of the flows, and the lack of pillows have been cited as evidence for a subaerial origin for the Catoctin volcanics (Reed, 1955). Pillow structures at this locality, however, indicate that the extrusion of the Catoctin was not solely subaerial.

SITE DESCRIPTION

In some parts of the Blue Ridge, the Catoctin Formation is separated from the underlying basement rocks by the clastic metasedimentary rocks of the Swift Run Formation. Where the Swift Run is absent, as at this locality, the Catoctin lies directly on the basement rocks, represented here by the Pedlar Formation. The Pedlar is one of several coarse-grained metamorphic and igneous rock units exposed in the core of the Catoctin–Blue Ridge Anticlinorium. Radiometric ages for these rocks generally fall in the 1.0 to 1.2 Ga range. Considerable textural variation exists within the rocks assigned to the Pedlar Formation although all have charnockitic affinities. The typical mineralogy of the Pedlar is that of hypersthene granodiorite or hypersthene-bearing quartz monzonite. At this site coarsely layered Pedlar is exposed on the Appalachian Trail approximately 100 ft (30 m) north of its juncture with the trail from the parking area (Fig. 2). This outcrop displays broad (6 to 8 in [15 to 20 cm]) bands of reddish iron staining in the Pedlar. The origin of this staining is uncertain. It may reflect pre-Catoctin weathering activity or it may have been caused by mineralizing fluids associated with Catoctin volcanism.

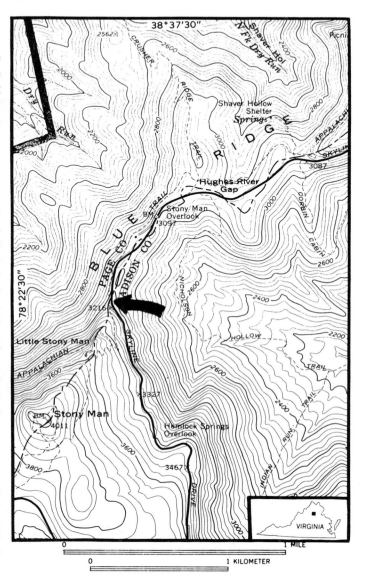

Figure 1. Location of Catoctin Formation outcrops detailed in Figure 2.

Massive, metamorphosed basalt flows characterize the overlying Catoctin Formation. The mineral composition of the metabasalts is typically albite, chlorite, epidote, actinolite, and opaque ores. The abundance of green minerals is reflected in the color of the rock and in the common reference to the Catoctin greenstones. Investigations by Reed and Morgan (1971) indicate that the basalts originally were of tholeiitic composition and were spilitized during Paleozoic metamorphism. The thickness of the Catoctin is uncertain, but has been estimated to be as much as 2,000 ft (6,096 m) in Shenandoah National Park (Gathright, 1976).

The age of the Catoctin has been somewhat problematic. Rankin and others (1969) reported an age of 820 Ma for the Catoctin. Their data were obtained from zircons collected at five localities in Pennsylvania, Virginia, and North Carolina. The samples were taken from presumed stratigraphically equivalent horizons at those sites. The 820 Ma age is in contradiction to data from granitic rocks that are cut by dikes of greenstone similar, if not identical to, Catoctin feeder dikes. Zircons from these granitic rocks yielded a U-Pb age of 730 Ma, thereby establishing a maximum age for the Catoctin (Lukert and Banks, 1984). This is also in accord with a 600 Ma Catoctin age obtained from Rb-Sr data by Mose and others (1985).

Although some authors have suggested that parts of the Catoctin may have been submarine flows (e.g., Bloomer and Bloomer, 1947; Brown, 1958), most investigators have followed Reed (1955), in ascribing a subaerial origin to the Catoctin volcanics. Such an origin is in accord with the originally tholeiitic composition of the lavas. Similarly, the common occurrence of columnar jointing and the presence of amygdaloidal and scoriaceous flows supports a subaerial origin. In addition, the lack of pillows was cited as evidence for subaerial extrusion.

The polygonal columns in the talus pile just above the parking area (Fig. 2) indicate terrestrial extrusion for at least part of the Catoctin in this area. The nature of the Pedlar–Catoctin contact at this locality is also in accord with subaerial extrusion. To observe the contact, walk south on the Appalachian Trail from the iron-stained Pedlar exposure about 75 ft (22.9 m) to the point where the trail begins to rise. At this point, move off the trail to the right. Careful examination of the outcrops within a few feet of the trail will reveal the contact between the vesicular, somewhat phyllitic metavolcanics and the coarse-grained basement rocks (Fig. 2). The basement rocks immediately beneath the metavolcanics are somewhat altered and display a reddish color. This alteration may reflect weathering of the basement rocks prior to Catoctin volcanism. In any case, the vesicles in the metavolcanics are additional evidence for subaerial extrusion.

It is the presence of pillow structures that gives this site its greatest significance. The pillows are small (approximately 12 by 5 in [30.5 by 12.7 cm]), somewhat flattened, and outlined by narrow, light tan borders. The presence of pillows is generally regarded as prima facie evidence of subaqueous extrusion. The pillows may be observed by continuing south from the contact zone about 75 ft (22.9 m) onto the first narrow ledge below the trail. (The width of the ledge is greatly exaggerated in Fig. 2.) The pillows are best displayed in the near vertical exposures between the ledge and the trail. Since the discovery of the pillows here, Gathright (personal communication, 1981) has indicated the presence of pillow-like structures in and near Swift Run Gap. The areal extent of pillow development has not been determined either at this locality or in the Swift Run Gap area. It seems likely that pillow development is not widespread or it would certainly have been noted in previous studies. If the occurrence of pillows is areally restricted, this would suggest extrusion of the lavas into small embayments in the preCatoctin surface. In any case, the

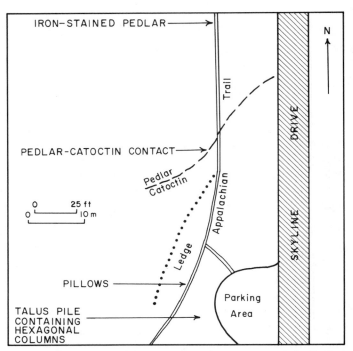

Figure 2. Sketch map of Catoctin Formation exposures along Skyline Drive, at mile 39.1, Madison County, Virginia.

presence of the pillows reveals that the extrusional environment of the Catoctin was not everywhere the same.

REFERENCES CITED

Bloomer, R. O., and Bloomer, R. R., 1947, The Catoctin Formation in Central Virginia: Journal of Geology, v. 55, p. 94–106.

Brown, W. R., 1958, Geology and mineral resources of the Lynchburg quadrangle, Virginia: Virginia Division of Mineral Resources Bulletin 74, 99 p.

Gathright, T. M., II, 1976, Geology of the Shenandoah National Park, Virginia: Virginia Division of Mineral Resources Bulletin 86, 93 p.

Lukert, M. T., and Banks, P. O., 1984, Geology and age of the Robertson River Pluton, in Bartholomew, M. J., ed., The Grenville event in the Appalachians and related topics: Geological Society of America Special Paper 194, p. 161–166.

Mose, D. G., Diecchio, R. J., DiGuiseppi, W. H., and Nagel, M.S., 1985, Confirmation of a latest Precambrian (600 m.y.) age for the Catoctin Formation in Virginia: Geological Society of America Abstracts with Programs, v. 17, p. 126.

Rankin, D. W., Stern, T. W., Reed, J. C., Jr., and Newell, M. F., 1969, Zircon ages of felsic volcanic rocks in the upper Precambrian of the Blue Ridge, central and southern Appalachians: Science, v. 166, p. 741–744.

Reed, J. C., Jr., 1955, Catoctin Formation near Luray, Virginia: Geological Society of America Bulletin, v. 66, p. 871–896.

Reed, J. C., Jr., and Morgan, B. A., 1971, Chemical alteration and spilitization of the Catoctin greenstones, Shenandoah National Park, Virginia: Journal of Geology, v. 79, p. 526–548.

Shores Complex and Mélange in the central Virginia Piedmont

William Randall Brown, Department of Geology, University of Kentucky, Lexington, Kentucky 40506

LOCATION

A 1 mi (1.6 km) section just east of Shores Station on the north side of the James River, Fluvanna County, Arvonia 7½-minute quadrangle (Fig. 1). Exposure is along a rough gravel Chesapeake and Ohio Railroad access road beside the tracks. It is a private road and no questions are asked of small groups, but large groups should get permission from: Chessie System Railroad, Division Office, P.O. Box 1254, Richmond, Virginia 23210, phone (804)-226-7400.

INTRODUCTION

A belt of intensely-deformed metagraywackes and metapelites as much as 5 mi (8 km) wide, locally containing zones of mélange, extends northeastward through the central Virginia Piedmont. It has been mapped from northeastern Appomattox County on the south, to where it disappears beneath the Culpeper Mesozoic basin in Fauquier County on the north, a distance of about 100 miles (161 km). The belt was first recognized where it crosses the James River and its local mélange character was first observed just east of Shores Station on this river (Brown, 1976). The term Shores Mélange has been used informally for those parts of the belt which contain blocks, and the rocks of the entire intensely-deformed belt have been named Shores Complex (Bland and Blackburn, 1980; Evans, 1984). The geologic site just southeast of Shores has been chosen because of its accessibility and good exposure of the mélange.

It has become apparent that the Shores Complex is part of a major zone of thrusting and obduction which has been traced from southeastern Pennsylvania, southwestward across Maryland, and two-thirds of the way across Virginia; and includes the Baltimore ophiolitic complex, Drake and Morgan's (1981) Piney Branch Complex and Potomac River allochthon, and possibly other allochthonous complexes. The whole belt is marked by a strong magnetic lineament (Zietz and others, 1977; Zietz and others, 1980). That part of the lineament in Virginia, referred to here as the Shores Lineament, is shown in Figure 2.

In central Virginia, the Shores Complex lies between Cambrian Chopawamsic volcanic-arc rocks on the east and the relatively deep-water rear-arc basinal sediments of the Evington Group and Hardware Metagraywacke to the west. Vibroseis reflection profiling across the central Virginia Piedmont shows the Shores Lineament to have the subsurface form of a low-angle overthrust (Glover and Costain, 1982; Wehr and Glover, 1985, Fig. 4D). It appears that the Shores Complex marks the major zone of movement (suture) between the rocks of the volcanic arc and the sediments of the rear-arc basin. This movement occurred chiefly or completely during closure of the rear-arc basin by arc-continent collision of the Taconic Orogeny.

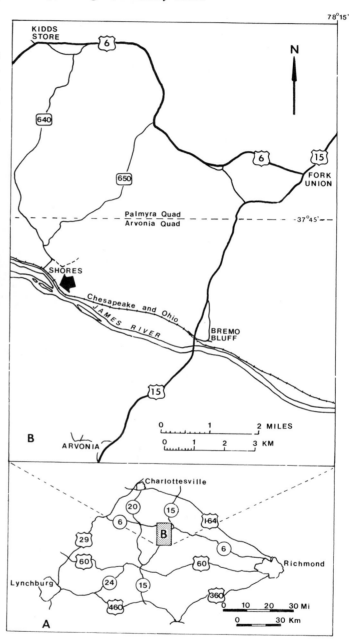

Figure 1. Location map.

The Shores Complex is part of a group of rocks that were mapped as Evington Group (?) by Smith and others (1964) and Brown (1969). Evans (1984), however, considers the complex to be distinct from the Evington Group and suggests that the name Evington be limited to those continent-border, off-shelf, west-derived sediments west of his Buck Island fault zone near the Scottsville Mesozoic basin. He also gave the name Hardware Metagraywacke to those apparently east-derived sediments be-

Figure 2. Index map showing the extent of the Shores magnetic lineament–mélange zone in Virginia and its relation to other major structures. Magnetics from Zietz and others (1977); eastern Piedmont in part from Farrar (1984); northern Piedmont from Pavlides (1981); Maryland from the Geological Map of Maryland (1968).

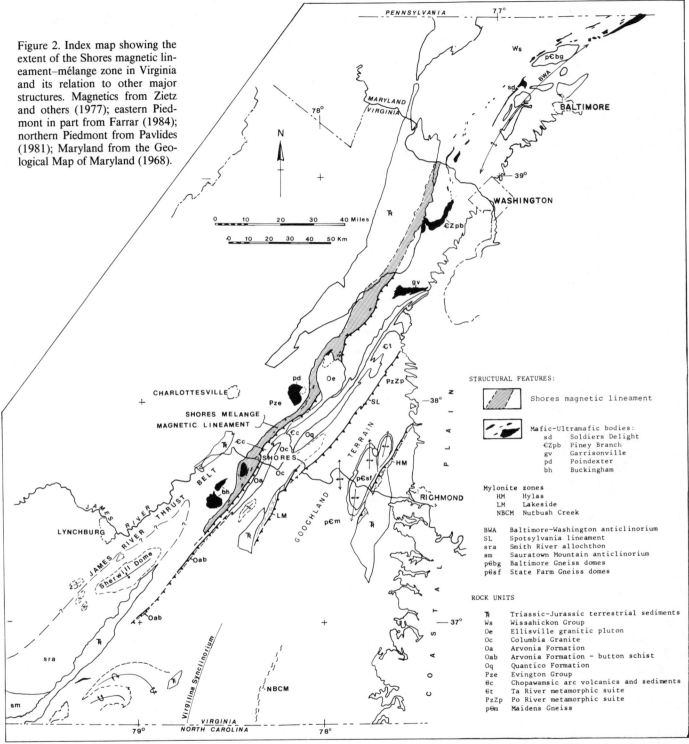

tween his Buck Island fault zone and the Shores Complex. These useful distinctions are followed in this paper (Fig. 3).

The Shores Complex consists chiefly of complexly and intricately polydeformed magnetite-rich metagraywacke, siliceous schist and phyllite, numerous enclosed greenstone sheets and lenses, and occasional metagabbroic and ultramafic masses. Some

of the latter may be measured in miles (km). The metasediments are predominantly in the greenschist facies, although Evans (1984) suggests that they were once of higher grade and have been retrograded. Zones of abundant fragments and blocks of greenstone, ultramafic, and metagraywacke (mélange) occur locally, at east of Shores and at Holiday Creek at the Buckingham-

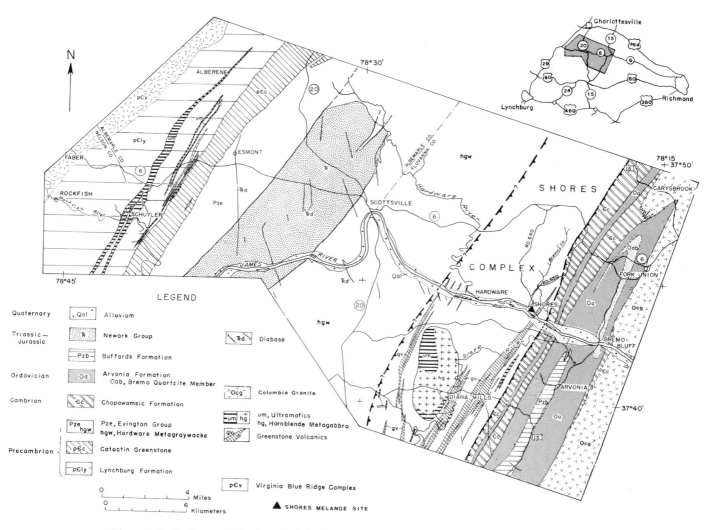

Figure 3. Geologic map of the Arvonia-Schuyler district in central Virginia showing the geologic setting of the Shores Complex in the vicinity of Shores.

Appomattox county line. Small masses of sheared ultramafic are widespread and brecciated soapstone occurs at Sams Creek 2 mi (3.2 km) west-northwest of Andersonville in Buckingham County. The largest mafic-ultramafic body definitely within the Shores Complex is the Diana Mills body 4 mi (6.4 km) southwest of Shores; it is 4.5 mi (7.2 km) long by 1.8 mi (2.9 km) wide and has been shown by gravity measurements to be rootless (see Fig. 3 and Graham, 1975). It consists largely of metagabbro but also includes a large percentage of serpentinite, some of which is intensely brecciated. Several large mafic-ultramafic bodies occur just west of the magnetic lineament and Shores zone of intricate deformation. These include two in Buckingham County (Ern, 1968) and one at Poindexter in Louisa County (Hopkins, 1960, and Fig. 2). It is not known if these have any direct relation to the Shores Complex.

The limits of the complex are imperfectly known; but in central and north-central Virginia they appear to more or less coincide with the Shores magnetic lineament (Fig. 2). At James

River the western limit to the zone of intense deformation, about 4 mi (6.4 km) west of Shores, essentially coincides with the edge of this magnetic high. Beyond this limit, which Evans (1984) interprets to be a fault, the intense deformation gives way abruptly to the much less deformed Hardware Metagraywacke, which at first dips steeply southeastward and is characterized by northeastward-plunging reverse drag folds. Beyond a narrow zone of these steep dips, foliation and bedding dip gently southeastward; but the rocks have a dominant spaced cleavage which is axial-planar to rare gently refolded nearly recumbent isoclinal folds.

Along James River southeast of Shores the complex extends about 1.0 mi (1.6 km) to where it is overthrust by Cambrian Chopawamsic volcanics (Fig. 3). Just east of this fault, in the Long Island and Arvonia synclines, the Chopawamsic Formation, mostly felsic and mafic volcanics, is overlain unconformably by the Late Ordovician Arvonia Slate. Further east the Arvonia is nonconformable upon the Columbia Granite (the granitic part of

Brown's [1969] Hatcher Complex), which intrudes the Chopawamsic and has been dated at 454 ± 9 Ma (Mose and Nagel, 1982).

Northeast of James River, the complex has been mapped in part by Pavlides who, in mapping the Mineral and Lahore 7½-minute Quadrangles about 31 mi (50 km) northeast of Shores, found that the stock-like Ellisville monzonitic pluton intrudes the Shores melange and includes at its north end a granitoidal olistolithic mass 5.6 mi (9 km) long which in turn, includes a large ultramafic mass from the melange (Brown and Pavlides, 1981). The Ellisville is somewhat foliated, but its intrusion disrupted the trend and foliation of the Shores Complex and, therefore, apparently postdates its major movement. A Rb/Sr whole-rock isochron age of 440 ± 8 Ma has been obtained for the Ellisville (Pavlides and others, 1982). From this it must be inferred that foliation development and major movement in the Shores Complex occurred before latest Ordovician time.

Southwest of Shores, the Complex has been traced about 28 mi (45 km) into northwestern Appomattox County where it and the magnetic lineament appear to terminate. It appears likely that the complex has the form of a large wedge or slice bordered by faults and overridden and cut off to the south by the eastern fault or some other thrust, possibly the post-metamorphic fault which has changed the Arvonia Slate into a button schist from southern Buckingham County southwestward (Fig. 2).

SITE DESCRIPTION

Rocks typical of much of the Shores Complex, including metagraywacke, greenstone metabasalt, ultramafic, trondhjemite, melange, and felsic Chopawamsic metavolcanics which have been thrust upon the complex from the east, are exposed in the river bluffs on the north side of the Chesapeake and Ohio gravel access road extending for 1 mi (1.6 km) between Shores Station and Bremo Creek to the southeast. The river bluffs are largely forest or brush covered, but most lithologies are readily seen along the access road. Units of interest and significance are designated by number on Figure 4 and the distance of each southeastward from Shores is indicated below.

Stop 1. 0.00–0.17 mi (0.00–0.27 km). Greenstone metabasalt with minor metagraywacke and quartz-mica schist; four small Mesozoic diabase dikes. Crops are mainly in the woods, but the greenstone is exposed along Virginia 640 just northeast of Shores. The rock is dark green, aphanitic or nearly so, and weakly to strongly foliated. Locally it is amygdaloidal with fillings of epidote. Banded portions may have been mafic tuff. The microscope shows original textures to be largely or wholly destroyed. Chief minerals are actinolite, epidote, chlorite, albite, and quartz with some magnetite, biotite, and clinozoisite. Bland (1978) and Davis (1977) showed that the major and trace element chemistry of these and other Shores greenstones places them in the ocean-floor basalt (OFB) classification of Pearce and Cann (1973) and Pearce (1975, 1976).

Stop 2. 0.17–0.29 mi (0.27–0.47 km). Metagraywacke with minor greenstone. The metagraywacke in the Shores Com-

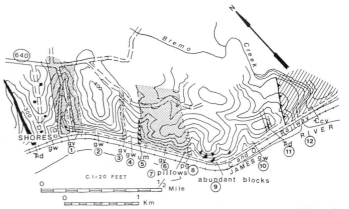

Figure 4. Map showing the rock units and Stops 1 through 12 at the Shores geologic site. ꝑd, Mesozoic diabase dike; gw, metagraywacke; gv, greenstone metabasalt; um, ultramafic; pg, plagiogranite (trondhjemite); Ccv, Chopawamsic volcanics.

plex varies appreciably from place to place, but it is generally medium gray to greenish gray weathering tan, fine-grained, and moderately to strongly foliated. Spaced cleavage is common and crenulation foliation is locally present. Much of the rock has been intricately deformed and consists of blebs, lumps, hooks, and rootless isoclines of quartz and quartz-feldspar segregations in a jumbled matrix of fine mica, chlorite, feldspar, and quartz. Bedding segments of brown-weathering calcareous sandstone occur locally; in places as sheared-off isoclines.

Stop 3. 0.29–0.31 mi (0.47–0.50 km). Greenstone, some yellow-green epidosite, thin laminated in part, possible flattened pillows.

Stop 4. 0.31–0.35 mi (0.50–0.56 km). Metagraywacke and schist like that at Stop 2.

Stop 5. 0.35 mi (0.56 km). Chlorite-actinolite ultramafic, 6.6 feet (2 m) thick block or lens. Mostly actinolite, medium-grained, anhedral to subhedral, randomly oriented with some tremolite and interstitial chlorite.

Stop 6. 0.35–0.52 mi (0.56–0.84 km). Greenstone similar to that at Stops 1 and 3. Tight refolded folds in thin laminae.

Stop 7. 0.50 mi (0.80 km). Pillows in greenstone (Fig. 5).

Stop 8. 0.52–0.54 mi (0.84–0.87 km). Plagiogranite (trondhjemite), white to medium gray, medium-grained, composed chiefly of quartz, albite, and minor epidote, muscovite, and tourmaline; chlorite and iron oxide occur in darker varieties. Albite is considerably altered with tiny muscovite inclusions. Granulated in part. Small angular inclusions of greenstone and ultramafic.

0.54–0.60 mi (0.87–0.96 km). Cover.

Stop 9. 0.60–0.77 mi (0.96–1.24 km). Mélange consisting of abundant angular blocks of greenstone metabasalt, some metagraywacke and ultramafic, in intricately deformed matrix of quartz-chlorite-mica schist or gneiss with blebs, hooks, lenses, and tiny isoclines of quartz-feldspar segregations (Fig. 6). Blocks range from a few inches (cm) to 49 feet (15 m) long.

0.77–0.80 mi (1.24–1.29 km). Cover.

Stop 10. 0.80–0.97 mi (1.29–1.56 km). Metagraywacke, contorted, some blocks and refolded rootless isoclinal folds.

Stop 11. 0.90 mi (1.45 km). Three vertical Mesozoic diabase dikes in metagraywacke. The westernmost, about 20 inches (50 cm) thick, has good horizontal columns. The easternmost and largest dike is about 33 feet (10 m) thick.

Stop 12. 0.97–1.0 mi (1.56–1.61 km). Chopawamsic felsic volcanic, lustrous light greenish gray, speckled with iron oxide, strongly sheared, finely porphyritic. Phenocrysts a millimeter or more long are albite with tiny inclusions of sericite and epidote minerals. Matrix is chiefly muscovite, quartz, albite, and epidote. Beta-form quartz phenocrysts characteristic of many felsic Chopawamsic volcanics are absent here. Bland (1978, p. 102–114), on the basis of trace element concentrations in Chopawamsic samples from the Quantico area of northern Virginia, and using the discrimination diagrams of Pearce and Cann (1973), concluded that the Chopawamsic volcanics represent low-K tholeiites of an island arc affinity. This conclusion, and the classification of the Shores greenstones as ocean-floor basalt

Figure 5. Pillow structures in greenstone metabasalt 0.5 mi (0.8 km) southeast of Shores.

(OFB), along with field relations and vibroseis reflection profile, have provided evidence for a Chopawamsic volcanic arc thrust against a Shores-Hardware-Evington rear arc basin.

Figure 6. Shores Mélange between 0.6 and 0.7 mi (1.0 and 1.1 km) southeast of Shores, Virginia. *A,* Block of laminated and weakly foliated epidote-calcite-quartz-chlorite rock (metabasalt ?) in highly contorted metagraywacke. *B,* Small metagraywacke block in intricately deformed gneissic matrix. *C,* Small fragments of greenstone (dark massive rock) in quartzose metagraywacke. *D,* Typical intricately-deformed mélange matrix.

REFERENCES CITED

Bland, A. E., 1978, Trace element geochemistry of volcanic sequences of Maryland, Virginia, and North Carolina and its bearing on the tectonic evolution of the central Appalachians [Ph.D. thesis]: University of Kentucky, Lexington, 328 p.

Bland, A. E., and Blackburn, W. H., 1980, Geochemical studies on the greenstones of the Atlantic Seaboard Volcanic Province, south-central Appalachians, in Wones, D. R., ed., Proceedings, Caledonides in the USA, International Geological Correlation Project 27, Caledonides Orogen: Virginia Polytechnic Institute and State University, Blacksburg, Virginia, Memoir 2, p. 263–270.

Brown, W. R., 1969, Geology of the Dillwyn quadrangle, Virginia: Virginia Division of Mineral Resources Report of Investigation 10, 77 p.

——1976, Tectonic melange (?) in the Arvonia Slate District of Virginia: (Abstract), Geological Society of America Abstracts with Programs, v. 8, no. 2, p. 142.

Brown, W. R., and Pavlides, Louis, 1981, Melange terrane in the central Virginia Piedmont: Geological Society of America Abstracts with Programs, v. 13, no. 7, p. 419.

Davis, P. A., 1977, Trace element model studies of Late Precambrian–Early Paleozoic greenstones of Virginia, Maryland, and Pennsylvania [Ph.D. thesis]: University of Kentucky, Lexington, 135 p.

Drake, A. A., Jr., and Morgan, B. A., 1981, The Piney Branch Complex—a metamorphosed fragment of the Central Appalachian Ophiolite in northern Virginia: American Journal of Science, v. 281, p. 484–508.

Ern, E. H., 1968, Geology of the Buckingham quadrangle, Virginia: Virginia Division of Mineral Resources Report of Investigations 15, 45 p.

Evans, Nicholas H., 1984, Latest Precambrian to Ordovician metamorphism and orogenesis in the Blue Ridge and western Piedmont, Virginia Appalachians [Ph. D. dissertation]: Virginia Polytechnic Institute and State University, Blacksburg, 313 p.

Farrar, Stewart S., 1984, The Goochland granulite terrane: remobilized Grenville basement in the eastern Virginia Piedmont: Geological Society of America Special Paper 194, p. 215–227.

Geological Map of Maryland, 1968, Maryland Geological Survey, Scale 1:250,000.

Glover, Lynn, III, and Costain, J. K., 1982, Vibroseis reflection seismic structure along the Blue Ridge and Piedmont James River profile, north central Virginia: (abstract), Geological Society of America Abstracts with Programs, v. 14, no. 7, p. 497.

——Costain, J. K., Coruth, C., and Farrar, S. S., 1983, The Eocambrian-Early Ordovician margin of North America and the Taconic suture in Virginia (abs.): Geological Society of America Abstracts with Programs, v. 15, no. 6, p. 582.

Graham, George M., 1975, An integrated geological and geophysical study of the Diana Mills pluton, Virginia [M.S. thesis]: University of Kentucky, Lexington, 49 p.

Hopkins, H. R., 1960, Geology of western Louisa County, Virginia: [Ph.D. dissertation], Cornell University, Ithaca, New York, 98 p.

Mose, D. G., and Nagel, M. S., 1982, Plutonic events in the Piedmont of Virginia: Southeastern Geology, v. 23, p. 25–39.

Pavlides, Louis, 1981, The central Virginia volcanic-plutonic belt: an island arc of Cambrian(?) age: U.S. Geological Survey Professional Paper 1231-A, A1–A34 p.

Pavlides, Louis, Arth, J. G., Daniels, D. L., and Stern, T. W., 1982, Island-arc, back-arc, and melange terranes of northern Virginia: tectonic, temporal, and regional relationships: Geological Society of America Abstracts with Programs, v. 14, no. 7, p. 584.

Pearce, J. A., 1975, Basalt geochemistry used to investigate past environments on Cyprus: Tectonophysics, v. 25, p. 41–67.

——1976, Statistical analysis of major element patterns in basalts: Journal of Petrology, v. 17, pt. 1, p. 15–43.

Pearce, J. A., and Cann, J. R., 1973, Tectonic setting of basic volcanic rocks determined using trace element analyses: Earth and Planetary Science Letters, v. 19, p. 290–300.

Smith, J. W., Milici, R. C., and Greenberg, S. S., 1964, Geology and mineral resources of Fluvanna County, Virginia: Virginia Division of Mineral Resources Bulletin 79, 62 p.

Wehr, Frederick, and Glover, Lynn, III, 1985, Stratigraphy and tectonics of the Virginia–North Carolina Blue Ridge: Evolution of a late Proterozoic–early Paleozoic hinge zone: Geological Society of America Bulletin, v. 96, p. 285–295.

Zietz, Isidore, Calver, J. L., Johnson, S. S., and Kirby, J. R., 1977, Aeromagnetic Map of Virginia: U.S. Geological Survey, Geophysical Investigation Map GP-915, scale 1:500,000, and Map GP-916, scale 1:1,000,000, in color.

Zietz, Isidore, Haworth, R. T., Williams, Harold, and Daniels, D. L., 1980, Magnetic anomaly map of the Appalachian orogen: Memorial University of Newfoundland Map no. 2a, scale 1:2,000,000.

Ophiolites(?) in the Lynchburg Group near Rocky Mount, Virigina

James F. Conley, Virginia Division of Mineral Resources, P.O. Box 3667, Charlottesville, Virginia 22903

LOCATION

The city of Rocky Mount, Virginia is located in Franklin County in the southwestern Piedmont of Virginia. It is situated on U.S. 220 about halfway between the cities of Roanoke and Martinsville (Fig. 1).

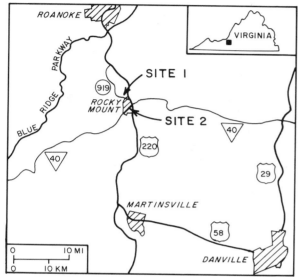

Figure 1. General location map of southwest Virginia showing the location of *Site 1* and *Site 2,* north and south of Rocky Mount.

INTRODUCTION

Metamorphosed ultramafic, gabbroic, and basaltic rocks interlayered with metasedimentary rocks crop out on a regional scale on the southwestern limb of the Blue Ridge anticlinorium. In the Rocky Mount area these metaigneous and metasedimentary rocks are at upper greenschist facies and metamorphic grade increases along strike to the southwest. The ultramafic and mafic rocks are found within a sequence of metasedimentary rocks that was originally named the Lynchburg Formation by Jonas (1927) and later elevated to group status by Furcron (1969). The lower unit of the Lynchburg Group in the Rocky Mount area is the Ashe Formation (Rankin, 1970), which is overlain by the Alligator Back Formation (Rankin and others, 1973).

Mafic and ultramafic metaigneous rocks are more commonly found in the Alligator Back than in the Ashe Formation. The Ashe Formation in the Rocky Mount area is primarily a layered biotite-plagioclase gneiss, whereas the Alligator Back contains metamorphosed ultramafic rocks, gabbros, and basalts and a metasedimentary sequence composed of interlayered metagraywackes, metagraywacke conglomerates, graphitic schists, pelitic schists, impure marbles, and quartzites. Some sequences of metaigneous rocks consist of ultramafic rocks overlain by meta-

gabbros that, in turn, are overlain by metabasalts; some consist of metagabbros overlain by metabasalts, whereas others contain only metabasalts. The metabasalts at the top of these sequences are overlain by metasedimentary rocks and may contain metasedimentary layers between flows. These sequences of metaigneous and metasedimentary rocks are found at several horizons in the Alligator Back and appear to represent ophiolites and their sedimentary cover rocks that have been repeated by thrusting (Conley, 1981). The variations in the stratigraphy of the metaigneous rocks suggest that some of the ophiolite sequences are fairly complete, whereas others are structurally truncated successions.

The ultramafic rocks are metamorphosed dunites and metapyroxenites. The metagabbros are generally coarser grained towards their centers and some are layered. Metabasalts are altered to foliated, fine-grained amphibolites. Evidence of baked zones and zones of thermal alteration are not observed along contacts between the metaigneous rocks and metasedimentary rocks. These contact zones would be expected to be preserved as relicts by later greenschist grade metamorphism if the metaigneous rocks were intrusive into the metasedimentary rocks.

SITE DESCRIPTIONS

Site 1, Grassy Hill. Exposed along Virginia 919 at the crest and along the southeastern flank of Grassy Hill just north of Rocky Mount (Fig. 2) are altered ultramafic rocks that comprise an elongate, northeast-trending stratiform rock mass that is approximately 3.7 mi (6 km) long and about 0.6 mi (1 km) in breadth. This rock mass is one of the three largest known ultramafic bodies in Virginia. The site displays the ultramafic core of this body, which to the southeast is overlain by metagabbro and metabasalts. This particular ultramafic body was mapped in detail and in reconnaissance by Michael B. McCollum (map in preparation). At the crest of Grassy Hill are exposures of altered metaperidotite and talc-tremolite schist. The rock may show faint compositional layering and is dark green. It is schistose to granular and is composed of talc, tremolite, chlorite, opaque minerals, and may contain relict olivine that is partially replaced by iddingsite, antigorite, and chlorite. Locally the body contains thin veins of fibrous anthophyllite.

Near the crest of Grassy Hill the ultramafic rocks are crisscrossed by numerous thin (3 ft. [1 m] thick) metapyroxenite dikes. These dikes are composed of tremolite, talc, chlorite, and opaque minerals. The rock contains numerous large (0.8–1.2 in [2–3 cm]) relict pyroxene phenocrysts that are partially to totally altered to uralite. McCollum has mapped a gabbro body overlying the ultramafic rocks on the southwestern end of this body and metabasalts are locally present above metagabbros on the southeast edge of the body, which may be observed on Virginia 919 on the southeastern flank of Grassy Hill at the bottom of the hill, 0.75 mi (1.2 km) east of Site 1.

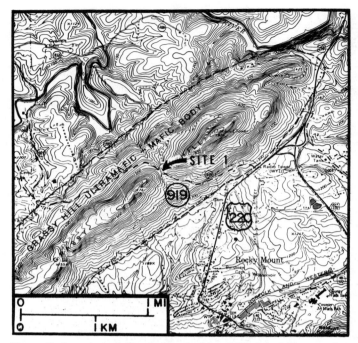

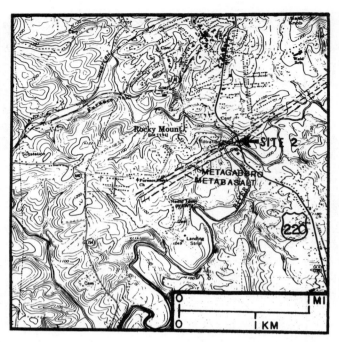

Figure 2. Location of Site 1: North side of Virginia 919, opposite parking area of garbage collection site at the crest of Grassy Hill, 0.8 mi (1.3 km) west of the intersection of Virginia 919 and U.S. 220 Business (North Main Street) at the northern city limits of Rocky Mount, Franklin County, Virginia, Boones Mill 7½-minute Quadrangle.

Figure 3. Location of Site 2: Road cut on east side of U.S. 220 Business (South Main Street) at the southern city limits of Rocky Mount, opposite the intersection of U.S. 220 and Virginia 640 (Scuffling Hill Road) southward along U.S. 220 Business to end of exposure on the north bank of the Pigg River.

Site 2, Southern City Limits of Rocky Mount. At this site, rock is exposed in a road cut on the east side of U.S. 220 Business from just north of the intersection of Virginia 640 (Scuffling Hill Road) southward to the north bank of the Pigg River (Fig. 3). Gabbro is exposed in the northwestern part of the road cut and to the southeast it is overlain by a steeply southeastward dipping sequence of metabasalt flows. The exposed contact between the metagabbro and overlying metabasalts is parallel to compositional layering in the metabasalts. This would suggest that the metabasalts poured out over the metagabbro rather than the metagabbro intruding the metabasalts. A white mica schist interlayer (metapelite bed) lies about 33 ft (10 m) above the base of the flows indicating that sedimentation was occurring during the time of vulcanism. What appears to be distorted relict flow banding and elongated masses that could be

poorly preserved pillow structures can be observed in the metabasalt flows.

The metagabbro is a mottled dark-green and white, coarse-grained, granoblastic rock composed of actinolite, albite, epidote, and chlorite with accessory amounts of quartz, zircon, and opaque minerals. Pyroxene (augite) may occur as relict phenocrysts that have been partially replaced by actinolite and chlorite.

Metabasalt flows that overlie the metagabbro are dark greenish-black or mottled black and white, fine- to medium-grained, schistose rocks. The metabasalts are composed predominantly of actinolite nematoblasts, oriented parallel to the plane of schistosity. Most actinolite crystals are randomly oriented on this plane whereas some may show a preferred lineation on the plane. Variable amounts of albite, chlorite, epidote, and quartz may be present in the rock.

REFERENCES CITED

Conley, J. F., 1981, Stratigraphic relationships between rocks of the Blue Ridge anticlinorium and the Smith River allochthon in the southwestern Virginia Piedmont, *in* Conley, J. F., Marr, J. D., Jr., and Berquist, C. R., Jr., eds., Thirteenth Annual Virginia Geologic Field Conference: Virginia Division of Mineral Resources, 34 p.

Furcron, A. S., 1969, Late Precambrian and early Paleozoic erosion-depositional sequences of northern and central Virginia: Georgia Geological Survey Bulletin 80, p. 57–88.

Jonas, A. I., 1927, Geologic reconnaissance in the Piedmont of Virginia: Geologi-

cal Society of America Bulletin, v. 38, p. 837–846.

Rankin, D. W., 1970, Stratigraphy and structure of Precambrian rocks in northwestern North Carolina, *in* Fisher, G. W., Pettijohn, F. J., and Reed, J. C., eds., Studies of Appalachian geology—central and southern: New York, Interscience Publishers, p. 227–245.

Rankin, D. W., Espenshade, G. H., and Shaw, K. W., 1973, Stratigraphy and structure of the metamorphic belt in northwestern Virginia—A study from the Blue Ridge across the Brevard zone to the Sauratown Mountains anticlinorium: American Journal of Science, v. 273, p. 1–40.

The Roanoke Rapids complex of the Eastern slate belt, Halifax and Northampton Counties, North Carolina

J. Wright Horton, Jr., 928 National Center, U.S. Geological Survey, Reston, Virginia 22092
Edward F. Stoddard, Department of Marine, Earth, and Atmospheric Sciences, North Carolina State University, Raleigh, North Carolina 27695

LOCATION

The site is located at Roanoke Rapids, North Carolina, along the tailrace below Roanoke Rapids dam on the Roanoke River, in the northeast quadrant of the Roanoke Rapids 7½-minute quadrangle (Fig. 1), 4.3 mi (7 km) south of the Virginia state line. This hydroelectric dam and the tailrace are owned by North Carolina Power. Most geologic features exposed along the tailrace can be observed by traversing the south side, which is open to the public "at your own risk." The north side of the tailrace and the adjacent riverbed north of it show similar features and can be reached, with permission, by walking across the lower part of the dam. Permission can be requested from officials at the North Carolina Power station at the north end of Oakwood Avenue (Fig. 1).

INTRODUCTION

The Eastern slate belt of North Carolina and southern Virginia is the easternmost terrane in the exposed southern Appalachian Piedmont. It consists of greenschist-facies to lower amphibolite-facies metavolcanic and metasedimentary rocks similar to those of the Carolina slate belt farther west. A preliminary Cambrian rubidium-strontium whole-rock age from felsic metavolcanic rocks near Princeton, N.C. (P. D. Fullagar, oral communication, 1984) indicates that some Eastern slate belt rocks are also similar in age to those of the Carolina slate belt. The stratigraphic relationships between the two slate belts, if any, are unknown. They are separated by the amphibolite-facies Raleigh belt, which includes high-grade equivalents of Eastern slate belt rocks as well as rocks believed by Farrar (1985a) to be a Grenville sequence.

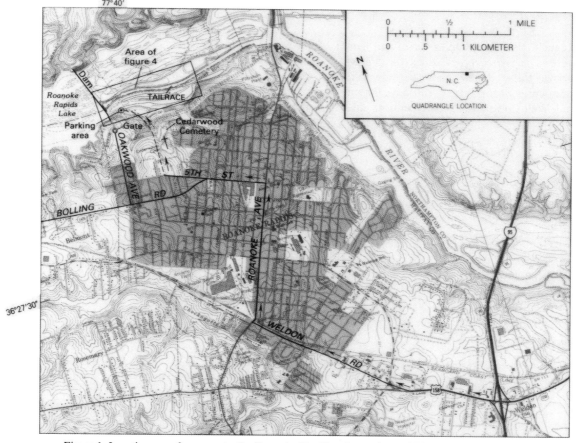

Figure 1. Location map for access to the Roanoke Rapids tailrace. Area of Figure 4 is outlined.

Fossils indicate that the Carolina slate belt is an exotic terrane that in Cambrian time was located away from North America (Secor and others, 1983). The Eastern slate belt, which lies farther east, is therefore another candidate for an accreted terrane. Recent tectonic models suggest that the two slate belts originated in a volcanic-arc setting associated with subduction (Rankin, 1975; Kish and Black, 1982; Farrar and others, 1983; Glover and others, 1983; Kite and Stoddard, 1984). These models vary considerably, however, postulating both east- and west-dipping subduction zones, possible multiple or compound arcs, and crustal environments ranging from oceanic to continental (Kish and Black, 1982). If the Halifax County mafic-ultramafic complex is an ophiolite as proposed by Kite and Stoddard (1984), it is likely that the Eastern slate belt originated at least partly in an oceanic setting. The geology of the Eastern slate belt has not been mapped in detail, and geologic reconnaissance maps by Parker (1968), Farrar (1985a), and Wilson (1981) differ considerably. Good bedrock exposures in the Eastern slate belt are sparse, even by Piedmont standards, but the tailrace below Roanoke Rapids dam is a spectacular exception. It reveals 1 mi (1.6 km) of nearly continuous fresh bedrock and is aligned roughly perpendicular to the regional north strike of cleavage and bedding. A variety of metavolcanic, metasedimentary, and intrusive rocks, spectacular folds, and other features of the Eastern slate belt are clearly exposed.

SITE DESCRIPTION

The Roanoke Rapids complex (informal name) of the Eastern slate belt is well exposed along the tailrace (Fig. 2) below Roanoke Rapids dam. To reach the tailrace from I-95, exit west onto U.S. 158 (Fig. 1) and drive 0.8 mi (1.3 km). Bear right on Weldon Road (sign to "business district") and

Figure 2. Tailrace below Roanoke Rapids dam, looking eastward from dam. Light-colored rock in foreground on both sides of tailrace is pink leucogranite. Darker rocks beyond the leucogranite are metamorphosed volcanic and volcanogenic sedimentary rocks.

proceed 1.3 mi (2.1 km). Turn right on Roanoke Avenue (North Carolina 48) and continue 1.2 mi (1.9 km) through downtown Roanoke Rapids. Turn left onto Fifth Street at a traffic light and continue 0.8 mi (1.3 km), avoiding Bolling Road, which forks to the left. Turn right onto a gravel road at the far end of Cedarwood Cemetery and proceed 0.6 mi (1 km), beyond the cemetery and beneath the powerlines, to a public parking area just outside the fence around the North Carolina Power station (Fig. 1). A trail leads northward 500 ft (150 m) to outcrops along the tailrace.

The Roanoke Rapids complex, informally named by Farrar (1985b), is the largest volcanic-plutonic complex in the Eastern slate belt (Farrar, 1985c). It consists of greenschist-facies volcanic

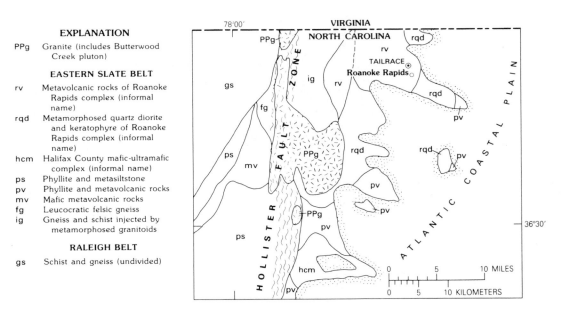

EXPLANATION

PPg   Granite (includes Butterwood
       Creek pluton)

EASTERN SLATE BELT

rv   Metavolcanic rocks of Roanoke
       Rapids complex (informal
       name)
rqd   Metamorphosed quartz diorite
       and keratophyre of Roanoke
       Rapids complex (informal
       name)
hcm   Halifax County mafic-ultramafic
       complex (informal name)
ps   Phyllite and metasiltstone
pv   Phyllite and metavolcanic rocks
mv   Mafic metavolcanic rocks
fg   Leucocratic felsic gneiss
ig   Gneiss and schist injected by
       metamorphosed granitoids

RALEIGH BELT

gs   Schist and gneiss (undivided)

Figure 3. Regional geologic map modified from Farrar (1985a).

and hypabyssal intrusive rocks. The main intrusive body, which ranges in composition from quartz diorite to trondhjemite, is south and east of Roanoke Rapids (Fig. 3) and may represent a center of volcanism (Farrar, 1985b). This intrusion is bordered on the north by a sequence of felsic metavolcanic rocks interlayered with volcanogenic metasedimentary rocks. These rocks constitute most of the section exposed along the tailrace below the Roanoke Rapids dam.

The tailrace (Fig. 4) exposes a metamorphosed sequence of volcaniclastic rocks and interlayered volcanic sandstones, conglomerates, and mudstones. Greenstone occurs locally as dikes (Fig. 5), sills, and possibly flows. Other metamorphosed igneous rocks include quartz diorite, quartz keratophyre, and leucogranite.

The metavolcanic rocks exposed along the tailrace are chiefly crystal and crystal-lithic metatuffs and lapilli metatuffs. They consist of plagioclase and quartz phenocrysts and lithic clasts in a matrix of sodic plagioclase, quartz, and chlorite. Common accessory minerals include epidote, calcite, white mica, and opaque minerals. The lapilli are felsic metaigneous rock, have sharp contacts with the groundmass, and may be as much as several inches (centimeters) in diameter. The crystal and lithic fragments typically have rounded corners, indicating that the original pyroclastic material has been reworked by sedimentary processes. Greenstone layers as much as a few ft (meters) thick, which are concordant and interstratified with the metamorphosed pyroclastic and sedimentary rocks, presumably originated as flows and/or sills. Similar greenstone also occurs as dikes (discussed below).

In some places, it is difficult to distinguish between the metamorphosed pyroclastic rocks, which commonly appear reworked, and the interlayered metasedimentary rocks derived from them. The metasedimentary rocks are chiefly volcanic metasandstones that grade into lesser amounts of volcanic metaconglomerate and metamudstone. The volcanic metasandstones include lithic (volcanic) metaarkose and minor feldspathic and lithic metagraywacke (>10–15% chlorite matrix). These volcanic metasandstones are composed of quartz (as much as 60%), plagioclase, white mica, calcite (locally as much as 20%), and accessory epidote and opaque minerals. Blue quartz is widespread. Lithic clasts, which may be as much as several inches (centimeters) in diameter, are subangular to subrounded and dominantly felsic. Clasts of pink metagranitoid, gray phyllitic metamudstone, and white quartzite have also been observed. Parker (1964, and written communication, 1985) observed graded bedding and scour-and-fill structure in these rocks 0.6 mi (1 km) east of the area shown in Figure 4 near the North Carolina 48 bridge (see Fig. 1). Subarkosic quartz-pebble metaconglomerate grades into matrix-supported orthoquartzitic metaconglomerate at Locations 11 and 12 (Fig. 4). Laminated metamudstone is composed of fine-grained plagioclase, white mica, chlorite, quartz, and calcite. Chert pebbles occur in layers of pebbly metamudstone at Location 4 (Fig. 4).

Quartz keratophyre is interlayered at some places with met-avolcanic and metasedimentary rocks (e.g. Location 2, Fig. 4). It is composed chiefly of plagioclase, quartz, and lesser amounts of chlorite and white mica. Accessory minerals include calcite, epidote, opaque minerals, and, locally, titanite. Oligoclase grains reach a maximum length of 3 mm, but most are no longer than 1 mm. Whether the quartz keratophyre originated as hypabyssal sills, extrusive flows, or both, is unclear. Epidote quartz diorite at Location 14 (Fig. 4) contains angular greenstone xenoliths and is clearly intrusive.

Greenstone dikes, which are generally fine grained, commonly porphyritic, and locally amygdaloidal (Location 5, Fig. 4), range from 0.4 in (1 cm) to several meters in thickness and cut across the older layered rocks. The greenstone is composed chiefly of plagioclase (now albite-oligoclase), chlorite, and epidote. Minor accessory minerals commonly include white mica, titanite, opaque minerals, quartz, and calcite. Actinolite pseudomorphs after hornblende were observed in one thin section. Local relict phenocrysts of plagioclase are strongly saussuritized, and their original compositions are uncertain. Flowage zoning is indicated by phenocrysts concentrated near the centers of some dikes (Location 6, Fig. 4). The foliation defined by fine-grained chlorite is more obvious in thin sections than in the field. Some of these dikes appear to cut across vertical folds (discussed below) in the country rock but locally have a weak cleavage that may be axial planar to such folds (Fig. 4).

A body of nonfoliated but highly altered pink leucogranite crops out near the western end of the tailrace (Location 1, Fig. 4) and in the riverbed to the north. The leucogranite is composed mainly of pink microcline, plagioclase, and quartz. Secondary accessory minerals include epidote, calcite, and sericitic white mica. Opaque minerals occur only in trace amounts. The leucogranite is pervasively fractured and locally brecciated on a microscopic scale, and the fractures are healed by a fine-grained mixture of quartz, plagioclase, epidote, and sericite. The apparent absence of cross-cutting greenstone dikes suggests, if exposures are adequate, that the leucogranite may be the youngest rock unit exposed here.

Rocks along the tailrace have been metamorphosed under lower greenschist-facies (chlorite zone) conditions and observed textures can be explained by a single progressive metamorphic event. Calcite and ankerite occur locally as fracture-filling material. Limits on the age of regional metamorphism in this part of the Eastern slate belt are generally lacking, but this metamorphism should predate the 286–293 Ma rubidium-strontium cooling ages of biotite from the eastern, undeformed side of the Butterwood Creek pluton (Russell and others, 1985) (see Fig. 3).

The Roanoke Rapids complex is within the Roanoke Rapids structural block of Farrar (1985a), which includes all rocks of the Eastern slate belt east of the Hollister mylonite zone (see Fig. 3). The sequence of regional deformation events proposed by Farrar (1985a) is based almost entirely on relationships west of the Hollister zone, and we have been unable to apply it in the Roanoke Rapids area.

The dominant and apparently earliest cleavage or schistosity

EXPLANATION

LOCATION OF FEATURE MENTIONED IN TEXT

1   Leucogranite
2   Quartz keratophyre, folds
3   2 cleavages (photo, figure 6), folds
4   Pebbly metamudstone
5   Greenstone dikes
6   Greenstone dikes
7   Roanoke Rapids folds
8   Ductile deformation zones, 2 fold sets, greenstone dike
9   Greenstone dike, 2 fold sets (photos, figures 5, 7)
10 Ductile shear zone (S_c?)
11 Conglomerate, Roanoke Rapids folds
12 Conglomerate, Roanoke Rapids folds
13 Folds
14 Quartz diorite and xenoliths
15 Sigmoidal boudinage along S_c

x·x·x ·········· Contact of pink leucogranite, dotted in water

PLANAR FEATURES*

Strike and dip of dominant cleavage, S_s (commonly parallel to layering, S_O)

88 △   Inclined

◇   Vertical

Strike and dip of slip cleavage or crenulation cleavage, S_c

73 ▲   Inclined

◆   Vertical

Strike and dip of axial surface of fold

80 ■   Inclined

■   Vertical

LINEAR FEATURES*

Axis of minor fold (R-Roanoke Rapids set; other sets unassigned)

z ➤ 87   Dextral ("Z" fold)

s ➤ 55   Sinistral ("S" fold)

➤ 85   Symmetrical ("M" fold)

*Where two symbols are combined, their intersection marks point of observation

```
0    200   400   600   800   1000 FEET
|----|----|----|----|----|
0        100       200      300 METERS
CONTOUR INTERVAL 2 FEET
```

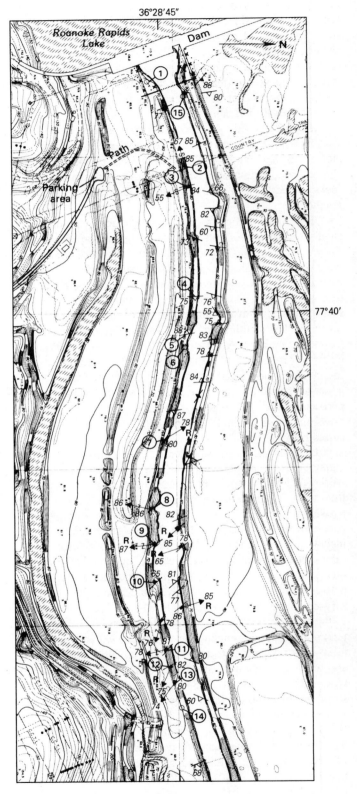

Figure 4. Detailed map of tailrace below Roanoke Rapids dam, showing geologic features and locations mentioned in text. Topographic base is reproduced with permission from City of Roanoke Rapids, Engineering Department.

Figure 6. Plan view showing relationship of S_s and S_c. Curvature of S_s indicates sinistral shear along the subvertical shear band cleavage, S_c, which strikes N.39°W. S_s strikes N.16°W. and dips 60°NE. at a point between the shear bands. Location 3, Figure 4. Top of photo is southwest; nickel coin indicates scale.

Figure 5. Plan view of greenstone dike (under nickel coin), cutting vertical folds in metavolcanic rocks that have subvertical hinge surfaces striking about N.10°W. Weak cleavage in the greenstone, however, is approximately axial planar to adjacent folds. The greenstone dike is therefore deformed, but less intensely than is the metavolcanic rock. Location 9, Figure 4. Top of photo is west.

exposed along the tailrace, here called S_s, strikes N.23°E. to N.32°W., has an average strike of about N.10°W., and dips almost everywhere more than 60°, both east and west. The primary bedding or layering, S_O, appears essentially parallel to this foliation at most locations along the tailrace, but local deviations have been observed. S_s commonly occurs both as a continuous cleavage, characterized by oriented mineral grains that are evenly distributed, and as a spaced cleavage that has planar to anastamosing micaceous seams a few millimeters to several centimeters apart. In a few places (Location 8, Fig. 4), strain is concentrated in ductile deformation zones as much as 66 ft (20 m) thick, which are parallel to this cleavage. S_s in these zones becomes phyllitic and more closely spaced and grades into almost a mylonitic fabric, but it has no obvious overprint texture or change in orientation. Fine-grained disseminated pyrite is abundant in layers parallel to the cleavage in these zones, causing a characteristic rusty weathering. The greenstone dikes, oriented at an angle to S_s in these zones, are not notably deformed or offset, and S_s is poorly defined in the greenstone (Location 8, Fig. 4).

Locally, a second subvertical cleavage strikes N.35°–60°W. This cleavage, here called S_c, is found primarily in areas where deformation associated with S_s has been strong and particularly in the ductile deformation zones parallel to S_s. Both cleavages are thought to have been produced by a single continuous deformation. We interpret S_c as a shear-band cleavage (White and others, 1980; Gapais and White, 1982; = C-surface of Berthé and others,

1979; Simpson and Schmid, 1983). Measured dihedral angles between S_c and S_s range from 21° to 44° but cluster at about 36°. S_s surfaces curve into S_c surfaces, and the sense of rotation is consistently that of sinistral shear (Fig. 6). S_c is most common as a spaced strain-slip cleavage, which locally grades into a crenulation cleavage associated with subvertical crenulation axes. These cleavages also appear to be axial planar to larger mesoscopic folds. A 2.7-in (7-cm) thick ductile shear zone at Location 10 (Fig. 4), which strikes N.46°W. and dips 65°SW., contains a parallel mylonitic foliation, is bounded by sinistral drag folds, and may be a local manifestation of S_c. Sigmoidal boudinage of competent layers has been produced in a few places by sinistral shear along S_c (Location 15, Fig. 4).

Several sets of vertical to steeply inclined folds are well exposed along the tailrace, but none fit readily into the series of regional fold generations proposed by Farrar (1985a). All of these folds deform the dominant cleavage, S_s, and layering, S_O, as well as weak subhorizontal crenulation lineations on these surfaces.

The most prominent and widespread system of mesoscopic folds consists of open-to-close, commonly angular, vertical folds, including chevron folds (Fig. 7) and crenulations, which have subvertical axial surfaces (dipping 80° to 90° NE. or SW.) that strike N.32°–65°W. and subvertical axes (plunging 80°-90°). These folds, here called the Roanoke Rapids folds, are well exposed at many locations along the tailrace (Locations 7, 8, and 11, Fig. 4). A strain-slip cleavage or crenulation cleavage, similar in form and orientation to the shear-band cleavage, S_c, is axial planar to these folds. This cleavage is found only locally in the hinge zones of these folds or in areas where such folds are abun-

Figure 7. Plan view of dextral, close, vertical chevron folds that typify the Roanoke Rapids folds. Subvertical hinge surfaces strike about N.35°W. Location 9, Figure 4. Top of photo is west; nickel coin indicates scale.

dant. The Roanoke Rapids folds are commonly symmetrical, but where they are asymmetrical, they typically have a dextral (clockwise or Z) pattern when viewed down the plunge. This asymmetry is therefore opposite to the asymmetry associated with

the shear-band cleavage, S_c, which otherwise appears to be axial planar to the folds. This difference may be related to the fact that ". . . the normal asymmetry of a group of folds is opposite to the sense of rotation displayed by the limb of the lower order fold on which the group is located" (Hansen, 1971, p. 24–25 and Figs. 11 and 12). Regional mapping of fold and shear-band asymmetries in the surrounding areas is needed in order to fully understand the relationships exposed along the tailrace.

Three minor sets of open-to-close, commonly angular (and locally chevron), vertical to steeply inclined folds, all of which have steep to subvertical axes, are less widespread at this locality. These sets of folds include (1) dextral to symmetrical folds characterized by axial surfaces that strike N.74°–78°W. (Locations 8 and 13, Fig. 4); (2) sinistral folds characterized by axial surfaces that strike N.4°–12°W. (Location 3, Fig. 4); and (3) sinistral folds characterized by axial surfaces that strike N.55°–65°E. (Location 2, Fig. 4). These minor fold sets have been observed only in small areas and may have only local significance. The steep to subvertical axes and axial surfaces, similar styles, and apparent absence of overprint relations among these fold sets and the Roanoke Rapids folds suggest that all of these folds are products of a single deformation event. The strike of the steep to subvertical cleavage and layering varies widely but systematically along the tailrace (Fig. 4); this variation suggests the presence of vertical folds, yet to be mapped, at a megascopic scale. The orientations of the fold sets and cleavages suggest the possibility of wrench faulting nearby—a hypothesis not yet tested by mapping of sufficient detail.

REFERENCES CITED

Berthé, D., Choukroune, P., and Jegouzo, P., 1979, Orthogneiss, mylonite and non-coaxial deformation of granites: the example of the South Armorican Shear Zone: Journal of Structural Geology, v. 1, no. 1, p. 31–42.

Farrar, S. S., 1985a, Tectonic evolution of the easternmost Piedmont, North Carolina: Geological Society of America Bulletin, v. 96, no. 3, p. 362–380.

—— 1985b, Stratigraphy of the northeastern North Carolina Piedmont: Southeastern Geology, v. 25, no. 3, p. 159–183.

—— 1985c, The southeastern Piedmont of Virginia: Progress in tectonic definition [abs.]: Geological Society of America Abstracts with Programs, v. 17, no. 2, p. 91.

Farrar, S. S., Glover, L., III, and Costain, J. K., 1983, An alternate tectonic model for the eastern Appalachian Piedmont suggested by a decollement between the Goochland terrane and the Carolina-Eastern slate belts [abs.]: Geological Society of America Abstracts with Programs, v. 15, no. 6, p. 570.

Gapais, D., and White, S. H., 1982, Ductile shear bands in a naturally deformed quartzite: Textures and Microstructures, v. 5, no. 1, p. 1–17.

Glover, L., III, Costain, J. K., Coruh, C., and Farrar, S. S., 1983, The Eocambrian–Early Ordovician margin of North America and the Taconic suture in Virginia [abs.]: Geological Society of America Abstracts with Programs, v. 15, no. 6, p. 582.

Hansen, E., 1971, Strain facies: New York, Springer-Verlag, 207 p.

Kish, S. A., and Black, W. W., 1982, The Carolna slate belt: Origin and evolution of an ancient volcanic arc; Introduction: Geological Society of America Special Paper 191, p. 93–98.

Kite, L. E., and Stoddard, E. F., 1984, The Halifax County complex: Oceanic lithosphere in the eastern North Carolina Piedmont: Geological Society of America Bulletin, v. 95, no. 4, p. 422–432.

Parker, J. M., III., 1964, The mile-long section of metamorphic rocks at Roanoke Rapids, North Carolina [abs.]: Elisha Mitchell Scientific Society Journal, v. 80, no. 2, p. 164.

—— 1968, Structure of easternmost North Carolina Piedmont: Southeastern Geology, v. 9, no. 3, p. 117–131.

Rankin, D. W., 1975, The continental margin of eastern North America in the southern Appalachians: The opening and closing of the proto-Atlantic Ocean: American Journal of Science, v. 275-A, p. 298–336.

Russell, G. S., Russell, C. W., and Farrar, S. S., 1985, Alleghanian deformation and metamorphism in the eastern North Carolina Piedmont: Geological Society of America Bulletin, v. 96, no. 3, p. 381–387.

Secor, D. T., Jr., Samson, S. L., Snoke, A. W., and Palmer, A. R., 1983, Confirmation of the Carolina slate belt as an exotic terrane: Science, v. 221, no. 4611, p. 649–651.

Simpson, C., and Schmid, S. M., 1983, An evaluation of criteria to deduce the sense of movement in sheared rocks: Geological Society of America Bulletin, v. 94, no. 11, p. 1281–1288.

White, S. H., Burrows, S. E., Carreras, J., Shaw, N. D., and Humphreys, F. J., 1980, On mylonites in ductile shear zones: Journal of Structural Geology, v. 2, no. 1-2, p. 175–188.

Wilson, W. F., 1981, Geology of Halifax County, N.C.: North Carolina Geological Survey Section, Open-File Map, scale 1:125,000.

The western edge of the Raleigh belt near Adam Mountain, Wake County, North Carolina

Edward F. Stoddard, *Department of MEAS, North Carolina State University, Raleigh, North Carolina 27695*
J. Wright Horton, Jr., *U.S. Geological Survey, 928 National Center, Reston, Virginia 22092*
Albert S. Wylie, Jr., *Chevron USA, P.O. Box 1150, Midland, Texas 79702*
David E. Blake, *Department of MEAS, North Carolina State University, Raleigh, North Carolina 27695*

LOCATION

This site consists of two stops, both located in the northeast quadrant of the Bayleaf 7½-minute Quadrangle, Wake County, North Carolina. The localities are immediately west of State Route 1005 (north of Bayleaf and south of North Carolina 98) and adjacent to the Upper and Lower Barton Creek arms of Falls Lake Reservoir on the Neuse River (Fig. 1).

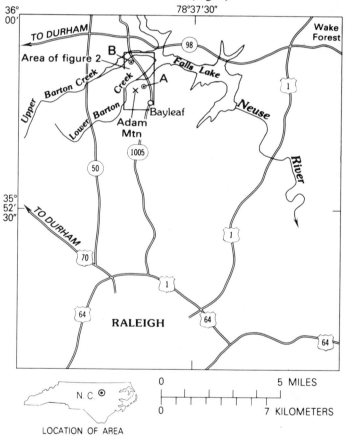

Figure 1. Location map showing field trip stops A and B.

INTRODUCTION

The Raleigh belt consists of amphibolite-facies gneisses and schists in the eastern Piedmont of North Carolina and southern Virginia. It is flanked by lower grade (mainly greenschist facies) metavolcanic and metasedimentary rocks of the Carolina slate belt on the west and Eastern slate belt on the east. Because the Carolina slate belt is considered to be a volcanic-arc terrane exotic to North America (Secor and others, 1983), terranes lying east of it are susceptible to a similasr interpretation. This field trip site displays rocks on both sides of a major lithotectonic boundary on the western edge of the Raleigh belt. Here a probable continental terrane of the Raleigh belt is separated from structurally higher oceanic terranes to the west (Falls Lake melange and Carolina slate belt) by a thrust fault that has been overprinted by late Paleozoic folding and regional metamorphism (Wylie and Stoddard, 1984; Horton and others, 1985).

The Falls Lake mélange consists of numerous pods and lenses of metamorphosed ultramafic (*u*) and mafic rocks enclosed in a matrix of biotite-muscovite schist (*ms*) (Fig. 2). These rocks have been variously ascribed to the Raleigh belt (Parker, 1978), to the Carolina slate belt (Wylie and Stoddard, 1984; Farrar, 1985), or to a separate mélange terrane dividing the two belts (Horton and others, 1985). The mélange lies east of low-grade metavolcanic rocks of the Carolina slate belt (Cary sequence of Parker, 1979) and west of felsic gneisses and schists of the Raleigh belt. The melange terrane, which is 2.5 to 5 mi (4 to 8 km) wide, extends from southern Wake County northward into Granville County (Carpenter, 1970; Horton and others, 1985; J. W. Horton, Jr., and D. E. Blake, unpublished mapping) and perhaps farther north. Ultramafic fragments in the melange have tectonic (and possibly sedimentary) contacts with the enclosing schist. The schist probably had a graywacke-mudstone protolith, and the ultramafic fragments, possibly dismembered ophiolite, may represent the oceanic crust of the Carolina slate belt as suggested by Stoddard and others (1982).

East of the mélange, the Raleigh belt is devoid of ultramafic rocks (Fig. 2). In that area, potassium-rich quartzofeldspathic gneiss (*fg*) may have felsic igneous and/or arkosic sedimentary protoliths of continental provenance, and graphitic schist (*gs*) may have originated in a restricted basin. The presence of these rock types as well as the regional gravity low in the area suggest a continental origin. Although it has been suggested that parts of the Raleigh belt may date to the Grenville (Farrar, 1984), the ages of these rocks are unknown.

Because of the extreme lithologic differences between the Falls Lake mélange and the Raleigh belt and the existence of a thin and discontinuous zone of thinly banded "ribbon" gneiss along the contact (near locations 2a, 2b, and 2c in Fig. 2), we interpret the contact between these two terranes as a thrust fault (Wylie and Stoddard, 1983; 1984; Horton and others 1985) that juxtaposes the ophiolitic mélange against the Raleigh belt conti-

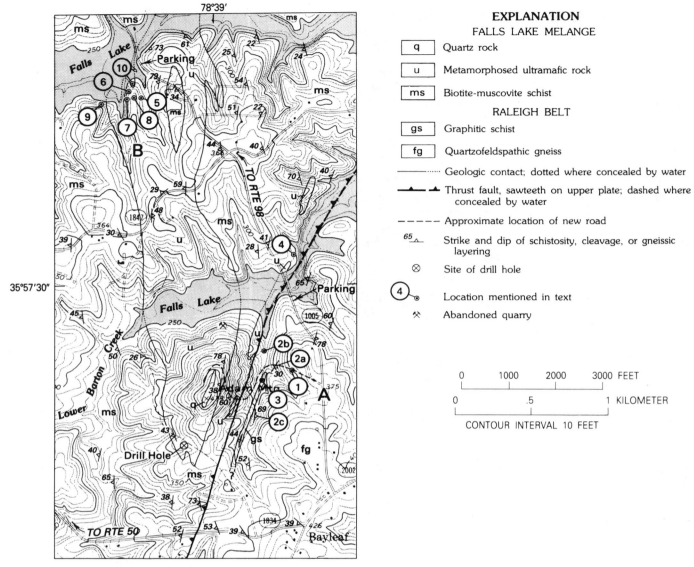

Figure 2. Geologic map of part of the Bayleaf 7½-minute Quadrangle showing the field trip area and localities mentioned in the text.

nental rocks. The age of the thrust is unknown, but upright folds and metamorphism superimposed on the rocks of both terranes are inferred to be Alleghanian in age (Wylie, 1984; Wylie and Stoddard, 1984). Another interpretation of the contact, favored by Parker (1979), is that it is an unconformity.

SITE DESCRIPTION

The creek at Stop A (Fig. 2) is just east of the contact between the lithologically dissimilar Falls Lake melange and Raleigh belt. Within and east of the creek are outcrops of quartzofeldspathic gneiss containing a persistent interlayer of graphitic schist. These rocks belong to the Raleigh belt. The area west of the creek is underlain by biotite-muscovite schist containing pods and lenses of metamorphosed ultramafic rocks, as well as amphibolite and hornblende gneiss. These rocks belong to the Falls

Lake mélange (Horton and others, 1985). The contact between the two assemblages of rocks is interpreted as a thrust fault whose moderate westward dip at this location is attributed to later folding. Consequently, the ultramafic-bearing schist of the mélange structurally overlies the Raleigh belt rocks (Fig. 2). The ridge at Stop B contains an exposure of a metamorphosed ultramafic body that includes a variety of rock types.

Stop A

Stop A (Figs. 1 and 2) is along a small creek that drains northward into Lower Barton Creek on the east side of Adam Mountain (elevation 478 ft; 146 m). A large part of this area is currently under residential development. Permission to examine exposures within the development, called Adam Mountain, and to gain access to the creek area via the new road should be requested from the security guard on duty. Park along State Route 1005 or, with permission, near the creek within the development's land.

***Quartzofeldspathic Gneiss* (fg).** The area east of the creek at Stop A is underlain predominantly by biotite-free or biotite-poor quartz-muscovite-plagioclase +/- microcline gneiss that is poorly foliated. Good examples have been excavated during the construction of a new road that extends westward from S.R. 1005 to Adam Mountain (Fig. 2, Location 1). Closer to the creek, the quartzofeldspathic gneiss is more biotite rich and has a stronger foliation. Excellent exposures in the creek (Fig. 2, Locations 2b and 2c) show two S-surfaces. First, thin gneissic compositional banding is defined by alternate layers rich and poor in biotite and opaque minerals; these layers are generally no thicker than 0.2 in. (5 mm). Second, a younger cleavage axial planar to gently plunging folds having south- and north-trending axes is defined by the parallel alignment of biotite flakes (Fig. 3). The gneissic banding has strikes and dips of N. 10°–31°E., 75°–82° SE. at Location 2b and N. 8° E., 69° SE. at Location 2c (Fig. 2). The younger cleavage at Location 2b strikes N. 8° W. to N. 20° E. and dips 32°–52° E.

***Graphitic Schist* (gs).** Several good exposures of graphitic schist occur in and near the creek in the southern part of the area (Fig. 2, Location 3). The schist is a thin but persistent unit that has been mapped southward for over 16 mi (25 km) by Parker (1979). However, because we have been unable to trace the graphitic schist north of Location 3 (Fig. 2), we suggest that the graphitic unit may be truncated by the inferred thrust. Near Adam Mountain, the schist is rich in kyanite; the two most common assemblages contain (in addition to muscovite, quartz, graphite, and tourmaline) kyanite+garnet+staurolite or kyanite+staurolite+biotite. In the graphitic schist, compositional layers less than 0.4 in. (1 mm) thick that are rich in graphite and aluminous minerals alternate with layers of similar thickness that are nearly pure quartz. Garnets are commonly elongate in the plane of layering. Many garnets appear to have been pulled apart and later annealed, having partial euhedral overgrowths. Staurolite contains abundant graphite inclusions, whereas kyanite porphyroblasts are free of inclusions. The layering and parallel foliation are overprinted by a crenulation cleavage that is defined by the alignment of muscovite flakes and along which some kyanite blades have grown. Although both surfaces strike approximately north at this locality, the compositional layering and parallel foliation dip steeply west, whereas the younger cleavage dips moderately west. Porphyroblasts of staurolite and kyanite are elongate parallel to the crenulation lineation formed by the intersection of these surfaces; the crenulation axes plunge about 20° south. Locally, an apparently earlier generation of kyanite is represented by folded or bent poikiloblasts parallel to the layering. Figure 4 displays some of the microstructures present in thin sections of graphitic schist.

***Biotite-muscovite Schist* (ms).** Biotite-muscovite schist of the Falls Lake melange underlies most of the region west of the creek at Stop A, but outcrops of it are scarce in the immediate

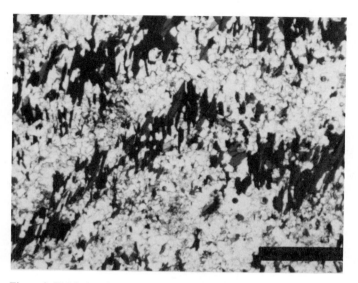

Figure 3. Thinly banded quartzofeldspathic gneiss from Stop A, Location 2b (see Fig. 2) showing two S-surfaces. A younger cleavage defined by biotite intersects the older gneissic banding at a high angle. The biotite cleavage is axial planar to gently plunging upright folds such as the one in the photograph.

vicinity of Adam Mountain. Here and elsewhere, this rock unit is represented most commonly by two-mica quartz-plagioclase +/- chlorite schist that locally contains garnet. Small exposures may be visible between ultramafic rocks in the drainage ditches along the new road on the east slope of Adam Mountain, but better exposures can be seen east of State Route 1005 and north of Lower Barton Creek (Fig. 2, Location 4).

***Stop B. Ultramafic Rocks* (u)**

Stop B (Figs. 1 and 2) is a low north-trending ridge approximately 1.2 mi (2 km) north of Stop A and just south of Upper Barton Creek. The ridge was cleared in 1979 during excavation for the reservoir and is on public land. Park on State Route 1005.

The biotite-muscovite schist unit contains a large number of lenses and pods of ultramafic rocks. One elongate ultramafic body underlies Adam Mountain and its northern and western slopes (Fig. 2), but is better exposed at Stop B, about 1.2 mi (2 km) north of Adam Mountain (Fig. 2). At Stop B, examples typical of nearly all the ultramafic lithologies present in Wake County can be found. The four predominant ultramafic rock types present are serpentinite (Fig. 2, Location 5), chlorite-actinolite schist (Fig. 2, Location 6), soapstone or talc schist (Fig. 2, Location 7), and hornblendite (Fig. 2, Location 8). The serpentinite is a pale green to white, massive, relatively dense rock containing very thin streaks of magnetite. Some asbestiform chrysotile can be found, but most of the serpentine is a fine-grained massive type (antigorite?). Olivine has been reported from serpentinite in a different body (Parker, 1979; Moye, 1981). Chromite occurs on Adam Mountain and was encountered as disseminated grains in serpentinite from an exploratory drill hole on the south end of the knob (Parker, 1979). Small pods of chromitite are present in other bodies in the area. Either chlorite or actinolite may predominate in the chlorite-actinolite schist; minor talc and magnetite octahedra are locally present. Soapstone contains some chlorite and actinolite, as well as a carbonate mineral whose former presence is indicated in nearly all instances by rhombohedral cavities in the rock. Both pure talc schist and soapstone occur at Stop B; clinopyroxene (relict?) is present but exceedingly sparse in the soapstone. One area of black hornblendite is present near the crest of the ridge (Fig. 2, Location 8); thin section analysis reveals that this rock consists mainly of blue-green amphibole, about 10 percent epidote, and minor garnet. Chemical analyses of serpentinite, chloritite, and soapstone from eight sites along the south side of Falls Lake (near a small abandoned quarry (Fig. 2)) are given by Moye (1981, Table 3). On the basis of these analyses as well as of petrography, Moye inferred that the protoliths of the Adam Mountain ultramafic body were dunite and harzburgite depleted in Al_2O_3, TiO_2, CaO, Na_2O, and K_2O. The crest of Adam Mountain is underlain by resistant quartz rock (q in Fig. 2) that is composed of fine-grained polycrystalline quartz crisscrossed by numerous veinlets of banded chalcedony and containing cavities lined with drusy quartz crystals. The origin of the quartz rock is unknown, but the same rock type also occurs nearby in other ultramafic bodies.

The western contact of the ultramafic body is constrained by a small outcrop of biotite-muscovite schist at the water line about 650 ft (200 m) west of the ridge crest (Fig. 2, Location 9). A reentrant of biotite-muscovite schist also appears to be present at the north end of the ridge (Fig. 2, Location 10). Contacts between the different rock types at Stop B are generally parallel to the undulatory schistosity, which varies considerably in orientation.

REFERENCES CITED

Carpenter, P. A. III, 1970, Geology of the Wilton area, Granville County, North Carolina [M.S. thesis]: Raleigh, North Carolina State University, 106 p.

Farrar, S. S., 1984, The Goochland granulite terrane: Remobilized Grenville basement in the eastern Virginia Piedmont: Geological Society of America Special Paper 194, p. 215–227.

——1985, Tectonic evolution of the easternmost Piedmont, North Carolina: Geological Society of America Bulletin, v. 96, no. 3, p. 362–380.

Horton, J. W., Jr., Stoddard, E. F., Wylie, A. S., Jr., Moye, R. J., Jr., and Blake, D. E., 1985, Melange terrane in the eastern Piedmont of North Carolina: Geological Society of America Abstracts with Programs, v. 17, no. 2, p. 96.

Moye, R. J., Jr., 1981, The Bayleaf mafic-ultramafic belt, Wake and Granville Counties, North Carolina [M.S. thesis]: Raleigh, North Carolina State University, 122 p.

Parker, J. M. III, 1978, Structure of the west flank of the Raleigh belt, North Carolina, in Snoke, A. W., ed., Geological investigations of the eastern Piedmont, southern Appalachians, Carolina Geological Society Field Trip Guidebook 1978: Columbia, South Carolina Geological Survey, p. 1–7.

——1979, Geology and mineral resources of Wake County: North Carolina Department of Natural Resources and Community Development, Geological Survey Section, Bulletin 86, 122 p.

Secor, D. T., Jr., Samson, S. L., Snoke, A. W., and Palmer, A. R., 1983, Confirmation of the Carolina slate belt as an exotic terrane: Science, v. 221, no. 4611, p. 649–650.

Stoddard, E. F., Moye, R. J., Kite, L. E., and Won, I. J., 1982, Eastern North

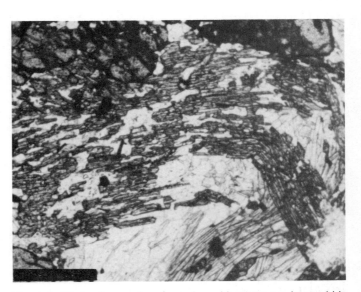

Figure 4. Photomicrographs of porphyroblast textures in graphitic schist of the Raleigh belt (Stop A, Location 3 in Fig. 2). (A) Garnet is flattened in the plane of compositional layering and parallel schistosity that is warped and crinkled in other parts of the same thin section. (B) Kyanite lies in the plane of compositional layering and is folded with it. Later kyanite (see A) cuts across the layering and appears to be parallel to the younger cleavage defined by biotite. Scale bars in A and B are 0.5 mm long.

Carolina ophiolite terranes: Setting, petrology, and implications: Geological Society of America Abstracts with Programs, v. 14, no. 1-2, p. 626.

Wylie, A. S., Jr., 1984, Structural and metamorphic geology of the Falls Lake area, Wake County, North Carolina [M.S. thesis]: Raleigh, North Carolina State University, 79 p.

Wylie, A. S., Jr., and Stoddard, E. F., 1983, Multiple deformation and polymetamorphism in northern Wake County, eastern North Carolina Piedmont: Geological Society of America Abstracts with Programs, v. 15, no. 2, p. 110.

——1984, The Falls Lake mylonite zone: A pre-Alleghanian lithotectonic boundary in the eastern North Carolina Piedmont: Geological Society of America Abstracts with Programs, v. 16, no. 6, p. 702.

Pilot Mountain, North Carolina

James Robert Butler, Department of Geology, University of North Carolina, Chapel Hill, North Carolina 27514

LOCATION

Pilot Mountain State Park is located just west of U.S. 52, 24 mi (38 km) north of Winston-Salem and 14 mi (22 km) south of Mount Airy, Surry County, Pinnacle 7½-minute quadrangle (Fig. 1). The Pilot Mountain State Park exit from U.S. 52 leads directly to the park entrance. The field trip narrative begins at the parking lot on the mountain top (Fig. 2, Location A). No rock collecting is allowed in the park without special written permit.

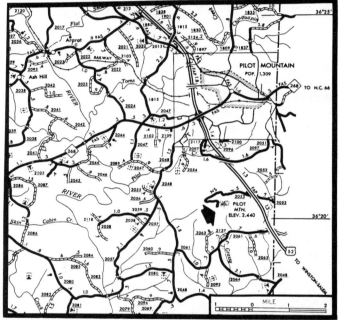

Figure 1. Highway map showing the location of Pilot Mountain.

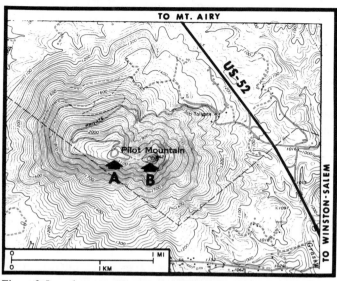

Figure 2. Location map, Pilot Mountain State Park, Pinnacle 7½-minute quadrangle, North Carolina.

INTRODUCTION

Pilot Mountain is a prominent isolated peak rising abruptly above the gently rolling Piedmont surface. The peak has an elevation of 2,421 ft (738 m) in contrast to the general Piedmont surface at about 1,000 ft (305 m). Pilot Mountain was a landmark for the early traders and settlers traveling into and across the Appalachians. It has several interesting geomorphic, regional geologic, and botanical features.

The mountain is one of the most conspicuous inselbergs (Kesel, 1974) in the Piedmont. Kesel (1974) used "inselberg" as a descriptive name for an isolated mountain that protrudes from a nearly flat plain, in preference to "monadnock," which has genetic connotations in the humid erosion cycle. Pilot Mountain is capped by a thick section of nearly horizontally bedded quartzite. The quartzite is flat-lying because it occurs near the crest of the Sauratown Mountains anticlinorium (Butler and Dunn, 1968).

The core of the anticlinorium is defined by extensive areas of quartzites, whose age and correlations are poorly known. One of these areas is Pilot Mountain; a larger mass includes Moores Knob and other peaks in and near Hanging Rock State Park about 12 mi (20 km) to the east (Espenshade and others, 1975).

Vegetation on Pilot Mountain includes many plants that are more typical of the Blue Ridge to the northwest than of the surrounding Piedmont. Examples are rhododendron, laurel, chestnut oak, pitch pine, and table mountain pine.

The Sauratown Mountains anticlinorium is a major tectonic feature in the Piedmont of North Carolina, but its detailed structure and regional significance are still poorly understood. Billion-year-old gneisses are exposed in the core, but age of the quartzite is not known; however, it is probably late Precambrian or early Paleozoic (Rankin and others, 1973). Bryant and Reed (1961) correlated the quartzite with the Cambrian Chilhowee quartzite farther west and proposed that the core of the anticlinorium was a tectonic window. Butler and Dunn (1968), Centini (1968), and Stirewalt and Dunn (1973) disputed the window interpretation on both stratigraphic and structural evidence, and proposed that Pilot Mountain was on the upright limb of a major recumbent, refolded fold. Simons and others (1982) rejuvenated the idea of a window in the Sauratown Mountains, but in a somewhat different position than that proposed by Bryant and Reed (1961). Recent geologic mapping in the western anticlinorium is producing changes in interpretations; Hatcher and others (1983) recognized both pre- and post-metamorphic thrusts and proposed that the Inner Piedmont-Sauratown Mountains boundary was a metamorphic gradient rather than a thrust fault as proposed by Espenshade and others (1975).

SITE DESCRIPTION

The parking lot at Pilot Mountain is situated on a gently northward-dipping slope of quartzite. Walk about 300 ft (90 m) east along the Little Pinnacle Trail to the overlook at the crest of Little Pinnacle (Fig. 2, Location A). The overlook provides good views of the layered quartzite on Big Pinnacle, the steep-sided main peak, and the surrounding region. Moores Knob and other peaks of the Sauratown Mountains are seen to the east. On a clear day, you can see buildings of Winston-Salem to the southeast and the Blue Ridge topographic front to the northwest. The Blue Ridge front is the sharp topographic break between the lower, flatter Piedmont surface and the higher, more rugged terrain of the Blue Ridge mountains. Stony Ridge is a prominent ridge 3 mi (5 km) southwest of Pilot Mountain; it is the type locality of the Stony Ridge fault zone, which extends for 43 mi (70 km) across the Piedmont just south of the axis of the Sauratown Mountains anticlinorium. The fault zone is marked by microbreccia and siliceous microbreccia, which typically stand up as ridges.

The Jomeokee Trail begins just north of Little Pinnacle, crosses the saddle, and circles Big Pinnacle (Fig. 2, Location B), affording good places to see features of the quartzite. The rocks are mainly clean quartzites, consisting of quartz with minor amounts of ilmenite, tourmaline, and other minerals, and schist, composed mainly of greenish phengite and quartz. Exposures along the trail show that the main schistosity is essentially parallel to bedding. Crenulation cleavage occurs in mica-rich layers, and cross-bedding in the quartzite. The axial trace of the anticlinorium passes through the saddle between Little Pinnacle and Big Pinnacle. The trail circling Big Pinnacle is rugged in some places, but generally follows the base of the quartzite and gives access to many of its interesting features. Cross-bedding is well exposed near the base of the quartzite on the southwestern side of Big Pinnacle. Schistosity and bedding are generally parallel, which is part of the evidence that the regional structure is a regional isoclinal fold. Crenulations and crenulation cleavage are locally well displayed. The quartzite lies with apparent conformity on biotite gneiss and schist; if the contact is a fault, any mylonite or breccia has been recrystallized by regional metamorphism. The structural features are discussed in more detail by Butler and Dunn (1968, p. 25–30) and by Stirewalt and Dunn (1973). Regional geology is shown on the map by Espenshade and others (1975).

Hanging Rock State Park, 10 mi (16 km) to the east of Pilot Mountain, also includes many quartzite and schist exposures. Most of the field trip stops in that area described by Butler and Dunn (1968) are still accessible.

REFERENCES CITED

Bryant, Bruce, and Reed, J. C., Jr., 1961, The Stokes and Surry Counties quartzite area, North Carolina—a window: U.S. Geological Survey Professional Paper 424D, p. D61–D63.

Butler, J. R., and Dunn, D. E., 1968, Geology of the Sauratown Mountains anticlinorium and vicinity, North Carolina: Southeastern Geology, Special Publication 1, p. 19–47.

Centini, B. A., 1968, Structural geology of the Hanging Rock area, Stokes County, North Carolina [Ph.D. thesis]: Chapel Hill, University of North Carolina, 57 p.

Espenshade, G. H., Rankin, D. W., Shaw, K. W., and Neuman, R. B., 1975, Geologic map of the east half of the Winston-Salem quadrangle, North Carolina-Virginia: U.S. Geological Survey Map I-709-B.

Hatcher, R. D., Jr., McConnell, K. I., and Heyn, Teunis, 1983, Preliminary results from detailed geologic mapping studies in the western Sauratown Mountains anticlinorium, North Carolina: Carolina Geological Society Field Trip Guidebook, p. 1–7.

Kesel, R. H., 1974, Inselbergs on the Piedmont of Virginia, North Carolina, and South Carolina: Types and characteristics: Southeastern Geology, v. 16, p. 1–30.

Rankin, D. W., Espenshade, G. H., and Shaw, K. W., 1973, Stratigraphy and structure of the metamorphic belt in northwestern North Carolina and southwestern Virginia: A study from the Blue Ridge across the Brevard Fault Zone to the Sauratown Mountains anticlinorium: American Journal of Science, v. 273–A, p. 1–40.

Simons, J. H., Bartley, J. M., and Butler, J. R., 1982, Regional significance of a newly recognized thrust at Danbury, Sauratown Mountains, North Carolina: Geological Society of America, Abstracts with Programs, v. 14, p. 618.

Stirewalt, G. L., and Dunn, D. E., 1973, Mesoscopic fabric and structural history of Brevard zone and adjacent rocks, North Carolina: Geological Society of America Bulletin, v. 84, p. 1629–1650.

Linville Falls fault at Linville Falls, North Carolina

Robert D. Hatcher, Jr., Department of Geology, University of South Carolina, Columbia, South Carolina 29208
J. Robert Butler, Department of Geology, University of North Carolina, Chapel Hill, North Carolina 27514

LOCATION

One half mi (1 km) south of the Linville Falls parking area off Blue Ridge Parkway east of the town of Linville Falls, North Carolina, Linville Falls 7½-minute Quadrangle, North Carolina (Fig. 1). This locality is within the Blue Ridge Parkway National Park. Consequently, the collectaing of samples and breaking of rocks is prohibited. The actual fault exposure is about 100 yd (90 m) above the falls at the end of the trail from the parking lot and is in the area indicated by a sign to not enter because of the dangerous rapids and the river. This area can be entered by walking carefully.

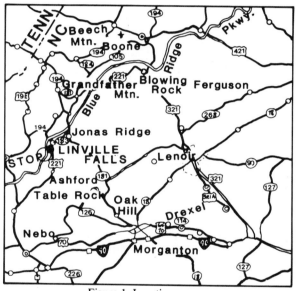

Figure 1. Location map.

INTRODUCTION

The Grandfather Mountain window is the largest structural window in the Blue Ridge. It is formed by the Linville Falls fault which separates medium-grade metamorphic basement and younger rocks of the Blue Ridge thrust sheets(s) from the lower grade (chlorite-biotite) metamorphic rocks of the window.

The structure inside the window is not simple. Basement and late Precambrian Grandfather Mountain Formation rocks were ductilely deformed, metamorphosed, and then faulted as they became part of this structure (Fig. 2). Even the rocks inside the window are thought to be allochthonous (Boyer and Elliott, 1982; Harris and others, 1981), since most reconstructions require the rocks inside the window to also be transported. Boyer and Elliott (1982) interpreted the footwall of the window as a duplex structure.

An intermediate thrust sheet, the Table Rock sheet, occurs in the southwest side of the window. It contains rocks of the Erwin Quartzite of the Chilhowee Group overlain by the Shady Dolomite (Bryant and Reed, 1970). Rocks of the Table Rock thrust sheet were isoclinally folded prior to being emplaced and the sequence is locally overturned near Woodlawn, a few miles south of Linville Falls. The Linville Falls fault lies immediately above the Table Rock thrust sheet and crenulation folds with the same orientations occur inside the window, in the Table Rock thrust sheet and in the hanging wall rocks, indicating post-thrusting emplacement of the crenulations.

The Linville Falls fault is thus a major fault in the Blue Ridge that probably formed during Alleghanian deformation. A Rb-Sr whole-rock age of 300 Ma was obtained from mylonite from the fault zone (Van Camp and Fullagar, 1982). It is not clear how the fault is related but it must emerge again as one of the major frontal Blue Ridge thrusts.

The purpose of this stop is to examine exposures of Erwin Quartzite of the Chilhowee Group (Lower Cambrian), the overlying Cranberry Gneiss (Elk Park Plutonic Group) basement rocks, and the fault zone that separates them. The Cranberry Gneiss is overthrust onto the Chilhowee at the type locality of the Linville Falls fault at Linville Falls.

SITE DESCRIPTION

Rocks exposed at Linville Falls belong to the Lower Cambrian Chilhowee Group, probably the Erwin Formation, which is the uppermost unit in the Chilhowee in this area (Bryant and Reed, 1970). Upon cursory examination, the sandstones appear to be only slightly deformed. However, further examination reveals they were polyphase-deformed and isoclinally folded prior to faulting and emplacement of later structures.

An intersecting set of linear structures is present at the overlook area above Linville Falls. One lineation could be a fold axis lineation related to the isoclinal recumbent folds that are present. The other lineation is a crenulation cleavage. These structures formed at right angles to one another. The earlier lineation (NW orientation) may actually be related to emplacement of the Blue Ridge thrust sheet and the Linville Falls fault or to isoclinal folding, which occurred prior to thrusting.

An excellent exposure of the Chilhowee Group rocks may be seen in the walls of Linville Gorge. Isoclinal recumbent folds in the Chilhowee sandstones may also be observed. A particularly good viewpoint is located just south of the falls which provides a panorama of Linville Falls (Fig. 3). The large block of sandstone that forms the barrier around which the river must flow contains a good exposure of a large isoclinal recumbent fold in Chilhowee sandstone.

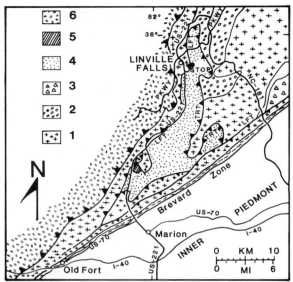

Figure 2. Map showing the major lithologic and tectonic units present in the vicinity of the Linville Falls locality (modified from Bryant and Reed, 1970). 1 - Basement (1.2-1.0 b.y.) gneisses. 2 - Late Precambrian Grandfather Mountain Formation. 3 - Late Precambrian Brown Mountain Granite. 4 - Lower Cambrian Chilhowee Group. 5 - Lower Cambrian Shady Dolomite. 6 - Late Precambrian (?) Tallulah Falls Formation and other rocks of the Hayesville thrust sheet. LFT - Linville Falls thrust. TRT - Table Rock thrust.

The crenulation cleavage present in the Chilhowee sandstones is a penetrative structure which passes from the rocks of the Chilhowee Group exposed beneath the Linville Falls fault in the Table Rock thrust sheet, through the fault zone, into the overlying rocks of the Cranberry Gneiss and into the higher thrust sheets of the Blue Ridge. The highest thrust sheet present about 2 mi (3 km) southwest of Linville Falls may be the Hayesville sheet (Hatcher, 1978). The crenulation cleavage that overprints all tectonic units may be the same generation crenulation cleavage identified originally by Butler (1973) in the Blue Ridge nearby. First descriptions of these rocks and the structures were made by Keith (1903, 1905, 1907) and more recently by Bryant and Reed (1970) in their work in the Grandfather Mountain window area.

The fault zone exposed above Linville Falls consists of a layer of finely laminated, fine-grained, chlorite-bearing porphy-

Figure 3. View of Linville Falls from overlook showing isoclinally folded Chilhowee (Erwin) sandstones. The Linville Falls fault is in the background but not visible above the upper falls at the top of the photograph.

roclastic blastomylonite which ranges up to 18 in (45 cm) thick. The blastomylonite has been deformed into open folds that are associated with the crenulation cleavage described above. The Cranberry gneiss is also mylonitic and retrograded from its original high metamorphic grade (granulite facies?) to greenschist facies assemblages. Rocks of the Chilhowee Group in the Table Rock thrust sheet in this area (Fig. 2) likewise are metamorphosed to the lower greenschist facies of regional metamorphism.

Bryant and Reed (1969) interpreted the northwest trending lineation mentioned above as a cataclastic lineation parallel to the transport direction of the major thrust sheets, and they recognized that many of the small fold axes also parallel the lineation. They interpreted this lineation as a result of rotation of fold axes into a direction of transport during thrusting. We recognize that this rotational process generally accompanies the formation of sheath folds, but no evidence exists to confirm that either of these lineations is related to the formation of sheath folds. The new lineation is probably formed during the early episode of isoclinal folding which produced lineation in parallelism with the fold axes. These could have simply been rotated *en masse* during thrusting and not transposed into the transport direction by inhomogeneous simple shear.

REFERENCES CITED

Boyer, S. E., and Elliott, David, 1982, Thrust systems: American Association of Petroleum Geologists Bulletin, v. 66, p. 1196–1230.

Bryant, Bruce, and Reed, J. C., Jr., 1969, Significance of lineation and minor folds near major thrust faults in the Southern Appalachians and the British and Norwegian Caledonides: Geological Magazine, v. 106, p. 412–429.

Bryant, Bruce, and Reed, J. C., Jr., 1970, Geology of the Grandfather Mountain window and vicinity, North Carolina and Tennessee: U.S. Geological Survey Professional Paper 615, 190 p.

Butler, J. R., 1973, Paleozoic deformation and metamorphism in part of the Blue Ridge thrust sheet, North Carolina: American Journal of Science, v. 273-A, p. 72–88.

Harris, L. D., Harris, A. G., de Witt, Wallace, Jr., and Bayer, K. C., 1981, Evalua-

tion of southern eastern overthrust belt beneath Blue Ridge-Piedmont thrust: American Association of Petroleum Geologists Bulletin, v. 65, p. 2497–2505.

Hatcher, R. D., Jr., 1978, Tectonics of the western Piedmont and Blue Ridge: review and speculation: American Journal of Science, v. 278, p. 276–304.

Keith, Arthur, 1903, Description of the Cranberry quadrangle (North Carolina-Tennessee): U.S. Geological Survey Geologic Atlas, Folio 90, 9 p.

Keith, Arthur, 1905, Description of the Mount Mitchell quadrangle (North Carolina-Tennessee): U.S. Geological Survey Geologic Atlas, Folio 124, 10 p.

Keith, Arthur, 1907, Description of the Roan Mountain quadrangle (Tennessee-North Carolina): U.S. Geological Survey Geologic Atlas, Folio 151, 12 p.

Van Camp, S. G., and Fullagar, P. D., 1982, Rb-Sr whole-rock ages of mylonites from the Blue Ridge and Brevard zone of North Carolina: Geological Society of America Abstracts with Programs, v. 14, p. 92–93.

The Concord gabbro-syenite complex, North Carolina

Harry Y. McSween, Jr., Department of Geological Sciences, University of Tennessee, Knoxville, Tennessee 37996

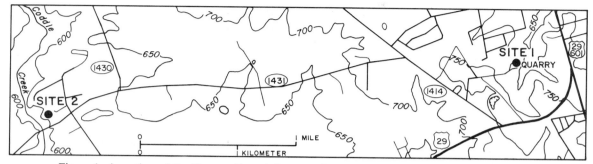

Figure 1. A section of the Kannapolis and Concord Quadrangles showing the locations of Site 1 (syenite) and Site 2 (gabbro). The contact between these lithologies follows approximately the 700-ft (212 m) contour in this area.

LOCATION

Two stops (Fig. 1) southwest of the town of Concord, Cabarrus County, North Carolina (Fig. 1). No special permission is needed for either stop. Site 1 is a quarry located in the southwestern corner of the Concord 7½-minute Quadrangle, and is accessible via Cavalier Court, an unpaved road connecting with Rock Hill Church Road (Rd. 1414) 0.2 mi (0.3 km) northwest of its intersection with U.S. 29. Cars and vans can be parked at the quarry (0.4 mi or 0.6 km down Cavalier Court), but buses should be parked near the radio tower. Site 2 is an exposure along Coddle Creek at Weddington Road in the Kannapolis quadrangle. A bridge crosses the creek 0.2 mi (0.3 km) southwest of the Weddington Road (Rd. 1431) and Concord Farms Road (Rd. 1432) intersection. Park at the turnout on the south side of the bridge, cross the fence on the opposite side of the road, and walk upstream on the right side of the creek for approximately 300 ft (90 m) to a small waterfall. The outcrop is a field of large gabbro boulders and is not visible from the road.

INTRODUCTION

Mafic intrusions are a major component of the bimodal (gabbro and granite) suite of Paleozoic postmetamorphic igneous rocks in the southern Appalachian Piedmont. The Concord complex is one of approximately 20 gabbroic intrusions exposed in an arcuate chain in the Charlotte and Carolina slate belts of North Carolina, South Carolina, and Georgia (McSween and others, 1984, and references therein). It covers an area of approximately 36 mi^2 (100 km^2) in the Concord, Concord SE, Kannapolis, and Harrisburg quadrangles (Fig. 2). This complex consists of a large oval-shaped gabbro stock enveloped by a discontinuous syenite ring dike of varying thickness. A map by Bell (1960) shows a 'parentheses' shape for the syenite, open at the northern and southern ends; however, Olsen and others (1983) found that the syenite is continuous across the northern perimeter, forming a

horseshoe-shaped outcrop pattern. A train of small satellite plugs of gabbro extend southward from the open end of the complex (only one of which is shown in Fig. 2; see Bell, 1960).

The complex is clearly delineated from surrounding rocks by its geophysical signature, and field trips to this locality would be enhanced by taking along appropriate geophysical maps. Aeromagnetic and aeroradioactivity surveys (Johnson and Bates, 1960) indicated a strong positive magnetic anomaly at the gabbro-syenite contact (Fig. 3) and anomalously high radioactivity centered over the syenite. The complex also has a pronounced positive gravity anomaly (Morgan and Mann, 1964). Olsen and others (1983) used the latter data to calculate a model for the geometry of the pluton at depth from a gravity profile along line A-B in Figures 2 and 3. Their model depicted a gabbro plug of approximately 20,000 ft (6 km) thickness and a steep-walled syenite ring dike with an outward dip of 78° on the west side and a vertical dip on the east side (Fig. 3).

Fullager (1971) reported a Rb-Sr crystallization age for the syenite of 404 ± 21 Ma (recalculated using the new Rb decay constant), and a Sm-Nd isochron for gabbro indicated an age of 407 ± 36 Ma (Olsen and others, 1983). The obvious association of these lithologies in space and time suggests that they may be related through magmatic differentiation. Initial Sr and Nd isotopic ratios for gabbro and syenite are virtually identical (Fullagar 1971; Olsen and others, 1983), and consistent with closed-system fractionation of mantle-derived magma without assimilation of crustal materials. Derivation of the syenite as a residual liquid from fractional crystallization of gabbroic magma was tested by Olsen and others (1983). They observed that the cumulus phases in most gabbros are either olivine + plagioclase or clinopyroxene + plagioclase, though all three cumulus minerals occur in a few samples. Using major and trace element chemistry, they found that syenite could be produced from a magma like that which formed the gabbro by fractionation of olivine + clinopyroxene +

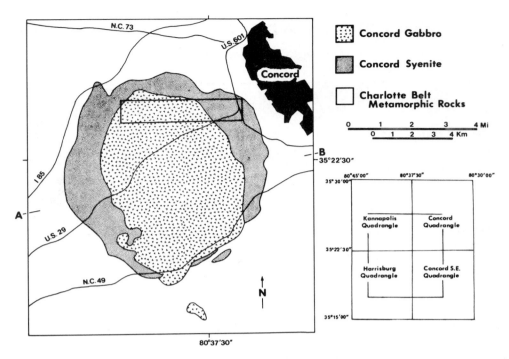

Figure 2. Geologic map of the Concord gabbro-syenite complex, after Olsen and others (1983). The area of Figure 1 containing the two sites is delineated by the rectangle just west of Concord.

plagioclase of the same compositions as the cumulus phases in the gabbro.

Olsen and others (1983) proposed the following model for the origin of the Concord complex. Gabbroic magma tapped from a larger body of similar magma at depth intruded first, possibly as a partly crystalline mush. Syenite could not have formed by fractionation of this mafic magma at the present erosional level, due to the absence of intermediate rock types, but it must have formed at depth by fractionation of olivine + clinopyroxene + plagioclase from a magma of similar composition (the source chamber from which the gabbro magma separated). A fractured region at the contact of the gabbro and country rocks was subsequently intruded by the syenitic magma. The ring dike structure did not form by subsidence of a central gabbroic block into syenitic magma, because the magnitude of the gravity anomaly is inconsistent with this geometry and because the width of the ring dike would require the steep-walled gabbro to sink many thousands of feet (kilometers) (Morgan and Mann, 1964).

McSween and others (1984) have argued that the Concord complex and related plutons represent the eroded core of a subduction-related magmatic arc. Whatever their tectonic setting, these intrusions represent an important Paleozoic igneous event in the Piedmont of the southern Appalachians. The Concord complex is the most thoroughly studied of these plutons, and its geometry and contact relations with surrounding rocks are readily inferred from its topographic (see below) and geophysical expression. Syenite, the ultimate differentiation product of these gabbroic magmas, is better exposed here than in other plutons.

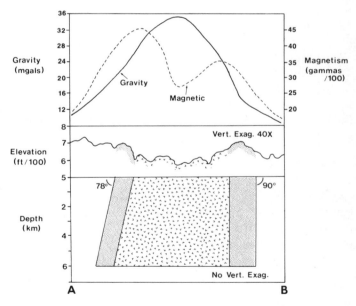

Figure 3. Geophysical and topographic traverses and geologic cross-section along line A–B in Figure 2. Symbols for gabbro and syenite are the same as in Figure 2. Gravity data were taken from the simple Bouguer anomaly map of Morgan and Mann (1964), and magnetic data were extracted from the aeromagnetic map of Johnson and Bates (1960). The cross-section was calculated from gravity data by Olsen and others (1983).

SITE DESCRIPTIONS

Two sites were selected to illustrate the petrologic variations within the Concord complex. Site 1 is an abandoned quarry in syenite. The quarry is now filled with water, but large excavated blocks of syenite remain. Elsewhere syenite outcrops consist of massive, weathered boulders in light-colored, feldspathic saprolite.

The syenite at Site 1 is a coarse-grained seriate porphyry containing megacrysts of perthite and antiperthite in a groundmass of microcline, sodic plagioclase, clinopyroxene, biotite, hornblende, quartz, magnetite, ilmenite, pyrite, zircon, monazite, and apatite. Some megacrysts are zoned, with perthitic cores surrounded by bands of plagioclase (now altered to clay minerals), followed again by perthitic rims. Mafic minerals are concentrated into clots interspersed between megacrysts. Modal analyses and mineral compositions for samples from Site 1 (SC-2) are given by Olsen and others (1983).

Site 2 consists of dark, spheroidal gabbro boulders and is a typical exposure for this lithology. Gabbro weathers to a distinctive, gray- or light brown-colored saprolite. The Concord gabbro is petrologically similar to others in the Piedmont. Samples are medium-grained meso- or orthocumulates consisting primarily of plagioclase, olivine, clinopyroxene, orthopyroxene, and hornblende, with accessory biotite, apatite, magnetite, ilmenite, sulfides (mostly pyrrhotite), and minor secondary minerals. Orthopyroxene and hornblende poikilitically enclose other phases. Layering has not been observed in the rocks at Site 2 or elsewhere in the pluton, and this, plus the limited Fe-enrichment trend in gabbro samples, is consistent with emplacement of this mafic unit (and many others in the Piedmont) as crystal mushes (McSween and others, 1984). Modal analyses and mineral compositions for samples from Site 2 (CG-6) are given by Olsen and others (1983).

In driving from one site to the other, the dramatic topographic difference, approximately 100 ft (30 m) of relief, between deeply eroded gabbro and the resistant syenite ridge is apparent. In Figure 1, the contact between these lithologies follows approximately the 700 ft (212 m) contour interval. Lithologic contacts between gabbro and syenite are not generally exposed but can be fixed rather closely, as illustrated by the topographic profile along line A-B (Fig. 3). Using this topography as a guide, further reconnaissance by car along U.S. 29 or other paved roads demonstrates the ring dike structure of the complex.

REFERENCES CITED

Bell, H., 1960, A synthesis of geologic work in the Concord area, North Carolina: U.S. Geological Survey Professional Paper 400B, p. 189–191.

Fullagar, P. D., 1971, Age and origin of plutonic intrusions in the Piedmont of the southern Appalachians: Geological Society of America Bulletin, v. 82, p. 2845–2862.

Johnson, R. W., and Bates, R. G., 1960, Aeromagnetic and aeroradioactivity survey of the Concord Quadrangle, North Carolina: U.S. Geological Survey Professional Paper 400B, p. 182–195.

McSween, H. Y., Jr., Sando, T. W., Clark, S. R., Harden, J. T., and Strange, E. A., 1984, The gabbro-metagabbro association in the southern Appalachian Piedmont: American Journal of Science, v. 294, p. 437–461.

Morgan, B. A., and Mann, V. I., 1964, Gravity studies in the Concord Quadrangle, North Carolina: Southeastern Geology, v. 5, p. 143–155.

Olsen, B. A., McSween, H. Y. Jr., and Sando, T. W., 1983, Petrogenesis of the Concord gabbro-syenite complex, North Carolina: American Mineralogist, v. 68, p. 315–333.

Carolina slate belt near Albemarle, North Carolina

James Robert Butler, Department of Geology, University of North Carolina, Chapel Hill, North Carolina 27514

LOCATION

Two regionally important stratigraphic units in the Carolina slate belt are exposed along North Carolina 24-27 on both sides of Lake Tillery (Pee Dee River), about 7 mi (11 km) east of Albemarle, Stanly and Montgomery Counties, Morrow Mountain 7½-minute quadrangle (Fig. 1). Site A (Fig. 2) is 1.7 miles (2.7 km) east of the Lake Tillery bridge and 0.6 mile (1 km) east of the intersection of North Carolina 24-27 and State Road 1150. At Site B (Fig. 2), mudstone of the Tillery Formation and a metagabbro sill are exposed along North Carolina 24-27 0.3 mi (0.5 km) west of the Lake Tillery bridge. Since traffic conditions at Site B are dangerous, extreme caution must be used when stopping.

INTRODUCTION

The Carolina slate belt includes volcanic and sedimentary rocks of Late Precambrian and possibly Cambrian age, metamorphosed to chlorite and biotite grade, that extend through the Piedmont from Georgia to Virginia. The segment in central North Carolina is probably the best known part of the belt. The Albemarle region includes type localities for several stratigraphic units. Two of the best exposures of the lower units and of metagabbro that intrudes them are described here. These localities are on the

Albemarle 15-minute quadrangle, which was mapped by Conley (1962). Conley and Bain (1965) established a stratigraphic section for the Carolina slate belt, with the Uwharrie Formation as the lowest unit and the Tillery Formation conformable above it (Fig. 3). The uppermost units in this part of the belt were the Morrow Mountain Rhyolite and Badin Greenstone of the Tater Top Group, which were interpreted to lie with angular unconformity above folded older units (Conley and Bain, 1965). Stromquist and Sundelius (1969) reinterpreted part of the middle and upper stratigraphic sequence. They considered the Tater Top Group to be interlayered with other units of the Albemarle Group, rather than unconformably overlying them; consequently, their stratigraphy is somewhat different than that of Conley and Bain (1965), as shown in Figure 3. Rocks in the Albemarle area are mildly deformed and metamorphosed; the dominant structures are open folds plunging southwest and the regional metamorphism is chlorite grade.

The Uwharrie and Tillery Formations are late Proterozoic to Early Cambrian in age. Rb/Sr whole-rock ages (Hills and Butler, 1969) and U/Pb zircon ages (Wright and Seiders, 1980) for rhyolites from the Uwharrie are 550 to 580 Ma. Fossils found so far in this part of the slate belt are all from units above the Tillery Formation and give ambiguous ages. St. Jean (1973) reported Early to Middle Cambrian trilobites from a locality southwest of Albemarle. Recently discovered metazoan fossils

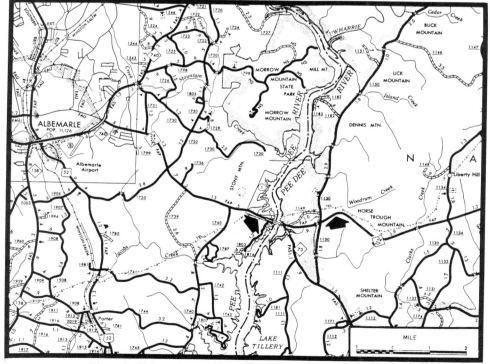

Figure 1. Highway map, showing the location of the field trip stops.

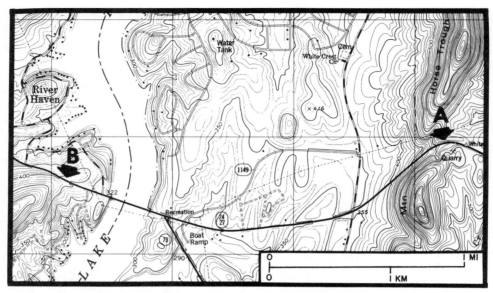

Figure 2. Location map, Morrow Mountain 7½-minute quadrangle, North Carolina.

from localities southwest of Albemarle are identified as a Late Proterozoic Ediacarian fauna (Gibson and others 1984), and the "trilobites" were reclassified as part of the Ediacarian fauna.

Butler and Ragland (1969) discussed petrology and chemistry of rocks in the Albemarle region. Their data included analyses of rhyolite from Site A and metagabbro from Site B. The rhyolite is a sodium-rich rhyolite typical of this part of the Carolina slate belt.

SITE DESCRIPTION

Site A: There is ample parking for several vehicles on the northern side of the highway. Rocks of the Uwharrie Formation are exposed along the highway just east of the parking area and in the large abandoned quarry south of the highway. The quarry is on private land and may be posted, especially during hunting season; if so, permission may be required for access. Site A exposes rhyolitic rocks that are near the top of the Uwharrie Formation. The site is on the western flank of the Troy anticlinorium and bedding dips moderately westward. Rhyolites of the Uwharrie Formation are more resistant than mudstones of the underlying Tillery Formation and form generally higher and more rugged topography; the Pee Dee River cut its valley into the less resistant Tillery mudstones.

The roadcut and quarry at Site A are in rhyolitic epiclastic conglomerates and sandstones and possibly tuffs. Rhyolitic clasts here are generally non-vesicular and extremely fine-grained; many are probably devitrified glass, as indicated by local occurrence of spherulites. Dipyramidal quartz crystals (low quartz paramorphs after high quartz) commonly occur as phenocrysts in the rhyolite and as clasts in reworked tuffaceous material.

There is some flow-banded rhyolite present in the outcrop, but no clear evidence of pumice or flattened pumice has been demonstrated which would indicate the presence of welded tuffs.

These rocks were interpreted as mainly reworked felsic volcanic debris by Gibson and Teeter (1984), with features such as matrix-supported and clast-supported conglomerates with well-rounded and angular clasts, and poorly sorted and well-sorted compositional layers with some cross bedding. The variety of features suggests both subaerial and subaqueous deposition, by streams on a fan-delta environment and by waves and/or currents on the submerged fan-delta front (Gibson and Teeter, 1984, p. 37).

In the quarry, a nearly vertical metabasaltic dike cuts the rhyolitic beds. The transition from Uwharrie to Tillery occurs just downhill from this site and is evident in roadcuts as a change from whitish to reddish soil and saprolite.

Site B: The parking area at Site B is near the crest of a hill underlain by a metagabbro sill (Conley, 1962, Plate 1). Traffic conditions at this stop are dangerous and extreme caution must be used. There is a parking turnout with space enough for about three vans or equivalent vehicles on the north side of the road, just west of the ridge crest. Approach the stop driving west, stay in the right lane of the highway, and carefully pull onto the parking area on the right (north) shoulder of the road. Walk back eastward along the outcrop, staying well onto the shoulder away from the highway and in the shallow drainage ditch. The metagabbro sills are primarily in the Tillery and overlying McManus or Cid Formations, and they form elongate ridges as much as several miles long. Northward from the parking area, Stony Mountain in the left foreground is also underlain by a metagabbro sill. Prominent hills farther north to the right of Stony Mountain and near Lake Tillery are Morrow Mountain, Sugarloaf Mountain, and Tater Top, all underlain by rhyolite. Conley (1962) interpreted these rhyolite bodies to be unconformable above the Tillery, but Stromquist and Sundelius (1969) considered them to be volcanic interbeds or accumulations of lava and volcanic debris around sites of eruption.

Walk eastward from the parking area. The first outcrops are

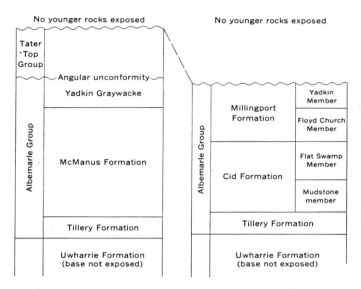

Figure 4. Photograph of contact of metagabbro sill (above) with mudstone (below). Point of hammer is just below the contact.

Figure 3. Stratigraphy of the Carolina slate belt, Albemarle area, North Carolina. Left column after Conley and Bain, 1965. Right column after Stromquist and Sundelius, 1969.

massive, medium-grained metagabbro with some xenoliths of mudstone and cross-cutting thin metabasaltic dikes. The original minerals of the metagabbro, presumably mostly pyroxene and plagioclase, are completely replaced by a low-grade metamorphic assemblage, mainly actinolite, epidote, chlorite, and albite. Pseudomorphs of actinolite after augite are conspicuous in much of the metagabbro.

Farther eastward, the lower contact of the sill with underlying laminated mudstone is well exposed (Fig. 4). The contact is parallel to bedding in the mudstone, which dips moderately westward. The mudstone is mostly laminated to thinly bedded, with some sandy layers locally showing grading. At several places in the outcrop, there are intraformational conglomerates, zones as much as 3 ft (2 m) thick of mudstone clasts in a muddy matrix, bounded above and below by unbroken beds of mudstone. Associated with these conglomerates are small folds and faults indicating layer-parallel movement. These disrupted beds are interpreted

to be soft-sediment deformation, caused by slumping of sediment on the seafloor shortly after bed deposition. The Tillery mudstone ("argillite" in many reports) was deposited in relatively quiet conditions below wave base. The rhyolitic volcanism typical of the Uwharrie had ceased and any highlands had been eroded, as no coarse rhyolitic debris is found in the Tillery. However, rhyolitic and other volcanic deposits are again abundant higher in the stratigraphic section.

Other field trip stops in this region are described by Stromquist and Conley (1959) and by Gibson and Teeter (1984). Morrow Mountain State Park is an interesting geological and archeological site; it is reached by traveling westward on North Carolina 24-27 and following signs to the park (Fig. 1). Flowbanded rhyolite is common on Morrow Mountain. The rhyolite there is homogeneous and fine-grained enough to have excellent conchoidal fracture. The rhyolite was used extensively by Indians to make arrowheads, spear points, and cutting tools of many kinds. The top of Morrow Mountain is covered by many tons of debris from prehistoric point-making. Collecting is not permitted in the park without special written permit.

REFERENCES CITED

Butler, J. R., and Ragland, P. C., 1969, Petrology and chemistry of meta-igneous rocks in the Albemarle area, North Carolina slate belt: American Journal of Science, v. 267, p. 700–726.

Conley, J. F., 1962, Geology of the Albemarle quadrangle, North Carolina: North Carolina Division of Mineral Resources, Bulletin 75, 26 p.

Conley, J. F., and Bain, G. L., 1965, Geology of the Carolina slate belt west of the Deep River-Wadesboro Triassic basin, North Carolina: Southeastern Geology, v. 6, p. 117–138.

Gibson, G. G., and Teeter, S. A., 1984, A stratigrapher's view of the Carolina slate belt, southcentral North Carolina: Carolina Geological Society, Field Trip, Guidebook, 43 p.

Gibson, G. G., Teeter, S. A., and Fedonkin, M. A., 1984, Ediacarian fossils from the Carolina slate belt, Stanly County, North Carolina: Geology, v. 12, p. 387–390.

Hills, F. A., and Butler, J. R., 1969, Rubidium-strontium dates for some rhyolites

from the Carolina slate belt of the North Carolina Piedmont: Geological Society of America Special Paper 121, p. 445.

St. Jean, Joseph, 1973, A new Cambrian trilobite from the Piedmont of North Carolina: American Journal of Science, v. 273-A, p. 196–216.

Stromquist, A. A., and Conley, J. F., 1959, Geology of the Albemarle and Denton quadrangles, North Carolina: Carolina Geological Society, Field Trip, Guidebook, 36 p.

Stromquist, A. A., and Sundelius, H. W., 1969, Stratigraphy of the Albemarle Group of the Carolina slate belt in central North Carolina: U.S. Geological Survey Bulletin 1274-B, 22 p.

Wright, J. E., and Seiders, V. M., 1980, Age of zircon from volcanic rocks of the central North Carolina Piedmont and tectonic implications for the Carolina volcanic slate belt: Geological Society of America Bulletin, Part 1, v. 91, p. 287-294.

The Kings Mountain belt and spodumene Pegmatite District, Cherokee and York Counties, South Carolina, and Cleveland County, North Carolina

J. Wright Horton, Jr., U.S. Geological Survey, 928 National Center, Reston, Virginia 22092
J. Robert Butler, Department of Geology, University of North Carolina, Chapel Hill, North Carolina 27514

LOCATION

The Kings Mountain belt and spodumene Pegmatite district are located in Cherokee and York Counties, South Carolina, and Cleveland County, North Carolina, south of Kings Mountain, North Carolina (Fig. 1). This area includes the most complete stratigraphic section, the most productive mineral deposits, and the most intensively studied parts of the belt.

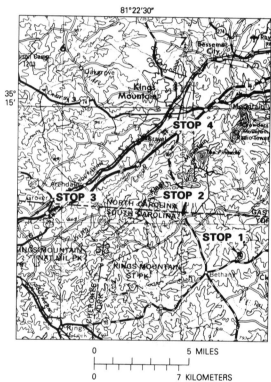

Figure 1. Part of the Charlotte 2-degree Quadrangle showing stop locations.

INTRODUCTION

The Kings Mountain belt, defined by King (1955), is characterized by metasedimentary and metavolcanic rocks that are now designated the Battleground and Blacksburg Formations (Horton, 1984). The rocks of the belt commonly have steep dips and form an area of greenschist-facies and/or epidote–amphibolite facies metamorphism. The Kings Mountain belt is about 110 mi (177 km) long and has a maximum width of about 12 mi (20 km)

(Fig. 2). It is bounded by amphibolite facies rocks of the Inner Piedmont belt on the west and of the Charlotte belt on the east.

The lower part of the Battleground Formation consists of recognizable metavolcanic rocks interlayered with quartz-sericite schist of uncertain origin (Horton, 1984). Metavolcanic facies include hornblende gneiss, feldspathic biotite gneiss, and schistose volcaniclastic rocks. These rocks grade laterally and vertically into quartz-sericite schist. This schist has a quartz content higher than that of normal igneous rocks, lacks volcanic textures, and probably formed from epiclastic or sedimentary material and hydrothermally altered volcanic material. The upper part of the Battleground Formation consists of quartz-sericite schist interbedded with kyanite (or sillimanite) quartzite, quartz-pebble metaconglomerate, spessartine-quartz rock, and quartzite. The Battleground Formation is intruded by metatonalite of Late Proterozoic age (Horton, 1984).

The Blacksburg Formation consists of sericite schist or phyllite with interlayered marble, micaceous quartzite, amphibolite, and minor calc-silicate rock. The sericite schist is commonly graphitic and contains more white mica but less quartz and plagioclase than does schist of the Battleground Formation. The Blacksburg Formation is predominantly metasedimentary in origin, but amphibolite lenses have basaltic compositions and may be metamorphosed sills or flows (Horton, 1984). The age of the Blacksburg Formation is unknown, and its stratigraphic relationship to the Battleground Formation, if any, is uncertain because of intervening faults.

As many as five episodes of folding and related deformation have been recognized in the Kings Mountain belt (Horton, 1981b). The pattern of rock units shown on Figure 3 is controlled largely by folds of the two earliest episodes, F_1 and F_2. The largest map-scale folds are the South Fork antiform and Cherokee Falls synform, which are interpreted as F_2 structures (Horton, 1981b). Folds are locally disrupted by tectonic slides or ductile faults that are roughly parallel to the regional schistosity (Butler, 1981; Horton, 1981b). Structures younger than F_2 are conspicuous in the major shear zones but sporadically distributed elsewhere and rarely affect the map pattern.

Ductile shear zones are located along both margins of the Kings Mountain belt and within it. The most significant of these, the Kings Mountain shear zone, separates the Kings Mountain belt from the Inner Piedmont belt to the northwest (Figs. 2 and 3). Rock units and metamorphic isograds on both sides of the shear zone are truncated against it (Horton, 1981b). The Kings Mountain belt is separated from the Charlotte belt to the south-

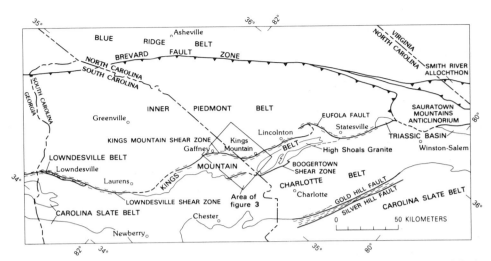

Figure 2. Generalized geologic setting of the Kings Mountain belt modified from Horton and Butler (1981). Area of Figure 3 is outlined.

east by the Boogertown shear zone in southernmost North Carolina, but structural breaks are lacking along this boundary in other areas (Fig. 2).

Spodumene pegmatite forms dikes in an elongate zone, conservatively at least 25 mi (40 km) long and 0.6 to 2 mi (1 to 3 km) wide, trending northeast from the South Carolina–North Carolina state line near Grover to a few miles (kilometers) east of Lincolnton, N.C. This zone, originally called the Carolina "tin-spodumene belt" (Kesler, 1942), is located near the southeast edge of the Inner Piedmont belt within 2.2 mi or 3.5 km (west) of the Kings Mountain shear zone. Dikes of spodumene pegmatite also are present in the shear zone (Horton, 1981a). All dikes of spodumene pegmatite in the area are also within 2 mi (3 km) (east) of the Cherryville Granite. Pegmatites that lack spodumene are found in and near the granite and locally in gradational contact with spodumene pegmatite (Kesler, 1961; 1976). The Cherryville Granite and both types of pegmatite are of Mississippian age (Kish, 1977; 1983). The Carolina tin-spodumene belt constitutes one of the largest developed reserves of lithium in the world (Evans, 1978). Additional field trip stop descriptions in the area are available in Horton and Butler (1977) and Horton and others (1981).

SITE DESCRIPTIONS

The first three stops show distinctive rock units of the Battleground and Blacksburg Formations. The fourth stop shows dikes of spodumene pegmatite in and adjacent to the Kings Mountain shear zone.

Stop 1: Kyanite Quartzite at Henry Knob

Henry Knob (Kings Mountain, N.C.-S.C., 7½-minute Quadrangle) is located in York County, S.C. (Fig. 1). From I-85 at Kings Mountain, N.C., exit onto North Carolina 161 (=South

Carolina 161) and drive south 7.9 mi (12.7 km); at Bethany (McGill's store) turn left (east) on South Carolina 55 and drive 1.8 mi (2.9 km); turn left (north) on Secondary Road 508 and drive 200 ft (60 m). Park at the base of Henry Knob (beyond the parking area with trash containers); walk up the dumps on the right and continue about 1,000 ft (300 m). Henry Knob is part of the Henry's Knob Recreational Facility, owned by the State of South Carolina. Visitors should check in with the resident caretaker; you may have to sign a liability waiver.

Henry Knob is a prominent monadnock of kyanite quartzite, one of the characteristic rock units of the Battleground Formation. It was extensively mined for kyanite from 1948 through 1969. A detailed geologic map is available from Espenshade and Potter (1960, Plate 10).

Kyanite quartzite (or kyanite-quartz rock) at Henry Knob forms conformable lenses as much as 165 ft (50 m) thick surrounded by quartz–white mica schist of the Battleground Formation. Espenshade and Potter (1960) reported an average mode (in volume percent) of 65% quartz, 28% kyanite, and 7% pyrite based on nine thin sections of kyanite quartzite. The abundance of pyrite at Henry Knob is much higher than that of most kyanite and sillimanite quartzite units of the Battleground Formation, which generally contain less than 1% pyrite. Rutile and white mica are common accessory minerals. Lazulite, barite (fluorescent), and topaz (fluorescent) have also been observed. The kyanite quartzite is medium gray and typically consists of kyanite crystals as long as 0.4 in (1 cm) in a matrix of fine- to coarse-grained quartz. Kyanite appears as blades parallel to the foliation and as random aggregates and radial clusters that cross the foliation. Quartz inclusions are abundant in the kyanite. Pyritic white mica schist interlayered with the kyanite quartzite contains minor amounts of kyanite, andalusite, staurolite, and tourmaline (Espenshade and Potter, 1960). Cordierite has also been reported from Henry Knob (H. H. Posey, oral communication, 1983). The mineralogy of Henry Knob is essentially a reflection of

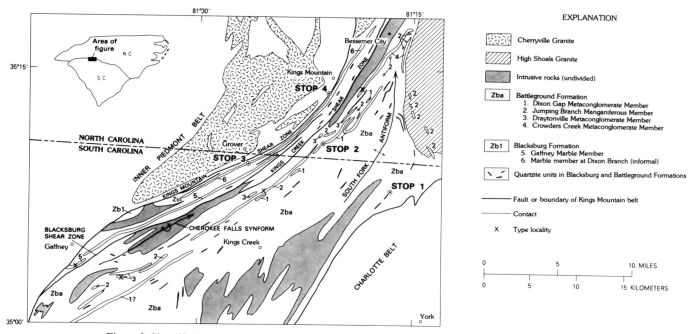

Figure 3. Simplified geologic map of the central part of the Kings Mountain belt modified from Horton (1984) showing field guide stop locations: 1, Henry Knob; 2, Dixon Gap; 3, Grover quarry; 4, Foote mine.

Alleghanian-age isochemical metamorphic conditions (Horton and Stern, 1983; Sutter and others, 1984).

Espenshade and Potter (1960) found evidence for hydrothermal leaching of volcanic rocks at similar deposits in other areas of the Piedmont but favored a sedimentary origin for high-alumina quartzites in the Battleground Formation. Kyanite quartzite in the vicinity of the Pinnacle (Fig. 1) grades laterally into quartz-pebble metaconglomerate that can be traced for many miles and therefore almost certainly had a sedimentary origin (Horton, 1981b). Other high-alumina quartzites may have formed by metamorphism of fine-grained silica and clay produced by hydrothermal alteration of volcanic or epiclastic material in hot springs or solfataras. The abundant pyrite, presence of rutile and lazulite, and thin, discontinuous lenses favor this latter interpretation at Henry Knob.

The compositional layering (S_0) and schistosity (S_1) at Henry Knob strike about N.30°E. and are nearly vertical. A weak crenulation cleavage (S_2?) is nearly parallel to S_1; crenulation-axis lineations plunge gently southwest where the two surfaces intersect (Horton, 1977, p. 79). A northwest-trending diabase dike about 50 ft (15 m) thick crosses the base of Henry Knob on the southwest side.

The mine at Henry Knob was a major domestic producer of kyanite during the 1950s and 1960s. Pyrite was recovered as a byproduct from 1960 until the mine became inactive in 1969. Henry Knob is now of interest not only to geologists but also to environmental scientists in other fields because of serious reclamation problems caused by the tailings, dumps, and acid mine water.

Stop 2: Conglomerate at Dixon Gap

Dixon Gap (Kings Mountain, N.C.-S.C., 7½-minute quadrangle) is located in Cleveland County, N.C. (Fig. 1). From Henry Knob, return to South Carolina 161 and head north; turn left (west) on paved road S-46-705 3.2 mi (5.2 km) north of Bethany and 0.3 mi (0.5 km) south of the North Carolina state line. Drive 0.2 mi (0.3 km) and turn right (northwest) on a gravel road. After 1.3 mi (2.1 km), the gravel road crosses the state line and becomes North Carolina Secondary Road 2245. Continue another 1.2 mi (1.9 km), passing the intersection with Secondary Road 2288. Park 660 ft (200 m) beyond that intersection on the far (northwest) end of the road cut at the crest of Kings Mountain ridge. A foot trail leads 1,000 ft (300 m) east-northeast (uphill) to large exposures of metaconglomerate.

The gravel road in South Carolina that becomes Secondary Road 2245 in North Carolina can be difficult in wet weather and may be impassable for buses and other large vehicles. In these cases, one can reach Dixon Gap from I-85 by taking the exit immediately to the southwest of the exit for North Carolina 161. Exit onto Dixon School Road (Secondary Road 2283); turn left (southwest) and follow it 1.1 mi (1.8 km) to Dixon Church; turn left (southeast) on Secondary Road 2245 and drive 1.5 mi (2.4 km) to the crest of Kings Mountain ridge at Dixon Gap.

Although privately owned, the land north of the road at Dixon Gap is not posted or fenced and has been open to hikers and picnickers for many years. Please respect the property and keep it clean so this accessibility can continue.

This is the type locality of the Dixon Gap Metaconglomerate Member, one of several quartz-pebble metaconglomerate units in the Battleground Formation. Chloritoid-bearing phyllite and sericitic quartzite or quartz-sericite schist of the Battleground Formation are also exposed.

The Dixon Gap Metaconglomerate Member of the Battleground Formation, named for this locality (Horton, 1984), forms this segment of Kings Mountain ridge. The quartz-pebble metaconglomerate is light gray, generally clast supported, and consists of 50–60% pebble-size clasts and 40–50% matrix. The pebbles are moderately sorted and have an average diameter of about 0.4 in (1 cm) although pebbles as long as several inches can be observed. They are mainly white quartz with smaller amounts of gray magnetite-hematite-quartz rock that contains 15–40% specular hematite and/or magnetite. Pebbles of fine-grained hornblende gneiss and quartz–white mica phyllite have been observed but are uncommon. Visual estimates from three thin sections indicate that the conglomerate at this locality has a matrix composition of about 50% quartz, 25% white mica, 10% chloritoid, 5% kyanite, and 10% opaque oxides. Quartz grains in the matrix are angular to rounded, poorly sorted, and have an average diameter of about 0.06 in (1.5 mm). The rock is pressure welded and cemented by recrystallized quartz. Locally, the metaconglomerate grades into layers of pebbly metasandstone or quartzite. Pebbles show no evidence of deformation or metamorphism prior to deposition. Pebbles of quartz and magnetite-hematite-quartz rock resemble layers from quartzite and ferruginous quartzite units within the Battleground Formation. The hornblende gneiss and phyllite pebbles are similar to metavolcanic and metasedimentary facies in the lower part of the Battleground Formation. The pebble compositions, angularity, and dirty matrix suggest an intraformational source (Horton, 1977; France and Brown, 1981). The coarse grain size indicates a high-energy depositional environment, possibly the littoral zone of an ancient shoreline as suggested by Horton and Butler (1977).

Bedding in the conglomerate strikes N.44°E. and dips 75°NW. Questionable graded beds suggest that the section becomes younger to the northwest (Horton and Butler, 1977). Pebbles are slightly flattened in the plane of regional schistosity (S_2), which is believed to be axial planar to the South Fork antiform, but the conglomerate at Dixon Gap is not strongly deformed. Flattened pebbles in another metaconglomerate, the Draytonville Metaconglomerate Member of the Battleground Formation, are used as strain indicators by Hatcher and Morgan (1981).

Chloritoid-bearing phyllite and micaceous quartzite of the Battleground Formation crop out along the road cut east of the conglomerate at Dixon Gap. The schistosity, S_2, which is axial planar to the South Fork antiform (Fig. 3), is slightly steeper than the bedding, S_0. Both surfaces dip steeply northwest and both strike northeast, with the bedding consistently to the east of the schistosity as expected here on the west limb of the South Fork antiform (Horton and Butler, 1977).

The road cuts at Dixon Gap contain a complete set of post-F_2 structural features (Horton, 1977) and are therefore a good place to examine the sequence of overprinting. Two sets of crenulations that plunge gently northeast have been observed on S_2. The weaker crenulation cleavage, S_3, related to F_3 crenulations, is commonly not visible because it is nearly parallel to S_2. Larger F_4 crenulations are associated with a prominent subhorizontal crenulation cleavage, S_4, which dips gently northeast. Nearly vertical kink planes, S_5, strike about N. 50° W. F_5 kink folds plunge steeply northwest where these kink planes intersect S_2. A more detailed analysis of the structure at Dixon Gap is available in Horton (1977, p. 85–88), and a table summarizing structural features in this part of the Kings Mountain belt is available in Schaeffer (1981, p. 75).

The bluffs of metaconglomerate near Dixon Gap provide a panoramic view to the northwest. About 0.9 mi (1.4 km) to the northwest is the valley of Kings Creek, which in this area is located approximately along the Kings Creek shear zone. This shear zone separates the Battleground Formation from the Blacksburg Formation on the far (northwest) side. The cluster of white open pits and tailings ponds 2.1 mi (3.4 km) to the northwest is the Moss mica mine operated by King Minerals, Inc. This mine in pegmatite and coarse-grained Cherryville Granite of the Inner Piedmont belt is described as a field trip stop by Horton and others (1981).

Stop 3: Marble at the Grover Quarry

The Grover quarry of Vulcan Materials Company (Grover, N.C.-S.C. 7½-minute Quadrangle) is located in Cherokee County, South Carolina (Fig. 1). From I-85, take the Grover, North Carolina, exit (0.5 mi or 0.8 km south of the N.C. state line) onto U.S. 29 and turn south. Just south of the U.S. 29 and I-85 interchange, turn left (northeast) onto the frontage road parallel to I-85. After 0.6 mi (1.0 km), this road turns to the right, away from I-85, and leads 0.5 mi (0.9 km) to the quarry. Advance permission, hard hats, and safety glasses are required to enter the quarry. Permission should be requested by contacting the Quarry Superintendent, Vulcan Materials Company, Blacksburg, SC 29702 (phone: 803-839-2344).

This is the type locality of the marble member at Dixon Branch, a major unit within the Blacksburg Formation. It may or may not be correlative with the Gaffney Marble Member (Horton, 1984). The following discussion is adapted from a field trip stop description by Horton and others (1981, p. 219–220).

The medium light-gray to medium dark-gray banded dolomitic marble at this quarry is typical of the marble member at Dixon Branch in most respects but appears more Mg-rich and more graphitic than average. Horton and others (1981, p. 220) reported a chemical analysis in weight percent (provided by Vulcan Materials Company) of 50.4% $CaCO_3$, 31.7% $MgCO_3$, 2.3% Fe and Al oxides, and 15.6% acid-insoluble material. This represents the commercial product that may include minor amounts of amphibolite mixed with the marble. The stone quarried here is used as agricultural lime and crushed stone.

The thickness of the marble at this locality is about 330 ft (100 m). The compositional layering and parallel foliation are oriented about N. 73°E., 45° NW. A mineral lineation plunges about 30° in a N. 50° E. direction. Mats of white tremolite needles are locally concentrated along foliation planes. Lenses of amphibolite and calc-silicate rock are present within the marble. The marble unit interfingers with medium-gray, fine-grained biotite-sericite schist on both sides. The metamorphic grade here is lower amphibolite facies.

Vulcan Materials Company shifted its operation from the older Blacksburg quarry to this site in 1979. The Blacksburg quarry, 1.2 mi (2 km) southwest of the Grover quarry, exposes a variety of folds, as well as late brittle faults, that have been analyzed in detail by Horton (1977). It was described as a field trip stop by Horton and Butler (1977) and remains accessible with permission, but is largely filled with water.

Stop 4: Spodumene Pegmatite and Host Rocks at the Foote Mine

The Foote mine (Kings Mountain 7½-minute Quadrangle) is on the south side of the town of Kings Mountain in Cleveland County, N.C. (Fig. 1). From I-85, exit north (toward Kings Mountain) on North Carolina 161 and abruptly turn left (southwest) onto the frontage road. This road parallels I-85 for 0.6 mi (1 km) and then curves right, leading to the mine office. Hard hats and safety glasses are required. Access is at the company's discretion. Permission for visits should be requested a month or more in advance by contacting the Operations Manager, Foote Mineral Company, P.O. Box 792, Kings Mountain, NC 28086 (phone: 704-739-2501).

This is one of two large open-pit mines in spodumene pegmatite of the "Carolina tin-spodumene belt." Unlike the Lithium Corporation of America mine, which is located entirely in the Inner Piedmont, the Foote mine straddles the boundary between the Kings Mountain and Inner Piedmont lithotectonic belts (Fig. 2). A geologic map of the Foote mine and vicinity is available in Horton and others (1981, Fig. 4).

Fine-grained phyllonitic schist of the Blacksburg Formation is exposed along the southeast wall of the pit; it is located within the Kings Mountain shear zone and represents the northwest edge of the Kings Mountain belt. The phyllonitic schist consists of muscovite, biotite, plagioclase, and quartz, with accessory garnet, staurolite, andalusite, tourmaline, and chlorite (Horton, 1977). The sharp contact between this rock and amphibolite of the Inner Piedmont belt, which underlies most of the pit, is well exposed. The amphibolite is dark gray to greenish black, fine grained, and thinly layered. It is composed of green hornblende, plagioclase, and quartz with small amounts of sphene, epidote, and other accessory minerals (Horton, 1977). The amphibolite contains layers of calc-silicate minerals and calcite as well as chloritic biotite schist. Pegmatite dikes intruded into schist are typically parallel to the steep schistosity; those intruded into amphibolite are more irregular and discordant. Amphibolite adjacent to the dikes is commonly brecciated (Kesler, 1961).

The spodumene pegmatite is generally homogeneous, although compositional zoning is found on a small scale. The aver-

age composition by weight is 20% spodumene, 32% quartz, 27% albite, 14% microcline, 6% muscovite, and 1% trace minerals (Kesler, 1961). Primary trace minerals include beryl (0.4%), manganapatite, zircon, ferrocolumbite, and cassiterite. Coarse-grained spodumene and microcline (generally less than 12 in [30 cm] long) and minor muscovite (generally less than 1 in [2 cm] long) are found in a granular matrix of fine- to medium-grained quartz and albite. Fractures in broken crystals of spodumene and microcline are filled with fine-grained albite and quartz. Border zones, which extend less than 3.3 ft (1 m) from the contacts, are composed of fine-grained albite and quartz and contain relatively little spodumene or microcline. Pods of coarse-grained zoned pegmatite are found locally within the more homogeneous pegmatite. The Rb-Sr whole-rock age of the pegmatite is 340 ± 5 Ma (Kish, 1983).

Amphibolite is altered to biotite schist within about 24 in (60 cm) of the pegmatite contacts (Kesler, 1961). Holmquistite, a lithium amphibole, is found in these alteration zones and is most abundant adjacent to the pegmatite. Kesler (1961) also noted partial replacement of biotite by chlorite and associated sulfide mineralization within about 70 ft (21 m) of the pegmatites.

Two generations of aplite dikes have been recognized (Horton and Simpson, 1978; Horton and others, 1981). The younger aplite dikes, unlike the older ones, contain trace amounts of spodumene, have chilled margins, and have biotite-rich contact aureoles in the amphibolite.

The structural geology of the Foote mine has been analyzed in detail (Horton, 1977; Horton and Simpson, 1978), and five fold episodes have been recognized (Horton and others, 1981). Intrafolial, isoclinal F_1 folds have an axial-plane foliation, S_1. F_2 folds are tight to isoclinal, plunge gently to moderately southwest and northeast, and have an axial-plane schistosity, S_2, oriented about N. 34° E., 58° NW (Horton and others, 1981). S_2 is the dominant foliation, but S_2 and S_1 are generally subparallel, and their intersection produces thin wedges and crenulations in the schist. S_3 is a steep, northwest-dipping slip cleavage associated with the Kings Mountain shear zone; related F_3 crenulations formed by the intersection of S_3 and S_2 plunge gently southwest. Spodumene pegmatite dikes are commonly parallel to S_3. Most are undeformed but some have a gneissoid foliation parallel to S_3 in the host rocks. Flowage folds within the pegmatite that do not affect the contacts may be related to emplacement of the dikes (Horton and others, 1981). F_4 kink folds and open folds plunge gently northeast and southwest and have axial surfaces that dip gently eastward (Horton and others, 1981). Open, upright cross folds and small open crenulations, F_5, plunge northwest. Two sets of nearly vertical faults, striking N. 15° E. and N 85° W., respectively, have been described (Horton and Butler, 1977; Horton and others, 1981). A more detailed summary of the structural geology is available from Horton and others (1981).

Approximately 100 mineral species that have been identified at the Foote mine are listed by Marble and Hanahan (1978). Metamorphic and hydrothermal minerals at wallrock contacts include holmquistite, biotite, tourmaline (schorl), pyrrhotite, garnet (spesssartine-grossular), ferroaxinite, epidote-clinozoisite, albite, and quartz. Secondary hydrothermal minerals in joints and vugs include apatite, fairfieldite, switzerite, roscherite, lithiophilite, lithiophosphate, eosphorite, vivianite, tetrawickmanite, bikitaite, eucryptite, bertrandite, bavenite, milarite, brannockite, eakerite, swinefordite, tin-bearing titanite, and rhodochrosite-siderite (White, 1981). Supergene minerals include birnessite, cryptomelane, gypsum, and hydrated phosphates of manganese

and iron (White, 1981). More than 30 phosphate minerals have been identified (White, 1981). The list of new mineral species first recognized and described from this mine includes three phosphate minerals (switzerite, kingsmountite, and an unnamed triclinic analog of roscherite) and three tin minerals (brannockite, eakerite, and tetrawickmanite) (White, 1981). Readers interested in the mineralogy of the Foote mine should refer to White (1981) and references therein.

REFERENCES CITED

Butler, J. R., 1981, Geology of the Blacksburg South quadrangle, South Carolina, *in* Horton, J. W., Jr., Butler, J. R., and Milton, D. J., eds., Geological investigations of the Kings Mountain belt and adjacent areas in the Carolinas (Carolina Geological Society Field Trip Guidebook 1981): Columbia, South Carolina Geological Survey, p. 65–71.

Espenshade, G. H., and Potter, D. B., 1960, Kyanite, sillimanite, and andalusite deposits of the Southeastern States: U.S. Geological Survey Professional Paper 336, 121 p.

Evans, R. K., 1978, Lithium reserves and resources: Energy, v. 3, p. 379–385.

France, N. A., and Brown, H. S., 1981, A petrographic study of Kings Mountain belt metaconglomerates, *in* Horton, J. W., Jr., Butler, J. R., and Milton, D. J., eds., Geological investigations of the Kings Mountain belt and adjacent areas in the Carolinas (Carolina Geological Society Field Trip Guidebook 1981): Columbia, South Carolina Geological Survey, p. 91–99.

Hatcher, R. D., Jr., and Morgan, B. K., 1981, Finite strain and regional implications of the deformed Draytonville metaconglomerate near Gaffney, South Carolina, *in* Horton, J. W., Jr., Butler, J. R., and Milton, D. J., eds., Geological investigations of the Kings Mountain belt and adjacent areas in the Carolinas (Carolina Geological Society Field Trip Guidebook 1981): Columbia, South Carolina Geological Survey, p. 100–109.

Horton, J. W., Jr., 1977, Geology of the Kings Mountain and Grover quadrangles, North and South Carolina [Ph.D. thesis]: Chapel Hill, University of North Carolina, 174 p.

——1981a, Shear zone between the Inner Piedmont and Kings Mountain belts in the Carolinas: Geology, v. 9, p. 28–33.

——1981b, Geologic map of the Kings Mountain belt between Gaffney, South Carolina, and Lincolnton, North Carolina, *in* Horton, J. W., Jr., Butler, J. R., and Milton, D. J., eds., Geological investigations of the Kings Mountain belt and adjacent areas in the Carolinas (Carolina Geological Society Field Trip Guidebook 1981): Columbia, South Carolina Geological Survey, p. 6–18.

——1984, Stratigraphic nomenclature in the Kings Mountain belt, North Carolina and South Carolina, *in* Stratigraphic Notes, 1983: U.S. Geological Survey Bulletin 1537-A, p. A59–A67.

Horton, J. W., Jr., and Butler, J. R., 1977, Guide to the geology of the Kings Mountain belt in the Kings Mountain area, North Carolina and South Carolina, *in* Burt, E. R., ed., Field guides for Geological Society of America, Southeastern Section Meeting, Winston-Salem, North Carolina: Raleigh, North Carolina Department of Natural and Economic Resources, p. 76–143.

——1981, Geology and mining history of the Kings Mountain belt—A summary and status report, *in* Horton, J. W., Jr., Butler, J. R., and Milton, D. J., eds., Geological investigations of the Kings Mountain belt and adjacent areas in the Carolinas (Carolina Geological Society Field Trip Guidebook 1981): Columbia, South Carolina Geological Survey, p. 194–212.

Horton, J. W., Jr., Butler, J. R., Schaeffer, M. F., Murphy, C. F., Connor, J. M., Milton, D. J., and Sharp, W. E., 1981, Field guide to the geology of the Kings Mountain belt between Gaffney, South Carolina, and Lincolnton, North Carolina, *in* Horton, J. W., Jr., Butler, J. R., and Milton, D. J., eds., Geological investigations of the Kings Mountain belt and adjacent areas in the Carolinas (Carolina Geological Society Field Trip Guidebook 1981): Columbia, South Carolina Geological Survey, p. 213–247.

Horton, J. W., Jr., and Simpson, M. G., 1978, Structural geology at the Foote Mine, Kings Mountain, N.C.: Geological Society of America Abstracts with Programs, v. 10, no. 4, p. 171–172.

Horton, J. W., Jr., and Stern, T. W., 1983, Late Paleozoic (Alleghanian) deformation, metamorphism, and syntectonic granite in the central Piedmont of the Southern Appalachians: Geological Society of America Abstracts with Programs, v. 15, no. 6, p. 599.

Kesler, T. L., 1942, The tin-spodumene belt of the Carolinas: U.S. Geological Survey Bulletin 936-J, p. 245–269.

——1961, Exploration of the Kings Mountain pegmatites: Mining Engineering, v. 13, p. 1062–1068.

——1976, Occurrence, development, and long range outlook of lithium-pegmatite ore in the Carolinas, *in* Vine, J. D., ed., Lithium resources and requirements by the year 2000: U.S. Geological Survey Professional Paper 1005, p. 45–50.

King, P. B., 1955, A geologic section across the southern Appalachians: An outline of the geology in the segment in Tennessee, North Carolina and South Carolina, *in* Russell, R. J., ed., Guides to southeastern geology: New York, Geological Society of America, p. 332–373.

Kish, S. A., 1977, Geochronology of plutonic activity in the Inner Piedmont and Kings Mountain belt in North Carolina, *in* Burt, E. R., ed., Field guides for Geological Society of America, Southeastern Section meeting, Winston-Salem, North Carolina: Raleigh, North Carolina Department of Natural and Economic Resources, p. 144–149.

——1983, A geochronological study of deformation and metamorphism in the Blue Ridge and Piedmont of the Carolinas [Ph.D. thesis]: Chapel Hill, University of North Carolina, 220 p.

Marble, Laura, and Hanahan, Jack, Jr., 1978, The Foote minerals: Rocks and Minerals, v. 53, no. 4, p. 158–173.

Schaeffer, M. F., 1981, Polyphase folding in a portion of the Kings Mountain belt, north-central South Carolina, *in* Horton, J. W., Jr., Butler, J. R., and Milton, D. J., eds., Geological investigations of the Kings Mountain belt and adjacent areas in the Carolinas (Carolina Geological Society Field Trip Guidebook 1981): Columbia, South Carolina Geological Survey, p. 72–90.

Sutter, J. F., Horton, J. W., Jr., and Kunk, M. J., 1984, Timing of Alleghanian metamorphism in the Kings Mountain belt of North Carolina and South Carolina: Geological Society of America Abstracts with Programs, v. 16, no. 3, p. 201.

White, J. S., 1981, Mineralogy of the Foote mine, Kings Mountain, North Carolina, *in* Horton, J. W., Jr., Butler, J. R., and Milton, D. J., eds., Geological investigations of the Kings Mountain belt and adjacent areas in the Carolinas (Carolina Geological Society Field Trip Guidebook 1981): Columbia, South Carolina Geological Survey, p. 39–48.

Diabase dike near Lancaster, South Carolina: The "Great Dyke of South Carolina"

James Robert Butler, Department of Geology, University of North Carolina, Chapel Hill, North Carolina 27514

LOCATION

The large diabase dike and adjacent granite that it intruded are well exposed in road cuts along US Highway 601, about midway between Lancaster and Pageland in the northeastern South Carolina Piedmont (Figs. 1 and 2), about 50 mi (80 km) southeast of Charlotte, North Carolina. The outcrop is about 4,000 ft (1,200 m) south of Flat Creek and 0.9 mi (1.5 km) northeast of the village of Midway (intersection of US 601 and South Carolina 903). There is ample parking on the grassy shoulder of the highway just north of the outcrop, large enough even for several buses; in wet weather, be careful of soggy ground.

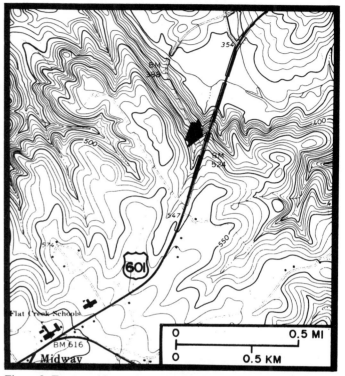

Figure 2. Topographic map showing location of the diabase dike outcrop. Taxahaw 7½-minute Quadrangle.

Figure 1. Road map showing locality.

INTRODUCTION

This large northwest-trending diabase dike is 1,123 ft (342 m) thick, which makes it one of the thickest diabase dikes in the eastern United States. It is part of a province of Mesozoic diabase dikes that extends through much of eastern North America (King, 1971; Ragland and others, 1983). In the southeastern United States, the dikes dominantly trend northwest, but there is also a group that trends almost north-south (Ragland and others, 1983). At this locality, the dike intrudes granite of the Pageland batholith, which is one pluton in a belt of late Paleozoic intrusions with ages of about 300 m.y. (Fullagar and Butler, 1979). The dike has contact-metamorphosed the adjacent granite. The dike is at the southeastern end of one of the most lengthy swarms in the Appalachian region; dikes extend along strike to the northwest of this locality for 125 mi (200 km) and one crosses the Brevard zone into the Grandfather Mountain window (King, 1971).

Steele (1971) chemically analyzed 95 samples from this road cut. The dike is a relatively homogeneous tholeiite with about 53 percent SiO_2, but there are slight modal and chemical variations that are symmetrical about the center of the dike. For example, alumina shows maxima of about 18 percent near each margin, but falls to 16 percent near the center. The dike contains 0.4 to 5.0 percent modal olivine that is extensively replaced by iddingsite (Steele, 1971). In spite of the modal olivine, chemically the dike samples are quartz normative (Steele and Ragland, 1976). This contrasts with most other dikes in the region, which are olivine tholeiites in both norms and modes, containing as much as 20 percent modal olivine (Butler and Howell, 1977). Steele and Ragland (1976) developed a model to explain the variations across the dike in this road cut by two pulses of similar tholeiitic magma, each undergoing some crystal fractionation. Additional information on dikes in this area is given by Butler and Howell (1977), Bell and others (1979), and Ragland and others (1983).

SITE DESCRIPTION

Proceeding southward along the road cut from the parking area, you first see spheroidal boulders of the Pageland granite. These are surrounded by saprolite of the granite. Granite near the dike contact is visibly changed by contact metamorphism into a darker rock that is partly recrystallized. A thin section of a sample taken 6 ft (2 m) from the contact contains augite, a mineral not observed in unmetamorphosed Pageland granite; therefore, the contact aureole probably reached pyroxene-hornfels facies of contact metamorphism.

The granite-dike contact appears as a distinct color change in the weathered rock, from light-colored saprolite of the granite to brownish-red saprolite of the diabase. This road cut was remarkably good when first made in 1966 and was dominantly composed of fresh rock; but weathering, soil creep, slumping, and plant growth are causing rapid deterioration. Proceeding south along the road cut, one passes the other contact of the dike, where the weathering is noticeably more intense. At the south end of the exposure, cross-bedded sands and gravels of the Middendorf Formation (Late Cretaceous) nonconformably overlie weathered granite. Rocks near the nonconformity have been heavily weathered twice, once in the Mesozoic before deposition of the Middendorf Formation and again in the Holocene.

The dike is thickest where it cuts the granite, as at this locality; the diabase narrows to the northwest and is only 115 ft (35 m) thick where it crosses into metavolcanic rocks 1.2 mi (2 km) from here (Butler and Howell, 1977). The dike can be traced only about 5 mi (8 km) northwest of here, but other dikes en echelon and parallel to it are numerous. The dike is covered by Coastal Plain sediments just southeast of this road cut, but it may continue for at least 9 mi (15 km) beneath the Coastal Plain, based on linear anomalies on aeromagnetic maps (Bell and others, 1979).

The Pageland granite is well exposed at Forty-Acre Rock, which is only 1.2 mi (2 km) north of this road cut. Forty-Acre Rock is a spectacular example of a "flatrock" type of exposure developed on the post-metamorphic granites near the Coastal Plain overlap. To reach Forty-Acre Rock, turn left (north) off US 601 just north of Flat Creek (intersection is visible from the parking area); go 2 mi (3 km) and turn left again onto an unpaved, dead-end road that leads to the exposure. Other field guides to this region are by Bell and others (1974) and by Hatcher and Butler (1979).

REFERENCES CITED

Bell, Henry, III, Books, K. G., Daniels, D. L., Huff, W. E., Jr., and Popenoe, Peter, 1979, Diabase dikes in the Haile-Brewer area, South Carolina, and their magnetic properties: U.S. Geological Survey Professional Paper 1123-C, 18 p.

Bell, Henry, III, Butler, J. R., Howell, D. E., and Wheeler, W. H., 1974, Geology of the Piedmont and Coastal Plain near Pageland, South Carolina, and Wadesboro, North Carolina: Carolina Geological Society Guidebook, 23 p.

Butler, J. R., and Howell, D. E., 1977, Geology of the Taxahaw quadrangle, Lancaster County, South Carolina: South Carolina Division of Geology, Geologic Notes, v. 20, p. 133–149.

Fulllagar, P. D., and Butler, J. R., 1979, 325 to 265 m.y.-old granitic plutons in the Piedmont of the Southeastern Appalachians: American Journal of Science, v. 279, p. 161–185.

Hatcher, R. D., Jr., and Butler, J. R. (compilers), 1979, Guidebook for Southern Appalachian Field Trip in the Carolinas, Tennessee, and northeastern Georgia: International Geological Correlation Program, Caledonide Orogen Project, 171 p.

King, P. B., 1971, Systematic pattern of Triassic dikes in the Appalachian region-Second Report: U.S. Geological Survey Professional Paper 750-D, p. D84–D88.

Ragland, P. C., Hatcher, R. D., Jr., and Whittington, David, 1983, Juxtaposed Mesozoic diabase dike sets from the Carolinas: A preliminary assessment: Geology, v. 11, p. 394–399.

Steele, K. F., 1971, Chemical variations parallel and perpendicular to strike in two Mesozoic dolerite dikes, North Carolina and South Carolina [Ph.D. thesis]: Chapel Hill, University of North Carolina, 203 p.

Steele, K. F., and Ragland, P. C., 1976, Model for the closed-system fractionation of a dike formed by two pulses of dolerite magma: Contributions to Mineralogy and Petrology, v. 57, p. 305–316.

Geology of the Haile Gold Mine, Lancaster County, South Carolina

Henry Bell, U.S. Geological Survey, Reston, Virginia 22092

LOCATION

Three miles east of Kershaw, Lancaster County, South Carolina, there are two groups of mine workings referred to collectively as the Haile Gold Mine. Their location on opposite sides of Haile Gold Mine Creek and the surrounding mine area is shown on Figure 1. The mine is easily accessible from U.S. 601, South Carolina 188, and unpaved county roads. There are several property owners; however, in early 1984 the property was operated by Mineral Mining Corporation, Kershaw, South Carolina.

INTRODUCTION

The Haile Mine, discovered about 1827, is the most important of the numerous gold mines in eastern South Carolina. Gold production of 296,000 ounces from both placer and lode deposits has followed the cyclic pattern characteristic of mines in the Appalachian piedmont. Beginning with the high grade placer deposits, gold mining progressed to large low grade lode deposits associated with strata-bound massive pyrite bodies and extensive hydrothermal alteration. Subsequent to the end of gold mining in 1942, the altered rocks have been mined for fillers and extenders for use in various industries. The quartz-sericite alteration of the felsic metavolcaniclastic host rocks and the essentially strata-bound habit of the gold ore bodies has implied to some an epithermal volcanogenic origin (Worthington and others, 1980; Spence and others, 1980). The old mine workings described in reports by Graton (1906), Pardee and Park (1948), Newton and others (1940) and others are now obliterated by waste from the mining of the industrial minerals back filled into the open pits and by development for renewed gold mining. New mining operations, however, continually produce new exposures and generate opportunities for new insights into the structure and origin of the ore deposit.

SITE DESCRIPTION

The mine is in lower Paleozoic metamorphic rocks of the Carolina slate belt which locally contain upper Middle Cambrian trilobites in central South Carolina (Samson and others, 1982). The ore deposits (Fig. 2) are in the upper part of a predominantly metavolcaniclastic unit, the Persimmon Fork Formation, originally rhyolitic or rhyodacitic in composition, where it is interlayered with thinly bedded metasedimentary rocks, probably the Richtex Formation (Bell, in preparation). Primary structures resembling spindle-shaped lapilli, bombs, and pumaceous fragments are locally so abundant that certain rocks are readily recognized as tuffs and volcanic breccias. Some of the thinly layered units show conspicuous primary sedimentary structures such as cross-bedding, disrupted and crumpled layering overlain and underlain by undisturbed beds, and slump structures. Layering and primary structures have been obliterated locally, however, in the most intensely altered rocks. Metamorphosed dikes probably similar in age to the volcaniclastic rocks occur in the

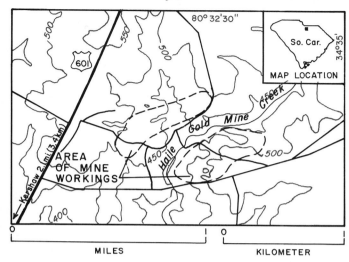

Figure 1. Index map showing location of the Haile Gold Mine workings, Kershaw 7½-minute Quadrangle, Lancaster County, South Carolina.

mine workings. They strike northeast, north, and northwest. Several encountered in drill core are diabase, but light colored saprolite indicates felsic varieties may also be present. More conspicuous are the numerous unmetamorphosed late Mesozoic diabase dikes which cut the older dikes and the volcaniclastic rocks. These are olivine-bearing tholeiitic diabase, which differ megascopically mainly in grain size, texture, abundance of olivine and feldspar phenocrysts, and thickness. In general, the diabase dikes in the Haile mine area have the conspicuously oriented trend of about N.45°W. that is characteristic of these dikes in this part of the southeastern piedmont. Although the strike changes locally, the dip is commonly vertical or nearly so.

Unconformably overlying all the metamorphic and igneous rocks are sedimentary rocks of the Coastal Plain. These consist of poorly indurated iron oxide stained cross-bedded sands, sandy clay lenses and gravels containing fragments of angular vein quartz, very-fine-grained mica schist, rounded white clay fragments of local origin and rounded quartz pebbles of exotic origin. The ages of these rocks are uncertain but probably Tertiary and Quaternary and they may include rocks of the Upper Cretaceous Middendorf Formation in thicker sections.

The mine is near the axis of an east-trending major regional anticlinal structure into which are intruded coarse-grained granite plutons of Pennsylvanian age. Superimposed on the major fold are smaller folds and numerous faults and abundant fractures. The foliation and cleavage cutting the volcaniclastic rocks is steeply dipping to the northwest at divergent angles, mostly between 50 and 70 degrees. Two or possibly three foliations all striking northeast at close angles can be distinguished locally. The most easily recognized faults, seen in hand specimens and in limited exposures in the mine workings, have displacements of a few feet or inches commonly with drag folds and breccias. There may be, however, less easily identified northeast-trending regional

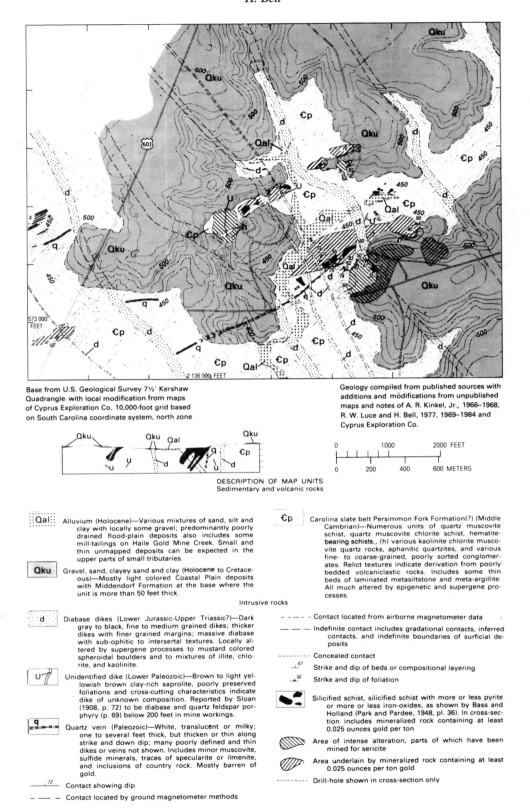

Base from U.S. Geological Survey 7½' Kershaw
Quadrangle with local modification from maps
of Cyprus Exploration Co. 10,000-foot grid based
on South Carolina coordinate system, north zone

Geology compiled from published sources with
additions and modifications from unpublished
maps and notes of A. R. Kinkel, Jr., 1966–1968,
R. W. Luce and H. Bell, 1977, 1969–1984 and
Cyprus Exploration Co.

0	1000	2000 FEET
0	200 400	600 METERS

DESCRIPTION OF MAP UNITS
Sedimentary and volcanic rocks

Qal   Alluvium (Holocene)—Various mixtures of sand, silt and clay with locally some gravel; predominantly poorly drained flood-plain deposits also includes some mill-tailings on Haile Gold Mine Creek. Small and thin unmapped deposits can be expected in the upper parts of small tributaries.

Qku   Gravel, sand, clayey sand and clay (Holocene to Cretaceous)—Mostly light colored Coastal Plain deposits with Middendorf Formation at the base where the unit is more than 50 feet thick.

Ɛp   Carolina slate belt Persimmon Fork Formation(?) (Middle Cambrian)—Numerous units of quartz muscovite schist, quartz muscovite chlorite schist, hematite-bearing schists,, (h) various kaolinite chlorite muscovite quartz rocks, aphanitic quartzites, and various fine- to coarse-grained, poorly sorted conglomerates. Relict textures indicate derivation from poorly bedded volcaniclastic rocks. Includes some thin beds of laminated metasiltstone and meta-argillite. All much altered by epigenetic and supergene processes.

Intrusive rocks

d   Diabase dikes (Lower Jurassic-Upper Triassic?)—Dark gray to black, fine to medium grained dikes; thicker dikes with finer grained margins; massive diabase with sub-ophitic to intersertal textures. Locally altered by supergene processes to mustard colored spheroidal boulders and to mixtures of illite, chlorite, and kaolinite.

U7   Unidentified dike (Lower Paleozoic)—Brown to light yellowish brown clay-rich saprolite, poorly preserved foliations and cross-cutting characteristics indicate dike of unknown composition. Reported by Sloan (1908, p. 72) to be diabase and quartz feldspar porphyry (p. 69) below 200 feet in mine workings.

q   Quartz vein (Paleozoic)—White, translucent or milky; one to several feet thick, but thicken or thin along strike and down dip; many poorly defined and thin dikes or veins not shown. Includes minor muscovite, sulfide minerals, traces of specularite or ilmenite, and inclusions of country rock. Mostly barren of gold.

72 Contact showing dip

– – – Contact located by ground magnetometer methods

– – – – Contact located from airborne magnetometer data

——— Indefinite contact includes gradational contacts, inferred contacts, and indefinite boundaries of surficial deposits

········· Concealed contact

67 Strike and dip of beds or compositional layering

50 Strike and dip of foliation

Silicified schist, silicified schist with more or less pyrite or more or less iron-oxides, as shown by Bass and Holland (Park and Pardee, 1948, pl. 36). In cross-section includes mineralized rock containing at least 0.025 ounces gold per ton

Area of intense alteration, parts of which have been mined for sericite

Area underlain by mineralized rock containing at least 0.025 ounces per ton gold

········· Drill-hole shown in cross-section only

Figure 2. Geologic map and section, Haile Gold Mine area, Lancaster County, South Carolina.

faulting in the mine area. Most Mesozoic diabase dikes cut the ore-bearing strata and older mafic dikes without apparent offset or displacement due to faulting, but some dikes were probably intruded into steeply dipping northwest-trending faults. Small displacements along faults into which diabase dikes were emplaced may account for mineralogic differences in the lodes on the opposite sides of some dikes as described by Sloan (1908). Aligned kinks in the map pattern of thicker dikes suggest that pre-dike east-northeast fracturing or faulting influenced the intrusion of these dikes where they cross the creek between the two areas of mine workings.

The area of intensely leached and silicified rocks around the Haile Mine extends beyond the limits of Figure 1 and 2 as much as 3.5 mi (5.6 km) in a westerly direction from the mine. It is at least 1 mi (1.6 km) wide, and it may be much larger. In the eastern part of the area extensions of the altered rocks may be concealed by sedimentary rocks of the Coastal Plain. The altered rocks extend below the deepest drilling (Spence and others, 1980) which is about 500 ft (152 m) and therefore extends deeper than the 100 ft (30.5 m) common for surficial weathering. Probably the area is not uniformly altered. Large parts are poorly exposed because of vegetation and soil cover.

The ore-bearing rocks are fine-grained quartz-muscovite schists, quartz-chlorite-muscovite schists, aphanitic quartzites, hematite-bearing schists, and various argillaceous kaolinite-chlorite-muscovite-quartz rocks that result from intense volcano-genic hydrothermal alteration processes and metamorphism on which is superimposed a younger chemical weathering (Bell, 1982). The rocks are predominantly light gray and light shades of greenish gray whereas the most altered are nearly white. The most intensely altered now consist of quartz, muscovite with traces of fuchsite locally, and pyrite. Less intensely altered rocks are richer in muscovite and kaolinite with minor mixed layer clays and rutile. These commonly retain much of their original volcaniclastic character. Some samples yield an x-ray diffraction pattern characteristic of dickite rather than well crystallized kaolin. Other rocks altered by syngenetic volcanic processes include dense light-colored cherty rocks with resinous luster and conchoidal fracture, and distinctive deep purplish red hematite-stained rock.

The intensity of the superimposed weathering is indicated by a diabase dike. Where exposed in mine workings near the top of an interfluvial divide and directly beneath Coastal Plain sedimentary rocks the dike is reduced by weathering processes almost beyond recognition to illite, chlorite, and kaolinite-rich saprolite; but in stream bottoms and mine workings 75 ft (22.9 m) below the interfluvial area it is largely unaltered.

In most of the rocks there is abundant pyrite or evidence of its former presence such as iron-oxide stained cavities or pseudomorphs of iron-oxides replacing pyrite crystals. Pyrite occurs as disseminations, as thin cross-cutting veins, and as layers of very-fine-grained crystals simulating bedding in thinly layered units. A body of massive pyrite, exposed in one of the mine pits, showed pyrite crystals ranging from very-fine- to medium-grained. The

upper few feet (meters) were reduced to massive gossan below overlying coastal plain rocks. Underlying the massive pyrite, a conglomerate having a light colored matrix contains very fine disseminated pyrite, and rock fragments including lens-shaped silica-rich cherty masses; irregular lens-shaped masses resembling fragments of massive pyrite and consisting of very-fine-grained aggregated dark pyrite; and angular flattened masses of light colored clay or sericite contrasting with the matrix. These last are interpreted to be fragments resulting from brecciation of altered volcanic rocks redeposited in a matrix of similar composition. The upper contact of this conglomerate is gradational into the massive pyrite by an increase of pyrite in the matrix. Locally a similar conglomerate unit overlying the massive pyrite grades into siliceous rock with lumpy texture and less pyrite.

Characteristically the volcaniclastic rocks of the slate belt are cut by abundant milky quartz dikes and veins, which produce vast amounts of surface float. In the Haile mine area some veins are up to several ft (m) thick and trend northeast or east and one dike, locally as much as 12 ft (3.7 m) thick, extends more than a thousand ft (305 m) along strike. Commonly these quartz veins are discontinuous on a scale of tens of ft (m) and pinch and swell along strike and dip. They include traces of muscovite, fragments of country rock, minor amounts of pyrite, and locally traces of specularite or ilmenite. They are mostly barren of gold. In the altered rocks of the mine area, very thin anastomosing and ptygmatic white quartz veinlets are widespread and abundant. Locally at least 6 veins per ft (30 cm) occur, ranging from less than an eighth of an inch to 10 in (2.5 cm to 25.4 cm) thick and striking mostly between 20° east and west of north. The thicker veins with a pinch and swell habit are thought to be a product of regional metamorphism of the felsic volcaniclastic rocks, but many of the thinner veins which are closely associated with intensely sericitized rocks may be a syngenetic coproduct of the intense sericitic alteration.

Lenticular masses of silicified and pyritized gold-bearing rocks, occurring in the mine, are postulated to be stratiform sinter-like deposits of hot spring origin (Spence and others, 1980). In the mine area, however, there are altered rocks that result from processes which extended into the interlayered and overlying thinly bedded metasedimentary rocks. This indicates that hydrothermal processes in the mine area were long-lived and that some hydrothermal alteration, probably fracture controlled, postdates the deposition of most of the volcanic and volcaniclastic rocks and continued after deposition of the overlying pelitic rocks had begun (Bell, in preparation). Zeolites, although widely distributed in the southern piedmont have not been identified in the Haile Mine probably because they are easily reduced to clay and other minerals by supergene processes. However, drill-core rich in carbonate minerals, probably calcite, were encountered in several parts of the mineralized area (Earl Jones, personal commun., 1979).

Both gold ore bodies and massive pyrite ore bodies occur in lodes which are the most intensely silicified and pyritized rock and so similar in details of lithology and mineralogy on both sides of Haile Gold Mine Creek that they are thought to be parts of a single deformed strata-bound deposit. The gold ore bodies are lenticular masses plunging northeast in the strata-bound lodes. The boundaries of the ore bodies are, in general, determined by assay values, but locally they are sharply demarcated along foliation planes, fractures, and possible faults (Graton, 1906; Sloan, 1908; and Newton and others, 1940). Typical ore-bodies range in thickness from about 40 to 100 ft (12.2 to 30.5 m) and extend down dip 300 to 475 ft (91.4 to 144.8 m) and along strike 200 to 600 ft (61.0 to 182.9 m). Massive pyrite ore bodies associated with the gold ore bodies occur mostly in the northeast extensions of the lodes at several stratigraphic levels. The massive pyrite bodies are generally low in gold values (Schrader, 1921).

The ore from the early placers along Haile Gold Mine Creek and the uppermost oxidized parts of the lodes contained visible native gold; some in large nuggets and with a fineness of .923 (Tuomey, 1844). Visible gold is reported along fractures or joint surfaces, and as thin flakes. Some of it is described as rusty, a result of iron-oxide coatings. In the most recently mined ore, the gold is submicroscopic although presumably in the native state and largely in pyrite-rich highly silicified rock. During examination of drill core samples using the scanning electron microscope (SEM), unidentified phases, or crystals, containing high amounts of tellurium were identified (Bell and Larson, 1983). Most tellurium-bearing crystals are either lead- or silver-tellurium, but some grains are gold-tellurium and gold-silver-tellurium. The tellurium-bearing grains all occur in pyrite and along fractures apparently confined to the pyrite crystals or crystal aggregates.

The grade of ore from open pits during the last years of mining averaged about 0.14 ounces per ton but varied widely (Newton and others, 1940, p. 4). The possibility that silver may be more abundant in unoxidized ore is suggested by the report of silver-rich bullion produced from the deepest parts of the mine (Spilsbury, 1882).

Pyrite is the most important sulfide, but molybdenite occurs in significant amounts in the gold ore. Other ore minerals known to occur in trace amounts include pyrrhotite, arsenopyrite, sphal-erite, and chalcopyrite. The pyrite occurs in several generations characterized by differences in grain size and habit. These include fine-grained "sub-massive" pyrite, disseminated euhedral pyrite such as commonly occurs in the gold ore, and large zoned cubes of pyrite which are less common but widespread. Sulfur isotope analyses of pyrite show a range in $d^{34}S$ values from 1.78 to 3.07 which suggested to LeHuray (1983) a strong magmatic sulfur component and an environment dominated by fresh water rather than Paleozoic seawater during formation of the pyrite.

Molybdenite is widespread in the ore zone and occurs as veinlets and disseminated grains; the gold ore averages about 70 ppm (molybdenum), but contains as much as 490 ppm molybdenum locally. Commonly the molybdenum content increases from the top to just below the bottom of the ore zone in both lodes, tending to support the conclusion that the lodes on the opposite sides of Haile Gold Mine Creek are parts of a single strata-bound deposit.

The stratabound massive sulfide ore-bodies seem clearly related to the origin of the lower Paleozoic rocks in which they occur. Features such as the conglomerate with fragments of very-fine-grained aggregated dark pyrite link the origin of these masses to their enclosing volcaniclastic rocks. The origin of the gold ore-bodies is less clear, but because they are also essentially strata-bound, they are considered to be syngenetic with their enclosing rocks and result from early Paleozoic volcanogenic processes. The altered rocks have been attributed to the effects of hot springs or fumeroles discharging in a subaerial environment. Such springs and fumeroles are a phenomena characteristic of the waning phases of volcanic activity commonly localized by faults and continuing to discharge hot water long after the volcanic activity ceased. At the Haile Mine the cross-cutting alteration in the overlying pelitic rocks may result from these hot springs and fumeroles. All the mining and exploration in the Haile mine area has been at relatively shallow depths and largely within the realm of surficial weathering. The presence of mineralized rock characteristic of deeper parts of hydrothermal systems has not been tested but the occurrence of tellurium minerals, dickite, and silver-rich ore suggests that these environments may exist somewhere in the vicinity.

REFERENCES CITED

Bell, Henry, in preparation, Geology of the Haile Mine, Lancaster County, South Carolina.

Bell, Henry, 1982, Strata-bound sulfide deposits, wall-rock alteration and associated tin-bearing minerals in the Carolina Slate Belt, South Carolina and Georgia: Economic Geology, v. 77, no. 2, p. 294–311.

Bell, Henry, and Larson, Richard R., 1983, A new occurrence of telluride minerals in South Carolina: Southeastern Geology, v. 24, no. 4, p. 189–194.

Graton, L. C., 1906, Reconnaissance of some gold and tin deposits of the southern Appalachians, with notes on the Dahlonega mines by Waldemar Lindgren: U.S. Geological Survey Bulletin 293, 134 p.

LeHuray, A. P., 1983, Lead- and sulfur-isotope systematics in some sulfide deposits of the Piedmont and Blue Ridge provinces of the southern Appalachians, U.S.A.: U.S. Geological Survey, Open-file report 84-112, 184 p.

Newton, E., Gregg, D. B., and Mosier, McHenry, 1940, Operations at the Haile Gold mine, Kershaw, S.C.: U.S. Bureau of Mines, Information Circular 7111, p. 1–42.

Pardee, J. T., and Park, C. F., Jr., 1948, Gold deposits of the southern Piedmont: U.S. Geological Survey Professional Paper 213, 156 p.

Samson, S. L., Secor, D. T., Snoke, A. W., and Palmer, A. R., 1982, Geological implications of recently discovered Middle Cambrian trilobites in the Carolina slate belt: Geological Society of America Abstracts with Programs, v. 14, no. 7, p. 607.

Schrader, Frank C., 1921, Pyrite at the Haile Mine, Kershaw, South Carolina with a note on Pyritization at the Brewer Mine, near Jefferson: U.S. Geological Survey Bulletin 725-F, p. 331–345.

Sloan, Earle, 1908, A catalogue of the mineral localities of South Carolina: South Carolina Geological Survey Bulletin, ser. 4, no. 2, 505 p., reprinted 1958 by South Carolina Division of Geology.

Spence, W. H., Worthington, J. E., Jones, E. M., and Kiff, I. T., 1980, Origin of gold mineralization at the Haile mine, Lancaster County, South Carolina: Mining Engineering, v. 32, no. 1, p. 70–73.

Spilsbury, E. G., 1882, in Report of the Director of the Mint upon the production of the precious metals in the United States: Washington, D.C. Government Printing Office, p. 638.

Tuomey, M., 1844, Report on the geological and agricultural survey of the State of South Carolina: Columbia, S.C., A. S. Johnston, State Printer, 63 p.

Worthington, J. E., Kiff, I. T., Jones, E. M., and Chapman, P. E., 1980, Applications of the hot springs or fumerolic model in prospecting for lode gold deposits: Mining Engineering, v. 32, no. 1, p. 73–79.

The Brevard fault zone at Rosman, Transylvania County, North Carolina

J. Wright Horton, Jr., 928 National Center, U.S. Geological Survey, Reston, Virginia 22092
J. Robert Butler, Department of Geology, University of North Carolina, Chapel Hill, North Carolina 27514

LOCATION

The Brevard fault zone is exposed near Rosman, Transylvania County, North Carolina (Fig. 1), in the southwestern quadrant of the Rosman 7½-minute Quadrangle. The stops are along public roads and are accessible without special permission. Busloads of approximately 50 people have visited these outcrops in the past.

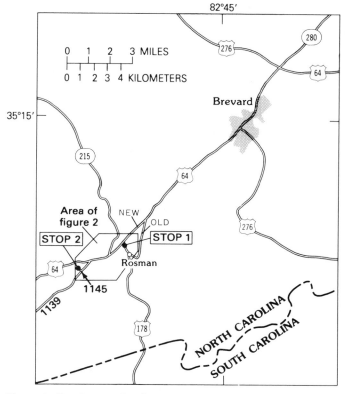

Figure 1. Road map showing stop locations near Rosman, North Carolina.

INTRODUCTION

The Brevard fault zone is widely recognized as one of the most fundamental structures of the southern Appalachian orogen. The type area of the zone is at Brevard, about 8 mi (13 km) northeast of Rosman (Keith, 1907).

The Brevard zone is a linear belt of mylonitic and cataclastic rocks that extends more than 370 mi (600 km) from the Gulf Coastal Plain onlap in Alabama almost to the North Carolina-Virginia border. It typically consists of mylonites and phyllonites in a zone 0.6–1.2 mi (1–2 km) wide, with a strong foliation that dips moderately to the southeast. Seismic reflection studies in northeastern Georgia indicate that the Brevard zone may be a splay from a major flat-lying decollement that underlies allochthonous crystalline rocks of the Blue Ridge and Inner Piedmont (Cook and others, 1983). Clark and others, (1978) conducted a seismic reflection study near Rosman with geophone spreads 4.3 and 5.6 mi (7 and 9 km) southeast of the Brevard zone. In addition to obtaining good reflections from the Brevard zone, Clark and others (1978) obtained large-amplitude reflections from a depth of approximately 3.7 mi (6 km), which they interpreted as the top of the Lower Cambrian Rome Formation, the contact between the Rome Formation and the underlying Shady Dolomite, and the base of the Lower Cambrian Shady Dolomite, all beneath allochthonous crystalline rocks.

In the Carolinas, the Brevard zone is the boundary between the Blue Ridge geologic province to the northwest and Piedmont geologic belts to the southeast. The Brevard zone is clearly a major boundary, as the distinctive Henderson Gneiss is the dominant lithology just southeast of the Brevard zone for more than 75 mi (120 km), but neither the Henderson Gneiss nor any rock of similar petrology and age is known northwest of the Brevard zone. The Henderson Gneiss is progressively mylonitized toward the Brevard zone (Horton, 1974). Rocks northwest of the Brevard zone are mainly gneisses and schists, including the Late Proterozoic and (or) early Paleozoic Tallulah Falls Formation, the Middle Proterozoic (Grenville-age) Toxaway Gneiss, and Paleozoic intrusions (Horton, 1982). Rocks of the Tallulah Falls Formation are also mylonitic in the Brevard zone.

Narrow units of phyllonite and mylonite within the Brevard zone are elongate parallel to the strike of the zone. Much of the Brevard zone phyllonite and some of the mylonite in the Rosman area are similar to and probably correlative with units of the Chauga River Formation as defined in South Carolina (Hatcher, 1969; Horton, 1982). The Rosman fault, a zone of extreme brecciation and disruption of layering within the Brevard zone that is clearly later than the mylonitization (Horton, 1974, 1980), juxtaposes these mylonitic and phyllonitic rocks of the Inner Piedmont (or Chauga) belt on the southeast against those of the Blue Ridge geologic province on the northwest (Horton, 1974, 1980, 1982). A distinctive tectonic melange, rarely more than 200 ft (60 m) thick, which contains exotic blocks of carbonate and other rocks of unknown affinity, is localized along the Rosman fault (Horton and Butler, 1976; Horton, 1980).

The numerous and varied interpretations of the Brevard zone are discussed by Roper and Justus (1973), Hatcher (1978),

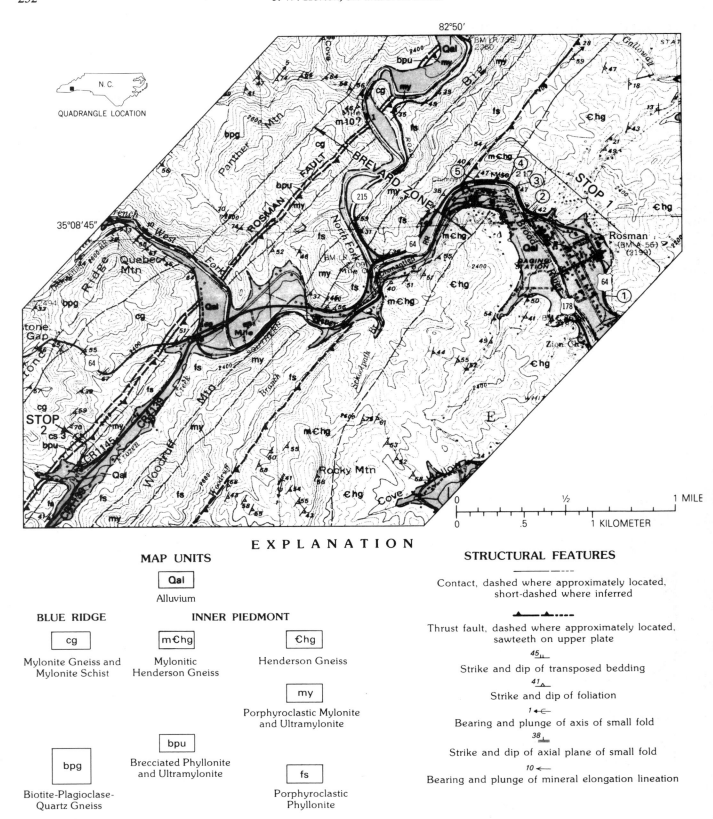

E X P L A N A T I O N

MAP UNITS

STRUCTURAL FEATURES

Qal

Alluvium

BLUE RIDGE

INNER PIEDMONT

cg

m€hg

€hg

Mylonite Gneiss and
Mylonite Schist

Mylonitic
Henderson Gneiss

Henderson Gneiss

my

Porphyroclastic Mylonite
and Ultramylonite

bpu

Brecciated Phyllonite
and Ultramylonite

bpg

fs

Biotite-Plagioclase-
Quartz Gneiss

Porphyroclastic
Phyllonite

Contact, dashed where approximately located,
short-dashed where inferred

Thrust fault, dashed where approximately located,
sawteeth on upper plate

45
Strike and dip of transposed bedding

41
Strike and dip of foliation

1
Bearing and plunge of axis of small fold

38
Strike and dip of axial plane of small fold

10
Bearing and plunge of mineral elongation lineation

Figure 2. Geologic map from Horton (1982) showing localities mentioned in the text.

Figure 3. Rock slabs illustrating progressive stages of mylonite development. Top Left: typical Henderson Gneiss collected 2.4 mi (3.9 km) southeast of the Brevard zone (southeast of Fig. 2); Top Center: strongly deformed Henderson Gneiss from locality 1 (Fig. 2); Top Right: protomylonite from locality 2. Bottom Left: porphyroclastic mylonite from locality 3; Bottom Center: ultramylonite from locality 5, showing cut surfaces approximately perpendicular to the foliation and mineral-elongation lineation; Bottom Right: cataclasite from Stop 2.

and Cook and others (1983). The regional geologic setting is shown by Hadley and Nelson (1971). A detailed geologic map of the Rosman quadrangle by Horton (1982) includes the area of this field guide. The geologic cross section in that report is partly constrained by the seismic reflection data of Clark and others (1978).

SITE DESCRIPTION

Exposures near Rosman are better than those near Brevard and are among the best anywhere along the Brevard zone. The Rosman section has been a focus of investigations by geologists, geochronologists, and geophysicists studying various aspects of the zone. Stops 1 and 2 illustrate different stages in the geological history of the Brevard zone, different modes of deformation, and evidence for reactivated faulting.

Stop 1: Development of Mylonite from the Henderson Gneiss

Stop 1 is a series of road cuts along the north side of old U.S. 64, where it parallels the French Broad River at Rosman (Figs. 1 and 2).

The Henderson Gneiss (Fig. 2, locality 1, and Fig. 3) is typically a medium-gray, biotite granite augen gneiss composed of microcline, oligoclase, quartz, biotite, and minor amounts of muscovite, epidote, and titanite. Exposures along the north side of old U.S. 64 at Rosman illustrate progressive stages in the development of mylonite from the Henderson Gneiss along the southeast edge of the Brevard fault zone. Although complicated by small-scale interfingering, a gradual textural progression is evident across strike from southeast to northwest, from strongly deformed Henderson Gneiss (Fig. 2, locality 1, and Fig. 3) to protomylonite (Fig. 2, locality 2, and Fig. 3) to porphyroclastic mylonite (Fig. 2, locality 3, and Fig. 3) to mylonite (Fig. 2, locality 4) to ultramylonite (Fig. 2, locality 5, and Fig. 3). Rock nomenclature in this field guide follows Sibson's (1977) textural classification of fault rocks.

The mylonitic rocks are yellowish gray, fine grained, and thinly laminated, and they contain round, grayish-pink porphyroclasts of microcline. The porphyroclasts, which are commonly elongate parallel to the mylonitic foliation, are enclosed by a matrix of fine-grained quartz, plagioclase, and micas with grains averaging 0.05 mm to 0.1 mm across (Horton, 1974).

As the rocks become increasingly mylonitic, the relict microcline porphyroclasts decrease in size and abundance (Fig. 3), and their roundness increases. The average grain size of the rock decreases, and textures become more equigranular. Thin laminae or ribbons of highly recrystallized, polygonal quartz become more prevalent, giving the rocks a streaky appearance. In thin section, the mylonites appear highly recrystallized, and undulatory extinction, although present, is not strong. As the intensity of mylonitization progresses, the number of crystals forming individual porphyroclasts increases and microcline twinning becomes less conspicuous in thin sections. The larger porphyroclasts are commonly brecciated, pulled apart, and rotated, the direction of rotation changing over small distances. Extension fractures in microcline are common along crystallographic cleavage planes at high angles to the mineral-elongation lineation, and these fractures are filled with fine-grained polygonal quartz and plagioclase where the fragments are separated. The rounding of porphyroclasts and the overall decrease in the percentage of microcline appear to be the results of chemical reactions along grain boundaries. Sheaths of fine-grained muscovite appear along the mylonitic foliation tangent to the microcline porphyroclasts. Within these sheaths, the microcline is rimmed and embayed by quartz, plagioclase, and myrmekite. Fine-grained muscovite increases in abundance as the amount of microcline diminishes. Medium-grained, dark-brown biotite decreases in abundance and eventually vanishes, as fine-grained, light-brown biotite appears. Chlorite is found only rarely in trace amounts as rims around garnet. Euhedral cubes of pyrite and hematite pseudomorphs after pyrite are sparsely disseminated, and the mylonitic foliation is not deformed around them. In spite of the mineralogical changes associated with mylonite formation, the overall major-element chemistry of the mylonitic rocks does not differ significantly from that of the Henderson Gneiss protolith (Bond, 1974).

The Henderson Gneiss has yielded U-Pb upper concordia intercept ages of zircons at about 600 Ma (Sinha and Glover, 1978) and 538 Ma (Odom and Fullagar, 1973) and a whole-rock Rb-Sr age of 524 ± 27 Ma (recalculated from Odom and Fullagar, 1973). Samples of mylonitized Henderson Gneiss from road cuts at this locality have been dated isotopically by Odom (1971), Odom and Fullagar (1973), Bond (1974), Bond and Fullagar (1974), and Sinha and Glover (1978). Zircons yield a U-Pb upper concordia intercept age of 456 Ma according to Sinha and Glover (1978), who interpret the intercept as evidence of mylonite formation during the Taconic orogeny. Other samples from these road cuts yield a whole-rock Rb-Sr isochron age of 348 ± 8 Ma according to Odom and Fullagar (1973, recalculated). They interpret this age as a time of mylonitic deformation and recrystallization in the Brevard zone but do not preclude additional times of movement earlier or later. Three samples of "mylonite" that are texturally transitional between the typical Henderson Gneiss and "blastomylonite" do not plot on either isochron (Odom, 1971). Bond (1974) and Bond and Fullagar (1974) found that samples of mylonitized Henderson Gneiss from this locality and others from Horseshoe, N.C., 21 mi (34 km) to

the northeast, could be plotted on a single isochron of 380 ± 15 Ma (recalculated). Sinha and others (1984) obtained a similar apparent Rb-Sr age of 390 Ma from slabs of a single block of mylonite collected near Rosman. Unlike previous investigators, however, Sinha and others (1984) interpret this apparent age as the chronologically meaningless slope of a mixing line between end members whose Rb-Sr systems are dominated by old porphyroclastic feldspars and new micas respectively.

Stonebraker (1973) determined K-Ar ages of whole rock samples and mica separates from eight traverses across the Brevard fault zone, including one at Rosman. All of these ages are about 300 Ma and show no discontinuity across the Brevard zone.

At least two fold generations are present in outcrops along this traverse. The mylonitic foliation is parallel to the axial surfaces of isoclinal folds, and a strong subhorizontal northeast-trending mineral-elongation lineation is parallel to the hinge lines of these folds. The mylonitic foliation is deformed around later open to tight upright folds, which have subvertical axial surfaces and also have subhorizontal, northeast-trending hinge lines. Kink bands locally deform the mylonitic foliation. Although a few small faults cut across the mylonites along this traverse, brittle textures are more prevalent at Stop 2.

Stop 2: The Rosman Fault

Stop 2 is an excavated hillside at a former landfill on County Road 1145 approximately 2.5 mi (4 km) west-southwest of Rosman (Figs. 1 and 2). From Stop 1, drive west on old U.S. 64, passing the intersection with Gloucester Road (North Carolina 215), and proceed another 0.9 mi (1.5 km). After crossing a bridge over the West Fork of the French Broad River, turn left (south) on County Road 1139. Drive 0.7 mi (1.2 km) and take the right (paved) fork onto County Road 1145. Follow the pavement 0.3 mi (0.5 km) until it ends at Stop 2 (see Fig. 2).

The Rosman fault, a narrow zone of relatively brittle cataclastic deformation within the Brevard zone that post-dates the formation of mylonite (Horton, 1980), is well exposed at this locality. Cataclasite fabric (Fig. 3, lower right) is concentrated in a narrow, continuous unit of semicohesive brecciated graphitic phyllonite and ultramylonite of Chauga belt affinity; this unit has been called a tectonic melange or broken formation (Horton, 1980, 1982). Numerous cross-cutting faults and associated drag folds of varied orientation range from microscopic to mesoscopic in scale, and narrow microbreccia zones are common. These features produce a chaotic appearance at outcrops (Fig. 4), but southeast-dipping reverse faults (listric thrust faults?) are the most pervasive (Horton, 1982). Lenses and pods of black amorphous carbon are common along these faults (Horton, 1980). Small bodies of quartzite are highly competent in contrast to the carbon-rich material. Keith (1907) showed a lens of marble cropping out at this location, but we found only a small amount of carbonate as fracture-filling material. All bodies of Brevard-zone marble observed or reported in the Rosman area are asso-

Figure 4. The Rosman fault at Stop 2.

ciated with this unit of brecciated phyllonite and ultramylonite along the Rosman fault. Fragments of mylonite and phyllonite from adjacent units are common. Such features indicate a phase of more brittle cataclastic deformation that post-dates the formation of mylonite.

The Rosman fault separates mylonitized rocks of the Chauga belt, or the northwest flank of the Inner Piedmont belt on the southeast, from those of the Blue Ridge belt on the northwest. It therefore is the major lithotectonic boundary within this segment of the Brevard zone. The unit of mylonite gneiss and mylo-

nite schist that borders the Rosman fault on the northwest within the Brevard zone at this locality (Fig. 2) belongs to the graywacke-schist member of the Tallulah Falls Formation (Horton, 1982). The average attitude of the Rosman fault, about N.35°E., 40°SE. in this area, is essentially parallel to the pre-existing mylonitic foliation in the Brevard zone. We have traced the Rosman fault in reconnaissance about 33 mi (53 km), from Fletcher, North Carolina, 27 mi (43 km) northeast of Rosman, southwestward to the Toxaway River, 6 mi (10 km) southeast of Rosman, near the South Carolina state line.

REFERENCES CITED

Bond, P. A., 1974, A sequence of development for the Henderson Augen Gneiss and its adjacent cataclastic rocks [M.S. thesis]: Chapel Hill, University of North Carolina, 53 p.

Bond, P. A., and Fullagar, P. D., 1974, Origin and age of the Henderson Augen Gneiss and associated cataclastic rocks in southwestern North Carolina [abs.]: Geological Society of America Abstracts with Programs, v. 6, no. 4, p. 336.

Clark, H. B., Costain, J. K., and Glover, L., III, 1978, Structural and seismic reflection studies of the Brevard ductile deformation zone near Rosman, North Carolina: American Journal of Science, v. 278, p. 419–441.

Cook, F. A., Brown, L. D., Kaufman, S., and Oliver, J. E., 1983, The COCORP seismic reflection traverse across the southern Appalachians: American Association of Petroleum Geologists, Studies in Geology 14, 61 p.

Hadley, J. B., and Nelson, A. E., 1971, Geologic map of the Knoxville quadrangle, North Carolina, Tennessee, and South Carolina: U.S. Geological Survey,

Miscellaneous Investigations Map I-654, scale 1:250,000.

Hatcher, R. D., Jr., 1969, Stratigraphy, petrology, and structure of the low rank belt and part of the Blue Ridge of northwesternmost South Carolina: South Carolina Division of Geology Geologic Notes, v. 13, no. 4, p. 105–141.

—— 1978, Tectonics of the western Piedmont and Blue Ridge, southern Appalachians: Review and speculation: American Journal of Science, v. 278, p. 276–304.

Horton, J. W., Jr., 1974, Geology of the Rosman area, Transylvania County, North Carolina [M.S. thesis]: Chapel Hill, University of North Carolina, 63 p.

—— 1980, Post-mylonitic brittle deformation in the Brevard fault zone, North Carolina [abs.], in Wones, D. R., ed., Proceedings, The Caledonides in the USA, IGCP Project 27—Caledonide Orogen, 1979 Meeting: Blacksburg, Virginia Polytechnic Institute and State University Memoir 2, p. A17.

—— 1982, Geologic map and mineral resources summary of the Rosman quadrangle, North Carolina: Raleigh, North Carolina Department of Natural Resources and Community Development, Geological Survey Section, GM 185-NE (scale 1:24,000), MRS 185-NE (4 p. text).

Horton, J. W., Jr., and Butler, J. R., 1976, Structure and petrology of the Brevard zone and adjacent areas in the Rosman, N.C., 7.5′ quadrangle [abs.]: Geological Society of America Abstracts with Programs, v. 8, no. 2, p. 199.

Keith, Arthur, 1907, Description of the Pisgah quadrangle: U.S. Geological Survey, Geologic Atlas Folio 147, 8 p.

Odom, A. L., 1971, A Rb-Sr isotopic study: implications regarding the age, origin, and evolution of a portion of the Southern Appalachians, western North Carolina, southwestern Virginia, and northwestern Tennessee [Ph.D. thesis]: Chapel Hill, University of North Carolina, 92 p.

Odom, A. L., and Fullagar, P. D., 1973, Geochronologic and tectonic relationships between the Inner Piedmont, Brevard zone and Blue Ridge belts, North Carolina: American Journal of Science, v. 273-A (Cooper volume), p. 133–149.

Roper, P. J., and Justus, P. S., 1973, Polytectonic evolution of the Brevard zone: American Journal of Science, v. 273-A (Cooper volume), p. 105–132.

Sibson, R. H., 1977, Fault rocks and fault mechanisms: Journal of the Geological Society of London, v. 133, p. 191–213.

Sinha, A. K., and Glover, L., III, 1978, U/Pb systematics of zircons during dynamic metamorphism: Contributions to Mineralogy and Petrology, v. 66, p. 305–310.

Sinha, A. K., Hewitt, D. A., Rimstidt, J. D., and Partin, B., 1984, Distribution of strain and chemical domains in mylonite: correlation with isotopic ages [abs.]: Geological Society of America Abstracts with Programs, v. 16, no. 6, p. 657.

Stonebraker, J.D., 1973, Potassium-argon geochronology of the Brevard fault zone, Southern Appalachians [Ph.D. thesis]: Tallahassee, Florida State University, 134 p.

Winding Stair Gap granulites: The thermal peak of Paleozoic metamorphism

B. Steven Absher and Harry Y. McSween, Jr., Department of Geological Sciences, University of Tennessee, Knoxville, Tennessee 37996

LOCATION

The Winding Stair Gap exposure is a 1,220-ft-(370 m) long road cut on U.S. 64 in Macon County, North Carolina. The outcrop is located in the northeastern part of the Rainbow Springs 7½-minute Quadrangle within the Nantahala National Forest, 10.5 mi (17.5 km) south of the U.S. 64 South and U.S. 441 intersection near Franklin, North Carolina (Fig. 1). Paved parking areas suitable for buses are available at the crest of Winding Stair Gap adjacent to the Appalachian Trail, and just north of the exposure at a scenic overlook. Access to the upper tiers (for specimen collecting) is via the northern end of the outcrop.

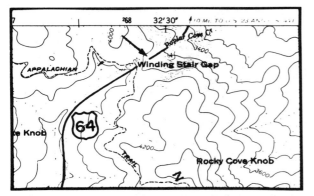

Figure 1. Extreme northern part of the Rainbow Springs 7½-minute Quadrangle showing the location of Winding Stair Gap.

INTRODUCTION

Winding Stair Gap is located along the thermal axis of Paleozoic metamorphism outlined by regional Barrovian isograd patterns (Force, 1976). It probably represents the best exposure of granulite facies rocks in the southern Appalachians. The Paleozoic age for these rocks is inferred from their position east of the Hayesville Thrust, which separates allochthonous Late Precambrian and Early Cambrian metasedimentary and possibly metavolcanic rocks from Grenville basement west of the fault (Hatcher and Butler, 1979). The units at Winding Stair Gap probably correlate with intervals in the Tallulah Falls Formation or Coweeta Group mapped at lower metamorphic grade elsewhere in the Hayesville thrust sheet. Peak metamorphism was synchronous with development of structures previously inferred to be Taconic in age (Hatcher and Butler, 1979; Absher and McSween, 1985).

Six lithologic units, including an orthopyroxenite of possible

ophiolitic origin, indicate the complex stratigraphy in an area of otherwise poor exposure. The presence of this wide range of rock compositions enables constraints to be placed on pressure and temperature of metamorphism. Mineral assemblages in this outcrop that are indicative of granulite facies metamorphism are summarized in Figure 2. The superposition of univariant reaction curves and mineral stability fields for the units at Winding Stair Gap on a P-T diagram (Fig. 3) allows an estimate of the pressure and temperature of peak metamorphism and an approximate retrograde path (Absher and McSween, 1985). The upper pressure boundary (reaction 1) is limited by the presence of orthopyroxene and plagioclase in the orthopyroxenite, biotite-garnet granulite, and biotite-hornblende granulite units. Clinopyroxene has been observed only in the calc-silicate gneiss, not in the mafic granulites. The lower pressure limit for the rocks at Winding Stair Gap is based on the absence of cordierite in the sillimanite schist unit, although the exact position of reaction curve 2 would vary depending on the Mg-Fe ratio of cordierite. Minimum temperatures for peak metamorphism are constrained above 700°C by the absence of primary muscovite according to reaction 3. The presence of sillimanite rather than kyanite as the stable aluminosilicate polymorph (reaction 4) also serves as a low temperature limit to peak metamorphism. Peak metamorphic conditions of 750-775°C and 6.5-7.0 kb, calculated from several exchange geothermometers and one geobarometer (Absher and McSween, 1985), coincide with those determined from mineral stability fields (Fig. 3).

The approximate retrograde metamorphic path at Winding

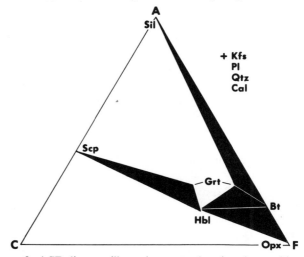

Figure 2. ACF diagram illustrating prograde mineral assemblages at Winding Stair Gap. These assemblages are consistent with granulite facies metamorphism.

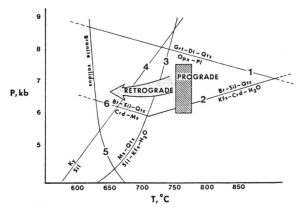

Figure 3. P-T diagram for prograde and retrograde metamorphic conditions at Winding Stair Gap, after Absher and McSween (1985). The lined rectangle represents data from three mineral exchange geothermometers and one geobarometer. Reactions are discussed in the text.

Stair Gap, indicated in Figure 3 by the arrow, has been determined largely by the mineral assemblages in the quartzo-feldspathic gneiss. The presence of primary muscovite and kyanite in non-foliated rocks of granitic composition indicates crystallization of these anatectic melts at temperatures between 700°C and 625°C (reactions 3, 4, 5). Quartzo-feldspathic gneiss samples containing muscovite + sillimanite and muscovite + kyanite suggest that crystallization of *in situ* melts occurred over a range of temperatures during cooling. The lower pressure range during cooling is restricted by the absence of cordierite in the sillimanite schist unit (reaction 6). Although no upper pressure limit during cooling has been determined, a pressure increase is not likely.

Winding Stair Gap thus represents granulite facies rocks formed at the culmination of Paleozoic metamorphism in the Blue Ridge. The isobaric cooling path during retrograde metamorphism suggests cooling was not due to uplift, but possibly resulted from crystallization of local intrusions which provided the heat source for granulite facies conditions (Absher and McSween, 1985). Migmatitic melts formed at Winding Stair Gap scavenged fluids, resulting in $P_{H_2O} < P_{Total}$. Upon cooling and crystallization, these melts liberated water which facilitated localized retrograde metamorphism. Deformation occurred in three phases before, during, and after peak metamorphism. The protoliths for these rocks were part of a sedimentary and volcanic sequence containing a sliver of ultramafic rock. Similar ultramafic bodies elsewhere in the Hayesville thrust sheet have been suggested to be ophiolitic in origin (e.g., McElhaney and McSween, 1983).

SITE DESCRIPTION

The Winding Stair Gap outcrop illustrates the granulite facies lithologies and deformational structures that characterize the thermal axis of Paleozoic metamorphism. The distributions of six lithologic units identified at Winding Stair Gap are shown on a cross-sectional map of the northern side of the exposure (Fig. 4a). Although both sides of the road cut contain the lithologic units and deformational features herein described, the northern side is shown because the three set back benches allow greater sampling access. Each tier is offset from the level below by approximately 16.5 ft (5 m). However, it is most convenient to observe these lithologies at road level.

Sillimanite schist at the northeastern end of the exposure is composed of biotite, garnet, sillimanite, plagioclase, K-feldspar, and quartz. A strong planar fabric produced by biotite and coarse-grained sillimanite needles is present. Monomineralic segregations of garnet are frequently confined to hinge zones of isoclinal folds and may represent restite from partial melting. Kyanite-bearing K-feldspar segregations occur in irregular, lenticular bodies, generally sigmoidally shaped. These lenses may have formed during migmatization.

Biotite-hornblende granulite is a fine- to medium-grained, dark green, granoblastic rock occurring as 3 to 16.5-ft- (1 to 5 m)-wide bands closely associated with sillimanite schist in isoclinal recumbent folds in the northeast part of the outcrop and as enclaves in quartzo-feldspathic gneiss to the southwest (Fig. 4a). This lithology is composed of hornblende, plagioclase, biotite, ilmenite, and quartz, with lesser amounts of garnet, orthopyroxene, or cummingtonite.

Quartzo-feldspathic gneiss, composed of quartz, plagioclase, K-feldspar, biotite, and pink almandine garnet, is partly migmatized and exhibits wide variability in textures and field relationships. One textural type is a well-foliated metatexite migmatite characterized by a conspicuous gneissic appearance of leucosome segregations and thinner biotite-garnet bands. Weakly foliated biotite- and garnet-bearing gneisses occur as discordant layers of variable thickness. Sillimanite is the aluminosilicate mineral in the metatexite migmatite and weakly foliated gneiss. Kyanite-bearing, non-foliated granofels of granitic to granodioritic composition cut metatexite migmatite and weakly foliated gneiss lithologies.

Biotite-garnet granulite is a dark gray, composite unit containing hornblende-biotite-orthopyroxene and biotite-garnet-hornblende assemblages. Coarse-grained biotite and poikiloblastic garnet up to 1.2-in (3 cm) diameter are conspicuous in the biotite-garnet-hornblende assemblage.

Calc-silicate gneiss occurs as zoned, lenticular bodies rarely over 3.3 ft (1 m) long. These lenses occur within quartzo-feldspathic gneiss surrounding the biotite-garnet granulite unit and probably represent a thin, boudinaged layer. Orange-colored garnet (grossular-almandine composition) and the abundance of light green minerals (hornblende, diopside, and clinozoisite) help differentiate the calc-silicate gneiss from the quartzo-feldspathic gneiss. Locations of boudins are shown by dots in Figure 4a; these are mapped only to a height of about 7 ft (2 m) above the base of each tier.

The orthopyroxenite is a 50 to 66-ft, (15 to 20 m) thick non-foliated unit occurring at the southwest end of the exposure. Plagioclase and quartz veins form a reticulated pattern through-

Winding Stair Gap

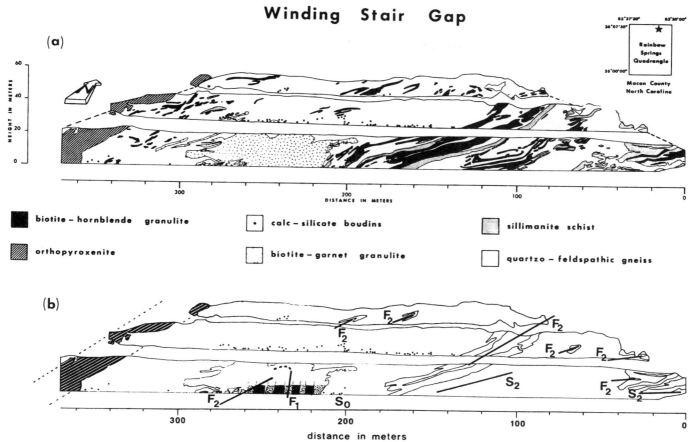

Figure 4. (a) Outcrop map of the northern side of Winding Stair Gap exposure showing six lithologic units. (b) Structural elements at Winding Stair Gap. Key lithologic units are outlined. The hornblende-biotite-orthopyroxene and biotite-garnet-hornblende assemblages defining S_0 in the biotite-garnet granulite unit are illustrated by shaded and dashed patterns, respectively.

out the orthopyroxenite. Highly weathered orthopyroxenite occurs at the southwest end of the second tier and is distinguished by its dark, reddish-brown coloration. Contact relations with other units are not clear because of complex folding and recrystallization.

Deformation occurred before, concurrent with, and after peak metamorphism. Visible structural features are illustrated in Figure 4b. The earliest structural element is compositional layering, S_0. A regular succession of hornblende-biotite-orthopyroxene and biotite-garnet-hornblende assemblages within the biotite-garnet granulite occurs at the base of the Winding Stair Gap exposure (Fig. 4b). Calc-silicate gneiss boudins completely surrounding the biotite-garnet granulite unit also indicate an original compositional layer. The orthopyroxenite unit does not fit readily into the proposed stratigraphy, suggesting that it may have been structurally emplaced. Boudinage of the calc-silicate layer and repetition of the hornblende-biotite-orthopyroxene and biotite-garnet-hornblende assemblages are the result of folding (F_1) prior to peak metamorphism (Fig. 4b). The F_1 fold axis is oriented N36°W, plunging 8°NW. No direct evidence of S_1 schistosity

accompanying F_1 has been observed, although intrafolial folds in the metatexite migmatite could represent S_0 or S_1.

The dominant schistosity of Winding Stair Gap, S_2, is defined by the parallel alignment of biotite, sillimanite, and hornblende and the development of gneissic banding which cuts across S_0 and F_1. At Winding Stair Gap, S_2 (Fig. 4b) is oriented approximately N45°W; 20 to 40°SW, but the orientation of S_2 in much of the surrounding area is to the NE, dipping 30 to 80°NW. The alignment of prograde minerals, such as sillimanite and biotite, into a pervasive schistosity indicates the deformation producing S_2 was contemporaneous with peak metamorphism. Isoclinal recumbent folds, F_2, are apparent in the sillimanite schist and biotite-hornblende granulite units, refolding of F_1 in the biotite-garnet granulite, folding of calc-silicate boudins, and folding of the orthopyroxenite (Fig. 4b).

Deformation following peak metamorphism is neither widespread nor pervasive. In the sillimanite schist, crenulations of biotite and sillimanite deform S_2 schistosity. Some K-feldspar lenses, containing retrograde kyanite and coarse-grained muscovite, are deformed into S-shaped bodies. Crenulations of retrograde biotite and cummingtonite replacing hypersthene in the orthopyroxenite followed peak metamorphism.

The earliest structural events, F_1 and S_1(?), predate development of the metatexite migmatite. F_2 and S_2, developed during peak metamorphism, are contemporaneous with migmatization. The weakly foliated gneiss probably crystallized late in the prograde metamorphic event or immediately following granulite facies metamorphism. The last deformation, F_3, is associated with nonfoliated granitic rocks and retrograde minerals. F_3 and the retrograde phases do not appear to represent a separate, unrelated metamorphic and deformational episode, however. The extent of retrograde metamorphism is very limited at Winding Stair Gap, and crenulations do not develop a penetrative fabric. The F_3

event probably represents a minor deformation feature produced as the rocks cooled from the metamorphic peak.

The folding events recognized in the Winding Stair Gap area are consistent with regional patterns outlined by Hatcher and Butler (1979). The earliest fold development at Winding Stair Gap possibly coincides with regional F_1, interpreted as Taconic structures formed prior to peak metamorphism. F_2 and S_2 at Winding Stair Gap may be equivalent to the F_2 folds and dominant S-surface interpreted as Taconic features developed near the thermal peak of metamorphism. The crenulation at Winding Stair Gap, F_3, does not directly correlate with the regional F_3 event of upright isoclinal to open folds, but may be related to a regional F_4 crenulation previously interpreted as Acadian.

REFERENCES CITED

Absher, B. S., and McSween, H. Y., 1985, Granulites at Winding Stair Gap, North Carolina: the thermal axis of Paleozoic metamorphism in the southern Appalachians: Geological Society of America Bulletin, v. 96, p. 588–599.

Force, E. R., 1976, Metamorphic source rocks of titanium placer deposits—a geochemical cycle: United States Geological Survey Professional Paper 959-B, 16 p.

Hatcher, R. D., Jr., and Butler, J. R., 1979, Guidebook for southern Appalachian field trip in the Carolinas, Tennessee, and northeastern Georgia: International Geological Correlation Program, Caledonide Orogen Program 27 Fieldtrip, Guidebook, 117 p.

McElhaney, M. S., and McSween, H. Y., 1983, Petrology of the Chunky Gal Mountain mafic-ultramafic complex, North Carolina: Geological Society of America Bulletin, v. 94, p. 855–874.

Chunky Gal Mountain and Glade Gap

Robert D. Hatcher, Jr., Department of Geology, University of South Carolina, Columbia, South Carolina 29208

LOCATION

Chunky Gal Mountain and Glade Gap are located adjacent to U.S. 64 6 mi (10 km) east of Shooting Creek, Clay County, North Carolina, Shooting Creek and Rainbow Springs 7½-minute Quadrangles (Fig. 1).

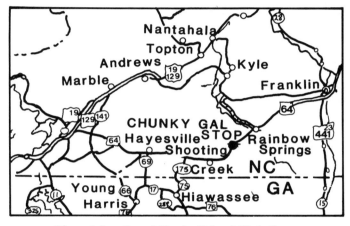

Figure 1. Location of Chunky Gal and Glade Gap.

INTRODUCTION

The central Blue Ridge of North Carolina consists of a medium to high grade assemblage of metasedimentary and meta-igneous rocks. Rocks of the western Blue Ridge consist of a sequence of upper Precambrian to Lower Cambrian clastic and carbonate rocks that were deposited on continental basement along the ancient continental margin of North America. Rocks of the eastern Blue Ridge consist of an assemblage of late Precambrian(?) metaclastic and metavolcanic rocks deposited on the outer continental margin, and possibly, in part, on oceanic crust, since continental basement is rare here. These rocks were metamorphosed and intruded by granitic plutons during the early to middle Paleozoic. The central Blue Ridge contains the boundary between these eastern and western assemblages: the Hayesville thrust (Fig. 2). The Hayesville thrust is a premetamorphic fault that has a sinuous outcrop trace, indicating it has been folded after emplacement. There are few, if any, large accessible exposures of the Hayesville fault, but the Chunky Gal Mountain–Glade Gap exposure contains structures representative of the structural style of the central Blue Ridge just east of the outcrop trace of the Hayesville fault. This exposure is important because of its size and the nature of the rocks and structures exposed.

Exposures at Chunky Gal Mountain and Glade Gap contain representative lithologies of the western part of the Hayesville thrust sheet just east of the trace of the Hayesville fault on the east flank of the Shooting Creek window (Fig. 2). The purpose of this stop is twofold: 1) to illustrate the structural style of this part of the central-eastern Blue Ridge and 2) to examine several of the representative lithologies of this part of the Blue Ridge. The stratigraphy in this part of the Blue Ridge consists of rocks of the Tallulah Falls Formation (deep water clastic-volcanic assemblage) overlain by the Coweeta Group of cleaner clastic metasedimentary rocks. The rocks exposed at this stop belong to the Coweeta Group and to a mafic complex that is part of the Buck Creek or Chunky Gal mafic-ultramafic complex described by McElhaney and McSween (1983).

SITE DESCRIPTION

Chunky Gal Mountain

The Chunky Gal Mountain exposure consists of a multiply folded assemblage of metasandstone and amphibolite with minor calc-silicate quartzite. At its northeastern end the exposure has been cut by several small brittle faults. The central part of the exposure is mylonitized, while the southern third has been cut by a major fault, the Chunky Gal Mountain fault, which emplaced amphibolite of the Buck Creek or Chunky Gal Mountain complex over clastic rocks of the Coweeta Group (Coleman River Formation metasandstones). Muscovite is still present indicating that the rocks are in the upper amphibolite facies, since sillimanite is abundant nearby.

The rocks consist of quartz-feldspar-biotite metasandstones that contain additional amounts of garnet and calc-silicate layers. In addition, they were intruded by several trondhjemite(?) dikes less than 4 in (10 cm) wide. Amphibolite at road level in the southern part of the cut consists of hornblende-plagioclase gneiss. The amphibolites appear less deformed than the adjacent biotite metasandstones (biotite gneiss) only because the amphibolites recrystallize and contain fewer strain markers than the adjacent rocks. Rocks of the Coweeta Group at the northern end of the exposure consist of layered metasandstones and minor schists with boudins of calc-silicate quartzite. The pressure was lowered immediately adjacent to the boudins during their formation and garnet-quartz formed pressure shadows parallel to layering. Layering in the metasandstones at Chunky Gal Mountain has probably been transposed several times through progressive deformation.

The contact exposed between the amphibolite and biotite metasandstone units must be a major fault. The name Chunky Gal Mountain fault was first applied to the fault at this exposure

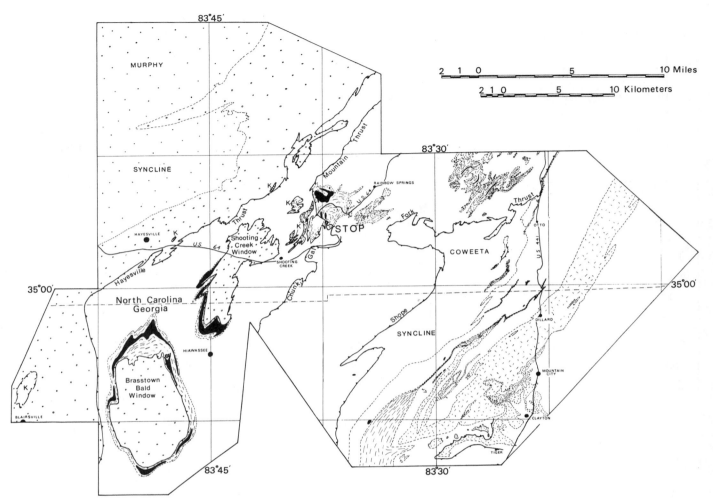

Figure 2. Generalized geologic map of part of the Blue Ridge of western North Carolina and northeast Georgia. The Rabun granite and related bodies are cross-hatched. Ultramafic bodies are black. Mafic bodies are lined. Great Smoky Group rocks are shown by circles. Murphy belt rocks are shown by dots. Tallulah Falls Formation and Coweeta Group rocks are unpatterned.

by Hatcher and Butler (1979). As one approaches this contact, the metasandstone becomes more gneissic and mylonitic. It is mylonitic for a distance of 130 to 150 ft (40 to 45.7 m) from the contact. The amphibolite in the hanging wall is not mylonitic.

Several ductile shear zones occur within the biotite gneiss unit, which indicate a down-to-the-west sense of shear. The Chunky Gal Mountain fault contact exhibits the same shear sense as rotated porphyroclasts within the mylonitic rocks, indicating the fault must be either an extensional normal fault or a folded thrust. Because of the overall character and style of the deformation illustrated in this exposure, I favor the latter interpretation. For purposes of discussion, the contact is considered a synmetamorphic thrust, perhaps a tectonic slide, yet the fabrics are clearly truncated by the fault contact. The mylonitic fabric in the biotite gneiss in the footwall has been rotated into parallelism with the

fault contact. However, the foliation in the amphibolite, which obviously must be earlier, appears truncated at the contact. Therefore, the dominant foliation in this area, probably S_2, is related to formation of the thrust in a sequence of progressively deformed rocks.

The deformation plan at Chunky Gal Mountain is dominated by the presence of an early foliation, which contains passive flow isoclinal and isoclinal recumbent folds that have been refolded by more upright passive flow folds, then by brittle faults (Fig. 3). The foliation in the amphibolite in the hanging wall may be an earlier foliation truncated by the fault.

Glade Gap

Rocks at Glade Gap consist mostly of amphibolite with

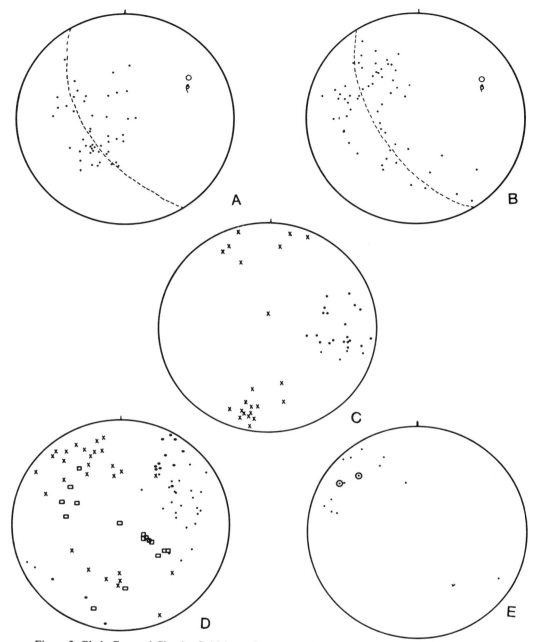

Figure 3. Glade Gap and Chunky Gal Mountain mesoscopic fabric data. A. 48 poles to S_1 surfaces in Domain I (βN56E, 31NE). B. 59 poles to S_1 in Domain II (βN55E, 29NE). C. 24 late flexural-slip folds and crenulation cleavage measurements. Dots—axes; squares—poles to axial surfaces. Domain II: dots—axes; x's—poles to axial surfaces. E. 15 poles to fault and shear planes (other than joints). Circled dots—poles to major fault surfaces.

some interlayered metasandstone and pelitic schist. These rocks were isoclinally folded and later refolded by upright folds. Veins of quartz and quartz-feldspar were emplaced at several different times. Some remain planar while others in similar orientations are folded. An excellent isoclinal recumbent fold occurs on the south side of the Glade Gap cut. This fold is composed of interlayered thin metasandstone and pelitic schist and is completely surrounded by amphibolite.

REFERENCES CITED

Hatcher, R. D., Jr., and Butler, J. R., 1979, Guidebook for southern Appalachian field trip in the Carolinas, Tennessee and northeastern Georgia: International Geological Correlation Program-Caledonide Orogen Project (27), 117 p.

McElhaney, M. S., and McSween, H. Y., Jr., 1983, Petrology of the Chunky Gal Mountain mafic-ultramafic complex, North Carolina: Geological Society of America Bulletin, v. 94, p. 855–874.

Ocoee Gorge; Appalachian Valley and Ridge to Blue Ridge transition

Robert D. Hatcher, Jr.,Department of Geology, University of South Carolina, Columbia, South Carolina 29208
Robert C. Milici, Virginia Division of Mineral Resources, Box 3667, Charlottesville, Virginia 22903

LOCATION

Ocoee Gorge is located along U.S. 64 in southeastern Tennessee in the western edge of the Blue Ridge (Fig. 1). Getting to this location is relatively easy, although parking conditions and access to the rocks vary considerably along the highway. Parking is readily available at Parksville Dam at the west end of the series of stops to be described. However, at Maddens Branch adequate parking is available for only one bus or four vans. The shoulders along the highway at both of these stops provide limited walking space, and the traffic moves very fast. Pay attention to the traffic at both Parksville Dam and Maddens Branch. Boyd Gap at the east end of the site contains adequate parking and plenty of room to walk and look at rocks.

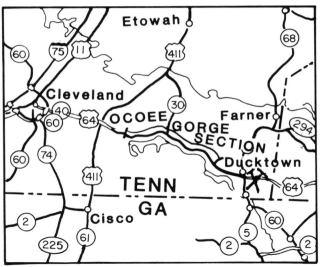

Figure 1. Location map.

INTRODUCTION

The transition from the southern Appalachian Valley and Ridge to the Blue Ridge involves major changes in both tectonic style and stratigraphic sequences (Fig. 2). In general, the rocks of the Valley and Ridge consist of a sequence of platform clastic and carbonate rocks (Lower Cambrian to Mississippian) that were mostly not metamorphosed and cleaved and were thrust faulted and folded during the Alleghanian orogeny. Rocks of the frontal Blue Ridge consist of upper Precambrian to Lower Cambrian clastic and minor carbonate rocks deposited from the ancient continental slope and rise to the outer continental platform. These rocks were folded, cleaved, and metamorphosed, except for the rocks in the frontal thrust sheet, during an earlier deformational-thermal event (Taconic?), then were later overthrust toward the west during the Alleghanian event.

The Ocoee Gorge is the type area for the Ocoee Series (Safford, 1856). This set of exposures contains almost complete sections of the upper parts of the Ocoee Series rocks in this area, including the Sandsuck Formation and Wilhite Slate of the Walden Creek Group and the upper part of the Great Smoky Group (Fig. 3). An argument can also be made in the gorge, but not at one of the stops described below, that the contact between the Great Smoky Group and Walden Creek Group is a comformable stratigraphic contact (J. O. Costello, pers. communication, 1985) and not a fault.

This series of exposures in Ocoee Gorge is important for several reasons to our understanding of southern Appalachian tectonics. It is a classic late Precambrian stratigraphic section and type area and affords the opportunity to observe in almost continuous exposure the transition from the foreland to the internal parts of an orogenic belt.

The Ocoee Gorge stops contain an assemblage of late Precambrian, Cambrian, and Ordovician sedimentary to metasedimentary rocks that range from unmetamorphosed at the west end (Parksville Dam) to biotite grade at Boyd Gap. The transition from subgreenschist grade (anchizone), brittly deformed Cambrian and late Precambrian rocks at Parksville Dam to very low grade but strongly cleaved rocks at Maddens Branch to strongly cleaved and polyphase deformed rocks at Boyd Gap is well illustrated in the section. It is also possible to observe the transition from probable Alleghanian deformation at the west edge of the Blue Ridge thrust sheet to Taconic deformation farther east in the Maddens Branch locality (Stop B) to polyphase Taconic deformation at Boyd Gap (Stop D). In addition, stratigraphic and compositional changes may be examined in the Chilhowee Group and Sandsuck Formation at the frontal thrust, the Walden Creek Group rocks at Maddens Branch, and the Boyd Gap Formation rocks of the Great Smoky Group at Boyd Gap (Fig. 1).

SITE DESCRIPTIONS

Stop A: Parksville Dam Locality

Stop A is located on U.S. 64 at Parksville Dam, Tennessee (Fig. 2). The rocks exposed in road cuts on the north side of the highway at Parksville Dam consist of the Lower Cambrian Chilhowee Group sandstones and shales along with rocks of the Sandsuck Formation of late Precambrian age. These rocks exhibit the effects of brittle deformation and essentially no metamorphism, but abundant fractures and small normal and thrust faults

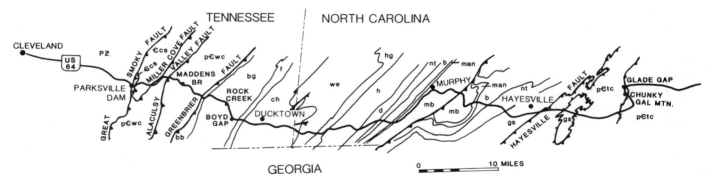

Figure 2. Geologic map showing locations of stops between Parksville Dam and Ducktown, Tennessee, in the Ocoee Gorge Appalachian Valley and Ridge to Blue Ridge transition, as well as the location of the Glade Gap/Chunky Gal Mountain stop farther east.

are present, indicating the close proximity to a major brittle fault zone. In the westernmost cut, rocks of the Knox Group in the footwall of the Great Smoky fault may be observed in the underbrush about 30 ft (10 m) from the highway. Rocks exposed here were considered by Milici (1978) to consist of a slice of Chilhowee Group rocks between rocks of the Knox and older rock units to the west and the late Precambrian Sandsuck Formation to the east. If the entire mountain is traversed, most of the stratigraphic sequence in the Chilhowee Group may be observed. However, there are better places to observe this sequence toward the northeast rather than attempting to climb the mountain at this stop.

It is likely that the rocks exposed at Parksville Dam belong

to the uppermost sandstone unit, the Hesse Sandstone, which consists of massive layers of quartz sandstone containing abundant *Scolithus* tubes and some shale beds. Some vestiges of cross-bedding may be present in a few places, but it is more likely that features that resemble cross-beds are Liesegang bands produced by oxidation of iron-rich unstable minerals and diffusion of iron oxide/hydroxide precipitates. Immediately southeast of the block of Hesse Sandstone is a mass of shale identified by Milici (1978) as Murray Shale, which is fault bounded as well as repeated by faulting in several blocks (Fig. 4). Farther southeast this fault block is bounded by the late Precambrian Sandsuck Formation. Perhaps because of the contrasting rheology of the different fault blocks produced by differing amounts of sandstone

SYSTEMS	SERIES	UNIT	THICKNESS (Meters)	DOMINANT ROCK TYPE(S)
ORDOVICIAN	Lower Ordovician	Knox Group	1000	Dolomite, (siliceous) limestone
CAMBRIAN	Middle and Upper Cambrian	Consauga Group	700	Shale, limestone
	Lower Cambrian	Rome Formation	600	Shale, limestone
		Shady Dolomite	400	Dolomite, shale
		Chilhowee Group Hesse Sandstone Murray Shale Nebo Sandstone Cochran Formation	1000	Sandstone, shale, conglomerate
PRECAMBRIAN	Ocoee Series	Walden Creek Group Sandsuck Formation Wilhite Formation Shields Formation	2800	Shale (slate), sandstone, conglomerate, limestone
		Great Smoky Group Boyd Gap Formation Farner Formation Copperhill Formation	7000	Graywacke, sandstone, shale

Figure 3. Stratigraphic column of rock units exposed in Ocoee Gorge.

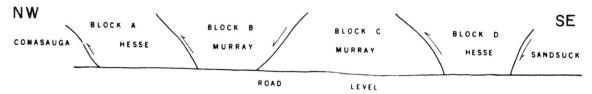

Figure 4. Sketch of structural units that compose the Chilhowee fault slice. Section is 0.3 mi (0.5 km) long.

and shale, some of the blocks remain undeformed while others appear complexly deformed. Those that contain the greatest amount of sandstone appear more intensely deformed. There is also a general positive relationship between proximity to the trace of the Great Smoky fault and the intensity of deformation. It is likely that all the brittle deformation in this exposure is the result of Alleghanian thin-skinned deformation.

Stop B: Cleavage in Walden Creek Group Slates at Maddens Branch

Stop B at Maddens Branch lies about 6.7 mi (11.2 km) east of Parksville Dam on U.S. 64 (Fig. 1). The rocks at Maddens Branch are thought to be correlative with the Walden Creek Group described farther northeast by Neuman and Nelson (1965). Hurst and Schlee (1962) considered the rocks at Maddens Branch to be the youngest rocks in the late Precambrian succession in this area. The rocks consist of greenish-gray slate with thin interlayers of fine-grained, tan, carbonate-rich siltstone along with moderate to thick beds of massive- to medium- or coarse-grained gray feldspathic sandstone. The standstone beds range from thicknesses of a few inches to 3 ft (1 m) or more. Ripple and scour marks are present on some bedding planes. Part of the section is composed of variegated sand and shale or siltstone layers ranging from greenish to reddish gray or to more buff colors.

The deformation at Maddens Branch is dominated by folds with amplitudes ranging from a few inches to 30 ft (10 m). Folds range from very tight to open and from passive slip in the slates of the Walden Creek Group to flexural slip in the more massive sandstones. Vergence of folds is toward the northwest. These folds appear to be part of one dominant episode of folding, which is probably older than the brittle deformation at Parksville Dam. Later, low amplitude open folds have not been observed.

Slaty cleavage is the dominant penetrative fabric element. Its orientation is very consistent (Fig. 5) and indicates very little post-cleavage folding or other deformational processes that would rotate the cleavage. Fold axial surfaces parallel cleavage. The slaty cleavage probably results from reorientation of grains with minimal pressure solution during lower greenschist facies metamorphism. However, pressure solution is dominant in some of the more carbonate-rich layers.

Study of silty carbonate and slate layers provides an opportunity to attempt to differentiate between primary depositional and later tectonic structures. Cleavage refraction is evident at boundaries between pelitic and silty carbonate layers. Upper surfaces of some carbonate beds were rippled. Their amplitudes were increased during flattening strain that accompanied the formation of slaty cleavage. This interpretation is based upon the observation that corresponding lower surfaces of many silty carbonate layers in the same positions in folds are not deformed but remain as bedding-parallel surfaces. Other silty carbonate layers exhibit equivalent layer boundary distortion into cuspate folds whose axial surfaces parallel the orientation of slaty cleavage planes. However, cleavage refracts toward the cuspate fold hinges away from the more arcuate hinges. On the upper sides of beds, cusps are synclines, arcuate hinges are anticlines.

Some silty carbonate layers are continuous, others are discontinuous. Continuous layers become folded surfaces during deformation; discontinuous layers become boudins. Formation of slaty cleavage appears to have enhanced the separation of initially continuous or slightly discontinuous layers. The discontinuous silty carbonate layers may have originally been connected by filamentous laminae that were destroyed during cleavage formation. They may have initially had a rippled configuration so that cleavage formation resulted in preservation of material in the crestal portions or ripples, but the thinned troughs were destroyed. Some boudin-shaped pods of silty carbonate are presently flattened into ellipses whose long axes are parallel to slaty cleavage planes. Other folded silty carbonate layers are, in part, strongly flattened and partially dismembered into the plane of the slaty cleavage, indicating that incipient transposition occurred during folding. Bedding serves primarily as a strain marker, particularly where beds are very thin.

Folds and slaty cleavage appear to have formed at the same time. Poles to fold axial surfaces and cleavage planes have the same orientation (Fig. 5). Holcombe (1973) has described the slaty cleavage as a domainal slaty cleavage that results from a concentration of layer silicates in zones parallel to cleavage planes and zones in which quartz and other nondimensionally oriented grains are concentrated. Cleavage is expressed as quartz-filled, gash fractures in metasandstone beds, perhaps related to layer parallel extension, fracturing, and quartz filling as pressure solution dissolved some of the quartz from the sandstone in adjacent layers. The orientation of the gashes in the sandstones is subparallel to slaty cleavage in adjacent slates but fans more about fold axes than the slaty cleavage, indicating perhaps that the formation of extension gashes occurred early on and that rotation occurred later. Numerous joints and a few small late brittle faults are also present at Maddens Branch. These all appear to be late structures.

One small fault near the west end of the cut dips steeply but irregularly crosses both bedding and cleavage. This structure has a striated movement surface upon which slickensides and fibers indicate dip-slip motion.

A small early thrust is present at the easternmost end of the exposure. It occurs in the slate just above the contact with the

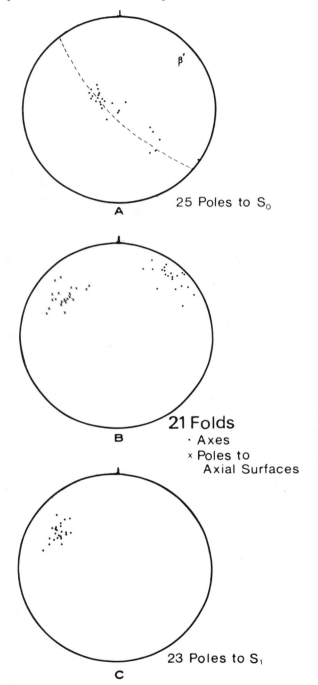

A 25 Poles to S₀

B 21 Folds
· Axes
× Poles to
  Axial Surfaces

C 23 Poles to S₁

Figure 5. Mesoscopic fabric data from Maddens Branch locality. A. 25 poles to bedding (S_0). The orientation of β is N52°E, 9°NE. B. 21 passive-slip folds. C. 23 poles to slaty cleavage (S_1). Note the correspondence between orientations of folds in B and the axis in A.

sandstones and displays the classic features of a decollement thrust. It starts and ends in incompetent layers and exhibits drag folds in the zone of maximum displacement. It could have formed early, during the cleavage-forming event, or even earlier as a diagenetic feature.

Metamorphic grade at Maddens Branch is in the Barrovian chlorite zone. X-ray diffraction study of slate chips indicates strong orientations of chlorite and muscovite parallel to the slaty cleavage, suggesting that recrystallization and formation of the slaty cleavage occurred coevally with the metamorphic thermal peak.

Stop C: Faulting in the Boyd Gap Formation at Rock Creek

Stop C is located about 8 mi (12 km) east of Stop B on U.S. 64 (Fig. 2). Turn into a paved parking area just beyond the beginning of the increased road grade. The biotite isograd and pyrite-pyrrhotite transition occur approximately at this point. Deformation of graywacke, sandstone, siltstones, and shales in the Boyd Gap Formation of the Great Smoky Group at Rock Creek resembles in several ways the deformation in broken formation zones observed in parts of the Valley and Ridge, such as at Dunlap, Tennessee (see Wilson and Wojtal, this volume). At Rock Creek, intense deformation is confined to the upper two-thirds of the exposure, generally above a basal fault zone (Fig. 6). The basal fault zone consists of about 6 ft (2 m) of complexly deformed silty shale. Several small breccia blocks up to 2 ft (60 cm) long and 1 ft (25 cm) thick are encapsulated within the shale. These blocks consist of coarse, quartzofeldspathic graywacke, conglomeratic graywacke, and fine silty graywacke. No irregular quartz veins or masses of quartz are common within this zone but are present very close by on the adjacent mountainsides.

The basal fault zone is overlain by about 20 ft (7 m) of silty graywacke. The unit is folded into a tightly appressed overturned syncline, which has an intricately folded and faulted eastern limb. Next, to the east along the outcrop, is a mass of complexly deformed silty shale and siltstone that contains irregular folded boudins of silty graywacke sandstone.

The easternmost part of the outcrop is dominated by a flat-topped anticline of silty graywacke. The western limb of the anticline is vertical to overturned, but the eastern limb is pinched off so that irregular pieces of silty graywacke trail behind the main mass, surrounded by deformed silty shale. Silty black shale in the core of the flat-topped anticline exhibits a poorly developed cleavage. Subhorizontal to folded bedding within the silty shale is well defined and is accentuated by the growth of quartz and pyrrhotite in thin irregular veins.

The subhorizontal graywacke is overlain by a fault that follows the bedding of overlying strata. The fault zone that extends upward from near road level at the eastern end of the outcrop is composed of fractured and folded shale and siltstone that contain irregular masses of coarser beds. Strata at the top of the outcrop are above the broken formation zone.

Exposures of the folded hanging-wall beds occur along U.S.

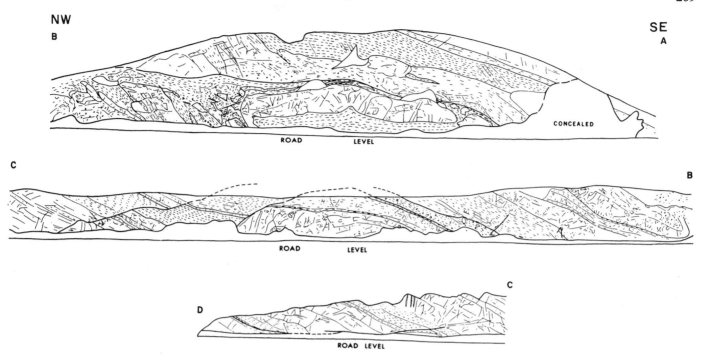

Figure 6. Diagram showing broken formation zone in the Boyd Gap Formation at Rock Creek.

64 east of the Rock Creek fault. The finer-grained beds exhibit a well-developed cleavage, and quartz-filled extension fractures are common. However, an extensive fracture zone like the one developed at Dunlap, Tennessee, is absent here.

Stop D: Boyd Gap[1]

Stop D is located about 2 mi (3.3 km) east of Stop C at Boyd Gap, the highest point on U.S. 64 in this area (Fig. 2). Numerous exposures of thick bedded metagraywacke and interbedded shale may be observed here, where the rocks have been folded into west-vergent modified concentric folds. This exposure is at the eastern end of the type section of the Boyd Gap Formation of the Great Smoky Group. The two major rock types of this unit, dark gray laminated slate or phyllite and metagraywacke, are about equally represented and constitute most of the exposed beds. A few of the beds are matrix-rich metagraywacke. The metagraywacke units are composed mainly of quartz and feldspar with subsidiary biotite, muscovite, quartz, and feldspar. Some of the large quartz grains have an opalescent blue color. Most noteworthy of the accessory minerals are graphite and iron sulfides, both indicative of strongly reducing conditions during deposition. Metagraywacke beds contain pyrrhotite. However, pyrite rather than pyrrhotite is present in the dark slate and matrix-rich metagraywackes, a situation explained by Merschat and Larson

(1972) as being the result of nonequilibrium caused by the much higher sulfur content of the dark rocks as compared to the metagraywackes. Polished section examination invariably reveals small blebs of chalcopyrite associated with the pyrrhotite. Occasionally, sphalerite and rare galena may also be observed here.

Bedding or initial stratification is prominent. Less common primary features to be observed here include graded betting, slump structures, flame structures, intraformational fragments in some metagraywackes, rare concretionary structures, and a few penecontemporaneous folds.

The dominant anisotropies are bedding (S_0) and slaty cleavage (S_1). Their geometric relations are most obvious in the fine-grained beds. Bedding is obviously folded but generally strikes about N45°E. The attitude of slaty cleavage is much more limited, with an average strike of about N30°E and dip 65° southeast. At a few places kinking, which may be an early stage in the development of crenulation cleavage (S_2), may be noted. An almost microscopic crenulation cleavage may be observed on some S_1 surfaces in the slates. Top of bed or facing directions, as determined by bedding cleavage relations, agree with the sedimentary top-bottom criteria. A series of prominent folds at Boyd Gap is asymmetric, similar to disharmonic folds of bedding with a subhorizontal plunge (Fig. 7). The slaty cleavage is approximately axial planar to these folds. Differences in competency have controlled the behavior of the individual layers in folds. The least competent slates are obviously thinner on the limbs than on the crests or troughs, demonstrating that flowage has occurred within individual beds. The more competent metagraywacke layers maintained fairly uniform thickness from limb to limb, although minor systematic variations indicate that they, too, are

---
[1]Authors' note: This stop was originally described by L. S. Wiener and C. E. Merschat and published in an earlier guidebook (Wiener and Merschat, 1978). It was intended that they be coauthors on the Ocoee Gorge stop, but their names were removed at the request of P. M. Brown of the North Carolina Geological Survey.

SE   NW

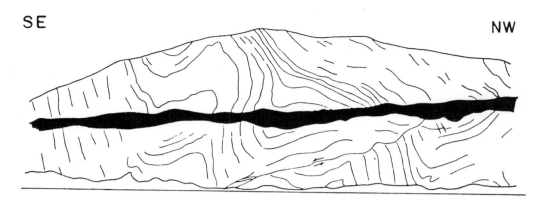

Figure 7. Sketch of the Boyd Gap locality showing northwest-vergent
folds in thick sandstone layers cut by a small fault. View is southwest
from the bench on the northeast side of the highway.

deformed by flow to some extent. The fold mechanisms in the least competent beds range between flexural flow and passive flow. Those in the more competent metagraywacke beds range between flexural slip and flexural flow.

Nearly north-trending, left lateral strike-slip cross faults are also found in this exposure. These faults displace all other features in the rocks and are clearly postmetamorphic brittle structures.

Slickensides on the fault surfaces are subhorizontal or plunge very gently southward. Displacement along any one of these faults is not extensive. By matching fold axes from one wall of the exposure to the other, we estimate the displacement to be about 150 ft (45 m). Faults with similar genetic relations have been noted in the mines and denuded area of the nearby Copper Basin. The greatest displacement known here is about 350 ft (100 m).

REFERENCES CITED

Holcombe, R. J., 1973, Mesoscopic and microscopic analysis of deformation and metamorphism near Ducktown, Tennessee [Ph.D. thesis]: Palo Alto, Stanford University, 225 p.

Hurst, V. J., and Schlee, J. S., 1962, Ocoee metasediments, north central Georgia (Geological Society of America, Southeastern Section Annual Meeting, Guidebook no. 3): Georgia Department of Mines, Mining and Geology, 28 p.

Merschat, C. E., and Larson, L. T., 1972, Disseminated sulphides in late Precambrian Ocoee rocks: Geological Society of America Abstracts with Programs, v. 4, p. 92–93.

Milici, R. C., ed., 1978, A structural transect in the southern Appalachians, *in* Field trip in the southern Appalachians: Tennessee Division of Geology

Department Report of Investigations 37, p. 5–52.

Neuman, R. B., and Nelson, W. H., 1965, Geology of the western Great Smoky Mountains, Tennessee: U.S. Geological Survey Professional Paper 349-D, 81 p.

Safford, J. M., 1856, A geological reconnaissance of the State of Tennessee: Nashville, Tennessee Geological Society, 1st Biennial Report State Geologist, 164 p.

Wiener, L. S., and Merschat, C. E., 1978, Structure of Boyd Gap, *in* Hatcher, R. D., Jr., Merschat, C. E., Milici, R. C., and Winer, L. S., Field Trip 1—A structural transect in the southern Appalachians, Tennessee and North Carolin, *in* Milici, R. C., ed., Field trips in the southern Appalachians: Tennessee Division of Geology Report of Investigations 37, p. 39–40.

The Cartersville fault at Carters Dam

Keith I. McConnell, Georgia Geologic Survey, 19 Martin Luther King, Jr., Drive, Atlanta, Georgia 30334

LOCATION

The Cartersville fault is exposed near the powerhouse at the base of Carters Dam located in southeastern Murray County, approximately 60 miles (96 km) north-northwest of Atlanta on the Oakman 7½-minute Quadrangle (Fig. 1). To reach Carters Dam, travel north from Cartersville on U.S. 411 until you see the signs for the Carters Dam powerhouse.

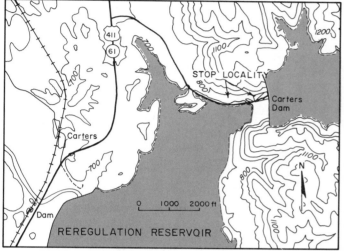

Figure 1. A part of the Oakman 7½-minute Quadrangle depicting the location of the Cartersville fault outcrop.

INTRODUCTION

Carters Dam is built along the scarp of the Cartersville fault which separates late Precambrian age rocks of the Blue Ridge Province to the east from Lower Cambrian age rocks of the Valley and Ridge Province to the west (Fig. 2). Immediately east of the fault and comprising the hanging wall are metamorphosed coarse clastics, argillites, and subordinate amphibolites of the Ocoee Supergroup. In the footwall of the Cartersville fault are shales and carbonates of the Conasauga Group.

The exposure at Carters Dam provides an excellent opportunity to observe the relationships between the Blue Ridge and Valley and Ridge Provinces as well as being the best exposure in Georgia of a fault (i.e., the Cartersville fault) that has been the subject of controversy for nearly a century. In particular, to the south near the town of Cartersville, the position of the trace of the Cartersville fault, and indeed, whether the fault even exists east of Cartersville, has been debated for 75 years. Visitors to the Carters Dam site can observe the style and intensity of deformation associated with the Cartersville fault. For those interested in the controversy surrounding the Cartersville fault, additional stops in the Cartersville area are recommended. Two Georgia Geological Society field trips (Bentley and others, 1966; Costello and others,

1982) have been held in the Cartersville area and have had the Cartersville fault problem as their major theme.

The Cartersville fault is the westernmost thrust sheet carrying Grenville basement in the hanging wall. Two external basement massifs, the Corbin Gneiss Complex and Fort Mountain Gneiss (McConnell and Costello, 1984), are present just east of the trace of the Cartersville fault (Fig. 3). As originally mapped by Hayes (1891), the Cartersville fault extended directly through the town of Cartersville and thrust rocks of the Ocoee series (pre-

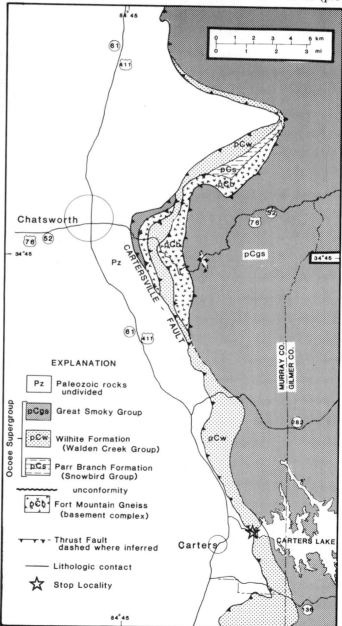

Figure 2. Generalized geologic map of the Carters Dam area.

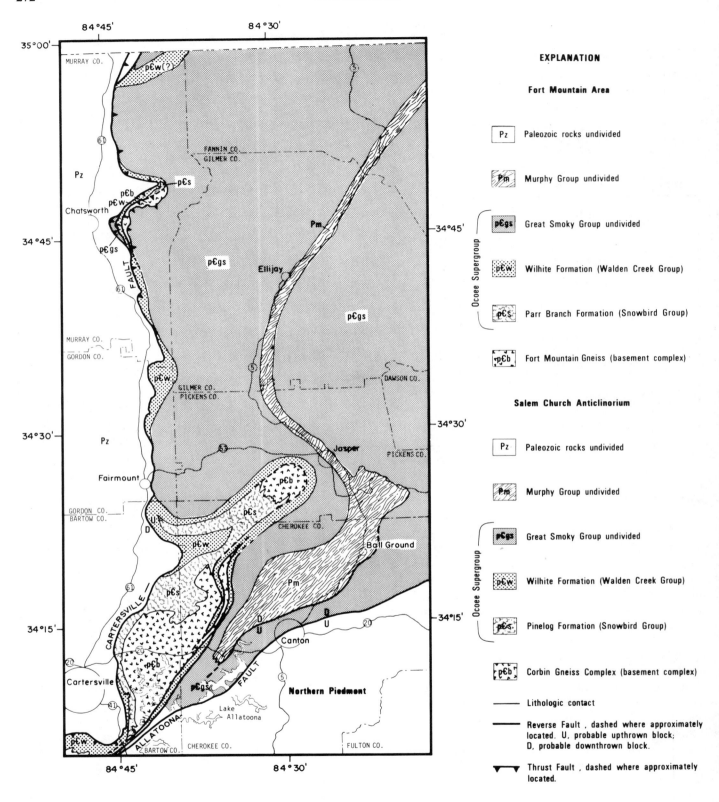

Figure 3. Generalized geologic map of the Blue Ridge in north-central Georgia (after McConnell and Costello, 1984).

sumed to be of Silurian age, on the basis of work by Hayes in east Tennessee) over other Paleozoic sediments. In a subsequent report, Hayes (1901) placed the fault a few miles to the east and changed the age of Ocoee rocks to Algonkian. According to Hayes (1901, p. 410), the fault was ". . . marked by a bed of breccia, a few inches or feet in thickness, and made up of comminuted fragments of the formations on either side." In 1944, Stose and Stose (1944) extended the Great Smoky fault of Keith (1927) southward, connecting it with the Cartersville fault; many subsequent workers (Cressler and others, 1979; McConnell and Costello, 1982) have followed this interpretation. However, Costello and others (1982) pointed out that in its type area the Great Smoky fault transports upper Precambrian and lower Paleozoic rocks over younger Paleozoic rocks. Rocks as young as Rome Formation (Cambrian) lie in the hanging wall of the Great Smoky fault (Neuman and Nelson, 1965). In contrast, the Cartersville fault brings Precambrian metamorphic rocks of the Blue Ridge into contact with lower Cambrian rocks. This fault relationship is similar to the relationship found across the traces of the Miller Cove, Sylco Creek, and Alaculsy Valley faults to the north; therefore, the fault present at this locality (Carters Dam) and separating the late Precambrian Ocoee Supergroup from the lower Cambrian sequence may not be the Great Smoky fault. If the Great Smoky fault is represented in Georgia it may be further to the west, possibly represented by klippen of Lower Cambrian formations lying west of the trace of the Cartersville fault (Costello and others, 1982).

The brittle deformation evident along the Cartersville fault is Alleghenian in age. The southwestern continuation of the Cartersville fault, southwest of Cartersville, thrusts rocks of the Talladega belt over rocks as young as the Mississippian-age Fort Payne Chert (Crawford, 1977). Subsequent to the emplacement of the Cartersville thrust sheet, the plane of the fault was deformed by late stage folding (F_3). Major reentrants in the Cartersville fault near Chatsworth and Cartersville (Fig. 3) are, at least in part, the result of these late stage folds. The trend of these folds is northwest-southeast and they plunge moderately to steeply to the southeast. These folds are open and are variously vergent to the southwest and northeast. In the Chatsworth area, F_3 folds are responsible for arching of units while further to the south, in the Cartersville area, F_3 folds interfere with earlier-formed, northeast trending structures to form a dome and basin pattern resulting in the separation of exposed masses of the Corbin Gneiss Complex. Also near Cartersville, where Ocoee Supergroup rocks trend into the Talladega belt, F_3 folds rotate rock units and earlier-formed planar structures from a northerly strike northeast of Emerson to a southwesterly strike south of Emerson (McConnell and Costello, 1982). Ocoee Supergroup rocks are, therefore, traceable around the reentrant in the Cartersville fault and at least a part of the Talladega belt is composed of Ocoee Supergroup rocks.

SITE DESCRIPTION

At Carters Dam, the steeply dipping Cartersville fault juxta-

Figure 4. Photograph of the brecciated zone associated with the Cartersville fault near the powerhouse of Carters Dam.

poses metasandstones, metaconglomerates, phyllites, and amphibolite of the Ocoee Supergroup against shales and limestones of the Conasauga Group. The fault is expressed by a zone of intensely sheared metasandstone and graphitic schist approximately seven feet (2.1 m) thick (Fig. 4). Shales of the Conasauga Group are exposed along a service road above the powerhouse and in the adjacent ditch. These shales are tan when fresh, but weather to a red-orange to pink color. The effects of compression, the development of small scale folds and associated cleavages, become increasingly evident in the Conasauga as the fault zone is approached from the west. The brecciated zone is characterized by graphitic schist and matrix-supported angular blocks of quartzite. At the Carters Dam overlook at the top of the dam, you are standing on top of the scarp of the Cartersville fault. On the horizon to the west, are the Armuchee Ridges (Clark and Zisa, 1976) of the Valley and Ridge Province. These ridges are capped by the Silurian age Red Mountain Formation that, in Alabama, contains economic deposits of iron ore.

REFERENCES CITED

Bentley, R. D., Fairley, W. M., Fields, H. H., Power, W. R., and Smith, J. W., 1966, The Cartersville fault problem: Georgia Geologic Survey Guidebook 4, 38 p.

Clark, W. Z., Jr., and Zisa, A. C., 1976, Physiographic map of Georgia: Georgia Geologic Survey Map, 1:2,000,000.

Costello, J. O., McConnell, K. I., and Power, W. R., 1982, Geology of late Precambrian and early Paleozoic rocks in and near the Cartersville District, Georgia: 17th Annual Field Trip, Georgia Geological Society: 40 p.

Crawford, T. J., 1977, Cartersville fault, *in* Chowns, T. M., ed., Stratigraphy and economic geology of Cambrian and Ordovician rocks in Bartow and Polk Counties: Georgia Geologic Survey Guidebook 16-A, 21 p.

Cressler, C. W., Blanchard, H. W., Jr., and Hester, W. G., 1979, Geohydrology of Bartow, Cherokee, and Forsyth Counties, Georgia: Georgia Geologic Survey Information Circular 50, 45 p.

Hayes, C. W., 1891, The overthrust faults of the southern Appalachians: Geological Society of America Bulletin, v. 2, p. 141–154.

—— 1901, Geological relations of the iron-ores in the Cartersville district, Georgia: American Institute of Mining Engineers Transactions, v. 30, p. 403–419.

Keith, A., 1927, Great Smoky overthrust: Geological Society of America Bulletin, v. 38, p. 154–155.

McConnell, K. I., and Costello, J. O., 1982, The relationship between the Talladega belt rocks and Ocoee Supergroup rocks near Cartersville, Georgia, *in* Bearce, D. N., Black, W. W., Kish, S. A., and Tull, J. F., eds., Tectonic studies in the Talladega and Carolina slate belts, southern Appalachian orogen: Geological Society of America Special Paper 191, p. 36–46.

——1984, Basement-cover rock relationships along the western edge of the Blue Ridge thrust sheet in Georgia, *in* Bartholomew, M. J., ed., The Grenville event in the Appalachians and related topics: Geological Society of America Special Paper 194, p. 263–279.

Neuman, R. B., and Nelson, W. H., 1965, Geology of the western Great Smoky Mountains, Tennessee: U.S. Geological Survey Professional Paper 349-D, 81 p.

Stose, G. W., and Stose, A. J., 1944, The Chilhowee Group and Ocoee series of the southern Appalachians: American Journal of Science, v. 242, p. 367–390, 401–416.

62

The Pumpkinvine Creek Formation at the type locality

Charlotte E. Abrams and Keith I. McConnell, Georgia Geologic Survey, 19 Martin Luther King, Jr., Drive, Atlanta, Georgia 30334

LOCATION

The type locality of the Pumpkinvine Creek Formation is located on I-75, approximately 32 mi (51 km) north of downtown Atlanta, and approximately 1.5 mi (2.4 km) north of where I-75 crosses Lake Allatoona (Fig. 1). Outcrop is readily apparent on both sides of the highway, but exposures of the Pumpkinvine Creek Formation are more extensive and less weathered on the east side of the highway. Do not attempt to cross the highway.

INTRODUCTION

The Pumpkinvine Creek Formation (McConnell, 1980; McConnell and Abrams, 1984) is the northwesternmost mafic metaigneous rock unit present in the northern Piedmont of Georgia. The rocks at this outcrop lie just to the southeast of the Blue Ridge–Piedmont terrane boundary (McConnell and Abrams, 1984). The boundary between rocks of the Pumpkinvine Creek Formation and the Blue Ridge (i.e., Ocoee Supergroup) in this area is the extension of the Hayesville fault (Hatcher and Odom, 1980), and is termed the Allatoona fault (McConnell and Abrams, 1984). Hatcher and Odom (1980) proposed that the Hayesville fault separates a dominantly volcanic, ultramafic, granite-bearing terrane on the southeast (i.e., the Pumpkinvine Creek Formation and northern Piedmont) from a nonvolcanic, basement bearing terrane on the northwest (i.e., Blue Ridge). In the area of this stop, a similar separation in terranes is evident.

The Pumpkinvine Creek Formation is of both stratigraphic and economic importance in the northern Piedmont. The Pumpkinvine Creek Formation and the associated Canton and Univeter Formations comprise what is termed the "Dahlonega gold belt." Numerous gold and massive sulfide mines and prospects are associated with the Pumpkinvine Creek Formation. Relict volcanic features and the presence of exhalative units lead Abrams and McConnell (1984) to recognize that base and precious metal deposits of the area are volcanogenic. Stratigraphically, the Pumpkinvine Creek Formation represents one of the few lithologically distinct and laterally continuous units in the northern Piedmont and, therefore, is a good stratigraphic marker unit.

The Pumpkinvine Creek Formation can be traced northeastward to near Dahlonega (German, in press). Southwestward, rocks of the Pumpkinvine Creek Formation have been mapped to the vicinity of New Georgia. Rocks of similar lithology and stratigraphic position can be traced southwestward to the Alabama-Georgia line. Based on chemical similarities and similar tectonic positions, McConnell (1980) proposed that the Hillibee Greenstone and Pumpkinvine Creek Formation were exposed parts of the same unit. Trace element data from the Pumpkinvine

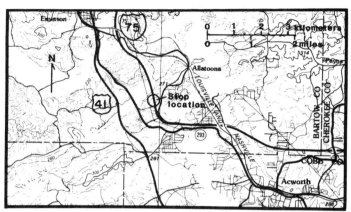

Figure 1. Location map. Location: east side of I-75, approximately 32 mi (51 km) north of downtown Atlanta, 21.5 mi (35 km) north of I-285, and approximately 1.5 mi (2.4 km) north of Lake Allatoona, Bartow County, Georgia; Cartersville 1:100,000.

Creek Formation and Hillabee Greenstone suggest that these mafic volcanic rocks are tholeiitic basalts and that they formed in a rift environment (McConnell, 1980). McConnell and Abrams (1984) interpreted these data to suggest that the Pumpkinvine Creek Formation formed in a back-arc basin.

SITE DESCRIPTION

McConnell (1980) designated the outcrop on I-75 and an outcrop along strike to the southwest on U.S. 41 as the type localities of the Pumpkinvine Creek Formation. Both exposures are easily accessible (Fig. 1) and contain the relict volcanic textures characteristic of this unit, but exposures are better along the Interstate highway.

The Pumpkinvine Creek Formation is composed predominantly of fine-grained amphibolite with intercalated sericite phyllite, and felsic gneiss (McConnell and Abrams, 1984). The felsic gneiss unit occurs both as thin (approximately 1 ft [30 cm]) interlayers within the main mass of amphibolite and as a separate mappable unit occupying the core of a regional fold defined by the two limbs composed of amphibolite (Fig. 2). The felsic gneiss is dacitic in composition and locally contains relict volcanic textures. This unit has been termed the Galts Ferry Gneiss member (Fig. 2) of the Pumpkinvine Creek Formation. Also included in the Pumpkinvine Creek Formation are thin, discontinuous lenses of sulfide and oxide facies banded iron formation.

At the I-75 location, the Pumpkinvine Creek Formation is composed primarily of chloritic amphibolite with layers of garnet-hornblende plagioclase gneiss (interpreted to be a crystal tuff, McConnell and Abrams, 1983). Mineral assemblages in the mafic members of the Pumpkinvine Creek Formation are

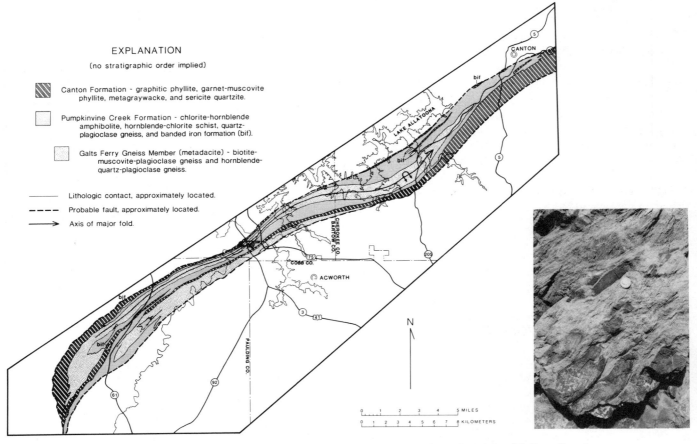

Figure 2. Geologic map of the Pumpkinvine Creek and Canton formations (after McConnell and Abrams, 1984). Insert: photograph of the deformed pillows in the Pumpkinvine Creek Formation at the type locality.

indicative of at least one episode of upper greenschist to lower amphibolite facies metamorphism (McConnell, 1980). The chloritic amphibolite is composed of epidote, chlorite, hornblende, quartz, and albite with accessory calcite and pyrite. At the southern end of the exposure is the contact between the Pumpkinvine Creek Formation and the structurally overlying Canton Formation.

Relict volcanic features characteristic of this unit are present at the I-75 exposure. Hurst and Jones (1973) recognized relict pillow structures in the amphibolite of the Pumpkinvine Creek Formation just to the west on U.S. 41. At the I-75 stop, relict pillows (Fig. 2) can be observed as elongate shapes with their long axis in the plane of the metamorphic foliation.

Also present are amygdules composed of radiating epidote clusters.

At least three fold events can be recognized in the rocks at this location. The metamorphic foliation formed coincident with first generation folding which transposed original layering. Isolated fold noses related to this fold generation are visible in rocks in this area. Second generation folds trend northeast and are vergent to the northwest as were first generation folds. Second generation folds define the regional outcrop pattern in the northern Piedmont. Third generation folds trend northwest-southeast and plunge moderately to the southeast. Both later fold generations have a well-developed crenulation cleavage (McConnell, 1980).

REFERENCES CITED

Abrams, C. E., and McConnell, K. I., 1984, Geologic setting of volcanogenic base and precious metal deposits of the west Georgia Piedmont: A multiply deformed terrain: Economic Geology, v. 79-7, p. 1521–1539.

German, J. M., in press, The geology of the northeastern portion of the Dahlonega gold belt: Georgia Geologic Survey Bulletin 100.

Hatcher, R. D., Jr., and Odom, A. L., 1980, Timing of thrusting in the southern Appalachians, USA: model for orogeny?: Journal of the Geological Society of London, v. 137, p. 321–327.

Hurst, V. J., and Jones, L. M., 1973, Origin of amphibolites in the Cartersville-

Villa Rica area, Georgia: Geological Society of America Bulletin, v. 84, p. 905–911.

McConnell, K. I., 1980, Origin and correlation of the Pumpkinvine Creek Formation: A new unit in the Piedmont of northern Georgia: Georgia Geologic Survey Information Circular 51, 19 p.

McConnell, K. I., and Abrams, C. E., 1983, Geology of the New Georgia Group and associated sulfide and gold deposits: West-central Georgia: Society of Economic Geologists Annual Meeting, Atlanta, 1983, Guidebook: 27 p.

——1984, Geology of the greater Atlantic region: Georgia Geologic Survey Bulletin 96, 127 p.

Corbin Gneiss Complex: Southernmost exposures of Grenville basement in the Appalachian Blue Ridge

John O. Costello, Department of Geology, University of South Carolina, Columbia, South Carolina 29208

LOCATION

Red Top Mountain State Park and vicinity, (1.3 mi or 2.1 km) east of Exit 123 (Red Top Mountain Road) on I-75, Bartow County, Allatoona Dam 7½-minute Quadrangle Georgia (Fig. 1). NOTE: Written permission must be obtained before samples can be gathered on State Park property. Write to: Chief Naturalist, Department of Natural Resources, Parks and Historic Sites Division, 270 Washington Street, S.W., Atlanta, Georgia 30334.

Alternatively, fresh samples can be gathered from roadcuts and outcrops approximately 5 mi (8 km) north of the site along the Georgia 20 right-of-way between Roland Springs Community and the Bartow/Cherokee County line, Allatoona Dam 7½-minute Quadrangle. To reach these exposures, take I-75 north from Red Top Mountain Road. Exit onto Georgia 20 (Exit 125) and proceed East.

INTRODUCTION

The Corbin Gneiss Complex is an assemblage of retrograded granulite facies orthogneisses and paragneiss that composes the southernmost exposures of Grenville basement in the Blue Ridge Geologic Province. Apart from this, the significance of the Corbin and similar basement terranes as a sediment source throughout the late Precambrian Ocoee Supergroup of the Blue Ridge and northeastern Talladega Belt in Georgia is widely recognized

(McConnell and Costello, 1984). Although distant from present exposures of basement, clastic detritus, possibly originating from a Grenville terrane, has also been reported in two Paleozoic sequences to the southwest of the Corbin Gneiss Complex. Sibley and Sears (1982) observed bluish quartz grains in a diamictite facies of the Middle Ordovician Rockmart Slate in western Georgia and speculated on a Proterozoic Y or Z source. Additionally, a diamictite facies of the Silurian (Harris and others, 1984) Lay Dam Formation of the Talladega Belt in Alabama contains clasts of various lithologies including granitoid rocks that yield U-Pb ages of 1.1 b.y., suggesting derivation from Grenville crust (Telle and others, 1979; Tull, 1982). The Corbin Gneiss Complex is the closest exposure of basement to the above rock units and it or a similar basement terrane may be the source.

SITE DESCRIPTION

Retrograded granulite facies gneisses that yield U-Pb ages ex-

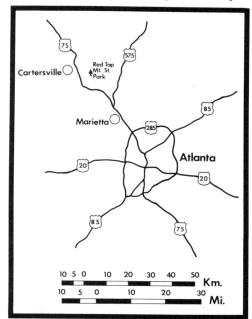

Figure 1. Locality maps for Red Top Mountain State Park (adapted from Allatoona Dam, Ga., 7½-minute Quadrangle).

ceeding 1 b.y. (Odom and others, 1973) form two external massifs of Grenville basement east of Cartersville, Georgia. These rocks were originally described as the Corbin and Salem Church Granites by Hayes (1900) and Bayley (1928) respectively. However, later studies have shown the massifs to be lithologically complex (Kesler, 1950; Martin, 1974; Costello, 1978) and they have been renamed the Corbin Gneiss Complex by McConnell and Costello (1984). The massifs consist of a suite of intrusive felsic to mafic orthogneisses and a meta-arkosic paragneiss country rock. The Corbin Gneiss Complex and its nonconformable Ocoee Supergroup mantle are extensively exposed within the core of the Salem Church anticlinorium in Bartow, Cherokee, and Pickens counties (see regional maps in McConnell and Costello, 1982, 1984).

A combination of accessibility and sufficiency of exposure makes Red Top Mountain State Park an attractive location to observe the Corbin Gneiss Complex. The site (Fig. 1) is located close to the southern border of the Corbin outcrop area (Fig. 2). Here, abundant boulders of two orthogneiss facies and less conspicuous exposures of paragneiss occur in the road cuts and lie scattered throughout the woods. More extensive exposures of the three lithologies ranging from fresh rock to saprolite occur along the lakeshore about the park and are accessible by foot or boat. Low reservoir levels and reduced foliage during winter insure maximum exposure.

Descriptions of rock units given herein are field descriptions and can be confirmed with the unaided eye or handlens. More elaborate petrographic descriptions are presented by Kesler (1950), Martin (1974), and McConnell and Costello (1984).

The most characteristic facies of the Corbin Gneiss Complex, and the one dominant within the site, is a quartz monzonitic to dioritic gneiss (McConnell and Costello, 1984) containing large, light-gray to white microcline crystals in a medium-grained groundmass of blue quartz, plagioclase, biotite, and garnet (*Yco* on Fig. 2). Field evidence and data presented by Kesler (1950) and McConnell and Costello (1984) indicate that while this facies is texturally uniform, the groundmass varies compositionally between an oligoclase-quartz-mica gneiss and an andesine-augite gneiss. This accounts for the bulk compositional range described above. Within the site the coarse-grained, spotted-looking gneiss is dominantly felsic in composition and generally, it is tectonically altered to an augen gneiss (Fig. 3). However, locally, the porphyritic texture of the protolith is well preserved.

Less prevalent are bodies of mafic orthogneiss and felsic paragneiss. The mafic facies of the Corbin is described as an andesine-augite gneiss by Kesler (1950) and a metagabbro by Crawford (in Cressler and others, 1979). Scattered bouldery outcrops displaying the dark greenish gray color and medium-grained texture of this variety of the Corbin (*Ycm* on Fig. 2) occupy the northwestern part of the site around the Red Top Mountain Park rental cabins (Fig. 1) and lie more widely separated elsewhere within the park. The dominant mineral in the mafic orthogneiss is plagioclase with smaller amounts of orthoclase, quartz, biotite, and pyrrhotite. Ilmenite is abundant as an accessory mineral and

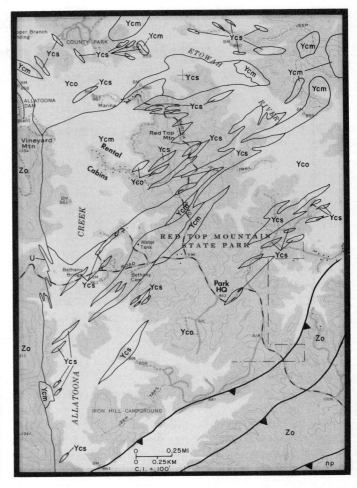

Figure 2. Generalized geologic map of Red Top Mountain State Park and vicinity (modified after Crawford in Cressler and others, 1979). Corbin Gneiss Complex is shown as: *Yco*, Porphyritic Facies (augen gneiss); *Ycm*, Andesine-augite Facies, *Ycs*, Meta-arkosic Paragneiss. Ocoee Supergroup undivided, *Zo*. Metavolcanic and meta-sedimentary rocks of the northern Piedmont, *np*. *U* marks the unconformity exposure.

wave-generated concentrations occur in sand along the lakeshore where this facies or its dark-red residuum is present.

Paragneiss consisting chiefly of quartz, feldspar, biotite, and garnet with varying amounts of graphite (*Ycs* on Fig. 2) occupies lenticular pods and belts within the park area. Fortuitous or not, most of the outcrop patterns in the park align with the regional orientation of the Salem Church anticlinorium (F_2) suggesting that structure, at least in part, controls orientation. Outcrops of the paragneiss are usually poorly preserved. Typical exposures contain a clay-rich, highly weathered rock that breaks easily into blocky fragments.

Contact relationships between the phases of the Corbin were established during Grenville orogenesis. At present, lithologic contacts are mostly obscured within the site by deep weathering, heavy vegetation, and Lake Allatoona. However, sharp crosscutting of paragneiss by coarse-grained felsic gneiss is locally observ-

Figure 3. Porphyritic facies of the Corbin Gneiss Complex displaying the characteristic augen texture.

able within the site (Fig. 4). One such exposure, accessible only when reservoir levels drop during winter months, lies at the western abutment of Bethany Bridge on Lake Allatoona (Fig. 1).

Boundary relationships between the orthogneiss phases are more complex. Where preserved, they can be either sharp or diffuse lending uncertainty to any interpretation of intrusive timing. However, basic xenoliths reported by Martin (1974) and some outcrop patterns shown on Figure 2 suggest that mafic Corbin is the older of the two orthogneiss facies at the site.

All facies of the Corbin Gneiss Complex are weakly to strongly foliated. Gneissic foliation is parallel to the fabric of extensive mylonite zones within the Corbin (Costello, 1978). Segments of lakeshore near the eastern boundary of the park contain exposures where a gradational transformation from augen texture to mylonite is observable. Relict Grenville fabrics may be present within the gneisses, but most tectonic structures are believed to be Paleozoic in age.

Mylonitic foliation in the Corbin Gneiss Complex is subparallel to the regional schistosity (S_1) of the late Precambrian Ocoee Supergroup (Costello and others, 1982; McConnell and Costello, 1984). Similar mylonites in Grenville age Fort Mountain Gneiss near Chatsworth, Georgia, to the north yield Rb-Sr ages of 368 ± 9 m.y. (Russell, 1976). These data suggest synchronicity of peak metamorphism and ductile deformation in the western Blue Ridge in Devonian time.

The Corbin Gneiss Complex is nonconformably overlain by greenschist facies (chlorite grade) rocks of the late Precambrian Ocoee Supergroup (*Zo* on Fig. 2). This relationship is locally structurally modified, generally appearing as an overturned contact to the west or a fault contact to the east (Fig. 2). See McConnell and Costello (1982, 1984) or Costello and others (1982) for a more complete structural synthesis. The nonconformity is visible (during winter months) at the location indicated by the *U* on Figure 2 on the west side of Lake Allatoona opposite the boat ramp at Bethany Bridge. A visit to the Corbin/Ocoee contact ties off an excursion nicely and removes any suspicions that the Corbin may be post-Ocoee.

Figure 4. Crosscutting relationship between the porphyritic facies (below) and the meta-arkosic paragneiss (above) of the Corbin Gneiss Complex.

The hills west of Lake Allatoona are held up by Ocoee metasandstones and metapelites. In this area, the Ocoee dips eastward beneath the Corbin Gneiss Complex on the overturned limb of a westward vergent anticline. West of Bethany Bridge and north of the road, massive light gray to tan metasandstone and interlayered dark gray phyllite form two conspicuous outcrops along the lakeshore. Approximately 50 ft (15 m) north of the northernmost outcrop, augen textured Corbin Gneiss lies in contact with the basal Ocoee conglomerate. The conglomerate contains rounded grains of blue quartz and microcline which are obviously derived from the adjacent coarse-grained gneiss.

The history of the Corbin Gneiss Complex is intricate. It has been a part of this dynamic planet for more than a billion years and much work remains to assemble the detailed aspects of its evolution. However, enough data have been gathered on the Corbin and similar basement rocks to allow the construction of a generalized chronicle:

1. Magma preserved as Corbin orthogneiss intrudes country rock of sedimentary origin prior to 1 Ga (Odom and others, 1973).

2. Post emplacement metamorphism elevates the Corbin terrane to granulite facies conditions (Martin, 1974) before 700 Ma ago (Dallmeyer, 1975).

3. Late Proterozoic extension of the North American craton rifts the Grenville orogen opening the proto–Atlantic Ocean (Iapetos). Basins (i.e., Ocoee) form and fill with immature, craton-derived sediments. The Corbin terrane, perhaps occupying a horst or half-graben structure close to or outboard of the late Proterozoic–early Paleozoic hinge zone (Wehr and Glover, 1985), is buried by Ocoee sediments. Succeeding Eocambrian–Early Cambrian clastics and carbonates record tectonic stabilization and early Paleozoic passive margin evolution.

4. Mid-Paleozoic ductile deformation (Russell, 1976) and metamorphism affects the western Blue Ridge of Georgia. The Corbin Gneiss Complex is retrograded, the Ocoee is prograded, and both are penetratively deformed.

5. The late Paleozoic Alleghanian orogeny affects large-scale lateral displacement of the western Blue Ridge. The Corbin and its Ocoee cover sequence are emplaced in their present structural position above Lower Cambrian shelf sedimentary rocks of the Cartersville District, Georgia.

6. The region is eroded to its present topography.

REFERENCES CITED

Bayley, W. S., 1928, Geology of the Tate quadrangle, Georgia: Georgia Geologic Survey Bulletin 43, 170 p.

Costello, J. O., 1978, Shear zones in the Corbin gneiss of Georgia, *in* short contributions to the geology of Georgia: Georgia Geologic Survey Bulletin 93, p. 32–37.

Costello, J. O., McConnell, K. I., and Power, W. R., 1982, Geology of late Precambrian and early Paleozoic rocks in and near the Cartersville District, Georgia: Georgia Geological Society 17th Annual Field Trip, Guidebook: 42 p.

Cressler, W. S., Blanchard, H. E., Jr., and Hester, W. G., 1979, Geohydrology of Bartow, Cherokee and Forsyth Counties, Georgia: Georgia Geologic Survey Information Circular 50, 45 p.

Dallmeyer, R. D., 1975, ^{40}Ar/^{39}Ar age spectra of biotite from Grenville basement in northwest Georgia: Geological Society of America Bulletin v. 86, p. 1704–1744.

Harris, A. G., Repetski, J. E., Tull, J. F., and Bearce, D. N., 1984, Early Paleozoic conodonts from the Talladega slate belt of the Alabama Appalachians: Geological Society of America Abstracts with Programs, v. 16, p. 143.

Hayes, C. W., 1900, Geological relations of the iron-ores in the Cartersville District, Georgia: American Institute of Mining Engineers Transactions, v. 30, p. 403–419.

Kesler, T. L., 1950, Geology and mineral deposits of the Cartersville District, Georgia: U.S. Geological Survey Professional Paper 224, 97 p.

Martin, B. F., Jr., 1974, The petrology of the Corbin Gneiss [M.S. thesis]: University of Georgia, Athens, 133 p.

McConnell, K. I., and Costello, J. O., 1982, Relationship between Talladega belt rocks and Ocoee Supergroup rocks near Cartersville, Georgia, *in,* Bearce, D. N., Black, W. W., Kish, S. A., and Tull, J. F., eds., Tectonic studies in the Talladega and Carolina slate belts, southern Appalachian orogen: Geological Society of America Special Paper 191, p. 19–30.

——1984, Basement-cover rock relationships along the western edge of the Blue Ridge thrust sheet in Georgia, *in,* Bartholomew, M. J., ed., The Grenville event in the Appalachians and related topics: Geological Society of America Special Paper 194, p. 263–280.

Odom, A. L., Kish, S. A., and Leggo, P. J., 1973, Extension of "Grenville basement" to the southern extremity of the Appalachians: U-Pb ages of zircons: Geological Society of America Abstracts with Programs, v. 5, p. 425.

Russell, G. S., 1976, Rb-Sr evidence from cataclastic rocks for Devonian faulting in the southern Appalachians: Geological Society of America Abstracts with Programs, v. 8, p. 1081.

Sibley, D. M., and Sears, J. W., 1982, The Taconic orogeny in the Georgia Valley and Ridge Province: Geological Society of America Abstracts with Programs, v. 14, p. 616–617.

Telle, W. R., Tull, J. F., and Russell, C. W., 1979, Tectonic significance of the bouldery facies of the Lay Dam Formation, Talladega slate belt, Chilton County, Alabama: Geological Society of America Abstracts with Programs, v. 11, p. 215.

Tull, J. F., 1982, Stratigraphic framework of the Talladega slate belt, Alabama Appalachians, *in,* Bearce, D. N., Black, W. W., Kish, S. A., and Tull, J. F., eds., Tectonic studies in the Talladega and Carolina slate belts, southern Appalachian orogen: Geological Society of America Special Paper 191, p. 3–18.

Wehr, F., and Glover, L., III, 1985, Stratigraphy and tectonics of the Virginia–North Carolina Blue Ridge: Evolution of a late Proterozoic–early Paleozoic hinge zone: Geological Society of America Bulletin, v. 96, p. 285–295.

The Brevard fault zone west of Atlanta, Georgia

Keith I. McConnell, Georgia Geologic Survey, 19 Martin Luther King Jr. Drive, Atlanta, Georgia 30334

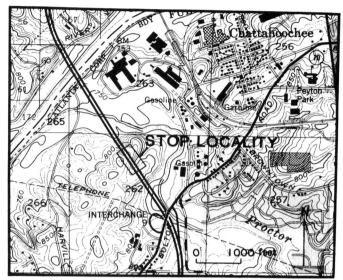

Figure 1. Part of the Northwest Atlanta 7½-minute Quadrangle showing the location of the Brevard zone outcrop.

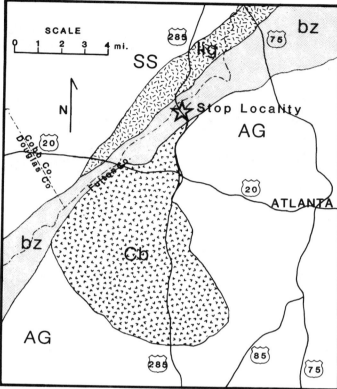

Figure 2. Generalized geologic map for the Brevard zone stop near Atlanta. SS=Sandy Springs Group; lig=Long Island Gneiss; bz=Brevard zone; Cb=Ben Hill Gneiss; AG=Atlanta Group.

LOCATION

One of the best exposures of the Brevard fault zone is located along I-285 approximately 12 mi (19 km) west of the center of Altanta, Georgia. This exposure is present just south of the Chattahoochee River (Fig. 1) on both sides of the Interstate. The outcrop on the east side of the Interstate is better for examining the textures and structures of the Brevard zone. Traffic along this section of Atlanta's Interstate system is very heavy, and extreme caution should be used in viewing this outcrop.

INTRODUCTION

The Brevard fault zone stretches for more than 375 mi (600 km) from south of the Virginia–North Carolina line to the Coastal Plain overlap in Alabama. Interpretations regarding the nature of movement and the extent of displacement associated with the Brevard fault zone vary. Some workers have suggested that faulting in the Brevard zone is characterized by left-lateral strike-slip faulting with a thrust component (Reed and others, 1970), while others have interpreted movement in the Brevard zone to be right-lateral strike-slip (Reed and Bryant, 1964) with a thrust component (Higgins 1966). Hatcher (1971, 1978a) interpreted the Brevard fault zone to be a reactivated backlimb thrust on the Blue Ridge thrust sheet and Burchfiel and Livingston (1967) have suggested that the Brevard is a linear root zone similar to alpine-type root zones. The many other interpretations proposed for the Brevard fault zone are too numerous to list in this guide. The reader is referred to summary articles (Roper and Justus, 1973; Hatcher,

1978a) for a more complete listing of the various interpretations. However, it can be said that in Georgia the Brevard fault zone separates two terranes that contain similar lithologies and stratigraphic sequences (Hatcher, 1978b; Kline, 1980; McConnell and Abrams, 1984). These similarities suggest that the Brevard fault zone may not be a major terrane boundary but may actually separate two parts of the same terrane.

In the Greater Atlanta Region, the Brevard fault zone (Fig. 2) is bounded on the northwest by the Sandy Springs Group (Higgins and McConnell, 1978). The Sandy Springs Group, a stratigraphic succession of dominantly metasedimentary rocks with interlayered metavolcanic rocks, is composed of three formations: a lower mixed gneiss-schist-amphibolite unit, the Powers Ferry Formation; a micaceous quartzite, the Chattahoochee Palisades Quartzite; and an aluminous schist, the Factory Shoals Formation (McConnell and Abrams, 1984). The Sandy Springs Group commonly has been interpreted to be equivalent to the Tallulah Falls Formation in northeast Georgia (Hatcher, 1974) and the Heard Group and Jackson's Gap Group in eastern Alabama (Hatcher, 1975; McConnell and Abrams, 1984). To the

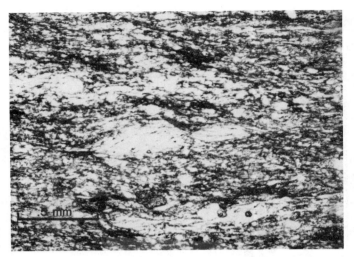

Figure 3. Photomicrograph of button schist texture from the Brevard zone.

southeast, the Brevard fault zone is bordered by rocks of the Atlanta Group (Higgins and Atkins, 1981). The Atlanta Group is a metasediment-dominated sequence with lesser amounts of interlayered metavolcanic rocks and is intruded by late Paleozoic granites. Numerous formations have been defined in the Atlanta Group; the reader is referred to Higgins and Atkins (1981) for a complete description of units. McConnell and Abrams (1984) have suggested that many of the rock units in the Atlanta Group are correlative with the Sandy Springs Group.

Crawford and Medlin (1973) denoted the Brevard fault zone by the presence of a well-developed secondary foliation that they termed a "cataclastic foliation." This "cataclastic" foliation (S_2) is axial-planar to second-generation folds (F_2) and is present in areas outside of the Brevard fault zone (McConnell and Abrams, 1978); therefore, the presence of a second foliation should not be a criterion for inclusion in the Brevard fault zone. In this report, the Brevard zone is limited to rocks that have undergone intense ductile deformation. Rocks present in the Brevard fault zone near Atlanta include protomylonite, mylonite, blastomylonite, button schist (Fig. 3), and phyllonite. Occurring in association with the phyllonites are muscovite aggregates (Higgins, 1966) and zones of flattened and poikiloblastic garnets (Abrams, 1983).

The Brevard fault zone marks a linear region of low-grade retrogressive metamorphism separating upper amphibolite facies rocks on either side. The post-metamorphic history of the Brevard fault zone in the Atlanta area is interpreted to involve little vertical displacement. The lack of major vertical displacement is suggested by the occurrence of equivalent lithologies and stratigraphic successions on either side of the Brevard fault zone. Also, if significant post-metamorphic vertical displacement had occurred

along the Brevard zone, metamorphic grade should be substantially different across the zone; however, this is not the case in the Greater Atlanta Region.

Stratigraphic control is another aspect of the Brevard fault zone in the Atlanta area. Hatcher (1975) indicated that the Brevard zone is stratigraphically controlled for at least part of its length and is bordered by several equivalent rock units (i.e., Heard Group, Sandy Springs Group, Tallulah Falls Formation, Ashe Formation) for most of its length. In the Atlanta area, the Brevard zone apparently lies within rocks of the Sandy Springs Group and its equivalents for most of its length. However, the absence in the Atlanta area of units defined as Chauga River Formation (a unit common along the southeastern border of the Brevard zone in northeastern Georgia; Hatcher, 1969) complicates the issue of stratigraphic control of the Brevard zone.

SITE DESCRIPTION

The exposure of the Brevard fault zone along I-285 is perhaps the best for demonstrating the intensity of ductile shearing, the variety of lithologies involved, and the subsequent overprint of late-stage brittle deformation. The I-285 exposure occurs just southwest of an area where similar lithologies and lithologic sequences are apparent on either side of the Brevard zone, thus establishing limits on the displacement.

The Brevard fault zone along the I-285 cut is approximately 1 mi (1.6 km) wide (Fig. 2). The Long Island Gneiss, a biotite-epidote-plagioclase gneiss (Higgins and McConnell, 1978), is in contact with the Brevard fault zone to the northwest; to the southeast, the Brevard grades, through decreasing intensity of mylonitization, into an undefined granitic rock (possibly the Ben Hill Granite). Textures ranging from protomylonite to mylonite can be observed in rocks exposed at this outcrop. Late-stage brittle features (e.g., small-scale normal faults and flinty crush zones) also are apparent. Mylonitic foliation strikes approximately N35°E and dips moderately to the southeast, approximately parallel to regional foliation in this area. At least three fold episodes are evident from close examination of the rocks at this stop. Isolated fold noses (F_1) within the predominant folaition attest to transposition of an earlier layering. F_2 folds are tight to isoclinal and trend approximately N25°E. A late, southwest-oriented warping is also evident in rocks at this exposure.

Other than structural features apparent at this exposure, the most striking feature is the compositional layering exhibited. Darker layers generally have a higher proportion of epidote, chlorite, graphite, tourmaline, and opaques than the lighter layers. The more leucocratic layers appear in thin section to be mylonitized granite or granite gneiss, whereas graphite in the darker rock suggests that it may have had a metasedimentary parent, possibly the Sandy Springs Group (McConnell and Costello, 1980).

REFERENCES CITED

Abrams, C. E., 1983, Geology of the Austell-Frolona antiform, northwestern Georgia Piedmont [M.S. thesis]: Athens, University of Georgia, 119 p.

Burchfiel, B. C., and Livingston, J. L., 1967, Brevard zone compared to Alpine root zones: American Journal of Science, v. 265, no. 4, p. 241–256.

Crawford, T. J., and Medlin, J. H., 1973, The western Georgia Piedmont between the Cartersville and Brevard fault zones: American Journal of Science, v. 273, p. 712–722.

Hatcher, R. D., Jr., 1969, Stratigraphy, petrology and structure of the low rank belt and part of the Blue Ridge of northwesternmost South Carolina: South Carolina Division of Geology, Geologic Notes, v. 13, p. 105–141.

—— 1971, Stratigraphic, petrologic, and structural evidence favoring a thrust solution of the Brevard problem: American Journal of Science, v. 270, p. 177–202.

—— 1974, An introduction to the Blue Ridge tectonic history of northeast Georgia: Georgia Geologic Survey Guidebook 13-A, 60 p.

—— 1975, Second Penrose field conference: The Brevard zone: Geology, v. 3, p. 149–152.,

—— 1978a, Tectonics of the western Piedmont and Blue Ridge, southern Appalachians: reveiw and speculation: American Journal of Science, v. 278, p. 276–304.

—— 1978b, The Alto allochthon: a major tectonic unit of the northeast Georgia Piedmont, *in* Short contributions to the geology of Georgia: Georgia Geologic Survey Bulletin 93, p. 83–86.

Higgins, M. W., 1966, Geology of the Brevard lineament near Atlanta, Georgia: Georgia Geologic Survey Bulletin 77, 49 p.

Higgins, M. W., and Atkins, R. L., 1981, The stratigraphy of the Piedmont southeast of the Brevard fault zone in the Atlanta, Georgia, area, *in* Wigley,
P. B., ed., Latest thinking on the stratigraphy of selected areas in Georgia: Georgia Geologic Survey Information Circular 54-A, p. 3–40.

Higgins, M. W., and McConnell, K. I., 1978, The Sandy Springs Group and related rocks in the Georgia Piedmont—nomenclature and stratigraphy, *in* Sohl, N. F., and Wright, W. B., eds., Changes in stratigraphic nomenclature by the U.S. Geological Survey, 1977: U.S. Geological Survey Bulletin 1457-A, p. A98–A105.

Kline, S. W., 1980, Sandy Springs sequence rocks southeast of the Brevard fault zone, near Atlanta, Georgia, and their bearing on the nature of the zone: Geological Society of America Abstracts with Programs, v. 12, no. 4, p. 181.

McConnell, K. I., and Abrams, C. E., 1978, Structural and lithologic controls of Sweetwater Creek in western Georgia, *in* Short contributions to the geology of Georgia: Georgia Geologic Survey Bulletin 93, p. 87–92.

—— 1984, Geology of the Greater Atlanta Region: Georgia Geologic Survey Bulletin 96, 127 p.

McConnell, K. I., and Costello, J. O., 1980, Guide to geology along a traverse through the Blue Ridge and Piedmont provinces of north Georgia, *in* Frey, R. W., ed., Excursions in southeastern geology: American Geological Institute, v. 1, p. 36–46.

Reed, J. C., Jr., and Bryant, B., 1964, Evidence for strike-slip faulting along the Brevard zone in North Carolina: Geological Society of America Bulletin, v. 75, p. 1177–1195.

Reed, J. C., Jr., Bryant, B., and Myers, W. B., 1970, The Brevard zone: a reinterpretation, *in* Fisher, G. W., Pettijohn, F. J., Reed, J. C., Jr., and Weaver, K. N., eds., Studies of Appalachian geology, central and southern: New York, Wiley Interscience, p. 261–269.

Roper, P. J., and Justus, P. S., 1973, Polytechnic evolution of the Brevard zone: American Journal of Science, v. 273-A, p. 105–132.

Structural and petrologic features of the Stone Mountain granite pluton, Georgia

Willard H. Grant, Geology Department, Emory University, Atlanta, Georgia 30322

LOCATION

Stone Mountain is located 14 mi (22.5 km) due east of the intersection of I-85 and I-75 in north Atlanta. The park is accessible from any interstate going to Atlanta by exiting on I-285 and following it until it intersects U.S. 78 (Stone Mountain Freeway) east of Atlanta. A modest fee is charged at the gate, where a map of the entire park is available. Since access to most of the field trip stops is in parts of the park not open to the general public, permission must be obtained. This can be done by going to Confederate Hall at the west end of the mountain and requesting permission at the administrative offices to visit the quarry on the east end of the mountain (East Quarry), Stop 3, and the extension of Old U.S. 78, Stop 2, that terminates in the lake (Figs. 1 and 2).

The Confederate Memorial, a carving of Lee, Jackson, and Davis, the steep north-facing slope of the mountain, is a readily visible landmark for orientation. Structural features are best observed on dark, wet, cold winter days when the clouds hang low on the mountain. All the light is diffuse and the wetness enhances structural details to the maximum. Mid-summer viewing, especially on bright sunny days, ranges from poor to impossible in the quarries.

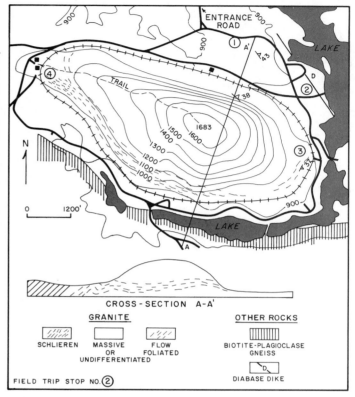

Figure 2. Geologic map of Stone Mountain and cross section based on flowage structures. Field-trip locations also noted.

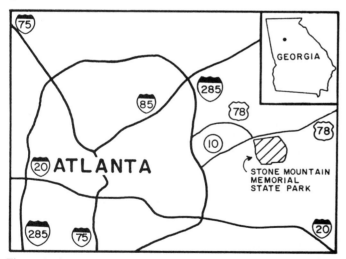

Figure 1. General location map (Stone Mountain 7½-minute Quadrangle) Stop locations are within the park and are circled numbers.

INTRODUCTION

Stone Mountain is a granite monadnock which rises approximately 780 ft (238 m) above the surrounding deeply weathered metamorphic rocks of the Piedmont. The monadnock is an asymmetrical dome with a nearly vertical slope on the north, a gentle slope on the west, and steep slopes on the east and south sides. The west slope is the only one open to the public.

The shape of the mountain is controlled by topographically subparallel dilation joints (Hopson 1958). This jointing forms lenticular sheets of various sizes which spall off the mountain. The dilation is caused by expansion of the granite due to removal of confining pressure resulting from the erosion of some 36,300 ft (11,000 m) of rock, over the past 71 m.y., (Dallmeyer, 1978). On the gentle slopes solution pits occur. These are probably formed by the dilation of the granite and associated weathering. The pits range in size from 6 in (15 cm) to a few tens of ft (m). They are elliptical in shape, and a few are circular. The pits are occupied by a number of plants, some of which are endemic. Dimorphas make the pits red in the spring, and the yellow confederate daisies (Fig. 3) occur in the fall. These plants and pits are of much interest to botanists. Dr. M. Burbanck, an associate in the Emory University Biology Department, has contributed greatly to their understanding.

The composition of the granite varies from 26 to 43 percent

Figure 3. Pit and dome structure on undisturbed granite pavement, Stop 1. The water filled pit contains no plants, the pit behind the dome contains grass, and the one in the distance, trees.

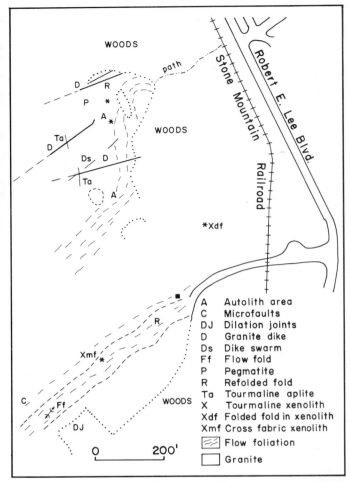

Figure 4. Map of the geologic features of the East Quarry. The map code is also used in other figures to aid in location.

oligoclase (An_{10} - An_{15}). Microcline ranges from 16 to 34 percent, quartz from 30 to 37 percent, muscovite about 9 percent, and biotite about 1 percent (Grant and others, 1980).

The granite intrudes both concordantly and discordantly into country rock composed primarily of biotite-plagioclase gneiss, interlayered with pods of amphibolite and minor mica schist. The rocks were regionally metamorphosed to above the sillimanite isograde. At the granite contact a slight grain-size enlargement occurs which is attributed to contact metamorphism. There is also some indication of contact metasomatism, which manifests itself as microcline porphyroblasts in the gneisses near the north granite contact (presently not exposed) (Grant 1962). Location is at the intersection north of U.S. 78 on Hugh Howell Road, Georgia 236.

Age determinations on the granite range from 281-325 Ma (Whitney and others, 1976; Dallmeyer, 1978; and Atkins and others, 1980). Contact and structural data indicate that the granite intrusion was late metamorphic. This is shown by crystal growth at contacts with pre-existing metamorphic mineral assemblages.

Planar structures in Stone Mountain include flow banding, flowage foliation and massive granite. Mica, mainly muscovite, defines these features. Xenoliths are mostly lens-shaped mica schist fragments, which show a strong orientation parallel to the flow structure. Biotite gneiss xenoliths also occur but are less common. Mapping of flowage structures around the south and east sides of the mountain suggests that the pluton is a rather thin sheet (Fig. 2). The intrusion of the sheet was controlled by a growing N65W fold system. This now parallels the long axis of the mountain. A pi diagram of xenolith orientations shows an S65E plunging axis. The N65W fold direction is one of those recognized by Atkins and Higgins (1978). It is also possible, using the same structural control, that the granite was intruded from below by northwest trending dikes, mentioned by Herrmann

(1954). The granite was intruded in a number of pulses rather than a single episode. This is supported by earlier granite autoliths contained in the main granite mass and later cross-cutting granite dikes. All these granites are of similar composition. The variations are recognized by structural or textural features.

Plots of lineations contained in xenoliths show plunge angles up to 30° to the southeast and bearings from N60E to S5E. Rotation of this data about 15° gives a lineation distribution covering the entire compass rose except the ranges between N5W to N65E and S5E to S65W. This distribution supports the idea of a growing, expanding magmatic intrusion between thin layers of simultaneously folding country rock. The addition of more magma breaks the layers in pieces (xenoliths) which spread out in sunburst fashion as the magma-filling blister expands.

Jointing is not well developed on the mountain; there is a northwest and a northeast striking set in the West Quarry area. This lack of extensive jointing may be partially responsible for the monadnock which rises uniquely in a milieu of deeply weathered granite which extends to both the north and east of the park area.

Figure 5. Granite dike (D) cross cutting massive granite in the East Quarry, Stop 3. White patches are dry granite, gray areas are wet.

Figure 6. Flow banding on the north edge of East Quarry (Fig. 8) where the woods patch intersects the quarry floor.

SITE DESCRIPTIONS

The locations of all stops are given on Figure 1.

Stop 1

Pavement area (across Robert E. Lee Boulevard) Southeast of the Stone Mountain Inn. This area is typical of undisturbed southeastern pavement exposures. It is characterized by solution pits and low domes (Fig. 3). Solution pits usually dominate these surfaces. The pits contain plants which are endemic to the granite outcrops. Studies of soils in similar pits show an increasing of clay mineral content with soil thickness. The greenish gray-to-black colors are due to lichen and fungi which tend to obscure the fabric details in the rock.

Stop 2 - Diabase Dike.

An early Mesozoic diabase dike crosscuts the granite, Figure 2. The two rocks contrast sharply in the color of their weathering products. The diabase color, mainly ochre, has a few streaks of red and a scattering of black spots of hydrated manganese oxides. The white granite saprolite has a bulk density of about 1.5 and is strongly kaolinized. Some microcline remains but none of the oligoclase. Muscovite remains apparently more or less unweathered. Quartz, the most resistant mineral, remains in abundance. The kaolin enriched B-soil horizon immediately above the saprolite is a pale brown color, indicating low iron content of the

granitic parent material. Also note that the granite fabric, in this case massive, is still recognizable. Moving west from the lake, core stones and saprocks, in various degrees of weatheredness, are exposed in the old road cut.

Stop 3 - East Quarry

Specific locations of features described are shown on Figure 4, where they are designated by letters, e.g., (A) for autoliths. The quarry has had several names in the past (King, Kellog) and is on the east end of the mountain. This area contains some excellent examples of granite dikes (Fig. 5), flowage (Fig. 6), xenoliths (Fig. 7), and evidence of the multiple intrusion of mineralogically similar granites. The earliest granite is in the foliated and flow-banded autoliths (A), the main granite phase is either flow-foliated, flow-banded (Fig. 6), or massive. The latest phase occurs in dikes (D) and thin 1.2-inches (3-cm-wide) dikes in swarms (Ds, Fig. 8), which crosscut the main granite. Flowage folds (Ff) (Fig. 9) suggest intrusion came from the east. A second flow fold on the south flank of the mountain suggests intrusion also came from the south. Both directions are consistent with the blister model. Two folded folds (R) occur; the northern most, morphologically resembles (Atkins and Higgins, 1978) an isocline coaxially refolded by a recumbent isocline (Fig. 10). This particular fold may be a large autolith, and is believed to be a "cast fold," that is a mold defined by a tectonic opening and filled by granite. The refolded fold, (R, Fig. 11) near the quarry entrance road is a type 2 interference fold (Ramsey, 1967). Occurring with this fold and at many other places on the mountain are the "cats paws," small rounded white patches with tourmaline in the center, probably of metasomatic origin. Pegmatites (P) crosscut the general fabric of the granite. Tourmaline aplites (Ta) are small veins with a quartz and feldspar ground mass and tourmaline concentration in the center. The aplites crosscut both the main granite mass and

Figure 7. Biotite-plagioclase gneiss xenolith (X) in massive Stone Mountain.

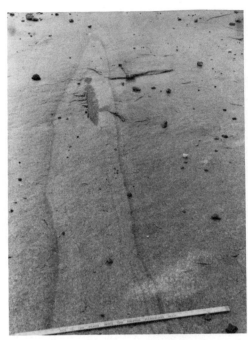

Figure 9. One end of a doubly plunging north-trending synform (Ff) on Figure 4.

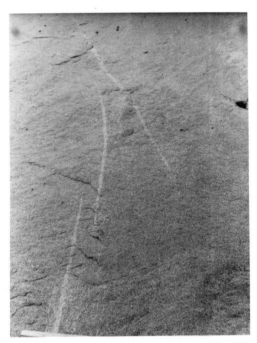

Figure 8. Dike swarms (Ds) on the East Quarry Map, Figure 4.

Figure 10. Refolded isocline (R) in the northern part of Figure 4.

the granite dikes (Fig. 12). The pegmatites, aplites, and quartz veins represent the last phase of magmatic activity. Xenoliths occur in two types: muscovite-biotite-garnet schist and biotite-plagioclase gneiss (Fig. 7). These rocks may contain tourmaline (Xt) of probable metasomatic origin and/or muscovite (Xmf) which crosscuts the pre-existing schistosity. Muscovite is probably the result of thermal metamorphism.

Microfaults (C) (Fig. 13) are small brecciated zones about 0.5 inch (1 cm) in width and a few feet (m) long. They appear in outcrop as long thin dark lines. One of these faults shows offset

cats-paws which along with their brecciated fabric suggests post-granite solidification shearing. Bleaching at the edges may indicate slight hydrothermal activity. These microfaults also occur at Mount Arabia, a migmatitic monadnock in South DeKalb County.

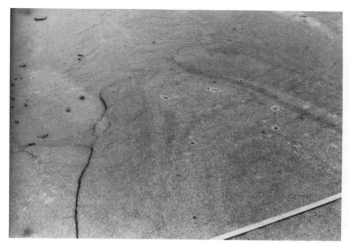

Figure 11. Interference flowage fold, possibly the result of shifting flow directions. Note the white tourmaline-centered spots locally called "catspaws."

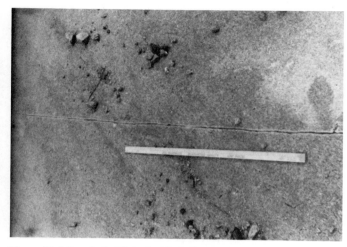

Figure 13. Microfault (C) is the black line bordered by a thin bleached zone; it runs across the photo at right angles to the yardstick.

Figure 12. Tourmaline aplite (Ta) cross cutting a granite dike (D) whose contact is shown in the lower left-hand corner.

Figure 14. Dilation jointing in the East Quarry (DJ). The quarry floor shows excellent flowage foliation.

Dilation or sheet joints are exposed at (DJ) (Fig. 14). A drill hole in the northern half of the northeast-trending quarry face shows a small offset which may be due to rock expansion.

Stop 4

West Quarry, near Confederate Hall and the area immediately east of the railroad. The major features to be seen in the

West Quarry are biotite schlieren and jointing. The schlieren appear to be the result of shearing out of xenoliths rather than magmatically formed concentrations of biotite. They are in some cases traceable into xenoliths and have the same mineral composition—biotite, garnet, quartz and varying amounts of feldspar and muscovite. On dark wet days pear-shaped flowage folds are visible approximately 100 ft (30 m) up the path from the railroad. Vertical jointing and solution pits are well exposed in this area.

REFERENCES CITED

Atkins, R. L., and Higgins, M. W., 1978, Relationship between Superimposed Folding and Geologic History in the Georgia Piedmont: Geological Society of America Abstracts with Programs, v. 10, p. 361.

Atkins, R. L., Higgins, M. W., and Gottfried, D., 1980, Geochemical Data Bearing on the Origin of Stone Mountain Granite, Panola Granite, and Lithonia Gneiss near Atlantia, Georgia: Geological Society of America Abstracts with Programs, v. 12, p. 170.

Dallmeyer, R. D., 1978, Ar^{40}/Ar^{39} Incremental Release Ages of Hornblende and

Biotite Across the Georgia Inner Piedmont: Their Bearing on Late Paleozoic-Early Mesozoic Tectonothermal History: American Journal of Science, v. 278, p. 124–149.

Grant, W. H., 1962, Field Excursion, Stone Mountain-Lithonia District, Georgia: Georgia Geological Survey Guidebook 2, p. 21.

Grant, W. H., Size, W. B., and O'Connor, B. J., 1980, Petrology and Structure of the Stone Mountain Granite and Mount Arabia Migmatite, Lithonia, Georgia: Geological Society of America 1980 Atlanta Field Trip no. 3, 17, p. 41–57.

Hermann, L. A., 1954, Geology of the Stone Mountain-Lithonia District, Georgia: Georgia Geological Survey Bulletin, v. 61, p. 139.

Hopson, C. A., 1958, Exfoliation and Weathering at Stone Mountain, Georgia: Georgia Mineralogical Newsletter, v. 11, p. 65–79.

Ramsey, J. G., 1967, Folding and Fracturing of Rocks: New York, McGraw-Hill Book Company, p. 568.

Whitney, J. A., Jones, L. M., and Walker, R. L., 1976, Age and Origin of the Stone Mountain Granite, Lithonia District, Georgia: Geological Society of America Bulletin, v. 87, p. 1067–1077.

The Bartletts Ferry and Goat Rock fault zones north of Columbus, Georgia

Thomas B. Hanley, *Department of Chemistry and Geology, Columbus College, Columbus, Georgia 31993*
James C. Redwine, *Southern Company Services, Inc., Birmingham, Alabama 35202*

LOCATIONS

The stops described herein are approximately 10 to 12 mi (16 to 20 km) north of Columbus, Georgia, on the east bank of the Chattahoochee River and on the west side of I-185 (Fig. 1).

INTRODUCTION

Long narrow zones of highly strained rocks characterized by brittle and ductile deformation are common in the crystalline parts of the Appalachian orogen of the U.S. and Canada and in its extensions, the Caledonian and Hercynian orogens of Europe. The Eastern Piedmont fault system (EPFS) consists of a number of such zones whose traces form an interlaced network extending from Alabama to Virginia (Hatcher and others, 1977). This text describes mylonitic rocks associated with the Bartletts Ferry fault and the Goat Rock fault, parts of the EPFS in west central Georgia and east central Alabama.

North of Columbus the EPFS consists of the Bartletts Ferry fault, the Goat Rock fault and the Towaliga fault (Fig. 2). The latter two were originally mapped by Crickmay (1933, 1952) on the basis of lithologic contrasts and on the presence of mylonitic rocks. Bentley (1969) describes the Bartletts Ferry fault in Alabama. Higgins (1971) extends the Bartletts Ferry fault into Georgia and includes examples from it and the Goat Rock fault in his paper on mylonite nomenclature and origin.

The Bartletts Ferry and Goat Rock faults occur on the south side of the Pine Mountain window (PMW). Evidence has been

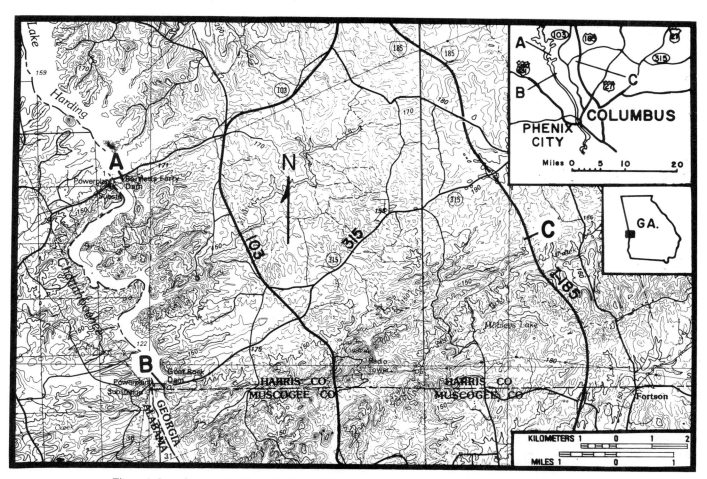

Figure 1. Location map for Stops A, B, and C. Base maps are parts of the Opelika, Alabama-Georgia, and Thomaston, Georgia, 1:100,000 metric scale 30 × 60-minute quadrangles.

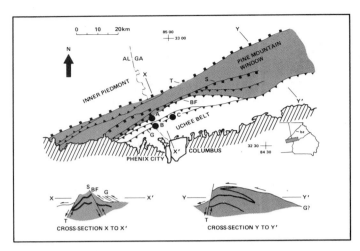

Figure 2. Geologic map of west-central Georgia and east-central Alabama showing the set of faults which locally makes up the Eastern Piedmont fault system. A, B, and C are stops described in the text. The Pine Mountain window (PMW) is gray. From north to south the faults are the Towaliga fault (T), Shiloh fault (S), Bartletts Ferry fault (BF), and the Goat Rock fault (G). Thrust faults are barbed on the overlying sheet and normal faults have tick marks on the down-thrown side. The position of the Brevard zone and the location of the mapped area are shown on the inset map. Two cross sections illustrate the structure of the rocks of the PMW. The black unit is the Hollis Quartzite, an important marker within the PMW. Map and cross sections are generalized from Sears and others (1981b). More detailed cross sections are in Sears and Cook (1984).

presented elsewhere (Clarke, 1952; Bentley and Neathery, 1970; Schamel and Bauer, 1980; Schamel and others, 1980; Holland, 1981; Sears and others, 1981a; Sears and Cook, 1984) that the rocks of the PMW are a stack of several large recumbent anticlines (nappes) involving metasediments of possible Upper Proterozoic-Lower Cambrian age with Grenville-age cores. These folds were formed when Africa collided with America in middle to late Paleozoic time and the Piedmont allochthon was emplaced onto the American continent (Cook and others, 1979, 1980).

The mylonites at the sites described herein are either the decollement zone of the allochthon or are splays from that zone. They involve a variety of lithologies and are multiply deformed. The types of structures are described according to cross cutting relationships seen at each site. Much of the structural complexity seen at these sites is probably due to the complexity of the mylonitization process.

Work at the eastern terminus of the PMW emphasizes the multiple nature of deformation along the Goat Rock and Towaliga faults and suggests strike-slip as well as thrust components in its history (Hatcher and others, 1981, 1985). The present arched structure of the PMW and the mylonites on its flanks may be due to uplift after emplacement of the allochthon or may be the result of the allochthon overriding a stack of nappes (Sears, 1980; Hatcher, 1984, Sears and others, 1981a, 1981b). Regional relationships are discussed further by Kish and others (1985).

We use a terminology developed by Wise and others (1984) for classifying fault-related rocks. As a broad category, mylonites are generally foliated rocks in which some degree of grain-size reduction is the result of ductile deformation. In protomylonites, more than 50 percent of the rock consists of megacrysts that survived the process of grain reduction. If survivor megacrysts make up from 10 to 50 percent of the rock it is a protomylonite. In ultramylonites, megacrysts make up less than 10 percent of the rock. All three types of mylonites are found in this area; however, orthomylonites dominate the stops described below.

STOP A: BARTLETTS FERRY DAM

Take Bartletts Ferry Road west from Georgia 103. Park at the end of Bartletts Ferry Road at Bartletts Ferry Dam. Outcrops described in this guide are on the east side of the Chattahoochee River just downstream from the dam. The dam may be crossed on foot.

The Bartletts Ferry fault is a broad zone of mylonite which dips to the southeast, separating the Uchee belt from the PMW. The description below is based on Holland (1981) and Schamel and others (1980).

Highly strained rocks extend along the east bank of the river between the dam and the raceway, which joins the river below the dam. The northern and central parts of the exposure consist of light and dark gray orthomylonites which are interlayered on a scale of up to 4 in (10 cm). The southern part consists of a uniform light-gray, mylonitic quartzofeldspathic gneiss with minor lenses of blastomylonite.

In thin section, the layered orthomylonite exhibits a mortar texture with ovoid porphyroclasts of strained or partially strained plagioclase and quartz embedded in a lenticular matrix of very fine-grained polygonized quartz, plagioclase, biotite and muscovite. This lenticular arrangement constitutes the mylonitic foliation in the rock. Fine-grained biotite disseminated in the matrix controls the color of the layered orthomylonite. Biotite also forms small distinct grains or very fine-grained clusters oriented parallel to the mylonitic foliation. Muscovite, in some cases, forms large, sigmoidal porphyroclasts. Garnet is present as euhedral to subhedral synkinematic grains.

Within the mylonite are pods and lenses of metapelite, amphibolite and calcsilicate. The metapelite occurs as lenses of coarse-grained phyllonite consisting of muscovite, biotite, quartz, graphite, garnet, sillimanite and pyrite in varying proportions and found in the central and southern parts of the exposure. The phyllonitic schistosity is kinked and disharmonically folded on a small scale within the lenses. Russell (1976) reports an Rb-Sr whole-rock age of 375+/−11 Ma (late Devonian) for these phyllonites.

Ovoid pods of amphibolite up to 7.5 in (20 cm) long and thin broken and smeared out rafts of calcsilicate were exposed by penstock construction. The amphibolite pods are flattened in the mylonitic foliation and in some cases have fine-scale layering cut by thin biotite envelopes. The calcsilicate is light green, sugary textured and consists of quartz, plagioclase and diopside.

Rocks of the adjacent Uchee belt and the PMW were probably interfolded and severely strained during the emplacement of the Piedmont allochthon to produce this assemblage of lithologies. The layered mylonites and the more felsic mylonites may have developed from the biotite and hornblende gneisses which dominate the Uchee belt. Amphibolite bodies of various sizes are common in the Uchee belt and probably supplied the amphibolite pods; a major one, the Hudson Rapids amphibolite, is exposed about 6 mi (10 km) to the east, just south of the Bartletts Ferry fault. Calcsilicate rocks and calcareous amphibolite are also components of the Uchee belt. The most reasonable source for the metapelite lenses would be the schists of the PMW, though metapelites are found in the Uchee belt as well.

A complex structural history is recorded in these rocks. The dominant planar structure in the outcrop is the south-dipping layering which is parallel to a mylonitic fabric. The mylonite is tightly to isoclinally folded with horizontal fold axes and axial surfaces oblique or nearly parallel to the earlier mylonitic foliation. Both the mylonite and its intense folding are attributed to the emplacement of the Piedmont allochthon. Localized small-scale faults, breccias and flexural folds are also present.

The mylonitic foliation is deformed into moderately tight to isoclinal folds which plunge slightly to the northeast. A sporadically developed axial-plane cleavage dips to the southeast. Where viewed from the southwest, these are S folds whose sense of overturning (vergence) is to the northwest. In extreme cases, regional S_2 is transposed parallel to itself, causing confusion as to the nature of the planar fabric and belying the complexity of its formation. These folds may have formed as stress trajectories shifted slightly during emplacement of the Piedmont allochthon.

The penetrative augen-studded lenticular orthomylonite foliation which dominates the outcrop is the regional S_2 mylonitic foliation referred to by Schamel and others (1980). It strikes northeast and dips moderately to the southeast. A subhorizontal, streaky lineation is composed of colinear elongate feldspar grains, small fold axes and the intersection of the mylonitic fabric with the slightly later transposition fabric which cuts across it.

Later deformation produced a variety of nonpenetrative structures. Their relationships to each other are unclear and assignment to specific deformational events is not attempted.

Widely spaced conjugate kinks and shears with offsets of tens of centimeters to several meters are found in the northern part of the outcrop. The structures are best developed in the steep northern limbs of larger earlier folds. The line of intersection formed by pairs of conjugate shear planes and the axes of large open folds associated with these shears plunge steeply to the south; the acute bisectrix of the shears is perpendicular to the lineation in the outcrop.

Narrow brecciated zones showing minor offset and cemented by a very fine-grained chloritized gouge are present in the walls of the penstock tailrace and in some construction blocks. Bleaching associated with these zones ex-

tends several centimeters into the mylonite.

Holland (1981), Schamel and others (1980) and Schamel and Bauer (1980) suggest that regional S_2, was produced by movement on the Bartletts Ferry fault and the emplacement of the Piedmont allochthon. The 375+/–11 Ma Rb-Sr date reported by Russell (1976, 1985) for the phyllonite lens may be a reasonable date for the formation of mylonitic foliation as suggested by Schamel and others (1981); however, this late Devonian date is earlier than the late Paleozoic date normally cited for the emplacement of the Piedmont allochthon as part of the Alleghanian orogeny (Cook and others, 1980; Hatcher and Williams, 1982; Windley, 1984). Mylonitic fabric appears to date from the earliest stages off the emplacement and/or assembly of the Piedmont allochthon prior to late Paleozoic time (Williams and Hatcher, 1982, 1983).

STOP B: GOAT ROCK DAM

Take Goat Rock Road west from Route 103. Park at the end of Goat Rock Road. Walk west on the old road to the dam and descend the metal stairway on the east side of dam down to the river level outcrops. Beware of slippery rocks and high water, especially in the spring. The dam cannot be crossed.

The rock types and structures exposed at Goat Rock Dam are similar to those at Bartletts Ferry. The outcrops are dominated by a mylonitic foliation that dips moderately to the southeast and strikes northeast. As at Bartletts Ferry, Holland (1981) and Schamel and others (1980) call the mylonitic fabric regional S_2 and relate it to the emplacement of the Piedmont allochthon.

Orthomylonite dominates the outcrops south of Goat Rock Dam. This is a medium to dark gray rock with a well-developed penetrative foliation. Augens of feldspar up to one cm long occur in a foliated fine-grained to aphanitic matrix. Foliation is defined by a combination of very fine lenticular layering and mineral orientation. It is parallel in most cases to a rough layering defined by variation in mineral content and developed on a scale of 4 to 8 in (10 to 20) cm. Plagioclase dominates these rocks both in the matrix and among the larger grains. Lesser amounts of microcline, quartz, biotite and muscovite also occur in both the matrix and as larger grains.

Thin zones of fine-grained to aphanitic ultramylonite cut across the orthomylonitic foliation. The fine-grained ultramylonites tend to be light in color while aphanitic ultramylonites range in color from pink with black streaks to green to black.

Relics of layered gneiss and amphibolite are found within the orthomylonite. These help unravel the sequence of events at this site.

Layered gneiss (Fig. 3) consists of alternating layers of light, strongly lineated gneiss one to 2 in (5 cm) thick and darker, less lineated biotite gneiss 4 to 8 in (10 to 20 cm) thick. Both lithologies are lineated orthomylonites consisting of ovoid feldspar porphyroclasts in a fine-grained quartz and mica matrix. Most of the porphyroclasts are slightly twinned plagioclase, some

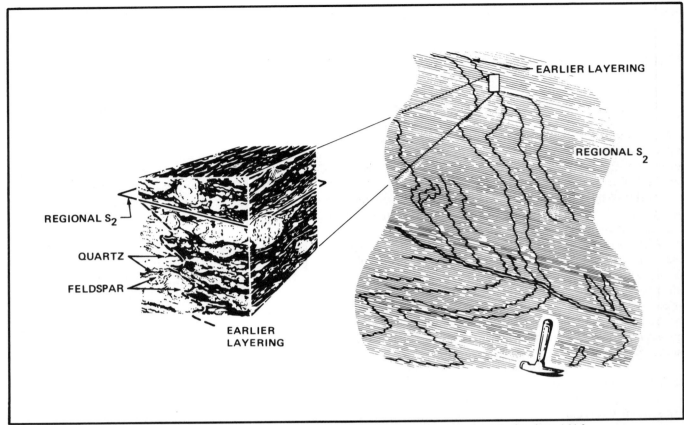

Figure 3. Sketch from photograph of relict mass of folded gneiss within an orthomylonite about 164 ft
(50 m) downstream from Goat Rock Dam, (Stop B). Insets show classic mylonitic textures from Higgins
(1971). Observer is looking east. Block diagram illustrates textural relations in sections cut parallel to
and at right angles to strike. Premylonitic layering was produced by an early nonmylonitic event.
Mylonitization produced a strongly lineated fabric consisting of quartz ribbons parallel to the axial
surfaces of the folds. Both the quartz ribbons and feldspar spindles are strongly elongated parallel to the
fold axes.

with exsolved K-feldspar or biotite inclusions. The fine-grained matrix consists of quartz, feldspar and muscovite. The darker layers contain more biotite.

Some of the structural details in the layered gneiss are shown in the inset block in Figure 3. The lineation in the gneiss is subparallel to the strike of the layering and is defined by both slightly flattened spindles of feldspar and quartz ribbons extending parallel to strike. Sections cut normal to the lineation show the mylonitic foliation discordant to the layering in the gneiss. The mylonitic foliation in the surrounding orthomylonites is represented in the layered gneiss by interbraided phyllonitic networks of muscovite and biotite and by the cross-sections of the quartz ribbons and the slight elongation of the feldspar spindles parallel to the phyllonitic foliation. These planar structures form axial-plane cleavage to folds in the gneiss.

A pod of well-foliated amphibolite surrounded by mylonitic gneiss crops out about 300 ft (100 m) farther south. It is approximately 3 ft (1 m) in diameter and consists of plagioclase, hornblende and minor opaques, along with retrograde amphibole and epidote.

Evidence for several episodes of deformation are present in this outcrop. Locally, nonmylonitic layering appears to be retained from the protolith of the mylonitic. This was folded and mylonitized during the emplacement of the Piedmont allochthon, producing the lineated mylonitic foliation, regional S_2, which dominates this site. Narrow zones of ultramylonite discordant to earlier planar mylonitic structures were also formed during this event. Closely spaced joints with epidote mineralization occur locally. Some of these have slickensided surfaces, indicating dip-slip movement. These and healed fractures seen in thin section are related to post-mylonitization brittle deformation.

STOP C: GOAT ROCK ORTHOMYLONITE ON I-185

Figure 4 illustrates the roadcut which is Stop C, 1.3 mi (2.2 km) south on I-185 from the Mulberry Grove Ga. 315 exit. The dominant lithology here is a fine- to very fine-grained orthomylonite to ultramylonite. Though dark at this exposure, this rock weathers to a silvery gray color. Numerous fractures are coated with sheaves of pink laumontite.

Mylonitic structure is well developed, tectonic inclusions are present, and porphyroclasts are abundant in this rock. The matrix of the mylonites is composed of quartz, biotite, and white mica. Darker rocks have matrices dominated by biotite and very fine-grained quartz. Lighter rocks have matrices in which quartz is a little coarser grained and biotite less abundant.

The porphyroclasts are generally untwinned to weakly twinned plagioclase with little evidence for recrystallization or brittle deformation. Some have a micaceous envelope. They are elliptical in cross section and without tails. In many cases, their long axes are oblique to the enclosing mylonitic fabric. K-feldspar porphyroclasts, much less common than plagioclase, have fine-grained k-feldspar and myrmekite at their margins. Muscovite is the other common porphyroclast. It occurs as streamlined sigmoidal books of various sizes.

A less common mylonitic lithology is one rich in aluminous minerals. This rock also has a well-developed mylonitic foliation. It contains garnet and sillimanite porphyroclasts dispersed through it in addition to plagioclase and muscovite porphyroclasts. The garnets are poikiloblastic angular to ellipsoidal and have been broken in many cases. The sillimanite has been boudinaged, folded and smeared out into fibrolitic tails.

Amphibolitic and quartzofeldspathic gneiss as well as pegmatoidal clasts are present in the mylonite. These have been streamlined in the foliation. The locations of some of these clasts are shown in Figure 4. In addition, a mass of calcareous amphibolitic gneiss is exposed on the natural surface above the road cut.

The mylonitic foliation in both dark mylonites and aluminous mylonites is cut by dark finer grained ultramylonite and is folded on a microscopic scale. This, plus the angular relationship between the mylonitic foliation and the ellipse axes of the porphyroclasts, indicate late stage deformation during the mylonitic event.

Northern, central and southern zones can be distinguished in the outcrop based on lithologies and structure. The southern zone is marked by large, south-verging disharmonic folds that plunge slightly to the east. Lack of axial planar structures indicates a flexural flow mechanism of folding.

The central part of the outcrop consists of dark orthomylonite in which the fluxion structure is well developed. It is separated from the southern part of the outcrop by a weathered narrow reentrant. This reentrant contains coarser muscovite porphyroclasts and appears to be more schistose than adjacent mylonites. This is interpreted to be a zone of slip.

The central zone grades into the northern zone which is characterized by less consistent dips of foliation, the presence of thin gray quartz rich layers and pods, a few thin greenish gray calcsilicate lenses, and lenses of earlier coarser mylonite. Two types of folds occur in the northern zone. The most obvious are s-shaped folds of the foliation (viewed from the east). These appear to reflect a shear couple with overlying material moving down to the southeast. Folds belonging to the other set tend to be tight to isoclinal. A sketch of one of these folds is shown on the outcrop sketch. The cross section seems to indicate flattening

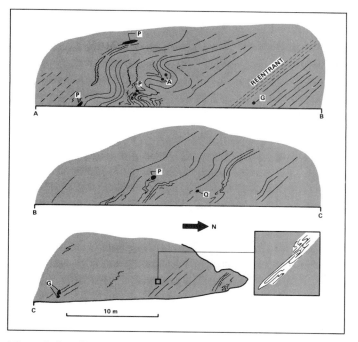

Figure 4. Goat Rock fault orthomylonite at Stop C. Observer is looking west. Pegmatite pods are indicated by P, gneiss pods by G, amphibolite pods by A and quartz pods by Q. Detail of flattened fold is shown by the inset at the north end of the outcrop. South-verging disharmonic folds are at the southern end of the exposure.

parallel to layering of an earlier fold and a component of down-dip movement of the overlying mass, though this sense is not consistent for all cases. The gray quartzites are markers for these folds.

Schamel and others (1980) suggest a structure consistent with the nappe model to explain the southward vergence of the large folds in the southern part of the outcrop. In this model (Fig. 5A), the down-dip extension of the Bartletts Ferry mylonitic foliation was folded by the development of a second thrust fault, the Goat Rock fault, above the Bartletts Ferry fault. This would have caused the overturning of the Bartletts Ferry fault mylonite in this vicinity, and the development of minor parasitic folds on the inside of the resultant synform. The large folds in the southern part of the outcrop would be these parasitic folds. The Goat Rock fault itself would be somewhere to the south of this outcrop.

There is another possible explanation for the south-verging folds. They may have resulted from the buckling of layers due to their sliding off a growing arch located to the north (the PMW structure). In this model (Fig. 5B), discrete surfaces of slip would develop parallel to and probably controlled by the existing fluxion structure. As the arching of the PMW area progressed, gravity instability would increase until the overlying material slid to the south. Some of the strain would have been accommodated by the development of the south-verging folds as mylonite piled up on the flank of the arch. Slip along specific planes parallel to the mylonitic foliation might have caused narrow zones of retrogression, as in the case of the narrow reentrant. Strain distributed

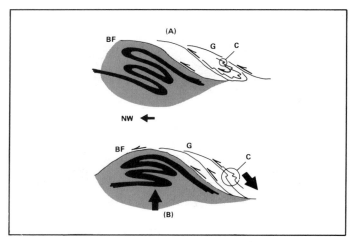

Figure 5. Alternate models for the formation of the disharmonic folds seen at the south end of the roadcut at Stop C. Model (A) (Schamel and others, 1980) suggests that the folds formed as parasites to an overturned fold; (B) suggests that the folds formed as overlying material slid off the rising Pine Mountain arch.

through the body of the rock may have caused the flattening and southern-side-down apparent sense of movement for folds seen in the northern end of the outcrop.

REFERENCES CITED

Bentley, R. D., 1969, Strike slip faults in Lee County, Alabama: Geological Society of America Abstracts with Programs, pt. 4, p. 5.

Bentley, R. D., and Neathery, T. L., 1970, Geology of the Brevard Fault Zone, and related rocks of the Inner Piedmont of Alabama: Alabama Geological Society Guidebook, 8th Annual Field Trip, p. 119.

Clarke, J. W., 1952, Geology and mineral resources of the Thomaston Quadrangle, Georgia: Bulletin 59, Georgia Geological Survey, 103 p.

Cook, F. A., Albaugh, D. S., Brown, L. D., Kaufman, S., Oliver, J. E., and Hatcher, R. D., Jr., 1979, Thin-skinned tectonics in the crystalline southern Appalachians: COCORP seismic reflection profiling of the Blue Ridge and Piedmont: Geology, v. 7, p. 563–567.

Cook, F. A., Brown, L. D., and Oliver, J. E., 1980, The Southern Appalachians and the growth of continents: Scientific American, v. 243, p. 156–168.

Crickmay, G. W., 1933, The occurrence of mylonites in the crystalline rocks of Georgia: American Journal of Science, v. 226, p. 161–177.

Crickmay, G. W., 1952, Geology of the crystalline rocks of Georgia: Bulletin 58, Georgia Geological Survey, 54 p.

Hatcher, R. D., Jr., 1984, Southern and central Appalachian basement massifs, in Bartholomew, M. J., ed., The Grenville Event in the Appalachians and Related Topics: Geological Society of America Special Paper 194, p. 149–154.

Hatcher, R. D., Jr., and Williams, H., 1982, Timing of large-scale displacements in the Appalachians: Geological Society of America Abstracts with Programs, v. 14, p. 24.

Hatcher, R. D., Jr., Howell, D. E., and Talwani, P., 1977, Eastern Piedmont fault system: Speculation on its extent: Geology, v. 5, p. 636–640.

Hatcher, R. D., Jr., Hooper, R. J., and Odom, A. L., 1981, Transition from the east end of the Pine Mountain belt into the Piedmont, central Georgia: Preliminary results: Geological Society of America Abstracts with Programs, v. 13, p. 9.

Hatcher, R. D., Jr., Hooper, R. J., Heyn, T., McConnell, K. I., and Costello, J. O., 1985, Geometric and time relationships of thrusts in the crystalline southern

Appalachians: Geological Society of America Abstracts with Programs, v. 17, p. 95.

Higgins, M. W., 1971, Cataclastic Rocks: U.S. Geological Survey Professional Paper 687, 97 p.

Holland, W. A., Jr., 1981, The Kinematic and Metamorphic Development of the Towaliga-Goat Rock Mylonites, Georgia-Alabama [Ph.D. thesis]: Tallahassee, Florida, The Florida State University, 116 p.

Kish, S. A., Hanley, T. B., and Schamel, S., 1985, Geology of Southwestern Piedmont of Georgia: Geological Society of America, Guidebook, 98th Annual Meeting, Florida State University, Tallahassee, 47 p.

Russell, G. S., 1976, Rb-Sr evidence from cataclastic rocks for Devonian faulting in the southern Appalachians: Geological Society of America Abstracts with Programs, v. 8, p. 1081.

——1985, Reconnaissance geochronological investigations in the Phenix City Gneiss and Bartletts Ferry mylonite zone, in Kish, S. A., Hanley, T. B., and Schamel, S., 1985, Geology of the Southwestern Piedmont of Georgia: Geological Society of America Guidebook, 98th Annual Meeting, Florida State University, Tallahassee, p. 9–11.

Schamel, S., and Bauer, D., 1980, Remobilized Grenville basement in the Pine Mountain window, in Wones, D. R., ed., The Caledonides in the U.S.A.: Blacksburg, Virginia, Department of Geological Sciences, Virginia Polytechnic Institute and State University, Memoir 2, p. 313–316.

Schamel, S., Hanley, T. B., and Sears, J. W., 1980, Geology of the Pine Mountain Window and adjacent terranes in the Piedmont Province of Alabama and Georgia: Geological Society of America Guidebook, 29th Annual Meeting Southeastern Section, Florida State University, Tallahassee, 69 p.

Sears, J. W., 1980, Nappe tectonics in the Pine Mountain window of the Alabama Piedmont: Geological Society of America Abstracts with Programs, v. 12, p. 208.

Sears, J. W., and Cook, R. B., 1984, An overview of the Grenville basement complex of the Pine Mountain window, Alabama and Georgia, in Bartholomew, M. J., ed., The Grenville Event in the Appalachians and Related Topics: Geological Society of America Special Paper 194, p. 281–287.

Sears, J. W., Cook, R. B., Jr., and Brown, D. E., 1981a, Tectonic evolution of the western part of the Pine Mountain window and the adjacent inner Piedmont province, in Sears, J. W., ed., Contrasts in tectonic style between the Piedmont Terrane and the Pine Mountain window: Alabama Geological Society 18th Annual Field Trip, p. 1–13.

Sears, J. W., Cook, R. B., Jr., Gilbert, O. E., Jr., Carrington, T. J., and Schamel, S., 1981b, Stratigraphy and structure of the Pine Mountain window in Georgia and Alabama, in Wigley, P. B., ed., Latest thinking on the stratigraphy of selected areas in Georgia: Georgia Geological Survey Information Circular 54-A, p. 41–53.

Williams, H., and Hatcher, R. D., Jr., 1982, Suspect terranes and accretionary history of the Appalachian orogen: Geology, v. 10, p. 530–536.

——1983, Appalachian suspect terranes, in Hatcher, R. D., Jr., Williams, H., and Zeitz, I., Contributions to the Tectonics and Geophysics of Mountain Chains: Geological Society of America Memoir 158, p. 33–53.

Windley, B. F., 1984, The Evolving Continents, 2nd edition: New York, John Wiley and Sons, 399 p.

Wise, D. U., Dunn, D. E., Engelder, J. T., Geiser, P. A., Hatcher, R. D., Jr., Kish, S. A., Odom, A. L., and Schamel, S., 1984, Fault related rocks: Suggestions for terminology: Geology, v. 12, p. 391–394.

ACKNOWLEDGMENTS

The authors thank Bill Corwin, Mark Ray, and Ken Boyd of SCS Engineering Publications Department for their assistance in drawing preparation. Georgia Power Company facilitated work on this field guide by allowing access to the Bartletts Ferry Dam site during a recent construction project. Dan Sundeen confirmed the presence of laumontite along fractures at Stop C. Paul Beyer and the Administration of Columbus College have encouraged continued geological research in the Columbus area.

Petrography and structural geology of Uchee Belt Rocks in Columbus Georgia, and Phenix City, Alabama

Thomas B. Hanley, *Department of Chemistry and Geology, Columbus College, Columbus, Georgia 31993*

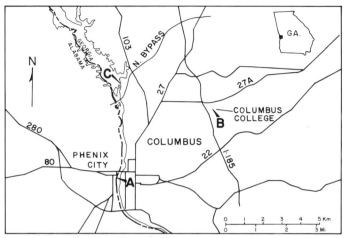

Figure 1. Map of the Columbus–Phenix City area with Stops A, B, and C indicated.

LOCATION

Uchee Belt rocks are well exposed in the vicinity of Columbus, Georgia. Stops A and B are on the banks of the Chattahoochee River in Phenix City, Alabama, and Stop C is located on the campus of Columbus College, Columbus, Georgia (Fig. 1).

INTRODUCTION

The Uchee belt is a large subdivision of the Piedmont of Georgia and Alabama. Bounded on the north by the Bartletts Ferry fault and extending beneath the Coastal Plain on the south (Bentley and others, 1985), its outcrops in the Columbus–Phenix City area are the most southern exposures of the Appalachian Piedmont. The Uchee belt consists of a sequence of hornblende gneiss, amphibolite and amphibolitic gneiss, gneissic metasediments, migmatite, and granitic to monzonitic gneiss. The Phenix City Gneiss, one of the main constituents of the Uchee belt, was named by Bentley and Neathery (1970) for exposures in the Chattahoochee River at Phenix City. The Rocky Shoals exposure of the Phenix City Gneiss (Stop A) shows the nature of a crenulation cleavage that merges with numerous shallow-dipping minor faults. These have been attributed by Schamel and others (1980) to proximity to the sole fault of the Piedmont allochthon. Lindsey Creek exposures (Stop B) are the most extensive of amphibolites in the immediate area, and exposures around Oliver Dam (Stop C) illustrate the lithologic diversity found in the Uchee belt.

SITE DESCRIPTION

Stop A: Rocky Shoals on the Chattahoochee

Approach Rocky Shoals from the west on U.S. 80 and 13th Street in Phenix City. Approach from the east by following U.S. 80 across the 14th Street Bridge from Columbus into Phenix City. Park in the parking lot at the right-angle turn in U.S. 80 south of the Fourteenth Street Bridge. Stop A (Fig. 2) lies below the parking lot along the river. Use extreme caution in descending the steep debris-strewn trail from the parking lot and in crossing the slippery rocks in the channel on the west side of the main rock pavement.

The nature of the Phenix City Gneiss and of two generations of penetrative structures that are developed in it are well displayed in the large pavement on the west side of the river below the Fieldcrest (once the Eagle-Phenix) dam. Most of the description and interpretation for this stop comes from Schamel and others (1980), most notably the map of the outcrop (Fig. 2). Additional information is presented in this guide on the nature of the tectonic surface they call "regional-S_2."

The pavement area is divisible into northern, central, and southern zones by differences in lithology and by structural discontinuities (Fig. 2). The southern zone differs from the other two by being richer in biotite and K-spar. Although gneissic banding in the southern zone is approximately parallel to banding in the central zone, there is a very streamlined amphibolite lens at the contact between the two, and deformation in the gneisses near the contact seems more intense. The central and northern zones are similar in lithology but are separated by a fault that truncates the layering in the central zone.

The Phenix City Gneiss is a thinly layered, coarse-grained, hornblende gneiss that contains amphibolite lenses and tabular masses and both concordant and discordant pegmatite veins and pods. The gneiss has a granoblastic texture and contains, as primary phases, varying quantities of andesine, quartz, hornblende, biotite, and minor apatite and opaques. Characteristically, K-spar is sparse to absent in the gneiss, though it and muscovite are abundant in the pegmatites. Epidote and sphene occur as sparse secondary overgrowths. Amphibolites in the gneiss are coarse grained.

The mineral content of the gneiss varies from dioritic and quartz dioritic to granodioritic. Its composition, layering, and the association with amphibolites suggest that the unit developed from a succession of andesitic volcanics and (or) volcanogenic sediments that were regionally metamorphosed to amphibolite facies.

Two types of layering occur at the Rocky Shoals locality. The dominant layering in the outcrop is developed on a scale of 0.25 in (0.5 cm), and is restricted to intrafolial folds. Since the thinner layering appears to be the earlier and is perhaps primary, it shall be called S_0. The delicate layering is more distinct and

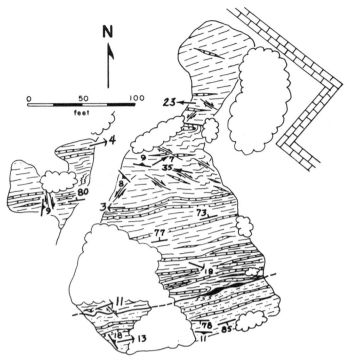

Figure 2. Geologic map of the Rocky Shoals outcrop of Phenix City Gneiss below the Fieldcrest Mill dam (once the Eagle-Phenix dam). Dashed-line pattern is amphibolitic gneiss; dot-dash pattern is biotite gneiss; amphibolite sliver (black) occurs along their mutual contact; dots only are pegmatites; sharp arrows and tick-marked strike-and-dip symbols are early folds (F_1) and lithologic layering ($S_0=S_1$) respectively; blunt arrows and barbed strike-and-dip symbols are later folds (F_2). Map redrafted from Schamel, et al. (1980).

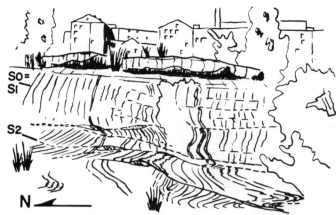

Figure 3. Sketch from photograph of the west side of the large Rocky Shoals outcrop. Observer is looking east. Folds (F_2) of steeply dipping layering ($S_0=S_1$) have shallow plunges and shallow axial surfaces. Slip surfaces (S_2) dip moderately to the south.

may be S_0 enhanced by metamorphic differentiation. Borders of the amphibolite masses are parallel to the coarser layering, as are the limbs of intrafolial isoclinal folds of the early layering. Shallow-dipping planar structures (S_2) and folds associated with them deform this layering on various scales. Some of these relationships are shown in Figure 3.

At least two periods of deformation (D_1 and D_2) are indicated by the relationships described above. D_1 produced the isoclinal folding (F_1) of S_0. These intrafolial folds are relatively uncommon and are difficult to measure because of the nature of the outcrop, but they appear to have variable orientations within S_1. They are called F_1, indicating that folding was roughly coeval with development of the steeply dipping layering; however, the F_1 folds may have developed earlier and then rotated into their present position by flattening in S_1. The weathered lenticular amphibolite may owe its form to such stretching and flattening.

The D_2 episode produced structures that are notable for their variability. The most widespread is a weakly developed, penetrative, subhorizontal planar structure, S_2, which is visible across the entire pavement exposure. Locally, S_2 is axial planar to small-scale crinkles and to concentric and modified-concentric F_2 folds. The F_2 folds plunge gently to the east-northeast. In other

places S_2 increases in intensity and merges with narrow zones of transposition or truncation of S_1. The limbs of some of the larger F_2 folds are tangential to some of these zones. These second-generation structures are well displayed in cross section on the west side of the large exposure (Fig. 3), at its southwest corner, and in the pavement exposure west of the channel.

Thin-section study of F_2 crinkles from the southern zone shows that S_0 and S_1 are defined, respectively, by layering and by the orientation of biotite grains parallel to layering. S_2 cleavage is developed as distinct zones of grain-size reduction and mica reorientation at approximately right angles to S_1.

Thin foliated granitic dikes occur within the narrow zones of transposition in which development of S_2 is intense. The foliation in the dikes is parallel to S_2. These thin granite bodies seem to be syntectonic because their emplacement was controlled by the zones of transposition and because they have a foliation that is parallel both to the zone and to S_2 in the surrounding rocks.

Stop B: Lindsey Creek Amphibolites

Bentley and others (1985) point out that amphibolites and hornblende gneiss are the characteristic lithologies of the Phenix City Gneiss. Good examples of these amphibolites are found in Lindsey Creek on the Columbus College campus (Fig. 1), which is southeast of the intersection of I-185 and Manchester Expressway (Alt. 27). Warm Springs Road is parallel to Manchester Expressway on its south side; East Lindsey Drive is parallel to I-185 on its east side. Take East Lindsey Drive south from Warm Springs Road. Drive through the brick gateway of Columbus College from East Lindsey Drive, cross Lindsey Creek, and park in the first parking lot on the left (north). Amphibolites and associated rocks are exposed in the creek bed (Fig. 4). Similar rocks are found along strike in Cooper Creek to the east.

The Lindsey Creek amphibolites consist of a variety of dark rocks. Hornblende schist and amphibolite are dominated by vary-

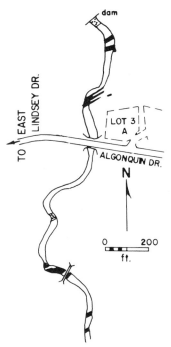

Figure 4. Location of outcrops in Lindsey Creek on the Columbus College campus. Solid pattern represents amphibolite and calcareous amphibolite; dots are massive grey biotite hornblende gneiss.

ing amounts of hornblende and plagioclase with quartz, diopside, epidote, calcite, sphene, and ilmenite as accessory minerals. Thin, leucocratic lenses ranging from 0.25 by 0.8 in (0.5 by 2 cm) to 0.5 by 4.0 in (1 by 10 cm) occur sporadically in the amphibolite. Thicker green lenses up to 1.5 in (4 cm) wide contain plagioclase, diopside, epidote, and scapolite. Diopside-bearing, dark, calcareous gneisses consisting of 2- to 5-in (5- to 25-cm) thick, gradationally colored layers occur with the amphibolites. Both the amphibolites and the calcareous gneisses contain green calcsilicate pods and lenses 1.5 to 2 in (3 to 5 cm) thick, consisting of plagioclase, diopside, quartz, garnet, hornblende, calcite, and sphene. In addition, veins of scapolite, vermicular epidote, hornblende, and grossularite cut the dark rocks. This suite is interpreted to be a sequence of volcanogenic sediments and perhaps volcanic rocks with a strong carbonate influence.

Biotite hornblende gneiss forms several massive, vaguely foliated outcrops in Lindsey Creek and along East Lindsey Drive north of the college entrance. The larger outcrops are light gray; however, smaller tabular masses intercalated with the amphibolites in the creek bed have a burnished orange color. These rocks are rich in quartz and plagioclase, are usually low in K-spar, and may be metaigneous associates of the amphibolites.

Structures developed in the amphibolites and other mafic rocks are different from those in the lighter gneisses. The mafic rocks tend to dip steeply to the north or south. Concentric folds associated with minor faults in the amphibolite are locally developed. Their plunges are shallow, though the dips of their axial surfaces vary from steep to shallow. In contrast, the lighter

gneisses generally dip to the north at moderate angles. The mafic rocks appear to have been caught up in a mass of more felsic material during deformation, yielding differently to stress due to their different competency. Foliations in the mafic rocks are either due to deformation (as in the case of the amphibolites and hornblende schists) or are relics of a sedimentary protolith (as in the case of the layered variety). At least some of the foliation in the felsic gneisses and the concentric folds in the amphibolites are interpreted to be related to the shallow dipping foliation, S_2, and the associated folding at Rocky Shoals related to D_2.

Stop C: Oliver Dam

Oliver Dam is west of the intersection of the Columbus North Bypass and River Road (Georgia 103) (see Fig. 1). The approach road and rocky pavement at the base of Oliver Dam provide good exposures of Uchee belt migmatitic gneisses. A great deal of rock is well exposed here. In addition, it seems reasonable to assume that the abundant boulders are local construction debris. The locality is reached by taking the turnoff to the municipal marina and boat ramp just north of the intersection of River Road (Georgia 103) and the North Bypass. Rather than turning down to the marina, continue parallel to the highway to the dam. Exposures along the road are always easily acccessible; however, the pavement below the dam can only be reached from the Georgia side by having Georgia Power unlock the gates for the tunnel through the dam or the walkway on top of the dam. This must be arranged ahead of time by calling the Georgia Power Company in Columbus (404-322-1661) and the dam operator (404-324-1760). The pavement is inaccessible during part of the spring due to high water. **WARNING:** the westernmost gate operates automatically; water in the river rises rapidly when the gates are opened!

Bentley and others (1985) use photographs of this outcrop to illustrate their description of Phenix City Gneiss but do not discuss the outcrop in detail. This site can be divided into two zones. The northern zone is structurally and lithologically complex and includes the rocks exposed along the approach road to the dam as well as the northern part of the pavement below the dam. A more homogeneous zone is present to the south. Cutting these rocks are quartz dikes and joints.

Traversing from north to south, exposures show calcareous amphibolite, biotite schist, amphibolite, gray gneiss, foliated pegmatite injection complexes, and porphyroclastic biotite gneisses, all of which show the effect of deformation and injection.

Oliver Dam Approach Road. This outcrop (Fig. 5) exhibits a coarse biotite schistosity that cuts across and wraps around masses of thinly layered amphibolite. Hornblende and plagioclase dominate the amphibolites. Masses that are more calcareous than others contain diopside, plagioclase, and significant calcite. Plate-shaped masses are as thin as 1 in (5 cm), have layering parallel to the edges of the plate, and are oriented parallel to the surrounding schistosity. Larger, blocky masses also show some degree of elon-

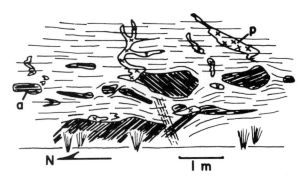

Figure 5. Sketch of road cut on the east side of the Oliver Dam service road. Observer is looking east. Amphibolitic masses (black), with internal foliation indicated, are enveloped and cut across by a later, shallow-dipping biotite schistosity (S_2?, dashes). Some of the amphibolite masses contain significant calcite. X-pattern indicates pegmatite. A zone of closely spaced fractures containing laumontite occurs at the bottom of the outcrop in the middle of the sketch.

gation parallel to the schistosity. Their internal layering, however, is variable from block to block and is cut obliquely by the schistosity in many places.

Schistosity is oriented approximately N55°W, 30°NE. This is different from the regional strike (ca. N60°E) but is roughly parallel to the orientation of the dominant foliation of the local gneisses.

Amphibolite blocks and slabs are interpreted to be fragments of a larger mass that was disrupted during development of the schistosity. Biotite seems to have grown in the zone of deformation due to an influx of potassium; the relationship of the pegmatites to both the amphibolite masses and schistosity suggests that pegmatite veins grew by diffusion during this time as well.

Oliver Dam Pavement. Gray gneiss is the dominant lithology on the pavement. It is generally a massive rock in which hornblende-rich layers are subordinate to light-gray diktyonitic migmatite. The neosomes are either interbraided or subparallel to one another and evenly spaced. Foliation in the paleosome in some cases trends subparallel to the adjacent leucosome, though in other cases paleosome layering is folded with axial surfaces oriented subparallel to the adjacent neosome. It is the neosome orientation in these rocks that defines their dominant foliation.

The darker, hornblende-rich layers are parallel to the dominant foliation in the lighter migmatitic part of the gneiss just described. These darker units contain tightly folded layers with axial surfaces that are parallel to the gneiss contact. In at least one instance, a pegmatite dike cuts the dark layer but ends abruptly (is sliced off?) at the lighter gneiss contact.

Also associated with the gray gneiss are large amphibolite gneiss masses. These are broken and injected by quartzofeldspathic pegmatite. Veins of quartz, carbonate, biotite, ilmenite, and pyrite occur in irregular chloritized fractures in some of the

darker masses. These amphibolite gneiss masses are best seen in place at the northeast end of the pavement on the raceway side of the cement wall. They are an extension of the amphibolite-dominated gneisses along the approach road.

Below the gray gneiss is the pegmatite injection complex. This is quite thick and consists of masses of various types of gneiss engulfed by foliated pegmatite. These pegmatite bodies are the dominant planar features in this part of the northern zone and are parallel to the foliation in the overlying gray gneiss. The engulfed masses of gneiss are distinctly layered and usually folded and in many instances are cut by dikes from the enveloping pegmatite.

Two types of masses have been recognized thus far in the injection complex. They are masses of amphibolite and masses of well-layered gneiss in which amphibolite layers are a minor lithology. Both types show the effects of severe deformation. Amphibolite masses occur either singly or in trains along foliation. The largest seen so far is nearly 3 ft (1 m) across. They tend to be flattened and in some cases are injected by quartzofeldspathic veins from the enveloping pegmatite. These have been interpreted as large clasts by Bentley and others (1985); however disruption of once more extensive amphibolite layers by deformation also seems plausible.

Masses of layered gneiss are extensively developed in the injection complex. They consist of a variety of well-defined lithologies. Garnet is present in some of the lighter layers, and thin amphibolite layers and lenses occur within these masses. One very large pointed boulder in the west-central part of the pavement illustrates the lithologies and their style of deformation.

Sheets of porphyroclastic rock 3 ft (1 m) or more thick occur near the south end of the northern zone. Concordant, foliated pegmatite is associated with these as are small trains of stretched and injected amphibolite masses. Bentley and others (1985) suggest that this assemblage of lithologies is a metaturbidite; however, because the attitude of these rocks is similar to the dominant planar structure in the rest of the northern zone rather than deformed like the metasedimentary gneissic masses, the interpretation that these are orthomylonites rather than metasediments is favored here. Rounding of the porphyroclasts and the fine grain size of the matrix are interpreted as being due to tectonic rather than sedimentary processes. Lineated phyllonitic to schistose schlieren are also found in these rocks.

REFERENCES CITED

Bentley, R. D., and Neathery, T. L., 1970, Geology of the Brevard Fault Zone, and related rocks of the Inner Piedmont of Alabama: Alabama Geological Society Guidebook, 8th Annual Field Trip, 119 p.

Bentley, R. D., Neathery, T. L., and Scott, J. C., 1985, Geology and mineral resources of Lee County, Alabama: Geological Survey of Alabama Bulletin 107, 110 p.

Schamel, S., Hanley, T. B., and Sears, J. W., 1980, Geology of the Pine Mountain Window and adjacent terranes in the Piedmont Province of Alabama and Georgia: Geological Society of America Southeastern Section Guidebook, 29th Annual Meeting, 69 p.

Mitchell Dam amphibolite along Alabama Highway 22, east-central Chilton and west-central Coosa Counties, Alabama

Karen F. Rheams, *Geological Survey of Alabama, Tuscaloosa, Alabama 35486*
David Allison and James F. Tull, *Department of Geology, Florida State University, Tallahassee, Florida 32306*

LOCATION

The Mitchell Dam Amphibolite is located along the Coosa River 9 miles (14.5 km) east of Clanton near the center of the Mitchell Dam 7½-minute quadrangle, Chilton and Coosa Counties, Alabama (Fig. 1). Two excellent outcrops illustrate the mineralogical, textural, and structural complexity of the amphibolite. Both exposures are easily accessible, located near the shoulder of Alabama 22.

INTRODUCTION

The Mitchell Dam Amphibolite is among the best exposures of medium to high metamorphic grade mafic rocks in the southern Appalachians. It is located at the southwestern limit of the exposed Appalachians in the Northern Alabama Piedmont, occurring only a few miles (km) from the overlap of the Gulf Coastal Plain. The amphibolite host is the Higgins Ferry Group (Neathery, 1975) of the Ashland Supergroup (Tull, 1978). This group contains a very thick, diverse sequence of graphitic, garnetiferous and manganiferous schists, quartzites, and concordant amphibolites units, of which the Mitchell Dam is structurally the uppermost. Near Mitchell Dam and to the southeast, the Higgins Ferry Group grades structurally upward into migmatitic gneiss, muscovite schist, and granitic plutons of the amphibolite-free Hatchet Creek Group (Neathery, 1975).

The Mitchell Dam Amphibolite is a sequence of medium- to coarse-grained, massive to thinly banded and foliated amphibolite in the Coosa lithotectonic block of the Northern Alabama Piedmont (Neathery and others, 1974; Tull, 1978; Rheams, 1982). The Coosa block is bounded on the northwest by greenschist facies middle Paleozoic metaclastic and metavolcanic rocks of the Talladega block; and on the southeast by late Precambrian amphibolite facies metasediments intruded by Ordovician and younger granitic plutons of the Tallapoosa block (Neathery and others, 1974). The boundaries of each block are delineated by large displacement faults. The amphibolite is composed predominantly of green to black hornblende and plagioclase (An_{10} to An_{50}) in almost equal proportions, with relict augite common (Kalk, 1972; Stow and others, 1984). The amphibolite units are surrounded by and intercalated with metasedimentary schists and gneisses of the Higgins Ferry Group (Neathery, 1975; Tull, 1978; Rheams, 1982). Chemical studies of the Mitchell Dam Amphibolite (Bloss, 1979; Schneider, 1980; Stow and others, 1984), and preserved igneous textures and minerals within the amphibolite (Bloss, 1979; Rheams, 1982), imply an igneous origin.

The chemistry of the Mitchell Dam Amphibolite has been

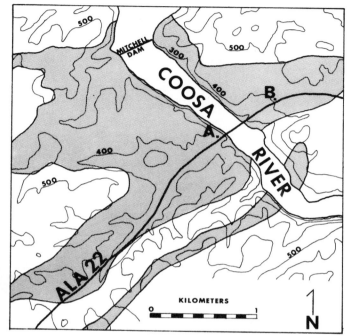

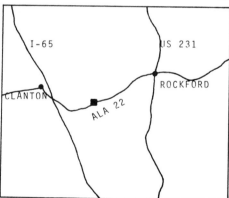

Figure 1. Location map of Mitchell Dam Amphibolite, Sites *A* and *B*. Amphibolite outcrop area shown in *dotted pattern* (from Mitchell Dam 7½-minute quadrangle, Chilton and Coosa Counties, Alabama).

discussed in depth by previous workers and indicates an igneous basaltic protolith with an ocean floor tectonic setting (Bloss, 1979; Schneider, 1980; Stow and others, 1984). Average major and trace element abundances and normative mineralogy for 48 Mitchell Dam Amphibolite samples are shown in Table 1. Other amphibolite bodies in the Northern Alabama Piedmont display similar chemistry and are located in similar stratigraphic position to the northeast along regional strike. An inferred age of late Precambrian is assigned to the amphibolite units of the Coosa

TABLE 1. AVERAGE MAJOR AND TRACE ELEMENT
ABUNDANCES AND NORMATIVE MINERALOGY FOR
48 MITCHELL DAM AMPHIBOLITE SAMPLES*

	Major Element Abundance (%)		Trace Element Abundance (ppm)			Normative Mineralogy	
SiO$_2$	49.70	(3.17)	Ba	62	(57)	Q	0.00
Al$_2$O$_3$	15.26	(1.58)	Co	39	(8.3)	Or	3.00
TiO$_2$	1.15	(0.38)	Cr	283	(167)	Ab	24.02
FeO	10.37	(1.64)	Cu	81	(60)	An	27.97
MgO	7.60	(2.74)	Li	19	(9.6)	Ne	0.00
CaO	10.99	(2.19)	Ni	142	(97)	Ol	11.51
Na$_2$O	2.80	(0.78)	Pb	17	(11)	Fo	6.24
K$_2$O	0.50	(0.56)	Sr	207	(144)	Fa	5.27
MnO	0.113	(0.054)	V	221	(47)	Di	22.44
P$_2$O$_5$	0.19	(0.05)	Zn	86	(24)	Wo	11.40
LOI	1.12	(0.51)				En	6.23
						Fs	4.82
Total	99.79					Hy	7.15
						En	4.05
						Fs	3.10
						Mt	1.69
						Il	2.21

*Stow et al., 1984.

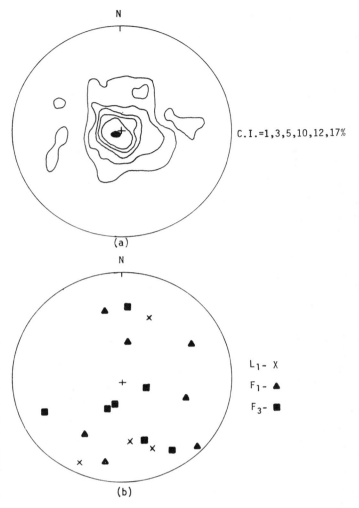

C.I.=1,3,5,10,12,17%

(a)

L$_1$- X
F$_1$- ▲
F$_3$- ■

(b)

Figure 2. Lower hemisphere projection of (a), 129 poles to S$_1$–foliation, Mitchell Dam Amphibolite; and (b), F$_1$ and F$_3$–fold axes of Mitchell Dam Amphibolite with L$_1$–mineral lineations.

block based upon their conformable relationship with the metasediment units (Tull, 1978).

The amphibolite and associated metasediments have four readily identifiable deformational phases resulting in metamorphic differentiation, development of foliation and lineation, intense folding, and transposition structures. The metamorphic grade of the area is the biotite-garnet-sillimanite-anatectic subfacies of the amphibolite facies. A retrogressive event postdates the peak of thermal metamorphism leading to actinolite, epidote, and sericite replacement assemblages.

Tight to isoclinal, often recumbent F$_1$–phase folds are prevalent throughout the Northern Alabama Piedmont (Tull, 1978). In the Mitchell Dam Amphibolite, F$_1$–folds are most easily identified by folding of the calc-silicate pods and/or pegmatite veins into a similar (style) geometry, while foliation (S$_1$) is defined by thin axial planar segregations of plagioclase and hornblende. These bands appear to be the product of metamorphic differentiation synchronous with the F$_1$–fold event.

Regionally, macroscopic and mesoscopic F$_2$–folds are rare and have not been recognized in the Mitchell Dam Amphibolite. They are, however, common in the surrounding metasediments on a microscopic scale. They are present as tight assymmetric crenulations or kinked micas accompanied by shearing of the S$_1$–foliation and flattening of the matrix around competent porphyroblasts (Tull, 1978).

F$_3$–phase folds fold S$_1$–foliation into tight-to-closed parallel geometries and, at a given outcrop, are generally colinear with F$_1$–folds. F$_3$–folds are rare in the amphibolite. F$_4$–cross-folds (1.2 mi or 1.9 km in wave length) are open, upright and megascopic in scale. The cross-folding creates complex interference patterns recognizable on maps of the metasedimentary stratigraphy. The Mitchell Dam Amphibolite is located near the culmination of an interference dome, accounting for the near

horizontal S$_1$–foliation and recumbent nature of F$_1$–folds. Stereogram plots of planar and linear structural elements (Fig. 2) indicate a general near horizontal orientation for the foliation. F$_1$–folds, F$_3$–folds, and mineral lineations (L$_1$), although coaxial at any given locality, are highly variable in orientation due to the complex interference of F$_4$–cross-folding and pre-existing structures.

If the interpretation that Higgins Ferry Group amphibolite units are metamorphosed basaltic flows, ash, and/or sills interlayered with late Precambrian sediments is correct, then the Higgins Ferry Group is at least a partial record of volcanic and sedimentary activity associated with late Precambrian rifting of a continental mass and the early formation of a passive continental margin. Subsequent Devonian through late Paleozoic closure of an ocean basin has caused metamorphism and large-scale transport of these sequences to their present position in one of several far-travelled thrust nappes (Tull, 1978).

The association of interlayered amphibolite and metasediments is consistent throughout much of the eastern Blue Ridge belt, the southwestern extension of which includes the high grade lthologies of the Northern Alabama Piedmont. The Eastern Blue Ridge belt extends from Alabama to Virginia. The amphibolites within this belt are likely representative of rift-related tholeiitic protoliths which formed along a once continuous, though no doubt irregular, rifted margin which was also accumulating vast quantities of fine-grained clastic sediments (Rankin, 1975; Thomas and others, 1980).

SITE DESCRIPTION

The amphibolite is exposed at two easily accessible stops adjacent to the Coosa River. Site A (Fig. 1) is located in a small abandoned quarry south of the dam on the west side of the river adjacent to Alabama 22. Site B is a large roadcut (Fig. 1) south of the dam on the east side of the Coosa River along Alabama 22.

Site A. Exposed in the quarry is a medium-grained, variably banded, foliated black amphibolite with a well defined hornblende lineation (L_{1m}) and a locally developed compositional layering/foliation intersection lineation (L_{li}). A striking characteristic of this exposure is the occurrence of ellipsoidal calc-silicate lenses or pods, completely enveloped by the foliation within the amphibolite (Fig. 3). These ellipsoidal pods are attenuated, isolated fold hinges. Light-colored stringers representing fold limbs thin and attenuate away from the thickened hinge area. Foliation (S_1) is axial planar to these early generation folds and a well developed hornblende mineral lineation is colinear with the fold axes. The morphology of the attenuated fold hinges indicates active transportation of compositional layering in response to extreme differential shear.

Whether the transported layering is of metamorphic, igneous, or sedimentary origin, or combinations thereof, is still open to debate (Kalk, 1972; Neathery, 1975; Tull and others, 1975; Bloss, 1979; Schneider, 1980; Rheams, 1982; Stow and others, 1984). However, recent mapping in this area has identified calc-silicate bearing metasedimentary units equivalent in mineralogical composition to the calc-silicate pods in the amphibolite. These thin but areally continuous units probably represent marl protoliths. The possibility therefore exists that the transposed calc-silicate compositional layers within the amphibolite originated as thin sediments interlayered with basaltic flows or ash.

Petrographically, the amphibolite exposed here is composed predominantly of dusky yellowish-green to pale yellowish-brown hornblende. Roughly 20 percent of the plagioclase (An_{10} to An_{30}) occurs as light-colored bands 0.06 to 0.5 in (1.5 to 12.7 mm) thick, while approximately 80 percent (An_{30} to An_{50}) occurs in the leucocratic pods, or disseminated in hornblende-rich domains. Phases in the calc-silicate pods are clinopyroxene, zoisite, epidote, grossular garnet, and calcic plagioclase. Relict pyroxene (augite), as opposed to the metamorphic pyroxene in calc-silicate pods, occurs as replacement relicts in hornblende cores. These relict pyroxenes are interpreted to have an igneous

Figure 3. Folded leucocratic pod, Site A, Mitchell Dam Amphibolite, Chilton County, Alabama.

origin (Kalk, 1972; Bloss, 1979; Rheams, 1982). Accessory minerals include epidote, actinolite, chlorite, garnet, white mica, quartz, and sphene. Epidote occurs principally as an alteration product of a plagioclase consuming reaction associated with a retrograde event.

Site B. Located in the roadcut just east of the Coosa River bridge are medium-grained, massive to thinly banded amphibolites. Early planar and linear fabric elements similar to those of Site A occur, and the mineralogy and petrographic relationships are essentially equivalent to those at Site A. Small, restricted areas at the eastern end and south of the roadcut at Site B contain retrograded amphibolites, with actinolite as the dominant amphibole. The amphibolite is medium- to coarse-grained and ranges in color from dusky green to greenish-black. Yellow-green to pale blue-green actinolite makes up 45 to 95 percent of the rock, with minor amounts of hornblende and biotite commonly present. Megascopic epidote group minerals are visible on cleavage surfaces.

Three low angle faults have been identified and described within the Mitchell Dam Amphibolite outcrop area (Bentley and Deininger, 1964; Neathery, 1975). The most spectacular of these faults occurs along much of the south-facing cut at Site B, and dips beneath the north-facing cut. A series of nearly horizontal, gently curving surfaces identifies the fault zone. Within the fault zone abundant chloritization and cataclastic material are typically present, along with isolated horses of relatively unaffected country rock. Excellent examples of concentric flexural slip drag folds occur in the footwall block 10 to 13.5 feet (3 to 4 m) in amplitude, and truncate against the fault zone. The sense of displacement indicated by the drag folds requires a component of normal slip in the history of fault displacement. The fault zone strikes N20°E and dips 10° to the south near the central part of the roadcut. Assymetrical, parasitic F_1–folds 1.5 to 3 feet (0.5 to 0.9 m) in wavelength defined by pegmatite veins are exposed in both walls of the outcrop offering excellent three-dimensional vantages of their orientation and morphology.

REFERENCES CITED

Bentley, R. D., and Deininger, R. W., 1964, Amphibolite exposures, *in* Deininger, R. W., Bentley, R. D., Carrington, T. J., Clarke, O. M., Jr., Power, W. R., and Simpson, T. A., Alabama Piedmont geology: Alabama Geological Society, 2nd Annual Field Trip, Guidebook, 64 p.

Bloss, Pamela, 1979, Geochemistry and petrogenesis of the Mitchell Dam Amphibolite, Chilton and Coosa Counties, Alabama [M.S. thesis]: University of Alabama, Tuscaloosa, 112 p.

Kalk, T. R., 1972, The petrology of the Mitchell Dam Amphibolite, Chilton and Coosa Counties, Alabama [M.S. thesis]: Memphis State University, Memphis, 66 p.

Neathery, T. L., 1975, Rock units in the high-rank belt of the Northern Alabama Piedmont, *in* Neathery, T. L., and Tull, J. F., eds., Geologic profiles of the northern Alabama Piedmont: Alabama Geological Society, 13th Annual Field Trip, Guidebook, 173 p.

Neathery, T. L., Bentley, R. D., Higgins, M. W., and Zietz, Isidore, 1974, Preliminary interpretation of aeromagnetic and aeroradioactivity maps of the crystalline rocks of Alabama: Part I, Geologic outline: Geological Society of America Abstracts with Programs, v. 6, p. 383.

Rankin, D. W., 1975, The continental margin of eastern North America in the southern Appalachians: The opening and closing of the proto-Atlantic Ocean: American Journal of Science, v. 275, p. 298–336.

Rheams, K. F., 1982, The petrography and structure of the Mitchell Dam Amphibolite and the surrounding Higgins Ferry Formation, Chilton and Coosa Counties, Alabama [M.S. thesis]: University of Alabama, Tuscaloosa, 124 p.

Schneider, H. I., 1980, Mineral, chemical and textural variations along s-surfaces in selected outcrops of amphibolite, Georgia and Alabama Piedmont [Ph.D. thesis]: Florida State University, Tallahassee, 110 p.

Stow, S. H., Neilson, M. J., and Neathery, T. L., 1984, Petrography, geochemistry, and tectonic significance of the amphibolites of the Alabama Piedmont: American Journal of Science, v. 284, p. 414–436.

Thomas, W. A., Tull, J. F., Bearce, D. N., Russell, G. S., and Odom, A. L., 1980, Geologic synthesis of the southernmost Appalachians, Alabama and Georgia, *in* Wones, D. R., ed., The Caledonides in the USA: Virginia Polytechnical Institute Department Geological Science Memoir 2, 329 p.

Tull, J. F., 1978, Structural development of the Alabama Piedmont province northwest of the Brevard Zone: American Journal of Science, v. 278, p. 442–460.

Tull, J. F., Deininger, R. W., and Neathery, T. L., 1975, Stops B–2A, B–2B, Mitchell Dam Amphibolite, *in* Neathery, T. L., and Tull, J. F., eds., Geologic profiles of the northern Alabama Piedmont, Alabama Geological Society, 13th Annual Field Trip, Guidebook, p. 158–161.

Olistostromal unit of the Silurian–Lower Devonian Lay Dam Formation, Talladega Slate Belt, Chilton County, Alabama

James F. Tull, Department of Geology, Florida State University, Tallahassee, Florida 32306
Whitney R. Telle, School of Mines and Energy Development, University of Alabama, Tuscaloosa, Alabama 35486

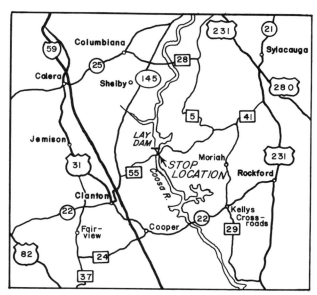

Figure 1. Location map of Lay Dam Formation olistostrome site, Coosa River, Clanton, Chilton County, Alabama.

LOCATION

One of the better exposures of an olistostromal sequence (metadiamictite facies) in the Lay Dam Formation lies immediately west of the Coosa River Bridge on Chilton County Road 55, approximately 12 mi (19 km) east of Jemison, Alabama (Fig. 1). To reach the site from the Lay Dam Exit on I-65, proceed north on Alabama 145 approximately 2 mi (3.2 km) to Chilton County Road 55. Turn right (east) and proceed approximately 6 mi (10 km) to the new Coosa River Bridge at Mim's Ferry. All access roads are improved and readily accessible. The stop is located in the SE¼ of Section 24, T 23 N, R 15 E, Chilton County, Alabama, and may be found on the Lay Dam, Alabama 7½-minute quadrangle. The metadiamictite is exposed on both sides of the highway. Good exposures of the metadiamictite also may be found on the east side of the bridge and at the old Mim's Ferry Landing south of the bridge.

INTRODUCTION

The Lay Dam Formation (Carrington, 1973) is a thick clastic sequence within the Talladega slate belt, a far-traveled lower greenschist facies thrust sheet which was emplaced upon the southern Appalachian foreland during the late Paleozoic Al-

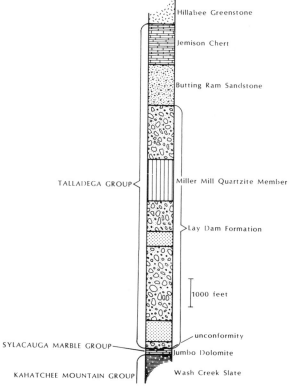

Figure 2. Generalized stratigraphy of the Talladega slate belt, western Chilton County, Alabama.

leghanian Orogeny (Tull, 1984). Rocks of the Talladega belt range in age from the Early Cambrian (Late Precambrian?) Kahatchee Mountain Group (Ocoee Supergroup to Chilhowee Group) to the Early Devonian Jemison Chert and Hillabee Greenstone (Fig. 2; Tull, 1982; Tull and Stow, 1979). The Talladega Group occurs above Lower Ordovician and older rocks of the Sylacauga Marble and Kahatchee Mountain Groups along a regional low-angle unconformity (Shaw, 1970). The Lay Dam Formation dominates the Talladega Group, occurring directly above the unconformity and represents a 13,200-ft thick (4000-m) clastic wedge that thins toward the craton. This site is representative of one of the thickest and most spectacular olistostromal deposits recorded in geologic literature. Boulders of metastable lithologies suspended in a fine-grained matrix indicate a period of extreme continental margin instability associated with the Acadian Orogeny following a period of Early Cambrian through Lower Ordovician passive margin sedimentation. Rapidly eroding highlands cored with Grenville basement lithologies is indicated as the source area of the clastic debris comprising

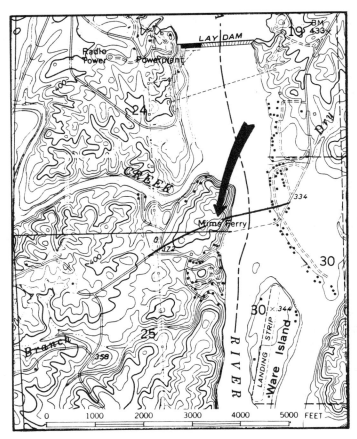

Figure 3. Site location of Lay Dam Olistostrome at the new Coosa River Bridge, Chilton County, Alabama, Lay Dam, Alabama 7½-minute quadrangle.

these metasediments and thus the Lay Dam Formation reveals much about the tectonic history of the southeastern Appalachians.

SITE INFORMATION

The Lay Dam Formation contains several recognizable sedimentary lithofacies including laminated and non-laminated phyllite, feldspathic metasandstone, metaconglomerate, metagreywacke, and a spectacular olistostromal metadiamictite facies. This stop provides an excellent exposure of one of the olistostromal units.

The metadiamictite facies represents an unusual and somewhat enigmatic sediment protolith which consisted of mudstone containing a polymictic assemblage of lithic clasts ranging in size from sand to boulders. The metadiamictite units are interbedded with the other lithofacies in the Lay Dam Formation on a scale varying from less than 3 ft to hundreds of feet (1 m to hundreds of m) and regionally intertongue in lenticular map pattern with these lithofacies. The major concentration of the metadiamictite facies lies between the Coosa River and I-65 in northeast Chilton County, where this facies comprises as

much as 60% of the Lay Dam Formation section (Telle, 1983; Fig. 4). In this area the metadiamictite facies is found locally both at the base, where it lies unconformably above metacarbonate and metaclastic rocks of the Sylacauga Marble and Kahatchee Mountain Groups, and at the top of the Lay Dam Formation, where it grades into the conglomeratic metasandstone of the overlying Butting Ram Quartzite (Fig. 2). Thus, diamictic sedimentation occurred intermittently throughout the deposition of the Lay Dam Formation in this region. Where the contacts between the diamictite facies and the psammitic or pelitic facies are exposed, both the upper and lower boundaries are generally planar and sharp, with no evident grading of the contact. A polymodal grain-size distribution in the metadiamictite results from the complete gradation from clay-size particles to boulders. The matrix, which we have arbitrarily defined as particles less than 0.08 in (2 mm), consists of randomly distributed silt to sand-size quartz, feldspar, and rock fragment grains chaotically intermixed with finer-grained white mica, chlorite, and carbonate. Sand grains are generally subangular to rounded. The matrix generally comprises 88% to 97% of the rock volume. At this exposure, the matrix ranges from 88% to 93% of the rock volume. The coarser clastic fraction thus forms only a small percentage of the total rock and it is matrix supported.

Clasts are distributed randomly in the matrix and consist of a wide variety of lithologies. These lithologies include sedimentary protoliths such as various carbonate rock types, pelites, cherts, and sandstones, as well as felsic plutonic and high-grade metamorphic rocks, all of which may be found at this site. The larger clasts generally range from subrounded to well-rounded, but are often flattened parallel to metamorphic foliation as are many of the carbonate clasts in this cut. Maximum clast size varies with locality, and at this stop a granite gneiss boulder was removed which was 5 ft (1.5 m) in diameter. Clasts larger than 8 in (20 cm) have been noted at most localities. The relative abundance of clast lithologies also varies with locality. Generally, granite and granitic gneiss are dominant by far, averaging over 80%, as they do at this stop. At a few localities, however, carbonate rock clasts (mostly fine- to medium-grained dolomitic marble) are dominant. Quartzite is also a common clast lithology and may represent 30% of the clasts locally. Pelitic clasts appear to be relatively uncommon, but may, in fact, be generally indistinguishable from the matrix because of the compositional similarities.

The gneissic metamorphic rock fragments have mineral assemblages indicative of upper amphibolite to granulite facies metamorphism which far exceeds that of the lower greenschist facies Talladega belt. Additionally, the metamorphic rock fragments display foliations discordant with the slaty cleavage developed within the matrix, and that are truncated at the boundaries of the clasts. Here at the Coosa River the degree of penetrative strain is high and has resulted in a moderate to high degree of flattening and stretching of even the felsic plutonic and gneissic clasts (Fig. 5). Vein quartz and carbonate strain shadows are well developed and trail away from the clast tips.

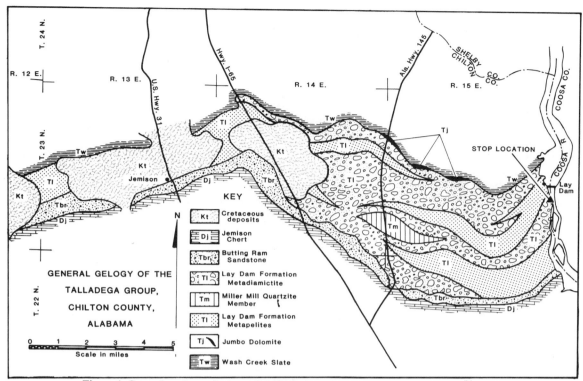

Figure 4. General geology of a part of the Talladega slate belt, Chilton County, Alabama.

Zircons extracted from a granite gneiss boulder at this locality (from the southwestern end of the northwest-facing cut) yield a U/Pb age of 1.1 billion years (Telle and others, 1979). This indicates that a nearby Grenville basement source was exposed during deposition of the diamictites. The clasts representing various sedimentary protoliths were probably derived from the sediment mantle of this basement source and generally can be matched well to units below the pre–Lay Dam Formation unconformity in the Kahatchee Mountain and Sylacauga Marble Groups of Early Cambrian through Early Ordovician Age.

Figure 5. Deformed gneissic boulders contained in olistostromal unit in the new Coosa River bridge site. Note foliation in boulders discordant with regional dynamothermal metamorphic foliation of matrix.

Granitic and carbonate sources generally yield very unstable fragments, but have yielded large, fresh clasts to the Lay Dam Formation. This implies very rapid mechanical erosion of a crystalline basement and its clastic/carbonate cover rocks. These factors point to rapid uplift and high relief, most likely along fault scarps proximal to the sedimentary basin margin. The large sizes of the crystalline and carbonate clasts also indicate the proximity of the source terrane and sediment transport by powerful torrential flow. Regional facies relationships, sediment thickness distributions, and clast size variations indicate that the source of the Lay Dam diamictite sediments lay to the east or southeast.

The metadiamictite facies is envisioned as a series of proximal subaqueous debris flows originating from an alluvial distributary system fed by vigorous mountain torrents on the adjacent land. Lithofacies associated with the metadiamictites represent components of the coalescing subaqueous fan system and delta front deposits. These deposits interfinger with the series of discrete debris flow deposits of the diamictons. The tectonic implication of this deposition is important—sediment accumulation was rapid and was concurrent with the rapid progressive uplift of a basement terrane bordering the southeast basin margin. Structurally induced debris flows, like the unit found at this stop, accompanied fan/delta growth on a tectonically formed slope.

The Lay Dam Formation grades upward into the Lower Devonian Butting Ram Quartzite (ridge on far horizon to the southeast) and locally overlies Lower Ordovician rocks (Harris and others, 1984). To the west of Jemison, the Lay Dam Formation consists of a rhythmically layered and locally graded sequence of weakly cleaved metasiltstone, metaclaystone, and metasandstone. This sequence interfingers with the diamictite facies east of Jemison. Approximately 3 mi (5 km) west of Jemison, the Lay Dam Formation metasiltstone has yielded conodant molds of post-Ordovician morphotype (Harris and others, 1984). Thus, the Lay Dam Formation in this region must range between Silurian and Early Devonian.

The Kahatchee Mountain and Sylacauga Marble Groups, which lie below the pre–Lay Dam Formation unconformity, are interpreted to represent a rifted margin to stable-carbonate-shelf facies which formed along the ancient passive continental margin of proto–North America during the latest Precambrian and lower Paleozoic (Tull, 1982). The regional unconformity at the base of the Talladega Group (which occurs a few hundred yards to the north of this exposure) marks a fundamental tectonic change in this continental margin—a time of uplift, erosion, collapse, and subsidence of the continental shelf. This activity was accompanied by differential uplift of sediment-mantled basement rocks to the east or southeast which shed debris, probably from exposed fault scarps, onto the fragmented shelf. A thick Silurian (?) to Lower Devonian clastic wedge was formed above the fragmented shelf which tapered toward the craton. These events were followed by a period of volcanic activity and regional dynamothermal metamorphism associated with the Acadian Orogeny, and subsequently by major thrusting and detachment from basement during the late Paleozoic Alleghanian Orogeny.

REFERENCES CITED

Carrington, T. J., ed., 1973, Talladega metamorphic front: Alabama Geological Society, 11th Annual Field Trip, Guidebook, p. 22–38.

Harris, A. G., Repetski, J. E., Tull, J. F., and Bearce, D. N., 1984, Early Paleozoic conodonts from the Talladega slate belt of the Alabama Appalachians—tectonic implications: Geological Society of America Abstracts with Programs, v. 16, p. 143.

Shaw, C. E., Jr., 1970, Age and stratigraphic significance of the Talladega Slate: Evidence of pre–Middle Ordovician tectonism in central Alabama: Southeastern Geology, v. 11, p. 253–267.

Telle, W. R., 1983, Geology of the Lay Dam Formation, Chilton County, Alabama [M.S. thesis]: University of Alabama, Tuscaloosa, 90 p.

Telle, W. R., Tull, J. F., and Russell, C. W., 1979, Tectonic significance of the bouldery facies of the Lay Dam Formation, Talladega slate belt, Chilton County, Alabama: Geological Society of America Abstracts with Programs, v. 11, no. 4, p. 215.

Tull, J. F., 1978, Structural Development of the Alabama Piedmont Northwest of the Brevard Zone: American Journal of Science, v. 278, p. 422–460.

—— 1982, Stratigraphic Framework of the Talladega slate belt, Alabama Appalachians, *in* Bearce, D. N., Black, W. W., Kish, S., and Tull, J. F., eds., Tectonic Studies in the Talladega and Carolina slate belts, Southern Appalachian Orogen: Geological Society of America Special Paper 191, p. 3–18.

—— 1984, Polyphase late Paleozoic deformation in the southeastern foreland and northwestern Piedmont of the Alabama Appalachians: Journal of Structural Geology, v. 6, p. 223–234.

Tull, J. F., and Stow, S. H., eds., 1979, The Hillabee metavolcanic complex and associated rock sequences: Alabama Geological Society, 17th Annual Field Trip, Guidebook, 64 p.

Mesozoic and Cenozoic compressional faulting along the Coastal Plain margin, Fredericksburg, Virginia*

Eugene K. Rader, Virginia Division of Mineral Resources, Charlottesville, Virginia 22903
Wayne L. Newell and Robert B. Mixon, U.S. Geological Survey, Reston, Virginia 22092

LOCATION

Exposures of the Stafford fault system are located south and southwest of Fredricksburg, Spotsylvania County, Virginia on or adjacent to U.S. 1 (Fig. 1).

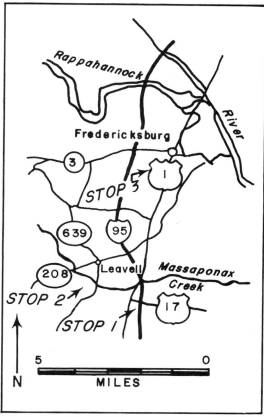

Figure 1. Road map of the Fredericksburg area showing the location of Stops 1, 2, and 3.

INTRODUCTION

Until recent years, most geologists have interpreted the Mesozoic and Cenozoic strata of the Atlantic Coastal Plain as a structurally simple, seaward-thickening sedimentary wedge. In general, the individual lithic units that compose the sedimentary section were believed to strike north and northeast parallel to the Coastal Plain margin and to dip gently and uniformly to the east and southeast toward the present-day Atlantic Coast. Early workers recognized major northwest-trending positive areas such

---

*Adapted from Mixon and Newell, 1978 and 1982.

as the Cape Fear arch and intervening broad downwarps such as the Salisbury basin and southeast Georgia embayment. Although minor faults of Cretaceous and post-Cretaceous age were observed at a few widely scattered localities (Cederstrom, 1945; Darton, 1950; White, 1952; Daniels and others, 1972), extensive zones of faulting such as those that characterize the Gulf Coastal Plain were not reported until the 1970s.

The first zone of faulting to be recognized in the Atlantic Coastal Plain was the Brandywine fault system in southern Maryland (Jacobeen, 1972). The Brandywine is a system of northeast-trending reverse faults defined in the subsurface mainly on the basis of deep borehole data and seismic reflection lines (Jacobeen, 1972). Field investigations along the Fall Line in northeastern Virginia and in South Carolina and Georgia have delineated other zones of reverse faults and related flexures (Mixon and Newell, 1977, 1978; O'Connor and Prowell, 1976; Prowell and O'Connor, 1978). In contrast to the Gulf Coast normal faults, these fault structures reflect the compressional stress regime which has prevailed in much of eastern North America during late Mesozoic and Cenozoic time (Brown and others, 1972; Sbar and Sykes, 1973; Zoback and Zoback, 1980).

This "cluster stop" is intended to familiarize participants with a part of the Stafford fault system, a series of "en echelon," northwest-dipping reverse faults along the Fall Line in northeastern Virginia (Fig. 1). Exposures of the thin sedimentary section at the inner edge of the Coastal Plain (Stops 1 and 2) and the Hazel Run fault are described (Stop 3).

SITE DESCRIPTIONS

Stop 1: Inner Coastal Plain Stratigraphic Section, Hicks Borrow Pit, Spotsylvania County, Va.

One of the better exposures of the thin Cretaceous and Tertiary units that constitutes the inner Coastal Plain section in this area is in the W. L. Hicks borrow pit (Fig. 2). The stratigraphic sections exposed in the borrow pit area include four of the six Coastal Plain formations used to delineate the Stafford fault system and to outline its deformational history.

The oldest unit exposed is the Potomac Formation (Fig. 3). Only the uppermost 10 to 15 ft (3 to 4.5 m) are exposed; however, data from a borehole at the edge of the pit indicate a total thickness of about 125 ft (37.9 m). The best exposure of the Potomac Formation is north of the narrow paved road which divides the borrow pit into two parts. The section includes, from bottom to top: 1) light-gray to yellowish-gray, trough-cross-bedded gravelly sand; 2) pale-gray to white, moderately well-sorted sand, characterized by very gently inclined bedding and

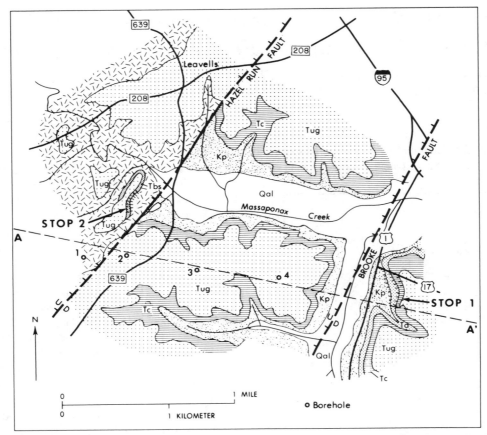

Figure 2. Generalized geologic map of Stop 1 and Stop 2 areas, Spotsylvania County. Kp, Potomac Formation; Ta, Aquia Formation; Tc, Calvert and (or) Choptank Formations; Tbs, burrowed sand; Tug, upland gravel; Qal, alluvium; Spotsylvania 7½-minute quadrangle. Modified from Mixon and Newell, 1982.
LOCATION OF STOPS 1 AND 2: STOP 1: East side of U.S. 1, 0.7 mile south of junction of U.S. 1 and I-95 south of Fredericksburg, Spotsylvania County, Spotsylvania 7½-minute quadrangle.
STOP 2: Proceed south of Leavells on Virginia 639 to first gravel road to west after crossing Massaponax Creek, pit 0.3 mile west of Virginia 639, Spotsylvania County, Spotsylvania 7½-minute quadrangle. Permission for entry should be obtained from Mr. W. L. Hicks of Fredericksburg for Stops 1 and 2.

laminae of black heavy minerals, grading upward to 3) greenish- and bluish-gray, yellowish-gray weathering sandy and clayey silt. The pale-gray sand contains two types of burrows. The upper-most Potomac beds are poorly exposed in the U.S. Highway 17 Bypass roadcuts north of the borrow pit. A well-preserved micro-flora in samples from a borehole, about 2 mi (3.3 km) south-southwest of the Hicks borrow pit, indicates that the Potomac beds in this area are Early Cretaceous in age (pollen zone I of Barremian(?) to early Albian age) and a correlation with the Patuxent Formation of Maryland.

The overlying lower and middle parts of the Aquia Forma-tion that crop out in the borrow pit consist of medium-gray, clayey and silty, fine- to very fine-grained, micaceous quartz sand containing some glauconite. The upper part of the Aquia is sim-ilar in texture and mineralogy but has weathered to a light yel-lowish gray. The base of the formation, exposed in a shallow gulley in the floor of the pit, is marked by a thin bed of quartz

pebbles, which are as much as 2 in (5 cm) in maximum dimension. Calcareous fossils, if once present, have been removed by leaching. The Aquia Formation here, and at other localities in updip areas of Caroline and Stafford counties, contain abundant coalified wood fragments and small logs as much as 3 ft (0.8 m) long. Trace fossils include noded *Ophiomorpha* burrows and very small vertical burrows filled with fecal pellets that may have been formed by marine worms. Correlations based on planktic fora-minifera at other sites indicate that the Aquia Formation is Paleo-cene in age (Loeblich and Tappan, 1957).

Lower(?) and middle Miocene sands and clay-silts of marine origin unconformably overlie the Aquia Formation in the high east wall of the W. L. Hicks pit. The basal stratum of the Miocene sequence in the Hicks borrow pit is poorly sorted, pebbly, and clayey sand bed ranging in thickness from 3 to 5 ft (1 to 1.4 m). This bed and the overlying blocky clay-silt, about 10 ft (3 m) thick, are conformably overlain at nearby localities by fine- to

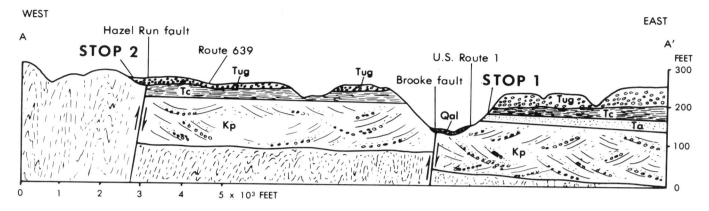

Figure 3. Cross section A-A' from vicinity of Leavells gravel pit (Stop 2) to Hicks borrow pit (Stop 1). Line of section shown on generalized geologic map. Kp, Potomac Formation; Ta, Aquia Formation; Tc, Calvert and (or) Choptank Formations; Tug, upland gravel; Qal, alluvium. Modified from Mixon and Newell, 1978.

very fine-grained sand, clayey and silty in part, which is more typical of Calvert and Choptank strata farther downdip. At the Hicks pit this sand section apparently has been removed by erosion, and upland gravel of Pliocene(?) age directly overlies the clay-silt unit (Fig. 3).

Outcrops of the clay-silt unit in the updip parts of the Coastal Plain in Stafford and Spotsylvania counties are generally highly leached and oxidized and contain few or no macrofossils. Unweathered material from boreholes in the Spotsylvania quadrangle contains diatoms and an abundant and well-preserved Neogene pollen assemblage dominated by oak. In addition, the presence of well-preserved specimens of the dinoflagellate *Palaeocystodinium golzowense* Alberti implies that the beds are no younger than middle Miocene (Lucy Edwards, oral commun., 1978).

In the vicinity of the Hicks borrow pit, the upland gravel is 20 to 40 ft (6 to 12 m) thick and is interbedded with minor amounts of clay and silt. The gravel consists mainly of pebbles of vein quartz, quartzite, chert, and metamorphosed plutonic and volcanic rocks. Pebbles of red mudstone and sandstone from the Culpeper Triassic basin are present but rare. The basal part of the unit includes some cobble- to boulder-size clasts. Trough cross-bedding is common. Sandy gravel grades upward into gravelly sand and then into silty and clayey sand, forming one or more fining-upward sequences.

Stop 2: Gravel Pit, Upland Deposits and Hazel Run Fault Near Leavells

The Leavells gravel pit is situated in northeastern Spotsylvania County at the extreme western edge of the highly dissected Coastal Plain upland (Fig. 2). The flat tops of the higher ridges in this area are apparently remnants of an extensive depositional surface of Pliocene(?) age associated with the directly underlying medium to coarse sand and gravel of the "upland gravels," here about 30 to 50 ft (9 to 15 m) thick. Near Leavells the depositional surface is about 250 to 265 ft (75.7 to 80 m) in altitude

but slopes gently eastward to about 240 to 250 ft (72.7 to 75.7 m) in the adjacent Guinea quadrangle.

To the west of the Pliocene(?) depositional surface, rolling hills underlain by saprolitized Piedmont crystalline rocks rise abruptly 50 ft (15 m) or more above the level of the flat-topped inner Coastal Plain ridges. This linear step in topography, Thornburg Scarp (Mixon, 1978), can be traced southward toward Richmond and northeastward toward Washington, D.C.

The Hazel Run fault extends from the edge of the Piedmont at Fraziers Gate and Leavells northeastward to Fredericksburg. Detailed mapping and boreholes at this locality show that the trace of the fault is between the gravel pit and the small pond in the valley adjacent to and east of the pit. Stratigraphic and structural relationships on each side of the fault are shown in the cross section (Fig. 3) extending from the vicinity of Leavells gravel pit (Stop 2) to the Hicks borrow pit on U.S. 1 (Stop 1).

The Piedmont crystalline rocks and at least five sedimentary rock units are exposed in the vicinity of the Leavells gravel pit. From oldest to youngest, the section includes: 1) the Potomac Formation, 2) the Calvert and (or) Choptank Formations, 3) gravelly sand containing abundant *Ophiomorpha* burrows, 4) high-level sandy gravel of fluvial origin (upland gravel), and 5) sandy and gravelly alluvium of Quaternary age which fills the Massaponax Creek valley.

Outcrops of crystalline rocks are found only on the relatively upthrown northwest side of Hazel Run fault. In the past, saprolitized gneiss and schist have been exposed in the small gullies that drain the gravel pit area. Fresh rock is exposed mainly along Massaponax Creek upstream from the fault. At one time, a borrow pit on the north side of Massaponax Creek displayed coarse high-level gravels containing saprolitized boulders resting on saprolite.

Coarse-grained, crossbedded arkosic sand typical of the Potomac Formation is exposed in the roadcut at the junction of Virginia 639 and the gravel road that leads westward to the pit. The borehole data indicate that the Potomac beds in this area

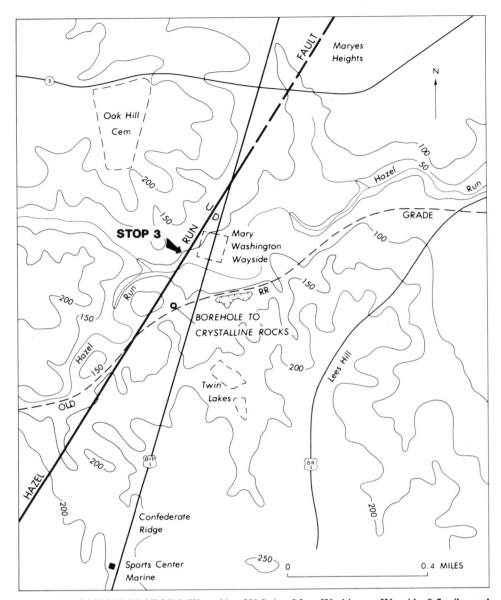

Figure 4. LOCATION OF STOP 3: West side of U.S. 1 at Mary Washington Wayside, 0.5 mile south of junction of U.S. 1 and Virginia 3 in Fredericksburg: Fredericksburg 7½-minute quadrangle. Modified from Mixon and Newell, 1982.

are about 120 to 130 ft (36.4 to 39.4 m) thick, approximately equal to the vertical separation on the Potomac Formation-crystalline rock contact across the fault.

Silty and clayey fine sand as much as 3 to 4 ft (0.9 to 1.2 m) in thickness underlies the floor of the gravel pit. This unit may represent the thin truncated edge of the Calvert Formation of the upthrown side of the fault.

A poorly to moderately well-sorted, medium- to coarse-grained quartz sand unit as much as 10 ft (3 m) thick is exposed in the lower walls of the gravel pit. Discontinuous stringers of gravel delineate trough cross-beds. Abundant rock fragments and feldspar reflect the Piedmont source terrane adjacent to the west.

Some beds contain abundant clay-lined, noded *Ophiomorpha* burrows which suggest a marginal-marine depositional environment. The contact between the burrowed sand unit and the underlying silty and clayey fine sand is covered by slumped material but is presumed to be unconformable.

The burrowed sand resembles the updip, marginal-marine facies of the Eastover Formation of late Miocene age (Ward and Blackwelder, 1980; Newell and Rader, 1982). Correlation with the Eastover has not been firmly established by detailed mapping between the rather isolated outcrops in the pit area and the fossiliferous sands of unquestioned Eastover age to the east and south.

The crossbedded sandy gravel exposed in the upper part of

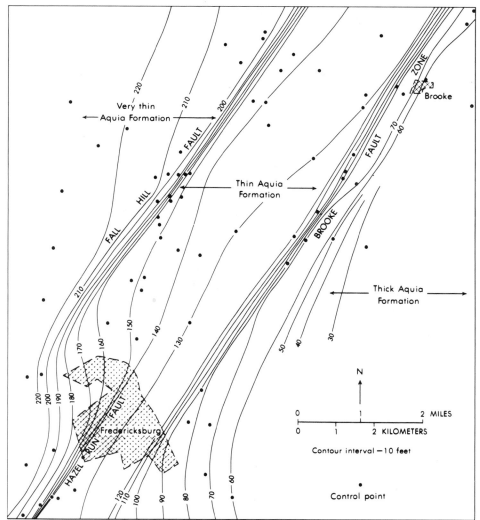

Figure 5. Structure contours on base of Aquia Formation, Fredericksburg area, showing flexuring and faulting of Paleocene beds coincident with Cretaceous fault trends. Structures delineate northeast-trending belts of thick, thin, and very thin Aquia strata. After Mixon and Newell, 1977.

the pit walls is thought to be of fluvial origin and appears to be separated from the underlying burrowed sand unit by a broadly undulating unconformity formed by stream scour. The gravel unit thickens to the east and merges with the extensive gravel sheet that underlies much of the Coastal Plain upland in the Fredericksburg area. These gravels are interpreted as upper-delta-plain deposits which, at the onset of marine regression in the middle(?) to late Pliocene, prograded southeastward across older lithic units of marginal marine and marine origin. The upland gravels are thought to be correlative, at least in part, with the upper regressive part of the Yorktown Formation in downdip areas to the east (Newell and Rader, 1982).

Stop 3: Hazel Run Fault, Mary Washington Wayside Park, Fredericksburg, Va.

The Hazel Run structure is naturally exposed at two

site located a short distance upstream from the U.S. 1 Bypass bridge over Hazel Run (Fig. 4). The fault plane of the main reverse fault has been excavated on the south side of Hazel Run in the steep bank near the abrupt change in stream gradient and the small waterfall about 500 ft (151.5 m) west-southwest of the highway bridge. At this locality, biotite augen gneiss of the Po River Metamorphic Suite (Pavlides, 1980) is thrust at a high angle over greenish-gray sand of the Potomac Formation. A borehole on the downthrown block, located on the abandoned Civil War era railroad just south of Hazel Run shows about 120 ft (36.4 m) of apparent vertical displacement of the Potomac Formation-crystalline rock contact. The marked change in stream gradient at this locality is caused by the juxtaposition of the resistant Piedmont crystalline rocks and the more easily eroded Coastal Plain beds. Additional outcrops showing the fault plane of the Hazel Run structure occur near the mouth of a narrow ravine on the north side of Hazel Run, about 300 ft (91 m) west of the picnic tables at Mary Washington Wayside roadside park.

Structure contouring and the borehole studies of the Hazel Run fault indicate about 60 ft (18.2 m) of vertical displacement of the base of the Aquia Formation versus about 120 ft (36.4 m) of displacement of the base of the Potomac Formation (Figure 5). Deformation of Coastal Plain beds of post-Aquia age along the Hazel Run structure is not entirely clear. Borehole and trenching studies of the fault about 1 mi (1.7 km) southwest of the Mary Washington Wayside suggests that the Calvert(?) Formation and the overlying upland gravels have not been appreciably warped or faulted along the main fault zone [Potomac Electric Power Company (PEPCO), 1976]. Thus, most of the movement

along the Hazel Run fault appears to have taken place in pre-Calvert time. However, a northwest-dipping reverse fault near the Sports Center Marine Store on U.S. 1 Bypass, about 1,500 ft (454.5 m) southeast of the Hazel Run fault trace, is reported to offset the base of the upland gravels about 18 in (45.7 cm) (PEPCO, 1973, 1976). This fault was not traceable upward through the upland gravel section, possibly because of the unconsolidated nature of the deposits at the time of faulting.

The increase in amount of displacement with depth on the Hazel Run fault and on other structures of the Stafford fault system indicates recurrent movement through much of Mesozoic and Cenozoic time.

REFERENCES CITED

Brown, P. M., Miller, J. A., and Swain, F. M., 1972, Structural and stratigraphic framework, and spatial distribution of permeability of the Atlantic Coastal Plain, North Carolina to New York: U.S. Geological Survey Professional Paper 796, 79 p.

Cederstrom, D. J., 1945, Geology and groundwater resources of the Coastal Plain in southeastern Virginia: Virginia Geological Survey Bulletin 63, 384 p.

Daniels, R. B., Gamble, E. E., Wheeler, W. H., and Holzhey, C. S., 1972, Some details of the surficial stratigraphy and geomorphology of the Coastal Plain between New Bern and Coats, North Carolina: Carolina Geological Society and Atlantic Coastal Plain Geological Association Annual Meeting and Field Trip Guidebook, 36 p.

Darton, N. H., 1950, Configuration of the bedrock surface of the District of Columbia and vicinity: U.S. Geology Survey Professional Paper 217, 42 p.

Jacobeen, F. H., Jr., 1972, Seismic evidence for high angle reverse faulting in the Coastal Plain of Prince Georges and Charles counties, Maryland: Maryland Geological Survey Information Circular 13, 21 p.

Loeblich, A. R., Jr., and Tappan, H. N., 1957, Correlation of the Gulf and Atlantic Coastal Plain Paleocene and lower Eocene formations by means of planktonic Foraminifera: Journal of Paleontology, v. 31, p. 1109–1137.

Mixon, R. B., 1978, The Thornburg scarp: a late Tertiary marine shoreline across the Stafford fault system *in* Mixon, R. B., and Newell, W. L., 1978, The faulted Coastal Plain margin at Fredericksburg, Virginia—Tenth Annual Virginia Geology Field Conference, October 13-14, 1978: Guidebook Reston, Virginia, 50 p.

Mixon, R. B., and Newell, W. L., 1977, Stafford fault system: Structures documenting Cretaceous and Tertiary deformation along the Fall Line in northeastern Virginia: Geology, v. 5, p. 437–440.

—— 1978, The faulted Coastal Plain margin at Fredericksburg, Virginia—Tenth Annual Virginia Geology Field Conference, October 13-14, 1978: Guidebook Reston, Virginia, 50 p., 12 figs.

—— 1982, Mesozoic and Cenozoic compressional faulting along the Atlantic Coastal Plain margin, Virginia *in* Lyttle, P. T. (ed.), Central Appalachian

Geology, NE-SE Geological Society of America, Field Trip Guidebooks: American Geological Institute, Falls Church, VA, p. 29–54.

Newell, W. L., and Rader, E. K., 1982, Tectonic control of cyclic sedimentation in the Chesapeake Group of Virginia and Maryland *in* Lyttle, P. T. (ed.), Central Appalachian Geology, NE-SE Geological Society of America, Field Trip Guidebooks: American Geological Institute, Falls Church, VA, p. 1–27.

O'Connor, B. J., and Prowell, D. C., 1976, The geology of the Belair fault zone and basement rocks of the Augusta, Ga. area: Georgia Geological Society Guidebook 16, p. 21–32.

Pavlides, Louis, 1980, Revised nomenclature and stratigraphic relationships of the Fredericksburg Complex and Quantico Formation of the Virginia Piedmont: U.S. Geological Survey Professional Paper 1146, 29 p.

Potomac Electric Power Company, 1973, Preliminary safety analysis report, Douglas Point Nuclear Generating Station, Units 1 and 2: Washington, D.C., v. 2, 150 p.

—— 1976, Geologic investigation of the Stafford fault zone: Washington, D.C., 53 p.

Prowell, D. C., and O'Connor, B. J., 1978, Belair fault zone: Evidence of Tertiary fault displacement in eastern Georgia: Geology, v. 6, p. 681–684.

Sbar, M. L., and Sykes, L. R., 1973, Contemporary compressive stress and seismicity in eastern North America: an example of intra-plate tectonics: Geological Society of America Bulletin, v. 84, p. 1861–1882.

Ward, L. W., and Blackwelder, B. W., 1980, Stratigraphic revision of upper Miocene and lower Pliocene beds of the Chesapeake Group, Middle Atlantic Coastal Plain: U.S. Geological Survey Bulletin 1482-D, 61 p.

White, W. A., 1952, Post-Cretaceous faults in Virginia and North Carolina: Geological Society of America Bulletin, v. 63, p. 745–748.

Zoback, M. L., and Zoback, M. D., 1980, State of stress in the conterminous United States: Journal of Geophysical Research, v. 85, no. B11, p. 6113–6156.

71

Depositional environments and paleogeography of the interglacial Flanner Beach Formation, Cape Lookout area, North Carolina

Robert B. Mixon, U.S. Geological Survey, Reston, Virginia 22092

LOCATION

The 2 Stops described here are in the Central Neuse River estuary in the outermost Atlantic Coastal Plain, Craven and Pamlico Counties, North Carolina (Fig. 1).

INTRODUCTION

Abundantly fossiliferous strata exposed in wave-cut cliffs along the Neuse River estuary in the outermost Atlantic Coastal Plain, North Carolina (Fig. 1) record a high stand of the sea during a Pleistocene interglaciation. These exposures provide an unusually good opportunity to study lithologies, sedimentary structures, faunal and floral assemblages, and facies relationships of a typical transgressive-marine sedimentary sequence. Sediments deposited in swamp, bay or sound, restricted-lagoon, barrier-island or barrier-spit, and nearshore-shelf environments have been identified. The lateral and vertical facies successions in the Pleistocene deposits and the associated relict coastal landforms delineate barrier-backbarrier complexes flanking the mouth of an ancestral Neuse River estuary.

As elsewhere along the Atlantic seaboard, the outer Coastal Plain near Cape Lookout is composed of a steplike series of broad terraces that roughly parallel the present Atlantic coastline and that decrease in altitude to the east and south toward the coast (Cooke, 1931; Colquhoun, 1969; Daniels and others, 1972). These terraces are underlain by unconsolidated sandy, silty, clayey, and shelly Pleistocene deposits ranging in thickness from a few ft (m) in updip areas to more than 60 ft (18 m) near the coast. The terrace deposits, in turn, unconformably overlie more consolidated Coastal Plain strata of older Cenozoic and Mesozoic ages.

The sedimentary facies of the terrace deposits, defined by lithology, sedimentary structures, and fossil assemblages, indicate that both the terrace deposits and the associated terrace surfaces formed in barrier, backbarrier, and nearshore-shelf environments during interglacial high stands of the sea. Thus, the more common Pleistocene landforms identified and mapped in the field area include broad, sandy shelf plains, barriers and barrier-spits (some with relict beach ridges and Carolina bays), backbarrier flats, and low, linear scarps which are interpreted as relict marine shorelines.

In the Cape Lookout area, the conspicuous, north-trending, east-facing Grantsboro scarp (Mixon and Pilkey, 1976) divides the outer Coastal Plain lowland into two parts (Fig. 2). East of the Grantsboro scarp, a gently seaward-sloping plain ranges in altitude from about 20 ft (6 m) at the toe of the scarp to sea level at Core Sound; this feature is believed to be an emerged shelf plain of late Pleistocene age (Mixon and Pilkey, 1976, p. 18). West of the Grantsboro scarp are older, more dissected Pleistocene depo-

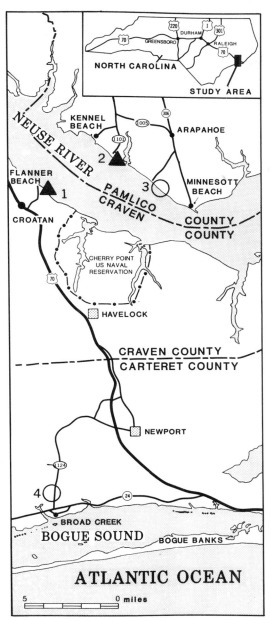

Figure 1. Map showing location of Stops *1* and *2* (triangles) and fossil localities (open circles) in nearshore-shelf facies at *3*, Smith Gut, and *4*, Broad Creek.

sitional surfaces of the relict Arapahoe and Newport barriers and associated backbarrier flats; altitudes of depositional surfaces range from about 25 to 40 ft (7.6 to 12 m; fig. 2). The barrier-backbarrier sequence west of the scarp constitutes the Flanner Beach Formation of middle or late Pleistocene age (DuBar and

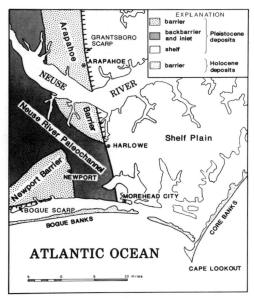

Figure 2. Map showing major paleogeographic features of outer Coastal Plain in the Cape Lookout area, North Carolina.

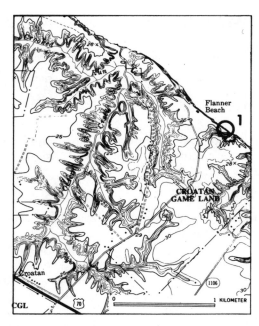

Figure 3. Map showing location and topography of Stop *1* at Flanner Beach, North Carolina (modified from Havelock 7½-minute quadrangle).

Solliday, 1963; DuBar and others, 1974; Mixon and Pilkey, 1976; McCartan and others, 1982). The well-sorted barrier sands and the shelly and muddy fine sands and clay-silts of the backbarrier facies of the Flanner Beach Formation are well exposed at both field trip stops.

SITE DESCRIPTIONS

Stop 1. Transgressive Pleistocene sequence in cliffs at Flanner Beach, Neuse River recreation area, Croatan National Forest, Craven County, North Carolina. The classic Flanner Beach locality is accessible by passenger car from Croatan and Havelock, North Carolina, via U.S. 70 and Route 1107. From the end of the Flanner Beach access road, walk downstream along the beach for about 0.3 mi (0.5 km) to the actively eroding cliff where the better exposures are located (Fig. 3).

The 28-ft-thick (8.5 m) Pleistocene section at this locality (Fig. 4), which is the type locality of the Flanner Beach Formation, records an advance of the sea across a coastal lowland similar to the low-relief, very gently seaward-inclined surface of the present-day North Carolina Coastal Plain. The lowermost part of the section, exposed at the downstream end of the cliff in the interval between low tide and 1 to 2 ft (30 to 61 cm) above the beach, consists of light-olive-gray to bluish-gray silty clay containing abundant rootlets and poorly developed argilans. This stratum is thought to be an organic soil (histosol?) formed at the surface of the emerged lowland. Just upstream from these outcrops, a brownish-gray peaty clay containing rooted cypress stumps as much as 8 ft (2.4 m) in diameter is present between the paleosol and the overlying marginal-marine deposits. In addition to cypress stumps, the peaty clay contains abundant fresh-water

diatoms and some marine or brackish-water species. The stumps and the mixed diatom assemblage suggest deposition near sea level in a fresh-water swamp close enough to the coast to be affected by occasional tidal inflow (Mansfield, 1928, p. 137). The warm-climate pollen assemblage in the peaty clay is dominated by pine, cypress, and oak (Whitehead and Davis, 1969). Both the rooted, olive-gray clay and the brownish-gray, peaty clay are part of a thin fluvial-paludal sequence that fills a shallow drainageway of the ancestral Neuse River (Mixon and Pilkey, 1976; Figs. 4, 7).

The disconformable contact between the cypress-stump beds and the overlying greenish- to bluish-gray, shelly sands of the transgressive phase of the Flanner Beach Formation is also well exposed near the downstream end of the cliff outcrops. A conspicuous lag deposit at the base of the shelly sands is composed of abundant large and small bivalves in a matrix of poorly sorted, fine to coarse sand containing scattered fine pebbles and fragments of wood (Fig. 5). Large, branching, sand- and shell-filled burrows penetrate the underlying organic-rich sediments to a depth of 12 in (30 cm) or more. The basal shell lag contains abundant disarticulated and articulated specimens of *Rangia cuneata,* a bivalve characteristic of very restricted environments such as brackish-water marshes. The *Rangia* occur together with *Dinocardium, Ensis, Anatina,* and other molluscan species that thrive in less restricted environments such as open bays and sounds or the shallow shelf. The mixing of species characteristic of different depositional environments could indicate an initial very rapid rise in sea level and a corresponding rapid landward shift of environments.

Above the basal lag deposits are 9 ft (2.7 m) of intensely bioturbated, muddy, fine to very fine, bluish-gray sand containing

Environment

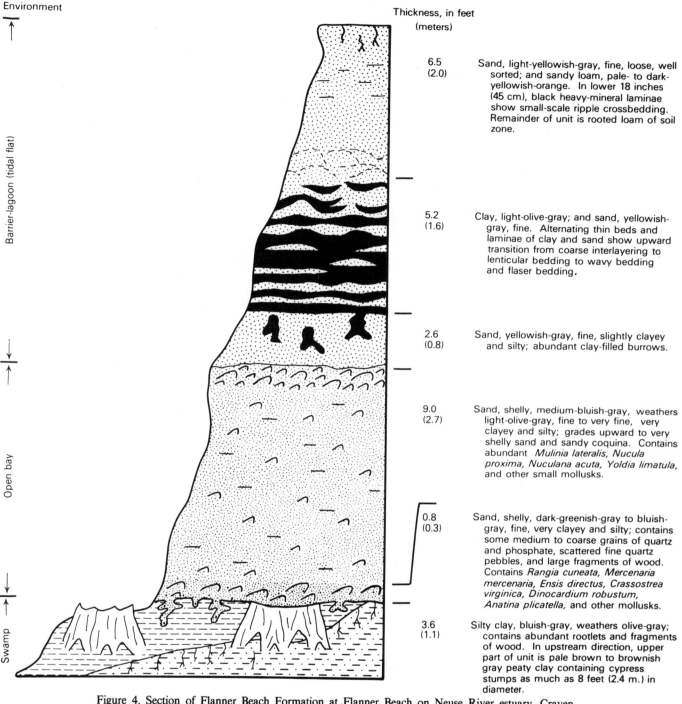

Barrier-lagoon (tidal flat)

Open bay

Swamp

Thickness, in feet
(meters)

6.5
(2.0)

Sand, light-yellowish-gray, fine, loose, well sorted; and sandy loam, pale- to dark-yellowish-orange. In lower 18 inches (45 cm), black heavy-mineral laminae show small-scale ripple crossbedding. Remainder of unit is rooted loam of soil zone.

5.2
(1.6)

Clay, light-olive-gray; and sand, yellowish-gray, fine. Alternating thin beds and laminae of clay and sand show upward transition from coarse interlayering to lenticular bedding to wavy bedding and flaser bedding.

2.6
(0.8)

Sand, yellowish-gray, fine, slightly clayey and silty; abundant clay-filled burrows.

9.0
(2.7)

Sand, shelly, medium-bluish-gray, weathers light-olive-gray, fine to very fine, very clayey and silty; grades upward to very shelly sand and sandy coquina. Contains abundant *Mulinia lateralis, Nucula proxima, Nuculana acuta, Yoldia limatula,* and other small mollusks.

0.8
(0.3)

Sand, shelly, dark-greenish-gray to bluish-gray, fine, very clayey and silty; contains some medium to coarse grains of quartz and phosphate, scattered fine quartz pebbles, and large fragments of wood. Contains *Rangia cuneata, Mercenaria mercenaria, Ensis directus, Crassostrea virginica, Dinocardium robustum, Anatina plicatella,* and other mollusks.

3.6
(1.1)

Silty clay, bluish-gray, weathers olive-gray; contains abundant rootlets and fragments of wood. In upstream direction, upper part of unit is pale brown to brownish gray peaty clay containing cypress stumps as much as 8 feet (2.4 m.) in diameter.

Figure 4. Section of Flanner Beach Formation at Flanner Beach on Neuse River estuary, Craven County, North Carolina. See Table 1 for complete listing of molluscan assemblage.

a fairly diverse assemblage of bivalves and gastropods (Mansfield, 1928; Fallow, 1973; Table 1). Juvenile forms of *Mulinia lateralis* are especially abundant. *Spisula solidissima* and *Donax roemeri protracta* that are characteristic of nearshore-shelf depositional environments and are present in fossil assemblages at the Smith Gut and Broad Creek localities near the eastern and southern edges of the Arapahoe and Newport barriers (see Table 1 and

Figs. 1, 2) are notably absent at this locality and at Stop 2. The overall aspect of the molluscan assemblage at this locality and the lack of bedding in the shelly sand suggest a low-energy, back-barrier depositional environment similar to the outer parts of Pamlico Sound and Core Sound today.

Conformably overlying the muddy and shelly sands are about 14 ft (4.3 m) of fine to very fine, yellowish-gray sand and

Figure 5. Photo showing burrowed contact between rooted, bluish-gray to olive-gray silty clay (below) and shelly fine sand facies of Flanner Beach Formation (above). Dime for scale.

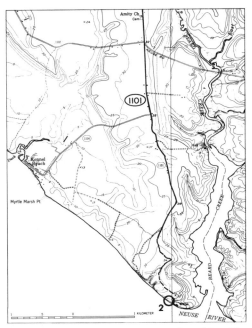

Figure 6. Map showing location and topography of Stop 2 at cliffs on Neuse River estuary west of Beard Creek, Pamlico County, North Carolina (modified from Upper Broad Creek and Arapahoe 7½-minute quadrangles).

TABLE 1. -- Fauna of Pleistocene fossil localities, Cape Lookout area, N.C. (modified from Mixon and Pilkey, 1976)

	Flanner Beach	Beard Creek	Smith Gut	Broad Creek
Gastropoda:				
Polinices duplicatus (Say). . . .	X	X	X	X
Tectonatica pusila (Say).	-	-	X	-
Crepidula fornicata (Linné). . .	-	-	X	-
"Nassa" trivittata (Say).	X	X	X	X
"N." acuta (Say).	X	X	X	X
Ilyanassa obsoleta (Say).	X	X	-	X
Busycon carica (Gmelin)	X	X	-	X
B. canaliculatum (Linné)	X	X	-	-
Oliva sayana Ravenel.	-	-	-	X
Eupleura caudata (Say).	X	-	-	X
Retusa canaliculata (Say)	X	X	X	-
Pelecypoda:				
Nucula proxima Say.	X	X	X	-
Nuculana acuta (Conrad)	X	X	X	-
Yoldia limatula Say	X	X	X	-
Anadara transvera (Say)	X	X	X	X
Cunearca incongrua (Say).	-	-	X	-
Lunarca ovalis (Bruguiere). . . .	X	X	-	X
Noetia ponderosa (Say).	X	X	-	-
Argopecten irradians (Lamarck). .	X	X	-	X
Crassostrea virginica (Gmelin). .	X	-	-	X
Anomia simplex (d'Orbigny). . . .	X	X	X	X
Carditamera floridana (Conrad). .	X	-	-	-
Venericardia tridentata	-	X	-	X
Cyclocardia borealis (Conrad) . .	-	-	-	-
Chama macerophyla Gmelin.	-	-	-	-
Dinocardium robustum (Solander) .	X	X	X	X
Mercenaria mercenaria (Linné). .	X	X	-	X
Chione cancellata (Linné). . . .	-	-	-	X
Gouldia cerina Adams.	-	-	X	-
Gemma gemma (Totten).	-	-	-	-
Transenella cubaniana (d'Orbigny)	-	-	X	-
Anatina (Raeta) plicatella (Lamarck) (=Labiosa canaliculata).	X	X	X	X
Mulinia lateralis (Say)	X	X	X	X
Spisula solidissima (Dillwyn) . .	-	-	X	X
Rangia cuneata (Gray)	X	-	-	X
Tellina alternata Say	X	X	-	-
T. (Angulus) agilis Stimpson. . .	X	X	X	-
Donax roemeri protracta (Conrad)	-	-	-	X
Abra aequalis (Say)	X	X	X	-
Ensis directus Conrad	X	X	X	-
Tagelus plebeius (Solander) . . .	-	-	-	X
Corbula contracta (Say)	-	X	X	X
Cyrtopleura costata (Linné). . .	-	X	-	X
Pandora trilineata Say.	X	X	X	-

light-olive-gray clay that constitute the relatively well-bedded and well-sorted upper part of the Flanner Beach Formation. The basal bed of this sequence is composed of 2.6 ft (0.8 m) of massively bedded fine sand containing abundant irregular, clay-filled burrows. Above the burrowed sand is about 5 ft (1.5 m) of interbedded, thinly bedded to laminated clay and fine sand. The middle part of this unit is dominantly clay containing a few very thin beds of sand and flat, isolated sand lenses (ripples) showing

foreset laminae. In the upper part of the unit, the lenticular-bedded clays and sands show an upward transition into wavy bedded sand and clay, and sand with clay flasers. Above the interbedded sand and clay, in the uppermost part of the cliff, is 6 to 7 ft (1.8 to 2.1 m) of loose, well-sorted fine sand and sandy loam of the soil zone. Near the base of the soil zone, black–heavy mineral laminae show small-scale crossbedding of current ripple origin. The rhythmic sand/clay bedding (including lenticular and flaser bedding), the small-scale current ripples, and the association with the underlying shelly sands of brackish-water origin suggest that the upper part of the Flanner Beach Formation at this

Environment

Thickness, in feet
(meters)

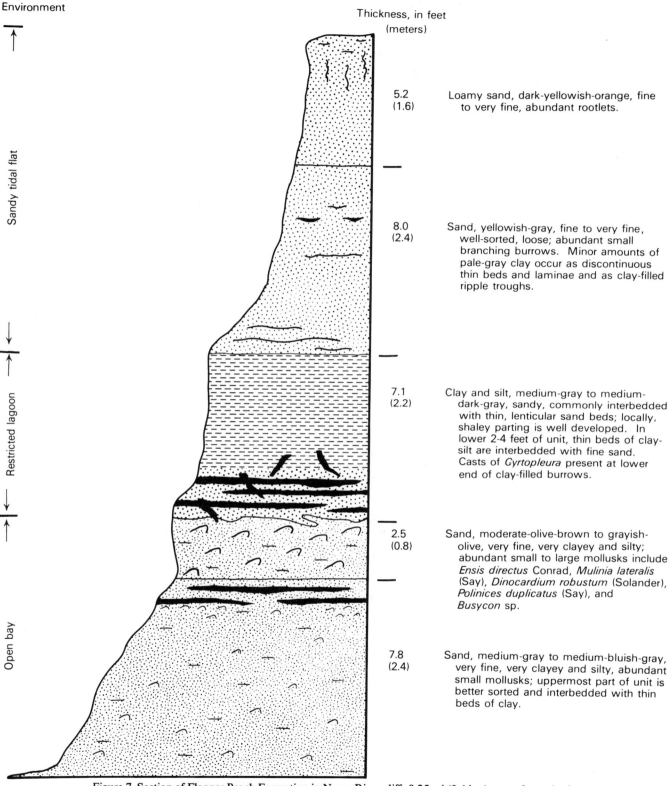

5.2
(1.6)
Loamy sand, dark-yellowish-orange, fine to very fine, abundant rootlets.

8.0
(2.4)
Sand, yellowish-gray, fine to very fine, well-sorted, loose; abundant small branching burrows. Minor amounts of pale-gray clay occur as discontinuous thin beds and laminae and as clay-filled ripple troughs.

7.1
(2.2)
Clay and silt, medium-gray to medium-dark-gray, sandy, commonly interbedded with thin, lenticular sand beds; locally, shaley parting is well developed. In lower 2-4 feet of unit, thin beds of clay-silt are interbedded with fine sand. Casts of *Cyrtopleura* present at lower end of clay-filled burrows.

2.5
(0.8)
Sand, moderate-olive-brown to grayish-olive, very fine, very clayey and silty; abundant small to large mollusks include *Ensis directus* Conrad, *Mulinia lateralis* (Say), *Dinocardium robustum* (Solander), *Polinices duplicatus* (Say), and *Busycon* sp.

7.8
(2.4)
Sand, medium-gray to medium-bluish-gray, very fine, very clayey and silty, abundant small mollusks; uppermost part of unit is better sorted and interbedded with thin beds of clay.

Sandy tidal flat

Restricted lagoon

Open bay

Figure 7. Section of Flanner Beach Formation in Neuse River cliffs 0.25 mi (0.4 km) west of mouth of Beard Creek, Pamlico County, North Carolina. See Table 1 for complete listing of molluscan assemblage.

locality was deposited in the intertidal or subtidal zone behind a barrier-island or barrier-spit system (Reineck and Singh, 1975).

Stop 2. Open-bay, restricted-lagoon, and barrier-sand facies of Flanner Beach Formation, Neuse River cliffs near mouth of Beard Creek, Pamlico County, North Carolina (Fig. 6). Stop 2 is on the north side of the estuary 0.25 mi (0.4 km) west of the mouth of Beard Creek and is accessible by passenger car from Arapahoe, North Carolina, via Routes 1005 and 1101. A dirt road extends from the end of Route 1101 south-southeastward through the woods for 0.9 miles (1.4 km) to the outcrops in cliffs overlooking the estuary (see Upper Broad Creek 7½-minute quadrangle). After rains, a 4 wheel-drive vehicle is needed for access to Stop 2. Stop 2 and other outcrops along the Neuse River estuary may also be visited by boat using a boat ramp at Ketner Heights on Slocum Creek near Havelock or public boat ramps at Kennel Beach and Minnesott Beach on the north side of the Neuse River estuary.

The thick section of the Flanner Beach Formation exposed in the wave-cut cliffs at Beard Creek (Fig. 7) is stratigraphically equivalent to the middle and upper parts of the section at Flanner Beach (see Stop 1). Although these sections are a relatively short distance apart (4 mi or 6.4 km) and are similar in overall aspect, some differences in lithology and sedimentary structures, reflecting differences in energy level and water depths, can be observed.

The beds of shelly fine sand that constitute the lower 10.3 ft (3.1 m) of the Beard Creek section are quite similar to the shelly sands at Flanner Beach and contain essentially the same brackish-water molluscan assemblage (see Table 1, Beard Creek locality). The ostracode assemblage in the Beard Creek beds, which is dominated by *"Haplocytheridae" setipunctata* (Brady) and *"H." bradyi* Stevenson, also suggests a restricted depositional environment (Mixon and Pilkey, 1976).

Overlying the shelly fine sand is a 7-ft (2.1 m)-thick, fining-upward sequence of fine sand and blocky clay-silt. This unit does not contain shell material. However, well-preserved casts of the large angel wing *Cyrtopleura* sp., a clam favoring soft substrates and quiet-water conditions, are found at the lower end of clay-filled burrows. The sparse fauna, the dominantly clay-silt lithology, and intertonguing relationships with the well-sorted, crossbedded, pebbly sands of the Arapahoe barrier that lies to the east, suggest deposition in the restricted, low-energy environment of a coastal lagoon. Eastward thinning of the clay-silt unit can be observed in the cliff exposures at this locality.

The upper part of the section at Beard Creek consists of about 13 ft (4.0 m) of fine, yellowish-gray sand and sandy loam containing isolated clay flasers and a few discontinuous beds of pale-gray clay. These well-sorted sediments appear to be the thin, finer grained, landward part of the Arapahoe barrier-sand deposits.

REFERENCES CITED

Colquhoun, D. J., 1969, Geomorphology of the lower coastal plain of South Carolina: South Carolina Division of Geology, MS-15, p. 1–36.

Cooke, C. W., 1931, Seven coastal terraces in the southeastern states: Washington Academy of Science Journal, v. 21, no. 21, p. 503–513.

Daniels, R. B., Gamble, E. E., Wheeler, W. H., and Holzhey, C. S., 1972, Carolina Geological Society and Atlantic Coastal Plain Geological Association Annual Meetings and Field Trip, Field Trip Guidebook, Oct. 7–8, 1972: 36 p.

DuBar, J. R., and Solliday, J. R., 1963, Stratigraphy of the Neogene deposits, lower Neuse estuary, North Carolina: Southeastern Geology, v. 4, no. 4, p. 213–233.

DuBar, J. R., Solliday, J. R., and Howard, J. F., 1974, Stratigraphy and morphology of Neogene deposits, Neuse River estuary, North Carolina, *in* Oaks, R. Q., Jr., and DuBar, J. R., eds., Post-Miocene stratigraphy, central and southern Atlantic Coastal Plain: Utah State University Press, Logan, p. 102–122.

Fallaw, W. C., 1973, Depositional environments of marine Pleistocene deposits in southeastern North Carolina: Geological Society of America Bulletin, v. 84,

no. 2, p. 257–267.

Mansfield, W. C., 1928, Notes on Pleistocene faunas from Maryland and Virginia and Pliocene and Pleistocene faunas from North Carolina: U.S. Geological Survey Professional Paper 150, p. 129–140.

McCartan, L., Owens, J. P., Blackwelder, B. W., Szabo, B. J., Belknap, D. F., Kriausakul, N., Mitterer, R. M., and Wehmiller, J. F., 1982, Comparison of amino acid racemization geochronometry with lithostratigraphy, biostratigraphy, uranium-series coral dating, and magnetostratigraphy in the Atlantic Coastal Plain of the southeastern United States: Quaternary Research, v. 18, p. 337–359.

Mixon, R. B., and Pilkey, O. H., 1976, Reconnaissance geology of the submerged and emerged Coastal Plain province, Cape Lookout area, North Carolina: U.S. Geological Survey Professional Paper 859, 45 p.

Reineck, H.-E., and Singh, I. B., 1975, Depositional sedimentary environments: New York, Springer-Verlag, 439 p.

Whitehead, D. R., and Davis, J. T., 1969, Pollen analysis of an organic clay from the interglacial Flanner Beach Formation, Craven County, North Carolina: Southeastern Geology, v. 10, p. 149–164.

The geology and physiography of the Orangeburg Scarp

Donald J. Colquhoun, Department of Geology, University of South Carolina, Columbia, South Carolina 29208

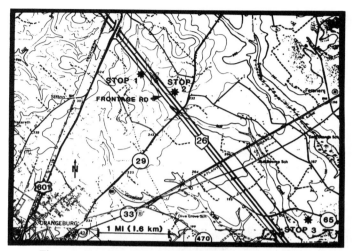

Figure 1. Location of Orangeburg Scarp, and Stops *1* to *3*.

LOCATION

The Orangeburg Scarp is located along and adjacent to I-26, Orangeburg, South Carolina, St. Matthews and Orangeburg 15-minute quadrangles and South Carolina, Department of Highways and Public Transportation, General Highway Map, Orangeburg County (Fig. 1).

INTRODUCTION

The Orangeburg Scarp (Pooser, 1965; Colquhoun, 1965; Citronelle Escarpment of Doering, 1960) is one of the oldest primary geomorphic landforms extant on the South Carolina coastal plain. It extends throughout the region aligned northeast, approximately paralleling the Atlantic Shoreline, and marks the division between the Upper and Middle Coastal Plain Physiographic subprovinces. The scarp is generally expressed in divide (interfluve) areas, by a seaward decrease in general surface elevation from about 350 ft (107 m) to 250 ft (76 m) occurring within 2 mi (3.2 km) from the northwest crest to the southeastern toe. The name is taken from the City of Orangeburg, South Carolina, which lies near the toe of the scarp (Fig. 1).

The scarp has resulted in this area from early–middle Pliocene (Campbell and others, 1975; Blackwelder and Ward, 1979), and possibly toward the south from early Miocene (pre-Seravallian and older) marine erosion (Colquhoun and others, 1983). As is the case for the younger seaward scarps (Surry, Suffolk, etc.), the Orangeburg Scarp results from ice-age sea level eustatic fluctuations. Warm interglacial (interstadial) conditions cause melting of glacial ice and consequent rise in sea level (Flint, 1971, p. 315–339).

Rising sea-level and consequent wave planation erodes the older underlying Tertiary and Cretaceous stratigraphic units whose position is dictated by the Cape Fear Arch and Southeast

Georgia Embayment tectonic trends. These Tertiary and Cretaceous units do not, therefore, parallel the present coast. Thus, toward the northeast in the vicinity of North Carolina, the Upper Coastal Plain consists of the basal Middendorf (Santonian and older) Formation, overlain by the Black Creek (Campanian) Formation. Toward the southwest in central South Carolina the latter is overlain by the Black Mingo Group (Paleocene) and successively toward the Georgia border by middle Eocene (Orangeburg Group), late Eocene, and Oligocene (?) (Barnwell Group). The uppermost stratigraphic unit occurring in the Upper Coastal Plain is the Citronelle Formation of Doering, locally termed the Upland Unit (Colquhoun and others, 1983).

Late Eocene, Oligocene, and early–middle Miocene strata lie seaward of the Orangeburg Scarp, underlying Pliocene and Pleistocene strata in the Middle and Lower Coastal Plains. The Eocene-Miocene units are widely separated from the scarp in central South Carolina, but more closely approach it toward the southwest. Within the Middle Coastal Plain the Duplin Formation overlies older middle Eocene units and is in turn overlain by alluvial fan sediments in the vicinity of the area shown in Figure 1. These alluvial sediments were mapped as Coharie Formation by Crooke (1936). As is the case for most Plio-Pleistocene units, these alluvial sediments are not universally present seaward of the scarp.

The Orangeburg Scarp is also delineated by contrasting soil groups on U.S. Department of Agriculture Soil Maps in areas where erosion has obliterated all topographic evidence. Part of the difference in soil series is probably related to age and climatic history of affected sediments—the land surface lying to the northwest being at least 4.3 Ma and probably much older, while that to the southeast being younger, probably between 4.3 Ma or older (the estimated age of the Duplin Formation) and 1.8 Ma (the estimated oldest age of the Wicomico Formation), which may be Waccamaw Formation equivalent or younger than the Waccamaw. The differing ages of the Upper and Middle Coastal Plain is also expressed in topography—the Upper Coastal Plain being more mature with well developed valleys and divides, and the Middle Coastal Plain generally developed as valleys with flat, extensive interfluves.

In summary, the Orangeburg Scarp is one of the oldest demonstrated primary geomorphic features extant on the Coastal Plain. It was formed, probably, by several episodes of erosion caused by eustatic sea level change and was last altered by shoreline processes in the Pliocene (Zanclian). The scarp is younger than the Upland Unit (pre-Yorktown) which overlies middle Eocene locally, late Eocene and possibly early Oligocene to the southwest. A significant drop in sea level occurred between the deposition of the Upland unit and the subsequent Duplin (Yorktown) submergence, and several Pliocene oscillations may have followed. The Citronelle and the alluvial fan units (Coharie

Formation) which overlie the Duplin Formation were emplaced in semi-arid climates.

SITE DESCRIPTIONS

Stop 1. Located on south side of I-26 on frontage road 1.3 mi (2.1 km) southeast of I-26 and U.S. 601 intersection (Fig. 1). Access via U.S. 601. Exposed on the south side of the road in a drainage ditch is the Upland Unit. The lithology consists of quartz and lithic pebbles and cobbles embedded in a clayey-sand matrix, which occasionally exhibits cross-bedding. The clay matrix is kaolinite. Angular kaolinite clasts, 2–4 mm in diameter, are also present indicating probable feldspar origin.

The unit was originally a feldspathic sand with cobbles; it compares favorably with the present semi-arid fluvial deposits in similar provenance settings, and not with existing Holocene fluvial deposits of the humid Southeastern United States. A comparison of the lithology expressed at Stop 1 with present Pleistocene fluvial sediments in a similar setting is available on the north side of U.S. 601, 1 mi (1.6 km) north of the Congaree River crossing. These borrow pits, 20 mi (32.2 km) northeast of Stop 1, contain significantly less kaolin, feldspar and cobbles.

A silicified log, thought to be cypress, 14 in (35.6 cm) in diameter, not in growth position, was collected at Stop 1. This is the only known fossil occurrence from the Upland Unit in the region. The upper delta plain fluvial Citronelle extends to elevations in excess of 320 ft (97 m) locally, and has been traced in auger holes for approximately 7 mi (11 km) with the trend of its channel aligned north-northeast. The channel terminates on the Escarpment face at approximately +265 ft (81 m) above mean sea level. The Citronelle lies upon green clayey sands of middle Eocene (Claiborne) age (Bernat, 1963; Pooser, 1965; Colquhoun, 1965).

Stop 2. Located on the north side of I-26. At end of frontage road walk 0.2 mi (0.3 km) northeast of I-26 and South Carolina 33 intersection (Fig. 1). Access via South Carolina 33. Exposed along the north wall of an old borrow pit are alluvial fan deposits emplaced at the toe of the escarpment. Basal coarse

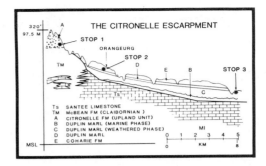

Figure 2. Stratigraphic cross-section along I-26.

gravel (pebble) sediments grade upwards to fine clayey sand. Regionally, the unit thins seaward of the scarp, the grain-size decreases, and the sorting factor becomes poorer (Bernat, 1963); all of which are characteristics of alluvial fan deposits in existing semi-arid regions. The alluvial sand unit lies locally on weathered and unweathered back-barrier marsh clays, but seaward upon well-sorted, fine-grained barrier island sands and then sandy, clayey shallow shelf sands and marls of the Duplin Formation (Yorktown of Blackwelder and Ward, 1979).

Stop 3. Located on the north side of Bull Swamp at dug pond about 330 ft (100 m) to right of County Road S-38-65, and 4.5 mi (7.2 km) south of intersection South Carolina 33 and County Road S-38-65 (Fig. 1). (Contact landowner in second house north of Bull Swamp on right for entry.) Spoil from a dug farm pond includes blue-grey Duplin (Yorktown)Formation with extensive macro and microfossils, listed in Pooser (1965) and Colquhoun (1965). A more extensive listing of Duplin macrofossils is found in Campbell and others (1975). Local auger drilling indicates the Duplin–Orangeburg Group (Siple and Pooser, 1975) contact drops with decreasing gradient from the toe of the escarpment to about 140 ft (43 m) above sea level in the subsurface. The Duplin Formation extends as high as about 164 ft (50 m) above sea level near the escarpment. The Duplin usually is weathered and completely underlies the alluvial fan unit, but four auger holes out of about fifty locally drilled encountered fresh material.

The relationship of the three stops to the local stratigraphy is indicated in Figure 2.

REFERENCES CITED

Bernat, P. E., 1963, Heavy mineral distribution in sediments in the vicinity of the Citronelle Escarpment between Orangeburg and St. Matthews South Carolina: [M.S. thesis], University of South Carolina, Columbia, 50 p.

Blackwelder, B. W., and Ward, L. W., 1979, Stratigraphic revision of the Pliocene deposits of North and South Carolina: South Carolina Geological Survey, Geologic Notes, v. 23, p. 33–43.

Campbell, L., Campbell, S., Colquhoun, D., and Ernissee, J., 1975, Plio-Pleistocene faunas of the central Carolina Coastal Plain: South Carolina Geological Survey, Geologic Notes, v. 19, p. 51–124.

Colquhoun, D. J., 1965, Terrace sediment complexes in central South Carolina: University of South Carolina, Department of Geology, Columbia, South Carolina, 62 p.

Colquhoun, D. J., Woollen, I. D., Van Nieuwenhuise, D. S., Padgett, G. G., Oldham, R. W., Boylan, D. C., Bishop, J. W., and Howell, P. D., 1983, Surface and subsurface stratigraphy, structure and aquifers of the South

Carolina Coastal Plain: South Carolina Department of Health and Environmental Control Report, Columbia, South Carolina, 78 p.

Cooke, C. W., 1936, Geology of the Coastal Plain of South Carolina: Geological Survey Bulletin 867, p. 132–133.

Doering, J. A., 1960, Quaternary surface formations of southern part of Atlantic Coastal Plain: Journal of Geology, v. 68, p. 182–202.

Flint, R. F., 1971, Glacial and Quaternary Geology: John Wiley and Sons, Inc., New York, New York, p. 315–339.

Pooser, W. K., 1965, Biostratigraphy of Cenozoic Ostracoda from South Carolina: Arthopoda, Art. 8, University of Kansas Publications, Lawrence, Kansas, 80 p.

Siple, G. E., and Pooser, W. K., 1975, Proposal of the name Orangeburg Group for outcropping beds of Eocene age in Orangeburg County and vicinity, South Carolina, *in* Cohee, G. P., and Wright, W. R., Changes in stratigraphic nomenclature by the U.S. Geological Survey, 1973: U.S. Geological Survey Bulletin 1395-A, p. A-55.

The Chandler Bridge Formation (Upper Oligocene) of the Charleston, South Carolina, region

Robert E. Weems, U.S. Geological Survey, Reston, Virginia 22092
Albert E. Sanders, The Charleston Museum, Charleston, South Carolina 29403

LOCATION

The Chandler Bridge Formation is exposed near the north-central border of the Ladson 7½-minute quadrangle (Fig. 1), 15 mi (24 km) northwest of Charleston, South Carolina. From I-26, take U.S. 78 exit (eastbound), proceed 1.9 mi (3.1 km) to U.S. 52 and turn left (north) on 52. Go 1.8 mi (2.9 km) to traffic light at intersection of U.S. 52 and U.S. 178, bear left onto 178 at traffic light (45 degrees left to northwest) and go 0.6 mi (1.0 km) to a pair of large overhead power lines. On far (northwest) side of power lines turn left (southwest) onto road paralleling power lines, go 0.7 mi (1.1 km) to ditch on left (southeast).

INTRODUCTION

The Oligocene is the most poorly represented epoch of the Cenozoic in the exposed Atlantic Coastal Plain north of Florida.

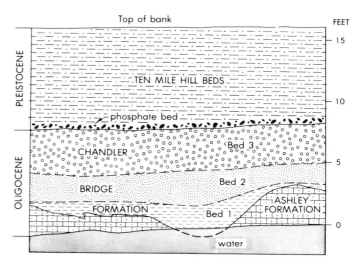

Figure 2. Detailed column of part of the ditch bank at the Chandler Bridge field locality.

The Charleston area, between the East Georgia Embayment to the southwest and the Cape Fear Arch to the northeast, is very unusual because at least two upper Oligocene stratigraphic units are present, the Ashley Formation of the Cooper Group and the Chandler Bridge Formation. A possible third Oligocene unit, the Edisto Formation, lies near the Oligocene-Miocene boundary and its proper epoch assignment is in dispute. Both the Ashley and Chandler Bridge are exposed at this site. A thin but visually prominent lag deposit of "Ashley River rock phosphate," extensively mined in the Charleston area in the late nineteenth and early twentieth centuries (Malde, 1959), is on top of the Chandler Bridge at this site.

SITE DESCRIPTION

Three discrete stratigraphic units are exposed along the banks of this drainage ditch (see Fig. 2). About midway down the vertical section is a layer of rounded to irregularly shaped rocks which represents the base of the Pleistocene part of the section. This layer is an example of the "Ashley phosphate rock" that was formerly mined in this region. At this locality, the Pleistocene overburden was too thick and the phosphate bed too thin to make mining economical. In some of the mined localities, individual blocks of rock phosphate weighed as much as 2,000 pounds (906 kg) (Leidy, 1877). Malde (1959) determined that the Charleston area phosphate rocks are predominantly composed of carbonate-fluorapatite.

The Pleistocene sands overlying the phosphate bed are

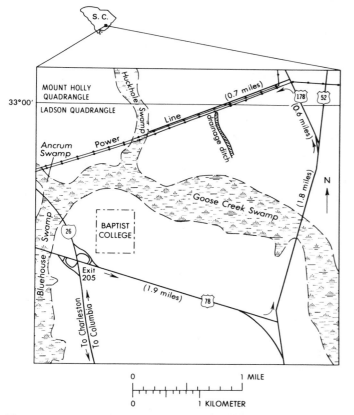

Figure 1. Location of outcrop of the Chandler Bridge Formation in drainage ditch near the north border of the Ladson 7½-minute quadrangle, Berkeley County, South Carolina.

medium- to fine-grained, clayey, and moderately deeply weathered with a soil profile dominated by orange to reddish-orange hues. Regional relationships (McCartan and others, 1984) suggest that this unit accumulated in an estuarine to freshwater environment located northwest (inland) from the Cainhoy barrier system of Colquhoun (1974). These beds (labelled Q3 in McCartan and others, 1984) have not been formally named, but they have been recently described as the Ten Mile Hill beds (see Weems and Lemon, 1984a, 1984b). The age of these beds (Cronin and others, 1981; McCartan and others, 1982; Szabo, 1985) is estimated to be about 200,000–240,000 years, which would make them equivalent to Stage 7 of Shackleton and Opdyke (1973). Their thickness at this locality is about 9 ft (2.7 m).

Below the phosphate rock bed is the Chandler Bridge Formation (Sanders and others, 1982). At this locality, the Chandler Bridge is about 5 to 8 ft (1.5 to 2.4 m) thick. At its base, the unit consists of medium- to dark-gray clay (Bed 1), which grades upward rapidly through light-gray, silty, fine-grained, quartz-phosphate sand containing abundant burrows (Bed 2) to a light-gray, medium-grained, massive, quartz-phosphate sand containing numerous lumps of phosphate and numerous flat, discoidal quartz pebbles (Bed 3). The basal clay bed (Bed 1), which at other localities is more typically a medium-yellowish-brown color, lies on a very irregular surface (Fig. 2). The basal contact shows numerous large pits and depressions, many of which are filled with carbonaceous clay. We suspect that these pits represent stump holes where trees were once rooted into the underlying unit. If this is true, then the largely unfossiliferous basal transgressive clay probably represents an estuarine to lagoonal environment of accumulation. Because the immediately overlying layer (Bed 2) is extensively bioturbated and contains porpoise and sea turtle skeletons, it must have accumulated in an environment which had water at least approaching normal salinity. This could represent an open marine shelf environment below wave base, but the stratigraphic position of this unit between an underlying lagoonal to estuarine unit and an overlying beach-front unit suggests that an open-bay environment of accumulation is more likely. This requires the fewest oscillations of sea level to bring together the observed stratigraphic sequence. The highest layer (Bed 3) is coarser grained than either of the lower two layers and commonly contains rounded to subangular 0.5 to 2 in (1 to 5 cm) phosphate pebbles and flat, polished, discoidal-shaped quartz pebbles 1 to 2 cm in longest diameter. At some localities, this bed contains porpoise skeletons which rest in the sediment parallel to each other. This mode of occurrence is highly reminiscent of the carcass distribution patterns that result from the stranding of schools of porpoises along modern coastlines (Sanders, 1980). The bed is too bioturbated to preserve bedding features, but all of the preserved characteristics of this unit suggest that it accumulated along a beach-face in shallow marine water. Therefore, the sediments of the Chandler Bridge Formation seem to record a shallow-marine transgression over a forest that was growing on the surface of the underlying, open-marine Ashley Formation. Only 8 ft (2.4 m) of section is preserved, but quite possibly

sediments which formed in deeper, more open-marine waters were once higher in this section but were stripped away by later erosion. The fossil cetacean assemblage is most similar to the cetacean faunas found in marine sediments of the upper Oligocene (middle Chattian) of Europe (Sanders, 1980; Sanders and others 1982).

The basal 2 ft (60 cm) of the ditch section is composed of the Ashley Formation (Ward and others, 1979; Weems and Lemon, 1984a, 1984b). This unit is a fine-grained quartzose and phosphatic calcarenite, containing abundant foraminifer tests and ostracode carapaces, which accumulated in a mid-shelf marine environment (Gohn and others, 1977; Hazel and others, 1977). Locally it may be more than 200 ft (65 m) thick. The Ashley is assigned to planktonic foraminifer biozone P20 and calcareous nannoplankton biozone NP24 of the upper Oligocene (lower Chattian; Hazel and others, 1977). Although both the Ashley and Chandler Bridge are late Oligocene, the cetacean faunas of the two units suggest that there is a significant age difference between them. All but one of the species, and all but two of the genera of the cetaceans in each unit are different from those in the other unit, although the Chandler Bridge species were readily derivable from the Ashley species. The sharp, undulating, probably once-forested contact between the two units also suggests that an interval of erosion and weathering could have taken place between the end of Ashley deposition and the beginning of Chandler Bridge deposition.

This locality is typical of the Chandler Bridge in every respect but one—it is exceptionally thick. The thickest known section of the Chandler Bridge (in auger hole) is 18 ft (5.5 m), but the most typical thickness is 1 to 2 ft (30 to 60 cm). This is because the relatively soft Chandler Bridge is usually preserved only low areas on the surface of the underlying Ashley Formation. The Ashley is quite resistant to erosion and tends to become case-hardened by recrystallization when exposed to freshwater. The repeated oceanic transgressions across the Charleston region in middle and late Cenozoic time have all tended to scour soft sediments down to the surface of the Ashley and then lay down a new blanket of sediment directly on the Ashley surface. During each regressive interval, when sea level was much lower than now, stream systems locally cut channels into the upper surface of the Ashley. Later, the next transgression filled these channels with sediment. Because these filled valleys are topographically low, later transgressions generally have not scoured all of the initially deposited sediments out of them. Thus middle to upper Cenozoic units around the Charleston region are most often preserved locally as thin patches in low areas on the surface of the Ashley and they are usually distributed in a mosaic pattern rather than in stacks, illustrating the principle of superposition in its simplest case. The present distribution of the Chandler Bridge is shown in Figure 3. Detailed stratigraphic analysis throughout the Charleston region indicates that many Miocene, Pliocene, and early to middle Pleistocene regional transgressions took place across the greater Charleston area. Each of these transgressions probably resulted in sediment deposition above the Chandler

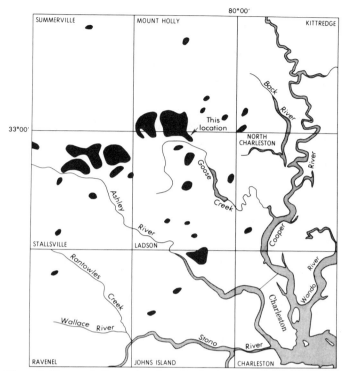

Figure 3. Known shallow subsurface distribution of the Chandler Bridge Formation (*black pattern*) in the region around Charleston, South Carolina. Arrow points to approximate location of the site discussed here. The 7½-minute quadrangles are identified and outlined.

Bridge at this site, but all of these other units were eroded or scoured off before the Pleistocene Ten Mile Hill beds were deposited directly on the Chandler Bridge about 200,000 years ago. Along this ditch, approximately 0.5 mi (1 km) southeast of the power line, the results of a later generation of this process can be seen. The land slopes down prominently (by outer Coastal Plain standards) and the visitor drops down from the lower (+ 35 ft [10.7 m] elevation) Talbot terrace of Colquhoun (1974), which is underlain by sediments constituting the Ten Mile Hill beds, onto his (+15 ft [4.6 m] elevation) Princess Anne terrace, which is underlain by the younger portions of the Wando Formation (McCartan and others 1980) equivalent to substages 5c and 5a of Shackleton and Opdyke (1973). On the basis of uranium/thorium ratios in contained corals, this part of the Wando has been dated elsewhere to be no older than about 85 thousand years (Cronin and others, 1981; McCartan and others, 1982; Szabo, 1985). Its relatively young age also is demonstrated in part by its poorly developed soil profile and its brownish to yellowish hues. At the southeast end of this ditch, the Ten Mile Hill beds have been scoured out and partly cannibalized to form the younger Wando beds. The phosphate lag bed at the base of the Wando is obviously full of reworked material, although a little of the bone and nodular phosphate were probably added in Wando time. We believe that this process of repeated scouring and cannibalizing of older material was very important in concentrating the phosphate rock into economically significant deposits.

Rock and nodular phosphate was mined extensively from the late nineteenth to the early twentieth century, and an excellent summary of the literature and production can be found in Malde (1959). The remaining phosphate deposits presently are not considered to be economically recoverable (Force and others, 1978). Although Sloan (1908) thought that the Oligocene and/or Miocene Edisto Formation was the source of the rock phosphate and Cooke (1936) ascribed its source to the "Hawthorn Formation," Malde considered the Ashley Formation to be the source of the mined phosphate material. The field relations at this locality show that rock and nodular phosphate are found directly above the Chandler Bridge Formation, which was unknown at the time of Malde's work, but does not appear as low in the section as the Ashley. Moreover, no case has been reported where rock phosphate still is imbedded within the top of the Ashley Formation where it had been phosphatized. All occurrences of rock phosphate directly on the Ashley have been erosionally isolated lag blocks found detached from the underlying Ashley sediments. Therefore the "Ashley River rock phosphate" must have been derived either from higher levels of the Chandler Bridge Formation, now eroded away, or from some other unit younger than the Chandler Bridge which is now entirely gone from this site. Nodular phosphate has been found at the contacts between most (but not all) of the stratigraphic units in the Charleston area, so the origins of this type of phosphate are probably diverse.

The Chandler Bridge is also notable because it contains abundant remains of vertebrates. Sharks, rays, teleost fish, turtles, crocodiles, birds, cetaceans, and sea cows are represented abundantly in these deposits. The cetacean fauna especially has made an important contribution to the elucidation of the evolutionary steps involved in transforming the primitive Eocene archeocetes into the fully marine odontocetes of the Miocene (Sanders, 1980). Isolated bones and teeth are abundant at this locality, but articulated skeletons are more rarely found. If visitors wish to see cleaned and prepared fossil specimens from this unit, it is possible to see some material on display at the Charlestown Museum at 360 Meeting Street in downtown Charleston. Only a small part of this material is on exhibit, but if arrangements are made well in advance with A. E. Sanders, more complete materials can be made available for perusal by groups of scientists, students, or interested individuals.

REFERENCES CITED

Colquhoun, D. J., 1974, Cyclic surficial stratigraphic units of the middle and lower coastal plains, central South Carolina *in* Oaks, R. Q., Jr., and DuBar, J. R., eds., Post-Miocene stratigraphy central and southern Atlantic Coastal Plain: Utah State University Press, Logan, p. 179–190.

Cooke, C. W., 1936, The geology of the coastal plain of South Carolina: U.S. Geological Survey Bulletin 867, p. 1–196.

Cronin, T. M., Szabo, B. J., Ager, T. A., Hazel, J. E., and Owens, J. P., 1981, Quaternary climates and sea levels of the U.S. Atlantic Coastal Plain: Science, v. 211, no. 4479, p. 233–240.

Force, E. R., Gohn, G. S., Force, L. M., and Higgins, B. B., 1978, Uranium and phosphate resources in the Cooper Formation of the Charleston region,

South Carolina: Geologic Notes, South Carolina Geological Survey, v. 22, no. 1, p. 17–31.

Gohn, G. S., Higgins, B. B., Smith, C. C., and Owens, J. P., 1977, Lithostratigraphy of the deep corehole (Clubhouse Crossroads corehole 1) near Charleston, South Carolina, *in* Rankin, D. W., ed., Studies related to the Charleston, South Carolina, earthquake of 1886; a preliminary report: U.S. Geological Survey Professional Paper 1028-E, p. 59–70.

Hazel, J. E., Bybell, L. M., Christopher, R. A., Frederiksen, N. O., May, F. E., McLean, D. M., Poore, R. Z., Smith, C. C., Sohl, N. F., Valentine, P. C., and Witmer, R. J., 1977, Biostratigraphy of the deep corehole (Clubhouse Crossroads corehole 1) near Charleston, South Carolina, *in* Rankin, D. W., ed., Studies related to the Charleston, South Carolina, earthquake of 1886; a preliminary report: U.S. Geological Survey Professional Paper 1028-F, p. 71–89.

Leidy, J. P., 1877, Description of vertebrate remains, chiefly from the phosphate beds of South Carolina: Journal of the Academy of Natural Sciences, Philadelphia, series 2, v. 8, pt. 3, article 9, p. 209–262.

Malde, H. E., 1959, Geology of the Charleston phosphate area, South Carolina: U.S. Geological Survey Bulletin 1079, 105 p.

McCartan, Lucy, Weems, R. E., and Lemon, E. M., Jr., 1980, The Wando Formation (Upper Pleistocene) in the Charleston, S.C., area, *in* Contributions to Stratigraphy: U.S. Geologic Survey Bulletin 1502-A, p. A110–A116.

McCartan, Lucy, Owens, J. P., Blackwelder, B. W., Szabo, B. J., Belknap, D. F., Kriausakui, N., Mitterer, R. H., and Wehmiller, J. F., 1982, Comparison of amino acid racemization geochronometry with lithostratigraphy, biostratigraphy, uranium-series coral dating, and magnetostratigraphy in the Atlantic Coastal Plain of the southeastern United States: Quaternary Research, v. 18, p. 337–359.

McCartan, Lucy, Lemon, E. M., Jr., and Weems, R. E., 1984, Geologic map of the area between Charleston and Orangeburg, South Carolina: U.S. Geological Survey Miscellaneous Investigations Series Map I-1472.

Sanders, A. E., 1980, Excavation of Oligocene marine fossil beds near Charleston, South Carolina: National Geographic Society Research Report, v. 12, p. 601–621.

Sanders, A. E., Weems, R. E., and Lemon, E. M., Jr., 1982, Chandler Bridge Formation—A new Oligocene stratigraphic unit in the lower Coastal Plain of South Carolina, *in* Contributions to Stratigraphy: U.S. Geological Survey Contributions to Stratigraphy Bulletin 1529-H, p. H105–H124.

Shackleton, N. J., and Opdyke, N. D., 1973, Oxygen isotope and paleomagnetic stratigraphy of Equatorial Pacific Core V28-238: Quaternary Research, v. 3, p. 39–55.

Sloan, Earle, 1908, Catalogue of the mineral localities of South Carolina: South Carolina Geological Survey Bulletin 2 (reprinted 1958), series 4, 505 p.

Szabo, B. J., 1985, Uranium-series dating of fossil corals from marine sediments of southeastern United States Atlantic Coastal Plain: Geological Society of America Bulletin, v. 96, p. 398–406.

Ward, L. W., Blackwelder, B. W., Gohn, G. S., and Poore, R. Z., 1979, Stratigraphic revision of Eocene, Oligocene, and lower Miocene formations of South Carolina: Geologic Notes, v. 23, no. 1, p. 2–32.

Weems, R. E., and Lemon, E. M., Jr., 1984a, Geologic Map of the Mount Holly quadrangle, Berkeley and Charleston Counties, South Carolina: U.S. Geological Survey Geological Quadrangle Map GQ-1579.

——1984b, Geologic Map of the Stallsville quadrangle, Dorchester and Charleston Counties, South Carolina: U.S. Geological Survey Geologic Quadrangle Map GQ-1581.

Stratigraphy and depositional environments of the kaolin belt in middle Georgia

Aubrey L. Long, R. L. Quintus-Bosz, and Ed L. Schrader*, J. M. Huber Corporation, Route 4, Huber, Macon, Georgia 31298

LOCATION

Two inactive kaolin mines about 6 mi (10 km) northeast of Huber, Georgia (Fig. 1), display the stratigraphy and depositional environments of the kaolin belt in middle Georgia. The northwest highwall of Huber's #26 mine and northeast highwall of Huber's #30 mine are about 2.0 mi (3.2 km) and 2.7 mi (4.3 km), respectively, northeast of Huber's mine shop, at latitude 32° 45′ and

Figure 2. Location of northwest highwall of Huber's *#26* mine and the northeast highwall of Huber's *#30* mine. In the southwest corner of the Dry Branch 7½-minute quadrangle and the northwest corner of the Marion 7½-minute quadrangle, Twiggs County, Georgia.

longitude 83°28′, in Twiggs County, Georgia (Fig. 2). The mines are in the southwest corner of the Dry Branch quadrangle and the northwest corner of the Marion quadrangle. These mines are located in the J. M. Huber Corporation active mine complex, which is approximately 11 mi (17.7 km) southeast of Macon, Georgia. Permission to enter should be obtained from the exploration office at Huber, Georgia, which is 4.0 mi (6.4 km) west of the mine shop, on the Warner Robins NE quadrangle.

INTRODUCTION

The kaolin-bearing sediments of the inner Coastal Plain in the southeastern United States constitute the world's leading source of high quality kaolin. Deposition of kaolinitic clays, especially during the Upper Cretaceous and Middle Eocene appears to have been widespread along a more than 300 mi (500 km) long belt, extending from Eufaula, Alabama, to Lexington County, west-central South Carolina (Patterson and Murray, 1975). However, commercial deposits of exceptionally pure kaolin that can be beneficiated to meet the strict quality specifications of the paper industry occur only between Macon, Georgia, and Aiken, South Carolina (Fig. 1), an area that has been referred to as the Georgia–South Carolina kaolin belt (Patterson and Murray, 1975; Schrader and others, 1983).

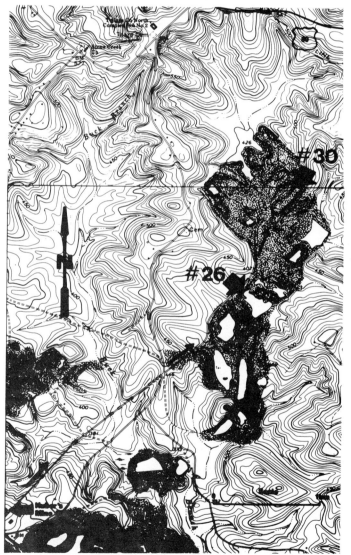

Figure 1. Location of Huber's mining operations in Twiggs County, Georgia, south of Macon.

*Present address: United Catalysts, Inc., P.O. Box 32370, Louisville, Kentucky 40232.

The area's generally low relief and dense vegetation make natural and roadside outcrops scarce and they usually provide only limited information. The kaolin mines, therefore, offer the best opportunity to study the early Coastal Plain sediments in the upper Coastal Plain. In the western part of the kaolin belt, especially near Huber, Georgia, where most kaolin is produced from Upper Cretaceous deposits, 50 to 100-ft-high (15 to 30 m) minewalls are not uncommon.

Starting at a layer of massive, Upper Cretaceous kaolin, the two suggested stops (Fig. 2) combine to expose an almost 100-ft (30 m) thick sequence of early Tertiary sediments. These sections clearly illustrate the diverse lithologies and typically complex sedimentary character generated by shifting environments during the initial stages of deposition along the inner Coastal Plain.

The pre-Jurassic igneous and metamorphic rocks of the Piedmont Province, which are exposed immediately north of the Fall Line (Fig. 1), constitute the basement of the inner Coastal Plain. Intense pre–Upper Cretaceous weathering and erosion of the Piedmont produced an uneven, southeastward sloping surface. Contour maps of the basement locally show a well developed network of southward-flowing streams (Herrick and Vorhis, 1963). Near Huber, Georgia, the basement dips to the southeast at a rate of approximately 58 ft per mi (11 m/km; Buie, 1980). Weathering of the upland Piedmont crystalline rocks, especially those of granitic composition, is considered by most investigators as the principal source of the kaolin on the Coastal Plain (Hurst and others, 1979).

Within the Georgia–South Carolina kaolin belt, a sequence of Upper Cretaceous beds constitute the oldest sedimentary rocks of the Coastal Plain. Forming a wedge-shaped unit, these beds overlie the pre-Jurassic basement with a major unconformity. Locally overlapped by younger sediments, the Upper Cretaceous is poorly exposed in the Huber area. The top of the Cretaceous was determined from drill data (Patterson and Buie, 1974) to dip to the southeast at a rate of about 35 ft per mi (7 m/km). Truncated updip by erosion, this unit is only a few feet (meters) thick near the Fall Line.

The Cretaceous units are typically made up of light colored crossbedded sands, clayey sands, and clay beds, considered to have been deposited in an upper deltaic environment. Although minor clay zones can be found randomly distributed throughout the Cretaceous, the larger and economically important kaolin deposits only occur in the uppermost part of the section. A pronounced undulating surface at the top of these kaolin bodies marks the contact with the overlying Tertiary sediments.

The Paleocene to Middle Eocene Huber Formation (Buie 1978) overlies unconformably the Cretaceous and, in places, also overlaps the crystalline basement. Like the underlying strata, the Huber sediments form a wedge-shaped unit that increases in thickness downdip, from 2 ft (0.6 m) near the Fall Line to 180 ft (55 m) near Tarversville, Georgia (Buie, 1978). From the abundant drill data available for this area, Buie and others (1979) determined that the top of the Huber Formation dips to the southeast at a rate of 15 ft per mi (3 m/km).

Much like the Cretaceous, the Huber is made up of mostly light colored, intricately crossbedded sands, clayey sands, sandy clays, and lenticular clay beds. Based on the sedimentary structures and the identification of brackish to marine fossils, an estuarine to marginal marine environment has been proposed for the Huber Formation (Tschudy and Patterson, 1975; Horstmann, 1983).

Most of the larger deposits of Tertiary kaolin are found distributed within a relatively narrow vertical interval at the top of the Huber Formation. Unlike the Cretaceous clay, the Tertiary kaolin deposits appear to have been subjected to extensive bioturbation, as evidenced by the presence of locally numerous, kaolin-filled tubular burrows (bryozoa?). The Tertiary kaolin is further distinguished from its Cretaceous equivalent by its higher content of dark heavy minerals, the presence of pisolitic textures, and irregular versus subconchoidal fracture (Buie, 1978).

An erosional unconformity also marks the contact between the Huber strata and the overlying marine sediments of the Late Eocene (Jackson) Barnwell Group. The sharply contrasting lithologies, fossil contents, stratification, and coloration of the sediments usually make this contact very conspicuous (Figs. 3, 4, and 5). The abundant fossil fauna of the Late Eocene sediments are characteristic of a littoral to neritic marine environment (Hurst and others, 1979).

In the middle Georgia area, the Barnwell Group is repre-

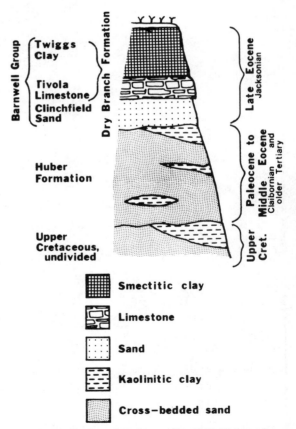

Figure 3. Generalized stratigraphic section of the Huber mine area.

Figure 4. Northeast highwall in Huber's mine #30 showing Huber Formation (Tc) and overlying units of the Barnwell Group.

sented by the Clinchfield Sand, Dry Branch Formation, and Tobacco Road Sand. The Tobacco Road Sand is not present in the two mine stops and only the Twiggs Clay and Tivola Limestone members of the Dry Branch Formation are represented. The fossil-rich Clinchfield is made up of extensively bioturbated, well sorted, medium-grained, calcareous sand. Vertically and laterally, these sands gradate into sandy clays and clays of the overlying Twiggs Clay. The Twiggs Clay forms large-scale lenticular deposits of dark gray, poorly to thin bedded clay and sandy clay. While kaolin is the principal and often only clay mineral in the Upper Cretaceous to early Tertiary sediments, montmorillonite is the dominant clay mineral in the Late Eocene deposits (Huddlestun and Hetrick, 1979). The maximum reported thickness for these sediments is 100 ft (30 m). The Tivola Limestone is made up of thin beds of light colored, loosely cemented, fossiliferous hash (bryozoa, pecten, echinoid). This unit usually occurs interfingered with gray-green sandy clay and silt of the lower Twiggs Clay.

SITE DESCRIPTION

The suggested stops, Huber mines #26 and #30 (Fig. 2), are both inactive kaolin mines and are located within the area designated by Buie (1978) as the type-locality for the Huber Formation. Together, the sections at both stops expose a relatively thick sequence of Upper Cretaceous to Late Eocene sediments, that typify the general stratigraphy, lithology, and depositional environments of the oldest Coastal Plain sediments in the middle Georgia area. Descending stratigraphically through these sections (Fig. 3), the units discussed below can be clearly distinguished.

Twiggs Clay

The highwall at Huber's #30 mine (Fig. 4) offers the best and most complete exposure of the Late Eocene sediments in this area. The Twiggs Clay, a member of the Dry Branch Formation, is predominantly a 25 ft (7.6 m) thick sequence of thin bedded and laminated, gray to dark gray clay and sandy clay. Oxidation of these sediments produces a very characteristic greenish gray to bluish gray coloration. A layer of light cream fossiliferous limestone marks the contact with the underlying Tivola Limestone.

Figure 5. Northwest highwall in Huber's mine #26. Ku = upper Cretaceous; Tc = Claibornian (Huber Formation); Tj = Jacksonian.

The lowermost section of the Twiggs Clay is made up of gray to dark gray, poorly bedded clay and sandy clay. The sediments appear to have been extensively reworked by organisms and contain abundant fossil remains; mostly molds of small gastropods and bivalves. Upwards, both the degree of bioturbation and the abundance of fossils decrease significantly and the very fine- to fine-grained sand can be observed forming distinct, very thin and discontinuous laminae within the clay. The uppermost section of exposed strata is made up entirely of thin bedded to laminated clay and contains foraminifers and coccoliths.

The lithology, stratification, and fossil fauna of the Twiggs Clay suggests that its deposition took place in an environment that changed from an initial near-shore to an offshore shelf environment.

Tivola Limestone

The Tivola Limestone consists of a sequence of interbedded fossiliferous limestone and clayey sand. In the Huber area, the generally 0.3 to 1.0-ft (9 to 30 cm) thick beds of limestone are made up of light cream, very porous and loosely cemented shell fragments, mainly of bryozoa, pecten and echinoid. Locally, these beds may merge into a single zone, up to 15 ft (4.6 m) thick. The dark gray clayey sand to sandy clay is very similar to the lowermost Twiggs Clay. The measured thickness of the Tivola Limestone is 20 ft (6.1 m).

The highly bioturbated clayey sediments and fossil contents of the limestone indicate that the Tivola Limestone must have been deposited in a near-shore, shallow shelf environment. Its gradational contact with the Twiggs Clay further suggests that both units probably were deposited during the same transgression.

Clinchfield Sand

Even though exposed at both sites, Huber #30 mine (Fig. 4) provides the best exposures of the Clinchfield Sand member of the Late Eocene Barnwell Group. It is represented by an approximately 15-ft (4.6 m) thick, uniform bed of yellow-brown, well sorted, medium-grained calcareous sand. Inorganic sedimentary structures are poorly preserved, suggesting extensive reworking by organisms. This unit contains abundant fossil remains, such as oyster shells, pecten, shark and ray teeth, fish vertebrae, and echinoids. The clay content in these sands increases significantly upward in the section, changing from sand to clayey sand. Where the Clinchfield Sand lies directly on a clay bed of the Huber Formation, large rip-up clay clasts are common. Drill and surface data indicate that the Clinchfield gradates laterally and vertically into the clayey sediments of the Twiggs. The fossils in these extensively bioturbated, well-sorted sands were deposited in a high energy, shoreline environment. The Clinchfield Sand is thought to represent the initial stage of the Late Eocene transgression that culminated with the deposition of the Twiggs Clay.

Huber Formation

In mine #26 (Fig. 5), the Huber Formation has a thickness of about 45 ft (31.7 m) from the sharp contact with the Barnwell Group above to the sharp contact with the Cretaceous sediments below. The most prominent feature of the Huber Formation is the kaolin bed at the top of the section. The kaolin beds in both mines are 3 to 10 ft (0.9 to 3.0 m) thick, but only mine #26 exposes the complete Huber section. A high water level in mine #30 conceals the lower part of the Huber and the Upper Cretaceous kaolin bed (Fig. 4). The Huber kaolin beds exhibit a hackly fracture and contain the tubular forms that resemble borings or bryozoa discussed by Buie (1978). The kaolinitic clay units usually grade downward into a kaolinitic sand, but some locations in both pits exhibit a sharp lower contact between the clay and sand. In the latter case, the lower contact is always very irregular.

In mine #26, the kaolin is underlain by a 25-ft (7.6 m) thick section of thin bedded, fine- to medium-grained, moderately well-sorted micaceous sand in a matrix of light gray to white kaolinitic clay. Both the sedimentary structures and lithologic composition observed in this sequence closely resemble those described by Horstmann (1983) for a tidal flat deposit.

Indentification of imprints and internal molds of a small mollusk, collected from the top of this unit, helped establish a Claiborne age for the Huber Formation (Buie and Fountain, 1967), previously considered part of the Cretaceous sedimentary sequence. In other areas of the kaolin belt, palynological evidence not only corroborated a Claiborne age (Scrudato and Bond, 1972; Tschudy and Patterson, 1975), but further indicated that deposition of the Huber strata may have started during the late Paleocene (Tschudy and Patterson, 1975).

Below the kaolinitic sand is a 15-ft (4.6 m) thick interval of yellowish-brown, medium- to very coarse-grained quartz sand. Well-preserved crossbedding, abundant dark minerals, and kaolin clasts, especially near its base, characterizes this unit. An abrupt contact separates this sand from the overlying strata of kaolinitic sand.

Upper Cretaceous

The lowermost unit in the section is the Cretaceous sediments. The Cretaceous that is exposed in mine #30 (Fig. 4) is an almost pure kaolin clay. This clay is white to cream in color and breaks with a conchoidal fracture. Drilling in this area indicates that this clay bed is as much as 49 ft (15 m) thick.

REFERENCES CITED

Buie, B. F., 1978, The Huber Formation of eastern central Georgia: Georgia Geological Survey Bulletin 93, p. 1–7.

Buie, B. F., 1980, Kaolin deposits and the Cretaceous-Tertiary boundary in east-central Georgia, in Frey, R. M., ed., Excursions in southeastern geology—Guidebook for Annual Meetings, Volume 2: American Geological Institute, Falls Church, p. 311–322.

Buie, B. F., and Fountain, R. C., 1967, Tertiary and Cretaceous age of the kaolin deposits in Georgia and South Carolina: (Abstract), Geological Society of America Special Paper 115, p. 465.

Buie, B. F., Hetrick, J. H., Patterson, S. H., and Neeley, C. L., 1979, Geology and industrial mineral resources of the Macon-Gordon kaolin district: U.S. Geological Survey Open-File Report 79-526, 36 p.

Herrick, S. M., and Vorhis, R. C., 1963, Subsurface geology of the Georgia Coastal Plain: Georgia Geological Survey Information Circular 25, 78 p.

Horstmann, K. H., 1983, Estuarine–tidal flat depositional model for the Paleocene-Eocene Huber Formation, east-central Georgia: [M.S. thesis], Duke University, Durham, 203 p.

Huddlestun, P. F., and Hetrick, J. H., 1979, The stratigraphy of the Barnwell Group of Georgia: Georgia Geological Survey Open File Report 80-1, 89 p.

Hurst, V. J., Kunkle, A. C., Smith, J. M., Pickering, S. M., Shaffer, M. E., Smith, R. P., Williamson, J. W., and Moody, W. E., 1979, Field conference on kaolin, bauxite and Fuller's earth: Clay Minerals Society, Annual Meeting, 107 p.

Patterson, S. H., and Buie, B. F., 1974, Kaolin of the Macon-Gordon area in Field conference on kaolin and Fuller's earth, November 14–16, 1974, Atlanta, Georgia: Georgia Geologic Survey for the Society of Economic Geologists, p. 1–21.

Patterson, S. H., and Murray, H. H., 1975, Clays, in Lefond, S. J., ed., Industrial minerals and rocks: American Institute of Mining, Metallurgical, and Petroleum Engineers, Baltimore, Maryland, p. 519–586.

Schrader, E. L., Long, A. L., Muir, C. H., Quintus-Bosz, R. L., and Stewart, H. C., 1983, General geology and operation of kaolin mining in the "southeastern clay belt": a perspective from Huber, Georgia, Field Trip Guide for 112th Annual American Institute of Mining, Metallurgical, and Petroleum Engineers' Meeting, Atlanta, Georgia: 25 p.

Scrudato, R. J., and Bond, T. A., 1972, Cretaceous-Tertiary boundary of east-central Georgia and west-central South Carolina: Southeastern Geology, Volume 14, p. 233–239.

Tschudy, R. H., and Patterson, S. H., 1975, Palynological evidence for Late Cretaceous, Paleocene and early and middle Miocene ages for strata in the kaolin belt, Central Georgia: U.S. Geological Survey, Journal of Research, v. 3, p. 437–445.

Trail Ridge and Okefenokee Swamp

Robert E. Carver, *Department of Geology, University of Georgia, Athens, Georgia 30602*
George A. Brook and Robert A. Hyatt, *Department of Geography, University of Georgia, Athens, Georgia 30602*

LOCATION

Trail Ridge is located in southeastern Georgia and northeastern Florida (Fig. 1). The ridge is a Pleistocene coastal terrace that forms the eastern margin of Okefenokee Swamp, a large freshwater swamp.

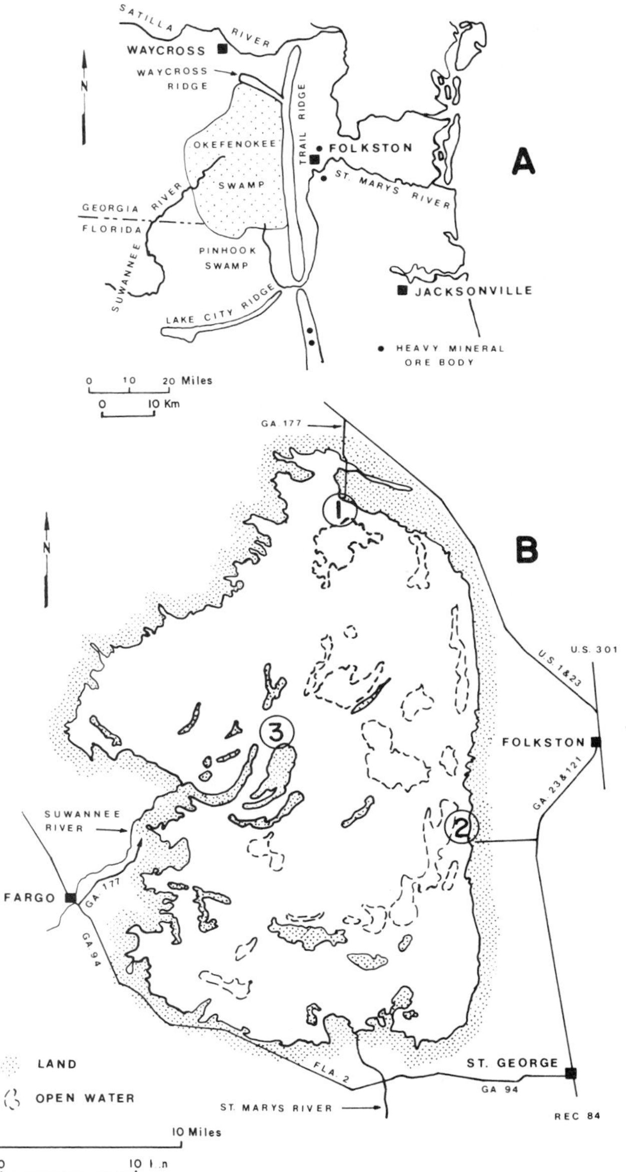

Trail Ridge is accessible by automobile where U.S. 1 and 23 crosses the Ridge about 20 mi (32 km) southeast of Waycross, Georgia. The road from Georgia 23 and 121 to the Suwannee Canal Recreation Area (Camp Cornelia) and Georgia 94 west of St. George, Georgia also cross Trail Ridge.

INTRODUCTION

Trail Ridge is one of the better developed Pleistocene Atlantic Coastal Terraces. It has been, and is, the site of several heavy mineral mining operations. Okefenokee Swamp is a freshwater peat swamp that is, in some respects, an analog of ancient coal swamps. The swamp is a National Wildlife Refuge and National Wilderness; and the swamp, swamp basin, and Trail Ridge complex have been the subject of extensive, ongoing biological, ecological, and geological studies. The current status of these extended studies is reported in Cohen and others (1984).

Trail Ridge is a coastal terrace that rises from an elevation of 80 ft (24 m) on the eastern side to about 175 ft (53 m) near the center, and slopes to the swamp level of 125 ft (38 m) on the western side (Fig. 2; Pirkle and Czel, 1983). It is one of the

Figure 1. (*A*) Location map of Trail Ridge and Okefenokee Swamp, after Pirkle and Czel (1983) and Pirkle and others (1977). (*B*) Detail map Okefenokee Swamp; showing: *1*, Okefenokee Swamp Park; *2*, Suwannee Canal Recreation Area; and *3*, Stephen C. Foster State Park, after U.S. Fish and Wildlife Service Leaflet RF 41590-1.

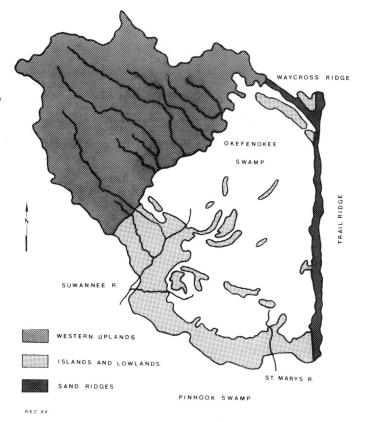

Figure 2. Features of the Okefenokee Swamp drainage basin.

most prominent Pleistocene coastal terraces of the southeast Atlantic Coastal Plain and forms the east edge of Okefenokee Swamp.

The swamp basin lies to the west of Trail Ridge. Uplands to the west of the swamp (Fig. 2) provide much of the water that flows to the swamp, although ground-water flow from Trail Ridge is a significant source. Water that is not lost by evapotranspiration drains from the swamp through the St. Marys River (to the Atlantic) and Suwannee River (to the Gulf of Mexico).

Several low islands occur within the swamp and the southern border of the swamp (Fig. 2) is a low, island-like ridge that separates Okefenokee and Pinhook Swamps; in fact, some workers include Pinhook Swamp in the Okefenokee Swamp complex. The swamp itself is heavily vegetated, with only a few areas of open water (called prairies). Cypress and shrubs dominate in the shallow water along the swamp margins and island shores, and water plants, especially water lilies, dominate in the deeper water. Masses of partly decayed organic matter accumulating on the bottom are commonly bouyed up by gases produced by decomposition and float to the surface. These floating islands, called batteries, are firm enough to walk on and are responsible for the Amerind name Okefenokee, "land of trembling earth."

Alligators and raccoons are the most numerous large animals in the swamp. Alligators normally ignore humans, but they are unpredictable and close approach should be avoided. Alligators do, however, attack and kill dogs that are in or near the water. Dogs should be left at home. Since the early 1960s rabies has been endemic in the raccoon population. Raccoons may be quite friendly, especially in areas where they are fed by visitors, but they should be avoided.

Climate

Because the climate of the area is important to some aspects of the geology of the area, a brief description of the climate follows.

The climate of southeastern Georgia and northeastern Florida is characterized by hot wet summers, warm dry falls, cool moist winters, and warm moist springs (Rykiel, 1977). The temperature-precipitation regime of the region falls in Koppen's Humid Subtropical (Cfa) climate type. The average January temperature is 53°F (11.7°C) and the average July temperature is 81°F (27.2°C). Annual precipitation averages 50 in (127 cm), with the greatest precipitation, in the form of convectional thunderstorms, during July.

Pre-Pleistocene Geology

The coastal terraces of southeastern Georgia and northeastern Florida are developed on predominantly Miocene sediments, with perhaps some Pliocene sediments overlying the Miocene or filling channels in Miocene sediments (Pirkle and Pirkle, 1984). Strata of Miocene (and Pliocene) age are thin and lenticular, individual units rarely extending more than a mile or two in any

TABLE 1. PLEISTOCENE COASTAL TERRACES OF THE GEORGIA AND FLORIDA ATLANTIC COAST*

Terrace	Elevation Above Mean Sea Level	
	Feet	Meters
Wicomico	95-100	30
Penholoway	70- 75	22
Talbot	40- 45	13
Pamlico	24	7.5
Princess Anne	13	4
Silver Bluff	4.5	1.5

*Elevations are for the associated shoreline in the central coastal area of Georgia from Hoyt and Hails (1974).

given direction, and consist of sand, clay, calcite, and dolomite mixed in varying proportion with few "mineralogically pure" units. The problem of correlation is compounded by the facts that outcrops are scarce and identifiable, diagnostic fossils are extremely rare, probably as a result of intense ground water leaching related to the humid subtropical climate. However, the preponderance of evidence suggests that the intermediate Pre-Pleistocene sediments are of near-shore to estuarine origin, and probably largely of Miocene age.

Coastal Terraces and Trail Ridge

Six coastal terraces of Pleistocene age approximately parallel the Atlantic Coast of Georgia and Florida. The terraces were constructed during Pleistocene interglacial high stands of the sea, and the highest terraces are the oldest. The six terraces, with the elevation above sea level of the associated shorelines, are given in Table 1. Hoyt and Hails (1974) hold that each terrace is a formation consisting of a barrier island facies, on the seaward side, and a marsh-lagoon facies on the inner side.

Much of the earlier literature was concerned with the difficulty of correlating terraces in different areas. Winker and Howard (1977) point out that the terraces in the Carolinas and northern Florida have been uplifted, confusing correlations made on the basis of elevation alone. Opdyke and others (1984) have explained that the uplift in Florida is due to isostatic rebound accompanying solution of limestone in the karst area of central Florida.

Trail Ridge in the Okefenokee Swamp area correlates with the Wicomico Terrace, the Wicomico Formation of Hoyt and Hails (1974), and the Trail Ridge Sequence of Winker and Howard (1977). Taking the terrace elevation as halfway between the toe and crest elevation, as calculated by Winker and Howard (1977), the terrace elevation at Okefenokee Swamp is about 122 ft (37.2 m), representing an uplift of at least 22 ft (6.7 m) (cf. Table 1) in this area.

Winker and Howard (1977), Pirkle and Pirkle (1984), and Pirkle (1984) all agree with Hoyt and Hails (1974) that Trail Ridge is an ancient beach ridge. This is consistent with the presence of heavy-mineral ore bodies along the southern part of the ridge (Fig. 1). Pirkle and Pirkle (1984) hold that easily transported fine sands were carried out to sea during the formation of Trail

Ridge, leaving coarser sands, enriched in heavy minerals, to form the body of the ridge. Heavy mineral deposits east of the ridge, according to Pirkle and Pirkle, formed during a pause in the regression that followed the formation of Trail Ridge. Intense ground-water leaching during the interglacial during which Trail Ridge formed, and succeeding interglacials that probably had climates similar to the present climate, leached iron from the ilmenite, the most desirable heavy mineral species, leaving it enriched in titanium.

The age of Trail Ridge is somewhat in doubt. It has been considered to be Miocene, Pliocene, and Pleistocene (See review in Opdyke and others, 1984), however, the preponderance of evidence appears to suggest a Pleistocene age (Pirkle and Czel, 1983).

Western Uplands

The western uplands (Fig. 2) are rather flat and slope gently to the Okefenokee Swamp. The major features of the area are "cypress domes." Cypress domes are actually shallow depressions that trap water during rainy periods. Cypress trees grow in these wet depressions and, because the centers of the depression stay wet longer than the edges, the trees in the center grow taller, producing a dome-shaped mass of vegetation with shrubby trees around the edges and tall cypress at the center.

Carbonates, intricately interbedded and interlensed with permeable sands and less permeable clays, lie close to the surface in many parts of the western uplands and Wadsworth and others (1983) suggest that the domes are sinkholes partly filled with originally overlying insoluble sediment. Pirkle and Pirkle (1984) also stress the importance of carbonate solution in the Okefenokee area, and present a figure, an air photo, that shows very numerous features described as solution depressions in the southwestern fringe of the swamp.

Okefenokee Swamp

The Okefenokee Swamp is developed in a basin of gently irregular topography. The swamp basin includes a number of low, arcuate ridges that are convex to the southeast and now occur as low islands and bars in the swamp (Cohen and others, 1984; Pirkle and Pirkle, 1984). The origin of these ridges is not clear at this writing.

About 7,000 years ago peat began accumulating in the swamp basin (Cohen and others, 1984). This is somewhat surprising because the basin, formed by Trail Ridge, is probably several hundred thousand years old. The uplift of the karst area south of the swamp, reported by Opdyke and others (1984) may well be responsible for the flooding of the basin, because the swamp drainage is essentially to the south.

Maximum peat accumulation is about 14.5 ft (4.4 m), in the troughs between the arcuate bars and islands mentioned above. Two types of peat account for 90 percent of the volume of peat in the swamp: *Taxodium* (cypress) dominant peat, which accumulates in the shallower parts of the swamp, and *Nymphea*

(water lily) dominant peat that accumulates in the deeper parts. (Cohen, Andrejko and others, 1984).

Prairies, open water areas in the swamp, probably are the result of fires during severe droughts; the fires burning through dry peat and producing temporary areas of deeper water. The fires, thought to be set by lightning, occur during the 20 to 25 year droughts that are characteristic of the region (Izlar, 1984).

Interesting components of the swamp vegetation are freshwater sponges, whose siliceous spicules are an important constituent of the peat; and bladderworts, carnivorous water plants that trap and digest very small insects and other animals to obtain nitrogen.

SITE DESCRIPTION

The margins of Okefenokee Swamp can be reached by automobile at Okefenokee Swamp Park, the Suwannee Canal Recreation Area, and Stephen C. Foster State Park. Access to the interior of the swamp is restricted, to some extent, but adequate. The Suwannee Canal Recreation Area features a 4,000-ft (1219 m) boardwalk into the swamp and an observation tower. Boat launching ramps, boat and canoe rentals, and guided tours are available at the three facilities mentioned above and listed below. Outboard motors on boats to be taken into the swamp are limited to a maximum of 10 horsepower. Camping facilities are available at Stephen C. Foster State Park. For further information, write or call: 1) Okefenokee Swamp Park, Waycross, Georgia 31501, 912-283-0583; 2) Concessioner, Suwannee Canal Recreation Area, Folkston, Georgia 31537, 912-496-7156; 3) Stephen C. Foster State Park, Fargo, Georgia 31631, 912-496-7509.

Canoe trips of 2 to 6 days duration, with camping on platforms in the swamp, can be arranged by permit from: Refuge Manager, U.S. Fish and Wildlife Service, Route 2, Box 338, Folkston, Georgia 31537, 912-496-3331. Because of danger from poisonous snakes and alligators, swimming in any part of the Swamp is prohibited.

Site 1. Okefenokee Swamp Park

If you are unable to visit all of the sites described, it would be best to skip this site and go directly to Site 2.

At the intersection of U.S. 1 and 23 and Georgia 177 (about 8 mi (12.9 km) south of Waycross, Georgia) turn south on Georgia 177 and proceed 5 mi (8.0 km) to Okefenokee Swamp Park. Admission to the park is $6.00 (1985). The park features natural history and historical exhibits, a boardwalk into the fringe of the swamp, and an observation tower. Georgia 177, from the intersection with U.S. 1 to the park, crosses Waycross Ridge, an east–west trending sand ridge (Fig. 2) that antedates and is truncated by Trail Ridge.

Site 2. Suwanee Canal Recreation Area

A road log to the Recreation Area follows:

0.0 mi (00 km). Entrance to Suwannee Canal Recreation Area. Turn right (west). Approximately 7.5 mi (12 km) S. of Folkston Georgia.

1.2 mi (1.9 km). Toe of Trail Ridge (75 ft [22.9 m] contour).

2.6 mi (4.2 km). Top of Trail Ridge (150 ft [45.7 m] contour).

2.8 mi (4.5 km). National Wildlife Refuge Boundary. The embankment on the left (to south) is spoil from the Suwannee Canal. In the 1890s an attempt was made to drain the Okefenokee swamp via the Suwannee Canal. The attempt failed, fortunately, because a channel could not be maintained, in the sand of Trail Ridge, below the ground water table. The ground water table under the center of Trail Ridge is higher than swamp level and the canal actually slopes toward, rather than away from the swamp. Swamp elevation is about 122 ft (37.2 m).

3.4 mi (5.5 km). Visitor area gate. Parking lot, Concessionaire (boat and canoe rentals, tours), Interpretive Center. Return to parking lot entrance. Entrance to Swamp Island Drive. This road, 3.7 mi (7.0 km) long, 15 mph (24 kph) speed limit, extends to Chesser Island in the swamp. The entrance gate is closed at 4:30 pm EST. At Chesser Island there are a typical swamp homestead, a low tower to observe a wildlife feeding area, a 4,000 ft (1,219.2 m) long boardwalk into the swamp, and an observation tower.

The boardwalk passes several "gator holes." These are pits in the swamp peat and moss that are excavated by alligators. The swamp vegetation can be observed at close range from the boardwalk, and the major swamp environments, cypress groves (houses), and open marsh (prairies), can be observed from the tower.

Site 3. Stephen C. Foster State Park

0.0 mi (km). St. George, Georgia. Proceed west on Georgia 94.

4.2 mi (6.8 km). Fire tower on left.

8.7 mi (14.0 km). Cypress domes on right. You are on the lowlands between Okefenokee and Pinhook swamps (Fig. 2). Numerous cypress domes occur in this area, but they are most obvious in cleared areas. I mention here and below a few domes that are in recently cut timberland that should be obvious for 10 years or longer.

11.2 mi (18.0 km). Manioc, Georgia.

11.6 mi (18.7 km). St. Mary's River, entering Florida, continue on Florida 2.

17.7 mi (28.5 km). Cypress dome on left.

26.2 mi (42.2 km). Cypress dome bisected by road.

26.8 mi (43.1 km). State Line, reenter Georgia, continue on Georgia 94.

34.9 mi (56.1 km). Intersection of Georgia 94, Georgia 177, and U.S. 441/Georgia 84. Turn right on to Georgia 177 toward Stephen C. Foster State Park.

45.6 mi (73.4 km). Okefenokee National Wildlife Refuge Boundary and gate.

51.4 mi (82.7 km). State Park gate.

52.1 mi (83.8 km). Boat basin parking lot, boat and canoe rentals, guided tours. A boardwalk along the fringes of the swamp begins on the far side of the boat basin. An interpretative center is nearby. Return to Highway intersection at mile 34.9 (56.1 km) Restart log for short trip to Suwannee River.

REFERENCES CITED

Cohen, A. D., Casagrande, D. J., Andrejka, M. J., and Best, G. R., eds., 1984. The Okefenokee Swamp: Its Natural History, Geology, and Geochemistry: Wetland Surveys, 4 Inca Lane, Los Alamos, New Mexico, 709 p.

Cohen, A. D., Andrejko, M. J., Spackman, W., and Corvinus, D., 1984, Peat deposits of the Okefenokee Swamp, in Cohen, A. D., Casagrande, D. J., Andrejka, M. J., and Best, G. R., eds., 1984, The Okefenokee Swamp: Its Natural History, Geology, and Geochemistry: Wetland Surveys, 4 Inca Lane, Los Alamos, New Mexico, p. 493–553.

Hoyt, J. H., and Hails, J. R., 1974, Pleistocene Stratigraphy of southeastern Georgia, in Oaks, R. Q., Jr., and Dubar, J. K., eds., Post-Miocene Stratigraphy Central and Southern Atlantic Coastal Plain: Utah State University Press, Logan, Utah, p. 191–205.

Izlar, R. L., 1984, Some comments on fire and climate in the Okefenokee swamp-marsh complex, in Cohen, A. D., Casagrande, P. J., Andrejka, M. L. and Best, G. R., eds., The Okefenokee Swamp: Its Natural History, Geology, and Geochemistry: Wetland Surveys, 4 Inca Lane, Los Alamos, New Mexico, p. 70–85.m

Opdyke, N. D., Spangler, D. A., Smith, D. L., Jones, D. S., and Lindquist, R.C., 1984, Origin of the epeirogenic uplift of Pliocene–Pleistocene beach ridges in Florida and development of the Florida karst: Geology, v. 12, p. 226–228.

Pirkle, E. C., Pirkle, W. A., and Yoho, W. H., 1977, The Green Cove Springs and

Boulougne heavy-mineral sand deposits of Florida: Economic Geology, v. 69, p. 1129–1137.

Pirkle, F. L., 1984, Environment of deposition of Trail Ridge sediments as determined from factor analysis, in Cohen, A. D., Casagrande, D. J., Andrejka, M. J., and Best, G. R., eds., The Okefenokee Swamp: Its Natural History, Geology, and Geochemistry: Wetland Surveys, 4 Inca Lane, Los Alamos, New Mexico, p. 629–650.

Pirkle, F. L., and Czel, L. J., 1983, Marine fossils from region of Trail Ridge, a Georgia–Florida landform: Southeastern Geology, v. 24, p. 31–38.

Pirkle, W. A., and Pirkle, E. C., 1984, Physiographic features and field relations of Trail Ridge in northern Florida and southeastern Georgia, in Cohen, A. D., Casagrande, D. J., Andrejka, M. J., and Best, G. R., eds., The Okefenokee Swamp: Its Natural History, Geology, and Geochemistry: Wetland Surveys, 4 Inca Lane, Los Alamos, New Mexico, p. 613–628.

Rykiel, E. J., Jr., 1977, The Okefenokee Swamp watershed: Water balance and nutrient budgets [Ph.D. thesis]: University of Georgia, Athens, 246 p.

Wadsworth, J. R., Jr., Brook, G. A., and Carver, R. E., 1983, Surface expression of heavily mantled intrastratal karst bordering Okefenokee Swamp, Georgia: Technical Papers of the 49th Annual Meeting of the American Society of Photogrammetry, p. 463–470.

Winker, C. D., and Howard, J. D., 1977, Correlation of tectonically deformed shorelines on the southern Atlantic coastal plain: Geology, v. 5, p. 123–127.

Devil's Mill Hopper, Alachua County, Florida

Thomas M. Scott, Florida Geological Survey, 903 W. Tennessee Street, Tallahassee, Florida 32304

LOCATION

Devil's Mill Hopper is a large, impressive solution collapse feature located in west-central Alachua County, Florida, northwest of Gainesville (Gainesville West 7½- minute quadrangle). The State of Florida has incorporated the Devil's Mill Hopper sinkhole and the land surrounding it into the Devil's Mill Hopper State Geological Site. Figure 1 shows the location of Devil's Mill Hopper in relation to I-75 and Gainesville.

Access to the sinkhole is limited to the designated trails and a boardwalk to the bottom of the sink. Please follow all the site regulations. Sampling permits may be obtained for research purposes by contacting the Bureau Chief, Bureau of Park Programs, Division of Recreation and Parks, Florida Department of Natural Resources, 3900 Commonwealth Boulevard, Tallahassee, Florida 32303.

INTRODUCTION

The Devil's Mill Hopper represents one of the few natural exposures of the Hawthorn Group where greater than 15 ft (4.6 m) of the section is exposed. Much of the initial research on the stratigraphy of north Florida is based on this outcrop and other outcrops in nearby sinkholes.

Devil's Mill Hopper has long been considered an important geologic feature in Florida due to the thickness of sediments exposed in the walls. It was first discussed in the geologic literature by Johnson (1888) and later by Dall and Harris (1892) as an important reference section for the Hawthorn Formation (now Group; Scott, in preparation). The Hawthorn Formation was named by Dall and Harris (1892) as the "Hawthorne beds" for the phosphatic sediments exposed in pits near Hawthorne east of

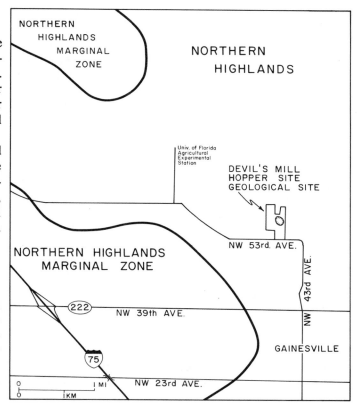

Figure 2. Location map of Devil's Mill Hopper state geological site showing physiographic zones, Gainesville 7½-minute quadrangle.

Gainesville in Alachua County, Florida. Since its inception the Hawthorne has been dropped, reinstated, redefined, and created continuing controversy. Scott (1983) provides a nomenclatural history of this unit. Currently, the Hawthorn Formation is being upgraded to group status (Scott, in preparation). Pirkle (1956) and Pirkle and others (1965) discussed the Mill Hopper as a "cotype" section with Brooks Sink in Bradford County for the Hawthorn Group. Besides being a reference section for the Hawthorn Group, the Devil's Mill Hopper is an example of an impressive karst feature.

The Devil's Mill Hopper, a deep, steep sided, nearly circular sinkhole, occurs near the southern terminus of the Northern Highlands physiographic province (White, 1970; Fig. 2). The Northern Highlands are separated from the limestone plain of the Western Valley by the Northern Highlands Marginal Zone (Williams and others, 1977). The Devil's Mill Hopper is located very near the boundary of the marginal zone and the highlands (Fig. 2). It is interesting to note that karst features are significantly more abundant and well developed in the marginal zone than in the highlands. However, none of the karst features in the marginal zone are as deep or steep sided as Devil's Mill Hopper. The

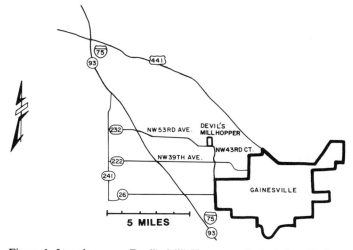

Figure 1. Location map, Devil's Mill Hopper geological site, Alachua County, Florida.

Devil's Mill Hopper is a rather anomalous karst feature in the Northern Highlands and the marginal zone since most karst depressions in these areas are relatively broad and shallow.

Stratigraphically, the Devil's Mill Hopper exposes post-Hawthorn sands, the entire section of the Hawthorn Group present in the area, and the top of the Ocala Group (the Crystal River Formation). The sections measured range from 98.5 ft (30 m; Puri and Vernon, 1964) to 116.8 ft (35.6 m; Williams, and others, 1977). The variability of the section thickness can be attributed to: (1) topography of the post-Hawthorn and upper Hawthorn surface and the underlying Ocala Group surface, (2) where the section was measured in the sink, and (3) problems related to slumping of the sediments. Sediment slump has obscured much of the section in most of the sink. It is interesting to note that this section which has been referred to as "type" Hawthorn actually is lithologically atypical when compared to the remainder of the Hawthorn in north Florida. The Devil's Mill Hopper Hawthorn section is very clayey and is representative of an areally limited phase of the group. This makes formational picks difficult. Formational identities discussed in this paper are tentative.

The Florida Geological Survey drilled a core next to the Mill Hopper in 1980. Figure 3 illustrates the stratigraphic section in the core along with the outcrop. This core, Mill Hopper #1 W-14641, is permanently stored at the Florida Geological Survey in Tallahassee and is available for inspection. Mill Hopper #1 compares very favorably with measured sections taken from various areas of the Devil's Mill Hopper.

The oldest sediments exposed in the sinkhole and encountered in the core are the limestones of the Ocala Group (Crystal River Formation) of Late Eocene (Jackson) age (Fig. 3). First named by Dall and Harris (1892), the Ocala Limestone was later raised to group status by Puri (1957). The limestones are wackestones containing coarse to fine bioclastic debris. The Crystal River Formation typically is very fossiliferous containing foraminifera, mollusks, bryozoans, and echinoids. Induration varies from soft to very hard where replaced by silica. These sediments can not be seen from the boardwalk. Bishop (pers. comm., 1984) indicates that the Crystal River Formation can be seen only in a cavity at the base of the sink hidden from view by overgrowth.

Immediately overlying the Crystal River Formation, unconformably, are the mixed phosphatic clastics and carbonates of the basal Hawthorn Group (the Penney Farms Formation). The Penney Farms Formation is named for the town of Penney Farms, Clay County, Florida where the Florida Geological Survey drilled the type core, W-13769, Harris #1 (Scott, in preparation). Approximately 10 ft (3 m) of the Penney Farms is present in the Mill Hopper core. It is estimated that about the same footage is present in the sink (Fig. 3). The Penney Farms Formation is Early Miocene (Early to Middle Aquitanian; Scott, in preparation).

Sediments tentatively assigned to the Marks Head Formation unconformably overlie the Penney Farms. The contact and most of this unit appear to be covered by slumped material in most of the outcrop. Huddlestun (pers. comm., 1984) reintro-

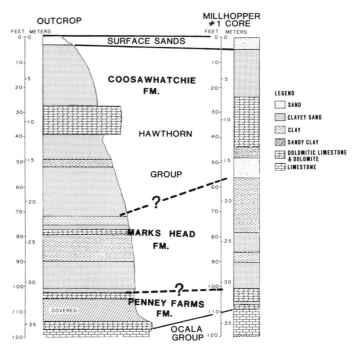

Figure 3. Comparison of Devil's Mill Hopper outcrop and Millhopper #1 core.

duced the Marks Head Formation for part of the Hawthorn Group in southeast Georgia. The present author has extended the usage into northeast Florida. The sediments assigned to the Marks Head in the Mill Hopper core and outcrop are predominantly phosphatic clays and sands with some carbonate lenses and clasts noted in outcrop. Approximately 30 ft (9.1 m) of Marks Head are present at this site. The Marks Head Formation is considered late Early Miocene (Lower Burdigalian) in age (Huddlestun, pers. comm., 1983).

Unconformably overlying the Marks Head is the Coosawhatchie Formation. Huddlestun (pers. comm., 1984) proposed this unit for southeast Georgia and the present author has extended it into northeast Florida. This unit consists of phosphatic clays, sands, and carbonates. Approximately 40 ft (12.2 m) of the Coowachatchie are present. The massive carbonate seen near the top of the Devils Mill Hopper is in the upper part of the Coosawhatchie. The Coosawhatchie Formation is Middle Miocene (early Serravallian) in age.

Phosphatic to very phosphatic clay sands and sandy clays are recognized at the top of the Coosawhatchie Formation. Brooks (1967) suggested that these sediments are reworked. He noted that they contain a Late Miocene invertebrate fauna with occasional zones containing Pliocene land vertebrates. These sediments are included in the Coosawhatchie until further studies can be completed. Approximately 14 ft (4.3 m) of this unit are present in the core and 24 ft (7.3 m) in the sink. This is overlain by 3 to 10 ft (0.9 to 3 m) of loose, gray to brownish quartz sand.

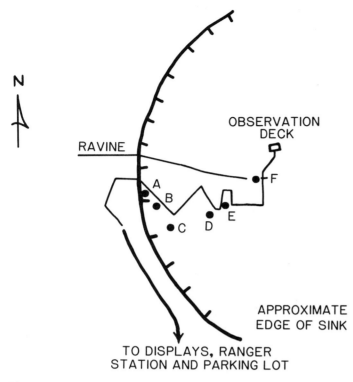

N

RAVINE

OBSERVATION DECK

A
B
C
D
E
F

APPROXIMATE EDGE OF SINK

TO DISPLAYS, RANGER STATION AND PARKING LOT

Figure 4. Schematic diagram of the west side of the Devil's Mill Hopper showing boardwalk into sink and points of interest, *A* to *F,* described in text.

SITE DESCRIPTION

Figure 4 is a schematic diagram of the boardwalk into the sink indicating where outcrops can be seen. A ravine enters the Mill Hopper from the west. The boardwalk roughly follows the south side of the ravine to the base of the sink. Where the boardwalk parallels the ravine near the top of the sink, blocks of carbonate can be seen in the stream. Occasional carbonate outcrops overlain by clayey, phosphatic quartz sand can also be seen. The clayey phosphatic sands crop out along the boardwalk at *A* on Figure 4. At *A* there is an outcrop approximately 4 ft (1.2 m) thick of the clayey phosphatic sand. This equates to the sands just above the limestone near the surface in Figure 3. At point *B* the limestone crops out next to the boardwalk.

Looking across the slope from *C* slump features are very common, obscuring a large part of the section. Several limestone blocks can be seen on the slumped slope. The limestone exposed above the slump and some of the sediments hidden by the slumped material are part of the Coosawhwatchie Formation. Just below the slumped slope sandy phosphatic clay and very clayey phosphatic sand crop out at point *D*.

The basal carbonate of the Hawthorn Group crops out at *E* and *F*. At *F* a small cavity can be seen. The light colored rock lying in the opening of the cavity is a carbonate cemented, slightly phosphatic, quartz sand. A small stream passes through the cav-

ity, then under the boardwalk, winding its way out into the bottom of the sink.

At the end of the boardwalk near the base of the Devil's Mill Hopper is a small observation deck from which you can enjoy the spectacular flora and tranquility of this site. It is also a good place to prepare for the climb to the top more than 100 ft (30.5 meters) above you.

Upon reaching the top of the Mill Hopper relax and then take a leisurely stroll around the top of the sink on the nature trail. Where the nature trail crosses the ravine which enters the sink from the west you can see the upper limestone exposed in the stream. Enjoy the trail.

REFERENCES CITED

Brooks, H. K., 1967, Miocene-Pliocene problems of Peninsular Florida, Southeastern Geological Society, 13th Annual Field Trip, Guidebook: p. 1–2.

Dall, W. H., and Harris, G. D., 1892, Correlation paper-Neocene: United States Geological Survey Bulletin, 84, p. 85–158.

Johnson, L. C., 1888, The Structure of Florida: American Journal of Science, 3rd Series, v. 36, p. 230–236.

Pirkle, E. C., 1956, The Hawthorn and Alachua formations of Alachua County, Florida: Florida Academy of Science, v. 19, no. 4, p. 197–240.

Pirkle, E. C., Yoho, W. J., and Allen, A. T., 1965, Hawthorn, Bone Valley and Citronelle sediments of Florida: Florida Academy of Science, v. 28, No. 1, p. 7–58.

Puri, H. S., 1957, Stratigraphy and zonation of the Ocala Group: Florida Geological Survey Bulletin 38, 248 p.

Puri, H. S., and Vernon, R. O., 1964, Summary of the geology of Florida and a Guidebook to the Classic Exposures: Florida Geological Survey, Special Publication No. 5 (revised), 312 p.

Scott, T. M., 1983, The Hawthorn Formation of Northeastern Florida: Florida Bureau of Geology, Report of Investigations 94, Pt. 1, p. 1–43.

Scott, T. M., in preparation, The lithostratigraphy of the Hawthorn Group in Florida: Florida Geological Survey Bulletin 59.

White, W. A., 1970, The geomorphology of the Florida peninsula: Florida Geological Survey Bulletin 51, 164 p.

Williams, K. E., Nicol, D., and Randazzo, A. F., 1977, The geology of the western part of Alachua County, Florida: Florida Bureau of Geology, Report of Investigations 85, 98 p.

The Central Florida Phosphate District

Thomas M. Scott, Florida Geological Survey, 903 W. Tennessee Street, Tallahassee, Florida 32305

LOCATION

The Central Florida Phosphate District is located in west-central Florida in parts of Polk, Hillsborough, Manatee, and Hardee counties (Fig. 1). The vast majority of phosphate mines occur within Polk County. 7½-minute topographic quadrangles covering the area of the phosphate district and the near-future southern expansion of the district include: Plant City, Lakeland, Auburndale, Nichols, Mulberry, Bartow, Keysville, Bradley Junction, Homeland, Fort Lonesome, Duette N. E., Baird, Bowling Green, Bereah, Duette, Fort Green, Wauchula and Griffins Corner.

INTRODUCTION

The active mines of the Central Florida Phosphate District

allow the visiting scientist a first-hand look at the economically important Miocene phosphorite deposits. Each mine section displays its own unique character resulting from the complex lithologic framework of the upper Hawthorn Group in this area. Often, the complexities of the Hawthorn section are evident in a single cut. Through the examination of mine sections from north to south and east to west in the district a visitor will be able to understand this complex Miocene section. Also interesting from the standpoint of the genesis of the phosphorites is the occurrence of multiple generations of phosphate, the phosphatized carbonate clasts, and the many episodes of reworking evident in the exposed sections.

SITE DESCRIPTION

Within the phosphate district there are approximately 15

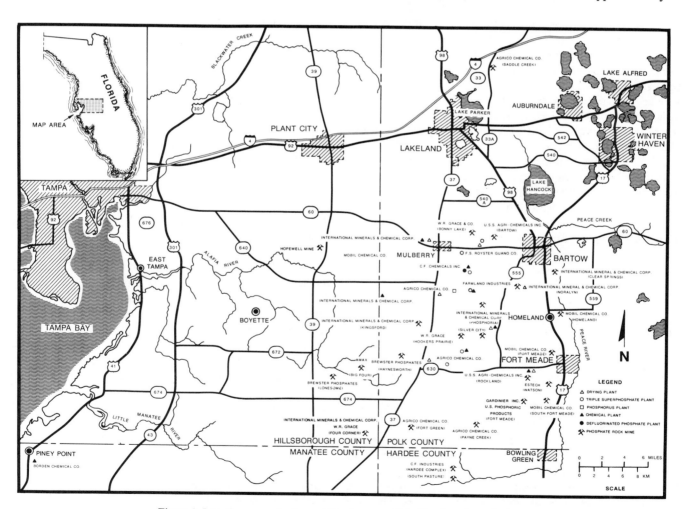

Figure 1. Location map of mines within the central Florida phosphate district.

TABLE 1. PHOSPHATE COMPANIES ACTIVE IN THE CENTRAL DISTRICT
IN 1985

Agrico Mining Company P.O. Box 1110 Mulberry, Florida 33860 (813) 428-1431	Gardinier, Inc. P.O. Box 937 Fort Meade, Florida 33841 (813) 285-8125

AMAX Chemical Corporation
Post Office Drawer 508
Bradley, Florida 33835
(813) 688-1130
	International Minerals and
	  Chemical Corporation (IMC)
	Clear Springs, Kingsford, and
	  Noralyn Mines
	P.O. Box 867
	Bartow, Florida 33830
	(813) 533-1121

Beker Phosphate Corporation P.O. Box 9034 Bradenton, Florida 33506 (813) 322-1311	
Brewster Phosphates Lonesome Plant Bradley,Florida 33835 (813) 634-5551	Mobile Chemical Company Ft. Meade and Nichols Mines P.O. Box 311 Nichols, Florida 33863 (813) 425-3011
Estech, Inc. Wauchula, Florida 33873 (813) 375-4321	USS Agri-Chemicals P.O. Box 867 Fort Meade, Florida 33841 (813) 285-8121
Silver City and Watson Mines P.O. Box 208 Bartow, Florida 33830 (813) 533-7164	W.R. Grace and Company P.O. Box 471 Bartow, Florida 33830 (813) 533-2171

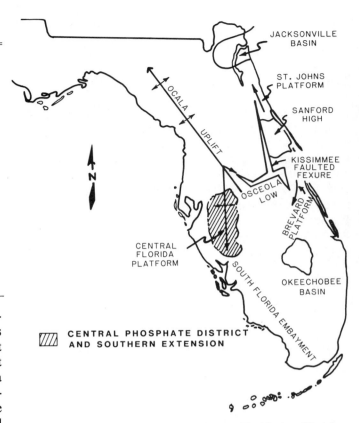

CENTRAL PHOSPHATE DISTRICT
AND SOUTHERN EXTENSION

Figure 2. Structural framework of peninsular Florida (modified from Hall, 1983).

companies operating 20 mines in 1985. It is not feasible to provide exact directions to each mine. Since each company requires visitors to obtain permission to visit the mines, it is important that a geologist contact the companies whose mines he wishes to visit and get directions to the company's main office. Typically, a company geologist or technician will meet the geologist and accompany him to the pits and provide information concerning the geology. A list of companies with mines is shown on Figure 1 and their addresses are provided in Table 1. Be sure to contact the companies you wish to visit prior to your trip so the proper arrangements can be made.

For years, the Central Florida Phosphate District has provided a large percentage of the United States phosphate production. Mining in the central district began in the late 1800's and has continued through the present. During this time mining has utilized lower grade materials as the higher grade deposits were depleted. During the early years mainly "river pebble" deposits were mined (deposits of phosphate concentrated by rivers and streams reworking the older sediments). As the "river pebble" deposits were exhausted, mining of "land pebble" deposits escalated. The present mines offer excellent opportunities for geologists to view the stratigraphy of the deposits and to collect a variety of phosphate minerals and vertebrate fossils. The phosphate bearing sediments in the central district have been discussed by numerous authors including Dall and Harris (1892), Matson and Clapp (1909), Cooke (1945), Cathcart (1950), Cathcart and McGreevy (1959), Altschuler and others (1964), Riggs and Freas (1965), Riggs (1967) and Hall (1983). The main phosphate producing beds were named the Bone Valley Gravel by Matson and Clapp (1909) and referred to later as the Bone Valley Formation. Current research by this author indicates that the Bone Valley should be downgraded to member status within the upper Hawthorn Group, Peace River Formation (Scott, in preparation). The Bone Valley grades downdip into the upper part of the Peace River Formation. The Peace River Formation (new name) con-

tains the phosphate reserves being prospected in the southern extension of the central district.

Regionally, the Central Florida Phosphate District lies on the Ocala Uplift on what Hall (1983) termed the Central Florida Platform (Fig. 2). The phosphate bearing section is generally thinnest in the northern part of the district and thickest downdip. More localized features such as the Lakeland and Winter Haven ridges affected the deposition of the Bone Valley Member (Altschuler and others, 1964; Fig. 3). The more localized controls have resulted in concentration of pebble phosphate on the subsurface expression of the Lakeland Ridge and its flanks, while in other peripheral areas the sand-sized phosphorite fraction is more abundant. Current mining is rapidly depleting the reserves of high grade ores in the central district. Most of the area containing concentrated pebble phosphorite has already been mined forcing the companies to mine the finer material and lower grade ores.

The Bone Valley typically consists of phosphorite gravel and sand incorporated in a matrix of quartz sand and clay. The percentages of these constituents varies widely both laterally and vertically within the section. Facies changes are both frequent and rapid. Phosphorite is almost always the most abundant single component in the Bone Valley. The phosphate gravel component is often the most common phosphorite size present. Locally, the interbedded nature of the fine and coarse (pebble) beds becomes quite complex. The phosphorite varies in color from white and

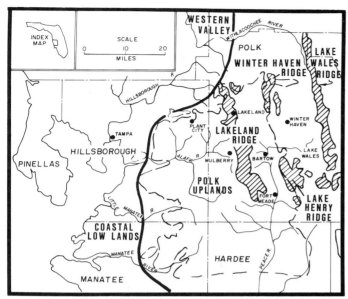

Figure 3. Map of west-central peninsular Florida showing limit of land pebble phosphate district (dashed line) and major topographic features. Shading indicates ridge areas above 150 feet (45.7 m) in elevation (modified from Altschuler and others, 1964).

tan to black. Typically the lighter colors are higher in the section. In the upper Bone Valley section the phosphorites are usually white, often earthy textured, due to leaching by ground water. This zone is often referred to as the "leached zone." Colors of the sediments of the Bone Valley range from blue greens, tans, and browns in the clays and "bed clays"; to olive grays, light grays, and brownish grays in the matrix; to light browns and white in the leached zone.

Throughout much of the phosphate district, the Bone Valley Member is underlain unconformably by the dolomites of the Arcadia Formation (new name) of the Hawthorn Group (Fig. 4; Scott, in preparation). The dolomites are sandy, sometimes clayey, phosphatic, and contain molds of mollusks. The upper part of the Arcadia Formation often contains clastic beds lithologically similar to those in the Peace River Formation. Colors vary from white to tan, yellowish, and gray. In the very southern part of the district the Bone Valley grades laterally into the upper Peace River Formation (Fig. 4). In some instances the Bone Valley may overlie beds of the Peace River Formation. The Peace River is typically a clayey, variably phosphatic, sometimes dolomitic quartz sand with scattered clay and carbonate beds. It is usually olive gray to light olive gray but may vary to lighter colors. The Peace River Formation is generally less phosphatic than the Bone Valley although zones of concentrated phosphorite do occur.

The Bone Valley Member is overlain unconformably by unconsolidated sands of Plio–Pleistocene age. These sands often appear to grade into the Bone Valley due to leaching of the upper Bone Valley by ground water. These sands, along with some of the leached zone are classified as overburden and are removed to expose the phosphate matrix.

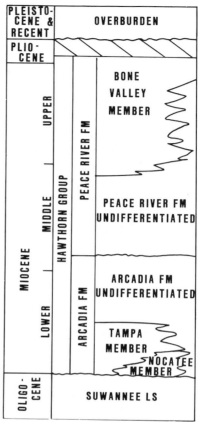

Figure 4. Stratigraphic column from the central Florida phosphate district.

Crissinger (in Webb and Crissinger, 1983), provides a detailed local stratigraphy of the district; ten units are recognized which recur throughout much of the district. Figure 5 graphically shows the relationship of these units within the central district. Unit 0 is the dolomites of the Arcadia Formation. Unit 1 consists of phosphatic, clayey quartz sands belonging to the Peace River Formation undifferentiated. Unit 2 is called the "bed clay." It is a sandy, silty, variably phosphatic clay that originally was thought to be a weathering product of the Arcadia dolomites. This author believes that these clays are the result of primary deposition, not weathering of the Arcadia Formation. Units 3, 4 and 5 are the clayey, phosphatic, sandy beds of the Bone Valley. Unit 3 is the phosphorite pebble unit of the Bone Valley and is well developed on the Lakeland Ridge. Unit 3 consists of abundant phosphate pebbles with a clay and sand matrix. Unit 4 is typically a sandy clay to clayey sand with abundant sand sized phosphorite and some pebble phosphorite. Unit 5 is generally a clay with sand sized phosphorite, variable amounts of pebble phosphorite and quartz sand. Unit 6 beds are channel deposits of reworked phosphorite. Unit 6 contains "button quartz" pebbles (fine to coarse gravel and occasionally cobble sized quartz clasts). Unit 7 is the "leached zone." Units 8 and 9 are overburden beds consisting of sands. Unit 10 is a modern phase reworking of the phosphate which commonly is referred to as "river pebble" phosphate. As

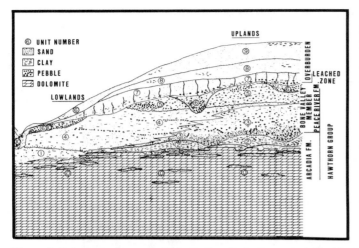

Figure 5. Schematic diagram of idealized units in the central Florida phosphate district (modified after Webb and Crissinger, 1983).

can be seen in Figure 5, all the units are not generally present at any single out crop, and they show some variations in vertical sequence.

The "leached zone" lies immediately above the phosphorite ore zone. The effects of leaching vary widely in this zone. The most intense leaching removes virtually all the phosphate mineral phases and silicates other than quartz. In less intensely altered zones and in zones of redeposition from groundwater, a number of secondary iron and aluminum phosphate minerals occur. These include beraunite, cacoxenite, crandallite, millisite, rock-bridgeite, vivianite, wavellite and others (Barwood and others, 1983).

By comparison the main phosphate mineral present in the un-leached portion of the phosphate matrix is a carbonate fluoro-apatite. One of the best areas to see the leached zone and associated minerals is at IMC's Clear Springs Mine along the Peace River southeast of Bartow.

The Florida Phosphate District is widely known for its abundant vertebrate remains. Both marine and terrestrial fossils are present throughout the district. Small shark teeth are virtually ubiquitous, being collected from all the mines. Large shark teeth are relatively common also. Other marine fossils include fish vertebrae, stingray tails, turtle plates, and sirenian ribs (which are abundant). The terrestrial fauna present is diverse including horse, camel, mastodon and many others. For more information see Olsen (1959) and Webb and Crissinger (1983).

The terrestrial vertebrates present in the phosphate bearing sediments provide an excellent opportunity to date the section. The oldest ages are from Unit 2 where a late Early to early Middle Miocene date has been determined (Webb and Crissinger, 1983). Within the Bone Valley Member, the age is late Middle to early Late Miocene while the reworked section falls in the latest Miocene to Early Pliocene (Webb and Crissinger, 1983). The overlying materials (units 8 and 10) fall into the Pleistocene.

It is obvious from the preceeding discussion that lithologic and stratigraphic specifics for any one particular site or mine can not be included here. However, company geologists will point out the Bone Valley Member and the Arcadia Formation (often called Hawthorn dolomites within the district).

REFERENCES CITED

Altschuler, Z. S., Cathcart, J. B., and Young, E. J., 1964, Geology and geochemistry of the Bone Valley Formation and its phosphate deposits, West-central Florida: Geological Society of America Field trip #6 1964 meeting, 68 p.

Barwood, H. L., Zelazny, L. W., Gricius, A., and Murochick, B. L., 1983, Mineralogy of the Central Florida Phosphate Mining District, in Southeastern Section Geological Society of America, Fieldtrip Guidebook: p. 83–89.

Cathcart, J. B., 1950, Notes on land pebble phosphate deposits of Florida: in Proceedings, Symposium on Mineral resources of the southeastern United States, University of Tennessee Press, p. 132–151.

Cathcart, J. B., and McGreevy, L. J., 1959, Results of Geologic exploration by core drilling, 1953, land pebble phosphate district, Florida: U.S. Geological Survey Bulletin 1046 K, 77 p.

Cooke, C. W., 1945, Geology of Florida: Florida Geological Survey, Bulletin 29, 339 p.

Dall, W. H., and G. D. Harris, 1892, Correlation Paper—Neocene: U.S. Geological Survey Bulletin 84, p. 87–158.

Hall, R. B., 1983, General Geology and stratigraphy of the southern extension of the Central Florida Phosphate District in Southeastern Section Geological Society of America, Fieldtrip Guidebook: p. 1.

Matson, G. C., and Clapp, F. G., 1909, A preliminary report of the geology of Florida with special reference to the stratigraphy: Florida Geological Survey 2nd Annual Report, p. 25–173.

Olsen, S. J., 1959, Fossil mammals of Florida: Florida Geological Survey, Special Publication 6, 74 p.

Riggs, S. R., 1967, Phosphorite stratigraphy, sedimentation and petrology of the Noralyn Mine, Central Florida Phosphate District [Ph.D. Thesis]: University of Montana, 268 p.

Riggs, S. R., and Freas, D. H., 1965, Stratigraphy and sedimentation of phosphorite in the Central Florida Phosphate District: Society of Mining Engineers, AIME Preprint #65H84, 17 p.

Scott, T. M., in preparation (to be published in late 1985), The lithostratigraphy of the Hawthorn Group in Florida: Florida Geological Survey Bulletin, 59 p.

Webb, S. D., and Crissinger, D. B., 1983, Stratigraphy and vertebrate paleontology of the central and southern phosphate districts of Florida, in Southeastern Section Geological Society of America, Fieldtrip Guidebook: p. 28–72.

The Everglades National Park

Paulette Bond, Florida Geological Survey, 903 W. Tennessee Street, Tallahassee, Florida 32304

LOCATION

The Everglades National Park is located at the southern extreme of peninsular Florida in Dade and Monroe Counties (Fig. 1). Florida 27 provides access from Homestead and Florida City to the southern part of the park and the locations discussed here.

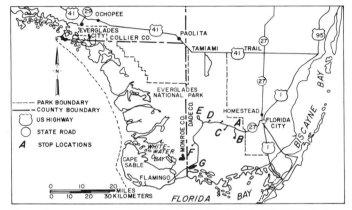

Figure 1. Location map of Everglades National Park. (A) Visitor's Center, Park Headquarters, and Entrance Station: (B) Anhinga Trail; (C) Long Pine Key; (D) Rock Reef Pass; (E) Pa-hay-okee Overlook; (F) Mahogany Hammock; (G) West Lake Trail.

INTRODUCTION

The Florida Everglades is an internationally recognized analogue for ancient coal-forming environments. The peats and peat-forming environments have been used to aid in reconstruction of paleoecologic and stratigraphic relationships in ancient coal seams as well as in attempts to ascertain swamp facies precursors for various coal lithotypes.

Before human intervention (primarily the excavation of an extensive system of drainage canals), the Kissimmee River flowed south to Lake Okeechobee and the lake level fluctuated sympathetically with natural inflow over a range of approximately 8 ft (2.4 m). Water levels in the Upper Everglades rose and fell with the level of Lake Okeechobee. Flood waters from Lake Okeechobee and local precipitation along the curved length of the Everglades combined to form a mass of water that flowed almost imperceptibly to the Gulf of Mexico (sheetflow of Parker, 1974). This chronic inundation allowed establishment of plant communities that produce the Everglades peats.

SITE DESCRIPTION

Anhinga Trail: A Freshwater Slough (B, Fig. 1)

Anhinga Trail traverses the northern part of Taylor Slough.

Taylor Slough, is similar to a river in that it delivers fresh water to several small bays bordering Florida Bay. The flow of water through sloughs, however, is slower than the flow through rivers; that feature, combined with the biological character and shallow nature of the sloughs, allows removal of pollutants and excess nutrients so that degradation of estuaries is alleviated (McPherson and others, 1976).

This stop focuses on a willow (*Salix caroliniana*) thicket that is characteristic of the relatively deeper waters of the slough. Ecologically, willow thickets serve as feeding, nesting, and roosting habitats for many herons, egrets, and other wading birds (Schomer and Drew, 1982).

The sediments that accumulate in Taylor Slough are primarily various types of peat and calcitic mud. A coring program in the slough (Gleason and others, 1974) showed the "deep" central trough to contain entirely peat. The well-drained flanks of the slough primarily contain calcitic mud with minor peat layers. The mouth of the slough, near Florida Bay, contains red mangrove (*Rhizophora mangle*) peat above calcitic mud suggesting a marine transgression (Gleason and others, 1974).

Pine Flatwoods Walking Trail at Long Pine Key (C, Fig. 1)

The pine forests here and elsewhere in the park occur on slightly elevated outcrops of the Miami Limestone. The elevation of the pinelands helps to protect them from the effects of flooding, and they are rarely inundated more than several weeks each year (McPherson and others, 1976). At this stop organic soils are thin or absent. They may, however, thicken in certain of the abundant solution holes that riddle the Miami Limestone in the vicinity of Stop C.

At this stop the forest undergrowth (understory) is relatively open. The upper branchy layer (canopy) of the pine forest is also open, and pine straw accumulates on the forest floor. These characteristics aid the initiation and spread of fires necessary to maintain the pine forest community (Schomer and Drew, 1982).

The dominant pineland species observed is Caribbean slash pine (*Pinus elliotti* var. *densa*). Periodic fires prevent tropical hardwoods from outcompeting the pines by checking the growth of hardwood seedlings and saplings. In addition, slash pine seeds need the stimulus of fire for germination. Mature pines are characterized by thick, moist bark that tends to protect them from milder fires (Schomer and Drew, 1982).

Rock Reef Pass (D, Fig. 1)

The "rock reefs" of the Everglades are long narrow ridges where the Miami Limestone is exposed at or near the surface. White (1970) describes Rock Reef as an "assemblage of structural lineaments that stand a little higher than the surrounding land because they have resisted solution better." Spackman and others (1964) note that they are unaware of any published information that justifies the term "rock reef."

Pa-hay-okee Overlook: Sawgrass Marsh (E, Fig. 1)

This stop is included to provide a view of the environment that most typifies the "true" Everglades. Sawgrass (*Mariscus jamaicensis*) is, in actuality, a sedge. Sawgrass marsh accounts for approximately 70 percent of the remaining Everglades (McPherson and others, 1976) and is also the most important peat-forming environment in south Florida (Spackman and others, 1964). Sawgrass marsh is generally higher in elevation than sloughs and lower than hardwood hammocks. The marsh is inundated to depths ranging from a "few inches to several feet." The optimum period for flooding is on the order of months (McPherson and others, 1976). In addition to sawgrass, muhley grass (*Muhlenberghia filipes*), beardgrass (*Andropogon* sp.), arrowhead (*Sagittaria lancifolia*), and many other true grasses grow here.

Spackman and others (1964) note the character of sawgrass is such that its peat is grossly amorphous and shows no woody, fibrous, or coarse granular texture. They further note that even though the leaf blades of sawgrass contain abundant silica in the sawtooth margin and covering some leaf veins, its peat does not show a concentration of silica.

Mahogany Hammock: Hardwood Hammock (F, Fig. 1)

The classic Everglades sawgrass marsh is punctuated by numerous small forested areas. Mahogany Hammock is an accessible example of a hardwood hammock. Rare paurotis palms (*Paurotis wrightii*) and massive mahoganys (*Swietenia mahogoni*) occur at this locality. Craighead (1974) notes that the distinctive dome-shaped profile of the hammock is functional in that it may deflect hurricane winds up and over the hammock itself. He theorizes that construction of the boardwalk through Mahogany Hammock left it vulnerable to rampant destruction during Hurricane Donna in 1960 (Craighead, 1974). Hardwood hammocks develop on slightly elevated areas that contain solution holes that are grouped together. The solution holes are frequently filled with peat, which may be deposited directly on bedrock or onto a layer of marl that lies directly above bedrock (Craighead, 1974; Spackman and others 1964).

West Lake Trail: Mangrove Environment (G, Fig. 1)

Stop G provides an opportunity to observe a mangrove community developed around a brackish-water lake. A short core taken from the eastern north shore of West Lake shows approximately 6 in (15 cm) of undifferentiated marl, nearly 2 ft (0.6 m) of red mangrove (*Rhizophora mangle*) root peat, and approximately 1 ft (0.3 m) of sawgrass (*Cladium jamaicensis*) peat (Gleason and others, 1974).

Along this trail red mangrove, white mangrove (*Laguncularia racemosa*), black mangrove (*Avicennia germinans*), and buttonwood (*Conocarpus erectus*) may be seen. Red mangrove, which dominates this site (Spackman and others, 1964), can grow in salt, brackish, or fresh water (McPherson and others, 1976). The black mangrove grows on surfaces that become submerged only at high tide (Wanless, 1974). The white mangrove also requires periodic saltwater inundation, while the buttonwood thrives on higher, drier ground (McPherson and others, 1976).

Florida's coastal mangrove forests act to buffer uplands from the effects of storms (McPherson and others, 1976). The coastal mangrove forest acts to break both wind and high tides. The mangroves also serve as baffles that aid in keeping inland waters free of suspended sediment and floating debris (McPherson and others, 1976).

REFERENCES CITED

Craighead, F. C., 1974, Hammocks of South Florida, *in* Gleason, P. J., ed., Environments of South Florida: present and past: Miami Geological Society, Memoir 2, p. 53–60.

Gleason, P. J., Cohen, A. D., Smith, W. G., Brooks, H. K., Stone, P. A., Goodrick, R. L., and Spackman, W., 1974, The environmental significance of Holocene sediments from the Everglades and saline tidal plain, *in* Gleason, P. J., ed., Environments of South Florida: present and past: Miami Geological Society, Memoir 2, p. 287–341.

McPherson, B. F., Hendrix, G. Y., Klein, H., and Tyus, H. M., 1976, The environment of south Florida, a summary report: U.S. Geological Survey Professional Paper 1011, 81 p.

Parker, G. G., 1974, Hydrology of the pre-drainage system of the Everglades in southern Florida, *in* Gleason, P. J., ed., Environments of South Florida: present and past: Miami Geological Society, Memoir 2, p. 18–27.

Schomer, N. S., and Drew, R. D., 1982, An ecological characterization of the lower Everglades, Florida Bay and the Florida Keys: Washington, D.C., U.S. Fish and Wildlife Service, Office of Biological Services, FWS/OBS-82/58.1, 246 p.

Spackman, W., Scholl, D. W., and Taft, W. H., 1964, Environments of coal formation: Geological Society of America, Annual Meeting, Field Guide, 67 p.

Wanless, H. R., 1974, Mangrove sedimentation in geological perspective *in* Gleason, P. J., ed., Environments of South Florida: present and past: Miami Geological Society Memoir 2, p. 190–200.

White, W. A., 1970, The Geomorphology of the Florida Peninsula: Florida Department of Natural Resources, Bureau of Geology, Geological Bulletin no. 51, 164 p.

Carbonate rock environments of south Florida

Paulette A. Bond, Florida Geological Survey, 903 W. Tennessee St., Tallahassee, Florida 32304

LOCATION

Representative exposures of Florida's Pleistocene carbonate rock environments are clustered at the southern tip of peninsular Florida (Fig. 1). The carbonate rocks can be examined onshore in Miami and on Windley Key, and offshore on Grecian Rocks and Rodriquez Bank, which are accessible by small boat. More precise location information is provided with the discussions for individual stops.

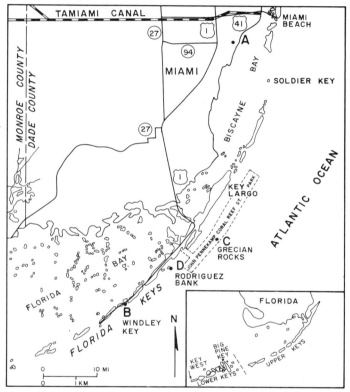

Figure 1. Generalized map showing the relative locations of sites comprising this cluster stop. Sites include: *A*, Oolitic member of the Miami Limestone; *B*, Key Largo Limestone at Windley Key quarry; *C*, the modern reef tract at Grecian Rocks; *D*, Rodriquez Bank, a modern carbonate mud mound.

INTRODUCTION

The carbonate rock-forming environments of south Florida have been studied extensively as modern analogues for ancient carbonate rocks. Because the organisms that form carbonate buildups change through time it is difficult to precisely compare modern environments with ancient ones. Heckel (1974) provides a thorough review of carbonate-forming environments with respect to carbonate buildups in terms of geographic as well as temporal distribution.

The oolitic member of the Miami limestone (Stop A; Fig. 1) is an example of a Pleistocene ooid shoal. Sedimentary structures preserved in the rock reflect the depositional processes that shaped this ancient shoal. Also, visible in the shoal is a diagenetic "overprint" characteristic of the Miami Limestone. Modern ooid shoals of the Bahamas were used extensively in the interpretation of this well-preserved ancient ooid shoal (Hoffmeister and others, 1967).

The Pleistocene Key Largo Limestone, an exquisitely preserved example of a coral reef, is closely related in space and time to the Miami Limestone. The quarry walls at Windley Key (Stop B; Fig. 1) provide a unique opportunity for viewing the internal structure of the ancient reef both parallel and perpendicular to its trend. Grecian Rocks (Stop C; Fig. 1), part of the modern Florida reef tract, may be compared and contrasted with the closely related Pleistocene reef at Windley Key.

Rodriquez Bank (Stop D: Fig. 1) is an example of a carbonate buildup that has maintained itself as a topographic feature even though it has no rigid organic framework. Unlike Grecian Rocks, carbonate sediments of Rodriquez Bank accumulate and are stabilized directly or indirectly by plants.

The carbonate rock-forming environments of south Florida are a natural field laboratory with ongoing research. The area is convenient to major population centers such as Miami, Florida, and to various small towns in the keys, making it relatively simple to arrange boat rental. Inasmuch as Grecian Rocks is located within Key Largo Coral Reef Marine Sanctuary and Rodriquez Bank is located within John Pennekamp Coral Reef State Park, urban development should not affect these areas.

STOP A: OOLITIC MEMBER OF THE MIAMI LIMESTONE

A good exposure of the oolitic member of the Miami Limestone is located in Coral Gables, Dade County, Florida, beneath the Le Jeune Avenue bridge over the Coral Gables Waterway (Fig. 2). The Oolitic member of the Late Pleistocene Miami Limestone occurs at the southern extreme of peninsular Florida. Major particle constituents of the oolitic facies include ooids, pellets (here used to refer to particles of ellipsoidal shape), and skeletal sand. Many of the aragonitic ooids and pellets are wholly or partly replaced by calcite. Where the oolite lies below the water table, the ooids and pellets have been dissolved and the rock is composed of spherical and ellipsoidal cavities bound together with calcite cement (Hoffmeister and others, 1967).

The oolite formed as a series of coalesced ooid shoals, originally separated by relict tidal channels (Hoffmeister and others, 1967) behind a seaward barrier bar (Halley and others, 1977;

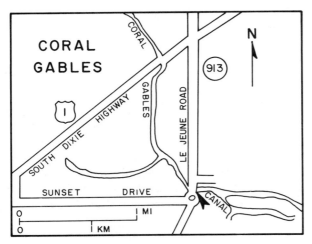

Figure 2. Arrow shows the location of the outcrop of Miami Limestone exposed in the Coral Gables canal.

Halley and Evans, 1983). The outcrop of the Miami Limestone at this stop exhibits characteristics of deposition in an active ooid shoal. Cross-bedding visible here resulted from periods of higher energy deposition (Halley and Evans, 1983). The average dip angle of the cross-beds is 20° and dip directions vary from N70°E to S60°E. The diverse directions of sediment transport are consistent with those seen in a modern ooid shoal complex such as that north of Andros Island, Bahamas.

Bioturbated layers separate cross-bedded sequences in this outcrop and represent periods of lower energy. In general, when a surface remains in a low-energy regime, the amount of bioturbation increases with time. The resultant stratigraphy will be disrupted roughly in proportion to the amount of time the bottom is subjected to bioturbation (Halley and Evans, 1983).

Close examination of the cross-beds and burrows reveals differential preservation due to variations in cementation (Halley and Evans, 1983). The cross-bed foresets are commonly preserved as alternating couplets of well cemented and poorly cemented layers apprpoximately 1 in (2 cm) thick. The well cemented layers are resistant to weathering. This mode of preservation is typical of the Miami Limestone regardless of elevation relative to sea level or dip angle. The finer-grained layers are frequently preferentially cemented. If alternate cross-beds are equal in grain size and degree of sorting, the layers that exhibit more compact grain packing tend to be better cemented. Halley and Evans (1983) suggest that when specific surface area is larger (finer grain size and closer packing) cementation is enhanced by an increased number of nucleation sites, lower permeability, and higher specific retention.

STOP B: THE KEY LARGO LIMESTONE

The Key Largo Limestone is exposed in the Windley Key quarry (Fig. 3), which is accessible by arrangement with the Miami Geological Society, P.O. Box 344156, Coral Gables, Florida 33114. The quarry is located on Windley Key just to the

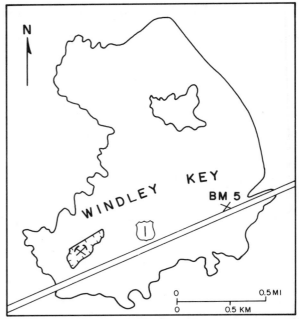

Figure 3. Location map of Windley Key quarry (traced from Plantation Key quadrangle, 7½-minute Series Orthophoto map).

north of U.S. Route 1. An alternate locality for the Key Largo Limestone is in the canal walls of the Key Largo Waterway; access to this locality is by small boat.

The Florida Keys are an arcuate strand of islands composed of Pleistocene Limestones. The Keys extend from Soldier Key at the north to Key West approximately 150 mi (240 km) south and west. They are frequently divided into two groups, the upper keys (extending from Soldier Key to Big Pine Key) and the lower keys (Big Pine Key to Key West; Fig. 1). The upper keys are composed of Key Largo Limestone and the lower keys of the Miami Limestone (examined in detail at stop A).

The Key Largo Limestone, which is remarkably homogeneous throughout the Florida Keys and is well exposed at the Windley Key quarry, is an excellent example of a Pleistocene reef limestone. It consists of a framework of reef-forming corals that act as a trap for fragmented skeletal materials. Virtually all of the coral species found today in the patch reefs of the Florida reef tract occur in the Key Largo Limestone (Hoffmeister and Multer, 1968). A panel from Windley Key quarry has been analyzed in detail by Stanley (1966). Figure 4 shows the location of this panel and details of the coral species within it. Stanley identified five main frame-building corals in the Key Largo Limestone: *Montastrea annularis, Porites astreoides, Diploria strigosa, D. clivosa,* and *D. labyrinthiformis.* These frame-builders make up approximately 30% of the formation's volume and *Montastrea annularis* comprises approximately half of that. The outcrops of the Key Largo Limestone exhibit no vertical or horizontal organic zonation.

The coral framework of the Key Largo Limestone is enclosed within a calcarenite that may be divided into a dominant

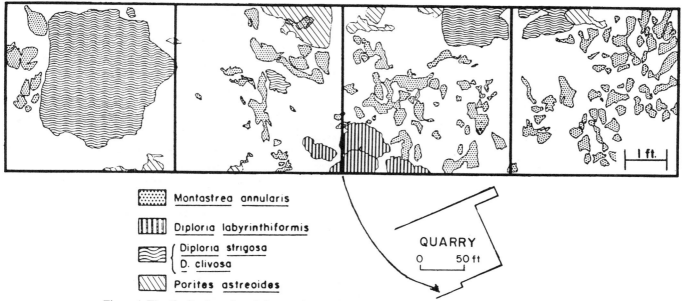

Figure 4. The distribution of corals in a section on the south wall of Windley Key quarry. At the extreme right is a tangential section through a branching *Montastrea* colony. At the extreme left is a large head of *Diploria* brain coral. *Porites astreoides* exhibits an encrusting habit (seen near the top of the section). (From Stanley, 1966)

poorly sorted facies and a subordinant well-sorted facies (Stanley, 1966). Both facies are poorly consolidated and friable. The poorly sorted calcarenite consists of *Halimeda* with lesser amounts of mollusks, corals, coralline algae, and foraminifera. The well-sorted calcarenite crops out as scattered patches of rock up to several yards (meters) across. It is devoid of organic framework material (Stanley, 1966), and is thought to represent a series of anastomosing channels that occurred between reef build-ups.

STOP C: GRECIAN ROCKS, A LIVING CORAL REEF

Grecian Rocks is located within Key Largo Coral Reef Marine Sanctuary (Fig. 5). It is accessible only by boat, and regular trips are scheduled by commercial dive boat operators on Key Largo. No collecting is permitted because this site lies within a National Marine Sanctuary.

Grecian Rocks, a living coral reef, provides an opportunity for comparison of a living coral reef with the Pleistocene reef examined at Windley Key quarry. The comparison must be qualified, however, because the Key Largo Limestone exposed at Windley Key represents a series of coalescing ancient patch reefs and thus cannot be directly compared to outer reefs, such as Grecian Rocks.

One of the most notable aspects of Grecian Rocks is the coral zonation initially described by Shinn (1963; Fig. 6). It consists of five zones: a rubble zone, a *Montastrea-Millepora* zone, an oriented *Acropora palmata* zone, an unoriented *A. palmata* zone, and a mixed *A. palmata–A. cervicornis* zone.

The rubble zone is comprised of coral rubble in a matrix of

Halimeda sand. It lies in 23 to 26 ft (7 to 8 m) of water. The rubble forms a substrate for holdfasts of gorgonians and other soft corals. Scattered large corals, primarily star corals (*Montastrea* sp.) and brain corals (*Colpophyllia* sp., *Siderastrea* sp., and *Diploria* sp.) inhabit the rubble zone (Shinn, 1980).

The *Montastrea-Millepora* zone is characterized by massive

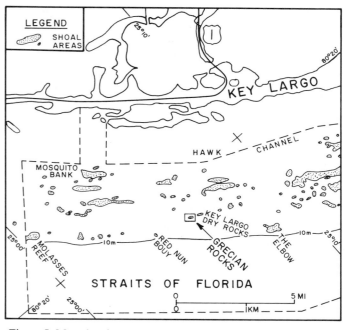

Figure 5. Map showing location of the Grecian Rocks reef within the Key Largo Coral Reef Preserve (*dashed line*; Shinn, 1980).

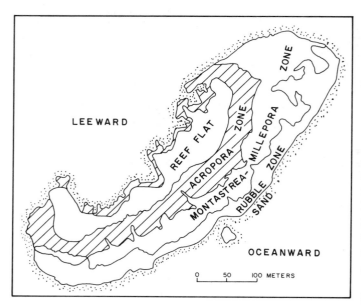

Figure 6. Map of Grecian Rocks showing major zonation of corals (modified from Shinn, 1980). The unoriented *Acropora palmata* zone is located within the reef flat; the mixed *Acropora palmata* zone is located leeward of the reef flat.

corals, mainly the stinging hydrocoral *Millepora*. It starts in 16 to 23 ft (5 to 7 m) of water, leeward of the rubble zone, and extends landward to a depth of 10 to 13 ft (3 to 4 m; Shinn, 1980).

The oriented *Acropora palmata* zone is the most spectacular of the coral zones. *Acropora palmata* in this zone occurs in water that ranges in depth from about 20 in to 13 ft (0.5 to 4 m). A well defined terrace marks the transition between the *Montastrea-Millepora* zone and the *Acropora* zone. The seaward edge of the *Acropora* zone is frequently a vertical cliff 3 to 6 ft (1 to 2 m) high. *Acropora palmata* has developed its orientation in response to wave attack. The zone takes the brunt of incoming sea-waves and swells and the orientation develops because *A. palmata* is subjected to predominantly unidirectional wave attack. Branches of *A. palmata* are oriented away from the direction of incoming waves; branches that protrude laterally are periodically broken away by storm waves (Shinn, 1963). Due to recent destruction of *A. palmata,* this zone may be locally unrecognizable.

The unoriented *A. palmata* zone (the reef flat) is the largest coral zone at Grecian Rocks. The coral here has grown upward to the spring low tide level and much of it is dead due to overcrowding. Along the leeward side of this zone, living *A. palmata* forms a strip that appears to be extending the reef flat. The boundary between the reef flat and the oriented *A. palmata* zone is transitional. In this zone occasional colonies of *A. palmata* are toppled onto their sides so that branches of the corals are exposed 12 to 16 in (30 or 40 cm) above the water at low tide. Some colonies have been completely overturned as a result of transport during hurricanes (Shinn, 1980).

The mixed *Acropora palmata-Acropora cervicornis* zone is

developed immediately landward of the reef flat and is characterized by delicate, unoriented *A. palamata* and thickets of *A. cervicornis,* which adapt well to the quiet water behind the reef. Scattered heads of the massive corals *Montastrea annularis, Diploria strigosa, Colcophyllia natans* occur on a bottom comprised of carbonate sand with interspersed storm-derived rubble.

The most spectacular geomorphic feature at Grecian Rocks is a series of channel-like grooves that are oriented approximately perpendicular to the trend of the reef tract (Shinn, 1980). The grooves and intervening spurs are a response to continual wave action. The spurs are coralline fingers that may reach approximately 30 ft (9 m) in height (Shinn, 1963). Because Grecian Rocks is located more than 0.6 mi (1 km) from the platform margin, the wave action is diminished in comparison with wave action at the outermost reefs so that the grooves, which cut both the *Montastrea-Millepora* zone and the oriented *Acropora palmata* zone, are somewhat reduced in size (Shinn, 1980).

STOP D: RODRIGUEZ BANK

Rodriguez Bank (Fig. 1) is accessible only by small boat. Specimen collection is permitted under the supervision of the State of Florida. The bank is an elongate flat-topped shoal that lies about 1 mi (1.6 km) east of Key Largo. Rodriguez Key, developed on Rodriguez Bank, is a mangrove-covered island 0.3 to 1 ft (.1 to .3 m) above mean low water (Turmel and Swanson, 1976).

Rodriguez Bank provides an example of an *in situ* buildup of skeletal carbonate sediment that may be used as a model for the development of a biostrome. Development of the bank was initiated at an embayment in the Pleistocene bedrock where lime mud collected in the absence of vigorous wave action (Turmel and Swanson, 1976). This bank (mud mound) has no rigid organic framework and plants are directly and/or indirectly responsible for stabilization of the sediment. The bank has maintained itself as a topographic feature in spite of hydrographic changes, and changes in biotic assemblages (Turmel and Swanson, 1976).

The surface of Rodriguez Bank and the areas that surround it contain a large population of marine plants and animals. Bottom sediments are produced by the fragmented calcareous skeletons of these organisms. Mangroves and turtle grass (both uncalcified) contribute organic matter to sediments while influencing growth.

The bank and its surrounding area may be divided into four zones based on the plants and/or animals that predominate (Turmel and Swanson, 1976, 1977): a mangrove zone, a grass and green algae zone, a *Goniolithon* zone, and a *Porites* zone. Figure 7 shows the areal distribution of these zones. The bands of *Goniolithon* and *Porites* are of particular note because they are well developed on the windward margin and are almost completely absent on the lee (west-facing) side. This distribution is a response to wave-action from the east, produced by prevailing trade winds. Sensitive corals require agitation and flushing, which wind-driven waves and tidal exchange produce. *Goniolithon* concentrates in

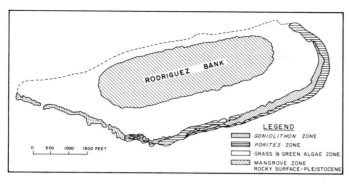

Figure 7. Map showing the surface zonation of dominant living organisms associated with Rodriquez Bank (Turmel and Swanson, 1976).

the shallowest water at the windward margin. This seems to be in response to the combination of intense light, agitation by waves at high water (keeping sand and silt in suspension), and periodic splashing by waves at low water, which prevents the zone from becoming completely dried out.

The mangrove zone constitutes the surface of Rodriquez Key, which is occupied by a dense forest of red mangroves (*Rhizophora mangle*), and is alternately exposed (low tide) and inundated (high tide).

The grass and green algae zone is comprised of turtle grass, *Thallassia testudinum,* and calcareous green algae of the family Codiaceae. This zone covers much of Rodriguez Bank and provides bottom cover in surrounding deeper waters. Turtle grass is a true grass with leaves, roots, rhizomes, and seeds. Its leaves are

coated with micro-organic slime that traps fine sediment from suspension. They also allow attachment and food-gathering of other plants and animals. The grass promotes deposition of suspended sediment and, where it is abundant, it stabilizes sediment in much the same way that sand dunes are stabilized by terrestrial grasses. Clay, silt, and sand sized particles on the bank are produced from the disaggregation of skeletons of the calcareous green algae *Halimeda tridens* and *Halimeda opuntia.* These are segmented, highly calcified plants that disintegrate after death. *Halimeda tridens* stands erect and branches while *Halimeda opuntia* forms cushion-like masses. The abundant unoriented pores of *Halimeda* plates allow segments to be broken easily as organisms nibble on uncalcified parts.

The *Goniolithon* zone is characterized by *Goniolithon strictum,* which is a branching form of calcareous red algae. Its branches are brittle and are made of high-magnesium calcite. Adjacent plants are intergrown with each other and completely cover the bottom in a dense growth that is wave resistant. When broken down, the brittle branches will yield rods (mainly granule size) greater than 0.1 in (2 mm) thick. The dense *Goniolithon* thickets host sponges, tunicates, laminated algal biscuits (oncolites), and green algae, as well as brittle stars, holothurians, molluscs, and crustaceans.

The *Porites* zone is characterized by *Porites divaricata,* which dominates a well-defined zone at the windward margin of the bank. The small branching finger coral colonies are intergrown and support a group of associated plants and animals much in the way of the *Goniolithon* thickets. *Porites* branches consist of porous aragonite and are fairly easily broken down to form gravel-size fragments.

REFERENCES CITED

Halley, R. B., and Evans, C. C., 1983, The Miami Limestone–A guide to selected outcrops and their interpretation: Miami Geological Society, Field Guide, 67 p.

Halley, R. B., Shinn, E. A., Hudson, J. H., and Lidz, B. H., 1977, Pleistocene barrier bar seaward of ooid shoal complex near Miami, Florida: American Association of Petroleum Geologists Bulletin, v. 61, p. 519–526.

Heckel, P. H., 1974, Carbonate buildups in the geologic record, *in* LaPorte, L. F., ed., Reefs in time and space: Society of Economic Paleontologists and Mineralogists Special Publication 18, p. 90–154.

Hoffmeister, J. E., and Multer, H. G., 1968, Geology and origin of the Florida Keys: Geological Society of America Bulletin, v. 79, p. 1487–1502.

Hoffmeister, J. E., Stockman, K. W., and Multer, H. G., 1967, Miami Limestone of Florida and its recent Bahamian counterpart: Geological Society of America Bulletin, v. 78, p. 175–190.

Multer, H. G., 1977, Field guide to some carbonate rock environments, Florida Keys and Western Bahamas: Dubuque, Kendall/Hunt Publishing Company, 415 p.

Shinn, E. A., 1963, Spur and groove formation on the Florida reef tract: Journal of Sedimentary Petrology, v. 33, p. 291–303.

——— 1980, Geological history of the Grecian Rocks, Key Largo Coral Reef Marine Sanctuary: Bulletin of Marine Science, v. 30, p. 646–656.

Stanley, S. M., 1966, Paleoecology and diagenesis of Key Largo Limestone, Florida: American Association of Petroleum Geologists Bulletin, v. 50, p. 1927–1947.

Turmel, R. J., and Swanson, R. G., 1976, The development of Rodriguez Bank, a Holocene mudbank in the Florida reef tract: Journal of Sedimentary Petrology, v. 46, p. 497–518.

——— 1977, The development of Rodriguez Bank, a recent carbonate mound, *in* Multer, H. G., Field guide to some carbonate rock environments, Florida Keys and Western Bahamas: Dubuque, Kendall/Hunt Publishing Company, p. 167–176.

St. Vincent Island (St. Vincent National Wildlife Refuge), Florida

Kenneth M. Campbell, Florida Geological Survey, 903 W. Tennessee Street, Tallahassee, Florida 32304

LOCATION

St. Vincent Island is a Holocene barrier island located in Franklin County, Florida, west-southwest of Apalachicola, Florida (Fig. 1). The island is located on the Apalachicola 2-degree quadrangle and on the Indian Pass and West Pass 7½-minute quadrangle maps.

The island is accessible only by boat. To obtain current access information, contact the Refuge Manager: St. Vincent National Wildlife Refuge, Post Office Box 347, Apalachicola, Florida 32320, (904) 653-8808.

INTRODUCTION

St. Vincent Island consists of an extensive, virtually undisturbed beach ridge plain of late Holocene age. The system of 12 beach ridge sets (approximately 180 individual ridges) provides a detailed window into late Holocene coastal and sea-level history.

Quaternary Stratigraphy

Undifferentiated Pleistocene. Schnable and Goodell (1968) discussed the Pleistocene and Holocene geology of the coastal area of the Apalachicola coast, including St. Vincent Island. Pleistocene sediments vary in thickness from approximately 120 ft (36.4 m) at Cape San Blas (west of St. Vincent Island) to less than 10 ft (3 m) to the east (Schnable and Goodell, 1968). Two of Schnable and Goodell's borings are located along the western portion of the lagoon side of St. Vincent Island. These borings show 22 to 25 ft (6.7 to 7.6 m) of Holocene sediment, but do not reach the base of Pleistocene sediment. Limestone is encountered in local water wells at 100 to 110 feet (30 to 33 m) below mean sea level (Schnable and Goodell, 1968). This would indicate approximately 75 to 90 ft (22.7 to 27.3 m) of Pleistocene sediment at St. Vincent Island.

Schnable and Goodell (1968) utilize a two-part breakdown of Pleistocene sediments and suggest that two depositional cycles are represented. Each depositional cycle is represented by a fining-upward sediment package separated from overlying and underlying units by unconformities.

The lower Pleistocene unit of Schnable and Goodell (1968) at St. Vincent Island consists of a coarse sand, overlain by silty clay and clay, in turn overlain by very clean fine quartz sand. The uppermost part of the lower unit consists of gray silty, clayey quartz sand or dark gray, quartz sandy, silty clay. Clays within the lower Pleistocene unit consist primarily of kaolinite with smaller quantities of montmorillonite (Schnable and Goodell, 1968).

The upper Pleistocene sequence (Schnable and Goodell, 1968) consists of a silty, fine to coarse quartz sand that grades

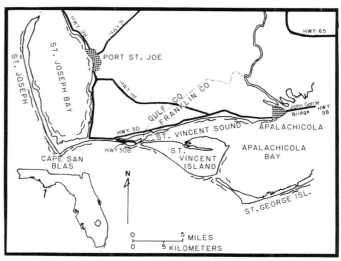

Figure 1. Location map, St. Vincent Island, Franklin Co., Fla.

upward to fine-grained sands and silty sands. The fine sands and silty sands are overlain by clayey sand and sandy clay.

Schmidt (1984) breaks Quaternary sediments into two lithologic packages, but does not attempt to differentiate between Pleistocene and Holocene sediments. The lower unit of Schmidt, onshore in the vicinity of St. Vincent Island, consists of massive clays with interspersed quartz sands that grade and thin northward into graded sands with some massive clays. Schmidt's upper unit consists of clean sands and clayey sands (Schmidt, 1984).

Holocene. Holocene sediments in the vicinity of St. Vincent Island consist primarily of the present-day barrier islands, spits and shoals, and sediments filling the old entrenched Apalachicola River Valley. Holocene sediments attain their greatest thickness (74 ft or 22.4 m) in the entrenched river valley. Outside the old Apalachicola River Valley, Holocene sediments range from approximately 10 to 25 ft (3 to 7.6 m) thick (Schnable and Goodell, 1968).

St. Vincent Island rests on basal Holocene sediments, which consist of clayey, fine-grained quartz sands, quartz sandy muds, and poorly sorted quartz sands. These Holocene sediments are deposited on an oxidized Pleistocene surface (Otvos, 1984). The island itself is constructed of well- to very well-sorted quartz sands, deposited under littoral or supratidal conditions. Many of the beach ridge swales and some of the ridges are blanketed with silts and clays. The first two ridges behind the present beach, and those immediately adjacent to Indian Pass, are dune decorated (Stapor, 1973).

Beach Ridge Development

Beach ridges are linear sand ridges that were deposited along the beach face in areas with a gentle offshore slope, low wave

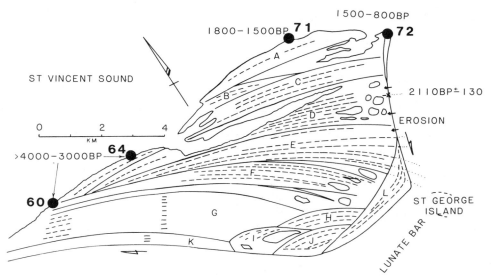

Figure 2. Diagrammatic sketch map of St. Vincent Island, Franklin County, Florida, showing the different sets of beach ridges (A, B, C, . . . L) that comprise this Holocene island. The numbers 60, 64, 71, and 72 locate critical Indian middens; these site numbers should begin with 8Fr to correspond with the site survey files of the Department of Anthropology, Florida State University. The C-14 date of 2110 BP locates the marine shell bed exposure; this shell bed stratigraphically overlies the clay/silt deposits that blanket beach-ridge set D. The open arrows indicate directions of present-day net longshore transport. Modified from Stapor and Tanner, 1977.

energy, and an abundant sand supply. Beach ridges form only on accreting beaches. Because they form on the beach face, they indicate the location and orientation of the coastline at the time they were formed.

Beach ridges are constructed by wave run-up in the swash zone by waves that fall in the long-term "maximum wave energy" category (Tanner and Stapor, 1972). The internal structure of beach ridges indicates that all or almost all of the ridge is constructed by wave run-up, with very little washover (Tanner and Stapor, 1972). Internal bedding is almost exclusively composed of planar, seaward-dipping cross beds. The ridge grows upward and seaward as material is deposited. Smaller ridges may be constructed by "fair weather waves" but will be destroyed when wave energy increases.

Beach Ridge Plain Development

Stapor (1973) describes several basic types of beach-ridge patterns, each of which is indicative of specific depositional environments. Stapor's categories are: (1) ridges convex seaward, occurring at the distal prograding tip of spits; (2) ridges concave seaward, which are being actively eroded perpendicular to their ridge and swale orientation; (3) ridges concave seaward deposited in embayed sections of the coast; and (4) ridges convex seaward, which represent net seaward growth, not lateral extention.

SITE DESCRIPTION

Category 3 beach ridges predominate on St. Vincent Island. Category 3 ridges that are essentially parallel indicate that sedi-

ment was transported from offshore without significant longshore transport. On St. Vincent, beach ridge sets A, B, C, F, G, H, J, and K (Fig. 2) are indicative of these conditions. Sets D and E (Fig. 2) are splayed to the east (increasing ridge crest spacing), a condition that indicates that westward longshore drift supplied the sand for their construction (Stapor, 1973). Beach ridges of set I (Fig. 2), although indistinct, appear to splay to the west, indicating a temporary reversal of drift direction (to the east).

Beach ridge sets H, I, J, and L (Fig. 2) were deposited in large part as a result of the migration of the West Pass lunate bar across the inlet (Stapor, 1973). The youngest ridge set, L, is of category 4. The sediment source for these ridges was sand delivered by easterly (primary) and southerly (secondary) transporting drift systems (Stapor, 1973).

Stapor and Tanner (1977) present a Holocene sea-level history developed from evidence preserved on St. Vincent Island and the adjacent mainland. The evidence utilized includes: (1) "average elevations of erosional scarps and beach ridge sets," (2) "elevation and distribution of cultural components of archeological sites," (3) "elevation of silt and clay beds containing marine shells," and (4) carbon-14 dating. Stapor and Tanner find evidence for four sea-level reversals.

Following a long period of sea-level rise beginning approximately 20,000 years before present (BP) and continuing through the early Holocene, sea level reached the highest Holocene sea-level stand approximately 5,000 BP (Stapor and Tanner, 1977). This sea-level stand is at approximately 5.0 ft (1.5 m) above present sea level. The evidence for this high Holocene sea-level stand consists of a wave-cut scarp cut into unconsolidated pre-Holocene sands on the mainland, shoreward of the present loca-

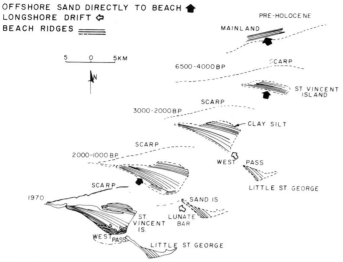

OFFSHORE SAND DIRECTLY TO BEACH ⬆
LONGSHORE DRIFT ⬅
BEACH RIDGES ═══

Figure 3. A detailed interpretation of the Holocene history of St. Vincent Island, Florida. Modified from Stapor, 1973.

tion of St. Vincent Island (Fig. 3). This appears to be the Silver Bluff position (Stapor and Tanner, 1977).

The first sea-level reversal documented by Stapor and Tanner (1977) occurred approximately 6,000 to 4,000 BP when sea-level rise changed to fall. The evidence consists of the topographically low silt- and clay-covered beach ridges on St. Vincent Island (Fig. 2, sets A-D; Fig. 3) located offshore of the topographically high (+5.0 ft or 1.5 m) scarp found on the mainland. St. Vincent Island beach ridge sets A-D were deposited at sea levels approximately 5.0 ft (1.5 m) lower than present, a sea-level fall of approximately 10 ft (3 m). Archaeological information provides the only present information regarding the time of formation of the mainland scarp, the St. Vincent beach ridges, and the sea-level reversal. Norwood occupation sites are present on both the mainland scarp and at sites 60 and 64 on St. Vincent Island (Fig. 2). This indicates that they were formed prior to 4,000 to 3,000 BP (Stapor and Tanner, 1977).

The second reversal (from fall to rise) occurred after the deposition of beach ridge sets A-D and prior to deposition of the marine shell bed that overlies the silts and clays that blanket set D. The marine shells are C-14 dated at 2,100 ± 130 BP (Fig. 2). The marine shells are exposed approximately 1.6 ft (49 cm) above present sea level, indicating a rise of about 6.5 ft (2 m) (Stapor and Tanner, 1977).

The third reversal (from rise to fall) occurred after deposition of the dated marine shells (2,110 ± 130 BP), but prior to the early Swift Creek (1,800–1,500 BP) occupation of archaeological site 71 (Fig. 2). This site is a midden that rests on silt and clay material. The contact is located in the middle of the present-day intertidal zone (Stapor and Tanner, 1977). The magnitude of sea-level fall is 6.5 to 8.2 ft (2 to 2.5 m) (Stapor and Tanner, 1977).

The last sea-level reversal documented by Stapor and Tanner (1977) occurred after the Weeden Island (1,500 to 800 BP)

occupation of site 72 (Fig. 2). The base of the shell midden at site 72 rests on silt and clay substrate 3.9 ft (1.2 m) below present sea level. Sea-level rise to the present position is 3.9 ft (1.2 m).

Based on elevation, there are two major groups of beach-ridge sets on St. Vincent Island. Sets A-D (Fig. 2) have crest elevations of about 3 ft (1 m). Sets E-L (Fig. 2) have crest elevations of approximately 8.2 ft (2.5 m) (excluding dune ridges and dune-decorated beach ridges that have elevations up to 16.4 ft; 5 m). Tanner and Stapor (1972) related beach-ridge height to wave height and concluded that fair weather waves are not responsible for beach-ridge formation. Stapor and Tanner (1977) state that most of the beach ridges on St. Vincent Island could be constructed by modern-day storm waves (3.3 to 4.9 ft or 1 to 1.5 m). The 3.3 ft (1 m) crest ridges (sets A-D) would require either less wave energy or lower sea level. As these ridges are blanketed with silt and clay, the lower sea level accounts for the difference in elevation.

Secondary deposits of alluvial silt and clay, marsh deposits, and shell beds have filled or partially filled many of the beach ridge swales. If these secondary deposits were removed, St. Vincent would be several islands, as the original surfaces of the swales are often below present-day sea level. This supports Stapor and Tanner's (1977) conclusion that mean sea level has for the past approximately 4,000 years fluctuated around a value that is less than 4.9 ft (1.5 m) below present mean sea-level position.

REFERENCES CITED

Otvos, E. G., 1984, Alternate interpretations of barrier island evolution, Apalachicola Coast, Northwest Florida: Litoralia, v. 1, no. 1, p. 9–21.

Schmidt, W., 1984, Neogene stratigraphy and geologic history of the Apalachicola Embayment, Florida: Fl. Geol. Survey, Bull. 58, 146 p.

Schnable, J. E., and Goodell, H. G., 1968, Pleistocene-recent stratigraphy evolution, and development of the Apalachicola Coast, Florida: Geol. Soc. America, Special Paper 112, 72 p.

Stapor, F. W., 1973, Coastal sand budgets and Holocene beach ridge plain development, northwest Florida [Ph.D. thesis]: Tallahassee, Florida State University, 221 p.

Stapor, F. W., and Tanner, W. F., 1977, Late Holocene mean sea level data from St. Vincent Island and the shape of the late Holocene mean sea level curve, *in* Proceedings, Coastal Sedimentology Symposium, Florida State University, Department of Geology, p. 35–68.

Tanner, W. F., and Stapor, F. W., 1972, Precise control of wave run-up in beach ridge construction, Zeitschrift fur Geomorphologie, v. 16, no. 4, p. 393–399.

Alum Bluff, Liberty County, Florida

Walt Schmidt, Florida Geological Survey, 903 W. Tennessee Street, Tallahassee, Florida 32304

LOCATION

Alum Bluff is located about two miles north-northwest of Bristol, in Liberty County, Florida. The bluff occurs along the east bank of the Apalachicola River in the NE¼ and SE¼ Sec.24,T.1N.,R.8W. (Fig. 1).

The Nature Conservancy owns the Bluff area and adjacent land. Their land is protected by fence and locked gates. Permission to enter can be obtained by contacting their Bristol Office at P.O. Box 789, Bristol, Florida 32321, or calling (904) 893-4153.

INTRODUCTION

The middle to late Tertiary deposits of the Florida Panhandle have been known since the late 1800s. They have been described on numerous occasions and have received the most attention from paleontologists attracted by their well preserved mollusk assemblages (predominantly Neogene). Alum Bluff was first described by Langdon in 1889 and is one of the best natural geologic exposures in Florida. Since that time other authors have described or named these units primarily based on their fossil assemblages. This location has been one of the primary outcrops used to establish the stratigraphic, paleoenvironmental and geologic history of the Florida Panhandle.

The sediments exposed at Alum Bluff include in ascending order: the Chipola Formation; the Hawthorn Formation; the Jackson Bluff Formation; an unnamed sandy clay interval; an unnamed clayey sand interval; and a unit composed of the Citronelle Formation and reworked "terrace sands" (Fig. 2). This section has been described often by geologists and paleontologists; a few of the more complete descriptions include: Dall and Harris (1892); Dall and Stanley-Brown (1894); Maury (1902); Matson and Clapp (1909); Gardner (1926–1944); Cooke (1945); Rainwater and others (1945); Puri (1953); Dubar and Beardsley (1961); Puri and Vernon (1964); Akers (1972); Banks and Hunter (1973); and Schmidt (1983).

SITE DESCRIPTION

Land access to the bluff area can be obtained by going 1.4 mi (2.2 km) north of the intersection of Florida 12 and 20 in Bristol, to the entrance of Skyland Ranch on the left (west). Travel this dirt road .2 mi (0.3 km) where the Nature Conservancy gate will be encountered. Pass gate and follow the dirt road 2.7 mi (4.3 km) to the Alum Bluff overlook (see Fig. 1). Road condition may vary depending on the amount of rain recently received. Normally the sands are very loose and high centers occur in the road. Pickup trucks and vans should have no trouble, however a small car with low clearance may have.

The bluff is quite steep and the fossiliferous horizons are

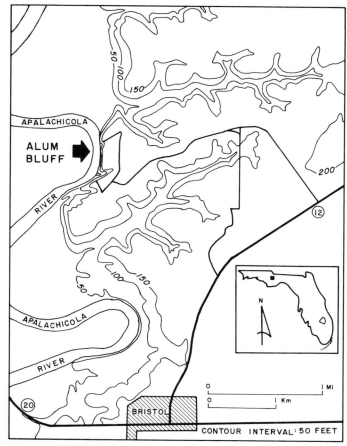

Figure 1. Location of Alum Bluff on the Apalachicola River, Liberty County, Florida. Bristol 7½-minute quadrangle (1945).

near the bottom. It is best to follow the bluff edge south until you reach a small drainage depression and descend from there. In summer the temperature and humidity can be very high. It is recommended to carry water along if you plan on being more than an hour or two.

The bluff can also be reached from the river by boat. There is a public launch about 1 mi (1.6 km) upstream from the Florida 20 bridge. This is accessible from Bristol on the east side of the river. The bluff is about 3 mi (4.8 km) up river from the boat launch.

Stratigraphy

Chipola Formation. The name Chipola Formation was suggested by Burns in 1889 (in Dall and Harris, 1892). He discovered and made large collections from shell beds exposed on the Chipola and Apalachicola Rivers. Dall and Stanley-Brown (1894) visited the area a few years later and called the formation the Chipola shell marl. Matson and Clapp (1909) included the

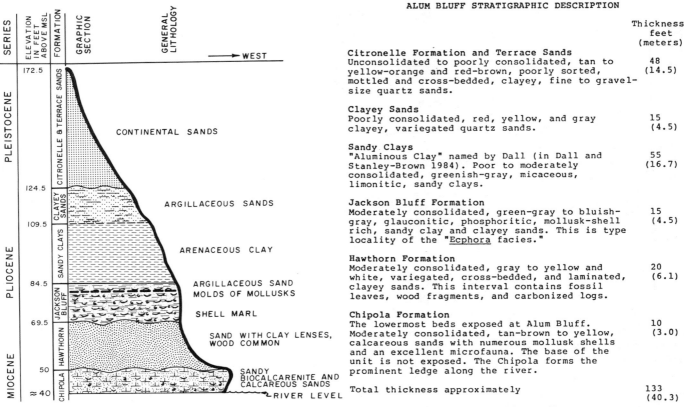

	Thickness feet (meters)
Citronelle Formation and Terrace Sands Unconsolidated to poorly consolidated, tan to yellow-orange and red-brown, poorly sorted, mottled and cross-bedded, clayey, fine to gravel-size quartz sands.	48 (14.5)
Clayey Sands Poorly consolidated, red, yellow, and gray clayey, variegated quartz sands.	15 (4.5)
Sandy Clays "Aluminous Clay" named by Dall (in Dall and Stanley-Brown 1984). Poor to moderately consolidated, greenish-gray, micaceous, limonitic, sandy clays.	55 (16.7)
Jackson Bluff Formation Moderately consolidated, green-gray to bluish-gray, glauconitic, phosphoritic, mollusk-shell rich, sandy clay and clayey sands. This is type locality of the "Ecphora facies."	15 (4.5)
Hawthorn Formation Moderately consolidated, gray to yellow and white, variegated, cross-bedded, and laminated, clayey sands. This interval contains fossil leaves, wood fragments, and carbonized logs.	20 (6.1)
Chipola Formation The lowermost beds exposed at Alum Bluff. Moderately consolidated, tan-brown to yellow, calcareous sands with numerous mollusk shells and an excellent microfauna. The base of the unit is not exposed. The Chipola forms the prominent ledge along the river.	10 (3.0)
Total thickness approximately	133 (40.3)

Figure 2. Geologic Section at Alum Bluff. (Revised after Dunbar and Beardsley, 1961, p. 165).

Chipola as a member in their Alum Bluff Formation, and Gardner (1926) later promoted the member to a formation. In 1953, Puri referred to the Chipola as a facies of the Alum Bluff Stage; then with Vernon (Puri and Vernon 1964), they recommended formation status.

Fossils identified from the Chipola Formation include mollusks, benthic and planktonic foraminifera, ostracods, corals, and calcareous nannofossils. Numerous authors have described the fauna. A summary of these references can be found in Schmidt and Clark (1980, p. 38). Planktonic foraminifera and calcareous nannofossils were used by Gibson (1967) and Akers (1972) to estimate the age of the sediments. It has been assigned to a Burdigalian age (late Early Miocene).

Hawthorn Formation. Dall and Harris (1892) in discussing the phosphoritic sediments being quarried near Hawthorne in Alachua County, Florida, assigned them the name "Hawthorne Beds." Matson and Clapp (1909) later designated this unit the Hawthorn Formation. Pirkle (1956; and Pirkle and others, 1965) designated the Devil's Millhopper Sink in Alachua County and Brooks Sink in Bradford County as cotype localities.

The Hawthorn Formation has been mapped in the Florida Panhandle by Cooke and Mossom (1929), Puri (1953), Vernon and Puri (1956), and others. It is considered Middle Miocene in age and thought to represent a deltaic or a pro-deltaic environment in the vicinity.

Jackson Bluff Formation. Puri and Vernon (1964) are credited with naming the Jackson Bluff Formation. They combined the *Ecphora* and *Cancellaria* biofacies because both are exposed at Jackson Bluff in Leon County, Florida. For a more complete historical record of nomenclatural changes leading up to current usage see Schmidt and Clark (1980, p. 41–58). They include references citing paleontological descriptions and age correlations. Fossils from the Jackson Bluff include: numerous mollusk shells, corals, ostracods, echinoids, bryozoans, barnacles, benthic and planktonic foraminifers, and calcareous nannofossils, among others. Based on planktonic foraminifera, it has been concluded that the Jackson Bluff was deposited in the Middle to earliest Late Pliocene (Akers, 1972).

Dubar and Beardsley (1961) reported on the paleoecology of the Jackson Bluff Formation deposits from the Alum Bluff site. They concluded that the sediments were deposited in the inner neritic zone in less than 8 fathoms (14.6 m) of normal open shelf marine water during a minor marine transgression.

Sandy Clays and Clayey Sands. These sediments are responsible for the name of this local outcrop. "From the efflorescence of ferrous sulphate arises the name *Alum* Bluff" (Langdon, 1889 p. 322). This interval has been called the "aluminous clay" by several authors. Dubar and Beardsley (1961) have suggested that this interval was deposited in stagnant, possibly lagoonal, conditions. There is an absence of marine fossils and the sulphurous odor and carbonaceous nature of the sediments also lead to this conclusion.

Citronelle Formation. The Citronelle Formation was named by Matson (1916, p. 168) for outcrops near the town of Citronelle, in southwestern Alabama. In northwestern Florida it consists of fluvial, cross-bedded sands, gravels and clays, and post-depositional limonite. In the Alum Bluff vicinity this unit consists of cross-bedded quartz sands, sandy more massive clay beds, clay pebbles, and brown to red, iron stained, bedded, clayey sands. The age of the Citronelle has been difficult to ascertain due to the lack of fossil material (Isphording and Flowers, 1983). More recent reworking and other similar deltaic deposits make correlation and differentiation time consuming and questionable. Recent workers have assigned a Late Pliocene–Pleistocene age to these sediments based on plant remains and stratigraphic position.

REFERENCES CITED

Akers, W. H., 1972, Planktonic Foraminifera and Biostratigraphy of Some Neogene Formations, Northern Florida and Atlantic Coastal Plain: Tulane Studies in Geology and Paleontology, v. 9, 140 p.

Banks, J. E., and Hunter, M. E., 1973, Post-Tampa, Pre-Chipola Sediments Exposed in Liberty, Gadsden, Leon, and Wakulla Counties, Florida: Transactions Gulf Coast Association of Geological Societies, v. 23, p. 355–363.

Cooke, C. W., 1945, Geology of Florida: Florida State Geological Survey Bulletin 29, 342 p.

Cooke, C. W., and Mossom, S., 1929, Geology of Florida: Florida State Geological Survey 20th Annual Report, 1927–1928, p. 29–27.

Dall, W. H., and Harris, G. D., 1892, The Neocene of North America: U.S. Geological Survey Bulletin 84, 349 p.

Dall, W. H., and Stanley-Brown, J., 1894, Cenozoic Geology along the Apalachicola River: Bulletin Geological Society of America, v. 5, p. 157.

DuBar, J. R., and Beardsley, D. W., 1961, Paleoecology of the Choctawhatchee Deposits (Late Miocene) at Alum Bluff, Florida: Southeastern Geology, v. 2, p. 155–189.

Gardner, Julia A., 1926–1944, The Molluscan Fauna of the Alum Bluff Group of Florida: U.S. Geological Survey Professional Paper 142, pts. 1-4, 1926; pt. 5, 1928; pt. 6, 1937; pt. 7, 1944, 709 p.

Gibson, T. G., 1967, Stratigraphy and Paleoenvironment of the Phosphate Miocene Strata of North Carolina: Geological Society of America Bulletin, v. 78, p. 631–650.

Isphording, W. C., and Flowers, G. C., 1983, Differentiation of Unfossiliferous Clastic Sediments: Solutions From the Southern Portion of the Alabama–Mississippi Coastal Plain: Tulane Studies in Geology and Paleontology, v. 17, No. 3, p. 59–83.

Langdon, D. W., Jr., 1889, Some Florida Miocene: American Journal Science 3rd Series, v. 38, p. 322.

Matson, G. C., 1916, The Pliocene Citronelle Formation of the Gulf Coastal Plain: U.S. Geological Survey Professional Paper 98, p. 167–192.

Matson, G. C., and Clapp, F. G., 1909, A Preliminary Report of the Geology of Florida, with Special Reference to the Stratigraphy: Florida State Geological Survey, Second Annual Report, p. 25–173.

Maury, C. J., 1902, A Comparison of the Oligocene of Western Europe and the Southern United States: Bulletin of American Paleontology, v. 3, No. 15, 94 p.

Pirkle, E. C., 1956, The Hawthorn and Alachua Formation of Alachua County, Florida Quarterly Journal of the Florida Academy of Science, v. 19, p. 197–240.

Pirkle, E. C., Yoho, W. H., and Allen, A. J., 1965, Hawthorn, Bone Valley, and Citronelle Sediments of Florida: Quarterly Journal of Florida Academy of Science, v. 28, no. 1, p. 7–47.

Puri, H. S., 1953, Contribution to the Study of the Miocene of the Florida Panhandle: Florida State Geological Survey Bulletin 36, 345 p.

Puri, H. S., and Vernon, R. O., 1964, Summary of the Geology of Florida and a Guidebook to the Classic Exposures: Florida State Geological Survey, Special Publication 5 Revised, 312 p.

Rainwater, E. H., Herring, D. G., and Ericson, D. B., 1945, Western Florida: Southeastern Geological Society, Third Field Trip, Nov. 9, and 10th, 1945, 93 p.

Schmidt, W., 1983, Neogene Stratigraphy and Geologic History Apalachicola Embayment, Florida [Ph.D. thesis]:, Florida State University, Tallahassee, 233 p.

Schmidt, W., and Clark, M. W., 1980, Geology of Bay County, Florida: Florida Bureau of Geology, Bulletin 57, 96 p.

Vernon, R. O., and Puri, H. S., 1956, A Summary of the Geology of Panhandle Florida and Guidebook to the Surface Exposures: Geological Society of American Southeastern Section Field Trip, March 24, 1945, 83 p.

Providence Canyons: The Grand Canyon of southwest Georgia

Arthur D. Donovan, Exxon Production Research Company, Houston, Texas 77252-2189
Juergen Reinhardt, U.S. Geological Survey, 928 National Center, Reston, Virginia 22092

LOCATION

Providence Canyons State Park is located in Stewart County, Georgia, in the Lumpkin SW 7½-minute quadrangle. The park entrance (Fig. 1) is located 0.15 mi (0.24 km) west of the intersection of Stewart County Road 23 and Georgia Road 39c.

The leader of any major field trip should advise the Park Superintendent of the purpose of the visit. No sampling or climbing of the canyon walls is allowed without the written permission of the Georgia Department of Natural Resources, 270 Washington Street, S.W., Atlanta, GA 30334.

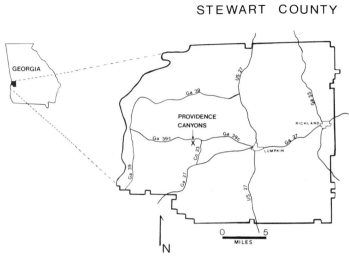

Figure 1. Location map of Providence Canyons State Park in southwestern Georgia.

INTRODUCTION

In Stewart and Quitman Counties, Georgia, spectacular exposures of the Providence Formation can be seen in numerous large gully exposures. These gullies are awesome, especially considering that they have developed as a result of poor land-use practices during the past 130 years. Providence Canyons, which is the type locality of the Providence Formation, is a complex of these gully exposures. Supplementary information on the stratigraphy of the area is available in Eargle (1955), Marsalis and Friddell (1975), and Reinhardt and Gibson (1980). Classic early stratigraphic studies in the eastern Gulf Coast by Stephenson (1911), Stephenson and Monroe (1938), and Monroe (1941) give valuable historical perspectives on the evolution of thinking on the entire Cretaceous section.

The Middle–Upper Maestrichtian Providence Formation is the youngest unit of the Upper Cretaceous Series in the Chatta-

hoochee River Valley region. This multilithologic, unconformity-bounded formation can be traced in a broad belt from central Alabama through the Chattahoochee River Valley and east to the Flint River in Georgia. Towards central Alabama, the Providence grades laterally into deposits mapped as the Prairie Bluff Chalk. The Providence Formation ranges in thickness from less than 100 ft (30 m) in parts of western Georgia to almost 300 ft (91 m) in eastern Alabama. The formation exhibits extraordinary areal variability in textures and composition along the outcrop belt, as well as in the shallow subsurface.

Donovan (1985) divided the Providence Formation into fine and coarse clastic facies. Fine clastic facies predominate in eastern Alabama, as well as in the shallow subsurface. Coarse clastic facies predominate along the outcrop belt in western Georgia. The Providence Formation (Fig. 2) contains two stacked facies sequences (informal lower and upper members) which are separated by a widespread regional flooding event. Each sequence displays an overall coarsening-upward or prograding nature, with a vertical (upward) and lateral (updip) change from fine clastic facies, dominated by horizontal burrows and a high diversity molluscan assemblage, to coarse clastic facies, dominated by vertical burrows and rare body fossils. The textural and faunal transitions in the Providence Formation indicate a downdip to updip change from offshore muds, dominated by infaunal deposit-feeders, to nearshore higher energy sands, dominated by infaunal suspension-feeders.

SITE DESCRIPTION

The large-scale geometry of the sedimentary units can best be traced by viewing the gully walls from the canyon rim. Details of a measured section can only be seen by careful inspection of fresh in-place gully walls. The Providence Canyons section was measured in Canyon #4, which may be reached by walking down into the canyons by footpaths located adjacent to Canyon #1, between Canyons #2 and 3, or between Canyons #3 and 4 (Fig. 3).

Special care should be exercised on the footpaths after rain, when the surfaces tend to be especially slippery. Caution near the canyon walls is advisable both along the canyon rim and at the base, since periodic rapid headward retreat of the canyons is still in progress.

The measured section at Providence Canyons State Park (Fig. 4) reveals a complex sequence of stacked, burrowed, laterally discontinuous, fining-upward cross-bedded sands. These sands can be divided into two stacked sequences by a distinctive

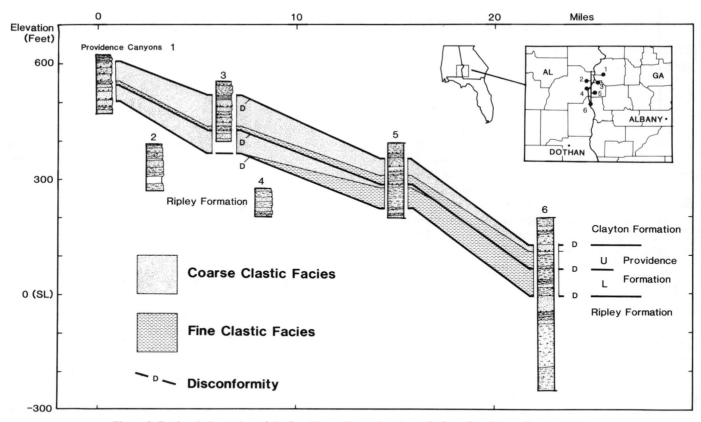

Figure 2. Regional dip section of the Providence Formation downdip from Providence Canyons State Park.

yellowish clayey sand horizon. This horizon (unit 12 on Fig. 4) is very continuous laterally and can be mapped throughout the canyons. Unit 12 is a pebbly sand at the base and grades upward into bioturbated clayey sand and relict-bedded sand and clay. These subfacies are common in the fine clastic facies of the Providence, which predominates downdip in the shallow subsurface and along the outcrop belt in eastern Alabama.

The base of unit 12 is interpreted as a transgressive disconformity within the Providence Formation, whereas the bulk of the unit appears to represent an updip tongue of Providence Formation fine clastic sediments. The regional significance of this break is demonstrated in Figure 2. The internal disconformity was used by Donovan (1985) to divide the Providence Formation into informal lower and upper members, both of which display a prograding vertical (upward) and lateral (downdip) nature.

The description that follows of the measured section corresponds to the units shown in Figure 4. Sedimentary units 1 through 6 are in the Ripley Formation; 7 through 15 constitute the Providence Formation; the remaining units, 16 through 18, represent highly weathered Clayton Formation (Paleocene) and Colluvium (Quaternary). Widespread regional disconformities, not textural changes, mark the boundaries of the Providence Formation. Cross-stratification in this report is divided into small-

scale [<4 in, (<10 cm)], medium-scale [4 to 12 in (10 to 30 cm)], and large-scale [>12 in, (>30 cm)].

PROVIDENCE CANYONS

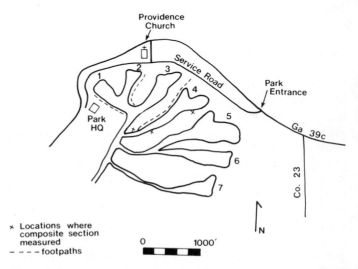

Figure 3. Sketch map of the digitate guillies comprising the Providence canyons. Location of the composite measured section is indicated along the walls of Canyon #4.

STRATIGRAPHY

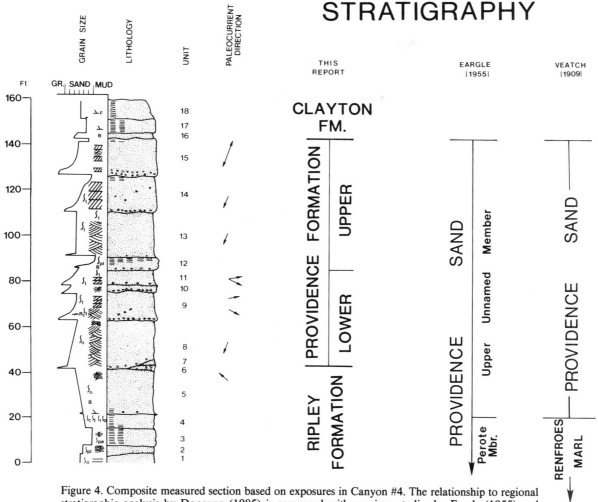

Figure 4. Composite measured section based on exposures in Canyon #4. The relationship to regional stratigraphic analysis by Donovan (1985) is compared with previous studies by Eargle (1955) and Veatch (1909).

Ripley Formation

Unit 1: 4 ft (1.2 m); Massive to cross-bedded sand; light gray; fine sand (quartz [q], muscovite [m], heavy minerals [hm], carbonaceous debris [c]); massive overall but one medium-scale cross-bed set at top; *Ophiomorpha.*

Unit 2: 4 ft. (1.2 m); Hummocky cross-stratified sand; dark yellowish orange; very fine to fine sand (q, m, hm, c), large-scale cross-stratification; *Planolites.*

Unit 3: 8 ft (2.4 m); Interbedded sand and clay; light olive gray to black; very fine to fine sand (q, m, hm, c)/silty clay; sands and clays are bedded on 2–4-in. (5–10-cm) scale, sands display planar laminations, clays contain sand partings and starved ripples; *Palaeophychus.*

Unit 4: 5 ft (1.5 m); Relict-bedded sand and clay; light-dark gray; very fine to fine sand (q, m, hm)/silty clay; locally contains sand-clay interbeds, sand component increases upward; *Ophiomorpha, Rosselia, Cylindricus, Thalassanoides.*

Unit 5: 19 ft (5.7 m); Coarsening-upward sand; moderate yellow; fine to medium sand (q, m, hm); changes from bioturbated to cross-stratified sands upward, medium-scale trough cross-stratified sets near top; *Ophiomorpha, Palaeophycus;* sharp and somewhat irregular base with local ironstone development.

Unit 6: 3 ft (0.9 m); Mottled sand; Grayish-pink, medium sand (q, m); red-stained root structures throughout, well developed clay clast lag in basal 4 in (10 cm), undulatory base.

Providence Formation

Unit 7: 2 ft (0.6 m); Pebbly sand; light brown; coarse sand/pb (q); massive with quartz pebbles scattered throughout and concentrated at base.

Unit 8: 18 ft (5.4 m); Fining-upward, cross-bedded sand; dusky purple; fine sand to pebbles (q, m, hm, clay clasts [cc]); dominated by large-scale trough cross-stratified sets, scattered small-scale sets at top; *Ophiomorpha;* scour base that truncates units 6 and 7 laterally.

Unit 9: 11 ft (3.2 m); Fining-upward sand; grayish-red purple; very fine sand to pebbles (q, m, hm, cc); contains large-scale tabular and trough sets displaying vertical decrease in bedform scale, sets near base up to 3 ft (1 m) high with small-scale sets superimposed; *Ophiomorpha;* distinct scour base with clay clast lag.

Unit 10: 4 ft (1.2 m); Fining-upward sand; multicolored; very fine sand to pebbles (q, m, hm, cc); medium-scale trough cross-stratified sands; *Ophiomorpha;* scour base with clay clast lag.

Unit 11: 7 ft (2.1 m); Fining-upward sand; pale blue; very fine sand to pebbles (q, m, hm, cc); medium-scale cross-stratified sands, upward decrease in bedform scale, upward increase in burrowing; *Ophiomorpha*; scour base with clay clast lag.

Unit 12: 6 ft (1.8 m); Muddy sand; pale yellowish-orange; clay to fine sand/pebbles (q, m, hm, cc); massive-bioturbated overall with scattered burrow fabrics near top, locally relict sand-clay interbeds near top; *Planolites*; pebble-sized quartz and clay clasts scattered near base.

Unit 13: 20 ft (6 m); Fining-upward sand; pale red-purple; fine sand to pebbles (q, m, hm, cc); large-scale trough cross-stratification, upward decrease in bedform scale, upward increase in burrowing; *Ophiomorpha, Thalassanoides,* well-developed clay clast lag at base.

Unit 14: 15 ft (4.5 m); Fining-upward sand; pale red-purple; fine sand to pebbles (q, m, hm, cc); large-scale tabular cross-stratification, upward decrease in bedform scale; *Ophiomorpha*; clay clast lag at base, upper 3 ft (0.9 m) of unit is a massive clayey sand.

Unit 15: 16 ft (4.8 m); Fining-upward sand; pale yellowish-orange; very

fine sand to pebbles (q, m, hm, cc); medium-scale tabular and trough cross-stratified sets scattered throughout; unit locally bioturbated, scattered distinct burrow fabrics *Ophiomorpha, Microcladichnus*; clay clast lag at base, upper 2 ft (0.6 m) of unit is a massive, rooted, clayey sand.

Clayton Formation

Unit 16: 3 ft (0.9 m); Pebbly sand; grayish-orange to pink; medium sand/pebbles (q, m, hm); massive-bioturbated; undulatory base with scattered quartz pebbles concentrated in basal 1 ft (0.3 m) of unit.

Unit 17: 6 ft (1.8 m); Very clayey sand; dark reddish brown; clay-granules (q); massive local ironstone development.

Colluvium

Unit 18: 8+ ft (2.4 m); Clayey sand; red; sand matrix with angular ironstone fragments.

REFERENCES CITED

Donovan, A. D., 1985, Sedimentology and stratigraphy of the Providence Formation [Ph.D. thesis]: Golden, Colorado School of Mines, 238 p.

Eargle, D. H., 1955, Stratigraphy of the outcropping Cretaceous rocks of Georgia: U.S. Geological Survey Bulletin 1014, 101 p.

Marsalis, W. E., and Friddell, M. S., 1975, A guide to selected Upper Cretaceous and lower Tertiary outcrops in the Lower Chattahoochee River Valley of Georgia: Georgia Geological Society Guidebook 15, 79 p. plus Appendix.

Monroe, W. H., 1941, Notes on deposits of Selma and Ripley age in Alabama: Alabama Geological Survey Bulletin 48, 150 p.

Reinhardt, J., and Gibson, T. G., 1980, Upper Cretaceous and Lower Tertiary Geology of the Chattahoochee River Valley, western Georgia and eastern

Alabama, *in* Frey, R. W., ed., Excursions in southeastern Geology, v. 11: Geological Society of America 1980 National Meeting, Atlanta, Georgia, American Geological Institute, p. 395–463.

Stephenson, L. W., 1911, Cretaceous (rocks of the Coastal Plain of Georgia), *in* Veatch, J. O., and Stephenson, L. W., Preliminary report on the geology of the Coastal Plain of Georgia: Georgia Geological Survey Bulletin 26, p. 66–216.

Stephenson, L. W., and Monroe, W. H., 1938, Stratigraphy of Upper Cretaceous series in Mississippi and Alabama: American Association of Petroleum Geologists Bulletin, v. 22, no. 12, p. 1639–1657.

Veatch, J. O., 1909, Second report on the clay deposits of Georgia: Georgia Geological Survey Bulletin 18, 453 p.

Sedimentary facies of the Upper Cretaceous Tuscaloosa Group in eastern Alabama

Juergen Reinhardt, U.S. Geological Survey, Mail Stop 925, Reston, Virginia 22092
L. W. Smith and D. T. King, Jr., Department of Geology, Auburn University, Auburn, Alabama 36849

LOCATION

These sedimentary facies are exposed in three, related stops in the eastern Gulf Coastal Plain in west-central Alabama, in the vicinity of Tuskegee eastward to Auburn. Specific stop locations are shown on the detailed topographic inset maps (Fig. 1). Except for one locality, the described exposures are along paved rights-of-way. No permission is required to view these outcrops and conventional transportation (car, bus, or van) is appropriate, although parking safely may be difficult along some of the secondary roads.

INTRODUCTION

Exposures of the Tuscaloosa Group in eastern Alabama exhibit sedimentary features characteristic of a broad alluvial plain. These sediments fringed the Appalachian orogen and comprised the coastal plain adjacent to the enlarging Gulf of Mexico Basin during early Late Cretaceous time. The sediments described herein are part of an extensive outcrop belt of Tuscaloosa deposits which can be traced from the southeastern margin of the Mississippi embayment through the type area in western Alabama and eastward to the Flint River in western Georgia, where it is over-

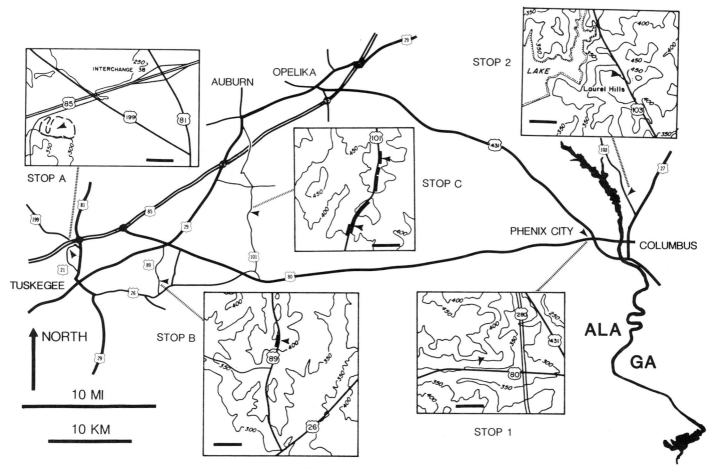

Figure 1. Location map of eastern Alabama showing the locations of Stops A, B, and C, and the supplementary Tuscaloosa stops. Inset A from Tuskegee 7½-minute quadrangle; inset B from Little Texas 7½-minute quadrangle; inset C from Auburn 7½-minute quadrangle. Supplementary stops are on the Phenix City and Fortson 7½-minute quadrangles. Bar scales on quadrangle insets equal approximately 1,000 ft (300 m).

lapped by younger Cretaceous units. Downdip, the Tuscaloosa is an important reservoir for water in western Georgia and most of Alabama. Further to the south and west, the deep Tuscaloosa stratum has become an increasingly significant source of oil and gas in Louisiana and Mississippi.

The stratigraphic and structural relationships of the Tuscaloosa Group have been extensively studied by Monroe and others (1946), Drennen (1953), and Eargle (1955). The Tuscaloosa in eastern Alabama lies unconformably on Piedmont rocks of the Uchee and Pine Mountain belts and generally dips to the south at 45 to 60 ft/mi (8.5 to 11.5 m/km). Locally, paleosols (fossil soils) preserved on the crystalline rock surface beneath the Tuscaloosa indicate a prolonged or intense period of subaerial weathering prior to the onset of terrigenous sedimentation (Freeman, unpublished data, 1981; Reinhardt and Sigleo, 1983; Sigleo and Reinhardt, 1985).

The texture and composition of the Tuscaloosa indicate that it is an immature sandstone locally derived from the adjacent Piedmont lithologies (Frazier, 1982). The depositional motif of the Tuscaloosa consists of superimposed fining-upwards sequences generally 10 to 20 ft (3 to 6 m) thick. The base of each sequence is marked by an erosional contact and the basal part of the sequence is commonly composed of a sandy conglomeratic facies. The middle part of the fining-upward sequence typically consists of trough and planar cross-stratified arkosic sand. The top of the sequence consists of a massive or mottled silty or sandy clay. Extensive paleosols within the clayey sediments of the Tuscaloosa have been described where the cycle top is not truncated by the overlying fining-upward sequence (Smith, 1984).

Local accumulations of leaves, silicified or lignitized wood, and networks of back-filled burrows (*Muensteria*) are common biogenic features of the Tuscaloosa in eastern Alabama. In contrast, the Tuscaloosa in western Alabama contains marine invertebrates in both surface exposures and in the shallow subsurface (Applin, 1964; Sohl, 1964). Palynomorphs in the Tuscaloosa in eastern Alabama strongly suggest an Eaglefordian or Late Cenomanian age (Christopher, 1980, 1982).

The primary sedimentary structures, the vertical and lateral facies relationships, and the various biogenic features in the updip Tuscaloosa section all indicate deposition in a continental environment, specifically in a sandy, low-sinuosity stream environment. Locally, broad, stable flood plains were well established. The crossbedded sand facies, which constitutes most of the Tuscaloosa section, was deposited in migrating bars and as channel fill deposits. The small amount of clay—representative of the flood plain—in most of the fining-upward sequences, and the lensoid geometry of many of the sand bodies suggest that lateral accretion and channel aggradation were the dominant processes in these stream systems (Reinhardt, 1980). Smith (1984) suggested that the Tuscaloosa streams were hybrid types, the product of coarse-grained meander belts. In areas that lacked fine overbank sediments, channels were probably relatively unconfined. These channels, and their associated sand bars, migrated laterally to produce broad sand sheets.

The geographically extensive fluvial facies of the Tuscaloosa in the eastern Gulf Coastal Plain suggests considerable relief and probable uplift in the southern Appalachians during the Cenomanian, a time of worldwide marine transgression. The aggradation of a major clastic wedge during this time interval profoundly affected the composition and supply of sediment to the Gulf Basin for the balance of the Cretaceous. The opportunity to study updip-to-downdip, as well as lateral, changes in the composition of the Tuscaloosa is outstanding in the highly dissected area of eastern Alabama where the coastal plain overlap can be traced for hundreds of miles along the fall line.

SITE DESCRIPTIONS

Three stops, which are outstanding representatives of the Tuscaloosa in Alabama, are described below. There are also abundant outcrops between the three, and supplementary stops are suggested at the end of this section.

Stop A: Floodplain Facies of the Tuscaloosa

Stop A is a series of borrow pits accessible only on foot; footing may be extremely treacherous following rain. Permission to visit can be obtained from: Andrew V. Sharpe, P.O. Box 2176, Auburn, AL 36830 (205/821-0952).

The section at Stop A contains an outstanding example of the bioturbated clay facies. The disconformable contact with the overlying channel sands is well exposed in the western part of the area. The site is a series of abandoned borrow pits adjacent to I-85. Although the pits have been abandoned for about ten years, the gullied exposures of the clay facies are fresh and support little or no vegetation in the disturbed area. A composite section of the Tuscaloosa sediments exposed at this site is presented in Figure 2.

The base of the section (unit 1) is composed of reddish brown clayey silt to silty clay, characterized by prominent subvertical joint sets and slickensided surfaces. In addition, the clay is riddled by horizontal to inclined, backfilled burrows, identified as the trace fossil *Muensteria* (Smith, 1984). Gray-to-green patches or mottles along fracture planes and associated with apparent root casts are thought to represent areas of iron reduction, either during pedogenesis or early diagenesis. Other features associated with pedogenesis are indurated nodules in the most iron-rich parts of the mottled clay, and hardened burrow fillings. These nodules are indurated rather than cemented, but their size and distribution in the sediment is reminiscent of "plinthites" in paleosols.

Unit 2 in the section (Fig. 2) is a fine-to-medium, micaceous, crossbedded arkosic sand. The quartz and potassium feldspar grains are subangular, poorly sorted, and stained to a very pale orange. The scale of the cross stratification decreases from medium- to small-scale with a corresponding decrease in overall grain size. Stacking and amalgamation of the sandy, channel fill sands are suggested by the intervening lens of silty clay and the

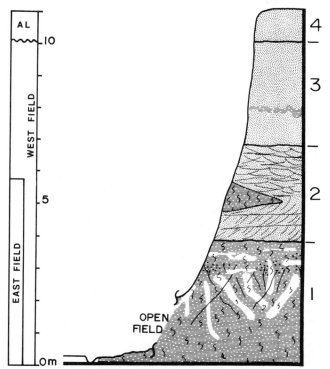

Figure 2. Measured section at Stop A; units, patterns, and symbols are described in text. The lower part of the section is better exposed in the eastern part, and the upper part of the section is better exposed in the western part of the locality. AL at the top of the section is considered a surficial unit of uncertain age and origin. Unconventional symbols within unit 1 include: black dots for hematite nodules and subvertical and horizontal blank patches for areas of reduced iron.

horizontal and vertical changes in grain size within unit 2. Unit 2 is the basal unit of a fining-upward sequence, but it is less representative of the sandy channel lithofacies than those seen in other Tuscaloosa sections.

Units 3 and 4 are probably post-Tuscaloosa deposits. Unit 3 is massive to locally planar laminated with a thin interval of convoluted lamination. The sedimentary structures and the composition of the sand (general lack of mica and little or no clay matrix) are unlike the Tuscaloosa sands in eastern Alabama. Both units 3 and 4 are provisionally categorized as surficial deposits, but detailed mapping to define the exact distribution and the constraints on age of the deposits has not yet been completed.

Stop B: Channel Sands and Pedogenesis

The measured section along the Macon County, Alabama 89 right-of-way is afforded by recent gullying of the Tuscaloosa and the capping alluvial terrace deposits. The section (Fig. 3) is composed of four fining-upward sequences in the Tuscaloosa, capped by the highest terrace associated with Chewacla Creek. The development of a paleosol at the top of the lowest fining-upward sequence is unusually complete in terms of preservation of root networks penetrating into the sandy substrate below the fine-grained alluvial soil.

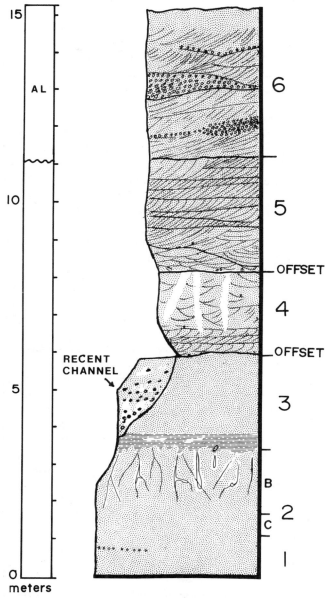

Figure 3. Measured section at Stop B; lateral offsets in the section correspond to short, northward traverses along Lee County Route 101. The lower part of the section may be poorly accessible due to recent gullying.

Unit 1 is a very pale orange, subangular, poorly sorted, fine-to-medium arkosic sand. The sand is massive with little variation in sediment size or composition, with the exception of thin concentrations of pebbles and granules in the lower part of the unit, and local enrichment of muscovite on poorly defined bedding surfaces.

Unit 2 is the part of the sequence most affected by Cretaceous pedogenic processes. The top of the paleosol consists of a 1 ft (0.3 m) interval of purple-to-ocher clay to sandy clay. The structure of the interval is massive and considered to be completely bioturbated. The sandy part of unit 2 (Fig. 2) has been divided into B and C, corresponding directly to the B and C horizons of the paleosol. The sandy B horizon is an indurated, bright brick red, massive sand. The interval contains some cemented nodules with concentric lamination (soil pisolites) and is cut by vertical and subvertical downward-branching fissures with diffuse, pale olive margins. The C horizon is a highly mottled zone with irregular patches of pale orange and bright red; it grades into the underlying, slightly stained sand.

Unit 3 is marked by a sharply erosional base. The lower part of the unit contains abundant clay clasts similar in color and composition to the clay in the top of unit 2. Overall, unit 3 is a fine-to-medium quartz sand with poorly defined trough crossbed sets. Adjacent to unit 3 is a local deposit interpreted as filling of a previously excavated gully (recent channel on Fig. 3).

Unit 4 is a trough crossbedded, fine-to-coarse arkosic sand; crystalline rock pebbles, intraformational clayey sand clasts, and coarse quartz grit occur throughout the lower part of the unit. The 7 ft (2 m)-thick unit is poorly sorted, miceous throughout, and fines upward. Reduced iron halos (root mottles) and clay infills of root casts are associated with the modern ground surface and are not related to a Cretaceous paleosol surface as in the underlying unit 2.

Further upsection, higher measured stratigraphic units were laterally offset slightly, because the lower part of the section was covered. The top of the Tuscaloosa section is similar in composition to the underlying sand body, but the style of bedform changes from predominantly trough to planar crossbeds.

The terrace deposits (unit 6) that cap the tops of the hills in this area are considerably coarser grained than the underlying Tuscaloosa sediments. The terrace deposits are deeply stained, contain no mica, and contain clay clasts and quartz pebbles with hematite rinds and coatings. The bedforms are small- to medium-scale trough crossbeds with abundant cut-and-fill structure throughout.

Stop C: Interstratified Channel Sands and Overbank Deposits

This stop is a series of roadcuts along Lee County, Route 101; the section (Fig. 4) is the composite of several closely spaced exposures along the north-trending road. Additional descriptions of this and other closely related stops are also available in Smith and King (1983). Although there are some similarities to the section described at Stop B, this stop affords a better opportunity to see the geometry and orientation of the channels and the local variation of the channel fills.

The lower part of the section is a 10 ft (3 m)-thick, medium grained arkosic sand (unit 1) with sparse polycrystalline quartz pebbles and clay intraclasts. The plane of the roadcut is parallel to the paleocurrent direction, giving the large trough crossbeds a planar, tabular appearance. Toward the top of the individual channel fill sequences, the bedforms are ripples; climbing ripples are well preserved. The sand grades upward into a bioturbated silty clay. The overall characteristics are like the clay facies de-

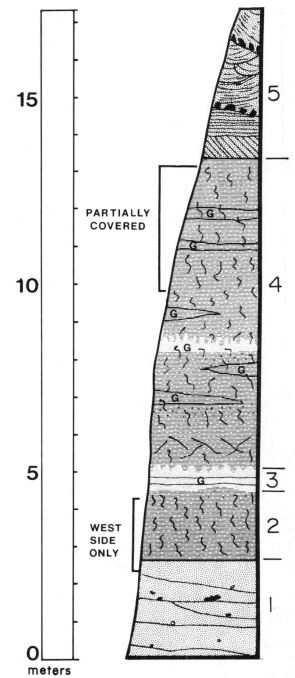

Figure 4. Composite measured section at Stop C; various parts of the section are exposed in a series of small roadcuts. Olive green sand to sandy clay intervals are marked G on the section.

scribed at Stop A, except that vertical fractures and slickensides are not prominent features at this location. Both the *Muensteria* ichnofacies and the incipient pedogenic features are as well developed here as at Stop A.

The middle of the section is interpreted as a thick sequence of overbank deposits dominated by the silty clay lithofacies. Two thin interbeds of gray-green silty sand are thought to be distal crevasse splay deposits, derived from breaching of channel levees. The sands contain an unusually large amount of biotite. The tops of the thin sand beds are bioturbated; considerable mixing of the sand and adjacent silty clay has taken place, causing the sand-clay contacts to be very diffuse.

Pedogenic features within unit 4 include inclined fracture surfaces with slickensides; these are thought to be craze planes, soft-sediment and compaction features of water-logged soils. Hardening and incipient cementation of the sediment through local concentration of iron oxides have produced nodular concretions ⅛ to ½ in (2 to 12 mm) in diameter.

Above a sharp and erosional contact, a highly channeled, medium-to-coarse arkosic sand exhibits some of the most spectacular and photogenic channel forms in the Tuscaloosa outcrop belt. Characteristic of this interval are granule- to boulder-sized clay intraclasts within the channel fill deposits, especially close to the bottom of the channel forms. Reduction rinds on many of the clay clasts strongly suggest that the changes in the oxidation state of the iron took place in the depositional environment. The abundance of overbank deposits with interbedded, highly channeled sands in this area suggests the dominance of stream systems more comparable to modern high bed-load meandering streams than most sand-dominated Tuscaloosa sections (Smith, 1984).

Supplementary Tuscaloosa Localities

Although Stops A, B, and C described in the previous section are some of the most instructive for understanding the nature of the basal Cretaceous sediments and sedimentation near the eastern margin of the Gulf Coastal Plain, there are many additional localities that are useful and reasonably accessible to large groups of geologists. Guidebooks of the Georgia Geological Society (Marsalis and Friddell, 1975; Reinhardt and Gibson, 1981), and the Alabama Geological Society (Carrington, 1983) contain the most useful stop descriptions of Tuscaloosa localities in the western Georgia and eastern Alabama area. Several of these localities were considered for this cluster stop, but were thought to expand the area of the stops too much from the three localities described (see Fig. 1).

Supplementary Stop 1 is a section of the Tuscaloosa 60 ft (20 m) thick, capped by a high terrace of the Chattahoochee River. It is located just west of the junction of U.S. 80 and 280 near Phenix City, AL. This section has been described in two of the guidebooks cited above and exhibits many of the features described at Stops B and C. The site is owned by John Franklin, c/o Casa Grande Apartments, Phenix City, AL. Permission is required for large groups; park next to the car wash that fronts the exposure.

Supplementary Stop 2 is on Georgia 103 at 54th Street in Columbus, GA. This locality is the finest exposure of the contact of the Tuscaloosa with the underlying crystalline rock and the associated Cretaceous paleosol (see Sigleo and Reinhardt, 1985; Reinhardt and Sigleo, 1983). A similar exposure of Tuscaloosa on weathered bedrock slightly further to the east is described in Marsalis and Friddell (1975).

REFERENCES CITED

Applin, E. R., 1964, A microfauna from the Coker Formation, Alabama: U.S. Geological Survey Bulletin 1160D, p. 65–70.

Carrington, T. J., ed., 1983, Current studies of Cretaceous formations in eastern Alabama and Columbus, Georgia: Alabama Geological Society, 20th Annual Field Trip, Guidebook, 52 p.

Christopher, R. A., 1980, Palynological evidence for assigning an Eaglefordian age (Late Cretaceous) to the Tuscaloosa Group of Alabama: Gulf Coast Section SEPM Research Conference Program and Abstracts, Houston TX, p. 12–13.

——— 1982, Palynostratigraphy of the basal Cretaceous units of the eastern Gulf and southern Atlantic Coastal Plains: *in* Arden, D. D., Beck, B. F., and Morrow, E., eds., Proceedings from the Second Symposium on the Geology of the Southern Coastal Plain, Georgia Geological Survey Circular 53, p. 10–23.

Drennen, C. W., 1953, Reclassification of the outcropping Tuscaloosa Group in Alabama: American Association of Petroleum Geologists Bulletin, v. 37, p. 522–538.

Eargle, D. H., 1955, Stratigraphy of the outcropping Cretaceous rocks of Georgia: U.S. Geological Survey Bulletin 1014, 101 p.

Frazier, W. J., 1982, Sedimentology and paleoenvironmental analysis of the Upper Cretaceous Tuscaloosa and Eutaw Formations in western Georgia: *in* Arden, D. D., Beck, B. F., and Morrow, E., eds., Proceedings from the Second Symposium on the Geology of the Southern Coastal Plain, Georgia Geological Survey Circular 53, p. 39–49.

Freeman, R. F., 1981, The interpretation of the occurrence and the chemical analysis of the Columbus laterite: Columbus College, Columbus, GA, 19 p. (unpublished manuscript).

Marsalis, W. E., and Friddell, M. S., 1975, A guide to selected Upper Cretaceous and lower Tertiary outcrops in the lower Chattahoochee River valley of Georgia: Georgia Geological Society, Guidebook 15, 79 p. + Appendix.

Monroe, W. H., Conant, L. C., and Eargle, D. H., 1946, Pre-Selma Upper Cretaceous stratigraphy of western Alabama: American Association of Petroleum Geologists Bulletin, v. 30, p. 590–607.

Reinhardt, Juergen, 1980, Tuscaloosa Formation (Cenomanian) from eastern Alabama to central Georgia: Its stratigraphic identity and sedimentology: Gulf Coast Section SEPM Research Conference Program and Abstracts, Houston TX, p. 25–26.

Reinhardt, J., and Gibson, T. G., 1981, Upper Cretaceous and lower Tertiary geology of the Chattahoochee River Valley, western Georgia and eastern Alabama: Georgia Geological Society, 16th Annual Field Trip, Guidebook, 88 p.

Reinhardt, J., and Sigleo, W. R., 1983, Mesozoic paleosols: Examples from the Chattahoochee River Valley, *in* Carrington, T. J., ed., Current studies of Cretaceous formations in eastern Alabama and Columbus, Georgia: Alabama Geological Society, 20th Annual Field Trip, Guidebook, p. 3–10.

Sigleo, W. R., and Reinhardt, J., 1985, Cretaceous paleosols from the eastern Gulf Coastal Plain: Southeastern Geology, v. 25, p. 213–223.

Smith, L. W., and King, D. T., 1983, The Tuscaloosa Formation: Fluvial facies and their relationship to the basement topography and changes in paleoslope, *in* Carrington, T. J., ed., Current studies of Cretaceous formations in eastern Alabama and Columbus, Georgia: Alabama Geological Society, 20th Annual Field Trip, Guidebook, p. 11–16.

Smith, L. W., 1984, Depositional setting and stratigraphy of the Tuscaloosa Formation, central Alabama to west-central Georgia [M.S. thesis]: Auburn University, Auburn, AL, 125 p.

Sohl, N. F., 1964, Pre-Selma larger invertebrate fossils from well-core samples in western Alabama: U.S. Geological Survey Bulletin 1160C, p. 55–64.

Cretaceous-Tertiary boundary southeast of Braggs, Lowndes County, Alabama

Charles W. Copeland, *Geological Survey of Alabama, Box O, University Station, Tuscaloosa, Alabama 35486*
Ernest A. Mancini, *Geological Survey of Alabama and University of Alabama, Tuscaloosa, Alabama 35486*

LOCATION

The Cretaceous-Tertiary boundary is exposed in a highway cut along the southwest side of Alabama 263, 4.6 mi (7.4 km) southeast of the intersection of Alabama 21 and 263 at Braggs, in the southwestern part of Lowndes County, Alabama. The exposure is 3.8 mi (6.1 km) northwest of the Lowndes–Butler County boundary in the NE¼ NW¼ sec. 22, T. 12 N., R. 13 E. (Fig. 1). The cuts along Alabama 263 were first opened by highway construction in 1968.

INTRODUCTION

This exposure of the contact between the Prairie Bluff Chalk (Late Cretaceous) and the Pine Barren Member of the Clayton Formation (early Paleocene) is the most accessible of the Cretaceous-Tertiary contacts in the Alabama Coastal Plain. The contact is exposed along the southwest side of the highway in the northern half of the cut for a linear distance of about 250 ft

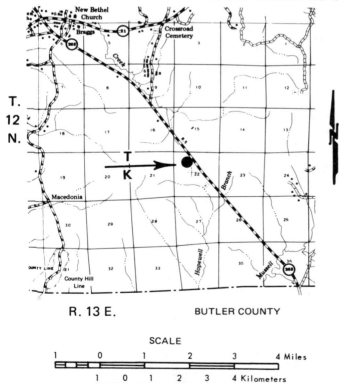

Figure 1. Location map to the Cretaceous-Tertiary contact, southeast of Braggs in the NE¼ NW¼ sec. 22, T. 12 N., R. 13 E., Lowndes County, Alabama

(76.2 m). The Cretaceous-Tertiary contact at the north end of the road cut is about 10 ft (3 m) above road level, and where measured, near where the Prairie Bluff dips below land surface, the contact is approximately 3 ft (0.9 m) above the center of the roadway.

The composite measured section accompanying this description (Fig. 2) includes beds from below road level in the ditch paralleling the outcrop and from two other points along the exposure. The sand bed, Bed 1 of Figure 2, occurs in the upper part of the Prairie Bluff, 9.4 ft (2.9 m) below the Cretaceous-Tertiary contact. The upper 9.4 ft (2.9 m) of the Prairie Bluff was measured at a point about 100 ft (30 m) south of the north end of the cut and the bed at the contact (Bed 3) and the overlying beds of the Pine Barren (Beds 4–14; 34 ft [10.4 m]) were measured and sampled at a point about 200 ft (60.9 m) from the north end of the roadcut at a place essentially bare of vegetation. The Prairie Bluff is approximately 100 ft (30 m) thick in western Lowndes County, and the Pine Barren Member of the Clayton Formation is about 115 ft (35.0 m) thick (R. W. Hedlund, Pers. comm., 1983). Summary descriptions of the Cretaceous formations exposed in the Alabama Coastal Plain are included in papers by Monroe (1941) and Copeland (1968). The stratigraphy and paleontology of the Paleocene and Eocene formations in Alabama are summarized in a monograph by Toulmin (1977).

DESCRIPTION

The contact between the Prairie Bluff Chalk and the Pine Barren Member of the Clayton is within an indurated limestone bed (Bed 3 of Fig. 2). Typical Cretaceous index foraminifers, *Rugoglobigerina rugosa* (Plummer) and *Globotruncana* sp. occur in Bed 2; the Paleocene species *Subbotina pseudobulloides* (Plummer) and *Subbotina trivialis* (Subbotina) occur in Bed 4. The non-indurated beds above and below the contact are similar in color and when cursorily examined seem to be of similar lithologic composition. However, when examined closely, the uppermost unconsolidated bed of the Prairie Bluff (Bed 2) is grayish-black, calcareous, silty and sandy, micaceous clay; and the lowest unconsolidated bed of the Pine Barren (Bed 4) is grayish-black (olive-gray when dry), very fine- to fine-grained, subangular, abundantly glauconitic, clayey, quartz sand. The lower beds of the Pine Barren also contain coarse to very coarse, subrounded, quartz grains that are frosted or clear and iron-stained in part. The large quartz grains are very similar to those occurring in Bed 3 and are not present in the underlying Prairie Bluff. Pelecypods typical of the Prairie Bluff such as *Exogyra*

Thickness
(feet)

Pine Barren Member of Clayton Formation

14. Clay, pale-yellowish-orange (10YR8/6) and medium-light-gray (N6), mottled, silty, calcareous, glauconitic, fossiliferous throughout with microfossils and molds and prints of shells; weathered to some extent.   13.7

13. Limestone, pale-yellowish-orange (10YR8/6) to dark-yellowish-orange (10YR6/6), weathered, glauconitic, silty, fossiliferous including leached molds and prints of mollusks.   0.5

12. Clay, dusky-yellow-green (5YR5/2) and greenish-gray (5G6/1) where weathered near the top of bed grading downward to clay, medium dark-gray (N4) silty, calcareous, glauconitic, micaceous, containing trace of carbonized wood fragments, fossiliferous including pectenids, venericards, microfossils, *Ostrea crenulimarginata* Gabb; pale-yellowish-orange (10YR 8/6) glauconitic, fossiliferous limestone cobbles occur as a discontinuous layer 3.4 feet above the base of the bed.   5.5

11. Limestone, light-olive-gray (5Y6/1), glauconitic, fossiliferous containing scattered medium, subrounded clear quartz grains; fossils include *Venericardia* sp. and other shells.   0.8

10. Clay, dark-gray (N3), silty, micaceous, calcareous, abundantly glauconitic (part of glauconite grains are internal molds of foraminifers) with very fine quartz sand and scattered medium to coarse, subrounded, clear and iron-stained quartz grains, fossiliferous, including shells and shell fragments, benthonic foraminifers and ostracodes.   2.4

9. Limestone, light-olive-gray (5Y6/1), glauconitic with scattered coarse, subrounded, iron-stained and clear quartz grains, fossiliferous, including shells, prints and molds of pelecypods and gastropods; specimens of *Ostrea pulaskensis* Harris are partially exposed on the upper surface of the bed.   1.3

8. Limestone, light-olive-gray (5Y6/1), glauconitic, argillaceous, with scattered coarse, subrounded, iron-stained and clear quartz grains, fossiliferous including fragments and shells of *Ostrea* sp.; upper and lower surfaces of bed are irregular.   1.1

7. Limestone, light-olive-gray (5Y6/1), glauconitic, argillaceous with scattered coarse and very coarse, subrounded, partly frosted, iron-stained and clear quartz grains, fossiliferous, including fragments and shells of *Ostrea pulaskensis* Harris; upper and lower surfaces of bed are irregular; limestone bed is 1.1 feet thick and is bounded by sandy clay partings 0.3 feet thick.   1.7

6. Sand, light-olive-gray (5Y5/2), clayey, very fine- to fine-grained, quartzose, abundantly glauconitic, containing coarse and very coarse subrounded, partly frosted, iron-stained and clear, quartz grains, fossiliferous including foraminifers, ostracodes and fragments of *Ostrea* sp.   1.6

5. Limestone, olive-gray (5Y4/1), argillaceous, glauconitic, containing small phosphate pebbles and scattered coarse and very coarse, subrounded, partly frosted, iron-stained and clear quartz grains, fossiliferous including thin pelecypod shells, probably *Ostrea* sp.; burrows are noticeable on relatively fresh surfaces; upper and lower surfaces of bed are irregular.   1.3

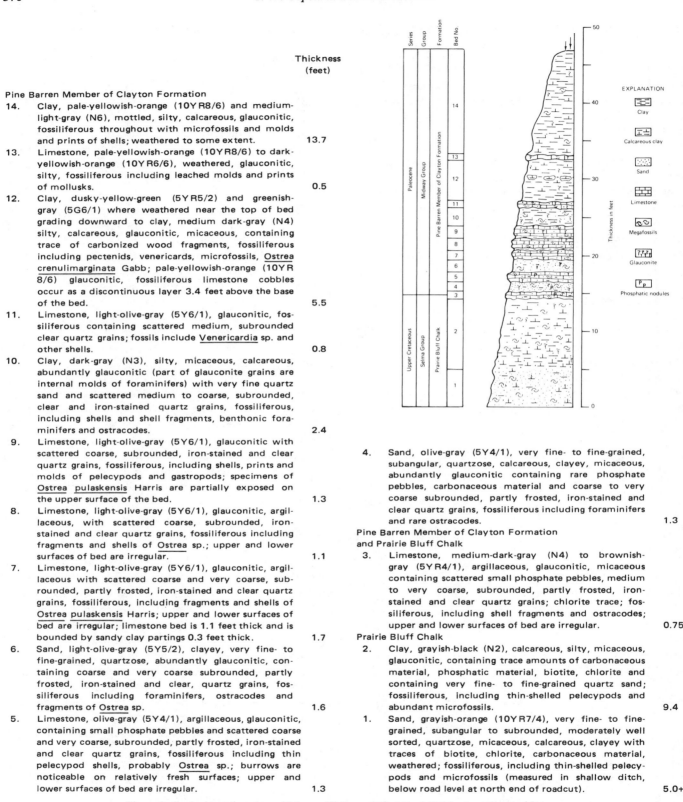

4. Sand, olive-gray (5Y4/1), very fine- to fine-grained, subangular, quartzose, calcareous, clayey, micaceous, abundantly glauconitic containing rare phosphate pebbles, carbonaceous material and coarse to very coarse subrounded, partly frosted, iron-stained and clear quartz grains, fossiliferous including foraminifers and rare ostracodes.   1.3

Pine Barren Member of Clayton Formation and Prairie Bluff Chalk

3. Limestone, medium-dark-gray (N4) to brownish-gray (5YR4/1), argillaceous, glauconitic, micaceous containing scattered small phosphate pebbles, medium to very coarse, subrounded, partly frosted, iron-stained and clear quartz grains; chlorite trace; fossiliferous, including shell fragments and ostracodes; upper and lower surfaces of bed are irregular.   0.75

Prairie Bluff Chalk

2. Clay, grayish-black (N2), calcareous, silty, micaceous, glauconitic, containing trace amounts of carbonaceous material, phosphatic material, biotite, chlorite and containing very fine- to fine-grained quartz sand; fossiliferous, including thin-shelled pelecypods and abundant microfossils.   9.4

1. Sand, grayish-orange (10YR7/4), very fine- to fine-grained, subangular to subrounded, moderately well sorted, quartzose, micaceous, calcareous, clayey with traces of biotite, chlorite, carbonaceous material, weathered; fossiliferous, including thin-shelled pelecypods and microfossils (measured in shallow ditch, below road level at north end of roadcut).   5.0+

Figure 2. Geologic section along Alabama Highway 263, 4.6 mi (7.4 km) southeast of Braggs.

costata Say and *Anomia argentaria* Morton were relatively common near the contact when the highway was first constructed but are now rare. Possibly they will be located more easily in the winter months. Beds 3, 4 and 5, at, and immediately above the contact, contain small dark-gray gravel-size phosphatic nodules. Biotite flakes and thin books of biotite up to 0.5 mm in size occur in the upper beds of the Prairie Bluff but were not noted in washed samples of the Pine Barren Member.

The Prairie Bluff in the ditch below the roadcut includes at least 5 ft (1.5 m) of very fine- to fine-grained, subangular to subrounded, moderately sorted, micaceous, fossiliferous quartz sand. The sand is probably a westward extension of a part of the Providence Sand, which occurs more prominently in central and eastern Lowndes County. The lower contact of the sand bed is covered by colluvium and its total thickness is not known; however, Smith (1978) measured 20 ft (6.1 m) of sand at this locality.

The sand bed at the base of the exposure is overlain by 9.4 ft (2.9 m) of grayish-black silty, sandy, calcareous, glauconitic, fossiliferous clay containing traces of carbonaceous material, phosphatic grains, chlorite and biotite. The fossil content of the Prairie Bluff clay beds is mainly pelecypod shells, shell fragments, and abundant microfossils.

The indurated bed containing the Cretaceous-Tertiary contact (Bed 3) is 9 in (23 cm) thick and consists of medium dark-gray to brownish-gray, glauconitic, fossiliferous, phosphatic limestone (sandy lime mudstone) containing phosphate pebbles, a trace of chlorite, and scattered medium to very coarse, subrounded, clear and partly stained quartz grains. Fossils noted in the bed include shell fragments and ostracodes. Bed 3 is the lowest, most prominent of the indurated beds occurring in the exposure.

Bed 4 at the base of the Pine Barren is 1.3 ft (39.6 cm) thick and consists of olive-gray, very fine- to fine-grained, abundantly glauconitic, fossiliferous quartz sand containing coarse and very coarse, partly clear and iron-stained quartz grains and dark-gray phosphatic pebbles. This basal sand is overlain by 1.3 ft (39.6 cm) of olive-gray, glauconitic, fossiliferous, clayey limestone (sandy lime mudstone) containing scattered coarse and very coarse subrounded quartz grains and phosphate pebbles (Bed 5). The upper and lower surfaces of the bed are undulatory and the thickness of the bed varies laterally. Extensive burrows are visible on freshly broken surfaces.

The lower limestone bed of the Pine Barren (Bed 5) is overlain by 1.6 ft (48.8 cm) of olive-gray, very fine- to fine-grained, moderately sorted, subangular, quartzose, calcareous, abundantly glauconitic, fossiliferous, clayey quartz sand that also contains coarse to very coarse subrounded, iron-stained and clear quartz grains similar to those in the underlying beds. This upper sand (Bed 6) contains fragments of *Ostrea* sp., other shells, foraminifers and ostracodes.

The overlying beds of the Pine Barren are mostly dark-gray, silty, micaceous, calcareous, sandy, glauconitic clay with thin interbeds of glauconitic, fossiliferous, partly sandy limestone. The clay beds weather to greenish gray and pale yellowish orange near the top of the exposure. Scattered iron-stained quartz grains gradually fine upwards and are not present above Bed 11. The clay content of the limestone beds also diminishes upward and glauconite and fossils are the prominent accessories.

BIOSTRATIGRAPHY

Studies of the Late Cretaceous and early Paleocene fossil assemblages of the Braggs or nearby localities are contained in reports by Cepek and others (1968), Worsley (1974), Toulmin (1977), and Smith (1978). Cepek and others (1968) studied the calcareous nannofossils and planktonic foraminifers, and Worsle (1974) and Thierstein (1981) examined samples from the Braggs localities and related extinctions of the Cretaceous forms to oceanic Cretaceous-Tertiary boundary events. The invertebrate faunas of the Pine Barren have been described from nearby outcrops in Butler County, Alabama, by Toulmin (1977). Smith (1978) identified a total of 82 species of ostracodes from the Prairie Bluff and Pine Barren localities at Braggs.

The age of the Prairie Bluff in the highway cuts southeast of Braggs has been discussed by Cepek and others (1968), Cepek and Hay (1969), Worsley (1974), and Thierstein (1981). The age of the formation in the Gulf and Atlantic Coastal Plain is included in papers by Smith (1975) and Smith and Mancini (1983). Cepek and others (1968) identified *Globotruncana gansseri* Bolli and several heterohelicid forms from the Braggs localities in the Prairie Bluff beginning 35 ft (10.7 m) above the Ripley contact and assigned the strata to the *Globotruncana gansseri* Subzone of Pessagno (1967). Cepek and others (1968) identified numerous nannofossils including *Nephrolithus frequens* Gorka from the Prairie Bluff, 5 ft (1.5 m) below the Cretaceous-Tertiary contact and considered *Nephrolithus frequens* to be an indicator of late Maestrichtian age. However, as pointed out in the papers by Smith (1975) and Smith and Mancini (1983), *Globotruncana gansseri* is of middle Maestrichtian age and *Nephrolithus frequens* occurs in the upper part of the *G. gansseri* planktonic foraminiferal zone.

All the papers referred to previously conclude that *Abathomphalus mayaroensis* (Bolli) and other species restricted to the *A. mayaroensis* Subzone of latest Maestrichtian age are not present in the Braggs locality. Furthermore, as pointed out by Pessagno (1967), Smith (1975), and Smith and Mancini (1983), the *Abathomphalus mayaroensis* Subzone is not presently known in outcrop within either the Gulf or Atlantic Coastal Plain.

Studies of the planktonic foraminifers from the Pine Barren Member of the Clayton discussed in papers by Cepek and others (1968) and Gibson and others (1982) indicate that the earliest known Paleocene planktonic forminiferal range zone, that of *Globigerina eugubina*, is not present at Braggs. Papers by Mancini (1981) and Gibson and others (1982) confirm that *Globigerina eugubina* Luterbacher and Premoli Silva is not known to occur in the lower Paleocene Clayton Formation in Alabama. The formations on either side of the Cretaceous-Tertiary contact exposed southeast of Braggs, Lowndes County, Alabama, do not include a record of the latest Cretaceous or earliest Paleocene.

Cepek and others (1968) identified several species of planktonic foraminifers including *Globoconusa daubjergensis* (Bronnimann) and *Subbotina triloculinoides* (Plummer) from the Pine Barren Member of the Clayton in the Braggs cuts and assigned the basal beds of the member to the *Globorotalia pseudobulloides* Zone of Bolli (1966). The assignment to the *Globorotalia pseudobulloides*

(= *Subbotina pseudobulloides*) Interval Zone was confirmed by Olsson (1970), Mancini (1981) and Gibson and others (1982) from studies of the Pine Barren in nearby Wilcox County. The upper beds of the Pine Barren Member and the overlying McBryde Limestone Member of the Clayton are included in the *Subbotina trinidadensis* Interval Zone as reported by Olsson (1970) and Mancini (1981).

REFERENCES CITED

Bolli, H. M., 1966, Zonation of Cretaceous to Pliocene marine sediments based on planktonic Foraminifera: Asociacion Venezolana, Geologia, Mineria y Petroleo Boletin Informacion, v. 9, p. 3–32.

Cepek, Pavel, and Hay, W. W., 1969, Calcareous nannoplankton and biostratigraphic subdivision of the Upper Cretaceous: Gulf Coast Association Geological Societies Transactions, v. 19, p. 323–336.

Cepek, Pavel, Hay, W. W., Masters, B. A., and Worsley, T. R., 1968, Calcareous plankton in samples from field trip stops, *in* Scott, J. C., Chm., Facies changes in the Selma Group in central and eastern Alabama: Alabama Geological Society Guidebook 6th Field Trip, 1968, p. 33–40.

Copeland, C. W., 1968, Facies changes in the Selma Group in central and eastern Alabama, *in* Scott, J. C., Chm., Facies changes in the Selma Group in central and eastern Alabama: Alabama Geological Society Guidebook 6th Field Trip, 1968, p. 2–26.

Gibson, T. G., Mancini, E. A., and Bybell, L. M., 1982, Paleocene to middle Eocene stratigraphy of Alabama: Gulf Coast Association Geological Societies Transactions, v. 32, p. 449–458.

Mancini, E. A., 1981, Lithostratigraphy and biostratigraphy of Paleocene subsurface strata in southwest Alabama: Gulf Coast Association Geological Societies Transactions, v. 31, p. 359–367.

Monroe, W. H., 1941, Notes on deposits of Selma and Ripley age in Alabama: Alabama Geological Survey Bulletin 48, 150 p.

Olsson, R. K., 1970, Planktonic Foraminifera from base of Tertiary, Millers Ferry, Alabama: Journal Paleontology, v. 44, p. 598–604.

Pessagno, E. A., Jr., 1967, Upper Cretaceous planktonic foraminifera from the

Western Gulf Coastal Plain: Paleontographica Americana, v. 15, no. 37, p. 242–445.

Smith, C. C., 1975, Upper Cretaceous calcareous nannoplankton zonation and stage boundaries: Gulf Coast Association Geological Societies Transactions, v. 25, p. 263–278.

Smith, C. C., and Mancini, E. A., 1983, Calcareous nannofossil and planktonic foraminiferal biostratigraphy, *in* Russell, E. E., and others, Upper Cretaceous lithostratigraphy and biostratigraphy in northeast Mississippi, southwest Tennessee and northwest Alabama, shelf chalks and coastal clastics: Society of Economic Paleontologists and Mineralogists, Gulf Coast Section, Spring field trip, April 7–9, 1983: p. 16–28.

Smith, J. K., 1978, Ostracoda of the Prairie Bluff Chalk, Upper Cretaceous, (Maestrichtian) and the Pine Barren Member of the Clayton Formation, Lower Paleocene, (Danian) from exposures along Alabama State Highway 263 in Lowndes County, Alabama: Gulf Coast Association Geological Societies Transactions, v. 28, p. 539–580.

Thierstein, H. R., 1981, Late Cretaceous nannoplankton and the change at the Cretaceous-Tertiary Boundary: Society of Economic Paleontologists and Mineralogists Special Publication, no. 32, p. 355–394.

Toulmin, L. D., 1977, Stratigraphic distribution of Paleocene and Eocene fossils in the eastern Gulf Coast region: Alabama Geological Survey Monograph 13, 602 p.

Worsley, Thomas, 1974, The Cretaceous-Tertiary boundary event in the ocean, *in* Hay, W. W., Studies in paleo-oceanography: Society of Economic Paleontologists and Mineralogists Special Publication, no. 20, p. 94–125.

St. Stephens Quarry (Lone Star Cement Company Quarry), St. Stephens, Washington County, Alabama, where a near complete Oligocene section, including the Eocene-Oligocene boundary, is exposed

Ernest A. Mancini, Geological Survey of Alabama and University of Alabama, Tuscaloosa, Alabama 35486
Charles W. Copeland, Geological Survey of Alabama, Box 0, University Station, Tuscaloosa, Alabama 35486

LOCATION

St. Stephens Quarry (Lone Star Cement Company Quarry) is located at St. Stephens Bluff on the west bank of the Tombigbee River, 2.2 mi (3.5 km) northeast of St. Stephens, Washington County, Alabama (Fig. 1).

INTRODUCTION

One of the most complete and continuously exposed marine Oligocene sections in North America is in the limestone quarry northeast of St. Stephens, Alabama (Fig. 1). The geologic section exposed at the quarry includes over 14 ft (4 m) of upper Eocene, Jackson Stage (Priabonian) strata and about 160 ft (49 m) of Oligocene, Vicksburg and Chickasawhay Stages

(Rupelian and Chattian) section. The Eocene-Oligocene contact is excellently preserved in the North Quarry.

The quarry is located near the southeastern nose of the Hatchetigbee anticline. The strata strike about N70°W and dip 2° to 2.5°SSW (Glawe, 1967).

The Eocene and Oligocene section exposed at the quarry consists of marginal marine to open outer shelf marine strata. The marls and limestones are highly fossiliferous with the clays being sparsely fossiliferous. The upper Eocene (Priabonian) Yazoo Clay and the Oligocene (Rupelian and Chattian) Red Bluff Clay/ Bumpnose Limestone, the Marianna Limestone, the Glendon Limestone Member and unnamed marl bed of the Byram Formation, and the Chickasawhay Limestone are exceptionally fossiliferous. Microfossil diversity is excellent throughout the section with the upper Eocene (Priabonian) Yazoo Clay and lower Olig-

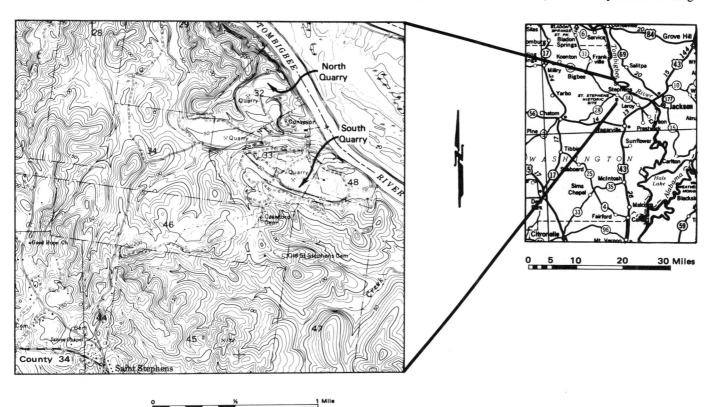

Figure 1. Location map of St. Stephens Quarry (Lone Star Cement Quarry), St. Stephens, Washington County, Alabama (portion of Saint Stephens 7½-minute quadrangle).

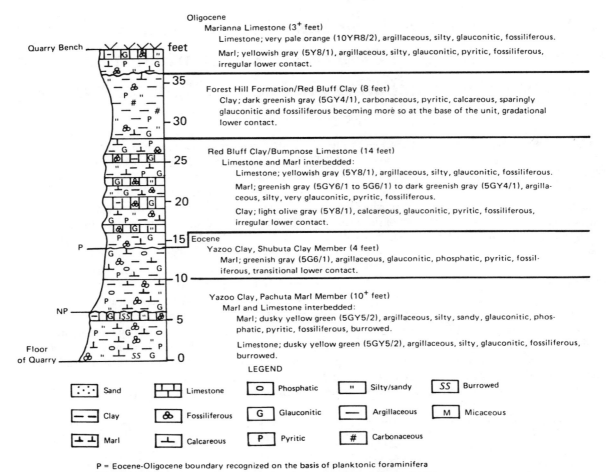

Figure 2. North Quarry measured Sec.32,T7N,R1W, Washington County, Alabama.

ocene (Rupelian) Red Bluff Clay/Bumpnose strata being most diverse. Macrofossils are best collected from the limestones, with the molluscan fauna present in the Chickasawhay the most diverse. See Cheetham (1963), Toulmin (1977), Dockery (1980, 1982), and MacNeil and Dockery (1984) for discussions of macrofossils. For discussions of ostracodes, see Huff (1970) and Hazel and others (1980); for foraminifera, see Bandy (1949), Deboo (1965), Poag (1966, 1972), Mancini (1979), and Waters (1983); and for calcareous nannoplankton, see Roth (1970), Bybell (1982), and Siesser (1983).

SITE DESCRIPTION

The route to the quarry involves turning right (northwest) off County Highway 34 at St. Stephens United Methodist Church and Cemetery onto a paved road and then turning right (north) on a gravel road 0.1 mile (0.2 km) from the junction of County Highway 34 and the paved road. The quarry gate is about 1.5 miles (2.4 km) from the junction of the paved road and the gravel road. The quarry includes parts of irregular sections 32, 33, 34 and 48, T. 7 N., R. 1 W., and on the Saint Stephens, Alabama, 7½-

minute Quadrangle (Fig. 1). The North Quarry measured section (sec. 32, T. 7 N., R. 1 W.), is about 0.25 mi (0.4 km) from the west bank of the Tombigbee River (Fig. 2). The Eocene-Oligocene contact is evident in the North Quarry. The South Quarry measured section (sec. 33, T. 7 N., R. 1 W.) is about 0.5 mi (0.8 km) from the west bank of the Tombigbee River (Fig. 3). About 140 ft (43 m) of Oligocene strata are exposed in the South Quarry. Presently, the quarry is not being mined for limestone; therefore, the exposures are starting to deteriorate.

The stratigraphic units exposed at St. Stephens Quarry in ascending order include the Pachuta Marl Member of the Yazoo Clay, the Shubuta Clay Member of the Yazoo Clay, the Red Bluff Clay/Bumpnose Limestone, the Forest Hill Formation/Red Bluff Clay, the Marianna Limestone, the Glendon Limestone Member of the Byram Formation, an unnamed marl member of the Byram Formation, the Bucatunna Clay Member of the Byram Formation, the Chickasawhay Limestone, and the Paynes Hammock Sand. See MacNeil (1944), Huddlestun (1965), May (1974), Toulmin (1977), and Coleman (1983) for a detailed lithologic description of these units.

The Yazoo Clay is a pale-olive calcareous, blocky clay at its

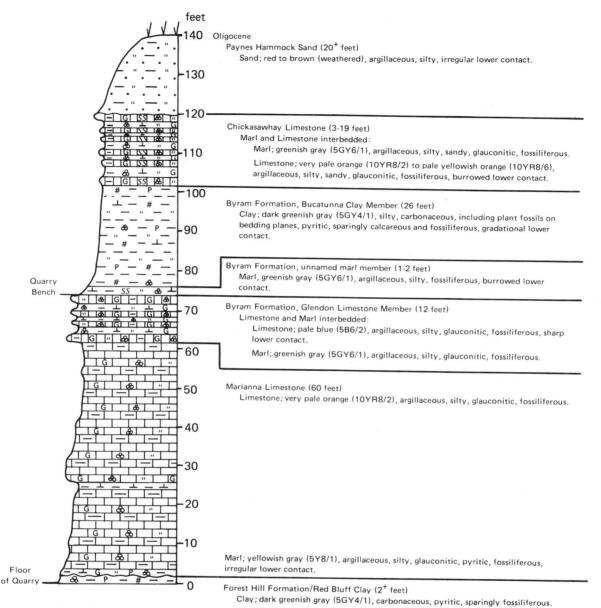

Figure 3. South Quarry measured section, Sec.33,T7N,R1W, Washington County, Alabama. See Figure 2 for lithologic legend.

type locality on the Yazoo River near Yazoo City, Yazoo County, Mississippi. It is about 200 ft (61 m) thick at this locality and thins eastward to about 72 ft (22 m) in Clarke County, Alabama (May, 1974; Toulmin, 1977). In western Alabama, the formation can be divided into four members (Murray, 1947). In ascending order, these include the North Twistwood Creek Clay, Cocoa Sand, Pachuta Marl, and Shubuta Clay Members. Only the Pachuta Marl and Shubuta Clay Members are exposed at St. Stephens Quarry (Fig. 2).

The Yazoo Clay grades laterally into the Crystal River Formation in eastern Clarke County and downdip in southern Alabama. The Crystal River includes 108 ft (32.9 m) of white, chalky, fossiliferous limestone at its type locality in the Crystal

River Rock Company Quarry, Citrus County, Florida (Moore, 1955; Toulmin, 1977).

The type locality for the Pachuta is on Pachuta Creek, southeast of Pachuta, Clarke County, Mississippi (Murray, 1947; May, 1974). The Pachuta consists of 10 ft (3 m) of greenish-gray, indurated, glauconitic, fossiliferous, argillaceous marl in its type area, southeast of Shubuta, Clarke County, Mississippi (Waters, 1983). At St. Stephens Quarry, the Pachuta includes 10+ ft (3+ m) of dusky yellow green (5GY5/2), argillaceous, silty, sandy, glauconitic, phosphatic, pyritic, fossiliferous, bioturbated marl interbedded with dusky yellow green (5GY5/2), argillaceous, silty, glauconitic, fossiliferous limestone. The lower contact of the Pachuta is not exposed in the quarry.

The Shubuta includes 90 ft (27.4 m) of grayish-olive green, blocky to massive, fossiliferous, calcareous clay along the Chickasawhay River, south of Shubuta and west of Hiwannee, including at its type locality as designated by Murray (1947), Clarke and Wayne Counties, Mississippi (May, 1974; Toulmin, 1977). At St. Stephens Quarry, the Shubuta thins to 4 ft (1.2 m) of greenish-gray (5G6/1), argillaceous, glauconitic, phosphatic, pyritic, fossiliferous, massive, blocky weathering marl. The lower contact of the Shubuta is gradational with the underlying Pachuta.

The Red Bluff Clay is a greenish-gray clay and marl interbedded with hard limestone ledges and chalky limestone in its type area on the Chickasawhay River, south of Shubuta, Wayne County, Mississippi (MacNeil, 1944). It thins eastward in Alabama from about 60 ft (18 m) in Choctaw County to about 20 ft (6 m) in Washington County (Copeland and Deboo, 1967). At St. Stephens Quarry, the 14 ft (4.3 m) of light olive gray (5Y8/1), calcareous, glauconitic, pyritic, fossiliferous clay and greenish gray (5GY6/1), argillaceous, silty, glauconitic marl interbedded with yellowish-gray (5Y8/1), argillaceous, silty, glauconitic, fossiliferous limestone overlying the Shubuta have been assigned by Glawe (1967) to the Red Bluff. Hazel and others (1980) place these beds in the Bumpnose Limestone. The Bumpnose consists of about 16 ft (5 m) of white, chalky limestone in its type locality area near Marianna, Jackson County, Florida (Moore, 1955). The lower contact of these strata is disconformable with the Shubuta and is marked by quartz, glauconite, phosphate grains, and shell hash.

The Forest Hill Sand consists of dark, thinly laminated sand and clay and sand at its type locality in Hinds County, Mississippi (MacNeil, 1944). It attains thicknesses in excess of 100 ft (30 m) in central and western Mississippi but thins rapidly to the east and pinches out in western Alabama (Copeland and Deboo, 1967). At St. Stephens Quarry, the 8 ft (2.4 m) of dark greenish-gray (5GY4/1), carbonaceous, pyritic, sparingly glauconitic, calcareous, and fossiliferous, laminated clay underlying the Marianna Limestone have been assigned by Glawe (1967) to the Forest Hill. Hazel and others (1980) place these beds in the Red Bluff. The lower contact of these strata is gradational with the underlying Red Bluff Clay/Bumpnose Limestone.

The Marianna Limestone is named for areas west of the Chipola River at Marianna, Jackson County, Florida, where it consists of white, chalky, fossiliferous limestone. The Marianna Limestone probably does not exceed 60 ft (18 m) in thickness and, in Mississippi and western Alabama, has a basal marl unit—the Mint Spring Marl Member (MacNeil, 1944). The Mint Spring is recognized in Mississippi as a formation. The type area is on Mint Spring Bayou, Warren County, Mississippi, where the Mint Spring is a greenish-gray, argillaceous to arenaceous, fossiliferous, glauconitic marl which has an average thickness of 3 to 6 ft (1 to 2 m) (May, 1974). At St. Stephens Quarry, the Marianna includes 60 ft (18.3 m) of very pale orange (10YR8/2), argillaceous, silty, glauconitic, fossiliferous limestone. A yellowish gray (5Y8/1), argillaceous, silty, glauconitic, pyritic, fossiliferous

marl occurs at the base of the Marianna at the quarry (Figs. 2 and 3). Hazel and others (1980) have designated this 1 to 2 ft (0.3 to 0.6 m) of marl the Mint Spring. The lower contact of the marl is burrowed.

The Byram Formation in Alabama includes, in ascending order, the Glendon Limestone Member, an unnamed marl member, and the Bucatunna Clay Member (Copeland and Deboo, 1967). At its type locality at Glendon Station, Washington County, Alabama, the Glendon consists of 12 ft (3.6 m) of gray, fossiliferous limestone (May, 1974). The Glendon includes 12 ft (3.6 m) of interbedded pale blue (5B6/2), argillaceous, silty, glauconitic, fossiliferous limestone and greenish-gray (5GY6/1), argillaceous, silty, glauconitic, fossiliferous marl at St. Stephens Quarry. The lower contact of the Glendon at the quarry is sharp.

In Mississippi, the Glendon and Bucatunna are elevated to formation status and the unnamed marl member, which is 3 to 12 ft (0.9 to 3.6 m) in thickness in Mississippi, is referred to as the Byram Formation. The Byram consists of greenish-gray, argillaceous to arenaceous, glauconitic, fossiliferous marl at its type locality along the west bank of the Pearl River, Hinds County, Mississippi (May, 1974). At St. Stephens Quarry, the unnamed member of the Byram consists of 1 to 2 ft (0.3 to 0.6 m) of greenish-gray (5GY6/1), argillaceous, silty, fossiliferous marl. The lower contact of this marl is burrowed at the quarry. This marl bed is designated the Byram Formation by Hazel and others (1980).

The type locality for the Bucatunna is along Bucatunna Creek, north of Denham Post Office, Wayne County, Mississippi. The Bucatunna consists of 29 to 102 ft (8.8 to 31.1 m) of dark gray, silty to arenaceous, micaceous, carbonaceous, fossiliferous clay (May, 1974). At St. Stephens Quarry, the Bucatunna includes 26 ft (7.9 m) of dark, greenish-gray (5GY4/1), silty, carbonaceous, pyritic, sparingly calcareous and fossiliferous clay. The lower contact of Bucatunna is gradational at the quarry.

The Chickasawhay Limestone includes 14 to 42 ft (4.3 to 12.8 m) of olive gray, argillaceous to arenaceous, fossiliferous limestone and bluish-green, fossiliferous marls and clays in Mississippi (May, 1974). The type locality of the Chickasawhay is near Waynesboro, Wayne County, Mississippi (Poag, 1972). The Chickasawhay at St. Stephens Quarry consists of 3 to 19 ft (0.9 to 5.8 m) of interbedded greenish gray (5GY6/1), argillaceous, silty, sandy, glauconitic, fossiliferous marl and very pale orange (10YR8/2) to pale yellowish orange (10YR8/6), argillaceous, silty, sandy, glauconitic, fossiliferous limestone. It has a burrowed lower contact at the quarry.

The type locality for the Paynes Hammock Sand is at Paynes Hammock Landing on a cutoff of the Tombigbee River, Clarke County, Alabama. At the type locality, the Paynes Hammock consists of 13 ft (3.9 m) of greenish, calcareous sand (MacNeil, 1944). At St. Stephens Quarry, the Paynes Hammock includes 20+ ft (6+ m) of weathered, red to brown, silty, argillaceous sand. The lower contact of the sand is irregular at the quarry.

PLANKTONIC FORAMINIFERAL BIOSTRATIGRAPHY

The Paleogene planktonic foraminiferal zonation utilized in this study was first established by Bolli (1957, 1966, 1972) and later modified by Stainforth and others (1975) and Stainforth and Lamb (1981). This zonation has been used widely as an accepted biostratigraphic standard for warm water areas of the world, including the Gulf Coastal Plain region.

The Pachuta Marl and Shubuta Clay Members of the Yazoo Clay at St. Stephens Quarry have been assigned by Mancini (1979) and Waters and Mancini (1982) to the upper Eocene (Priabonian) *Globorotalia cerroazulensis* (s.l.) Interval Zone (Fig. 4). *Hantkenina alabamensis* Cushman, *Globorotalia cerroazulensis cerroazulensis* (Cole) and *Globorotalia cerroazulensis cocoaensis* Cushman occur in the Pachuta and Shubuta (Mancini, 1979). The last occurrences of *Hantkenina alabamensis* and subspecies of *Globorotalia cerroazulensis* that are considered autochthonous are in the Shubuta (Deboo, 1965; Mancini, 1979; Waters and Mancini, 1982).

The Red Bluff Clay/Bumpnose Limestone, Forest Hill Formation/Red Bluff Clay, Marianna Limestone and lower part of the Byram Formation at St. Stephens Quarry have been assigned by Mancini (1979), Hazel and others (1980), and Stainforth and Lamb (1981) to the lower Oligocene (Rupelian) *Pseudohastigerina micra* Interval Zone. *Pseudohastigerina micra* (Cole) and *Globigerina ampliapertura* Bolli occur in the Red Bluff Clay/Bumpnose Limestone (Mancini, 1979) and Marianna Limestone (Stainforth and Lamb, 1981) at the quarry. Hazel and others (1980) place the Glendon Limestone Member of the Byram Formation from a corehole drilled along the Yazoo River, Warren County, Mississippi, in the *Pseudohastigerina micra* Interval Zone based on the occurrence of *Globigerina ampliapertura* and *Globigerina galavisi* Bermudez (= *Globigerina eocaena* Gumbel). The Byram (unnamed marl member of the Byram Formation at St. Stephens Quarry) has a probable assignment to the Oligocene (Chattian) *Globigerina ampliapertura* Interval Zone according to Hazel and others (1980). They identified *Globigerina ampliapertura* from the Byram from a corehole drilled along the Yazoo River, Warren County, Mississippi, while *Globigerina eocoena* was not observed in this marl.

The upper part of the Byram Formation and the lower part of the Chickasawhay Formation have been assigned to the Oligocene (Chattian) *Globoratalia opima opima* Range Zone (Poag, 1972; Hazel and others, 1980). *Globoratalia opima opima* Bolli, *Globigerina ciperoensis* Bolli, and *Globigerina angulisuturalis* Bolli have been reported from the lower Chickasawhay by Poag (1972).

The upper part of the Chickasawhay Formation and the Paynes Hammock Sand have been assigned by Poag (1972) to the Oligocene (Chattian) *Globigerina ciperoensis* Interval Zone. According to Poag (1972), the highest stratigraphic occurrence of *Globorotalia opima opima* is in the lower Chickasawhay beds. He (Poag, 1966; 1972) has reported *Globigerina ciperoensis, Globi-*

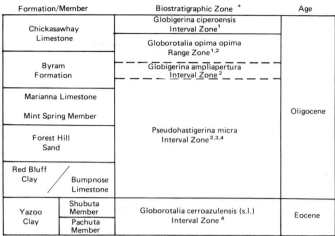

Figure 4. Upper Eocene and Oligocene planktonic foraminiferal biostratigraphy for southwestern Alabama.

gerina angulisuturalis, and *Globoquadrina globularis* Bermudez (= *Globoquadrina altispira globularis* Bermudez) from the Paynes Hammock.

At St. Stephens Quarry, the Eocene-Oligocene boundary, based on the vertical distribution of planktonic foraminifera, occurs at or near the top of the Shubuta Clay Member of the Yazoo Clay (Fig. 2). The Eocene-Oligocene boundary is recognized worldwide on the basis of the change in planktonic foraminiferal assemblages across this datum. Uppermost Eocene marine strata generally contain a diverse assemblage consisting of *Hantkenina, Cribrohantkenina, Globorotalia cerroazulensis* subspecies, and *Globigerina* species having a large test (Stainforth and others, 1975). Lowermost Oligocene marine strata are usually characterized by a less diverse assemblage predominated by *Globigerina ampliapertura, Globigerina gortanii* (Borsetti), *Globorotalia increbescens* (Bandy), and *Pseudohastigerina* species. The Eocene-Oligocene boundary is drawn worldwide at the top of the *Globorotalia cerroazulensis* (s.l.) Interval Zone (Stainforth and others, 1975), which at St. Stephens Quarry closely approximates the contact of the Shubuta Clay Member of the Yazoo Clay with the Red Bluff Clay/Bumpnose Limestone.

Based on calcareous nannoplankton, the Eocene-Oligocene boundary at St. Stephens Quarry could be placed at the disconformity (Fig. 2) in the Pachuta Marl Member of the Yazoo Clay (Bybell, 1982). The last occurrences of *Discoaster barbadiensis* Tan Sin Hok, *Discoaster saipanensis* Bramlette and Riedel, and *Reticulofenestra reticulata* (Gartner and Smith) are reported by Bybell (1982) at this datum. Cheetham (1957; 1963) using primarily cheilostome bryozoans also placed the Eocene-Oligocene boundary near the top of the Pachuta Marl Member in southwestern Alabama. Based on macrofossils, Huddlestun and Toulmin (1965) defined the boundary at the top of the Shubuta Clay Member of the Yazoo Clay.

Therefore, the Eocene-Oligocene boundary at St. Stephens Quarry based on planktonic foraminifera is 8 ft (2.4 m) higher in the section than where Bybell (1982) could place the boundary based on calcareous nannoplankton. This difference in elevation in placement of the Eocene-Oligocene boundary is not unusual. In fact, as reported by Gartner (1971), Stainforth and Lamb (1981),

Poore and others (1982), and Snyder and others (1984), the boundary as recognized by planktonic foraminifera occurs at a higher stratigraphic elevation in coreholes drilled in the Gulf of Mexico and the Atlantic Ocean than it would be if based on calcareous nannoplankton in these coreholes.

REFERENCES CITED

Bandy, O. L., 1949, Eocene and Oligocene foraminifera from Little Stave Creek, Clarke County, Alabama: Bulletins American Paleontology, v. 32, no. 131, 210 p.

Bolli, H. M., 1957, The genera *Globigerina* and *Globorotalia* in the Paleocene-Lower Eocene Lizard Springs Formation of Trinidad, B.W.I.: U.S. National Museum Bulletin 215, p. 61–81.

—— 1966, Zonation of Cretaceous to Pliocene marine sediments based on planktonic foraminifera: Asociacion de Venezolana Geología, Minería y Petróleo Boletin de Informacion, v. 9, p. 3–32.

—— 1972, The genus *Globigerinatheka* Bronnimann: Journal of Foraminiferal Research, v. 2, p. 109–136.

Bybell, L. M., 1982, Late Eocene to early Oligocene calcareous nannofossils in Alabama and Mississippi: Gulf Coast Association of Geological Societies Transactions, v. 32, p. 295–302.

Cheetham, A. H., 1957, Eocene-Oligocene boundary, eastern Gulf region: Gulf Coast Association of Geological Societies Transactions, v. 7, p. 89–97.

—— 1963, Late Eocene zoogeography of the eastern Gulf Coast region: Geological Society of America Memoir 91, 113 p.

Coleman, J. L., Jr., 1983, The Vicksburg Group carbonates—a look at Gulf Coast Paleogene carbonate banks: Gulf Coast Association of Geological Societies Transactions, v. 33, p. 257–268.

Copeland, C. W., and Deboo, P. B., 1967, Summary of upper Paleogene and lower Neogene stratigraphy of Alabama, *in* Jones, D. E., ed., Geology of the Coastal Plain of Alabama: Geological Society of America 80th Annual Meeting, New Orleans, Louisiana, Alabama Geological Society Guidebook: University, Alabama, p. 44–46.

Deboo, P. B., 1965, Biostratigraphic correlation of the type Shubuta Member of the Yazoo Clay and Red Bluff Clay with their equivalents in southwestern Alabama: Alabama Geological Survey Bulletin 80, 84 p.

Dockery, D. T., III, 1980, The invertebrate macropaleontology of the Clarke County, Mississippi area: Mississippi Department Natural Resources, Bureau of Geology, Bulletin 122, 387 p.

—— 1982, Lower Oligocene Bivalvia of the Vicksburg Group in Mississippi: Mississippi Department of Natural Resources, Bureau of Geology, Bulletin 123, 261 p.

Gartner, S., Jr., 1971, Calcareous nannofossils from the JOIDES Blake Plateau cores and revision of Paleogene nannofossil zonation: Tulane Studies Geology Paleontology, v. 8, no. 3, p. 101–121.

Glawe, L. N., 1967, Stop 16—Lone Star Cement Company Quarry at St. Stephens Bluff on Tombigbee River, 2.2 miles northeast of St. Stephens, Washington County, Alabama, *in* Jones, D. E., ed., Geology of the Coastal Plain of Alabama: Geological Society of America 80th Annual Meeting, New Orleans, Louisiana, Alabama Geological Society Guidebook, University, Alabama: p. 107–113.

Hazel, J. E., Mumma, M. D., and Huff, W. J., 1980, Ostracode biostratigraphy of the lower Oligocene (Vicksburgian) of Mississippi and Alabama: Gulf Coast Association of Geological Societies Transactions, v. 30, p. 361–401.

Huddlestun, P. F., 1965, Correlation of upper Eocene and lower Oligocene strata between the Sepulga, Conecuh, and Choctawhatchee Rivers of southern Alabama [M.S. thesis]: Florida State University, Tallahassee, 101 p.

Huddlestun, P. F., and Toulmin, L. D., 1965, Upper Eocene–lower Oligocene stratigraphy and paleontology in Alabama: Gulf Coast Association of Geological Societies Transactions, v. 15, p. 155–159.

Huff, W. J., 1970, The Jackson Eocene Ostracoda of Mississippi: Mississippi Geological, Economic, and Topographical Survey Bulletin 114, 289 p.

MacNeil, F. S., 1944, Oligocene stratigraphy of southeastern United States: American Association of Petroleum Geologists Bulletin, v. 28, p. 1313–1354.

MacNeil, F. S., and Dockery, D. T., III, 1984, Lower Oligocene Gastropoda, Scaphopoda, and Cephalopoda of the Vicksburg Group in Mississippi: Mississippi Department of Natural Resources, Bureau of Geology, Bulletin 124, 415 p.

Mancini, E. A., 1979, Eocene-Oligocene boundary in southwest Alabama: Gulf Coast Association of Geological Societies Transactions, v. 29, p. 282–286.

May, J. H., 1974, Wayne County geology and mineral resources: Mississippi Geological, Economic, and Topographical Survey Bulletin 117, 293 p.

Moore, W. E., 1955, Geology of Jackson County, Florida: Florida Geological Survey, Geological Bulletin 37, 101 p.

Murray, G. E., 1947, Cenozoic deposits of central Gulf Coastal Plain: American Association of Petroleum Geologists Bulletin, v. 31, p. 1825–1850.

Murray, G. E., 1963, North Twistwood Creek Clay: Corrected name for North Creek Clay, *in* DeVries, D. A., Jasper County mineral resources: Mississippi Geological, Economic, and Topographical Survey Bulletin 95, p. 97–100.

Poag, C. W., 1966, Paynes Hammock (lower Miocene?) foraminifera of Alabama and Mississippi: Micropaleontology, v. 12, p. 393–440.

Poag, C. W., 1972, Planktonic foraminifers of the Chickasawhay Formation, United States Gulf Coast: Micropaleontology, v. 18, p. 257–277.

Poore, R. Z., Tauxe, L., Percival, S. F., Jr., and LaBrecque, J. L., 1982, Late Eocene–Oligocene magnetostratigraphy and biostratigraphy at South Atlantic DSDP site 522: Geology, v. 10, p. 508–511.

Roth, P. H., 1970, Oligocene calcareous nannoplankton biostratigraphy: Eclogae Geologicae Helvetiae, v. 63, p. 799–881.

Siesser, W. G., 1983, Paleogene calcareous nannoplankton biostratigraphy: Mississippi, Alabama and Tennessee: Mississippi Department Natural Resources, Bureau of Geology, Bulletin 125, 61 p.

Snyder, S. W., Muller, C., and Miller, K. G., 1984, Eocene-Oligocene boundary: biostratigraphic recognition and gradual paleoceanographic change at DSDP site 549: Geology, v. 12, p. 112–115.

Stainforth, R. M., and Lamb, J. L., 1981, An evaluation of planktonic foraminiferal zonation of the Oligocene: Kansas University, Paleontological Contributions, Paper 104, 42 p.

Stainforth, R. M., Lamb, J. L., Luterbacher, H., Beard, J. H., and Jeffords, R. M., 1975, Cenozoic planktonic foraminiferal zonation and characteristics of index forms: Kansas University, Paleontological Contributions, Article 62, 425 p.

Toulmin, L. D., 1977, Stratigraphic distribution of Paleocene and Eocene fossils in the eastern Gulf Coast Region: Alabama Geological Survey, Monograph 13, 602 p.

Waters, L. A., 1983, Correlation of upper Eocene and lower Oligocene strata in Mississippi and Alabama [M.S. thesis]: University of Alabama, University, 169 p.

Waters, L. A., and Mancini, E. A., 1982, Lithostratigraphy and biostratigraphy of upper Eocene and lower Oligocene strata in southwest Alabama and southeast Mississippi: Gulf Coast Association of Geological Societies Transactions, v. 32, p. 303–307.

Dobys Bluff tongue of the Kosciusko Formation and the Archusa Marl Member of the Cook Mountain Formation at Dobys Bluff on the Chickasawhay River, Clarke County, Mississippi

David T. Dockery III, *Mississippi Bureau of Geology, P.O. Box 5348, Jackson, Mississippi 39216*

LOCATION

The Dobys Bluff Tongue of the Kosciusko Formation and the Archusa Marl Member of the Cook Mountain Formation are exposed on the east bank of the Chickasawhay River just below a large bend in the river south of Quitman, Mississippi, at the center of the north line of the NW¼SW½NW¼, Sec.18,T.2N., R.16E., Clarke County, Quitman 7½-minute quadrangle (Fig. 1). The exposure is accessible by boat or taking Mississippi 514 east from Quitman, turning south 0.75 mi (1.3 km) outside of the city limits onto a paved road leading to the Archusa Creek Water Park camp ground, continuing south past the camp ground and past Shiloh Church to a dirt road in Sec. 18 that veers southwest, and taking this road west along the ridgetop (avoiding logging trails leading in other directions) to a high bluff on the Chickasawhay River.

INTRODUCTION

Dobys Bluff provides excellent exposures of the Dobys Bluff marine tongue of the dominantly deltaic Kosciusko Formation, and of the Archusa Marl, a widespread marker unit in the subsurface of southern Mississippi. Exposures of this quality are unusual in the Gulf Coastal Plain area.

The marine shelf sediments of the Lisbon Formation (Middle Eocene) in Alabama grade westward along the eastern flank of the Mississippi Embayment into a deltaic depositional sequence. Formations within this sequence in ascending order include: (1) Winona Formation (marine shelf), (2) Zilpha Formation (marine shelf and prodelta), (3) Kosciusko Formation (deltaic), (4) Cook Mountain Formation (marine shelf and prodelta), and (5) Cockfield Formation (deltaic). The cyclical deposition recorded by these formations shows two marine transgressive events each followed by an episode of delta progradation. Associated with the latter marine transgression are the widespread carbonate deposits of the Cook Mountain Formation.

Thomas (1942) named the carbonate interval of the Cook Mountain Formation in Clarke County, Mississippi, the Archusa Marl Member. This unit is equivalent to the highly fossiliferous *Cubitostrea sellaeformis* (Conrad) zone in the upper Lisbon Formation of Alabama. To the northwest along the outcrop belt in central Mississippi (Newton County), the unit grades into fossiliferous sands of the Potterchitto Member of the Cook Mountain Formation.

The Archusa Marl Member has a restricted surface exposure in eastern Mississippi and a broad subsurface distribution in the southern part of the state. In the subsurface of southern Missis-

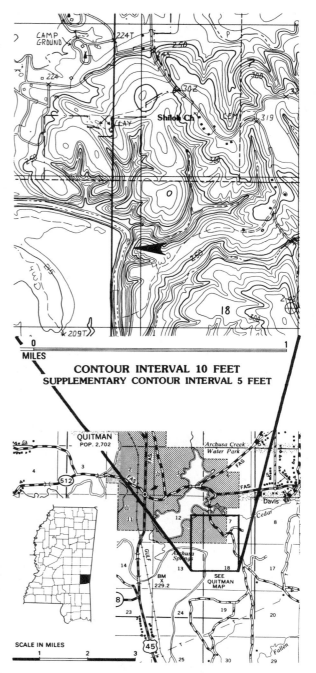

CONTOUR INTERVAL 10 FEET
SUPPLEMENTARY CONTOUR INTERVAL 5 FEET

Figure 1. Location map of the Kosciusko–Cook Mountain (Claiborne Group) outcrop at Dobys Bluff. Map is a reproduced part of the Quitman 7½-minute quadrangle. Arrow indicates the locality.

sippi, the Archusa Marl Member is referred to by the petroleum industry as the Cook Mountain Limestone or the *Camerina* Limestone. The latter name is based on the presence of the foraminifer *Camerina* [= *Nummulites*] *barkeri* (Gravell and Hanna). The Cook Mountain Limestone generally ranges between 60 ft (18.2 m) and 100 ft (30.3 m) thick in the subsurface of south-central Mississippi but thickens locally to over 200 ft (60.6 m) in a carbonate bank complex situated along the northern flank of the Wiggins Uplift in southern Mississippi. Along this structure, carbonate shelf mudstones and wackestones grade into the grainstones of the carbonate bank. An isolith map of the Cook Mountain Limestone is given in Dockery (1976, Fig. 6, p. 22).

The Archusa Marl Member is best exposed at Dobys Bluff where a 50-ft (15.2 m) thick section of this unit is continuously exposed for almost three hundred feet. Dobys Bluff is only a short distance downstream and to the east of the Archusa type

locality, which is on the south bank of the Chickasawhay River below the old U.S. 45 bridge south of Quitman. Another significant locality near Dobys Bluff is a site that produced one of the few Eocene land mammal finds in the Southeastern United States. At this site on the Chickasawhay River, a titanothere skull and jaw fragment were found in carbonate matrix only a foot (30 cm) above a bed containing *Cubitostrea sellaeformis*. Gazin and Sullivan (1942) named a new genus and species from these remains, *Notiotitanops mississippiensis*. Several rib fragments of a small rhinoceros were reported by Dockery (1980) from the base of Dobys Bluff in the Kosciusko Formation.

The upper part of the Kosciusko Formation exposed at Dobys Bluff differs from the predominantly deltaic facies of that formation in consisting of fossiliferous marine sands and clays. This unit was named the Dobys Bluff Tongue of the Kosciusko Formation by Dockery (1980). It contains a diverse molluscan

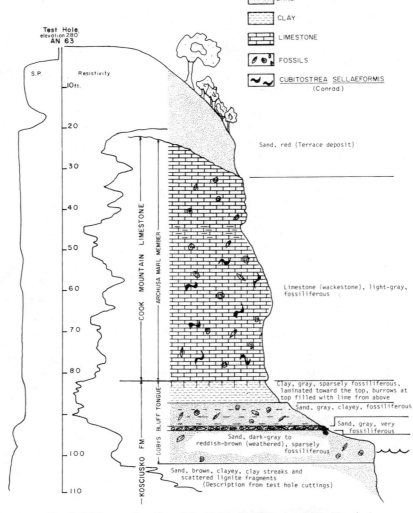

Figure 2. Measured section of the Dobys Bluff Tongue of the Kosciusko Formation and the Archusa Marl Member of the Cook Mountain Formation at Dobys Bluff.

Figure 3. Dobys Bluff Tongue of the Kosciusko Formation and the Archusa Marl Member of the Cook Mountain Formation at Dobys Bluff on the Chickasawhay River in Clarke County, Mississippi. Arrow indicates the Kosciusko–Cook Mountain contact.

fauna which differs in part from that of the overlying Archusa Marl Member. The marine invertebrate faunas of both the Dobys Bluff Tongue and the Archusa Marl Member are illustrated in Dockery (1980).

SITE DESCRIPTION

Dobys Bluff is an 80-ft (24.2 m) high bluff on the east bank of the Chickasawhay River (Fig. 2). Three stratigraphic units are exposed in this bluff for a distance of about 300 ft (90.9 m). These include in ascending order: (1) the Dobys Bluff Tongue of the Kosciusko Formation, (2) the Archusa Marl Member of the Cook Mountain Formation, and (3) Pleistocene Chickasawhay River terrace sand. The contacts between each of these units are disconformable.

The Dobys Bluff Tongue of the Kosciusko Formation is about 20 ft (6.1 m) thick at Dobys Bluff (although only about 18 ft (5.5 m) of the unit are exposed at low river level). It consists of fossiliferous marine sand in the lower 16 ft (4.8 m), and laminated, silty, fossiliferous, marine clay in the upper 4 ft (1.2 m). The latter clay section is fairly continuous in the subsurface and provides a means for recognizing the Cook Mountain-

Kosciusko contact on electric logs of wells (see electric log in Fig. 2). The upper contact of the Dobys Bluff Tongue with the Archusa Marl is sharp and can be readily seen from a distance (Figs. 3 and 4). Carbonate filled burrows extend below the contact into the clays of the Dobys Bluff Tongue and reworked, burrowed, clay lenses occur in the basal part of the Archusa Marl Member.

The invertebrate fauna of the Dobys Bluff Tongue is moderately well preserved. However, many of the larger specimens are distorted due to compaction. Mollusks comprise the majority of the invertebrate species, with the bivalve *Venericardia (Venericor) densata* Conrad and the gastropod *Calyptraphorus velatus nodovelatus* Palmer being common elements of the macrofauna. A concentration of molluscan shells occurs in a thin bed in the middle part of the Dobys Bluff Tongue.

The Archusa Marl Member of the Cook Mountain Formation consists of a 50-ft (15 m) thick section of massive carbonate wackestone. Fossil mollusk shells comprise the larger grains of the wackestone and rest in a carbonate mud matrix. *Cubitostrea sellaeformis* is common in the middle part of this unit where an oyster biostrome containing large articulated shells of this species forms a resistant ledge. Mollusks with aragonitic shells are also

Figure 4. Close view of the Kosciusko–Cook Mountain contact at Dobys Bluff. Arrow indicates the contact. The length of pick next to arrow is 26 in (10 cm).

well preserved in the carbonate matrix. Preservation of aragonitic shells is rare in other limestone units of the Gulf Coastal Plain, as the shells are usually leached from the calcareous matrix leaving only molds and casts. Though mollusks comprise the majority of the invertebrate species, valves and pedicles of the goose-neck barnacle *Euscalpellum eocenense* (Meyer) are moderately common.

Pleistocene terrace sand of the Chickasawhay River disconformably overlies the Archusa Marl Member. This sand is about 20 ft (6.1 m) in thickness and continues southward along the river's course overlying progressively younger Eocene and Oligocene units.

REFERENCES CITED

Dockery, D. T., III, 1976, Depositional systems in the upper Claiborne and lower Jackson Groups (Eocene) of Mississippi [M.S. thesis]: University of Mississippi, Oxford, 109 p.
——— 1980, The invertebrate macropaleontology of the Clarke County, Mississippi, area: Mississippi Bureau of Geology Bulletin 122, 387 p., 82 pl.
Gazin, C. L., and Sullivan, J. M., 1942, A new Titanothere from the Eocene of Mississippi, with notes on the correlation between the marine Eocene of the Gulf Coastal Plain and continental Eocene of the Rocky Mountain region: Smithsonian Miscellaneous Collections, v. 101, p. 1–13, 2 Figs.
Thomas, E. P., 1942, The Claiborne: Mississippi Geological Survey Bulletin 48, 96 p., 25 figs., 1 pl.

The Bashi-Tallahatta section at Mt. Barton, Meridian, Mississippi

David T. Dockery III, *Mississippi Bureau of Geology, P.O. Box 5348, Jackson, Mississippi 39216*

LOCATION

Mt. Barton is a prominent hill south of Meridian, Mississippi, in the SE¼ SE¼, sec. 24, T. 6 N., R. 15 E., Lauderdale County, Meridian South 7½-minute quadrangle (Fig. 1). The section at Mt. Barton is readily accessible by taking the 31st Street South exit off of I-20 at Meridian. Stop 1 of the section is at a field of boulders lining the right side of the exit ramp. Stops 2 and 3 are at Mt. Barton off of the first dirt road to the left from 31st Street South.

INTRODUCTION

Mt. Barton is an outlier of the Tallahatta Cuesta, a prominent cuesta along the eastern flank of the Mississippi Embayment extending from southern Alabama to north-central Mississippi. An excellent view of this cuesta to the west–northwest across the Okatibbee Creek flood plain is available from above stop 3 on Mt. Barton's west side (Fig. 2). Another fine view point is from the Tallahatta Cuesta along U.S. 45 south of Meridian. Excellent displays of a variety of Coastal Plain depositional environments and typical Coastal Plain geomorphology can be seen at this site.

The section in the vicinity of Mt. Barton was first described by Foster (1940), who included the upper part of the Tuscahoma Formation as presently defined within the Bashi Formation. This section illustrates a cyclical depositional sequence which begins with a marine transgression in the Bashi Formation, is followed by a regressive sequence consisting of the estuarine sediments of the Hatchetigbee Formation and the fluvial sediments of the Meridian Sand, and is concluded by a transgressive marine unit, the Basic City Shale Member of the Tallahatta Formation (Fig. 3). The Bashi and Hatchetigbee Formations are included in the Wilcox Group (Lower Eocene) and the Meridian Sand and Tallahatta Formation are included in the Claiborne Group (Lower-Middle Eocene). Age determinations for the Bashi, Hatchetigbee, and Tallahatta Formations based on calcareous nannoplankton studies are given in Siesser (1983) and Bybell and Gibson (1985).

SITE DESCRIPTION

The Mt. Barton section can be seen in a series of 3 stops. Stop 1 includes the Bashi Formation, Stop 2 includes the Hatchetigbee Formation, and Stop 3 includes the Meridian Sand and Basic City Shale Member of the Tallahatta Formation.

Stop 1

Stop 1 is on the right side of the 31st Street South exit from the I-20 east lane and is the larger of two boulder fields flanking

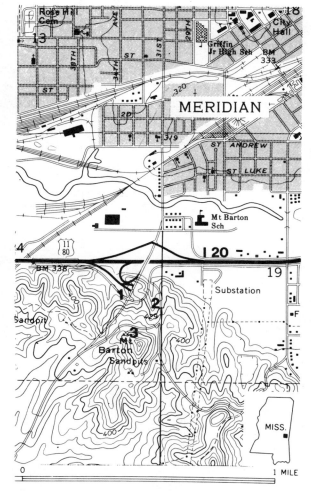

Figure 1. Location map of the Wilcox-Claiborne section at Mt. Barton south of Meridian, Lauderdale County, Mississippi. Meridian South 7½-minute quadrangle. Numbers 1-3 give locations for the 3 stops discussed in text.

each side of 31st Street South (Fig. 4). These car-size boulders are large sandstone concretions excavated from the Bashi Formation during the construction of I-20. This formation is now covered in the vicinity of the I-20 and 31st Street interchange.

The Bashi Formation is the only abundantly fossiliferous marine unit in the Wilcox Group in Mississippi and represents the largest marine transgression into the northern Gulf during that time. Fossil mollusks are abundant in the bases of the Bashi concretions (as they originally occurred in situ) at Stop 1, but are difficult to collect from the matrix. Common molluscan species include the large bivalves *Venericardia (Venericor) bashiplata* Gardner and Bowles and *Ostrea brevifronta* Dockery and the gastropods *Cornulina minax compressa* Dockery, *Pseudoliva san-*

Figure 2. View of Meridian, Okatibbee Creek flood plain, and the distant Tallahatta Cuesta from a sandstone ledge of the Tallahatta Formation on Mt. Barton.

Figure 4. Boulder-size concretions from the Bashi Formation placed along the exit ramp of 31st Street south of I-20 by the Mississippi Highway Department.

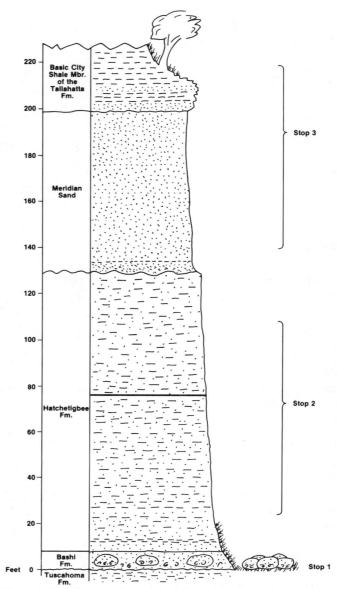

Figure 3. Composite section of the Bashi-Hatchetigbee-Meridian-Basic City sequence at Mt. Barton, for Stops 1-3.

tander Gardner, and *Bullia calluspira* Dockery. These species have thick shells and were well adapted for the high energy near-shore environments in which they lived. Many interesting vertebrate fossils also occur in the Bashi Formation in the Meridian area including an abundance of shark and ray teeth and less commonly alligator teeth and snake vertebrae. The Bashi Formation is about 8 ft (2.4 m) thick in the vicinity of Mt. Barton. Fossil invertebrates from this unit are illustrated in Dockery (1980, pl. 1-10).

Stop 2

Stop 2 is along a dirt road on the north side of Mt. Barton. After leaving Stop 1, it is accessible by traveling a short distance south on 31st Street, parking on the first dirt road to the left (which turns into an impassable trail), and walking around the western flank of Mt. Barton to a dirt road on the north side. The Hatchetigbee Formation is exposed in roadcuts along this road and contains a 121-ft (36.7 m) thick section of thinly bedded

Figure 5. Hatchetigbee Formation in a roadcut on the north slope of Mt. Barton.

Figure 6. Meridian Sand and Tallahatta Formation at a sand pit on the west side of Mt. Barton. Arrow indicates the Meridian-Tallahatta contact.

alternating sands and clays (Fig. 5). A 6-in (15.2 cm) thick lignite bed occurs in the middle part. Sand beds in the sequence vary from less than an inch to a few inches (cm) in thickness, and some sands show cross-laminations. These sediments are probably estuarine and deltaic in origin.

The nonfossiliferous estuarine sediments of the Hatchetigbee Formation grade westward into the thick undifferentiated deltaic sequences of the Wilcox Group. To the southeast, they grade into the alternating marine and estuarine deposits of the type locality at Hatchetigbee Bluff on the Tombigbee River in Washington County, Alabama. Gibson (1982) defined the Hatchetigbee Formation as nonmarine based on his observations in Georgia and eastern Alabama. To maintain this definition, he assigned all marine beds of the Hatchetigbee type section to the Bashi Formation and the intervening nonfossiliferous beds to the Hatchetigbee Formation. Dockery, Copeland, and Huddlestun (1984) pointed out that these marine beds were included in the original description of the Hatchetigbee as given by Smith and Johnson (1887) who considered the Hatchetigbee to contain both marine and nonmarine facies. Thus the Bashi is here restricted to include only the basal (and most extensive) marine unit above the Tuscahoma Formation. This unit represents the initial Lower Eocene transgression into the northern Gulf (Siesser, 1983). The overlying

Hatchetigbee Formation is a regressive sequence, which is non-fossiliferous in updip exposures but which grades downdip into fossiliferous marine strata.

Stop 3

Stop 3 is a sand pit on the west side of Mt. Barton (Fig. 6). Here a 70-ft (21.3 m) thick sequence of the Meridian Sand is well exposed. This unit consists of prominently cross-bedded sand and is bounded by disconformable contacts above and below. The lower contact shows a more prominent break in lithology and also shows moderate relief due to channeling.

There is disagreement concerning the stratigraphic placement of the Meridian Sand. Lowe (1933) named this unit and placed it in the Claiborne Group, citing evidence of a regional unconformity at its base. This placement has been continued by the Mississippi Bureau of Geology and the United States Geological Survey. Other workers have placed the Meridian Sand in the Wilcox Group, arguing that the top of the Meridian is a more mappable datum in the subsurface. Because of the similarities between the Meridian and Wilcox sands, the base of the Meridian is difficult if not impossible to establish on electric logs of wells. Also, even though the fluvially eroded lower contact of the Meridian is seen as a prominent disconformity on the outcrop, the fluvial sediments of the Meridian Sand in Mississippi are a continuation of the marine regression that began in the Hatchetigbee Formation.

Duplantis (1975) described the depositional environment of the Meridian Sand in Mississippi as a coarse-grained meander-belt fluvial system that developed in response to an increase in depositional slope. On the other hand, Wermund (1965) described the depositional environment of the Meridian in Alabama and part of Mississippi as a neritic bar. The fluvial environment proposed by Duplantis (1975) best fits the Meridian Sand sequence exposed at Mt. Barton.

Disconformably overlying the Meridian Sand at Stop 3 are the near-shore marine sediments of the Tallahatta Formation. This formation consists of two members in Mississippi, which in ascending order are the Basic City Shale and the Neshoba Sand. The Neshoba Sand Member is best developed in north-central Mississippi and pinches out in the vicinity of Meridian. Mt. Barton is capped by the lower 59 ft (17.9 m) of the Basic City Shale Member. Here it consists of claystone, sand, and some

quartzitic sandstones. This unit was originally named the buhr-stone, referring to its hard, siliceously cemented claystones and quartzitic sandstones, which were locally used as millstones. Clay sequences in the Basic City are soft in the subsurface, but become lithified when exposed above the water table. These lithified clays along with sand units form the erosion-resistant cap on the Tallahatta Cuesta. Claytone units of the Basic City Shale Member contain large amounts of cristobalite, a silicate that probably formed from the alteration of volcanic ash. Unweathered clays of this member contain high percentages of the zeolite mineral clinoptilolite, also a likely alteration product of volcanic ash.

The Basic City Shale was deposited in the nearshore environments of a transgressive marine shelf. Much of the detrital sediment that accumulated on this shallow marine shelf consisted of wind-blown volcanic ash. This ash weathered in the marine environment to clinoptilolite and cristobalite and formed a fine-grained bottom sediment. Fossil marine mollusks occasionally occur as molds in claystone or sandstone units. A diagnostic fossil that occurs infrequently in the claystone units is the large bivalve *Anodontia? augustana* Gardner.

REFERENCES CITED

Bybell, L. M., and Gibson, T. G., 1985, The Eocene Tallahatta Formation of Alabama and Georgia: its lithostratigraphy, biostratigraphy, and bearing on the age of the Claibornian Stage: U.S. Geological Survey Bulletin 1615, 20 p., 2 pl.

Dockery, D. T., III, 1980, The invertebrate macropaleontology of the Clarke County, Mississippi, area: Mississippi Bureau of Geology Bulletin 122, 387 p., 82 pl.

Dockery, D. T., III, Copeland, C. W., Jr., and Huddlestun, P. F., 1984, Reply to a revision of the Hatchetigbee and Bashi formations: Mississippi Geology, v. 4, no. 3, p. 11–15.

Duplantis, M. J., 1975, Depositional systems in the Midway and Wilcox Groups (Paleocene-Lower Eocene), North Mississippi [M.S. thesis]: Oxford, University of Mississippi, 87 p.

Foster, V. M., 1940, Lauderdale County mineral resources; Geology by V. M. Foster, and Tests by T. E. McCutcheon: Mississippi Geological Survey Bulletin 41, 246 p.

Gibson, T. G., 1982, Revision of the Hatchetigbee and Bashi Formations (Lower Eocene) in the Eastern Gulf Coastal Plain: U.S. Geological Survey Bulletin 1529-H, p. H33-41.

Lowe, E. N., 1933, Coastal Plain stratigraphy of Mississippi, Part 1, Midway and Wilcox Groups: Mississippi Geological Survey Bulletin 25, 125 p.

Siesser, W. G., 1983, Paleogene calcareous nannoplankton biostratigraphy: Mississippi, Alabama and Tennessee: Mississippi Bureau of Geology Bulletin 125, 61 p., 37 fig.

Smith, E. A., and Johnson, L. C., 1887, Tertiary and Cretaceous strata of the Tuscaloosa, Tombigbee, and Alabama Rivers: U.S. Geological Survey Bulletin 43, 189 p.

Wermund, E. G., 1965, Cross-bedding in the Meridian Sand: Sedimentology, v. 5, no. 1, p. 69–79.

88

Shelf marls and chalks in the Marine Section of the Upper Cretaceous: Mississippi

Ernest E. Russell, Department of Geology, Mississippi State University, Mississippi State, Mississippi 39762

LOCATION

Shelf marls and chalks in the marine section of the Upper Cretaceous of Mississippi are exposed west of Tombigbee River in western Lowndes County and eastern Oktibbeha County, east-central Mississippi. An additional optional stop is located in northern Noxubee County (Figure 1).

The four stops require approximately 40 mi (66.7 km) of driving mostly on gravel roads. The first two outcrops require short walks (5 to 10 minutes). The outcrops are steep and they are slick when wet.

INTRODUCTION

Marine strata of Campanian-Maestrichtian age crop out in a wide belt in northeast Mississippi. More than 825 ft (250 m) thick and deposited during several transgressive-regressive cycles, they afford an excellent opportunity to study shelf muds (marls and chalks) and transitional zone sediments (glauconitic sands

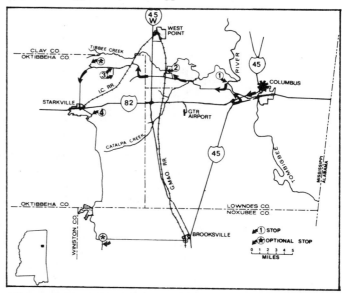

Figure 1. Generalized map showing approximate locations of stops described in text. Detailed maps accompany stop descriptions.

and calcareous clays) of late Cretaceous age. Since outcrops are rare and widely separated in the soft easily weathered coastal plains outcrop belt, sections in the southern part of this belt have been selected that are most representative of the principal marine lithofacies. The sections are keyed to a composite geologic column (Fig. 2). The sediments in all outcrops contain excellent microfaunas and floras and have representative megafaunas.

Regional Lithostratigraphy

Sediments in the composite geologic section (Fig. 2) were deposited on what appears to have been an open shelf during three transgressive and regressive cycles (Fig.3). The Tombigbee Sand and Mooreville Marl with the Arcola Limestone Member, the lowest sequence in the composite section, represent the earliest cycle. They are succeeded by the Demopolis-lower Ripley sequence which represents a major transgressive-regressive cycle. The uppermost Ripley Formation probably represents a truncated transgressive cycle. The Prairie Bluff Formation lies unconformably on the Ripley Formation and represents a transgressive phase with no regressive counterpart.

Marls which dominate the section represent shelf muds composed primarily of nannofossils with an admixture of more than 25 percent clay. Two thick beds of impure chalk which occur in the middle and upper part of the Demopolis Formation contain less than 25 percent clay. The calcareous sands and clays in the Ripley and the Tombigbee Sand were deposited in the

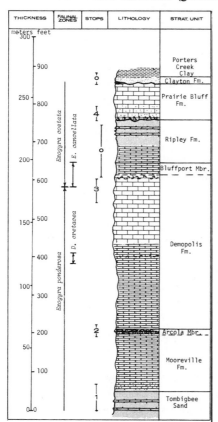

Figure 2. Composite geologic column at latitude of Columbus, Mississippi showing thickness of stratigraphic units and stratigraphic interval to be seen at each stop.

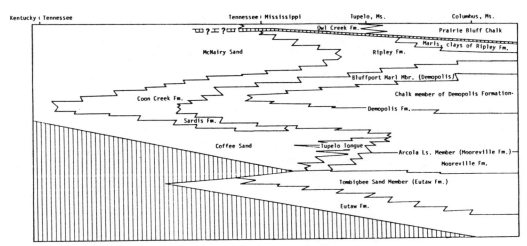

Figure 3. Diagrammatic geologic cross-section showing lithostratigraphic relationships in the post-McShan Upper Cretaceous Gulfian Series of the eastern Mississippi Embayment. Datum: Demopolis Formation. Thickness highly generalized.

transition zone. Northward in Mississippi and eastward in Alabama the shelf facies grade into shoreline facies.

Megafossils, abundant and diverse in the transition zone sediments of the Tombigbee Sand and Ripley Formation, are restricted in species in most of the marl-chalk lithofacies. Two important biostratigraphic Faunal Zones occur in this interval (Fig. 2): The *Exogyra ponderosa* Faunal Zone and the *E. costata* Faunal Zone. The *E. cancellata* Subzone of the latter zone occupies a thin stratigraphic interval at the top of the Demopolis Formation.

Structural strike in this area, a few degrees east of north with a dip in the order of 30 ft/mi (5 m/km) to the west, is related to downwarping of the Mississippi Embayment. Depositional strike appears to have been more or less east-west.

Unweathered outcrops are rare and widely separated in the coastal plains. The chalk, especially the marls and the loose clayey sands, weather rapidly into deep clay soils in the warm subtropical climate.

SITE DESCRIPTION

All outcrops discussed herein have been described in road logs in the Upper Cretaceous Field Trip Guide of the 1983 Annual Meeting of the Geological Society of America in New Orleans, Louisiana (E. E. Russell and others, 1983). An additional reference is Stephenson and Monroe (1940).

Stop 1: Tombigbee Sand-Mooreville Marl

Plymouth Bluff is located on the Tombigbee River (Fig. 4 (latitude 33°31′05″N; longitude 88°30′00″W). The bluff, exposing more than 80 feet (25 m) of transgressive marine sands and shelf muds, is owned by the Mississippi University for Women (MUW), Columbus, Mississippi. Permission for access

can be obtained by calling the Security Department on the MUW campus. The Stop can be reached by car in dry weather.

The bluffs, now cut off by the Tennessee-Tombigbee Waterway, have great historic interest. E. W. Hilgard (1860, p. 74) measured and described the bluffs in his report on the geology and agriculture of Mississippi. The appendix listed 46 species of fossils from collections by Dr. W. Spillman of Columbus, Mississippi. Stephenson and Monroe (1940, p. 72–73) also measured and described the bluff.

The bluff exposes 49 ft (15 m) of fossiliferous, calcareous, thick-bedded sand with five ledge-forming sandstones above low water levels, all of the Tombigbee Sand Member of the Eutaw Formation (Fig. 5). The third sandstone unit above the river forms a wide persistent bench with numerous casts of *Inoceramus* sp., occasional *Texanites* species, and, rarely, *Exogyra ponderosa* Roemer in attached position. The sands grade upward, through an interval of 6.5 to 10 ft (2 to 3 m), into the overlying Mooreville Formation of the Selma Group. The contact has been chosen where the sand distinctly becomes a marl. Stephenson and Monroe (1940, p. 73) chose the contact below the marl at a layer of phosphatic steinkern and determined that the contact was unconformable. The layer of phosphatic steinkern is not present at most other localities, and there is no physical or biologic evidence for an unconformity at the contact. The Tombigbee Sand appears to be transitional between the high-energy, shallow-water sands of the underlying Eutaw Formation and the shelf muds of the Mooreville Formation. With more than three meters of massive-bedded, blocky marl, the Mooreville Formation underlies the top of the bluff.

One of the most fossiliferous Tombigbee Sand outcrops, the entire interval contains a rich fauna of planktonic forams and an unusually well preserved flora of nannoplankton.

The plantonic forams and the nannoplankton in the bluff indicate a very Early Campanian Age for the Tombigbee Sand as well as the Mooreville Formation. For more information about

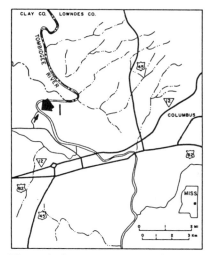

Figure 4. Stop 1, approximately 6 mi (10 km) west of Columbus, Mississippi. Access on dirt road off paved county road.

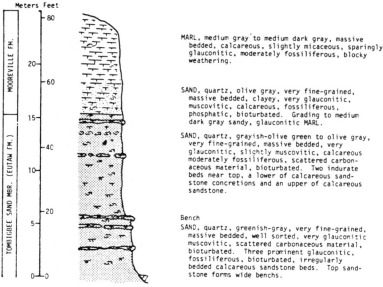

Figure 5. Composite geologic section at Plymouth Bluff, Lowndes County, Mississippi: Cluster Stop 1. Transgressive sequence from fossiliferous, massive-bedded transition zone sands of the Tombidgee Sand into the shelf muds of the Mooreville Formation. Contact is just above the uppermost indurated layer. Very large rudistids and ammonites have been collected from this bluff.

the biostratigraphy at this outcrop the reader is referred to Smith and Mancini (1983, p. 19).

Stop 2: Mooreville-Demopolis Formations

The Mooreville and Demopolis Formation are exposed on a low bluff on the south bank of Tibbee Creek below the GM & O Railroad bridge in Clay County, Mississippi (Fig. 6) (West Point 15-minute quadrangle; latitude 33°32′17″N, longitude 88°37′50″W). The outcrop area is reached by car and then by walking along a railroad track about 1,000 ft (300 m) to a railroad bridge. The outcrop extends for a considerable distance east and west of the bridge along the cut bank.

The best exposures are west of the bridge (Fig. 7) where more than 23 ft (7 m) of fossiliferous marl with discontinuous beds of limestone nodules are present in the cut.

The nodular limestone calcisphere wackestones and packstones of the Arcola Limestone Member mark the top of the Mooreville Formation (Fig. 7). The overlying marls are the basal beds of the Demopolis Formation. The contact between the Arcola Limestone Member and the Demopolis Formation was interpreted by Stephenson and Monroe (1940, p. 102) as being unconformable because of the presence of the phosphatic nodules and molds. Several horizons of phosphatic nodules and small limestone nodules occur above the upper limestone bed. At the type locality on the Alabama River there are 3 beds of nodular limestone. If there is an unconformity it is blended. Lins, et al. (1977) suggested that the calcisphere wackestones and packstones resulted from abrupt algal blooms caused by changes on a shallow shelf due to repeated cessation or "shutdowns" of upwelling. The limestone beds can be traced northward to Lee County,

Mississippi, to a point where they disappear abruptly in the Mooreville Formation as it grades laterally into the basal sands of the Tupelo Tongue of the Coffee Formation.

The marls are typical of the lithologies in the Mooreville and basal Demopolis Formations. Calcium carbonate mainly in the form of coccoliths make up 60-70 percent of the sediments and smectitic clays compose the bulk of the remainder.

The outcrop is in the *Exogyra ponderosa* Assemblage Zone and excellent *E. ponderosa erraticostata* Stephenson can be collected. There are rich calcareous nannoplankton floras and planktonic foraminifera faunas. The outcrop is in the upper part of the *Calculites Ovalis* Zone (NP19a) and the middle to upper part of the *Globotruncana elevata* Subzone (Smith and Mancini, 1982, p. 21). Therefore, it can be assigned a late early Campanian Age.

Stop 3: Demopolis Formation

A total of more than 65 ft (20 m) of typical massive-bedded, dense, impure, fossiliferous chalk in the upper Demopolis Formation is exposed in the two cuts (Figs. 8 and 9) on the West Point, 7½-minute quadrangle on Alexander Schoolhouse road (latitude 33°32′23″N, longitude 88°44′32″W). The outcrop can be reached by car even in wet weather.

The uppermost regressive Bluffport Marl Member of the Demopolis Formation is poorly exposed, as usual, at the top of the hill. In the southern cuts the impure chalk beds, with 70-80 percent calcium carbonate, are interbedded with marl beds. All are strongly bioturbated. The ichnofossil *Pseudobilobites* is present on this outcrop. Thin lenses of fossil hash lag form thin indurate lenses. These chalk beds are persistent along the outcrop and produce a characteristic "kick" on an electric log in the subsurface.

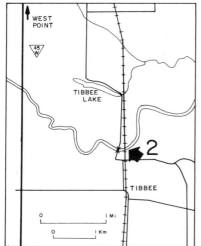

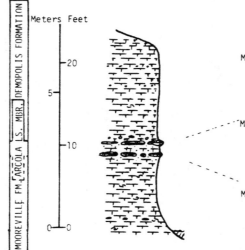

Figure 6. Stop 2, approximately 4 mi (6 km) south of West Point, Mississippi. Access to outcrop in south bank of Tibbee Creek is along railroad about 200 m north of gravel road.

MARL, grayish-olive, massive-bedded, weathers blocky, glauconitic, fossiliferous slightly muscovitic, moderately burrowed. Two horizons of phosphatized internal molds.

MARL, grayish olive to light olive gray massive-bedding, weathers blocky, glauconitic, fossiliferous slightly muscovitic, moderately fossiliferous, heavily bioturbated. Interbedded with two beds of LIMESTONE, yellowish-gray, slightly glauconitic, nodular, heavily burrowed.

MARL, dark greenish-gray to medium gray, massive-bedding, blocky fracture, phosphatic nodules, silty, glauconitic, slightly muscovitic, moderately fossiliferous, bioturbated. Upper two feet heavily burrowed and bioturbated, very glauconitic and phosphatic. Thin discontinuous lag bed of oyster hash.

Figure 7. Geologic section at Stop 2, just west of GM & O Railroad bridge on Tibbee Creek, Clay County, Mississippi. Outcrop exposes contact between the Mooreville Marl and overlying marls of the basal Demopolis Formation. The two thin nodular calcisphere wackestone and packstone beds are in the Arcola Limestone Member of the Mooreville Formation and mark the top of the formation. *Exogyra ponderosa erraticostata* Stephenson occurs in the beds below the Arcola Limestone.

The chalks are distinct from the marls in that they tend to fracture conchoidally, to weather more slowly, and to have very thin soils developed on them. "Bald" spots are common in the chalk belt. The chalk texture begins to become apparent when the calcium carbonate/clay ratio is higher than 75 percent.

The base of the *Exogyra cancellata* Subzone in the base of the *E. costata* Assemblage Zone occurs in the south cut where *E. cancella* Stephenson and *Pycnodonta mutabilis* (Morton) can be collected. The nannoplankton *Tetralithus aculeus, T. trifidus,* and *E. eximius* in the basal beds of the cut are absent in the upper beds. *Globotruncana calcarata* has not been found, but *G. ventricosa* is present in the basal beds of the cut; therefore, these beds are assignable to the *G. calcarata* Zonule of latest Campanian Age; whereas, the uppermost 36 ft (11 m) of the section can be assigned to the *Rugotruncana subcircumnodifer* Subzone of earliest Maestrichtian Age (Smith and Mancini, 1982, p. 24, 25).

Optional Stop: Ripley Formation

The most complete section of the Ripley Formation in this area is located on Rock Hill about 5 mi (8.3 km) north of Starkville, Oktibbeha County, Mississippi (Fig. 8), Pheba, Mississippi Topographic Quadrangle (latitude 33°32′32″N; longitude 88°47′50″W). It is described in two Cretaceous field trip guides (Russell and others, 1982; Russell and others, 1983); however, it is on private land and the owner is not always agreeable to its use. It is recommended that interested parties contact the Geology Department at Mississippi State University for information. Other Ripley outcrops in this area are described by Stephenson and Monroe (1940). The Ripley Formation can be seen in several outcrops east of the campus of Mississippi State University.

In this area the Ripley Formation, about 100 ft (30 m) thick, consists of a regressive basal calcareous clay that grades upwards into calcareous fossiliferous sands and an upper transgressive sand that grades upward into calcareous clays. A short distance to the south the middle sand tongue is absent.

Sands in the Ripley Formation underlie a persistent high ground separating the rolling prairies developed on the marls and chalks, from the low lying flatwoods, developed on the Tertiary Porters Creek Clay.

The Ripley Formation contains an excellent fauna of invertebrates, as well as vertebrates, and are in the *Exogyra Costata* Zone. There is an excellent nannoplankton flora (see Risatti, 1973; Smith and Mancini, 1983) which is assignable to the *Tranolithus phacelosus* Zone, *Reinhardites levis* Zone, and the basal part of the *Arkhangelskiella cymbiformis* Zone. The planktonic foraminifera (Clark, 1980, Smith and Mancini, 1983) fauna in the basal beds belong to the *Rugotruncana sub-circummodifer* Subzone whereas the uppermost beds are in the *G. aegyptiaca* Zonule of the *G. gansseri* Subzone. The nannoplankton as well as the planktonic foraminifera are related to those in European beds of early and middle Maestrichtian Age.

Stop 4: Ripley-Prairie Bluff Formations

A west facing cut bank east of the parking lot behind McKee Dormitory on east side Mississippi State University campus (Fig. 10), Starkville 7½-minute Quadrangle (latitude 33°37′17″N; longitude 88°32′04″W), is accessible to vehicle traffic.

The Stop exposes the uppermost Ripley clayey marls, the Ripley-Prairie Bluff unconformity and the basal sandy chalks of the Prairie Bluff Formation (Fig. 11). More than 10 meters of section is present.

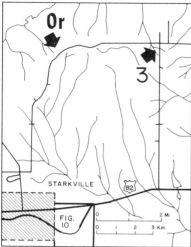

Figure 8. Stop 3, approximately 5.5 mi (9 km) northeast of Starkville, Mississippi. Arrow marked *Or* indicates optional stop in the Ripley Formation.

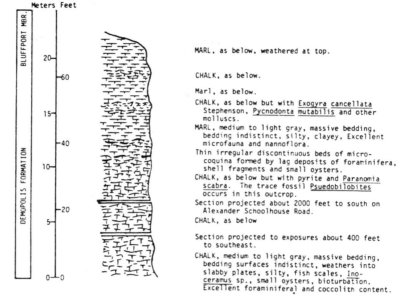

MARL, as below, weathered at top.

CHALK, as below.

Marl, as below.

CHALK, as below but with <u>Exogyra cancellata</u> Stephenson, <u>Pycnodonta mutabilis</u> and other molluscs.

MARL, medium to light gray, massive bedding, bedding indistinct, silty, clayey. Excellent microfauna and nannoflora.

Thin irregular discontinuous beds of micro-coquina formed by lag deposits of foraminifera, shell fragments and small oysters.

CHALK, as below but with pyrite and <u>Paranomia scabra</u>. The trace fossil <u>Psuedobilobites</u> occurs in this outcrop.

Section projected about 2000 feet to south on Alexander Schoolhouse Road.

CHALK, as below

Section projected to exposures about 400 feet to southeast.

CHALK, medium to light gray, massive bedding, bedding surfaces indistinct, weathers into slabby plates, silty, fish scales, <u>Ino-ceramus</u> sp., small oysters, bioturbation. Excellent foraminiferal and coccolith content.

Figure 9. Composite geologic section at Stop 3 on Alexander Schoolhouse Road, Oktibbeha County, Mississippi. Dense, impure chalk in ditches at north end of outcrop typical of upper Demopolis Formation. Bioturbated impure chalks interbedded with marls about 0.5 mi (0.8 km) to south are typical of uppermost Demopolis Formation where it grades upward into Bluffport Marl Member which is covered at top of hill. Thin-shelled oysters, *Pycnodonta mutabilis* (Morton) and *Exogyra cancellata* Stephenson occur in upper beds of outcrop.

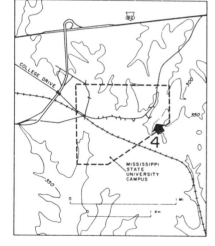

Figure 10. Stop 4, on east side of Mississippi State University campus in parking lot east of McKee Dormitory.

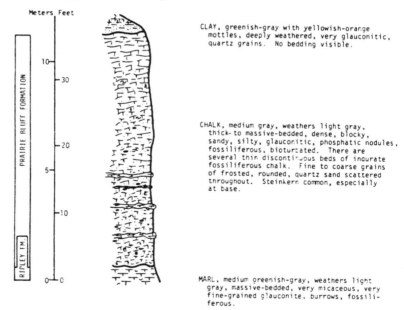

CLAY, greenish-gray with yellowish-orange mottles, deeply weathered, very glauconitic, quartz grains. No bedding visible.

CHALK, medium gray, weathers light gray, thick- to massive-bedded, dense, blocky, sandy, silty, glauconitic, phosphatic nodules, fossiliferous, bioturbated. There are several thin discontinuous beds of indurate fossiliferous chalk. Fine to coarse grains of frosted, rounded, quartz sand scattered throughout. Steinkern common, especially at base.

MARL, medium greenish-gray, weathers light gray, massive-bedded, very micaceous, very fine-grained glauconite, burrows, fossili-ferous.

Figure 11. Geologic section at Stop 4. Base of outcrop at north end, in marls and clays of the Ripley Formation.

Gullies at the north end of the parking lot expose about one meter of greenish-gray, massive-bedded, micaceous, clayey marl in the Ripley Formation. Fossils in the marl are thin-shelled and original shell material. Quartz sand and glauconite grains are very-fine-grained.

The slopes above the Ripley have been cut into more than nine meters of impure, medium gray, dense, fossiliferous, sandy chalk with several thin persistent indurated chalk beds. Fossil "steinkern," common in the basal 26 to 33 ft (8-10 m) of the Prairie Bluff Formation, are well developed at this outcrop. Phosphatic nodules, some steinkern, are disseminated throughout the beds and in several discontinuous layers. Fine- to coarse-grained, rounded, frosted quartz grains are common from the base of the formation upward in the outcrop (coarse-grained quartz is rare in the Cretaceous beds of this area). Steinkern, phosphatic nodules, and frosted quartz grains mark the base of the Prairie Bluff Formation and, with other features, indicate the unconformable relationships of the contact. There is no obvious faunal break at the Ripley-Prairie Bluff contact. About 60 mi (100 km) to the north the chalks grade into sands of the Owl Creek Formation.

There is a rich fauna, much of it molds and casts and most showing signs of post-depositional transport. The fauna includes *E. costata Saye, Sphenodiscus* sp., the small planktonic crinoid *Applinocrinus cretaceous* (Bather), and the large arenaceous foraminifera *Lituola taylorensis*. The latter two can be collected a few meters above the base of the formation.

The Prairie Bluff nannoflora is excellent, but solution features and calcite overgrowths are common. The Ripley flora is well preserved, diverse, and abundant. The outcrop is above the range of *Tetralithus nitidus* and *T. trifidus*.

Clark (1980) described a diverse and abundant fauna of planktonic foraminifera from the Ripley Formation in this vicinity, including *Globotruncana aegyptiaeca, G. gansseri, G. trinidadensis, Psuedoguembelina kempensis, Psuedo-textularia deformis, Heterohelix glabrans,* etc., which in the absence of *Globotruncana bulloides, G. fornicata, G. lapparenti, G. linneana* and *G. rosetta* indicate the fauna is similar to that of the *G. aegyptiaca* Zonule of the *G. gansseri* Subzone and, more or less, equivalent to strata in Europe of middle Maestrichtian Age (Smith and Mancini, 1982).

Optional Stop: Cretaceous-Tertiary Contact

Well developed outcrops exposing the unconformity between the upper Prairie Bluff Chalk and fossiliferous Tertiary Clayton Formation sediments are rare and difficult to access in this area. The nearest accessible outcrop exposing the Cretaceous-Tertiary contact is described by Russell and others (1983, p. 69–71) and is about 18 mi (30 km) south of Stop 4 (Fig. 12). It is located on the Lynn Creek Topographic Quadrangle; Noxubee County, Mississippi in Section 22, T 16 N, R 15, E (latitude 33°13′27″N; longitude 88°44′54″N).

At the Lynn Creek locality about 6.5 ft (2 m) of dense gray Prairie Bluff Creek is overlain unconformably by about 3.3 ft (1 m) of coarse fossiliferous sand in a channel, which in turn is overlain by more than 23 ft (7 m) of fossiliferous marl and calcareous clay of the Clayton and Porters Creek Formations. It appears that faunas of uppermost Maestrichtian Age and earliest Danian Age are absent (Smith and Mancini, 1982, p. 26).

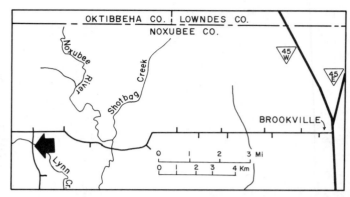

Figure 12. Optional stop (arrow) on gravel road approximately 10 mi (16 km) west of junction of U.S. 45E and 45W at Brookville, Mississippi.

REFERENCES CITED

Clark, M. S., 1980, Planktonic foraminifera of the Ripley Formation (Maestrichtian) Mississippi [M.S. thesis]: Starkville, Mississippi State University, 203 p.

Hilgard, E. W., 1860, Report on the geology and agriculture of the state of Mississippi: Mississippi Geological and Agricultural Survey Board, 391 p.

Lins, T. W., Johnson, F. E., Keady, D. M., and Russell, E. E., 1977, The Arcola Limestone: A Cretaceous calcisphere wackestone and grainstone: Geological Society of America, Abstracts with Programs, v. 9, no. 2, p. 159.

Risatti, J. B., 1973, Nannoplankton biostratigraphy of the Upper Bluffport Marl-Lower Prairie Bluff Chalk interval (Upper Cretaceous), in Mississippi, *in* Smith, L. A. and Hardenbol, J., eds., Proceedings of Symposium on Calcareous Nannofossils: Gulf Coast Section, Society of Economic Paleontologists and Mineralogists, Houston, Texas, p. 8–57.

Russell, E. E., Keady, D. M., Mancini, E. E., and Smith, C. C., 1982, Upper Cretaceous in the Lower Mississippi Embayment of Tennessee and Mississippi: Lithostratigraphy and Biostratigraphy: Field Trip Guidebook for 1982 Annual Meeting of Geological Society of America, New Orleans, Louisiana, 49 p.

Russell, E. E., Keady, D. M., Mancini, E. E., and Smith, C. C., 1983, Upper Cretaceous Lithostratigraphy and Biostratigraphy in Northeast Mississippi, Southwest Alabama, and Northwest Alabama, Shelf Chalks and Clastics: Field Trip Guidebook for Spring Field Trip of Gulf Coast Section, Society of Economic Paleontologists and Mineralogists, 72 p.

Smith, C. C., and Mancini, E. E., 1982, Biostratigraphy, *in* Russell, E. E., *et al.,* Upper Cretaceous in the Lower Mississippi Embayment of Tennessee and Mississippi: Lithostratigraphy and Biostratigraphy: Field Trip Guidebook for 1982 Annual Meeting of Geological Society of America, New Orleans, Louisiana, p. 15–26.

Smith, C. C., and Mancini, E. E., 1983, Calcareous Nannofossil and Planktonic Foraminiferal Biostratigraphy, *in* Russell, E. E., *et al.,* Upper Cretaceous Lithostratigraphy and Biostratigraphy in Northeast Mississippi, Southwest Alabama, and Northwest Alabama, Shelf Chalks and Clastics: Field Trip Guidebook for 1983 Spring Field Trip of Gulf Coast Section, Society of Economic Paleontologists and Mineralogists, p. 16–28.

Stephenson, L. W., and Monroe, W. H., 1940, Upper Cretaceous deposits: Mississippi Geological Survey Bulletin, 40, 296 p.

Coffee Landing, Hardin County, Tennessee: Type locality of the Coffee Sand

Ernest E. Russell, Department of Geology, Mississippi State University, Mississippi State, Mississippi 39762

LOCALITY AND ACCESSIBILITY INFORMATION

The type locality of the Coffee Sand is a series of bluffs on the Tennessee River north of Coffee Landing, Hardin County, Tennessee. The westernmost bluff is located at Lat. 35°16′28″ N; Long. 88°17′32″ W on the Milledgeville, Tennessee 7½-minute quadrangle (see index map, Fig. 1). The site is accessible by car and it is only a short walk along the river from Coffee Landing to one of the low bluffs. However, during high water the bluffs may not be approachable from Coffee Landing along the river's edge. All the bluffs are accessible by boat, and the bluff tops are only a short distance from paved roads although it is necessary to cross through woods. The bluffs are in an active cutbank of the Tennessee River, thus are very steep and unstable.

Figure 2. View of western bluffs from Coffee Landing on Tennessee River, during low water.

Therefore, caution is recommended, especially during wet weather.

INTRODUCTION

The spectacular Chalk Bluffs (Fig. 2) along the Tennessee River near Coffee Landing, were first described by Judge John Haywood in 1823. Later, Safford (1864, p. 362) designated them as the type locality of the Coffee Sand. The name came from a nearby steamboat landing—Coffee Landing. Safford (1869, p. 412) described the bluffs as follows:

(4) *On Top:* gravel and ferruginous conglomerate.

(3) *Sands,* with thin laminae of slaty clay; much like No. 1 below   10 ft (3 m)

(2) *Slaty Clay,* with but little sand; contains fragments of wood and leaves   20 ft (6.1 m)

(1) *Gray and Yellow Sands,* interstratified with numerous thin laminae and some thicker layers of slaty clay; strata of sand occasionally from three to six ft, without clay. Leaves, in fragments, and pieces of lignitic wood abundant. Projecting from the mass are the ends of two large trunks, their bark converted into lignite and their wood silicified. Contains pyrite and yields proto-salts of iron and ferruginous waters. Extending to the water's edge   65 ft (19.8 m)

One of the earliest described type localities in the Upper Cretaceous of western Tennessee, it represents the lagoonal lithofacies of a transgressive sea during Demopolis times. The Coffee Sands overlap Eutaw and Tuscaloosa sediments in western Tennessee. The bluffs present one of the most extensive outcrops of shallow water, nearshore, Upper Cretaceous clastics exposed in western Tennessee. Some of the bluffs are more than 100 ft (30 m) high with continuous exposures of several hundred feet (meters).

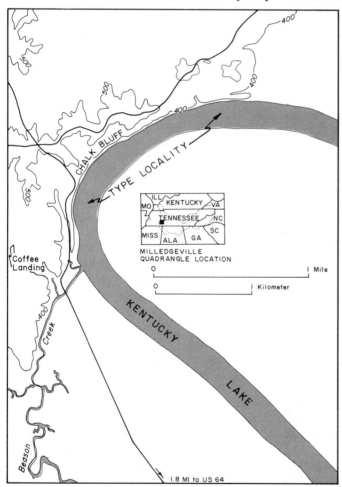

Figure 1. Location map showing type locality of the Coffee Formation northwest of Savannah in Hardin County, Tennessee. Milledgeville, Tennessee, 7½-minute quadrangle.

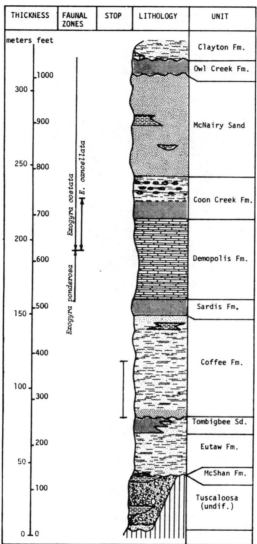

Figure 3. Composite geologic section of the Upper Cretaceous formations in western Tennessee. The bar under STOP column shows the stratigraphic position of the locality. At the latitude of Coffee Landing the Tuscaloosa and McShan Formations have been overlapped and the Tombigbee Sand Member of the Eutwa Formation has graded into nearshore facies.

SITE DESCRIPTION

The type locality of the Coffee Formation lies on the eastern edge of the Mississippi Embayment structure and the sediments dip gently to the west at about 27 ft/mi (5 m/km) The strike is about N10°E.

The outcrop is near the base of the Upper Cretaceous in this area and Paleozoic carbonates crop out a short distance to the east and southeast. Glauconitic Eutaw sands and clays crop out below the Coffee Formation in the bluffs. Hilltops to the northwest are capped by the Sardis Formation and outliers of the Demopolis Formation (Fig. 3). Thick gravel terrace deposits asso-

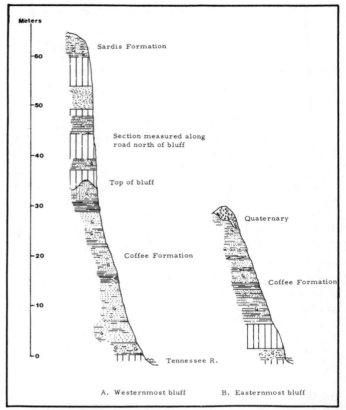

Figure 4. Measured geologic sections downstream from Coffee Landing, Hardin County, Tennessee. The westernmost section, located 0.5 mi (1 km) north of Coffee Landing, is not the westernmost bluff, but is the highest in that area. The upper part of the section was measured to the top of the ridge about 0.5 mi (0.8 km) to the northwest.

ciated with the Tennessee River crop out along the valley edge to the west and southwest, and dissected remnants of terrace deposits cap the hill north of Coffee Landing. They are part of an extensive Tertiary–Quaternary Tennessee River terrace system. The Cretaceous outcrop is shown on the West Sheet of the Geologic Map of Tennessee (Hardeman, 1966) and is described by Russell (1975, p. A24–A26).

Some ten bluffs are located on the northern cutbank of Tennessee River where it has less than 0.6 mi (1 km) to cut before being captured by Chalk Creek. Extensive flood plains lie south and east of the bluffs. The bluffs range in height from about 65 ft (20 m) to more than 100 ft (30 m). Two of the bluffs have been measured and described and are shown on Figure 4. The westernmost bluff is about 2,625 ft (800 m) north of Coffee Landing, and cannot be reached by walking along the river. It must be reached by boat or a perilous descent of the bluff from a paved road at the top of the outcrop. However, a smaller lower bluff about 1,000 ft (300 m) north of Coffee Landing exposes similar beds. The easternmost bluff (Fig. 4) is about 1.5 mi (2. 5 mi (2.5 km) northeast of or nearly 1.9 mi (3 km) downstream from Coffee Landing and is difficult to gain access to, but in most respects is similar to the other bluffs. Lithologically, all of the bluffs are similar and show the lenticular relationships of the sands and clays, which are typical of the Coffee Formation in Tennessee and northern Mississippi.

The first bluff north of Coffee Landing (about 0.3 mi [0.5 km]) exposes about 50 ft (15 m) of section at low water. The basal beds consist of several ft (m) of gray to greenish-gray glauconitic, micaceous quartz sand with gray clay partings, and large waterworn pieces of carbonized wood, some with teredo boring, are common. These beds have been assigned to the Eutaw Formation and crop out at the base of several bluffs. The basal graveliferrous sands of the Coffee Formation with *Ophiomorpha nodosa* and the deeply weathered surface on the Eutaw, typically present at the contact in this area are absent and the unconformity is not apparent. The basal sands and the thick clay lens at the top of the bluff are typical of the middle Coffee Formation. The yellowish-orange chert gravels at the top of the bluff are high level terrace deposits of the Tennessee River.

The next bluff (Fig. 4) downstream (about 0.6 mi [1 km] north of Coffee Landing) is very steep and exposes nearly 105 ft (32 m) of section. The upper 98 ft (30 m) of bluff consists of lithologies typical of the Coffee Formation. Sands dominate but 3 thick clay shale lenses crop out in this bluff.

The sands are light colored but may weather to light yellowish-orange or contain enough lignified plant remains to be brownish-gray. They are very-fine-grained and well sorted quartz with minor amounts of very-fine-grained light-colored glauconite and white mica. The sands exhibit various types of bedding, but, in general, they tend to be lenticular. Individual lenses may be bounded by wavey bedded clays and/or lignified plant remains. Flaser bedding is present, and small ripples are common.

When the western bluff (Fig. 4) was last measured in 1974 there were three lenses of grayish-brown clay exposed. The largest was more than 16 ft (5 m), and the others less than 10 ft (3 m) thick. Similar lenses occur in all the bluffs. They are lenticular and may stretch for more than 328 ft (100 m) before thinning to a feather's edge.

The clays (Fig. 5) are laminated to thin-bedded and weather to fissile fragments. Due to the large amount of comminuted lignified plant remains they are often chocolate brown in color. Typically the partings are near silt-size clean, clear, quartz, fine mica flakes and glauconite. The middle clay bed lies in a channel.

Fossils consist mainly of comminuted lignified plant remains concentrated mostly along bedding planes in the sands and clays. Large fragments of carbonized wood are present in the glauconitic sands and clays at the base of the bluff. Burrows as well as evidence of bioturbation are not conspicuous. Berry (1919, p. 35) described 13 genera and 16 species of plants from the Coffee Landing outcrops, one of which included *Halymenites major* Lesquereux (*Ophiomorpha nodosa* Lundgren) now recognized as a burrow. Bruce Wade (in Cockerell, 1916, p. 98–99) found fossil resin (amber) at Coffee Landing which contained an anterior wing, part of one antenna, and part of a leg of the insect *Dolophilus* (?) *praemissus* Cockerell.

The Coffee Formation overlies the Eutaw Formation and is

Figure 5. Laminated to thin-bedded clays interbedded with sand and lignified plant material. First bluff north of Coffee Landing, Tennessee.

overlain by the Sardis and Demopolis formations. Therefore, it is within the *Exogyra ponderosa* Faunal Zone, and is of Campanian age.

REFERENCES CITED

Berry, E. Wilbur, 1919, Upper Cretaceous floras of the eastern Gulf region in Tennessee, Alabama and Georgia: U.S. Geol. Survey Prof. Paper 112, 177 p.

Cockerell, T.D.A., 1916, Some American fossil insects: Proceedings U.S. National Museum, v. 51, p. 89–106.

Hardeman, W. D., 1966, Geologic Map of Tennessee, West Street: Tennessee Division of Geology.

Haywood, John, 1823, The natural and aboriginal history of Tennessee up to the first settlements therein by the white people in the year 1768: Reprinted Jackson, Tenn., 1959, 438 p.

Russell, E. E., 1975, Stratigraphy of the outcropping of the Upper Cretaceous in Western Tennessee, *in* Russell, E. E. and Parks, W. S., Stratigraphy of the Outcropping Upper Cretaceous, Paleocene, and Lower Eocene in Western Tennessee: Tennessee Division Geology Bulletin 75, p. A1–A65.

Safford, J. M., 1864, On the Cretaceous and superior formations of West Tennessee: American Journal of Science, 2nd ser., v. 37, p. 360–372.

—— 1869, Geology of Tennessee: Nashville, 550 p.

Earthquake-related features in the Reelfoot Lake area of the New Madrid Fault Zone in northwest Tennessee

Ernest W. Blythe, Jr., *The University of Tennessee, Martin, Tennessee 38238*

LOCATION

The Reelfoot Lake area is accessible from the east by I-40 and Tennessee 22; from the south by U.S. 51 and Tennessee 78; from the west by I-55 and I-155 and Tennessee 78; and from the north by the Purchase Parkway, U.S. 51-45W, and Tennessee 22 (Fig. 1).

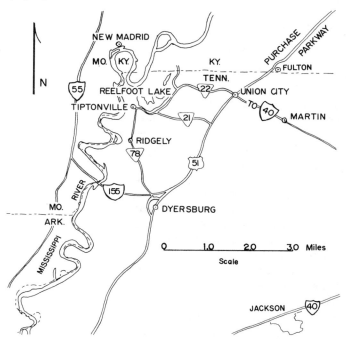

Figure 1. Map showing routes to the Reelfoot Lake area.

INTRODUCTION

Geologic features related to neotectonism in the New Madrid Fault Zone are clustered near Reelfoot Lake in northwestern Tennessee. The New Madrid earthquakes of 1811 and 1812 are regarded as the greatest earthquakes in United States history. Historical records indicate that fissures opened, through which water and sand squirted onto the surface (sand blows), and large landslides and surface faults occurred. Parts of Reelfoot Lake were formed due to subsidence accompanied by uplift in the nearby Tiptonville dome area during these earthquakes. Most of these features are now obliterated or obscure. Several notable features are visible, including Reelfoot Lake and its unique ecosystem.

Reelfoot scarp, which forms the eastern boundary of the Tiptonville dome and borders part of the western shore of the lake, shows evidence of having been actively uplifted on at least

three occasions during the past two thousand years (Russ, 1982). Both large and small landslides can be seen along the bluffs where dipping strata are deformed and are lower structurally than correlative strata in the Loess Hills Bluffs (Blythe, and others, 1975). Many of these landslides were probably triggered by earthquakes. Lines of sand blows; a prominent offset in the Loess Hills Bluffs, Ridgely Ridge, and the adjoining Tiptonville dome; and also the northeast-to-southwest seismic trend are suggestive of faults that have been active in the recent past.

Reelfoot Lake is on the eastern edge of the Mississippi River floodplain and partially occupies remnants of oxbow lakes that are abandoned channels of the Mississippi River (Fig. 2). The 115-ft (35-m) high Loess Hill Bluffs trend northeast-southwest at the eastern edge of Reelfoot Lake and form the boundary between the floodplain and the West Tennessee (loess) Plain to the east.

The stratigraphy of the Mississippi River floodplain in the vicinity of Reelfoot Lake consists of recent alluvial vertical and horizontal accretion sediments with clay plugs filling the oxbow

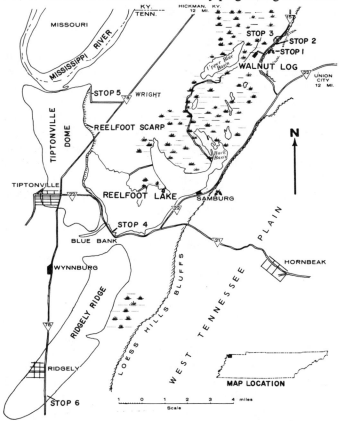

Figure 2. Generalized map of the Reelfoot Lake field trip area. (Location of the Tiptonville Dome and Ridgely Ridge is after Stearns 1979).

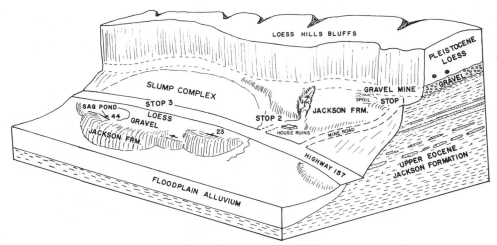

Figure 3. A block diagram of the location and geologic setting of Stops 1, 2, and 3.

lakes. Recent excavations (September 1984) for sewer lines near the southwest end of the lake revealed 3 to 6 ft (1 to 2 m) of vertical accretion silts and clays overlying another 6 to 10 ft (2 to 3 m) of horizontal accretion light gray, fine- to medium-grained sands with trough crossbed sets with maximum thicknesses of 8 to 12 in (20–30 cm). Sand blows on the floodplain south of Ridgely are composed of sand with similar composition.

The Loess Hills Bluffs are capped by as much as 86 ft (26 m) of Pleistocene loess composed mainly of silt with some clay intervals that are buried soil profiles. Locally terrestrial gastropods and calcareous concretions (loess dolls or loess kindchen) are abundant. Mastodon fossils have been excavated from the loess during gravel mining operations near Samburg. Pliocene-Pleistocene sandy gravel ranging from 0 to 16.5 ft (0 to 5 m) thick underlies the loess. It consists chiefly of brown to reddish-brown chert pebbles with chert and quartz sand. Thin beds of clay and sand with plant rootlets occur within the gravel in some locations. The top 33 to 66 ft (10 to 20 m) of the Upper Eocene Jackson Formation is found at the base of the bluffs. This formation is chiefly clay and claystone with beds and lenses of quartz sand and sandstone. Some of the clay and claystone contains abundant plant debris. The sandstones are former stream channels with a predominantly east to west transport direction as indicated by the orientation of cross-stratification. The stratigraphy of the Pleistocene loess and Plio-Pleistocene sandy gravel can be observed in several abandoned gravel strip mines along the bluffs in the vicinity of Samburg.

SITE DESCRIPTION

Locations recommended for inspection of bluff stratigraphy and also slumping along the bluffs are in the vicinity of Walnut Log. Stops 1, 2, and 3 are shown on Figure 2. Figure 3 is a generalized block diagram that shows more detail near these three stops. One of the best views of Reelfoot Lake is at Stop 4 on the south side of the lake. Stops 5 and 6 are at features associated with uplifted areas near Reelfoot Lake.

Stop 1: Gravel Mine

The Pleistocene loess and Plio-Pleistocene sandy gravel that overlie the Upper Eocene Jackson Formation are exposed in an abandoned strip mine just off Tennessee 157 about 0.3 mi (0.6 km) north of where the road goes west to Walnut Log. The mine road makes an acute angle to the southeast off Tennessee 157 and upslope into the mine. About 10 ft (3 m) of gravel, with well rounded chert and quartz pebbles and cobbles and fine to coarse sand, is exposed at the base of the outcrop. The sandy gravel is overlain by approximately 33 ft (10 m) of loess. Terrestrial gastropods are found about 6 ft (2 m) above the base of the loess.

Stop 2: Jackson Formation

Approximately 300 ft (100 m) north of where the mine road intersects Tennessee 157 and behind the ruins of the house on the east side of the road is an exposure of the Jackson Formation. About 20 ft (6 m) of the Jackson Formation crops out in the banks of a gully that cuts east into the bluff. Grayish-green massive claystone contains thin lenses of fine-grained sandstone and is overlain by a yellowish-gray medium-grained sandstone layer 8 in (20 cm) thick, which in turn is overlain by laminated clay predominantly yellowish gray with sand lenses near the base. The Jackson Formation is unconformably overlain by Plio-Pleistocene sandy gravel. The bottom 10 in (25 cm) of the sandy gravel is lithified with ferruginous cement.

Stop 3: Slump Features

Fuller (1912) recognized landslides along the Chickasaw Bluffs in his compilation of earthquake effects in this region. Many slumps and other types of mass movement are still recognizable along the western edge of the Loess Hills Bluffs. Two-tenths of a mile (0.3 km) north of Stop 2, Tennessee 157 crosses a large complex of slumps (see Figs. 2 and 3).

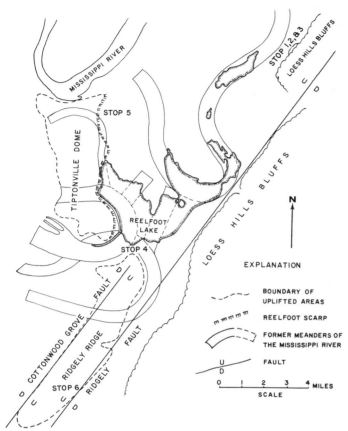

Figure 4. Map showing the relationship of Reelfoot Lake to former channels of the Mississippi River (after Russ, 1982 and Crone, and others, 1981) and to the structural features in the area (after Hamilton and Zoback, 1982).

Here the Jackson Formation crops out in a road cut on the west side of Tennessee 157. These strata dip 23 degrees to the southeast toward the bluff and are capped by sandy gravel. Nearer the bluff, across the road (east), loess that normally caps the bluff is in place at least 52 ft (16 m) below its position in the stratigraphic sequence of the bluffs. These strata have apparently slumped to this position, perhaps during the New Madrid series of earthquakes or during earlier quakes in the area. In the field and woods to the north and northeast of the Jackson outcrop are a series of dipping blocks containing the stratigraphic sequence of the Jackson Formation, sandy gravel, and loess. The first one is only a few feet (meters) across and has a vertical dip. The largest slump block in the complex has strata dipping at an angle of 44 degrees to the southeast. This block also has a small sagpond about 66 ft (20 m) in diameter associated with it. This slump complex is about 2000 ft (600 m) long and 650 ft (200 m) wide with the long axis parallel to the line of the bluffs.

Stop 4: The Reelfoot Lake Basin

Much of Reelfoot Lake occupies a complex of former channels of the Mississippi River. These former meanders of the Mis-

sissippi River are shown in Figure 4, which is modified after Crone and others (1981). Glenn (1933) suggested that the area subsided by rapid compaction of especially porous recent sediments as they were shaken by the 1811 and 1812 earthquakes. Fisk (1944) and Krinitzsky (1950) suggested that faulting is in part responsible for the structurally low area occupied by the lake. Krinitzsky inferred a normal fault from drill data on the west side of the lake beneath the Reelfoot scarp with a throw of 40 ft (10 m) and down on the lake side. Recent work by Stearns (1979) and Russ (1982) confirm as much as 10 ft (3 m) of uplift along the Reelfoot scarp and can account for much of this vertical offset as normal faulting in Holocene alluvium. The observations were made from closely spaced drilling across the scarp and from observations in an exploratory trench excavated by the U.S.G.S. across the scarp. Russ and others (1978) mapped a normal fault with as much as 10 ft (3 m) dip slip and also confirmed the main structure of the Reelfoot scarp as a monocline. Sexton and others (1982) conducted a high-resolution seismic-reflection survey that revealed no faults exhibiting significant offset of Tertiary and Paleozoic reflectors and suggesting monoclinal flecturing as a possible explanation for the scarp structure. They also concluded that if the Reelfoot scarp is associated with faulting, then the fault must be farther west of their localized study.

Various portions of the lake occupying the Reelfoot basin are visible from several places along the shoreline. One of the best locations for viewing the Blue Basin, the largest of the open bodies of water in this swamp and marsh complex, is from the Tennessee State Park picnic area near Blue Bank. This is also the deepest part of the lake. Maps showing the bottom elevations place water depths at about 19 ft (6 m) just north of this vantage point and from 3 to 10 ft (1 to 3 m) deep over most of Blue Basin.

Stop 5: The Tiptonville Dome and Reelfoot Scarp

The Tiptonville dome is the portion of the Lake County uplift bounded by the Reelfoot basin on the east (see Fig. 4). The boundary is defined by the north-to-south trending Reelfoot scarp mentioned earlier. The Tiptonville dome is a topographic high elevated about 20 ft (6 m) above the surrounding floodplain. Russ (1982) cited evidence for a tectonic origin for uplifted areas in Lake County and environs. (1) The surface of the uplifted area is significantly higher than any naturally occurring landform in the modern meander belt. (2) The original channel and levee gradients have been significantly warped, some to the extent that the slope of the original river-flow direction has been reversed. (3) The scarp on the east side of the dome has a predominantly convex slope, which would be consistent with an uplift origin. (4) Vertical offset of former channels of the Mississippi River is continuous from the Tiptonville dome across the Reelfoot scarp into the Reelfoot Basin. (5) Evidence for a monoclinal flexure is revealed in the exploratory trench excavated across the scarp (Russ and others, 1978). (6) Seismic reflection profiles show prominent reflectors to be arched upward beneath the areas that are topographically higher than the Reelfoot basin area.

The Reelfoot scarp at the eastern edge of the Tiptonville dome can be viewed by driving to the northeast along Tennessee 78 from its intersection with Tennessee 21-22 (see Fig. 2). After driving along Tennessee 78 for about 2.8 mi (4.5 km), the highway descends Reelfoot scarp. To the right the scarp slopes off to the western edge of Reelfoot Lake. To the left the scarp angles to the north. About 2 mi (3 km) farther to the northeast along Tennessee 78, turn to the left at the place indicated as Wright on Figure 2. Go west across the railroad tracks and continue west until the road has a right-angle turn toward the north. At Stop 5 you are standing at the level of the Reelfoot basin. Looking to the west you can see the topographic scarp and the higher level of the Tiptonville dome. To the left (south) the scarp can be seen as a more or less continuous feature. To the right the Reelfoot scarp ends where the levee built by the Corps of Engineers join it.

Stop 6: Sand Blows Associated with Ridgely Ridge and Bounding Faults

South of the Reelfoot basin is an uplifted area known as Ridgely Ridge. Ridgely Ridge is bounded on the west by the Cottonwood Grove fault and on the east by the Ridgely fault (see Figs. 2 and 4). Hamilton and Zoback (1982) confirmed these faults from seismic-reflection profiles taken across Ridgely Ridge by Geophysical Services Inc. for the U.S.G.S. They estimate the offset on the Cottonwood Grove fault to be about 260 ft (80 m)

of post–Middle Eocene movement. The Cottonwood Grove fault strikes N.40°E, and the northwest side is down.

The Ridgely fault is parallel to the Cottonwood Grove fault and bounds Ridgely Ridge on the southeast. The Ridgely fault is down to the southeast. Blythe and others (1975) suggest that the Ridgely fault extends to the northeast along the eastern margin of the Reelfoot basin roughly parallel to the bluffs and continues into the Loess Hills. Field evidence indicates that Upper Eocene strata are relatively high structurally in the Loess Hills Bluffs northwest of where the fault crosses into the Loess Hills. The Ridgely fault is believed to be responsible for the prominent offset in the line of bluffs to the northeast of Reelfoot Lake. Sand blows are abundant in the area south of Ridgely and are associated with Ridgely Ridge and the Ridgely fault. The sand blows are believed to have formed due to liquifaction during the New Madrid Earthquakes of 1811–1812. The sand blows are best viewed from the air but may be viewed in fields near Tennessee 78.

Two rather large sand blows can be observed on the left of Tennessee 78 0.2 mi (0.3 km) south of the Obion County line or 12 mi (20 km) south of the junction of Tennessee 78 and Tennessee 21-22 at Tiptonville (see Fig. 2). These sand blows are elongate from northeast to southwest and are approximately 60 by 100 ft (20 by 30 m) across. The sand is fine to coarse grained and is about 6 to 8 in (15 to 20 cm) thick overlying vertical accretion layers of silt and clay.

REFERENCES CITED

Blythe, E. W., Jr., McCutchen, W. T., and Stearns, R. G., 1975, Geology of Reelfoot Lake vicinity, *in* Stearns, R. G., ed., Field trips in West Tennessee: Tennessee Division of Geology, Report of Investigations no. 36, p. 64–75.

Crone, A. J., Russ, D. P., and Blythe, E. W., Jr., 1981, Earthquakes and related features of the Mississippi River Valley field trip guide, *in* Blythe, E. W., Jr., ed., Symposium and field trip guides: The University of Tennessee at Martin, Geosciences and Physics Department, 20 p.

Fisk, H. N., 1944, Geological investigation of the alluvial valley of the lower Mississippi River: Vicksburg, Miss., Mississippi River Commission, U.S. Corps of Engineers; 1 v. text 78 p. and 1 v. maps.

Fuller, M. L., 1912, The New Madrid earthquake: U.S. Geological Survey Bulletin 494, 119 p. (Reprint edition by Ramfre Press, Cape Girardeau, Mo.).

Glenn, L. C., 1933, The geography and geology of Reelfoot Lake: Tennessee, Academy of Science Journal, v. 8, p. 3–12.

Hamilton, R. M., and Zoback, M. D., 1982, Tectonic features of the New Madrid seismic zone from seismic reflection profiles, *in* McKeown, F. A., and Pakiser, L. C., eds., Investigations of the New Madrid, Missouri, earthquake region: U.S. Geological Survey Professional Paper 1236-F, p. 55–77.

Krinitzsky, E. L., 1950, Geological investigation of faulting in the lower Mississippi valley: Vicksburg, Miss., Waterways Experiment Station, Corps of Engineers, U.S. Army Technical Memorandum no. 3-311, 91 p.

Russ, D. P., 1982, Style and significance of surface deformation in the vicinity of New Madrid, Missouri, *in* McKeown, F. A., and Pakiser, L. C., eds., Investigations of the New Madrid, Missouri, earthquake region: U.S. Geological Survey Professional Paper 1236-H, p. 95–114.

Russ, D. P., Stearns, R. G., and Herd, D. G., 1978, Map of exploratory trench across Reelfoot scarp, northwestern Tennessee: U.S. Geological Survey Miscellaneous Field Studies Map MF-985, scale 1:39.4.

Sexton, J. L., Frey, E. P., and Malicki, D., 1982, High-resolution seismic-reflection surveying on Reelfoot Scarp, northwestern Tennessee, *in* McKeown, F. A., and Pakiser, L. C., eds., Investigations of the New Madrid, Missouri, earthquake region: U.S. Geological Survey Professional Paper 1236-J, p. 137–148.

Stearns, R. G., 1979, Recent vertical movements of the land surface in the Lake County Uplift and Reelfoot Lake Basin areas, Tennessee, Missouri, and Kentucky: U.S. Nuclear Regulatory Commission, NUREG/CR-0874, 37 p.

The Cockfield (Claiborne Group), Moodys Branch and Yazoo (Jackson Group) Formations at the Riverside Park locality in Jackson, Mississippi

David T. Dockery III, Mississippi Bureau of Geology, P.O. Box 5348, Jackson, Mississippi 39216

LOCATION

The Riverside Park locality is a steep-sided ravine along the eastern side of a golf course in northeast Jackson, Mississippi, in the NE¼ NW¼ NW¼, sec. 36, T. 6 N., R. 1 E., Hinds County, Jackson 7½-minute quadrangle (Fig. 1). It is readily accessible by taking the Lakeland Dr. east exit off of I-55 at Jackson, then turning south on Highland Dr. at the first traffic light, and continuing to the intersection with Riverside Dr. and Riverside Park Circle. At this intersection take Riverside Park Circle east, and after the circle turns north, stop in the parking lot on the right. This lot is next to the entrance of a nature trail and uphill from a golf course. The locality can be reached by walking east-northeast for a short distance over a part of the golf course to a ravine below a small pond.

INTRODUCTION

The Riverside Park locality has been long known as a collecting site for marine invertebrate fossils from the Jackson Group and is locally known as "Fossil Gulch." It is the alternate type locality for the basal unit of the Jackson Group, the Moodys Branch Formation (Moore and others, 1965). The type locality of this formation on Moodys Branch just southeast of the corner of Poplar Blvd. and Peachtree St. in Jackson is now partly covered with fill. At present, the Riverside Park locality is the only site in the Jackson area where a complete section of the Moodys Branch Formation can be seen. It is also an important stratotype for the diverse marine microfossils and macrofossils of the Jackson Group, a standard rock unit for the Gulf and Atlantic Coastal Plain.

Conrad (1856) established the terms Claiborne, Jackson, and Vicksburg as formal stratigraphic units within the Gulf Coast Eocene and placed them in their proper sequence. Presently, the Claiborne and Jackson Groups are recognized as Eocene, and the Vicksburg Group is recognized as Oligocene in age. The Jackson division was named for Jackson, Mississippi, where Conrad obtained a collection of fossil mollusks, which he recognized as intermediate in age between the Claiborne and Vicksburg faunas. Drawings of the Jackson mollusks were published in Wailes' *Report on the Agriculture and Geology of Mississippi* (1854, p. 289, pls. 14-17). This work was the first publication of the newly formed Mississippi Geological and Agricultural Survey. According to Wailes' (1854) description of the fossiliferous beds at Jackson and as indicated by the fossils, Conrad's collection came from the Moodys Branch Formation.

The Cockfield (Claiborne Group) and Moodys Branch (Jackson Group) Formations are the oldest units exposed at the crest of the Jackson Dome, a structural high above a buried

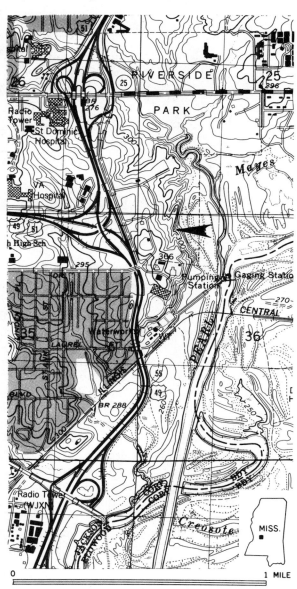

CONTOUR INTERVAL 10 FEET
DOTTED LINES REPRESENT 5-FOOT CONTOURS

Figure 1. Location map of the Cockfield (Claiborne Group), Moodys Branch, and Yazoo (Jackson Group) Formations outcrop at Riverside Park, Jackson 7½-minute quadrangle. Arrow indicates the locality.

Figure 2. Cockfield (Claiborne Group), Moodys Branch and Yazoo (Jackson Group) Formations at the Riverside Park locality. The lower arrow indicates the Cockfield-Moodys Branch contact, and the upper arrow indicates the Moodys Branch-Yazoo contact. Photograph taken by Perry Nations on March 11, 1965.

Cretaceous volcano and reef complex (Harrelson, 1981). These strata have a short outcrop belt along the Pearl River's western valley wall and associated tributaries, one of which is the Riverside Park locality. This outcrop belt is confined within the Jackson city limits. The Pearl River crosses the main outcrop belt of the Cockfield and Moodys Branch Formations 25 mi (41.2 km) northeast (by straight line) of Jackson. Exposures of the Moodys Branch Formation at Jackson have produced the most diverse and best preserved marine faunas of the Jackson Group as well as of any equivalent North American upper Middle Eocene unit. The molluscan species of this fauna are illustrated in Dockery (1977). Though the large species are not as well preserved as at other localities in Jackson, the Riverside Park locality does have a high diversity of small and medium-sized molluscan species.

SITE DESCRIPTION

The Riverside Park locality is a steep-sided ravine that is about 60 ft (18.2 m) long in an east-west direction and that has about 33 ft (10 m) of relief from its mouth to its western rim (this relief and the thickness of the exposed section vary depending on the degree of slumping). This locality, shown in Figure 2 as photographed in 1965, had a stable slope for a number of years until a pond was dug above the upper slope. Now there is large scale slumping along the upper end of the ravine; however, the section is readily visible and probably will continue to be so for some time. The exposed section includes part of the Cockfield Formation in the lower part, the Moodys Branch Formation exposed in the vertical walls, and part of the overlying Yazoo Formation exposed on the upper slope. The variations in the thicknesses of the Cockfield and Moodys Branch intervals are due to relief on the disconformable contacts between these units.

The Cockfield Formation consists of fluvial and deltaic sands and silty clays (Dockery, 1976), and is a major aquifer in the Jackson area as well as in other areas across central Mississippi. At the Riverside Park locality, the Cockfield Formation consists of 7 to 10 ft (2 to 3 m) of dark-gray silty clays and sands and a lignite bed in its lower part. The upper part of the formation is riddled with burrows extending downward from the

Moodys Branch Formation and consists largely of silty clay. These sediments were deposited in delta plain and estuarine environments associated with subsiding Cockfield delta systems. The Cockfield–Moodys Branch contact is erosional and has 3 or more ft (1+ m) of relief at the Riverside Park locality.

The Moodys Branch Formation consists of 10 to 13 ft (3 to 3.9 m) of glauconitic, fossiliferous sands deposited as a destructional shelf sand facies above subsiding delta systems. This unit is the basal marine transgressive unit of the Jackson Group, and is part of a transgressive sequence that extends up the Mississippi Embayment into Arkansas. Two facies, a northern and southern terrigenous facies, were recognized by Dockery (1977) from outcrop exposures of the Moodys Branch Formation and equivalents. The northern terrigenous facies contains the nearshore shelf sediments of the upper Mississippi Embayment and includes the Moodys Branch Formation in Yazoo County, Mississippi, and the White Bluff Formation in Arkansas. The southern terrigenous facies contains offshore shelf sediments and includes exposures of the Moodys Branch Formation in Clarke County, Mississippi; Jackson, Mississippi; and Montgomery, Louisiana.

The lower Yazoo Formation exposed at the Riverside Park locality consists of 10 or more ft (3+ m) of glauconitic, fossiliferous, blocky clay, which was deposited in offshore, quiet, marine waters. According to Siesser (1983), this clay contains a well preserved calcareous nannoplankton flora; however, the chalky and fragile invertebrate macrofossils are not so well preserved. The Moodys Branch-Yazoo contact is gradational.

This slumping and instability of the Yazoo Formation on the upper slope of the Riverside Park locality illustrates the problems encountered in the Jackson area and across a significant part of central Mississippi with the montmorillonitic clays that comprise this formation. These problems are apparent to travelers on I-20 and I-55 when they encounter undulating stretches of highway as they approach Jackson.

Several studies have been made of the Jackson Group section at the Riverside Park locality. Elder and Hansen (1981) studied the macrofossil assemblages at this locality and recognized three successive environments in the Moodys Branch Formation, which in ascending order are: inner shelf, inner middle shelf, and outer middle shelf. The succession of these environments indicates increasing water depth and distance from the shoreline. Siesser (1983) studied the calcareous nannoplankton at the Riverside Park locality and placed the Moodys Branch Formation in Martini's calcareous nannoplankton Zone NP 17 and the lower Yazoo Formation in NP18. The NP17-NP18 boundary approximates the Middle-Upper Eocene boundary according to Bignot and Cavelier (1981). Several workers have published on the well known molluscan faunas of the Moodys Branch Formation. Descriptions and illustrations of these faunas can be found in Harris and Palmer (1946–1947), Toulmin (1977), and Dockery (1977, 1980).

REFERENCES CITED

Bignot, G., and Cavelier, C., 1981, Table 1, *in* Pomerol, C., ed., Stratotypes of Paleogene Stages: Du Bulletin d'Information des Geologues du Bassin de Paris, Mem. no. 2, 301 p.

Conrad, T. A., 1856, Observations on the Eocene deposits of Jackson, Mississippi, with descriptions of thirty-four new species of shells and corals: Academy of Natural Sciences of Philadelphia, Proceedings 1855, v. 7, p. 257–263.

Dockery, D. T., III, 1976, Depositional systems in the upper Claiborne and lower Jackson Groups (Eocene) of Mississippi [M.S. thesis]: Oxford, University of Mississippi, 109 p.

—— 1977, Mollusca of the Moodys Branch Formation, Mississippi: Mississippi Geological Survey Bulletin 120, 212 p., 28 pl.

—— 1980, The invertebrate macropaleontology of the Clarke County, Mississippi, area: Mississippi Bureau of Geology Bulletin 122, 387 p., 82 pl.

Elder, S. R., and Hansen, T. A., 1981, Macrofossil assemblages of the Moodys Branch Formation (Upper Eocene), Louisiana and Mississippi: Mississippi Geology, v. 2, no. 1, p. 6–11.

Harrelson, D. W., 1981, Igneous rocks of the Jackson Dome, Hinds-Rankin Counties, Mississippi: Mississippi Geology, v. 1, no. 4, p. 7–13.

Harris, G. D., and Palmer, K.V.W., 1946–1947, The Mollusca of the Jackson Eocene of the Mississippi embayment (Sabine River to the Alabama River): Bulletins of American Paleontology, v. 30, no. 117, 563 p., 65 pl.

Moore, W. H., Bicker, A. R., Jr., McCutcheon, T. E., and Parks, W. S., 1965, Hinds County geology and mineral resources: Mississippi Geological Survey Bulletin 105, 244 p.

Siesser, W. G., 1983, Paleogene calcareous nannoplankton biostratigraphy: Mississippi, Alabama and Tennessee: Mississippi Bureau of Geology Bulletin 125, 61 p., 37 fig.

Toulmin, L. D., 1977, Stratigraphic distribution of Paleocene and Eocene fossils in the eastern Gulf Coast region: Geological Survey of Alabama Monograph 13, 602 p., 72 pl.

Wailes, B.L.C., 1854, Report on the agriculture and geology of Mississippi: E. Barksdale, State Printer, Jackson, 371 p.

The Vicksburg Group at the Haynes Bluff Quarry

David T. Dockery III, Mississippi Bureau of Geology, P.O. Box 5348, Jackson, Mississippi 39216

LOCATION

The Vicksburg Group is exposed in a rock quarry on east side of Mississippi 3 north of Redwood, Mississippi, in the NW¼, sec. 26, T. 18 N., R. 4 E., Warren County, Onward 15-minute Quadrangle (Fig. 1). Readily accessible by taking I-20 to the U.S. 61 bypass east of Vicksburg, traveling north on U.S. 61 bypass to Redwood, then continuing north on Mississippi 3 to dirt road on right opposite the office building of the inactive Mississippi Valley Portland Cement Company, and then following the dirt road by foot to the northernmost of two inactive quarries.

INTRODUCTION

The Haynes Bluff quarry provides the best exposures of the widespread Lower Oligocene Vicksburg Group in the type area of the group. The lower Oligocene marine fossils in the vicinity of Haynes Bluff, once known as Walnut Hills, are possibly the first fossils reported from Mississippi. This area was readily accessible by river boat and was first noted by the French naturalist Charles A. Lesueur in 1828 while in route to New Orleans on a flatboat. Lesueur made an excellent collection of fossils from the Haynes Bluff area during a stopover at Vicksburg. He illustrated these fossils the next year in a series of 12 beautifully drawn plates, which he never published. These plates were recently published in Dockery (1982a, 1982b).

Conrad (1848a, 1848b), who had seen part of Lesueur's collection at the Academy of Natural Sciences of Philadelphia, later collected fossils from the Vicksburg area and described 105 new species. In 1856, after studying the fossiliferous beds at Jackson, Mississippi, Conrad recognized the Vicksburg, Jackson, and Claiborne as formal stratigraphic units within the Gulf Coast Eocene and placed them in their proper sequence. Later in 1866, he placed the Vicksburg as Oligocene in age and the Jackson and Claiborne respectively as Upper and Middle Eocene. These age assignments were based on molluscan faunas and have remained largely the same up until the present, a period of almost 120 years.

The Vicksburg Group in Mississippi contains the best preserved and most diverse Lower Oligocene molluscan faunas in the Southeastern United States. Lower Oligocene marine sediments in the eastern Gulf and Atlantic Coastal Plains are highly calcareous, and aragonitic shells are absent due to leaching. West of the Mississippi River, the marine units of the Vicksburg Group grade into deltaic and fluvial clastic facies. The molluscan fauna of the Vicksburg Group, including the Red Bluff Formation in eastern Mississippi, is described and illustrated in Dockery (1982b) and MacNeil and Dockery (1984). These publications cite 555 species including subspecies, varieties, and unnamed forms.

The Vicksburg Group in the Vicksburg area includes the following formations in ascending order: (1) Forest Hill Formation, (2) Mint Spring Formation, (3) Marianna Limestone, (4) Glendon Limestone, (5) Byram Formation, and (6) Bucatunna Formation. This sequence records a marine regression in the Forest Hill Formation, a marine transgression that began with the Mint Spring Formation and climaxed during deposition of the middle Glendon Limestone, and a second marine regression that began in the upper Glendon Limestone and continued into the Bucatunna Formation. In eastern Mississippi, the basal deltaic sediments of the Forest Hill Formation grade into the fossiliferous clays and glauconitic sands of the Red Bluff Formation, the oldest marine unit of the Vicksburg Group.

All seven formations of the Vicksburg Group in Mississippi contain a marine fauna in some part if not throughout their sequence. However, three of these units, the Red Bluff, Mint Spring, and Byram Formations, contain a notable diversity of well preserved invertebrate species. In the Haynes Bluff quarry, a representative macrofauna and microfauna of the Vicksburg Group can be collected from the Mint Spring and Byram Formations. Care must be taken when collecting fossils adjacent to the old quarry road. This road is littered with Recent oyster and *Rangia* shells, which were used at one time to upgrade the carbonate content of the sediments in the manufacture of cement.

SITE DESCRIPTION

Marine units of the Vicksburg Group are well exposed in the Haynes Bluff quarry, a quarry in which the Glendon Limestone was mined for the manufacturing of cement. A measured section of this quarry taken from Fisher and Ward (1984) is given in Figure 2. The lower two units of the quarry sequence, the Mint Spring Formation and Marianna Limestone, are exposed in a steep-sided ditch that drains the quarry on its southwest corner and runs by the quarry access road near its junction with the quarry floor. Here the Mint Spring Formation is a calcareous, glauconitic, fossiliferous sand with a thickness of about 14 ft (4.3 m).. The thickness of the Mint Spring Formation is variable in the Vicksburg area, ranging from 5 to 25 ft (1.5 to 7.6 m) (Mellen, 1941). This variation is due to the relief of the lower disconformable contact with the Forest Hill Formation, a fluvial-deltaic unit with some fossiliferous estuarine zones below the upper contact (Dockery, 1982b). The Mint Spring Formation is a transgressive marine, near-shore shelf deposit that is strongly bioturbated and massive in appearance.

Capping the Mint Spring Formation along the upper edge of the ditch wall adjacent to the quarry floor is a foot-thick soft limestone unit attributed to the Marianna Limestone. This limestone averages about 30 ft (9.1 m) in thickness in eastern and central Mississippi but thins westward in the vicinity of Vicksburg.

Siesser (1983) noted that the Mint Spring Formation and

Marianna Limestone were diachronous along their outcrop belt in Mississippi. In eastern Mississippi, he placed these units in

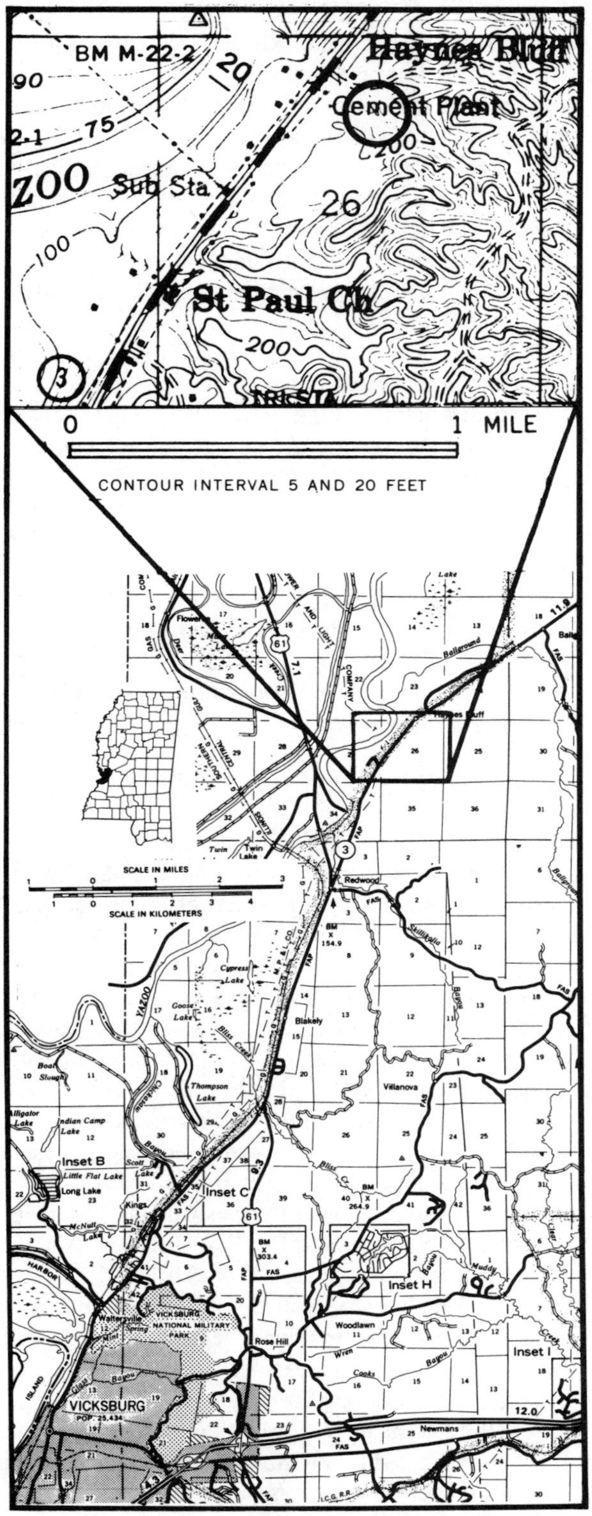

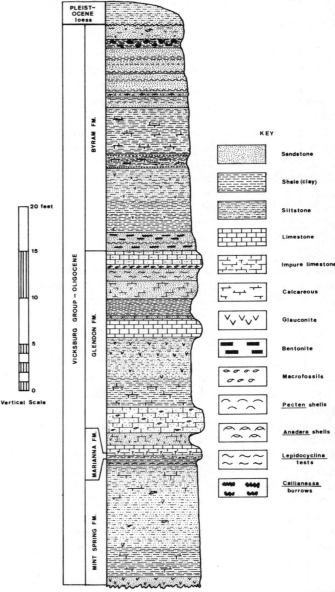

Figure 1. Location map of the Haynes Bluff quarry. Map is a 2 to 1 enlargement of the southeast part of the Onward 15-minute quadrangle. Circle indicates the locality.

Martini's calcareous nannoplankton zone NP21, and in central and western Mississippi, he placed them in NP22. The Mint Spring Formation and Marianna Limestone can be paleontologically distinguished in the field from the overlying Glendon Limestone and Byram Formation by the occurrence of the bivalve *Pecten poulsoni* Morton. This *Pecten* is characterized by the strongly inflated umbo of the right valve and by its U-shaped or V-shaped unilirate ribs.

The Glendon Limestone at the Haynes Bluff quarry is a 22-ft (6.7 m) thick sequence of limestone ledges interbedded with terrigenous sands and clays. Figure 3 is a photograph of the eastern quarry wall taken on June 12, 1976, when the quarry was still active and shows a fresh exposure of the Glendon Limestone and the overlying Byram Formation. The lower ledge of the

Figure 2. Measured section of the Haynes Bluff quarry taken from Fisher and Ward (1984).

Figure 3. Photograph of the eastern wall of the northern quarry at Haynes Bluff taken on June 12, 1976, when the quarry was still active. The Glendon Limestone forms the quarry wall above the lower floor, and the Byram Formation forms a wall above the upper floor. The resistant sand unit at the top of the Byram Formation is overlain by loess.

Glendon is a sparsely fossiliferous, bioturbated lime mudstone that forms the quarry floor, while the upper ledge is a hard bioclastic calcarenite and calcirudite. Molds of aragonitic molluscan shells, calcitic *Lepidocyclina supera* (Conrad) tests, and *Pecten byramensis* Gardner shells are abundant in the upper limestone ledge. The latter species is a guide fossil to both the Glendon Limestone and Byram Formation. It can be readily distinguished from *Pecten poulsoni* of the Mint Spring Formation and Marianna Limestone by its square-shouldered trilirate ribs. The Glendon Limestone maintains a rather uniform thickness of about 30 ft (9.1 m) across the state.

Conformably overlying the Glendon Limestone are the fossiliferous, glauconitic sands and clays of the Byram Formation. Unlike the Mint Spring Formation, the Byram has a variable sedimentary sequence consisting of bedded sands and clays. Figure 3 shows a prominent sand unit above a 13-ft (3.9 m) thick interval of sandy clay in the upper quarry wall. This sand unit is very fossiliferous and contains an abundance of the bivalve *Scapharca lesueuri* Dall; the lower clay interval is only sparsely fossiliferous. Differences in the near-shore shelf facies of the Mint Spring and Byram Formations are due to the deposition of the former in a transgressive, destructional shelf environment, while the latter was deposited in an aggrading, regressive shelf environment. Sparsely fossiliferous clay beds in the Byram probably represent periods of rapid sediment deposition.

A paleoenvironmental reconstruction of the Haynes Bluff quarry section based on a study of foraminiferal communities by Fisher and Ward (1984) indicates the greatest water depth, and thus the greatest extent of the marine transgression, to have occurred during deposition of the middle Glendon Limestone. They determined the following paleoenvironments within this section: (1) Mint Spring - hypersaline to normal marine marshes, (2) Marianna - shallow inner shelf, (3) Glendon - deep inner shelf, (4) Byram - normal bay/lagoon to hyposaline lagoon.

Disconformably overlying the Byram Formation is Pleistocene loess. This loess is badly gullied in the upper part of the quarry, and fossil land snails eroded from it are commonly mixed with Oligocene marine shells at lower levels. Snowden and Priddy (1968) radiocarbon dated land snail shells from the base and top of a nearby loess sequence at a gravel pit close to Redwood. They obtained age dates of 20,500 ± 600 and 18,200 ± 500 years B.P., thus correlating this unit with the Peorian loess of the Wisconsin glacial stage.

REFERENCES CITED

Conrad, T. A., 1848a, Observations on the Eocene Formation, and description of one hundred and five new fossils of that period, from the vicinity of Vicksburg, Mississippi, with an Appendix: Academy of Natural Sciences of Philadelphia Proceedings 1847, v. 3, p. 280–299.

—— 1848b, Observations on the Eocene formation, and description of one hundred and five new fossils of that period, from the vicinity of Vicksburg, Mississippi, with an Appendix: Academy of National Sciences of Philadelphia Journal, 2nd series, v. 1, p. 111–134, pl. 11–14 (plates reprinted in Dockery 1982b, Appendix I).

—— 1856, Observations on the Eocene deposits of Jackson, Mississippi, with descriptions of thirty-four new species of shells and corals: Academy of Natural Sciences of Philadelphia Proceedings 1855, v. 7, p. 257–263.

—— 1866, Checklist of invertebrate fossils of North America. Eocene and Oligocene: Smithsonian Miscellaneous Collections, v. 7, no. 200, p. 1–41.

Dockery, D. T., III, 1982a, Lesueur's Walnut Hills fossil shells: Mississippi Geology, v. 3, no. 3, p. 7–13.

—— 1982b, Lower Oligocene Bivalvia of the Vicksburg Group in Mississippi:

Mississippi Bureau of Geology Bulletin 123, 261 p., 62 pl.

Fisher, R., and Ward, K., 1984, Paleoenvironmental reconstruction of the Vicksburg Group (Oligocene), Warren County, Mississippi: Mississippi Geology, v. 4, no. 3, p. 1–9.

MacNeil, F. S., and Dockery, D. T., III, 1984, Lower Oligocene Gastropoda, Scaphopoda, and Cephalopoda of the Vicksburg Group in Mississippi: Mississippi Bureau of Geology Bulletin 124, 415 p., 72 pl.

Mellen, F. F., 1941, Geology, p. 1–88 *in* Mellen, F. F., McCutcheon, T. E., and Livingston, M. R., Warren County Mineral Resources: Mississippi Geological Survey Bulletin 43, 140 p.

Siesser, W. G., 1983, Paleogene calcareous nannoplankton biostratigraphy: Mississippi, Alabama and Tennessee: Mississippi Bureau of Geology Bulletin 125, 61 p., 37 fig.

Snowden, J. O., Jr., and Priddy, R. R., 1968, Geology of Mississippi loess, *in* Loess investigations in Mississippi: Mississippi Geological, Economic and Topographic Survey Bulletin 111, p. 12–203.

Holocene fluvial landforms and depositional environments of the Lower Mississippi Valley

Roger T. Saucier, U.S. Army Engineer Waterways Experiment Station, Vicksburg, Mississippi 39180

LOCATION

Holocene fluvial land forms are exposed along a 46-mi (74 km) circuitous route west of Natchez, Mississippi, in Concordia Parish, northeastern Louisiana. The route encompasses U.S. 84 and Louisiana 129, 565, and 15 through the communities of Ferriday, Stacy, Monterey, and Deer Park (Fig. 1).

Road shoulders are adequate for safe temporary parking at al locations. All lands beyond highway rights-of-way are privately owned. Best observation is in winter to early spring due to an absence of foliage and crops.

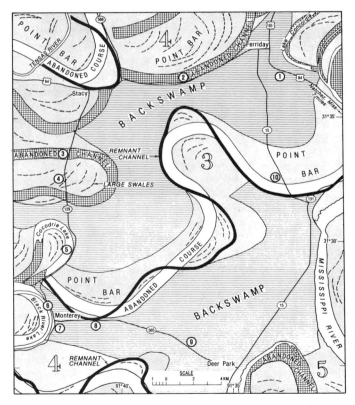

Figure 1. Map of Holocene fluvial landforms and depositional environments of the lower Mississippi valley showing the locations of stops 1-10.

INTRODUCTION

Most areas of Quaternary fluvial depositional environments and landforms have large areal extent, small relief, and few significant exposures, hence direct study is difficult. This is especially the case in the extensive 18,500 mi^2 (48,000 km^2) Holocene floodplain of the Mississippi River in the Lower Mississippi Valley. In this region, typical landforms are measured in plan in terms of miles, but only in feet in cross-section. So that fluvial landforms and environments characteristic of this large region, more than twice the size of New Jersey, can be observed and comprehended in terms of their inherent variability and postdepositional modification, this cluster stop was formulated. Considering the relatively short travel distances between stops in terms of the size of the alluvial valley, it is a unique opportunity.

Five meander belts have been formed by the Mississippi River since it ceased deposition of glacial outwash and converted from a braided stream pattern about 9,000 years ago (Saucier 1974). Because of progressive valley aggradation, there are widely varying degrees of veneering of the older meander belts by younger alluvium. Consequently, there are significant differences in fluvial landform morphology as a function of regional postdepositional history. Additionally, the life span of some individual landforms/environments is sufficiently long (hundreds to thousands of years) that age is important in morphology and lithology. For example, as a function of the degree of sediment filling over time, abandoned channels (cutoffs) variously occur as oxbow lakes, arcuate swampy depressions, or completely filled features with subtle topographic, hydrologic, and pedologic expression. This cluster stop includes features representing a wide range of age and morphologic variation.

The three Holocene Mississippi River meander belts visible at this cluster stop are designated 3, 4, and 5 (Fig. 1) (Saucier 1974). They represent different lengths of occupancy as well as different ages. Meander Belt 3 dates from about 4,700 to 6,000 B.P., a span of 1,300 years. During this time, the river channel meandered little, creating few cutoffs. Abandonment of the meander belt was slow, with meandering continuing during declining discharge. Final flow was confined to a small remnant channel that sometimes is within but often beyond the position of the full-flow channel. Meander Belt 4 dates from about 2,700 to 4,800 B.P., a period of 2,100 years. This longer active life resulted in a wider meander belt with numerous abandoned channels. The remnant channel is now occupied by the Ouachita River rather than local drainage. This stream is underfit in the relict channel and has not meandered significantly. Meander Belt 5 formed about 2,800 years B.P. and has been occupied continuously by the Mississippi River since then. It closely resembles Meander Belt 4 in number of cutoffs.

The three meander belts are separated by broad, flat floodbasin or backswamp areas. These are characterized by progressive vertical sediment accumulation in contrast to meander belts where lateral sedimentation on point bars has prevailed.

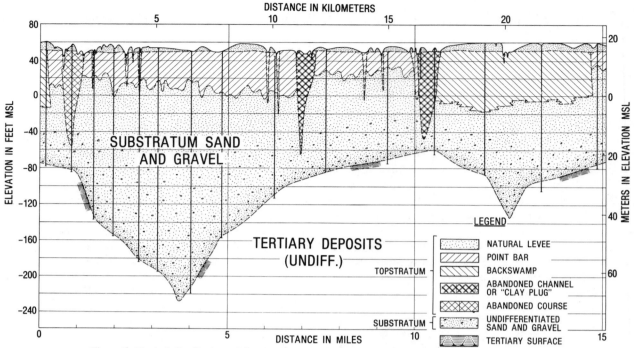

Figure 2. Typical distribution of Quaternary deposits and environments as determined by borings.

Features visible at the cluster stop locations are important elements in the present landscape. They are also indicative of shallow subsurface conditions. Figure 2, a cross section trending northwest from near Ferriday, is typical (Saucier 1967). Varying from about 100 to 300 ft (30 to 90 m) in thickness, the alluvial sequence overlies an erosional unconformity on Tertiary-age formations. The lower part of the substratum is Pleistocene in age (Saucier 1974) while the topstratum and upper part of the substratum (the upper 100 ft (30 m) of the sequence) is Holocene. Topstratum deposits consist primarily of clays, silts, and fine sands while the substratum is almost entirely sands and gravels.

SITE DESCRIPTION

The entire cluster stop area is relatively low in elevation (33 to 66 ft [10 to 20 m] above mean sea level) with low relief (averaging 10 to 16 ft [3 to 5 m]). Through-flowing streams have gradients of only inches per mile as do the present and abandoned meander belts. Small basin-drainage features are often stagnant and may experience reversals in flow depending on local precipitation patterns. Because of artificial flood-control levees and channelization, once extensive overbank flooding is now rare.

Particular attention should be paid to local and regional variations in land use (e.g., agriculture, settlements) and vegetation community composition at each location since these often are the only visible manifestations of alluvial landforms. There is a declining but still strong coincidence of towns, roads, farms, and other cultural features with natural levees and higher elevations. Until several decades ago, backswamp areas were uninterrupted forest tracts, but recently have been cleared for agriculture.

Abandoned channels and courses, remaining backswamp areas, and topographic depressions such as point bar swales are still largely wooded with bald cypress and water tupelo dominant. Disturbed areas and once-cleared low areas often support black willow, button bush and water locust. More completely filled channels support a mixed deciduous hardwood community like that of uncleared natural levees, consisting of oaks, elms, pecan, maples, hackberry, and sycamore.

Stop No. 1. Natural Levee Ridge

This stop is along U.S. 84 at a point 1.3 mi (2.1 km) southeast of its junction with U.S. 65 in Ferriday.

To the south and southwest of the highways, a gently sloping natural levee ridge flanks an inactive Mississippi River channel. The channel, now occupied by Lake Concordia, is 0.5 mi (0.8 km) north of the highway (Fig. 1). Although hard to discern, the natural levee slopes about 8 ft per mi (2.5 m/km) to the south.

A typical Mississippi River natural levee ridge is about 2 mi (3 km) wide (on each side of the channel) and has a crest elevation about 15 ft (4.5 m) above backswamp level. It is composed of firm to stiff, mottled gray and brown, well-oxidized silty and sandy clays. Since natural levees are formed by incremental sedimentation from overbank sheet flow, they occur peripheral to meander loops and may overlie a variety of fluvial landforms/environments.

Stop No. 2. Swamp-filled Abandoned Channel

This stop is situated on U.S. 84 a distance of 4.0 mi (6.4 km) west of its junction with U.S. 65 in Ferriday.

U.S. 84, following the crest of a natural levee ridge, flanks the 3,000-ft-wide (900-m) lower arm of a large, mostly swamp-filled abandoned channel in Meander Belt 4. A 10- to 15-foot high (3- to 4.5-m) scarp immediately north of the highway marks the south edge and former bankline of the channel.

Borings indicate the channel fill to be about 100 ft (30 m) thick and to consist of uniformly dark gray, soft, high-water-content organic clays and silts. Most of the fill was deposited shortly after cutoff. However, some sedimentation is still occurring since the channel is a depression and a link in a tortuous interior basin drainage system. Cross Bayou and Turtle Lake, the water bodies visible in the channel, are parts of this drainage system.

Stop No. 3. Completely Filled Abandoned Channel

This stop is along Louisiana 129 about 3.0 mi (4.8 km) south of its junction with U.S. 84 at the community of Stacy.

The abandoned channel at this stop is discernible only because of the presence of Boggy Bayou, which flows in a broad arcuate pattern along the outer limit of the abandoned channel, and Float Bayou, which flows in a similar pattern along the inner margin. Although this channel is about the same age as the one at Stop 2, initial and especially subsequent sedimentation were sufficient to nearly fill the feature. Surface elevations in the upper arm of the abandoned channel are only 5 ft (1.5 m) or less lower than the adjacent natural levee ridges. Near-surface soils, topography, and vegetation in the channel resemble those of a natural levee, but there is no regional slope.

Beneath the firm and silty surficial natural levee-like soils, much softer, wetter, and finer-grained channel fill deposits should be present. In a typical abandoned channel, channel fill deposits are thickest and finer-grained farthest from the point of cutoff. The upper and lower arms for several miles often have a wedge of silty and sandy fill representing river bedload rather than only suspended load as a sediment source.

Stop No. 4. Point Bar Accretion (Meander Scroll) Topography

This stop is situated along Louisiana 129 about 4.2 mi (6.8 km) south of the junction of this highway with U.S. 84 at the community of Stacy.

Point bar ridges and swales, manifestations of lateral accretion by a meandering river, are the most characteristic topography of the Lower Mississippi Valley. However, they are seldom as well developed and preserved as they are at this stop.

Linear to arcuate point bar ridges, composed of relatively light colored (tans and browns) silts and silty sands, alternate with parallel to subparallel swales filled with relatively dark colored (grays and browns) clays and silty clays. There are from 5 to 15 ridges and swales per mi (3 to 9 per km) in a typical sequence, and the relief from ridge crest to swale bottom is usually 5 to 10 feet (1.5 to 3 m) In the subsurface, fine-grained swale deposits and

the upper silty deposits of ridges extend to a depth of 20 to 30 feet (6 to 9 m), and grade downward into a relatively "clean" or clay-free substratum composed of sands and sands and gravels.

In a typical point bar sequence in a natural state, the ridges supported a natural levee-like vegetation community while the larger and deeper swales were swamp-filled depressions or contained shallow linear lakes such as those visible at this location. However, the present lakes are artificial impoundments maintained for agriculture and aesthetics.

At this stop, the trend of the ridges and swales is northeast (Fig. 1) or across the highway at nearly a 45° degree angle. Each ridge marks a relict bankline of a channel that was migrating to the east-southeast. Eventually it was cut off to create the abandoned channel visible at Stop 3.

Stop No. 5. Oxbow Lake

This stop is along Louisiana 129 about 3.9 mi (6.3 km) north of the junction of this highway with Louisiana 565 at the community of Monterey.

Cocodrie Lake, visible to the west of the highway, is rather small but otherwise is a classical Mississippi River oxbow lake. It is 2,000 to 3,000 ft (600 to 900 m) wide, 30 to 50 ft (8 to 15 m) deep, and about 4 mi (6.5 km) long while the total abandoned channel is about 8 mi (13 km) long. In a typical manner, the lake is asymmetrically situated in the abandoned channel, having more of its area in the lower arm than the upper arm. This reflects the pattern of post-cutoff sedimentation, especially the formation of sand wedges in the arms near the point of cutoff.

This oxbow lake maintains a connection with Black River Lake (Fig. 1) and interior basin drainage between Meander Belts 4 and 5 and thus experiences backwater flooding from the Ouachita and Red Rivers. Consequently, although sedimentation rates are low, the lake is filling with clays and silts and is slowly becoming shallower and smaller.

Stop No. 6. Modern Basin Drainage in Abandoned Course

This stop is situated along Louisiana 129 about 0.6 mi (0.9 km) north of its junction with Louisiana 565 in the community of Monterey.

Black River Lake, to the west of the highway, is an abandoned Mississippi River course occupied by a regional drainage system, the Ouachita River. When Meander Belt 4 was active, the Ouachita River was a tributary and it has been an underfit stream in the channel since abandonment. Occupations of segments of abandoned courses by streams of this type is common in the Lower Mississippi Valley. It happens most frequently when tributaries, flowing between meander belts or between a meander belt and the valley wall, are forced to breach natural levee ridges and enter abandoned channels because of meander belt coalescence or meander belt impingement against a valley wall.

Black River Lake experiences appreciable seasonal stage variations, responding both to floods in the Ouachita basin (from the north) and backwater flooding from the Red and Mississippi Rivers (from the south. The channel, several tens of feet deep, is now morphologically adjusted to the discharge of the Ouachita River; however, the meander radii are still those of the much larger Mississippi River. Channel fill deposits beneath the present lake consist of several ft of organic-rich clays and silts.

Stop No. 7. Crevasse Splay on Natural Levee

This stop occurs along Louisiana 129 about 0.2 mi (0.3 km) south of the junction of this highway with Louisiana 565 in the community of Monterey.

An inactive crevasse channel, manifested by a 10- to 16 ft-deep (3- to 5-m) by 100-ft (30-m) wide shallow, linear channel, crosses the highway at this location. It is associated with the Mississippi River course just west of the highway and formed when floodwater, normally topping the natural levee as sheet flow, locally became concentrated and channelized.

The crevasse is an erosional feature that conveyed floodwater through the crest of the natural levee into the adjacent backswamp. A short distance to the east, the channel shallows to ground level on the natural levee flank. This point is the apex of a crevasse splay, an alluvial-fan-like depositional feature covering an area of 0.5 mi^2 (1.3 km^2). The splay is indistinguishable on the ground but visible on aerial photos by a radial pattern of channel scars.

Crevasse splay deposits consist of up to 5 ft (1.5 m) of clays, silts, and fine sands. Crevasse splays have active lives of a few tens of years and coalesce with other splays or sheet flow deposits as natural levee ridges grow. Crevasse channels either fill with sediment as a meander belt is abandoned or remain as open channels such as this one; some function to convey tributary flow into relict channels and courses.

Stop No. 8. Remnant Channel of Abandoned Course

This stop is just north of Louisiana 565 about 1.7 mi (2.7 km) east of its junction with Louisiana 129 at the community of Monterey.

The line of trees north of the highway indicates the location of Plouden Bayou, a very small drainage channel marking the position of final flow in Meander Belt 3. It is a nearly sediment filled sluggish slough only a few tens of ft wide and several ft deep.

A remnant channel in an abandoned course cannot be identified as such just by its site-specific morphology (e.g., depth or channel configuration). One must consider planform and areal extent. Remnant channels have much higher length-to-width ratios and much lower sinuosities than similar features of other origin. For example, Plouden Bayou and its extensions are at most 325 ft (100 m) wide but are continuous for about 20 mi (30 km) in a broadly sinuous planform.

Deposits attributable to Plouden Bayou per se consist of only a few tens of feet of highly organic clays and silts. Underly-

ing Mississippi River abandoned course deposits are mostly loose clayey silts and sands. Abandoned course deposits, coarser than those in abandoned channels, reflect typically slower rates of abandonment and hence less abrupt reductions in river discharge.

Stop No. 9. Backswamp (Floodbasin Area)

This stop is along Louisiana 565 about 6.1 mi (9.8 km) east of its junction with Louisiana 120 at the community of Monterey.

The relatively low and extremely flat terrain visible in all directions is a large tract of backswamp located between Meander Belts 3 and 4 to the north, west, and south, and Meander Belt 5 to the east (Fig. 1). Bayou Cocodrie provides an outlet for local precipitation and backwater flooding.

Until 20 years ago, all cleared land in this area was uninterrupted and largely inaccessible swamp forest. Since then, hundreds of square miles have been cleared for agriculture now made possible in marginal areas by better drainage, modern farming practices and equipment, and economics.

Backswamp deposits at this stop consist of 80 to 100 ft (25 to 30 m) of mostly clays and organic clays laid down incrementally over thousands of years by floodwaters. When wet, the clays are dark gray and highly plastic. When dry, they tend to shrink and crack near the surface and sometimes disintegrate into pellet-sized particles locally referred to as "buckshot clays."

Stop No. 10. Obscure Abandoned Course

This stop is along Louisiana 15 about 6.8 mi (10.9 km) south of the junction of this highway with U.S. 85 in Ferriday.

The abandoned Mississippi River course and remnant channel at this stop are the same ones present at Stop 8; however, here they are largely obscured by a veneer of natural levee from the present meander belt. Such veneering is common in the Lower Mississippi Valley and complicates landform mapping and determining alluvial stratigraphy and chronology. This location illustrates the necessity in alluvial geology to always supplement field work with aerial photography and other remote sensing products as well as subsurface investigations by borings and geophysical methods.

The topography at this stop reveals a subdued, linear, undulating relief trending roughly east-west. When crops are in the fields, this relief is reflected only by field drainage lines. The larger ditches, following natural depressions, mark the locations of the remnant channel and the outer limits of the abandoned course.

Because of the veneer of natural levee deposits, abandoned course fill deposits occur at a depth of 10 ft (3 m) or more at this stop. The surficial natural levee deposits, while predominantly clayey, are much firmer and less organic than the underlying ones.

REFERENCES CITED

Saucier, R. T., 1967, Geological investigation of the Boeuf–Tensas basin, Lower Mississippi Valley: U.S. Army Engineer Waterways Experiment Station Technical Report E-757, 54 p.

——— , 1974, Quaternary geology of the Lower Mississippi Valley: Arkansas Archeological Survey Research Series No. 6, 26 p.

Domes of the North Louisiana Salt Basin

Joseph D. Martinez, Consulting Geologist, Plaquemine, Louisiana 70764

LOCATION

Nineteen salt domes (Fig. 1) are located in the North Louisiana Salt Basin, which lies between the Sabine Uplift and the Monroe Uplift. U.S. 167 from Alexandria to Ruston, Louisiana, provides access to this area. Winnfield, Vacherie, and Rayburns domes are precisely located on Figures 2, 5, and 7 in this paper. For greater detail on others see Kupfer (1975) and Anderson and others (1973).

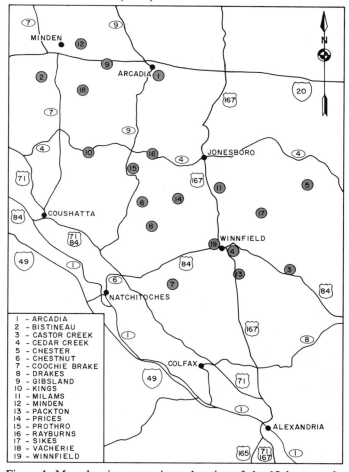

Figure 1. Map showing approximate location of the 19 known salt domes in the North Louisiana Salt Basin.

1 – ARCADIA
2 – BISTINEAU
3 – CASTOR CREEK
4 – CEDAR CREEK
5 – CHESTER
6 – CHESTNUT
7 – COOCHIE BRAKE
8 – DRAKES
9 – GIBSLAND
10 – KINGS
11 – MILAMS
12 – MINDEN
13 – PACKTON
14 – PRICES
15 – PROTHRO
16 – RAYBURNS
17 – SIKES
18 – VACHERIE
19 – WINNFIELD

INTRODUCTION

The salt domes in the coastal region either have little or no surface expression or are characterized by substantial positive topographic relief, as in the five islands described by Kupfer (this volume). In contrast, and reflecting a different growth history, the interior domes are generally overlain by topographic depressions over the central part of the dome, sometimes with uplift on the periphery (Martinez, 1980). Although such a depression occurs over part of Jefferson Island dome, one of the five islands, evidence of subsidence is not apparent over other coastal domes. With a few exceptions most of the domes in the North Louisiana Basin are neither precisely located nor well delineated (if at all) in the subsurface. Economic development has resulted in a varying degree of understanding of the subsurface geology of several of these domes. The Minden dome is associated with a substantial oil field and is therefore reasonably well known. This is also true of the Winnfield dome from which caprock has been extensively quarried for both limestone and anhydrite and from which salt has been mined by room and pillar methods (the mine is now flooded). To a much lesser extent, geologic information has been developed for the Drakes, Arcadia, and Gibsland domes, which are utilized for petroleum storage.

A great amount of geologic data and knowledge has been acquired in the last decade for the Vacherie and Rayburns salt domes in connection with investigations of the potential utility of these and other Gulf Coast salt domes for the isolation of high-level radioactive waste by the U.S. Department of Energy, its predecessors, and its contractors.

Because of the extraordinary effort by the Department of Energy to evaluate the potential of the Vacherie and Rayburns domes and the large amount of information that has been amassed regarding their characteristics and geology, they have been selected as two of three sites designated as specific field stops in this section. The other site selected is the Winnfield Dome, which offers the visitor interesting and instructive perspectives as a result of large quarrying operations in the caprock and because of published information concerning the interior of the salt stock derived from salt-mining operations.

The Coochie Brake Dome, although not pinpointed for a visit in this guide, has a well-developed depression overlying it that is expressed as a brake or swamp. It is being considered for development as a State Preservation Area (Caplinger, 1980) and therefore may be of future interest.

SITE DESCRIPTION

Winnfield Dome

The location of Winnfield Dome is shown in Figure 2 from Hoy and others (1962), which uses an idealized geologic map as a base and to which highway numbers have been added. The site is reached by traveling 3.7 mi (6 km) west on U.S. 84 from its intersection with U.S. 167 about 1 mi (1.6 km) east of the town of Winnfield. An air photograph of the caprock quarry is shown in Figure 3. Permission to visit the site should be secured from the Winn-Rock quarry operators at the location. A northwest-southeast geologic section from Belchic (1960) is shown in Figure 4. Huner (1939) described the Winnfield Dome as having a low interior depression about 1 mi (1.6 km) long and

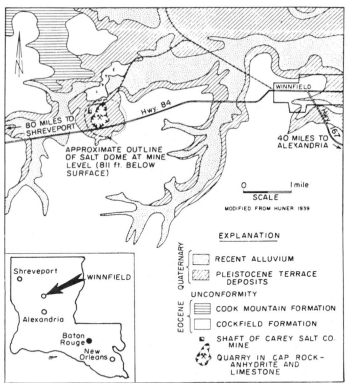

Figure 2. Location of Winnfield Salt Dome. From Hoy and others (1962).

0.5 mi (0.8 km) wide surrounded by a fairly conspicuous broken rim of Tertiary strata on all sides except the north. He described early observations of the outcrops at this locality but called attention to the fact that quarrying operations had (even then) greatly changed the nature of outcrops and the drainage patterns as well as the topography. Huner placed the dome within the area of Cockfield outcrop and observed that the Cook Mountain Formation outcrops on the encircling rim. According to him, exposures of Sparta Sand, the Cane River formation, and the Sabine Group are found on the inner flanks and in the low interior part of the dome. Most of the area occupied by caprock is overlain by Pleis-

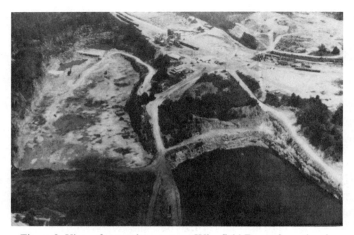

Figure 3. View of caprock quarry on Winnfield Dome from the air.

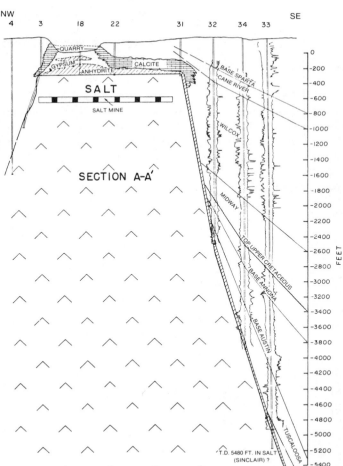

Figure 4. Northwest-southeast geologic section of Winnfield Salt Dome. From Belchic (1960).

tocene sediments. Where caprock has been quarried this is, of course, now also removed.

The salt stock is shown on Figure 2 to be nearly circular in shape and approximately 1 mi (1.6 km) in diameter (Hoy and others, 1962). Kupfer (1976) considered the base of the average salt dome in the North Louisiana Basin to be 18,000 to 28,000 ft (5,500 to 8,500 m) below sea level. Thus it seems reasonable to assume the vertical dimension of the Winnfield Dome to be in the range of 4 to 5 mi (6.4 to 8.0 km). As shown on Figure 4, caprock, in part exposed naturally at the surface, has been further revealed by quarrying operations. The complete sequence found in salt-dome caprock was present prior to removal by quarrying operations. Twenty to 75 ft (6 to 23 m) of limestone constituted the uppermost caprock zone, which was underlain by probably 20 to 50 ft (6 to 15 m) of gypsum. The lowermost caprock zone is anhydrite, which ranges in thickness from a few feet to more than 300 ft (90 m) (Hoy and others, 1962). All of the limestone has been removed with the possible exceptions of a few large blocks and stray fragments on unexcavated parts of the area. Belchic (1960) describes the calcite as normally a hard fractured crystalline white to variously gray rock with sharply defined black to light-gray bands varying in width from a few millimeters to several inches. These bands are reported to be frequently parallel, variously oriented, sometimes continuous for several feet but elsewhere deformed, twisted, broken, or brecciated. Solution cavities that are lined with crystals of calcite aragonite and pyrite are common. A wide variety of accessory minerals has been identified. The gypsum and anhydrite were also described by Belchic. The gypsum was characterized as being light- to medium-gray, granular or sugary in appearance, massive, and rarely fractured. Most, if not all, of this massive gypsum has been removed, but some gypsum is found associated with the anhydrite. The anhydrite is dark gray, massive, and granular to crystalline according to Belchic (1960). The anhydrite directly overlies the salt, the upper surface of which is essentially flat. Anhydrite grains were found to occur as an unconsolidated "sand" at the contact in a number of wells. Large solution cavities filled with water characterized the anhydrite-salt contact. The nature of this contact is clear evidence of dissolution.

The genesis of caprock through residual accumulation of anhydrite (Goldman, 1933, 1952; Taylor, 1938) seems well established. See Murray (1966) and Martinez (1974, 1978) for short and updated reviews.

A room-and-pillar salt mine at 811 ft (247.2 m) below the surface was operated from 1932 until 1965, at which time it was flooded and abandoned (Martinez and Wilcox, 1976). The internal structure of the salt and other features revealed by these mining efforts is well described by Hoy and others (1962). Greater attention has been given to the caprock in this narrative because of the excellent quarry exposures.

Rayburns Dome

Rayburns Dome is located in Bienville Parish, Louisiana, approximately 12 mi (19 km) west of the town of Jonesboro (Fig. 1). It may be reached by traveling west from Jonesboro for 12 mi (19.3 km) to the town of Friendship on Louisiana 4 and continuing 2 mi (3.2 km) beyond Friendship to the junction of Louisiana 4 with Louisiana 155, where beds on the southeastern part of the dome are upturned (Fig. 5).

The surface features of Rayburns Dome are summarized as follows in Law Engineering Testing Company (1981). A circular topographic low over the dome is surrounded by hills with 60 to 100 ft (18 to 30 m) of relief. The resistant ridge, which almost encircles the central depression, is formed of exposed rocks of the upper Eocene, Sparta Formation. Lower Tertiary and upper Cretaceous inliers are exposed within the central depression, although often covered by alluvium. The lower Tertiary units include the Cane River Formation and the Wilcox and Midway Groups.

Following Louisiana 4 to the northwest provides a traverse across the central part of the dome as shown on the geologic map of Figure 5. About 0.6 mi (1 km) from the junction of Louisiana 4 and Louisiana 155 an old lime quarry situated on the west side of Louisiana 4 exposes upper Cretaceous and Midway limes and

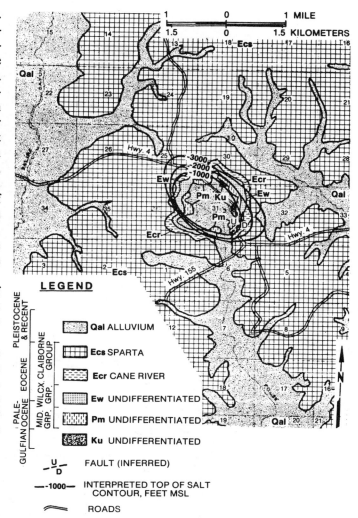

Figure 5. Geologic map of Rayburns Salt Dome. From Law Engineering Testing Company (1981).

marls, which are strongly dipping and faulted. Salt flats are prominently visible. It has been reported that the reworked pebbly contact of Tertiary on Cretaceous is also exposed. Another 0.6 mi (1 km) to the northwest on Louisiana 4, there is a dirt road on the left that leads to a nearby Department of Energy well site. Here a deep corehole in the salt was drilled under contract with the Institute for Environmental Studies of Louisiana State University for the U.S. Department of Energy to a depth of 5,008 ft (1,526.4 m). The top of the caprock was reached at 97.7 ft (29.8 m) below the surface. The top of the salt was intersected at 127.7 ft (38.98 m) below the surface. Thirty ft (9.1 m) of caprock was cored with a recovery of 22%; 2,542.0 ft (774.8 m) of salt was recovered below the base of the caprock with a recovery of 98%. Nominal core diameter was 4 in. (10.2 cm). These cores are currently stored at Louisiana State University (LSU) in Baton Rouge for the U.S. Department of Energy under quality assurance. A report containing a complete description of their lithol-

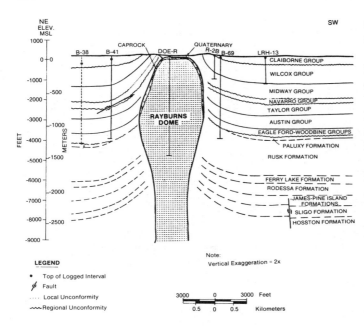

ogy has been prepared by Nance and Wilcox (1979). Hawkins (1978) prepared an engineering report of the drilling program.

A northeast-southwest section (Fig. 6) through Rayburns Dome was prepared by Law Engineering Testing Company (1981) for the National Waste Terminal Storage Program of the U.S. Department of Energy. The location of the LSU deep core-hole is indicated on Figure 6. Other well locations and geologic details are given in the section, which is drawn along the shortest horizontal dimension of the dome.

The results of numerous shallow drillings and coreholes to investigate the Quaternary geology over and surrounding the dome have been analyzed and summarized by Kolb (1977, 1978, 1979). His contributions to the structure and stratigraphy over Rayburns Dome are voluminous and significant, but space limitations preclude a thorough discussion here.

Vacherie Dome

Vacherie Dome straddles the boundary of Bienville and Webster Parishes in Louisiana, approximately 30 mi (48 km) east of Shreveport and about 10 mi (16 km) slightly east of south

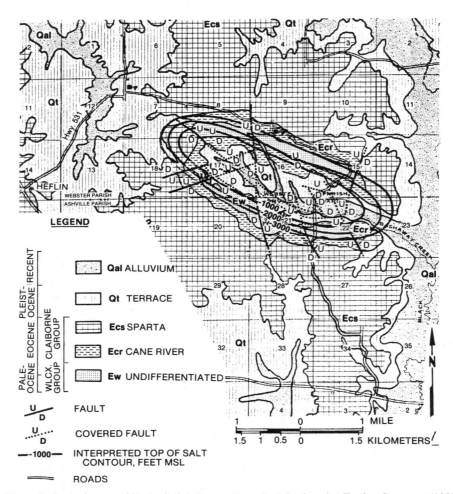

Figure 7. Geologic map of Vacherie Salt Dome. From Law Engineering Testing Company (1981).

from Minden (Fig. 1). The town of Heflin is situated about 10 mi (16 km) south of Minden. The Vacherie Dome can be reached by driving 2.4 mi (3.9 km) northeast on Louisiana 531 from Heflin, at which point a T junction is encountered where Louisiana 531 makes a 90° turn to the north (Fig. 7). Within 0.1 mi (0.16 km), at Central School, turn sharply to the right (east) and travel 2.4 mi (3.9 km) to a T junction at Galilee School, which is on the north side of the dome.

A right turn at this junction will take the visitor over the central part of the dome in about 1 mi (1.6 km). Other unimproved roads originating from the vicinity of Galilee School provide routes near and over the dome. The geologic map (Fig. 7) of Vacherie Dome and the surrounding area shows the road from Central School to Galilee School reaching and traversing the Cane River outcrop. Traveling south from the junction at Galilee School one crosses Wilcox and again Cane River and then Quaternary Terrace before reaching the alluvium of Bashaway Creek.

Bashaway Creek drains a central depression that overlies the dome. This low area is surrounded by hills with up to 200 ft (61 m) of relief (Law Engineering Testing Company, 1981). It is reported in that document that the resistant ridge partially encircling the central depression is formed by the upper Eocene Sparta Formation. Cane River and Wilcox inliers are identified as forming concentric rings within the depression.

A northeast-southwest section across the short axis of this elliptical shaped dome (Law Engineering Testing Company, 1981) is shown on Figure 8. The deep hole in the salt indicated on this section was drilled under contract with the Institute for Environmental Studies of Louisiana State University for the U.S. Department of Energy to a depth of 5,023 ft (1,531 m). The top of the caprock was intersected at 532 ft (162.2 m) below the surface. The top of the salt was reached at 805 ft (245.4 m) below the surface. Two hundred seventy-three ft (83.2 m) of caprock was cored with a recovery of 98%; 2,523 ft (769 m) of salt was cored with a recovery of 91%. Nominal core diameter was 4 in. (10.2 cm). These cores are also being stored at Louisiana State University in Baton Rouge for the U.S. Department of Energy under

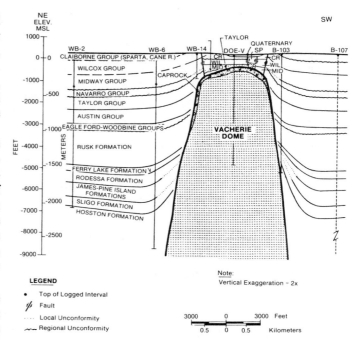

Figure 8. Northeast-southwest geologic section of Vacherie Salt Dome. From Law Engineering Testing Company (1981).

quality assurance. A report containing a complete description of their lithology has been prepared by Nance and others (1979). Hawkins (1978) prepared an engineering report of the drilling program.

Although the visible geology at Vacherie Dome is subtle and not particularly exciting, the dome is of much interest because of its early selection as one of the candidate sites for a potential repository for high-level nuclear waste. Its designation in this regard has been revised, and at this time recommendations are that it be eliminated from the reduced list of potential sites. It may, however, be proposed later for the second national repository. Much work has been done on this dome, and a large volume of literature has developed from this effort.

REFERENCES CITED

Anderson, R. E., Eargle, D. H., and Davis, B. O., 1973, Geologic and hydrologic summary of salt domes in Gulf Coast region of Texas, Louisiana, Mississippi, and Alabama: U.S. Geological Survey Open-File Report 4339-2, 294 p.

Belchic, H. C., 1960, The Winnfield Salt Dome, Winn Parish, Louisiana, *in* Shreveport Geological Society 1960 Spring field trip guidebook: Interior Salt Domes and Tertiary Stratigraphy of North Louisiana, p. 29–41.

Caplinger, Charles, 1980, Coochie Brake State Preservation Area, Winn Parish, La., Master Plan.

Goldman, M. I., 1933, Origin of the anhydrite caprock of American salt domes: U.S. Geological Survey Professional Paper 175, p. 83–114.

—— 1952, Deformation, metamorphism and mineralization in gypsum-anhydrite caprock, Sulphur Salt Dome, Louisiana: Geological Society of America Memoir 40, 160 p.

Hawkins, M. F., Jr., 1978, An engineering report of the core holes of Vacherie

and Rayburn's salt domes—north Louisiana salt dome basin: Prepared for the Office of Waste Isolation, Sub E-512-02500 (originally ORNL-Sub 7362), Institute for Environmental Studies, Louisiana State University, Baton Rouge, 361 p.

Hoy, R. B., Foose, R. M., and O'Neill, B. J., Jr., 1962, Structure of Winnfield Salt Dome, Winn Parish, Louisiana: American Association of Petroleum Geologists Bulletin, v. 46, p. 1444–1459.

Huner, J., 1939, Geology of Caldwell and Winn Parishes, Louisiana: State of Louisiana, Dept. of Conservation, Geological Bulletin no. 15, 356 p.

Kolb, C. R., 1977, *in* Martinez, J. D., and others, An investigation of the utility of Gulf Coast salt domes for the storage or disposal of radioactive wastes: Prepared for the Office of Waste Isolation, ORNL, Y/OWI/SUB/4112/37, Institute for Environmental Studies, Louisiana State University, Baton Rouge, 475 p.

—— 1978, *in* Martinez, J. D., and others, An investigation of the utility of Gulf Coast salt domes for the storage or disposal of radioactive wastes: Prepared for the Department of Energy, EW-78-C-05-5941/53, Institute for Environmental Studies, Louisiana State University, Baton Rouge, 310 p.

—— 1979, *in* Martinez, J. D., and others, An investigation of the utility of Gulf Coast salt domes for the storage or disposal of radioactive wastes: Prepared for the U.S. Department of Energy, E511-02500-A-1, Institute for Environmental Studies, Louisiana State University, Baton Rouge, 572 p.

Kupfer, D. H., 1975, *in* Martinez, J. D., and others, An investigation of the utility of Gulf Coast salt domes for the storage or disposal of radioactive wastes: Prepared for Oak Ridge National Laboratory, ORNL, ORNL-Sub-4112-10, Institute for Environmental Studies, Louisiana State University, Baton Rouge, p. 15–78.

—— 1976, *in* Martinez, J. D., and others, An investigation of the utility of Gulf Coast salt domes for the storage or disposal of radioactive wastes: Prepared for the Office of Waste Isolation, ORNL, ORNL-Sub-4112-25, Institute for Environmental Studies, Louisiana State University, Baton Rouge, 329 p.

Law Engineering Testing Company, 1981, Geologic area characterization, Gulf Coast salt domes project, Volumes III A & B, North Louisiana Study Area: Prepared for Battelle Memorial Institute, Office of Nuclear Waste Isolation.

Martinez, J. D., 1974, Tectonic behavior of evaporites, *in* Coogan, A. H., ed., Proceedings, 4th Symposium on Salt, Vol. I: Northern Ohio Geological Society and Kent State University, Cleveland, Ohio, p. 155–167.

—— 1978, Salt dome caprock—a record of geologic processes, *in* Coogan, A. H., and Hauber, Lukas, eds., Proceedings, 5th Symposium on Salt, Vol. 1: Northern Ohio Geological society, p. 143–151.

—— 1980, The nature and evolution of salt domes and their caprock: State of Louisiana, Department of Natural Resources, Geological Pamphlet no. 6, 16 p.

Martinez, J. D., and Wilcox, R. E., 1976, *in* Martinez, J. D., and others, An investigation of the utility of Gulf Coast salt domes for the storage or disposal of radioactive wastes: Prepared for the Office of Waste Isolation, ORNL, ORNL-Sub-4112-25, Institute for Environmental Studies, Louisiana State University, Baton Rouge, 329 p.

Murray, G. E., 1966, Salt structures of Gulf of Mexico Basin—A review: American Association of Petroleum Geologists Bulletin, v. 50, p. 439–478.

Nance, Damian, and Wilcox, R. E., 1979, Lithology of the Rayburn's dome salt core: Prepared for U.S. Department of Energy, E511-02500-3, Institute for Environmental Studies, Louisiana State University, Baton Rouge, 307 p.

Nance, Damian, Rovik, John, and Wilcox, R. E., 1979, Lithology of the Vacherie salt dome core: Prepared for U.S. Department of Energy, E511-02500-5, Institute for Environmental Studies, Louisiana State University, Baton Rouge, 343 p.

Taylor, R. E., 1938, Origin of the caprock of Louisiana salt domes: State of Louisiana, Department Conservation, Geological Bulletin no. 11, 191 p.

The Florida Parishes of southeastern Louisiana

Whitney J. Autin,, Louisiana Geological Survey, Box G, University Station, Baton Rouge, Louisiana 70893
J. J. Alford, Department of Geography, Western Illinois University, Macomb, Illinois 61455
Bobby J. Miller,Department of Agronomy, Louisiana Agricultural Experiment Station, Agricultural Center, Louisiana State University, Baton Rouge, Louisiana 70803
Robert P. Self, Department of Earth Sciences, Nicholls State University, Thibodaux, Louisiana 70310

LOCATION

The Florida Parishes of southeastern Louisiana (Fig. 1) are bounded on the west by the Mississippi River, on the east by the Pearl River, on the north by the Louisiana-Mississippi state line, and on the south by the Mississippi River deltaic plain. Principal cities include Baton Rouge and New Orleans.

INTRODUCTION

A region of geological diversity, the Florida Parishes of southeastern Louisiana (Fig. 1), provide an important key to understanding the Quaternary development of the northern Gulf of Mexico. Physiographic subsections of the region include the Loess Hills, the Pine Meadows, and the Southern Pine Hills (Thornbury, 1965). Elevations in the region range from near sea level along the coastal lakes in the southern part to more than 360 ft (110 m) near the Louisiana-Mississippi state line. Local relief varies from 25 ft (7.6 m) in the southern part of the area to over 200 ft (61 m) in the north. The degree of stream dissection and development of tributary networks progressively increases to the

north. The pattern displayed by summit elevations of major divides, relative drainage network development, and breaks in the slope of the upland surface indicates the presence of three major topographic belts with an approximate east-west trend.

The surface of lowest elevation approximately coincides with the Pine Meadows and the surface of highest elevation approximately coincides with the Southern Pine Hills. Between these surfaces lies a topographic belt of intermediate elevation and development characteristics. Upland landforms in the region are termed, in order of increasing elevation, the Prairie Terrace, the Intermediate Terrace, and the High Terrace (Fig. 2; Snead and McCulloh, 1984).

The origin and development of landscapes in the Florida Parishes have been controversial subjects for several years. The area consists of three upland lithostratigraphic units: the Citronelle Formation, the Prairie Formation, and loess. A braided stream sequence of coalescing alluvial fans is the generally recognized depositional environment of the Citronelle Formation (Self, 1983; Smith and Meylan, 1983). The distribution of the deposit and its relationship to regional topographic and stratigraphic features establish that it is older than the Prairie Formation. Map-

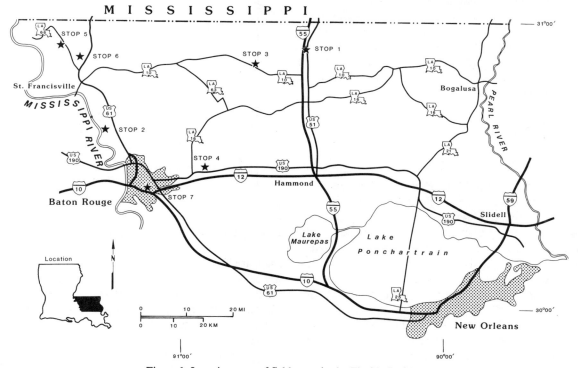

Figure 1. Location map of field stops in the Florida Parishes.

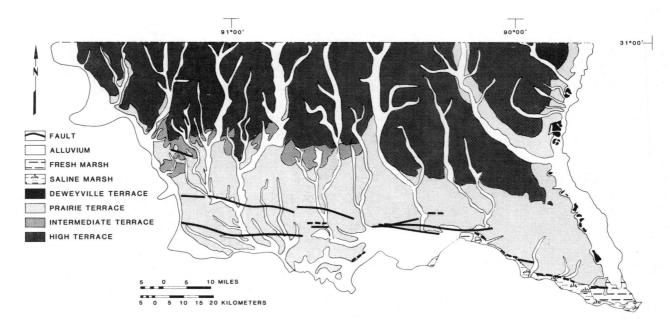

Figure 2. Geology of the Florida Parishes (from Snead and McCulloh, 1984).

ping of the Prairie Formation (Port Hickey Terrace) by Fisk (1938, 1944) is essentially correct, and the interpretation of its depositional environment as a meandering-channel alluvial plain has been recognized (Durham and others, 1967). Loess deposits of variable thickness blanket the landscape in a wedge-shaped fashion, with thicker deposits found along the western edge of the study area and a progressive thinning to the east (Miller and others, 1982). Sediments in alluvial landscape positions have been correlated with respective terrace levels along streams in the Tunica Hills. Geological structures such as the Baton Rouge fault zone have bearing on the distribution of physiographic and lithologic features in the Florida Parishes.

SITE DESCRIPTIONS

Stop 1.

The gravel pit at Kentwood (Fig. 3) provides a representative example of the lithology and character of the Citronelle Formation. Texture is an immature trimodal mix of sand, gravel, and clay, with sand being the most common component and clay the least common. Typical lithologies include muddy, gravelly sands, sandy gravels, and muddy sands. Bedding is highly lenticular and individual beds cannot be traced beyond an immediate exposure. Channels, cut and fill structures, scour surfaces, and cross-bedding are common.

Citronelle gravel fractions are subround tripolitic honey-colored cherts with mean grain sizes in the middle pebble range. Cobbles and boulders are absent or extremely rare. Very well rounded quartz granules are common and rip-up fragments of

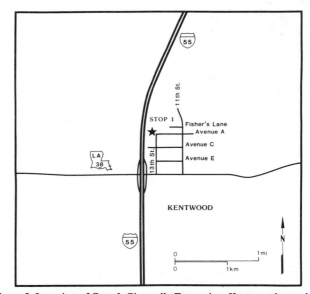

Figure 3. Location of *Stop 1,* Citronelle Formation, Kentwood gravel pit, end of 13th Street in Kentwood, S½, sec. 41, T. 1 S., R. 7 E., Tangipahoa Parish, Kentwood 7½-minute quadrangle.

shale are locally present. Citronelle sands are quite variable, but are most commonly subangular to round, moderately sorted, medium-grained quartz arenites with minor chert (Self, 1984). The sands contain a hematitic clay matrix which gives the unit its reddish color. Regional heavy mineral studies suggest an Appalachian mountain source (Rosen, 1969). Cullinan (1969) noted some quartz grains with sutured polycrystalline extinction which may suggest a crystalline Appalachian source. Textural and com-

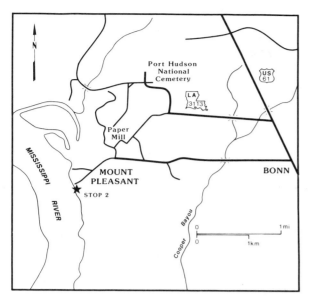

Figure 4. Location of *Stop 2,* Prairie Formation, Mt. Pleasant bluff, along the Mississippi River, W½, sec. 41, T. 5 S., R. 2 W., East Baton Rouge Parish, Port Hudson 7½-minute Quadrangle. Property owned by Pennington Oil Co., Baton Rouge, Louisiana.

positional data indicate multiple sources and considerable recycling.

Sedimentary structures, lithology, and sediment body geometry suggest that the Citronelle Formation probably represents a series of coalescing braided stream and alluvial fan deposits that blanketed the Florida Parishes and southwestern Mississippi in Plio-Pleistocene time. Three types of graveliferous deposits are recognized: (1) Massive gravel deposits contain interstitial sand and very little interbedded sand. These basal fan gravels and thick longitudinal bars formed in braided streams that flowed across the fans. (2) Channel-fill gravels develop near the heads of fans, with deposition and erosion in channels occurring almost simultaneously. (3) Interbedded sand and gravel deposits are cross-bedded with sharp contacts. Type 3 deposits are a complex series of longitudinal, transverse, and lobate bars and slack-water deposits of braided streams. The gravel pit at Kentwood is representative of this type of deposit. Plio-Pleistocene drainage patterns are believed to have controlled the distribution of Citronelle deposits. The formation may have developed in response to Plio-Pleistocene uplift to the northeast.

Stop 2

The Prairie Formation generally has meager exposures in the Florida Parishes, except for local bluffs where streams impinge along valley walls. The best exposure in the region is at Mt. Pleasant bluff, near the Port Hickey (Prairie) section of Brown (1938). This bluff is located on the Mississippi River about 20 mi (32.2 km) north of Baton Rouge (Fig. 4) where a section of loess overlies the upper part of the Prairie Formation. At Mt. Pleasant bluff, the Prairie sediments consist of point bar,

natural levee, and abandoned channel-fill environments. Modern stratigraphic characterization of this site is in progress (Autin and others, 1985).

In the Florida Parishes, the Prairie Formation crops out at elevations less than 75 ft (23 m) in the east and at elevations less than 125 ft (38 m) in the west. The Prairie Formation unconformably rests on the Citronelle Formation and dips beneath the Holocene Mississippi River delta plain to the south. Its maximum thickness may approach 600 ft (183 m) in the region (Parsons, 1967). The deposit is typically described as a fining-upward sequence, with a basal unit of coarse sand and gravel, a middle unit of medium sand, and an upper unit of laminated silt, clay, and fine sand (Fisk, 1944). Test borings (Autin, 1984, unpub. data) indicate that this description is applicable to the area.

The Prairie Formation is mapped as a coast-trending unit across south Louisiana, but also extends into major alluvial valleys as a fluvial deposit. The unit was initially defined by Fisk (1938) as both an alluvial terrace and a formation developed in the late Pleistocene. The upper part of the unit is generally believed to be equivalent to the Beaumont Formation of Texas and various coastal terraces of western Florida.

Stops 3 and 4.

A multiple set of loess deposits have been identified in the Mississippi River valley. The upper loess is correlated with Peoria Loess of the Upper Mississippi valley on the basis of stratigraphic position, lateral continuity, soil characteristics, and radiocarbon dates (Otvos, 1975). No attempt has been made to correlate pre–Peoria Loess in southeastern Louisiana with loess identified in the Upper Mississippi valley. Stratigraphic position, lithologic characteristics, and the development and morphology of its geosols are analogous to Loveland Loess as identified by Wascher and others, (1947) and Leighton and Willman (1950).

Geographic and landscape distribution patterns and characteristics of loess in the Florida Parishes result from its continuous west to east thinning pattern, erosion, and continuous pedogenic processes during and since deposition. Pre–Peoria Loess occupies parts of the land surface throughout the areas identified as Intermediate and High terraces (Snead and McCulloh, 1984). Pre-Peoria and Peoria loess cover the entire uplands in the western part of the region, with a combined thickness in excess of 19 ft (6 m). Erosion has removed all traces of loess from the landscape in some areas. In the direction of thinning loess, the underlying geosol is first evident at the surface on lower slope positions of the steepest landscapes. The underlying geosol eventually becomes exposed at progressively higher elevations and on less sloping landscapes. This relationship is evident along roadcuts in central St. Helena Parish (Fig. 5).

Erosion of loess has been less severe on the younger and more gently sloping Prairie Terrace where only Peoria Loess has been identified. At Grays Creek, loess is thin enough to contain an admixture of the underlying material throughout (Fig. 6). Here,

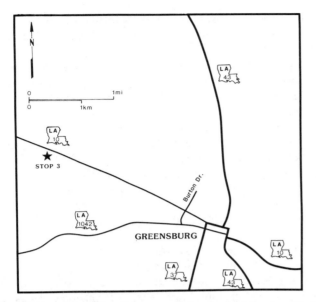

Figure 5. Location of *Stop 3,* loess overlying a geosol developed in the Citronelle Formation, T. 2 S., R. 4–5 E., St. Helena Parish, Greensburg 7½-minute and Felixville 15-minute quadrangles. Map shows various roadcuts along Louisiana 10.

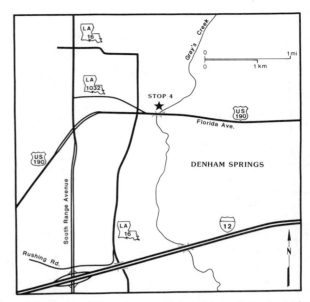

Figure 6. Location of *Stop 4,* loess overlying a geosol developed in the Prairie Formation, Grays Creek at U.S. 190, east of Denham Springs, S½ sec. 48, T. 6 S., R. 2 E., Livingston Parish, Denham Springs 7½-minute Quadrangle.

a silty mantle covers the terrace and forms a continuum with the thick loess farther west. Continuous loess deposits cover the Prairie terrace west of Tangipahoa Parish. Maximum loess thickness is about 13 ft (4.0 m) near the Mississippi River floodplain.

Sections containing unweathered loess are restricted to the western part of West Feliciana Parish. A fresh, moist exposure of unweathered loess reveals a massive but friable yellowish brown (10YR 5/4) to light yellowish brown (10YR 6/4) calcareous silt. Secondary carbonate concretions and pulmonate gastropod shells are present in some places. Unweathered loess has less than 10 percent clay (<2.0 micron), 85 to 95 percent silt (2–50 micron), and less than 10 percent sand (50–200 micron). The sand present is almost entirely very fine sand (50–100 micron). A distinctive distribution of particles within the silt-size fraction is characteristic of loess (Lewis and others, 1984).

Weathering, soil development, and mixing zones near loess-nonloess contacts have resulted in characteristics that differ significantly from those of unweathered loess. Pedogenic processes during and after loess deposition have resulted in a diffuse, gradational mixing zone between the loess and the underlying deposits. With increasing depth, the mixing zone contains increasing quantities of incorporated material from the underlying geosol. The thinnest mixing zones are beneath thick loess deposits near the Mississippi River floodplain where accumulation rates were most rapid. Thickness of the mixing zone is inversely related to loess thickness (Spicer, 1969).

Stops 5 and 6

Situated in the northwest corner of the Florida Parishes are the Tunica Hills, a dissected upland that encompasses most of

West Feliciana Parish and extends into southwestern Mississippi. This area provides one of the richest sources for Quaternary plant fossils in the lower Mississippi valley. Although fossils have been collected from several sites along these streams, the best known localities are at Percy bluff and Wilcox bluff (Fig. 7). The area has been extensively studied (Fisk, 1938; Delcourt and Delcourt, 1977; Otvos, 1978, 1980, 1981; Alford and others, 1983) however, there is no general agreement as to the relative or geochronometric ages of the fossiliferous units.

Alford and others (1983) identified stringers of sand, occasional pebbles, and the lack of primary carbonates at several T1 exposures indicating that the terrace containing the Percy bluff deposits (T1) did not contain in situ loess. Organic samples collected from four sites along the T1 terrace yielded dates ranging from about 15,000 to 3,000 years B.P. The dates showed good stratigraphic agreement since the older dates were from the base while the younger were from the upper part of the fill. Thus, the low terrace (T1) is too young to carry a Pleistocene loess. Wilcox bluff (T2) was also examined and Alford and others (1983) were able to collect well preserved wood for radiocarbon dating. Two dates (A-2006, A-2007) yielded dates of >38,000 years B.P. These dates establish that the flora and fill at Wilcox bluff are considerably older than at Percy bluff.

This antiquity does not necessarily mean that Wilcox bluff (T2) is Sangamonian. The samples were collected from near the base of the 49-ft (15 m) thick fill and the upper units could be significantly younger. It is noteworthy that Otvos (1980, 1981) has collected a number of samples from fossiliferous units that are Farmdalian (26,000–31,000 years B.P.). Quite possibly, the base of T2 is >38,000 years B.P. while the upper units were deposited during the Farmdale Substage.

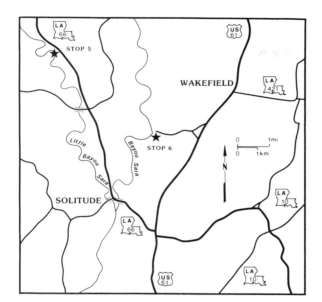

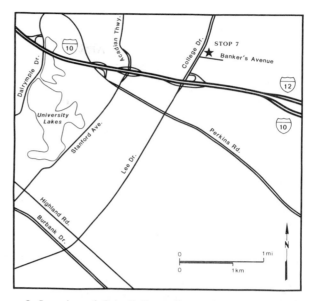

Figure 7. Location of *Stop 5,* Tunica Hills terrace, Percy bluff (T1), along Little Bayou Sara, W½, sec. 85, T. 1 S., R. 3 W.; Location of *Stop 6,* Wilcox bluff (T2), along Bayou Sara, W½, sec. 73, T. 2 S., R. 3 W., West Feliciana Parish, Weyanoke 7½-minute quadrangle.

Figure 8. Location of *Stop 7,* Baton Rouge fault zone, on Bankers Avenue in Baton Rouge, N½, sec. 94, T. 7 S., R. 1 E., East Baton Rouge Parish, Baton Rouge West 7½-minute quadrangle.

Soil and loess stratigraphy support this possibility. It has been recognized along the Mississippi valley that the Wisconsin Stage is represented by two loess deposits (Leighton and Willman, 1950). The Peoria Loess was deposited between 22,000 and 12,000 years B.P., while the older Roxana Silt was deposited between 30,000 years B.P. and perhaps as much as 75,000 years B.P. (Willman and Frye, 1970; McKay, 1980). If Wilcox bluff (T2) is Sangamonian, it should carry two loesses. However, T2 has only a single loess, indicating that this surface was not available for the early Wisconsinan loess fall.

Available evidence indicates that there are at least two fluviatile terraces flanking the streams draining the Tunica Hills. The youngest, containing the Percy bluff fossils, ranges from Woodfordian to Holocene. Terrace 2, containing the Wilcox bluff flora, is probably mid-Wisconsinan.

Stop 7

The Baton Rouge fault zone is the best documented structure in the Florida Parishes. Durham and Peeples (1956) correlated the surface expression of the fault with its subsurface displacement and documented the existence of a growth fault with about 20 ft (6 m) of displacement at the land surface and a dip of nearly 70 degrees to the south. At a depth of approximately 10,000 ft (3,000 m), the displacement increases to hundreds of feet (meters) and the dip reduces to 50 degrees or less. Surface expression of the Baton Rouge fault zone extends from the Baton Rouge area to the east along the north shore of Lake Pontchartrain (Fig. 2). The physiographic character of the fault can be observed in Baton Rouge (Fig. 8).

REFERENCES CITED

Alford, J. J., Kolb, C. R., and Holmes, J. C., Jr., 1983, Terrace stratigraphy in the Tunica Hills of Louisiana: Quaternary Research, v. 19, p. 55–63.

Autin, W. J., 1984, unpublished data.

Autin, W. J., Davison, A. T., Miller, B. J., and Lindfors-Kearns, F. E., 1985, Use of Mt. Pleasant bluff in correlation of the Prairie Formation, southeastern Louisiana: Geological Society of America Abstracts with Programs, v. 17, no. 3, p. 150.

Brown, C. A., 1938, The flora of Pleistocene deposits in the western Florida Parishes, West Feliciana Parish, and East Baton Rouge Parish, Louisiana *in* Fisk, H. N., Richards, H. G., Brown, C. A., and Steere, W. C., Contributions to the Pleistocene history of the Florida Parishes of Louisiana: Louisiana Department of Conservation, Geological Survey Bulletin 12, p. 59–96.

Cullinan, T. A., 1969, Contribution to the geology of Washington and St. Tammany parishes, Louisiana [Ph.D. thesis]: Tulane University, New Orleans, 287 p.

Delcourt, P. A., and Delcourt, H. R., 1977, The Tunica Hills, Louisiana-Mississippi: Late glacial locality for spruce and deciduous forest species: Quaternary Research, v. 7, p. 218–237.

Durham, Jr., C. O., Moore, C. H., Jr., and Parsons, B. E., 1967, An agnostic view of the terraces: *in* Field Trip Guidebook: Mississippi Alluvial Valley and Terraces, Geological Society of America Annual Meeting: p. 1–22.

Durham, Jr., C. O., and Peeples, E. M., 1956, Pleistocene fault zone in southeastern Louisiana: Transactions of the Gulf Coast Association of Geological Societies, v. 6, p. 65–66.

Fisk, H. N., 1938, Pleistocene exposures in western Florida Parishes, Louisiana *in* Fisk, H. N., Richards, H. G., Brown, C. A., and Steere, W. C., Contributions to the Pleistocene history of the Florida Parishes of Louisiana: Louisiana Department of Conservation, Geological Survey Bulletin 12, p. 3–25.

——1944, Geological investigation of the alluvial valley of the Lower Mississippi River: Mississippi River Commission, Vicksburg, 78 p.

Leighton, M. M., and Willman, H. B., 1950, Loess formations of the Mississippi Valley: Journal of Geology, v. 58, p. 559–623.

Lewis, G. C., Fosberg, M. A., Falen, A. L., and Miller, B. J., 1984, Identification of loess by particle size distribution using the Coulter Counter TA II: Soil Science, v. 137, p. 172–76.

McKay, E. D., 1980, Wisconsin loess stratigraphy: Illinois State Geological Survey Guidebook Series 13, p. 95–108.

Miller, B. J., Lewis, G. C., Alford, J. J., and Day, W. J., 1982, Loess parent materials in Louisiana and the Lower Mississippi valley: unpublished Field Trip Guidebook, 86 p.

Otvos, Jr., E. G., 1975, Southern limits of Pleistocene loess, Mississippi Valley: Southeastern Geology, v. 17, p. 27–38.

—— 1978, Comment on "The Tunica Hills, Louisiana-Mississippi: Late glacial locality for spruce and deciduous forest species" by P. A. Delcourt and H. R. Delcourt: Quaternary Research, v. 9, p. 250–52.

—— 1980, Age of Tunica Hills (Louisiana-Mississippi) Quaternary fossiliferous creek deposits; problems of radiocarbon dates and intermediate valley terraces in coastal plains: Quaternary Research, v. 13, p. 80–92.

—— 1981, Discussion of "Age of Tunica Hills (Louisiana-Mississippi) Quaternary fossiliferous creek deposits: Problems of radiocarbon dates and intermediate valley terraces in coastal plains by E. G. Otvos, Jr." Reply: Quaternary Research, v. 15, p. 367–70.

Parsons, B. E., 1967, Geological factors influencing recharge to the Baton Rouge ground-water system, with emphasis on the Citronelle Formation: [M.S. Thesis], Louisiana State University, Baton Rouge, 75 p.

Rosen, N. C., 1969, Heavy minerals and size analysis of the Citronelle Formation of the Gulf Coastal Plain: Journal of Sedimentary Petrology, v. 39, no. 4, p. 1552–65.

Self, R. P., 1983, Petrologic variation in Pliocene to Quaternary gravels of southeastern Louisiana: Transactions of the Gulf Coast Association of Geological Societies, v. 33, p. 407–15.

—— 1984, Plio-Pleistocene drainage patterns and their influence on sedimentation patterns in southeast Louisiana: Geological Society of America Abstracts with Programs, v. 16, no. 3, p. 194.

Smith, M. L., and Meylan, M., 1983, Red Bluff, Marion county, Mississippi: A Citronelle braided stream deposit: Transactions of the Gulf Coast Association of Geological Societies, v. 33, p. 419–32.

Snead, J. I., and McCulloh, R. P., comps., 1984, Geologic Map of Louisiana, 1984: Louisiana Geological Survey, Baton Rouge, 1:500,000.

Spicer, B. E., 1969, Characteristics of the loess deposits and soils in East and West Feliciana parishes, Louisiana [M.S. thesis]: Louisiana State University, Baton Rouge, 69 p.

Thornbury, W. D., 1965, Regional Geomorphology of the United States: John Wiley and Sons, Inc., New York, 609 p.

Wascher, H. L., Humbert, R. P., and Cady, J. G., 1947, Loess in the southern Mississippi Valley: Identification and distribution of the loess sheet: Soil Science Society of America Proceedings, v. 12, p. 389–99.

Willman, H. B., and Frye, J. C., 1970, Pleistocene stratigraphy of Illinois: Illinois State Geological Survey Bulletin Series 94, 204 p.

Louisiana Chenier Plain

John T. Wells, Institute of Marine Sciences, University of North Carolina at Chapel Hill, Morehead City, North Carolina 28557

LOCATION

The Chenier Plain extends along the Louisiana Gulf Coast west from Vermilion Bay to Sabine Lake. The region is accessible via Louisiana 27 south from Lake Charles and US 167 south of Lafayette, Louisiana to the coast road, Louisiana 82 (Figs. 1, 2).

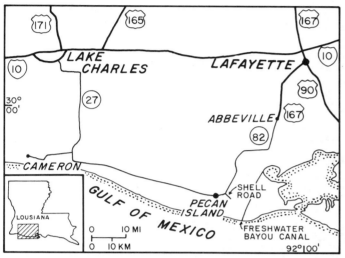

Figure 1. Location map for Louisana Chenier Plain stops.

INTRODUCTION

The Louisiana Chenier Plain is a 120 mi by 12 mi (200 km by 20 km) section of coast that extends west from Vermilion Bay, Louisiana, to the eastern margin of the Texas coast (Fig. 2). Its smooth form and relatively uncomplicated east-west orientation sets the chenier plain apart from the indented bays, barrier islands, and eroding headlands that characterize the Mississippi Delta Plain to the east. In cross section, the chenier plain is distinctive as a seaward-thickening wedge of muddy sediments; in plan view it is easily identified by the bundles of shore-parallel ridges of sand and shell hash that rest on the low-lying marsh and mudflat surfaces.

According to the early work of Russell and Howe (1935), the Louisiana Chenier Plain evolved during the Holocene as a sequence of prograding mudflats that were intermittently partially reworked into sand/shell chenier ridges. Pulses of sediment from the Mississippi River, transported by longshore currents, were responsible for the various stages of progradation. At times when the Mississippi River introduced sediment in the vicinity of the present chenier plain, the shoreline shifted seaward by processes of mudflat growth; during periods when its course took the discharge farther east, sediment influx to the chenier plain was low and wave attack was able to slow or halt the advance (Gould and

McFarlan, 1959). Cheniers formed during these latter periods and now stand as "islands" in the marsh. Five individual field sites within the chenier plain provide excellent examples of low-energy beaches and relict chenier ridges.

SITE DESCRIPTIONS

Stops 1 to 4 are in the eastern margin of the chenier plain at the mouth of Freshwater Bayou Canal which is accessible only by boat. Stop 5 is in the central chenier plain and is accessible by automobile.

Stop 1: Truncated Chenier Ridge and Perched Beach. Approximately 650 years ago, Chenier au Tigre was an active beach on the eastern margin of the Louisiana Chenier Plain. Isolated by mudflat growth, these beach deposits became stranded as the shoreline rapidly prograded seaward. Today, through subsequent shoreline retreat, Chenier au Tigre at the location of Stop 1 is a chenier ridge truncated by marine processes. Sediments that were landlocked in this relict chenier can now be seen again entering the dynamic beach environment from which they were previously derived.

In traveling east by boat from Freshwater Bayou Canal, Chenier au Tigre can easily be identified as the only ridge rising above the marsh surface that intercepts the shoreline at an oblique angle (Fig. 3). Chenier au Tigre and the other chenier ridges to the west provide a habitat for majestic oak trees from which, over the years, the name chenier has been derived (chene is the French word for oak). At an elevation of 6.5 ft (2 m) above the marsh surface, Chenier au Tigre in the early 1900s was the site of a small community of French Acadians who made a livelihood in the marshes of southwest Louisiana.

The beach deposits of Chenier au Tigre consist mainly of sand and shells with only minor amounts of silt and clay (Fig. 4). The shells usually occur as distinct layers, several inches (cm) thick, that have been concentrated by wave-sorting processes into lag deposits on the Gulf side of the ridge (Coleman, 1966). Typically, sediments on the backslope are finer in grain size. On windy days, waves can be observed eroding sediments from the chenier at its intersection with the coast, scattering the sediments across the beach face, and transporting the chenier to the west. On calmer days, the gentle wave swash sorts the particles but transports only the silts and clays.

Chenier au Tigre was formed along a stable section of coast, under conditions of overall low-wave and tide energy, and from sediment that was derived from both longshore and offshore sources. Just as part of the longshore contribution may have come from older updrift cheniers, the truncation of Chenier au Tigre now provides a source of sediment for the growth of new cheniers farther west. Although Chenier au Tigre looks like a beach ridge, it differs by being anchored in shoreface deposits that

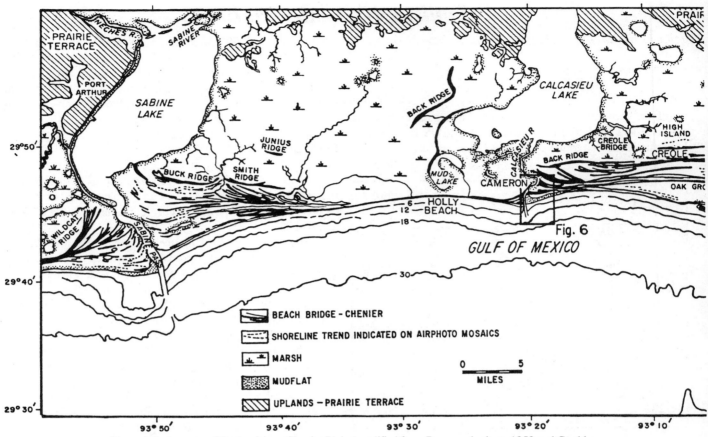

Figure 2. Index map of the Louisiana Chenier Plain (modified from Byrne and others, 1959 and Gould and McFarlan, 1959). The chenier plain extends from the western margin of Vermilion Bay on the central Louisiana coast (longitude 92°00′) to the eastern flank of the Pleistocene Trinity Delta in Texas (longitude 94°00′). Approximately 108 mi (180 km) of this 120 mi (200 km) stretch of coast can be traveled by car along Louisiana 82 and Texas 87. Further detail can be found on the Port Arthur 1:250,000 Topographic-Bathymetric Sheet NH 15-8, published by the U.S. Geological Survey. Enclosed areas show locations for individual Stops 1-4 (Fig. 3) and Stop 5 (Fig. 6).

are fine-grained. Thus, the primary difference between chenier ridges and the marsh/mudflat that surrounds them is 1) the lack of any appreciable sand or pure shell units in the marsh/mudflat sequences, and 2) a significantly higher percentage of fine sediment in these sequences (Price, 1955). However, in terms of processes, chenier ridges and beach ridges are nearly identical. Each is built by waves with the coarsest ridges resulting from storm activity.

The tendency to include as chenier ridges a wide variety of geomorphic features, including ancient oyster reefs, small spits, aeolian deposits, and natural levees, has lead Otvos and Price (1979) to suggest placing restrictions on use of the terms "chenier" and "chenier plain." They proposed applying the term chenier plain only in situations where at least two subparallel chenier ridges were separated by a muddy progradational unit.

The beach at Chenier au Tigre was fronted by a 3-to-6-ft (1-to-2-m)-thick mudflat in the late 1940s and early 1950s. Since that time, the muds have been dispersed, most likely transported

to the west, and the overall pattern on this stretch of coast is one of retreat. According to Morgan and others (1953), the beach at Chenier au Tigre has eroded more than 800 ft (240 m) in the past 100 years. Unpublished data taken at Louisiana State University (G. K. Kemp) shows a 40-ft (12-m) loss of shoreline between December 1980 and May 1983.

The marsh just west of Chenier au Tigre shows a good example of how this erosion proceeds. Here no continuous beach exists. The crenulated shoreline is a series of small wave-eroded embayments and protruding marsh headlands. Much of the marsh surface close to shore has been exhumed by prolonged flooding or by the physical impact of waves. The marsh/water boundary is formed by a steep scarp 1.5 ft (0.5 m) high. Fine-grained sediments eroded from the marsh are temporarily stored offshore as soft deposits a few cm thick.

Stop 2: Perched Beach. This stop is a perched beach where storms have forced a veneer of sand and shell hash onto the marsh surface. This type of shoreline is most common in the

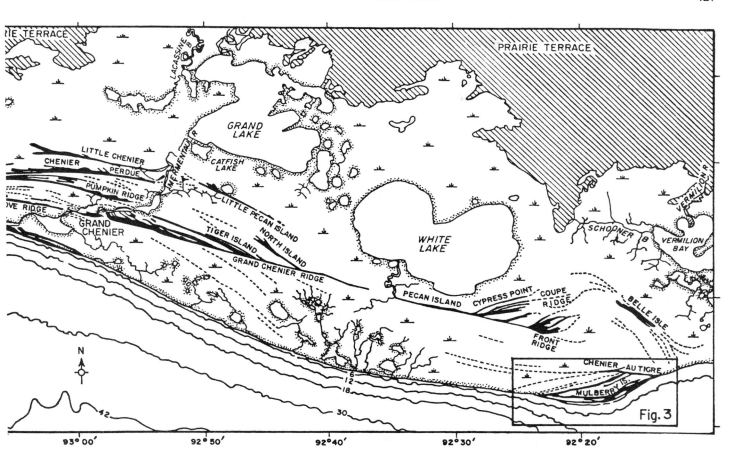

eastern section of the chenier plain. The perched beach can be recognized by the exhumed marsh cropping out in the surf zone. Rarely are the sands over 3 ft (1 m) thick and, near the waterline, they may be so thin that it is difficult to walk across them without sinking into the muds below. Mudcracks and mudballs are common features on the lower beach face at this stop.

Kaczorowski and Gernant (1980) have referred to the perched beach at this location as a Type I shoreline, one of three chenier-plain shoreline types to which Wells and Kemp (1982) added a fourth, the truncated or reactivated chenier shoreline of the previous stop. Beall (1968) identified the perched beach shoreline as a "transitional beach," formed under conditions of moderately high sediment supply and average coastal energy.

The shoreline in the vicinity of Mulberry Ridge at Stop 2 is part of a 36-mi (60-km) section that has been building for the last 600 years. Between 1100 and 600 years BP, the site of rapid coastal progradation shifted from the western back to the eastern chenier plain. From the eroding marsh west of Stop 1 to the western tip of Mulberry Ridge, the shoreline advanced by 1300 ft (400 m) during the 24-yr period between 1927 and 1951. Beach profiles taken from December 1980 to May 1983 show that, at least in the short term, this shoreline has been stable (G. P. Kemp, personal communication).

Stops 3 and 4: Active Mudflat. Collectively, Stops 3 and 4 are the best example of active mudflat growth on the Louisiana Chenier Plain and one of the best examples in North America. These stops show at two locations the processes by which sediment accumulates into mudflats, the same processes responsible for building virtually all of the chenier plain, a 12-mi (20-km)-wide wedge of mud that is 23-ft (7-m) thick. As shown in Figure 3, Stop 3 is the trailing (updrift) edge and Stop 4 is the leading (downdrift) edge of a large mudflat; however, recent and continuous translation of the entire mudflat to the west makes these locations inexact. Therefore, the following discussion can be applied to any field stop along approximately 6 mi (10 km) of coast beginning 0.6 mi (1 km) west of Freshwater Bayou Canal.

Fine-grained sediments building this mudflat are derived from the Atchafalaya River to the east. Carried as suspended sediment in the Atchafalaya mud stream (Wells and Kemp, 1981), the silts and clays are deposited in the nearshore zone as tidal mudflats (Fig. 5). Composition of the sediment in the mud stream is the same as that in the lower Atchafalaya River. Median particle diameters are 2-6 microns and the predominant clay minerals are Montmorillonite, Illite, and Kaolinite in the ratio 3:1:1. Estimates by Wells and Kemp indicate that only a small fraction of the greater than $175 \times 10^7 ft^3/yr$ ($50 \times 10^6 m^3/yr$) of fine-grained sediment that is transported to the eastern chenier plain ends up as intertidal mudflats. The remainder may be dispersed across the inner shelf as a blanket of fines.

Comparison of color infrared photographs, orthophoto-

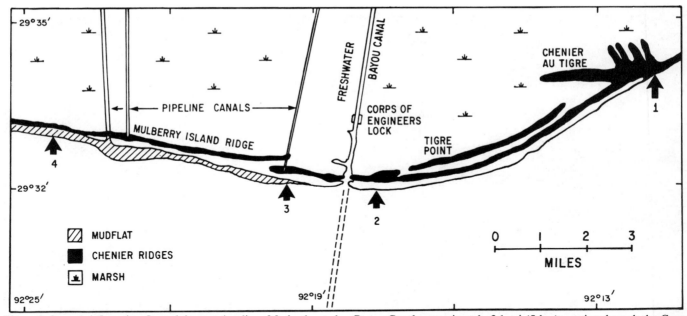

Figure 3. Stops 1-4. Location: Launch boat at Acadiana Marina located at the end of the shell road that is east of Louisiana 82 where it turns sharply west towards the town of Pecan Island. Travel south by boat approximately 1.8 mi (3 km) down Louisiana Fur Company Canal to a Y-shaped juncture, then east approximately 2.4 mi (4 km) to the intersection with Freshwater Bayou Canal. Travel south in Freshwater Bayou Canal approximately 3.1 mi (5 km), passing through the Corps of Engineers Lock and out to the Gulf (an additional 1.3 mi; 2 km). From the mouth of Freshwater Bayou Canal all field sites can be reached by boat and Stops 2 and 3 can be reached by walking. The best time to make these field stops is at low tide.

quads, and results of aerial and ground reconnaissance reveal that, during a 12-yr period from 1969-1981, 1) an increasing length of shoreline on the eastern chenier plain was fronted by mudflats, and 2) the locus of sedimentation was shifting to the west from its original location east of Freshwater Bayou Canal (Wells and Kemp, 1981). Most of the mudflats were discontinuous and separated by an eroding marsh shoreline similar to that just west of Stop 1. In 1984 these individual tidal flats merged into the single continuous mudflat dipicted in Figure 3. As first suggested by Morgan and others (1953), renewed mudflat growth

was a result of an increase in sediment discharge from the Atchafalaya River.

At low tide, 320-980 ft (100-300 m) of intertidal exposure (measured perpendicular to the shoreline) can be seen along most of the mudflat. These muds often fill in the embayments along the crenulated and previously eroded marsh shoreline. The upper 3-6 ft (1-2 m) are composed of a gel-like fluid mud similar in consistency to yogurt. Soft muds continue seaward beyond the low-tide line where patches of fluid mud have been noted from sampling and bottom probes to a distance of 3 mi (5 km) offshore.

Mudflats are capable of shifting rapidly and sometimes unexpectedly. As such, the chenier plain mudflats form temporary and transitory storehouses of littoral sediments that lead to extensive coastal progradation only upon their stabilization. Research in progress (G. P. Kemp and J. T. Wells) suggests, somewhat surprisingly, that chenier plain mudflats are deposited during the highest energy conditions this coast experiences on a regular basis, i.e., during winter cold fronts. The mudflat at Stop 4 prograded approximately 250 ft (75 m) between February 1981 and May 1983, all during the elevated water level associated with cold-front passage.

Perhaps the most important aspect of mudflat growth is that fluid muds attenuate incoming wave energy, leading to less wave attack at the shoreline itself (Wells and Roberts, 1981). This, in turn, creates conditions favorable for even further sedimentation. From shore or in a small boat, this property of muddy sediments is evident; breaking waves between Stops 3 and 4 are all but

Figure 4. Truncated chenier ridge at Stop 1.

Figure 5. Mudflat (at low tide) fronting the marsh at Stop 3.

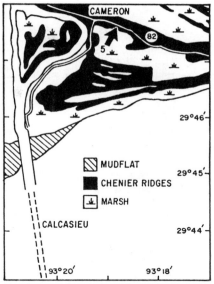

Figure 6. Index map to location of central chenier plain near mouth of Calcasieu Pass. Cameron Construction Pit (Stop 5) is located just off Louisiana 82, 11 mi (18 km) west of Creole, Louisiana.

absent. The mudflats forming here are thus the first stage in the feedback loop between coastal energy and shoreline response. As sedimentation continues over the next several hundred years, new mudflats will continue to merge with existing mudflats and, at its peak of development, the shoreline will become literally choked with fine-grained sediment. The final step will be stabilization by colonizing marsh grasses.

Stop 5: Cameron Construction Pit. The Cameron Construction Pit provides perhaps the best exposures in Louisiana of chenier-ridge stratigraphy and sedimentary structures. This stop is located within a broad fan of very young chenier ridges, less than 600 years old, that form the western tip of the extensive Oak Grove- Grand Chenier- Pecan Island trend (Fig. 2). The drive along Route 82 from Stop 4 follows this prominent trend which represents the 1100-year-old shoreline (Gould and McFarlan, 1959). The town of Cameron, just to the west, is a low-lying coastal community built on the high ground of these nested chenier ridges (Fig. 6). Proximity to the Gulf and subsidence of these now-inactive ridges have resulted in severe storm damage to the town of Cameron from Hurricane Audrey in 1957 and Hurricane Carla in 1961.

Figure 7A shows a N-S section across the chenier plain approximately 12 mi (20 km) east of Stop 5. The most prominent ridge is the 10-ft (3-m) high Oak Grove chenier that forms the stable substrate for Route 82. Each chenier was built on shallow, foreshore Gulf-bottom sediments that ranged from sand to silty clay. The ridges were isolated by subsequent progradation through the growth of mudflats and marsh (Fig. 7B). According to Gould and McFarlan (1959), only on the landward side do the chenier sediments extend over the adjacent marsh. Thus the earlier idea proposed by Russell and Howe (1935) that cheniers are beaches driven across vast expanses of marsh appears invalid. Cheniers are more likely developed as accretionary ridges along segments of the shore that are relatively stable. Relict beaches preserved as cheniers are often characterized by a smooth arcuate seaward front and a landward margin that is irregular from washover deposition.

Although a variety of structures and environments can be observed at Stop 5, the most notable are the interbedded sands and silts that have been altered by overwash processes (Fig. 7C). During storms, normal beach processes are interrupted by high tides and large waves that transport the coarser sediments inland as washover fans. These deposits can be recognized by abrupt changes in lithology and by steeply inclined beds that dip landward. The fact that the coarsest sediments in the cheniers (except for the shell) are fine sands reflects the nature of the sediment available for transport rather than an inability of winter storms and hurricanes to transport coarser particles.

The shell layers, typically a few inches (cm) thick, consist primarily of the surf clam *Mulinia,* fragments of the oyster *Crassostrea,* and the brackish clam *Rangia.* The layers of silt and the thin clay drapes are a result of rapidly changing conditions; a continuous supply of muds from the east, even though diminished during periods of chenier-building, provides an opportunity for incorporation of fine-grained sediments in the cheniers. Kaczorowski and Gernant (1980) have shown that layers of sand and shell account for approximately 75 percent (by volume) of a 18-ft (5.5-m) vertical sections with the remaining 25 percent being clay and mixed-clay units. The upper 3 ft (1 m) of section consists of fine aeolian sands and dark, rooted soils from colonization by marsh grasses.

The configuration of chenier ridges in the immediate vicinity of this stop suggests that Calcasieu Lake was once an estuary with free connection to the Gulf of Mexico. Successive shorelines, each building farther seaward, eventually closed the estuary by processes of spit growth. Gould and McFarlan (1959) speculated that Sabine Lake to the west and Grand Lake to the east may have experienced similar histories.

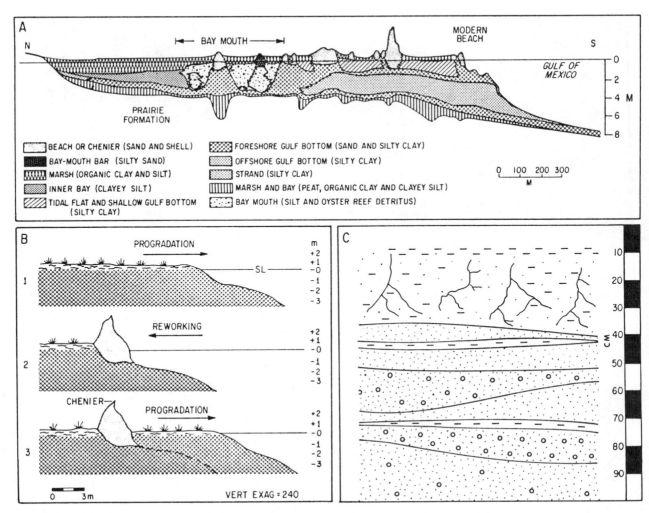

Figure 7A, 7B, 7C. Cross section of Louisiana Chenier Plain near Stop 5 (from Gould and McFarlan, 1959) showing stratigraphic relationship between chenier ridges and underlying sediments (A), model (from Hoyt, 1969) for chenier ridge development (B), and typical section of chenier ridge at Cameron Construction Pit showing inclined beds of sand and shell with thin layers, capped by root-burrowed soil profile (C).

REFERENCES CITED

Beall, A. D., Jr., 1968, Sedimentary processes operative along the western Louisiana shoreline: Journal of Sedimentary Petrology, v. 38, p. 869–877.

Byrne, J. V., LeRoy, D. O., and Riley, C. M., 1959, The Chenier Plain and its stratigraphy, southwestern Louisiana: Gulf Coast Association of Geological Societies, Transactions, v. 9, p. 237–260.

Coleman, J. M., 1966, Recent coastal sedimentation: Central Louisiana coast: Louisiana State University, Coastal Studies Institute #17, p. 73.

Gould, H. R., and McFarlan, E., Jr., 1959, Geological history of the chenier plain, southwestern Louisiana: Gulf Coast Association of Geological Societies, Transactions, v. 9, p. 261–270.

Hoyt, J. H., 1969, Chenier versus barrier, genetic and stratigraphic distinction: American Association of Petroleum Geologists Bulletin, v. 53, p. 299–306.

Kaczorowski, R. T., and Gernant, R. E., 1980, Stratigraphy and coastal processes of the Louisiana Chenier Plain: Gulf Coast Association of Geological Societies, Field Guide, p. 72.

Morgan, J. P., and Larimore, P. B., 1957, Changes in the Louisiana shoreline: Gulf Coast Association of Geological Societies, Transactions, v. 7, p. 303–310.

Morgan, J. P., Van Lopik, J. R., and Nichols, L. G., 1953, Occurrence and development of mudflats along the western Louisiana coast: Louisiana State University, Coastal Studies Institute, Technical Report 2, p. 34.

Otvos, E. G., and Price, W. A., 1979, Problems of chenier genesis and terminology—An overview: Marine Geology, v. 31, p. 251–263.

Price, W. A., 1955, Environment and formation of the chenier plain: Quaternaria, v. 2, p. 75–86.

Russell, R. J., and Howe, H. V., 1935, Cheniers of southwestern Louisiana: Geographical Review, v. 25, p. 449–461.

Wells, J. T., and Kemp, G. P., 1981, Atchafalaya mud stream and recent mudflat progradation: Louisiana chenier plain: Gulf Coast Association of Geological Societies, Transactions, v. 31, p. 409–416.

—— 1982, Mudflat and marsh progradation along Louisiana's Chenier Plain: A natural reversal in coastal erosion, *in* Boesch, D. F., Proceedings, Conference on Coastal Erosion and Wetland Modification in Louisiana, p. 39–50.

Wells, J. T., and Roberts, H. H., 1981, Fluid mud dynamics and shoreline stabilization: Louisiana Chenier Plain: Proceedings, 17th International Coastal Engineering Conference, p. 1382–1401.

Physiography of Louisiana Salt Domes (The Five Islands)

Donald H. Kupfer, Emeritus Professor, Department of Geology, Louisiana State University, Baton Rouge, Louisiana 70803

LOCATION

The Five Islands are a line of topographic mounds parallel to and southwest of U.S. Highway 90 between New Iberia and Morgan City, Louisiana (Fig. 1). They are 25 to 60 mi (42 to 100 km) from I-10 at Lafayette, Louisiana. Belle Isle is only accessible by boat: the others are 5 to 18 mi (8.3 to 30 km) by rural roads, from U.S. 90. All are privately owned, and the public is no longer admitted to any of the associated salt mines.

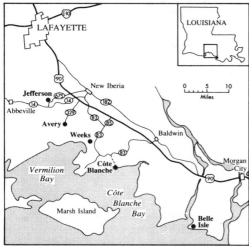

Figure 1. Location Map: The Five Islands.

INTRODUCTION

The Five Islands, arched up by rising salt stocks (diapirism), are the only large topographic hills in the swamp, marsh, and coastal plains of southern Louisiana. These islands were a favored habitation for prehistoric Indians and later European settlers. In addition, they contain the only active salt mines in Louisiana, and the largest salt mines in southeastern USA. Weeks Island has the only Strategic Petroleum Reserve that lies in a former salt mine. The flanks of each salt diapir are prolific petroleum reservoirs; several fields per dome. Two of the islands have scenic gardens. However, geologic outcrops on the islands are unspectacular to nonexistant.

About 120 salt domes have been located in onshore Louisiana, and over 25 have salt within 1,000 ft (300 m) of the surface (Howe and Moresi, 1931; Murray, 1961, tables p. 269–270; Halbouty, 1979, tables p. 50–52). Except for the older domes of north Louisiana (see Martinez, this guidebook), few of these have any surface expression.

In south Louisiana, one or two domes have a subtle radial drainage pattern, visible only after one knows where they are. The others were recognized as a result of gravity surveys and oil exploration activity. Only the Five Island salt domes (Jefferson, Avery, Weeks, Cote Blanche, and Belle Isle; Fig. 1) are prominent topographic features. Although not true islands, their topographic

relief (75-171 ft, 23-52 m) above a predominantly swampy terrain makes them "islands of dryness" and a favored habitation site. The Five Islands (and two deep-seated salt-controlled structures) are aligned along a N 42° W trend and are each 7.3 mi (11.8 km) apart (Fig. 1).

A salt *dome* is a domal (or anticlinal) sedimentary structure with dimensions measured in mi (tens of km) and a much smaller salt *stock* comprises its core. Upward movement of the low-density plastic salt arched the overlying sediments and then pierced them forming a diapir (for detailed definitions see Braunstein and O'Brien, 1968; Atwater and Forman, 1959). Louisiana onshore stocks are about 1+ mi (2+ km) in diameter. The tops are within 3 mi (5 km) of the surface, and most are a half to a quarter of that. In north Louisiana the salt base is between 15,000 and 29,000 ft (4,500 to 9,000 m); in south Louisiana the original salt horizon is much deeper, as much as 60,000 ft (18,000 m), but probably devoid of salt.

In rising from the original Jurassic Louann salt layer (1,500 to 5,000 ft [500 to 1,500 m] thick) to its present position, a salt stock typically passes through three active phases (pillow, bell, stock; Fig. 2) and two passive ones (compaction, erosion). The *pillow* phase is characterized by mainly horizontal movement as the salt migrates toward a center or ridge and doubles to triples in thickness, meanwhile draining nearly all the salt from adjacent withdrawal basins or rim synclines (Trusheim, 1960). Examples

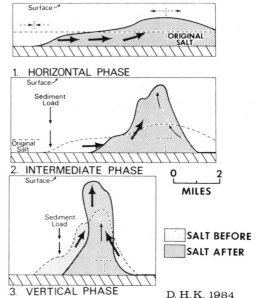

Figure 2. The three active phases of salt dome growth. (1) The salt moves from the original uniform bed into a pillow two or three times thicker and about 13 mi (20 km) in diameter. (2) The pillow becomes bell-shaped, diapirism may start, and an extensive dome is formed in the surrounding sediments. (3) Diapiric salt moves into a stock, a vertical column of salt 1+ mi (2+ km) in diameter, and the rim syncline converges on it; the salt nears the surface and may extrude.

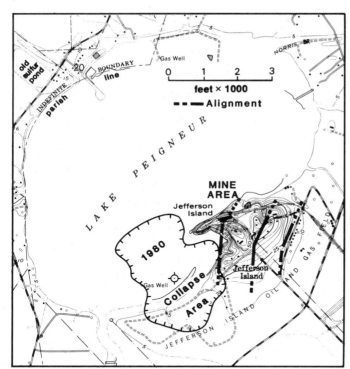

Figure 3. Jefferson Island and Lake Peigneur, showing the location of the alignments (lineaments), area of lake-floor collapse of November 1980, and approximate position of Texaco well. (From Delcambre 7½-minute Quadrangle).

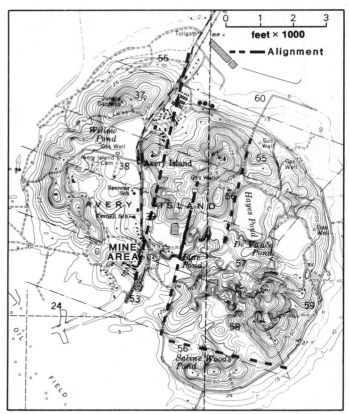

Figure 4. Avery Island, showing significant topographic valley-alignments and an apparent right-lateral offset of the "island." The Tabasco-pepper fields lie between the 5- and 20-ft (1.5 to 6 m) contours on the northeast and east sides of the island. The Blue Pond is a man-induced collapse lake; the Civil War salt mining was in this area. The other ponds are natural salt-collapse lakes. (From Delcambre 7½-minute Quadrangle).

of this phase are common under the continental slope area of the Gulf of Mexico (young, still forming) and along the internal northern edges of the salt basin (old, residual).

Phase two, the *bell,* is a transition from predominantly horizontal to predominantly vertical movement of the salt. The salt shapes are highly irregular and commonly are controlled by local faulting, sedimentary irregularities, and in particular by depositional patterns. Extreme variations in size, shape, and control exist (examples: coastal shelf of Louisiana, northern Spain, and the Dead Sea).

The third phase, the *stock,* also has many variations, but commonly is characterized by a vertical cylinder of salt at the core of a domal sedimentary structure. This is the phase of most of the onshore Gulf Coastal salt domes, although some, like the Five Islands, may have matured to stability (next phase).

After active salt movement has ceased, the stock (in the first passive phase) appears to continue to rise as the less-consolidated sediments around it compact. Most *compaction* occurs where the sediments are youngest and thickest, generally in the last rim synclines to form, although it may occur anywhere (random distribution of thick areas on isopach maps). Because rim synclines converge onto the stock with time, the last rim synclines lie against the stock, and so the salt stock appears to emerge, but not the whole dome. The Five Islands are just entering this stage, and the "island" part exactly overlies the salt stock; the "marsh line" (break from flat marsh to drained slope) clearly outlines the edge of the diapir (including any shale diapirism).

The final stage is regional uplift and *erosion* (second passive phase), exposing the internal structure of the dome and, if no ground-water aquifers are present, even the salt stock. Examples from the interior salt province demonstrate that active salt movement ceased about the end of the Cretaceous; compaction movement ceased about middle Tertiary; and erosion has dominated for at least 30 million years (see Kupfer, 1977, for north Louisiana; Seni and Jackson, 1983 for northeast Texas). As a result, these interior domes are breached and the cores are commonly topographic lows. In south-central Mississippi, the compaction movement ceased in the Plio-Pleistocene, but erosion has not yet become dominant.

At the Five Islands only the areas directly over the salt stocks are "uplifted," not the whole domes; the topographic mounds are only about 2 mi (3 km) in diameter; whereas the structural "domes" are 5 to 7 times that diameter, and are sinking. The numerous small lakes and other closed depressions (Figs. 3, 4, 5) indicate that salt solution and caprock formation are still active processes.

Alignments are shown on all three topographic maps (Figs. 3, 4, 5). These may represent faults, solution collapse features, or shear zones related to spines of differential salt movement (Kupfer, 1976, 1980). With minor exceptions, the only surface evidence is topographic (i.e., lineaments). Only those lineaments that influence the top-of-salt surface are shown, and these probably extend into the internal structure of the salt.

SITE DESCRIPTION

The physiographic interest and accessibility of the Five Islands decreases from north to south. The northern two islands (Jefferson and Avery) have good roads and allow public access to scenic gardens (fee charged). The others are generally closed to public access and local inquiry is suggested. For a detailed guidebook see Kupfer & Morgan (1976).

Jefferson Island (Fig. 3)

Jefferson Island (Iberia Parish, Delcambre 7½-minute Quadrangle) is the northernmost of the Five Islands, accessible via Louisiana 675 which runs west for 5 mi (8 km) from U.S. 90 to the dome and then south to Louisiana 14 (Fig. 1). Unlike the other four, Jefferson Island is not in a swamp, but lies on the dry Pleistocene Prairie Terrace, which is very flat and agricultural. A large lake, Lake Peigneur, covers most of the salt stock and is underlain by caprock at a depth of 800 ft (250 m), suggesting that it is a natural salt-collapse lake. Sulfur was mined in the 1930's by the Frasch process (O'Donnell, 1935) and some of the jetties in the lake still remain. Sulfur slag can be seen on the northwest shore where the plant was located.

Only a small topographic hill (55 ft or 20 m above the terrace level) is present. It is the site of the Joseph Jefferson (American mid-19th century actor) mansion and the old salt-mine buildings of Diamond Crystal Salt Company. The hill represents the latest spine of salt movement (Kupfer, 1976, 1980). Highly faulted Pleistocene sands up to 600 ft (200 m) thick are overlain by 5 to 100 ft (2 to 30 m) of clays; both are tilted and arched by the spine; the top-of-salt is at sea level. The clay/sand contact can be seen in a wavecut cliff on the westernmost point of the peninsula extending into Lake Peigneur.

In November 1980, a Texaco well was drilled into or adjacent to the salt mine. The mine flooded, draining Lake Peigneur and causing a large collapse structure (Autin, 1984). Most features are now under water, but an arcuate 'bite' into the otherwise oval-shaped shore line and minor collapse features can be seen from the Live Oak Gardens (admission fee). Texaco operates the oil and gas field, which lies principally on the south side of the stock and surrounds the small hill.

Several N 10-30° E lineaments, first recognized on airphotos, cross the hill area and are marked by seep springs, scarplets, and saddles. The relationship of these to the underlying salt is not clear, but they appear to be related to spines-of-salt motion and represent "lows" in the buried top-surface of the salt.

Avery Island (Fig. 4)

Avery Island (Iberia Parish, Delcambre 7½-minute Quadrangle) is second of the Five Islands to the southeast, 7 mi (12 km) south of U.S. 90 on Louisiana 329. It is the site of the Avery Island Jungle Gardens, McIlhenny Tabasco fields and bottling

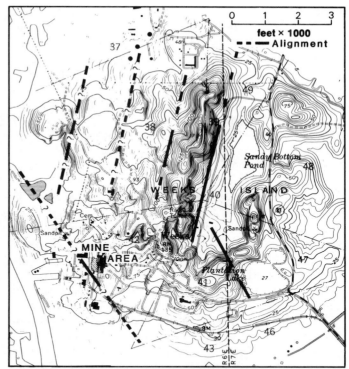

Figure 5. Weeks Island. The mine buildings are to the southwest and the Strategic Petroleum Reserve (SPR) buildings (new, not shown) are in the south-center of the island. Canals on the north (not shown) and west service the oil wells. The east-central low area and the scarp to the west of it are related to structure within the salt. The lakes are natural collapse lakes. (From Weeks 7½-minute Quadrangle).

plant, private estates, numerous oil operations on all flanks, several natural collapse-lakes, Indian archeological sites, Civil War salt-mining operations, and the International Salt Company operations (with barge canals connecting to the Intracoastal Canal). The highest point is 157 ft (48 m) above sea level. There is much to see here, but the island is privately owned and access (even after paying a small fee) is restricted to a small area along the north side. The Jungle Gardens (fee) and Tabasco bottling plant (free) are in this northern area, as are several very deep petroleum wells that penetrate for hundreds of meters through the salt overhang into the sediments below. Petroleum wells on the west side of the island are drilled from barges in canals; those on the east from land rigs.

The salt occurs in several spines, mostly without caprock, and much of it near the surface (15 to 75 ft or 5 to 20 m). Alignments are on valleys that overlie known boundary zones between spines-of-salt movement; some show displaced lithologies at the surface. It is not known if there is any right-lateral offset in the salt that is equivalent to the physiographic offset shown in Figure 4.

The above-salt geology is complex, with numerous fault blocks separating sequences of Pleistocene sands and clays. One small sand quarry (500 ft [150 m] NE of Blue Pond) shows horizontal to vertical bedding, contortion, and numerous small faults.

Weeks Island (fig. 5)

Weeks Island (Iberia Parish, Weeks 7½-minute Quadrangle) is the highest (171 ft or 52 m) of the Five Islands, and has the second oldest salt production (Myles Salt, now Morton Salt Co.). It is also the only *mine* taken over by the Strategic Petroleum Reserve (SPR) for the storage of oil (the other reserves are in solution-mined cavities). Salt mining continues to the north and west of the SPR cavity (NW center of map, Fig. 5) and may eventually continue below it. The main oil production is from the north flank, which is also the "overhang" flank (youngest spine). An archeological site is at water level on the northwest edge of the island, but it is inaccessible.

Some roads are open to the public, mainly Louisiana 83 (loop road from New Iberia to Baldwin, Louisiana), which crosses the east flank of the mound, but the salt mining and SPR areas are closed. Several natural-collapse lakes can be seen along this east side, with nesting egrets in season. From the higher points (the highest are not accessible without permission), one can see the small rises in vegetation that mark the Avery and Cote Blanche "islands" 7 mi (12 km) away to the NE and SW.

The surface geology of Weeks Island is unmapped, but consists of faulted Pleistocene sands and clays, which can be seen in several gullies (as east of BM 135, Fig. 5). A marked topographic ridge trends northerly across the center of the "island", bounded by a prominent escarpment on the east. This scarp corresponds to a major, sediment-bearing boundary zone in the salt, and marks the eastern limit of the older (now SPR) workings. The brine wells of the Morton Salt Company are to the east of this zone (NW of Sandy Bottom Pond) and in a separate spine of salt.

Other

Cote Blanche (St. Mary Parish, Kemper and Marone Point 7½-minute Quadrangles) is a somewhat smaller "island" (maximum elevation 97 ft [30 m]), on the north shore of Cote Blanche Bay. It can be reached by a 1.5-mi long (2.5 km) gravel road that extends south from Louisiana 83 at a point 13 mi (22 km) west of Baldwin, Louisiana, and 5 mi (8 km) southeast of Weeks Island. A small cable-drawn ferry crosses the Intracoastal Canal to the privately owned property. On the western flank of the mound are buildings and two air shafts for salt mining (Domtar Chemical Company of Canada), and associated barge canals. Most of the salt stock is to the south under the water of the bay and numerous petroleum platforms can be seen offshore. On the shoreline are 40-ft (10 m) high sea cliffs, probably the only ones in Louisiana; inaccessible except by foot across swampy ground. A few years ago these sea cliffs exposed several Pleistocene rollover, growth-faults with displacements of about 2 ft (60 cm) or less, but storm waves and slumping commonly obscure them.

Belle Isle (St. Mary Parish, Belle Isle 7½-minute Quadrangle) is the southernmost of the Five Islands, accessible only by boat. It is on the south shore of Wax Lake Bayou, near its entrance to Atchafalaya Bay. The highest point (78 ft [24 m]) is north of a circular lake that is underlain by caprock, probably a natural salt-solution collapse feature. Remains of the circa 1900 mining attempt have virtually disappeared. The 1960-1983 salt-mining operations by Cargill, Inc. have been discontinued. A large, shallow sand quarry is present to the north of the Cargill shafts. Sun Oil Company operates a gas-separation station on the south flank.

ANNOTATED BIBLIOGRAPHY

Autin, Whitney, 1984, Observations and Significance of Sinkhole Development at Jefferson Island: Louisiana Geological Survey, Geological Pamphlet No. 7, 75 p. (History of the drilling disaster, morphology and geology of island, and a general study of collapse features.)

Atwater, G. I., and Forman, M. J., 1959, Nature and growth of southern Louisiana salt domes and its effect on petroleum accumulation: American Association of Petroleum Geologists Bulletin, v. 43, p. 2592-2622. (Origin and evolution of salt domes and the associated petroleum-bearing structures.)

Braunstein, Jules, and O'Brien, G. D., 1968, ed., Diapirism and Diapirs, a Symposium: American Association of Petroleum Geologists, Memoir 8, 444 p. (Includes a comprehensive discussion and bibliography on diapirism.)

Halbouty, M. T., 1979, Salt Domes, Gulf Region, United States and Mexico: 2nd edition, Gulf Publishing Co., Houston, Texas, 529 p. (Textbook and sourcebook, complete bibliography on Gulf Coast salt domes.)

Howe, H. V., and Moresi, C. K., 1931, Geology of Iberia Parish: Louisiana Department of Conservation, Geological Bulletin 1, 187 p. (Surface geology of the Five Island area, comprehensive annotated bibliography on Louisiana salt domes.)

Kupfer, D. H., 1976, Shear Zones inside Gulf Coast salt stocks help to delineate spines of movement: American Association of Petroleum Geologists Bulletin, v. 60, p. 1434-1447. (Earlier, but more available version of Kupfer, 1980.)

Kupfer, D. H., 1977, Times and rates of salt movement in north Louisiana, *in* Martinez, J. D., and Thoms, R. L., 1977, Salt Dome Utilization and Environmental Considerations; a Symposium: Institute of Environmental Studies, Louisiana State University, Baton Rouge, Louisiana, 423 p. (Many aspects of salt dome geology), p. 119-134. (Rate statistics, Vacherie Salt Dome, LA.)

Kupfer, D. H., 1980, Problems associated with anomalous zones in Louisiana salt stocks, USA, *in* 5th Symposium On Salt: Northern Ohio Geological Society, Cleveland, Ohio, v. 1, p. 119-134. (Recognition of spines of motion, especially at Five Island salt domes; see also Kupfer 1976.)

Kupfer, D. H., and Morgan, J. P., 1976, Louisiana Delta Plain and Salt Domes: New Orleans Geological Society, New Orleans, Louisiana, 68 p. (Comprehensive guidebook from New Orleans to Weeks and Avery Islands; summary of wetland morphology and internal structure of salt, maps.)

Murray, G. E., 1961, Geology of the Atlantic and Gulf Coastal Province of North America: Harper, Inc., New York, 692 p. (Standard textbook with excellent Chapter V on Gulf Coast salt domes.)

O'Donnell, Lawrence, 1935, Jefferson Island salt dome, Iberia Parish, Louisiana: American Association of Petroleum Geologists Bulletin, v. 19, p. 1602-1644. (Subsurface geology, sulfur production, and caprock.)

Seni, S. J., and Jackson, M.P.A., 1983, Evolution of salt structures, east Texas diapir province: American Association of Petroleum Geologists Bulletin, v. 67, Part I, p. 1219-1244; Part II, p. 1245-1274. (Recent study on Texas salt domes.)

Trusheim, F., 1960, Mechanism of salt migration in northern Germany: American Association of Petroleum Geologists Bulletin, v. 44, p. 1519-1540. (Detailed study of the genesis of north German salt domes.)

Selected depositional environments of the Mississippi River deltaic plain

Harry H. Roberts, Coastal Studies Institute, Louisiana State University, Baton Rouge, Louisiana 70803

LOCATION

Depositional environments of the Mississippi River deltaic plain can be studied in a series of exposures in the Barataria Basin (upper deltaic plain) and in the Balize delta lobe (lower deltaic plain). Stops 1 and 2 are in Barataria Basin, which is an interdistributary basin defined by alluvial ridges of the modern Mississippi River on the east and Bayou Lafourche on the west (Fig. 1).

Stops designated for the lower delta are accessible only by boats suitable for safely navigating the river from Venice to the mouth of South Pass.

INTRODUCTION

The Mississippi, like other major river systems, consists of four primary components: drainage basin, alluvial valley, deltaic plain, and receiving basin. The stops described herein relate to selected depositional environments and the landforms, sedimentary facies, and other geological products within the plain of the Mississippi River delta complex. Previous studies (McIntire, 1958; Kolb and Van Lopik, 1958; Frazier, 1967) have demonstrated that the vast Holocene deltaic plain ($>$11,120 mi[2]; 17,900 km[2]) is the product of deposition associated with multiple changes in the river's course over the past 6,000 to 8,000 years. This fundamental process of switching the locus of deposition has resulted in the interfingering and overlapping of major delta lobes, like the modern Balize or bird-foot delta. The complexities of Louisiana's deltaic plain and coastline reflect the stages of deterioration associated with abandonment of delta lobes. During Holocene times, the Mississippi River has carried a very fine-grained sediment load to its low-energy receiving basin. Sediments arriving at the Gulf consist primarily of clay, silt, and fine sand. A dominance of clay-rich sediment (approximately 70%) results in significant dewatering and sediment compaction upon burial. This local process, coupled with substantial regional subsidence, causes most sedimentary facies to be quickly preserved in the subsurface. However, these changes, as well as those initiated by deposition, create a predictable sequence of sedimentary facies and associated landforms. A major characteristic of Mississippi River sedimentation is the orderly repetition of depositional events. The cyclic sedimentation concept provides a framework for organizing the complex environmental relationships and facies distributions resulting from delta building.

Recognizable cycles are present on many scales, both spatial and temporal. These vary in magnitude from the large overlapping lobes, like the modern Balize lobe, with thicknesses of 45 feet (14 m) to approximately 600 feet (1983 m) and areal extents of up to 2,510 mi[2], (4,040 km[2]) to small crevasses that are similar in shape but only fill restricted interdistributary bays.

As a product of major cyclic depositional events, the present deltaic plain is partitioned into interdistributary basins separated by slightly higher alluvial ridges. These lateral boundaries of the basins are linear strips of elevated land composed of active and abandoned natural levees. Although geological and vegetation maps of south Louisiana clearly define both basins and alluvial ridges (Saucier, 1974; Chabreck and Linscombe, 1978), these features are so large that it is difficult to comprehend them by ground-level observations. However, each major depositional setting of the deltaic plain possesses a suite of smaller scale characteristics that can easily be seen in the field. The Field Stops described herein are designed to concentrate on these observable characteristics with the understanding that they fit into a larger framework that may be observed only from maps, air photos, or observation on an overflight.

SITE DESCRIPTIONS

Upper Deltaic Plain

Interdistributary basins are composed of various depositional environments, primarily swamps, marshes, lakes, and lake deltas. Basins narrow upstream, where they are heavily influenced by fresh water, and open toward the Gulf, where salt water intrudes into the marshes and lakes. Sediments are introduced from the seaward end by daily or twice-daily tidal fluctuations. Upstream, parts of the basin (upper deltaic plain) receive fine-grained sediments during floods. Prior to the building of artificial levees, this annual influx could be quite significant and was one important process of basin filling.

Highly organic deposits of Barataria Basin's freshwater swamps result from collection of logs, leaves, branches, and aquatic plant debris on the swamp floor. In many ways these swamps are similar depositional settings to their ancient counterparts in the coal fields of the eastern United States. Stop 1 is designed to point out the salient features of these important highly organic environments of the upper deltaic plain (Fig. 1).

Woody swamp vegetation is replaced toward the Gulf by vast expanses of freshwater marsh. Peats, organic oozes, and humus are important deposits of these relatively stagnant, chemically reducing environments. Organic marsh deposits cover broad areas and essentially blanket abandoned delta lobes. Regional lignitic correlation markers commonly associated with subsurface deltas are equivalent to these modern freshwater marsh deposits. Stop 2 highlights a special kind of freshwater marsh, a flotant. This marsh is floating on a column of water that varies with the flood cycle. Thick organic deposits that have wide lateral continuity can potentially develop in these settings.

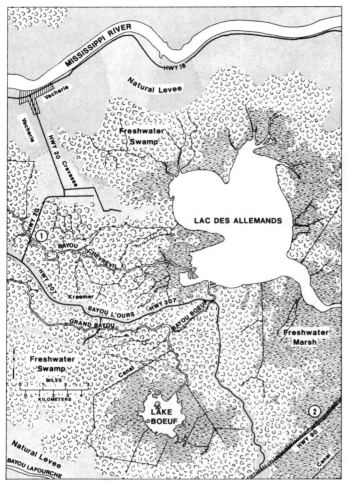

Figure 1. Map of the northern, freshwater part of the Barataria Basin showing the distribution of major environments (swamps and marshes) as well as the locations of two geological stops in the upper deltaic plain. Stop 1 is located near Bayou Chevreuil in the backswamp, and Stop 2 is located in the floating freshwater marsh found on the southern flank of Lac des Allemands.

Stops 1 and 2 are 22.5 mi (36.2 km) apart by paved highway and are easily accessible from either Baton Rouge, Thibodaux, or New Orleans.

Stops 1 and 2 are areas of very low relief (average elevations of approximately 2 ft [61 cm] above mean sea level) and are in the northern part of the basin, characterized by freshwater swamps and marshes. Sluggish bayous meander through these areas and usually empty into one of the basin's many lakes. These small streams are essentially stagnant during the low-water part of the yearly hydrologic cycle. The two sites are located along Louisiana 20 and 90 and have areas for parking vehicles. Features described in the following site descriptions are best seen in the late spring or early fall, when temperatures are most comfortable and the deciduous trees still have their leaves.

Stop 1 - Backswamp

From the intersection of Louisiana 18, which parallels the

Mississippi River, with Louisiana 20 at Vacherie, Louisiana, it is 7.8 mi (13 km) on Louisiana 20 to Stop 1. Louisiana 20 has been built down the axis of the Vacherie Crevasse, which originated from the Mississippi River and built into Lac des Allemands (Fig. 1). The change from cultivated fields to adjacent forested areas marks the transition from the coarse material (silts/sands) of the topographically higher crevasse deposits to the clay-rich backswamp. As Louisiana 20 turns to the southwest, one quickly leaves the crevasse surface and enters the backswamp, where Stop 1 is found at the intersection of Louisiana 20 with Bayou Chevreuil. North of the Bayou Chevreuil bridge a large shell-covered parking lot is located to the left.

The swamp site is approximately 250 yards (230 m) north of the parking lot. There is easy access to the swamp from the shoulder of the road. During most of the year, water covers the swamp floor, and at peak floods as much as 3 to 5 ft (1 to 2 m) of water can be present. In late spring and early summer, water levels are at their highest. However, except in the highest floods, this swamp locality is accessible and rarely has water depths of 2 ft (0.6 m). In fact, this site is a well-drained swamp, which means that the swamp floor is exposed to the atmosphere during the low-water part of the hydrologic cycle (September-November). In contrast, there are similar environments that remain wet throughout the year. Poorly drained swamps are where organic sediments are preserved from the degradational effects of exposure (oxidation) and are the best analogs to the woody coal-forming environments of the past. Unfortunately, poorly drained swamps are in remote sections of interdistributary basins and are not accessible by road.

High organic productivity, low rates of terrigenous sediment input (mostly clays and silty clays), and intense geochemical activity (early diagenesis) are characteristics of both well-drained and poorly drained swamps. However, in a well-drained situation, oxidizing and reducing conditions alternate during the sedimentary sequence. Reducing conditions are typical of poorly drained swamps. Note that at Site 1, as well as in poorly drained swamps, water-tolerant woody vegetation such as cypress (*Taxodium distichum*), tupelo gum (*Nyssa aquatica*), and swamp maple (*Acer rubrum*) are common, along with numerous other species of aquatic vegetation (Fig. 2). Dark, highly organic clays and peats accumulate in the poorly drained swamps, while steel-gray clays containing numerous root burrows and oxidizing diagenetic products such as carbonate and iron oxide nodules are the end products of well-drained settings. Pyrite, vivianite, and other minerals that form in reducing environments are most common in poorly drained swamp sediments (Coleman, 1966). Average accumulation rates as determined from C^{14} dates are about 1 ft/century (30 cm/century).

At Stop 1 a long-handled shovel is useful for digging into the swamp floor. A highly organic layer of recently deposited leaf litter, twigs, etc., is usually present above the oxidized light-gray clay. Note the homogeneous texture of the sediments, lack of stratification, and thorough burrowing, especially by roots. Plant material is eventually eliminated by oxidation, and faunal remains are leached during water table fluctuation, leaving the clay-rich terrigenous sediment as the final product of this depositional environment. Even when viewed on high-resolution X-ray radiographs, these sediments are generally devoid of faunal debris, but have early diagenetic products in the first couple of feet of deposition (<200 years).

A typical pattern of cyclic sedimentation in interdistributary basins such as Barataria Basin and Atchafalaya Basin, to the west, is for well-drained swamps to subside under low sediment-input conditions, causing water depths to increase until poorly drained conditions prevail. Eventually, water levels in the poorly drained swamp deepen to the point that woody vegetation is replaced by the open water of a lacustrine environment. Given sufficient time, a lacustrine delta will fill the lake (e.g., fed from a crevasse on the main river bordering the basin) and the cycle will start again, with the lake delta subsiding until a well-drained swamp forms stratigraphically above it. These basins are filled with sediments that represent variations on this theme. Stop 1 is an excellent site at which to observe the swamp-environment phase.

Stop 2 - Floating Fresh-Water Marsh (Flotant)

This Stop provides easy access to a vast expanse of floating freshwater marsh north of Highway 90 (Fig. 1). The site is located 3.5 mi (5.6 km) east along Highway 90 (toward New Orleans) after its intersection with Highway 20, which passes through the community of Kraemer, Louisiana. Adequate parking is available on the north shoulder of Highway 90 and across from the site at an EXXON station and bar. The site is actually adjacent to old Highway 90 and the railroad that parallels it (Fig. 1). From the access road (closed with a pipe gate) approximately 200 yards (185 m) east on old Highway 90 there is a small boardwalk from the road to the railroad track. On the north side of the track is the floating marsh, with trees of the back-swamp on the horizon (Fig. 3).

Figure 2. The swamp floor at Stop 1 covered with water (approximately 1-2 feet; 0.5 m deep). Note the numerous cypress knees and the complete covering of the water surface with duck weed (*Lemna minor*), a small floating aquatic plant. This picture was taken in the fall before the trees dropped leaves to the swamp floor.

Figure 3. The floating freshwater marsh (flotant) along Highway 90 south of Lac des Allemands in Barataria Basin. Maiden cane (*Panicum hemitomon*) and Roseau cane (*Phragmites communis*) are the two grasses typical of this environmental setting. Trees of the backswamp are visible in the background.

Freshwater marshes are important and widespread sedimentary environments of the Mississippi River's upper deltaic plain (Fisk, 1958; Frazier and Osanik, 1969). At Stop 2, near Lac des Allemands, there is an abrupt transition from swamp forest to the rather featureless grassy carpet of the fresh marsh (Fig. 1). This environment extends south to the landward limit of tidal influence, which is marked by brackish water marsh.

The freshwater marshland (Stop 2) is characterized by four observable features: (1) the dominant grass, maiden cane (*Panicum hemitomon*); (2) standing water of less than 5 percent salinity; (3) a thick and durable marsh-grass root mat under which are sediments ranging from peat to homogeneous steel-gray clay; and (4) a marsh that is actually floating (flotant).

Flotant is formed when dense mats of roots and other types of organic detritus are separated from the terrigenous sediment substrate beneath the marsh. These living rafts are up to 2 ft (60 cm) thick and certainly have proved strong enough to support a large party of field trip participants. Swiftly spreading flotant commonly invades open water bodies, and it can fill sizable lakes on time scales of a few tens of years.

As a result of the high primary productivity of marsh grasses, large quantities of organic matter are trapped in this environment. Extensive freshwater peats up to 12 ft (3.5 m) thick exist in Barataria Basin (Kosters and Bailey, 1983) and thus have the potential to be good stratigraphic markers (e.g., lignites in the Tertiary Wilcox Fm) in the subsurface.

It is advisable to take a long-handled shovel to Stop 2 for use in cutting through the floating root mat. Note the fine-grained terrigenous sediments beneath the mat and close proximity to swamp and lake environments.

Lower Deltaic Plain

Stops 3-5 are located in the modern Balize lobe of the

Mississippi River delta complex where fresh- and salt-water depositional environments interfinger (Fig. 4). The modern bird-foot delta has an area of approximately 300 mi^2 (185 km^2) and is less than 600 years old. It is the product of rapid deposition of primarily fine-grained sediments in a low-energy marine setting. These conditions allow constructive fluvial processes to dominate over the physical processes (waves and currents) of the receiving basin. As a result, the delta progrades basinward as a network of elongate distributaries. Within this general constructional framework, there is a cyclic repetition of initiation, growth, and abandonment of subdeltas that tends to fill the interdistributary bays and other shallow-water regions flanking the main river.

Even though they are of limited thickness, one of the major facies associated with many deltas, including the Mississippi, is the subdelta or crevasse splay. Deposits associated with these features have broad aerial extent and account for most of the marshland adjacent to major distributaries. These miniature deltas within the modern Balize lobe are initiated, built, abandoned, and subside to an open-water bay again in about 100 to 150 years (Coleman and Gagliano, 1964). Stop 3 is designed to point out the characteristics of this important sedimentary system in the lower delta (Fig. 4). The field trip participant will have the opportunity to investigate depositional components of two crevasse splays that are small enough to observe and comprehend at ground level. Although these kinds of systems are relatively thin (10 to 20 ft; 3 to 6 m), continuing subsidence and cyclic sedimentation result in the stacking of one subdelta on top of another.

Stops 4 and 5 concentrate on a rapidly prograding distributary and the diapiric mudlump islands that result from deformation of prodelta clays under the advancing subaerial delta (Fig. 5). These significant features are associated with the most dynamic depositional setting of the delta. As distributaries like South Pass prograde into a receiving basin, they deposit thick sand bodies that become essentially encased in a matrix of fine-grained sediments. Ancient counterparts to these sand bodies have been among the most productive reservoirs in the northern Gulf Coast oil and gas province.

In the modern delta it is easy to see the products of sediment deformation, as distributary-mouth-bar sands are deposited over thick accumulations of highly organic and watered prodelta clays. Stress applied by rapid sand deposition on low-shear-strength clays results in sediment instability and deformation. The end product of these conditions is an abnormally thickened distributary-mouth-bar sand and an underlying clay unit that is displaced vertically to form a feature known as a mudlump. These diapirs commonly become subaerial and form the highest topography in the lower delta. Shelf and prodelta clays are generally exposed in the mudlumps, thus indicating a vertical displacement of up to 400 feet (120 m) for these deposits. Because mudlump formation is associated with advancement of the distributary mouth bar, new mudlumps are constantly appearing, and change in these river-mouth features is the rule rather than

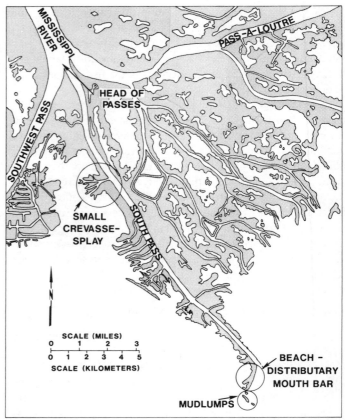

Figure 4. Location map of field trip stops along the South Pass distributary. Stop 3 is a small crevasse splay. Stop 4 is the South Pass beach-distributary mouth bar. Stop 5 focuses on the mudlump islands.

the exception. Mudlumps represent only one of many deformational features on the delta front that are initiated because of rapid deposition.

From New Orleans, Louisiana 23 runs directly to Venice, where there is a launch facility adequate for any trailerable boat. This boat launch can be found 0.25 mi (0.4 km) left at the end of Louisiana 23 where it meets with an access road to the docking areas. Local boats are available for hire in Venice.

Stop 3 - Natural Levee/Crevasse Splay

From Head of Passes (Fig. 4), it is approximately 3.8 mi (6.1 km) down South Pass to Stop 3. This depositional system is one of several small crevasse splays that have broken the narrow natural levee of South Pass and built into East Bay (Fig. 5). Like other crevasse splays of the delta, these have a well-developed channel network and coarse sediment lobes that are not vegetated. Much of the vegetation at Stop 3 is Roseau cane (*Phragmites communis*), a common fresh- to brackish-water marsh plant. The main distributary channel of this system is deep enough to accommodate a small motor boat. It is therefore possible to transit the entire crevasse splay to the shallow bar now forming at the termination of the distributary channel.

After entering the crevasse channel, a stop near the break to look at the natural levee deposits of South Pass would be instructive. Note that the natural levees of this major pass are very narrow. At this point on the river, water elevations change only slightly between flood and nonflood conditions. Up the river near New Orleans, however, the river surface may vary 16 to 18 ft (5 to 5.5 m) over the annual hydrologic cycle. Therefore, under natural conditions, natural levee deposits are thicker and more widespread upriver than near the Gulf. Natural levee deposits of fine sand and silt at Stop 3 are colonized by both marsh plants, on the permanently wet parts, and willows, on the levee crests. Sediments show the effects of oxidation on the highest parts of the levee, and all levee deposits display a variety of sedimentary structures, including extensive evidence of plant/animal burrowing.

Proceeding down the main crevasse channel, note the steep banks of the channel, which are stabilized by extensive root systems of Roseau cane. Toward the seaward end of the distributary, the Roseau cane gives way to smaller aquatic plants, which have recently colonized newly emergent parts of the prograding system. Digging through the sediments of these areas will reveal cycles of sedimentation related to the annual floods. Most of the sediment on these slightly subaerial lobes is in the silt range. Looking toward East Bay down the axis of the distributary, one can see logs and other debris caught on the shallow bar around which the channel is trying to split. This stop offers a variety of sedimentary settings, but if the environments of the South Pass distributary mouth are to be observed, Stop 3 should be alotted no more than one hour.

Stop 4 - Beach/Distributary-Mouth Bar

At the mouth of South Pass there is an abandoned Coast Guard station that has a small-boat dock, but the dock is frequently damaged by storms and it may be necessary to beach the

Figure 5. Low, oblique aerial view of one of several small crevasse splays that occur along the western margin of South Pass. Depositional units of this system are emphasized by the high-standing marsh vegetation, Roseau cane (*Phragmites communis*).

Figure 6. Aerial view of South Pass beach and associated mudlumps. Note the alignment of a previous beach that was active in the 1960s and early 1970s. The black material on the front of the beach represents an accumulation of woody fragments known as "coffee grounds." These organics are deposited with the bar sands after each flood. The beach has accreted to three mudlumps. (Photograph taken September 1981).

boat just upriver from it. Stop 4 is approximately 11 river miles (18 km) from Head of Passes (Fig. 4) and is easily accessible from the river.

The focus of this locality is the beach, which represents a subaerial expression of the distributary mouth bar. Although studies of the subaqueous bar have indicated a seaward progradation of more than 200 ft per year (60 m/yr) the mouth of South Pass has been stabilized by Corps of Engineers jetties. Consequently the beach is building seaward, with one end fixed at South Pass (Fig. 6).

The distributary mouth bar (DMB) is the coarsest sediment body (fine sand-silt) being formed in the lower delta. South Pass beach is simply the wave-reworked top of this much larger feature. However, the beach contains the primary low-angle foreshore bedding and other sedimentary structures as well as bioturbation features common to many other beaches. If the beach is observed after the spring flood and before the winter storms, it is likely that thick deposits of broken woody organics will be found deposited with the beach sand. These organics are locally known as "coffee grounds" (Fig. 6). During the late fall and winter they are covered by beach sand or reworked to the back of the beach. Digging on the bay side usually uncovers thick organic units beneath clean fine sands and silts. The organic units are modern analogs to thin "transported" coals that can be observed in the Pennsylvanian rocks of the eastern U.S. coal measures.

Under nondeformed conditions the DMB is 60 to 80 ft (18 to 25 m) thick. However, when weighting by bar sands displaces underlying prodelta and shelf clays, the sand body can abnormally thicken to as much as 400 ft (125 m). This is the case at South Pass, and as a consequence, mud diapirs are squeezed up through and in front of the prograding bar sands. These mudlumps are near the western end of South Pass beach, and during some years the beach actually incorporates a few of these features (Fig. 6). However, beach and mudlump morphology is in such a dynamic state of change that it is impossible to predict relationships between the two for any given season.

Stop 5 - Mudlump

Several mudlump islands occasionally become linked to the South Pass beach at its western end (Fig. 6). If this is the case when the locality is visited, dock the boat in the river (see Stop 4 for details) and walk the beach to observe the mudlumps. However, if the beach has been eroded at its western end, as is common in winter, it may be necessary to visit the mudlumps by boat. Unfortunately, sea state can make this trip difficult in a small boat. Favorable weather and sea conditions must exist before this trip is attempted.

Morgan and others (1963) explained the process of loading of the distributary mouth bar primarily during floods and the subsequent development of mud diapirs. They based their conclusions on borings taken at South Pass. References to these unusual features at the mouths of Mississippi River distributaries date back to the 1700s. With the building of the South Pass jetties by Captain James B. Eads in the 1870s, detailed studies of the river mouth area provided new data about the dynamic nature of these features. Reasons for their appearance as well as disappearance remained problematic until the Morgan and others (1963) studies.

Mudlumps still pose many interesting questions. Sediments that comprise these diapirs, for example, do not display structures that indicate flowage. Original bedding is usually preserved, even though the sediments are mostly clay and have been vertically displaced several hundred feet. Inspection of the youngest mudlumps indicates sediments of the upper delta front, which contain thin, bedded silts, while the oldest mudlumps expose rather homogeneous clays of a deeper depositional setting. Field observation will reveal that fracturing and faulting are common on all mudlump exposures.

Degassing and dewatering of the sediments are part of the mudlump formation process. Walking over the mudlumps will allow one to discover active mud vents, which look like miniature volcanoes belching clay-rich water and methane gas. These features are most active during and directly after flood season when, because of sediment loading, mudlump growth is at its peak.

When visiting the mudlumps, a shovel and spatula are helpful for cleaning surfaces to be inspected for sedimentary structures and other details of the sedimentary record. Because some horizons exposed by the diapiric process contain abundant microfossils, sampling equipment may be helpful. After finishing at the mudlumps, allow approximately one and a half hours to make the trip back upriver to Venice.

REFERENCES CITED

Chabreck, R., and Linscombe, G., 1978, Vegetative type map of the Louisiana coastal marshes: Louisiana Dept. of Wildlife and Fisheries, New Orleans, La.

Coleman, J. M., 1966, Ecological changes in a massive freshwater clay sequence: Transactions Gulf Coast Association Geological Societies, v. 16, p. 159–174.

Coleman, J. M., and Gagliano, S. M., 1964, Cyclic sedimentation in the Mississippi River deltaic plain: Gulf Coast Association of Geological Societies, v. 14, p. 67–80.

Fisk, H. N., 1958, Recent Mississippi River sedimentation and peat accumulation, in (van Aelst, E., ed.) Congres pour l'avancement d'etudes de stratigraphic et de geologie du carfonifiere, 4th, Heerlen: Maastrict, Netherlands, Compte Rendu, v. 1, p. 187–199.

Franzier, D. E. and Osanik, A., 1969, Recent peat deposits—Louisiana coastal plain. Geological Society of America Special Paper 114, p. 63–85.

Kolb, C. R., and Van Lopik, J. R., 1958, Geology of the Mississippi River deltaic plain, southeastern Louisiana: U.S. Army Corps of Engineers, Waterways Experiment Station Technical Reports 3-483 and 3-484, Vicksburg, Ms.

Kosters, E. C., and Bailey, A., 1983, Characteristics of peat deposits in the Mississippi River delta plain: Transition Gulf Coast Association of Geological Societies, v. 33, p. 311–325.

McIntire, W. G., 1958, Prehistoric Indian settlements of the changing Mississippi River delta: Louisiana State University, Coastal Studies Institute Series 1, 128 p.

Morgan, J. P., Coleman, J. M., and Gagliano, S. M., 1963, Mudlumps at the mouth of South Pass, Mississippi River: sedimentology, paleontology, structure, origin, and relation to deltaic processes: Louisiana State University Press, Baton Rouge, 190 p.

Saucier, R. T., 1974, Quaternary geology of the lower Mississippi Valley (map): U.S. Army Engineer Waterways Experiment Station, Corps of Engineers, Vicksburg, Ms.

Overflights of the lower deltaic plain of the Mississippi River delta complex

Harry H. Roberts, *Coastal Studies Institute, Louisiana State University, Baton Rouge, Louisiana 70803*

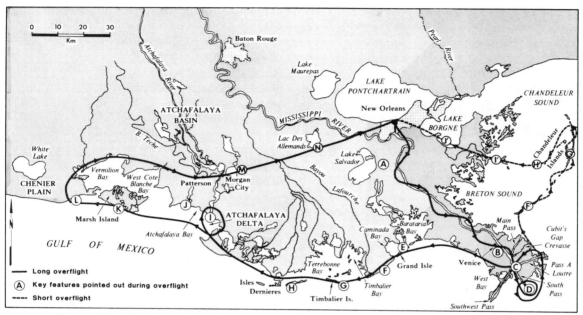

Figure 1. Map of south Louisiana showing the overflight routes along the coasts and over the environments of the lower deltaic plain.

LOCATION

Two overflight routes of the Mississippi River Delta complex augment or substitute for actual observation of depositional settings in the field (see Roberts, this Guidebook) (Fig. 1). Both overflights start and end at Lakefront Airport in New Orleans and incorporate depositional environments of the modern river, especially the Balize delta lobe. Arrangements can be made at Lakefront Airport to rent a light aircraft for the purpose of an overflight.

INTRODUCTION

Because of the immense scale of many features that compose the Mississippi River delta complex and relative inaccessibility of some depositional features, especially in the lower deltaic plain, an overflight can be very enlightening. The long route, Overflight 1, goes to the west after leaving the modern delta and incorporates coastal environments as well as the new Atchafalaya delta. This route takes about 3 to 4 hours in a Cessna 172 or equivalent aircraft that will accommodate a pilot and three passengers. The short route, Overflight 2, goes to the east and passes over the Chandeleur island arc, back-barrier shell islands, and marshlands of St. Bernard Parish. This flight pattern can be completed in about 2 to 3 hours.

OVERFLIGHT 1—RIVER FEATURES AND FLANKING ENVIRONMENTS

Between New Orleans and the town of Venice, in the modern Balize delta lobe, there are numerous features of the fluvial landscape that are best observed from the air. As the plane leaves Lakefront Airport, Lake Pontchartrain and another shallow marginal basin, Lake Borgne, are visible to the north and east of the river. Low sediment input and high biologic productivity are typical of these systems. Abundant brackish water clam shells (*Rangia cuneata*) are dredged from these lakes and used as roadbed material.

River Flank Environments (Site A)

As the flight follows the river toward the Gulf, the first major area of interest, Site A (Fig. 1), is just south of the river's last large meander, where its course straightens. The straight course appears to be the result of cutting into resistant marine clays roughly from New Orleans south, which minimizes lateral migration. At this point along the river's course (Site A) natural levees are over a mile wide and represent the highest topographic surface for housing sites, industry, and agriculture. Levees decrease in width southward in response to a decrease in the amplitude of the river's water level from low water to flood conditions.

Figure 2. A low-altitude aerial view of the Mississippi River looking south from Venice (upper right). Note the crevasse channels that break away from the river. These channels were the delivery systems for large bay-fill sequences that developed adjacent to the river course. The channel to the left was initiated in 1874, resulting in the Baptiste Collette subdelta. At the town of Venice a break occurred in 1838 that developed into the West Bay subdelta.

Figure 3. Oblique aerial view of the beach at South Pass (in background) and associated mudlumps (attached to end of beach).

Artificial levees now protect this important strip of natural levee land on both the bay and the river sides.

Looking to the west at Site A one views the vast fresh to brackish marshlands and lakes of a large interdistributary basin, Barataria Basin, which formed between the levees of the Lafourche river course (2,000-500 years B.P.) and similar deposits of the modern Mississippi River. Woody swamp vegetation in the northern parts of this basin quickly give way to fresh, brackish, and saline marshes as one progresses toward the Gulf. The lake margins to the west of Site A are characterized by marshes that are actually floating (flotants). Peats to 10 ft (3 m) thick are forming in parts of the Barataria Basin visible from this vantage point (Kosters and Bailey, 1983). Generally, peat quality decreases toward the Gulf as tidal influence pumps undesirable particulates and chemical constituents into the marsh sediments.

To the east of the river at Site A extensive marshlands cap the surface of a once extensive delta lobe, the St. Bernard delta (3,500-1,000 years B.P.), which built to the southeast. Abandoned distributary networks and their slightly elevated natural levees are evident on the marsh surface. The slight changes in relief associated with these features are accentuated by variations in the plant communities. It is not uncommon to see old distributary courses delineated by tree lines (usually willows). Southward the marshes of St. Bernard Parish give way to the open water of Chandeleur Sound.

Subdeltas (Site B)

Artificial levees have been constructed along the river to the town of Venice, Site B (Fig. 1). At the end of the artificial levee at Venice, a large channel has developed through the Mississippi's

natural levee and has built a distributary network to the southwest into West Bay. Across the river a similar channel has built a distributary system that has produced a subdelta to the east of the river. Figure 2 shows these two channels that have diverted Mississippi River water and sediment to shallow flanking bays since the 1800s. Looking downriver from Venice, the subdelta on the left or east is Baptiste Collette, and the one on the right is West Bay. Both subdeltas are in a late stage of deterioration. Note that in the West Bay subdelta only willow-lined levees of the once active distributaries are exposed. At its peak, marshlands of the West Bay subdelta probably occupied an area of nearly 100 mi^2. (160 km^2). Open water between distributaries of these two subdeltas indicates that subsidence and land loss have occurred.

These wedges of relatively coarse sediment on which the marsh is established range from a few feet to about 30 ft (9 m) in thickness. They have typical delta shapes, but have a much shorter "lifespan" than a major delta lobe. Formation and abandonment of major delta lobes occur on a time scale of approximately 1,000 years, while subdeltas are initiated, built, abandoned, and subside to an open-water bay again in slightly over 100 years. Within the time span of a major delta lobe like the bird-foot delta, several bay-fill sequences may be deposited in the same general location, creating a stacked arrangement in the subsurface as a product of cyclic depositional events (Coleman and Gagliano, 1964).

Major Distributaries (Site C)

At Site C (Figure 1), near "Head of Passes," south of Venice, the Mississippi River splits into three major distributaries: Pass a Loutre (east), South Pass, and Southwest Pass. Until recently both Southwest Pass and South Pass were regularly dredged to accommodate shipping, primarily to the ports of New Orleans and Baton Rouge. Now only Southwest Pass is maintained for ship traffic. Jetties have been built at the juncture of South and

Figure 4. View looking east of Grand Isle. Grand Terre is separated from Grand Isle by Barataria Pass, a tidal pass nearly 100 ft (30 m) deep.

Southwest passes to prevent shoaling and eventual closure of these passes. These structures are also present at the distributary mouths to minimize shoaling as the river builds a distributary mouth bar.

Looking down South Pass from Site C, numerous small subdeltas (crevasse splays) have resulted from small channels that have developed through the South Pass levees by natural processes. Large subdeltas like the Garden Island Bay complex, between South Pass and Pass a Loutre, contain many smaller crevasse splay features. As can be seen from Site C, as well as along the flightline to the end of South Pass, some particularly well-formed but small crevasse splays occur along the western margin of this distributary. The subaerial parts of these small features are dramatically highlighted by the high-standing (8 to 10 ft) (2.4 to 3.0 m) marsh plant *Phragmites communis.* Its thick root mats tend to stabilize the newly deposited sediment.

Distributary Mouth Features (Site D)

At Site D (Fig. 1) several distinct features can be observed near the mouth of South Pass, including a turbid river water plume, a beach, and numerous mudlump islands and shoals (Fig. 3). During the spring and early summer, when the river is in flood, tremendous volumes of sediment are transported, primarily as suspended load, to the delta-front environments. A sediment-laden plume can usually be seen extending from the South Pass distributary. The plume is usually deflected to the west by dominant east to west currents in the northern Gulf. Plume boundaries are generally very sharp, with turbid brackish-fresh water sometimes adjacent to the blue salt water of the Gulf.

South Pass distributary beach extends to the west as a product of the dominant drift and direction of wave attack. Dark material can be observed on the well-sorted, quartz-rich beach and behind it. This material represents transported woody organic fragments ("coffee grounds"). The beach itself represents the wave-reworked top of the distributary mouth bar, a large, linear sand body that has prograded southward from Head of Passes. Because of rapid subsidence in the lower delta, only the most recent beaches remain as subaerial units. Deposition of sand at the river mouth during floods creates sediment loading of underlying prodelta clays (Morgan and others, 1963). These rapidly deposited and highly incompetent fine-grained sediments, over which the coarser facies have prograded, are diapirically intruded into the distributary mouth bar sands, resulting in river-mouth islands (mudlumps). New mudlumps continue to appear as the progradation process proceeds. Sediments deposited in water as much as 300 to 400 ft (90 to 120 m) deep are now exposed at the surface in the oldest South Pass mudlumps. The youngest mudlumps expose alternating clays, silts, and thin sands of the distal bar. Mudlump formation is but one type of sediment deformation and mass movement common to the delta front environments. However, mudlumps are the only deformation features that are subaerial. As one flies over a mudlump, it is apparent that stratigraphy of the deposits remains intact (little soft sediment deformation), although faulting and tilting are common (Fig. 3).

Western Transgressive Coasts (Sites E, F, G, and H)

If the long overflight route is taken (Fig. 1), areas west of the modern delta will be observed. The modern Balize delta represents an active regressive phase of the Mississippi River delta complex. West of this lobe are the remnants of former delta lobes in various stages of deterioration. General coastline complexity and barrier morphology reflect the relative ages of coastal sectors. Holocene barrier islands, beaches, and spits associated with the western coasts represent the reworked remains of sediments once rapidly deposited in delta-building episodes. Rapid deposition and seaward progradation, now occurring in the modern Balize lobe, are followed by subsidence and marine reworking once the delivery system is abandoned for a more favorable site.

Site E (Fig. 1) gives a view of Grand Isle and Grand Terre (Fig. 4), which are barrier islands built by sediments eroded from the late Lafourche delta lobe (~1,000 years old). The thin veneer of sand represented by these islands makes them extremely vulnerable to storm damage and other forms of natural coastal change. Note the numerous manmade structures (groins, etc.) on Grand Isle designed to trap and stabilize beach and back-barrier sediment.

Since the Lafourche delta lobe preceded the modern Plaquemines and Balize lobes, it is in the first stages of deterioration. Site F (Fig. 1) gives a view of the Caminada-Moreau coastal area, which represents a rapidly eroding but emergent part of the Lafourche delta that functions as a sediment source for flanking barrier islands such as Grand Isle and Grand Terre. Boyd and Penland (1981) describe this site as Louisiana's major erosional headland where sediments are dispersed to both the east and west flanking coasts. Note that marshlands are being rapidly eroded in this area, sometimes leaving remnants of root networks in the surf zone.

Westward at Site G (Fig. 1), the Timbalier Islands represent flanking barriers on the west side of the Caminada-Moreau headland. Frequent hurricane impacts along the Louisiana coast have resulted in distinct washover features and rapid coastal retreat at this site. Note the accretionary ridges in East Timbalier Island, which indicate rapid westward spit growth. Along this flanking barrier island coast, structures associated with man's effort to stabilize these ever-changing features are clearly visible.

Site H (Fig. 1), west of the Timbalier Islands, is another group of barriers, Isles Dernieres, that have developed from erosion of an early lobe of the Lafourche delta. Note that these barriers are completely detached from the coastal marshlands. In this regard they are similar to the Chandeleur barrier island arc to the east of the modern delta. Penland and others (1985) show that barrier detachment has been a rapid process over the last 100 years. Storm washover features at Site H indicate the importance of this process to landward migration of these barriers. Trees visible on the back sides of the central barriers are largely mangroves. Their roots can be seen in the surf zone, indicating shoreline retreat. Both flood- and ebb-tide deltas are visible at the tidal passes.

Atchafalaya Delta (Site I)

After a westward flight from the Lafourche barriers along the coastal marshlands of central Louisiana, a large, shallow water body, Atchafalaya Bay, is encountered. At Site I (Fig. 1) it is apparent that rapid coastal progradation is underway. The numerous shoals and vegetated islands in this area are associated with the newly formed Atchafalaya delta (van Heerden and Roberts, 1980).

Although the Atchafalaya River has been diverting water and sediment from the Mississippi for several hundred years, it was not until the early 1950s that sediments (primarily clays) in significant quantities reached the coast (Fisk, 1952). Prior to this time the lakes and swamps of the Atchafalaya Basin functioned as natural sediment traps. By the late 1940s and early 1950s, the basin had filled to a point where sediments passed through to the coast. At this time the subaqueous phase of delta building was initiated, and an area characterized for centuries by land loss and coastal retreat started a progradational episode. Subaqueous growth continued until 1973, when the abnormally high flood of that year deposited near the river mouth a number of small shoals that were exposed at low tide. Appearance of these sand-rich shoals marked the beginning of the subaerial phase of delta growth (Roberts and others, 1980). Although flow down the Atchafalaya is controlled to 30 percent of that of the Mississippi, between 1973 and 1977 nearly 10 mi^2 (16.1 km^2) of new land was added to Atchafalaya Bay as a product of delta building from the Lower Atchafalaya River Outlet (Fig. 5).

Once shoals build to near the low-tide level, aquatic vegetation colonizes and helps to stabilize the sediment. Observations at Site I (Fig. 1) show that emergent parts provide substrates for marsh plants and willows, all of which trap sediment and protect

Figure 5. Photomosaic of the new Atchafalaya River delta taken after the 1976 flood, just three years after initiation of subaerial delta growth following the 1973 flood.

Figure 6. The Chandeleur Island arc looking toward the south. Note the large washover lobes and channels that have developed from the impact of storm overwash processes. *Spartina patens* and salt-tolerant mangrove trees have colonized the back side of the barrier.

against erosion. Note the same island morphologies, which indicate that small lobes fuse to form large emergent components of the delta (van Heerden and others, 1981). A similar small delta is also forming at the Wax Lake Outlet, Site J (Fig. 1), an artificial channel cut in 1942 from Six Mile Lake, north of Atchafalaya Bay.

Downdrift Coasts (Sites K and L)

As a product of sediment input to the coast by the Atchafalaya River, downdrift areas are also undergoing rapid changes. Site K (Fig. 1) is along the seaward coast of Marsh Island, where once-viable oyster reefs are now being heavily impacted by Atchafalaya River sediment. Significant increases in suspended sediment concentrations opposite and downdrift (west) of Atchafalaya Bay have literally smothered these oyster beds since the 1950s. Note the high angle of orientation of these reefs to the Marsh Island coast. It is thought that the reefs are seated on levees of subsurface distributaries associated with an early Holocene delta lobe (Coleman, 1966).

Site L (Fig. 1) is along the eastern margin of the Chenier Plain near Chenier Au Tigre where downdrift effects of Atchafalaya sedimentation have caused progradation of mudflats (Morgan and Larimore, 1957). Note the newly emergent mudflats in front of the latest chenier ridge. Marsh plants rapidly colonize these mudflats and help protect them from erosion. Continued growth of the Atchafalaya River delta should eventually be accompanied by reactivation of coastal progradation along the seaward margin of the entire Chenier Plain.

Swamps of the Atchafalaya Basin (Site M)

Returning to New Orleans from the Chenier Plain, the flight path crosses the southern part of the Atchafalaya Basin. Site M (Fig. 1) is north of Morgan City and the levees of Bayou Teche. Note the dense tree cover (dominantly cypress, tupelo gum, and swamp maple) and standing water. These highly organic environments can produce thick woody peats where influx of terrigenous sediment is minimal.

Numerous small lakes are also present in the southern basin. Swamps, lakes, and lake deltas are the most important depositional environments of the Atchafalaya interdistributary basin.

OVERFLIGHT 2—TRANSGRESSIVE BARRIERS

Breton and Chandeleur Barrier Islands (Sites F' and G')

Figure 1 shows an overflight route to the east of the modern delta that is somewhat shorter than the proposed western route. Both overflights start in New Orleans and proceed downriver to the bird-foot delta. After observing salient features of this actively building delta, Sites A-D, a course toward Breton Sound - Chandeleur Sound should be taken if the eastern overflight option is selected. Separating these shallow water bodies from the open

Figure 7. Highly irregular shell islands have developed behind the Chandeleur Island arc where productivity of calcareous shell-producing organisms is high.

Gulf is a chain of barrier islands, the Breton and Chandeleur groups, Sites F' and G' (Fig. 1). They represent the reworked distal margin of the St. Bernard delta lobe, which was constructed roughly 4,000 to 1,500 years B.P.

Site F' gives a view of Breton Island to the west and Grand Gosier Island to the east. These low topographic barrier remnants have been eroded and drastically modified by recent hurricanes. Spitlike accretion units are visible in emergent parts of Breton Island, and in both barriers, washover lobes are clearly visible. Eroded marsh remnants outcrop in the surf zone, indicating rapid retreat.

Site G' is in the approximate center of the Chandeleur barrier island arc (Fig. 6). Salt-tolerant grass (*Spartina patens*) and stunted black mangrove trees form much of the vegetative cover of the barrier islands. Their ability to trap sediments has helped establish a line of low eolian dunes toward the northern end of the Chandeleur arc. Dunes help direct overwash to interdune channels, thus reducing the overall retreat rate.

Quartz-rich sands of the Chandeleur barrier islands vary from thin accumulations at the southeast end (less than 12 ft [3.6 m] thick) to thicknesses in excess of 30 ft (9 m) at the northern extreme. The longshore drift of sediment set up by the angle of dominant wave approach favors a south to north sediment transport system. Subsidence and storm overwash drive the landward migration of these islands. Numerous cuts through the arc and large back-barrier overwash lobes are evidence of the importance of storm-related overwash processes (Fig. 6). As these islands transgress toward the St. Bernard marshes, a thin sheet of shell-rich sand is being left on the continental shelf.

Shell Islands (Site H')

Because of the updrift location of Chandeleur Sound from the active Mississippi River distributaries, sediment input is minimal and therefore carbonate shell debris becomes a significant part of the sediment. Oysters and other bivalves are abundant throughout the Chandeleur-Breton Sound region. Locally, shell debris is reworked into coquina islands (Fig. 7). Site H', Free Mason Islands, landward of the Chandeleur island arc, demonstrates the complicated surface morphology of these low-relief islands. Complicated spitlike accretion ridges are the dominant morphological features. Spit growth is driven by both violent summer storms (mostly from the south) and high wave action from the north accompanying winter cold-front passages.

Deteriorating St. Bernard Marshes (Site I')

The route from the Free Mason Islands to New Orleans crosses open water of Chandeleur Sound before encountering the St. Bernard marshes. The distal ends of these marshlands at Site I' are highly fragmented and in a late stage of deterioration. This region was formerly occupied by the St. Bernard delta lobe, and the present marshes represent a modern organic covering on that rapidly subsiding surface. Linear trends in the marsh represent the levees of once-active distributaries that are now in the subsurface.

Lake Borgne (Site J')

A large marginal delta lake, Lake Borgne, is intersected (Site J') on the flightline before reaching New Orleans. Like Lakes Pontchartrain and Maurepas to the north, this shallow water body is on the updrift side of the delta and receives only minimal terrigenous sedimentation. Being surrounded by marshlands, which are a source of the organic detritus on which many marine organisms feed, this lake is a highly productive fishery, especially for shrimp. As in the case of Chandeleur Sound, bottom sediments of Lake Borgne are rich in shell debris.

REFERENCES CITED

Boyd, R., and Penland, S., 1981, Washover of deltaic barriers on the Louisiana coast: Transactions, Gulf Coast Association of Geological Societies, v. 31, p. 243–248.

Coleman, J. M., 1966, Ecological changes in a massive freshwater clay sequence: Transactions, Gulf Coast Association Geological Societies, v. 16, p. 159–174.

Coleman, J. M., and Gagliano, S. M., 1964, Cyclic sedimentation in the Mississippi River deltaic plain: Transactions, Gulf Coast Association Geological Societies, v. 14, p. 67–80.

Fisk, H. N., 1952, Geological investigation of the Atchafalaya Basin and problems of Mississippi River diversion: U.S. Army Corps of Engineers, Mississippi River Comm., Vicksburg, Miss., p. 1–145.

Kosters, E. C., and Bailey, A., 1983, Characteristics of peat deposits in the Mississippi River delta plain: Transactions, Gulf Coast Association Geological Societies, v. 33, p. 311–325.

Morgan, J. P., and Larimore, P. B., 1957, Changes in the Louisiana shoreline: Transactions, Gulf Coast Association Geological Societies, v. 7, p. 303–310.

Morgan, J. P., Coleman, J. M., and Gagliano, S. M., 1963, Mudlumps at the mouth of South Pass, Mississippi River: sedimentology, paleontology, structure, origin, and relation to deltaic processes: Coastal Studies Inst., Louisiana State Univ., Baton Rouge, Coastal Studies Series No. 10, 190 p.

Penland, S., Suter, J.R., and Boyd, R., 1985, Barrier island arcs along abandoned Mississippi River deltas: Marine Geology, v. 63, p. 197–233.

Roberts, H. H., Adams, R.D., and Cunningham, R.H.W., 1980, Evolution of sand-dominant subaerial phase, Atchafalaya delta: American Association Petroleum Geologists Bulletin, v. 64, p. 264–279.

van Heerden, I. L., and Roberts, H. H., 1980, The Atchafalaya Delta - Louisiana's new prograding coast: Transactions, Gulf Coast Association Geological Societies, v. 30, p. 497–506.

van Heerden, I. L., Roberts, H. H., and Wells, J. T., 1981, Evolution and morphology of sedimentary environments, Atchafalaya Delta, Louisiana: Transactions, Gulf Coast Association Geological Societies, v. 31, p. 399–408.

The Bayou Lafourche delta, Mississippi River delta plain, Louisiana

Shea Penland, *Louisiana Geological Survey, Baton Rouge, Louisiana 70893*
William Ritchie, *University of Aberdeen, Aberdeen, AB9 2UF Scotland*
Ron Boyd, *Dalhousie University, Halifax, Nova Scotia B3H 3J5 Canada*
Robert G. Gerdes, *Louisiana State University, Baton Rouge, Louisiana 70893*
John R. Suter, *Louisiana Geological Survey, Baton Rouge, Louisiana 70893*

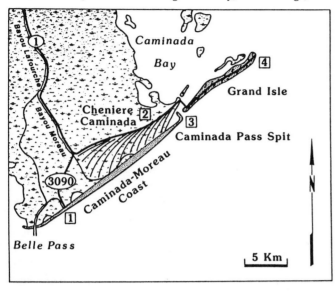

LEGEND

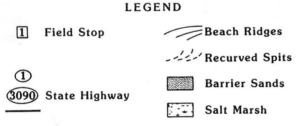

① Field Stop	⌒⌒ Beach Ridges
	⌒⌒ Recurved Spits
①	▦ Barrier Sands
③⓪⑨⓪ State Highway	▦ Salt Marsh

Figure 1. Map showing the field stop locations on the Bayou Lafourche delta cluster stop in coastal Louisiana.

LOCATION

In southeast Louisiana, State Highway 1 provides the only direct access to the shoreline of the delta plain along the natural levee of Bayou Lafourche through Cheniere Caminada to Grand Isle (Fig. 1). Abandoned approximately 300 years ago, the Bayou Lafourche delta represents the last major distributary of the larger Lafourche delta complex to be built. The Bayou Lafourche landscape contains some of the best examples of regressive and transgressive delta cycle landforms to be found within the entire Mississippi River delta plain. This cluster stop is designed to examine these features, particularly the regressive distributaries and beach ridges of Bayou Lafourche as well as the transgressive Caminada-Moreau coast, Caminada Pass spit, and Grand Isle flanking barrier island.

INTRODUCTION

Built by the largest river on the North American continent,

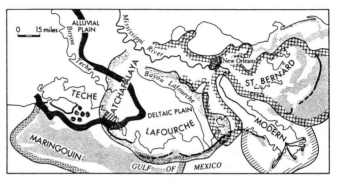

DELTA COMPLEX	AGE (YEARS BP)
Maringouin	7250-6200
Teche	5700-3900
St. Bernard	4600-1800
Lafourche	3500-400
Modern	Active
Atchafalaya	Active

Figure 2. The Mississippi River has built its delta plain by the classic delta switching process (Frazier, 1967). The Holocene delta plain consists of six delta complexes: two are currently active, the Atchafalaya and Modern, and four are abandoned, the Maringouin, Teche, St. Bernard, and Lafourche, and in various stages of transgressive barrier development.

the Mississippi delta plain is the depositional product of the classic delta switching process (Scruton, 1960), a fundamental sedimentological model taught to all earth scientists. This dynamic process consists of a regressive phase of deltaic sedimentation followed by a transgressive phase of coastal barrier sedimentation, collectively termed the delta cycle (Scruton, 1960). The Mississippi River delta plain is comprised of six major Holocene delta complexes which have been sequentially built at approximately 1000-year intervals (Fisk, 1944; Kolb and Van Lopik, 1958; Frazier, 1967). The Modern and Atchafalaya delta complexes are currently active in the regressive phase with major distributaries building south of Venice and Morgan City, respectively. The remaining four delta complexes are abandoned and are in various stages of transgressive barrier development (Frazier, 1967; Penland and others, 1981; Penland and Boyd, 1985).

Deposition of the Lafourche delta complex began around 3,300 years ago and actively continued up to 300 years ago (Frazier, 1967). The Lafourche delta complex completed the development of the Holocene subaerial delta plain by infilling the interdistributary area between the transgressive Teche and St. Bernard delta complexes. The deposition of the Lafourche delta complex was partly contemporaneous with the St. Bernard complex 1,000-3,000 years ago (Fig. 2). Other delta complexes also overlapped in time, indicating that the Mississippi flow was

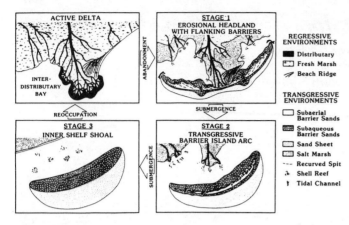

Figure 3. The genesis and evolution of transgressive Mississippi deltas is best summarized in a three stage model from Penland and Boyd (1981). This model begins with Stage 1 - *Erosional Headland and Flanking Barriers*. Next is Stage 2 - *Transgressive Barrier Island Arcs*. The sequence ends with Stage 3 -*Inner Shelf Shoals*.

shared between several concurrent delta lobes and even delta complexes, rather than a single distributary channel. Transgression of the Lafourche delta complex has produced the Isles Dernieres barrier island arc and the Bayou Lafourche barrier shoreline.

The genesis and evolution of transgressive Mississippi deltas can be summarized in a three stage model beginning with Stage 1, *Erosional Headland and Flanking Barriers* (Fig. 3). Here regressive deltaic sand deposits are reworked by the retreating shoreface and dispersed laterally by longshore transport into contiguous flanking barriers enclosing restricted interdistributary bays. Next comes Stage 2, *Transgressive Barrier Island Arcs,* where submergence of the erosional headland generates an intradeltaic

lagoon separating the barrier island arc from the retreating mainland. The sequence ends with Stage 3, *Inner Shelf Shoals.* The retreating barrier island arc is unable to keep pace with relative sea level rise or the retreating mainland and subsides below sea level. Following submergence, the barrier island arc continues to be reworked as a sandy shoal on the inner continental shelf. The Bayou Lafourche barrier shoreline is an example of a Stage 1 transgressive barrier.

BAYOU LAFOURCHE BARRIER SHORELINE

The Bayou Lafourche barrier shoreline consists of an erosional headland fronted by the Caminada-Moreau coast and two nearly symmetrical sets of flanking barriers: Caminada Pass Spit and Grand Isle to the east and East Timbalier Island and Timbalier Island to the west (Fig. 4). The Caminada-Moreau coast is composed of a low barrier beach in the form of a thin continuous washover flat approximately 3 ft (1 m) above mean sea level. Salt marsh has replaced fresh water marsh and is accumulating landward of the beach and can be seen outcropping in the surf zone. The eastern half of the erosional headland consists of a beach ridge plain composed of sediments probably derived from the older Bayou Blue and Bayou Robinson erosional headlands further east (Ritchie 1972; Gerdes 1985). Shoreface retreat is actively reworking the distributary sand bodies of Bayou Lafourche and Bayou Moreau and the beach ridge plain of Cheniere Caminada. The dominant wave approach direction to the Caminada-Moreau coast is from the southeast. This, together with the convex shoreline, produces a longshore sediment transport divergence from the central erosional headland. Sand eroded from the Caminada-Moreau coast shoreface accumulates in flanking barriers both to the east at Caminada

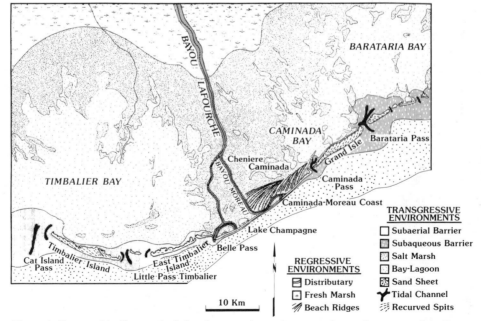

Figure 4. Geomorphic diagram depicting the regressive and transgressive environments of the Bayou Lafourche delta.

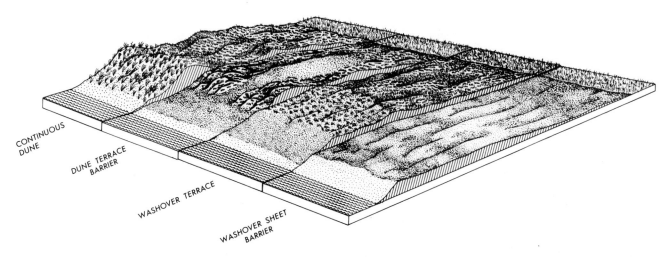

Figure 5. Coastal landforms along the Bayou Lafourche barrier shoreline range from washover flats to higher terraces to continuous dunes where the sediment supply is sufficient.

Spit and Grand Isle, and to the west at East Timbalier Island and Timbalier Island. Away from the central erosional headland, increasing downdrift sediment abundance leads to the development of small washover fans and low, hummocky dune fields which eventually coalesce further downdrift to form a higher, more continuous washover terrace and a foredune ridge (Fig. 5). Downdrift flanking barrier islands migrate laterally, in the direction of longshore sediment transport, by erosion of the updrift end and accretion downdrift. Washover sheets and multiple shallow breaches are common on the updrift or erosional ends of these islands. Downdrift, longshore bars become more prominently developed in the nearshore zone, attaching to the beachface toward the end of the system. In these downdrift zones, lateral building of recurved spits is taking place. Recurved spit morphology formed during the growth of both Timbalier Island and Grande Isle indicates the importance of an updrift sand source in the Bayou Lafourche erosional headland. Figure 6 illustrates a

historical map comparison for the years 1887 and 1978. Note the rapid retreat of the erosional headland in the vicinity of Bayou Lafourche and Bay Marchand, and the degree of lateral migration that can be observed at Timbalier Island. A similar pattern of shoreline change is seen at Grand Isle.

A stratigraphic strike section through the Bayou Lafourche barrier shoreline is shown in Figure 7. Along the Caminada-Moreau coast, distributaries associated with Bayou Lafourche and Bayou Moreau are exposed on the shoreface. Between Bayou Moreau and Caminada Pass spit, beach-ridge deposits associated with Cheniere Caminada are also exposed. The Bayou Lafourche barrier shoreline sand body thickens downdrift from the central headland. The mainland beaches are 3-6 ft (1-2 m) thick and overlie brackish and salt marsh deposits. At its point of attach-

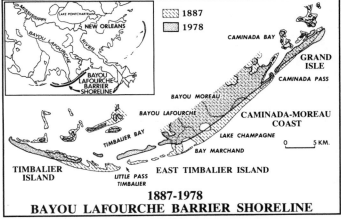

Figure 6. The coastal changes observed in the Bayou Lafourche delta and barrier shoreline are depicted based on a map comparison between 1887 and 1978. The pattern of shoreline change observed is representative of Stage 1, barrier shorelines (map source: National Ocean Survey).

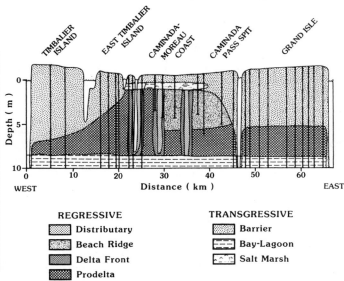

Figure 7. Stratigraphic strike section through the Bayou Lafourche delta illustrates the relationship between the regressive and transgressive delta cycle components.

ment to the headland, the Caminada Pass spit is 3-6 ft (1-2 m) thick, increasing to 16 ft (5 m) at Caminada Pass. Grand Isle is a 13 to 16 ft (4-5 m) thick flanking barrier island sand body, overlying an older, barrier sand body believed to be associated with one of the earlier Teche deltas. At the point of attachment to the central headland, East Timbalier Island is 3-6 ft (1-2 m) thick, reaching a maximum thickness of 16-19 ft (5-6 m) towards Cat Island Pass.

During regressive deltaic sedimentation, active delta complexes build distributary headlands which are separated by interdistributary bays. During transgression, sand moving alongshore from the erosional headland source into flanking barriers builds across the mouth of the interdistributary bays. Land loss leads to an increase in bay volume and continually increasing tidal prism, creating an environment suitable for tidal inlet generation. Tidal inlets are formed during storm events and especially during hurricanes, when elongated flanking barrier spits are breached by overwash processes. The increasing tidal prism of the restricted interdistributary bay is sufficient to maintain permanent fluid exchange through the barrier spit, resulting in the formation of a flanking barrier island and the production of flood and ebb tidal deltas. The result of flanking barrier-island growth and tidal-inlet generation is to produce a restricted interdistributary bay with intermediate salinities. Bioturbated muds accumulate in this environment, often accompanied by prolific oyster-reef growth (Coleman and Gagliano, 1964). Current examples of restricted interdistributary bays in the Bayou Lafourche delta are Caminada Bay and Timbalier Bay which are connected to the Gulf of Mexico through Caminada Pass and Little Pass Timbalier, respectively.

Figure 8. Oblique aerial photograph of the Caminada-Moreau coast along the Bayou Lafourche erosional headland (Field Stop 1).

STOP 1—CAMINADA-MOREAU COAST
SITE DESCRIPTION

The first field site is located at the end of Louisiana 3090, 3.5 mi (6.0 km) south of its intersection with Louisiana 1 (Fig. 1). The site is situated between Belle Pass and Louisiana 3090 covering 3.4 mi (5.3 km) and can be accessed by an automobile at low-tide.

The Belle Pass area represents the western boundary of the Bayou Lafourche erosional headland and Caminada-Moreau coast (Fig. 8). Bayou Lafourche enters the Gulf of Mexico here between stone jetties approximately 3,300 ft (1000 m) long. Because Belle Pass is the main distributary of the Lafourche delta, the coastline in this region projects seaward into the Gulf of Mexico. This headland has resulted in a concentration of wave energy, producing the highest rates of coastal erosion in Louisiana, if not the entire United States. Between 1887 and 1978, the coastline retreated 1.8 mi (3.1 km) at an average rate of 61 ft per year (18.6 m/yr) (Penland and Boyd 1985). The highest rates of erosion, 115 to 150 ft per year (35 to 45 m/yr) occurred during the period between the damming of Bayou Lafourche at Donaldsville in 1904 and the construction of the first jetties at Belle Pass in 1934. Further jetty and groin construction between 1945 and

1958 intercepted littoral drift from the Caminada-Moreau erosional headland westward to the Timbalier Islands. By 1969, sediments accumulating at the jetty had reversed shoreline erosion and initiated a period of gradual accretion at rates of 3.3 to 17 ft per year (1 to 5 m/yr). The westward direction of littoral drift and the resulting contrast between accretion east of the jetties and severe erosion west of the jetties can be seen from the 3,300 ft (1,000 m) of coastal offset found west of Belle Pass.

Between Louisiana 3090 and Belle Pass lies a transition region from slow coastal accretion at the jetties to severe coastal erosion where Louisiana 3090 intersects the coastline. East from Belle Pass there is a zone of low, hummocky dunes followed by washover flats. Behind the low dunes lies the remnant of Bay Marchand. This former interdistributary bay has been reduced in width from over 1.8 mi (3 km) in 1885 to less than 330 ft (100 m) in 1982.

STOP 2—CHENIERE CAMINADA

The second field site is located 7.1 mi (12 km) east of the intersection of Louisiana 1 and 3090 (Fig. 2). The site consists of a series of borrow pits located on the southside of Louisiana 1 which can be accessed by automobile.

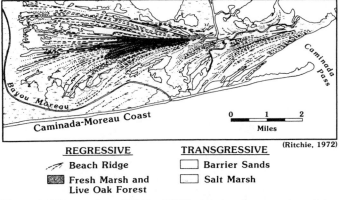

Figure 9. Diagram from Ritchie (1972) showing the geometry of the Cheniere Caminada regressive beach-ridge plain.

Regressive beach-ridge plains such as Cheniere Caminada are common depositional features within the Mississippi River delta plain in southeastern Louisiana. Cheniere Caminada is the largest beach-ridge plain to be found today (Fig. 9). It consists of more than 70 subparallel, seaward-flaring beach ridges in an arcuate fan-shaped configuration along the eastern levees of the Bayou Lafourche distributaries (Fig. 10). Ridges are primarily vegetated by *Quercus virginiana,* sp. and intra-ridge swales by *Spartina patens,* sp. Radiocarbon dates indicate beach-ridge building began approximately 720 years ago, when the Bayou Lafourche distributaries built seaward of the older, retreating Terrebonne shoreline and intercepted westward longshore sediment transport, resulting in the progradation of Cheniere Caminada. Near the fan apex, beach ridges are 23-26 ft (7-8 m) thick and thin westward to 6.6-10 ft (2-3 m) agains the levees of Bayou Moreau. A typical beach ridge vertical sequence coarsens upward, with shoreface silty sands overlain by a thin cap of beach, washover, and aeolian sands. Beach ridge progradation in this area ceased approximately 300 years ago with the abandonment of Bayou Lafourche.

Figure 10. Oblique aerial photograph of the regressive Cheniere Caminada beach-ridge plain (Field Stop 2).

STOP 3—CAMINADA PASS SPIT

The third field stop is located 6.8 mi (11 km) east of the intersection of Louisiana 1 and 3090 and accesses the coast via Elmer's Road (right turn off Louisiana Highway 1) which ends at the shoreline. The area is accessible to the east at low-tide by automobile.

Traveling east along the Caminada Pass spit, it is a distance of 5 mi (8 km) to the western margin of Caminada Pass (Fig. 11). This transition is accompanied by an increase in sediment abundance in the surf zone. Rates of shoreline change vary from the west to the east, from 16 ft (5 m) of erosion where the spit attaches to the headland to near stability or slight accretion adjacent to Caminada Pass. In addition, dune height, dune and vegetation stability, and the development of multiple offshore bars all increase east towards Caminada Pass. Channel and fan washover features give way to a continuous, vegetated dune crest of 5 to 10-ft (1.5- to 3-m) elevation (Fig. 5).

Figure 11. Oblique aerial photograph of the Caminada Pass recurved spit (Field Stop 3).

Towards the east end of the spit, abundant sediment has resulted in the development of a beach ridge landward of the present foredune ridge, similar to those of Grand Isle. Close by are several major spit breaches associated with major storms, such as Hurricane Flossie in 1956. These breaches were subsequently filled during post-storm recovery. This spit-breaching process is responsible for the generation of tidal inlets, such as Caminada Pass, and barrier islands, such as Grand Isle, which are visible from the eastern end of Caminada Spit. Also visible is the well-developed ebb-tidal delta seaward from Caminada Pass. Artificial jetty construction on the Grand Isle side of the pass has halted further lateral island migration.

STOP 4—GRAND ISLE FLANKING BARRIER ISLAND

The fourth field site is located at the public observation deck on the Grand Isle East State Park. The park is located at the end of Louisiana 1, 6.3 mi (10 km) east of the Caminada Pass bridge.

Grand Isle is the easternmost flanking barrier island of the Bayou Lafourche barrier shoreline. From the observation deck can be seen to the east Barataria Pass, and Fort Livingston on Grand Terre. To the south see the ebb-tidal delta of Barataria Pass; to the north, ridge and swale topography produced by a recurved spit building; and to the west the Grand Isle beach nourishment project. At Grand Isle, the characteristic flanking barrier, island pattern of updrift erosion and downdrift accretion occurs, similar to that observed at Timbalier Island. Prior to 1972, Grand Isle had historically eroded on its western end at Caminada Pass, and accreted downdrift on its eastern end adjacent to Barataria Pass. With construction of the single jetty on the western shore of Caminada Pass, erosion along western Grand Isle ceased and minor accretion began, averaging approximately 16.5 ft per year (5 m/yr). Along the central shoreline of Grand Isle, erosion rates of less than 16.5 ft per year (5 m/yr) are common. Farther downdrift, toward Barataria Pass, this erosional trend again turns to accretion of 16.5 to 33 ft per year (5 to 10 m/yr). Prior to jetty construction at Barataria Pass in 1958, the eastern end of Grand Isle accreted 10 to 20 ft per year (3 to 6 m/yr), typical of the downdrift end of a flanking barrier island. After 1958, sedimentation in this region accelerated, producing accretion rates in excess of 33 ft per year (10 m/yr) due to the construction of a jetty on the east end of Grand Isle. This field stop is a zone of sediment accumulation and recurved spit formation toward the east into Barataria Pass. In the surf zone, multiple nearshore bars can be seen attached on their updrift ends to the beachface. These bars migrate around the end of Grand Isle across a wide platform and weld to the active zone of recurved spit development, illustrating the process by which flanking barrier islands build downdrift away from their erosional headland sand source.

Figure 12. Oblique aerial photograph of Grand Isle (Field Stop 4).

REFERENCES CITED

Coleman, J. M., and Cagliano, S. M., 1964, Cyclic sedimentation in the Mississippi River deltaic plain: Gulf Coast Association of Geological Societies Transactions, v. 14, p. 67–82.

Fisk, H. N., 1944, Geological investigation of the alluvial valley of the lower Mississippi River: War Department, U.S. Army Corps of Engineers, p. 78.

Frazier, D. E., 1967, Recent deposits of the Mississippi River, their development and chronology: Gulf Coast Association of Geological Societies Transactions, v. 17, p. 287–311.

Gerdes, R. G., 1985, The Caminada-Moreau Beach Ridge Plain, in Penland, S., and Boyd, R., eds., Transgressive Depositional Environments of the Mississippi River Delta Plain, Louisiana Geological Survey, Guidebook Series No.

3, p. 127–140.

Kolb, C. R., and Van Lopik, J. R., 1958, Geology of the Mississippi River deltaic plain, southeastern Louisiana: Vicksburg, Mississippi, U.S. Army Corps of Engineers Waterways Experiment Station, Technical Report no. 3-483, p. 120.

Penland, S., and Boyd, R., 1981, Shoreline changes on the Louisiana barrier coast: Institute of Electrical and Electronic Engineers Oceans, v. 81, p. 209–219.

Penland, S., Boyd, R., Nummedal, D., and Roberts, H., 1981, Deltaic barrier development on the Louisiana coast: Supplement to Transactions, Gulf Coast Association of Geological Societies, v. 31, p. 471–476.

Penland, S., and Boyd, R., 1985, Transgressive Depositional Environments of the Mississippi River Delta Plain: Louisiana Geological Guidebook Series No. 3, p. 233.

Ritchie, W., 1972, A preliminary study of the distribution and morphology of the Caminada-Moreau sand ridges: Southeastern Geology, v. 14, p. 113–126.

Scruton, F. P., 1960, Delta building and the deltaic sequence, in Shepard, F. P., Phleger, F. B., and van Andel, T. H., Recent Sediments, northwest Gulf of Mexico: American Association of Petroleum Geologists Symposium Volume, p. 82–102.

Index

Vacherie Dome, 416–417
Valley and Ridge province, 69–72, 97–100,
    105–108, 110–122, 149–152, 185–190,
    201–206, 265–270
vertebrates, 325, 342
Virginia, 73–78
    Botetourt County, 105–108
    central, 207–208
    central, Fluvanna County, 209–214
    Eagle Rock Gorge, 105–108
    Fincastle Valley, 105–108
    Franklin County, Rocky Mount, 215–216
    Harpers Ferry water gap, 201–206
    Highland County, 97–100
    Lee County, 123–125

    southwestern, 113–118
    southwestern, Pepper, 119–122
    Spotsylvania County, Fredericksburg, 309–314
volcanism, 207–208

waterfalls, 72
water gaps, 69–72, 159–162, 201–206
Waverly Arch, 37–41
weathering, 328
Weeks salt dome, 431
West Virginia, Batoff Creek section, 109–112
    eastern, 113–118
West Virginia
    eastern, Pendleton County, 91–96
    Grant County, 69–72
    Harpers Ferry water gap, 201–206

    northwestern
        Ritchie County, 55–58
        Willow Island 7½-minute Quadrangle, 55–58
        Wood County, 55–58
    Pendleton County, Judy Gap, 85–90
    Pocahontas County, 101–104
    Raleigh County, 109–112
    Randolph County, Elkins, 79–83
    south-central, 59–68
Whiteoak Mountain synclinorium, 149–152
Wills Mountain anticline, 69–72
Winding Stair Gap, 257–260
windows, 123–125, 229–230
Winnfield Dome, 413–414

xenoliths, 286

Typeset by WESType Publishing Services, Inc., Boulder, Colorado
Printed in U.S.A. by Malloy Lithographing, Inc., Ann Arbor, Michigan